IOHANNES WALLIS D·D
GEOMETRIAE PROF· SAVILIANVS OXONIAE
REGALIS SOCIETATIS LONDINI SODALIS·

John Wallis c.1668

Engraving after W. Fairthorne by Giovanni Battista Cipriani (1727–85)

In Memoriam
Adolf Prag
(1906–2004)

The Correspondence of John Wallis

Volume III
(October 1668–1671)

Editors

PHILIP BEELEY
CHRISTOPH J. SCRIBA

OXFORD
UNIVERSITY PRESS

Great Clarendon Street, Oxford, OX2 6DP,
United Kingdom

Oxford University Press is a department of the University of Oxford.
It furthers the University's objective of excellence in research, scholarship,
and education by publishing worldwide. Oxford is a registered trade mark of
Oxford University Press in the UK and in certain other countries

© Philip Beeley and Christoph J. Scriba 2012

The moral rights of the authors have been asserted

First published in 2012

Impression: 1

All rights reserved. No part of this publication may be reproduced, stored in
a retrieval system, or transmitted, in any form or by any means, without the
prior permission in writing of Oxford University Press, or as expressly permitted
by law, by licence or under terms agreed with the appropriate reprographics
rights organization. Enquiries concerning reproduction outside the scope of the
above should be sent to the Rights Department, Oxford University Press, at the
address above

You must not circulate this work in any other form
and you must impose this same condition on any acquirer

British Library Cataloguing in Publication Data
Data available

Library of Congress Cataloging in Publication Data
Library of Congress Control Number: 2004270866

ISBN 978–0–19–856947–3

Printed and bound by
CPI Group (UK) Ltd, Croydon, CR0 4YY

PREFACE

The period covered by this third volume of *The Correspondence of John Wallis (1616–1703)* was one of remarkable political stability and economic prosperity. To an unprecedented degree, England was at peace with itself and with its neighbours. Scientific activity also prospered with the Royal Society and especially its industrious secretary Henry Oldenburg at the core. The institutional role of the Royal Society reflected London's increasing importance as a centre of intellectual and material commerce, of coffee houses and meeting places, of trade and shipping, of government and court, of finance, and of a postal service spreading out like so many arteries to the provinces.

The Oxford–London axis is of central importance to the present volume. Oldenburg regularly sent Wallis news of the weekly meetings of the Royal Society in London, to which Wallis replied as a matter of course. Through the medium of correspondence, the Savilian professor was a key participant in Royal Society debates such as those on the laws of motion or the theory of tides. With ordinary post between the English capital and the university city usually requiring at most one day, Wallis was able to make his presence felt, even when not in London.

The very nature of intellectual commerce and scientific debates conducted over distance poses difficulties when compiling an edition such as this. Many letters sent to Oldenburg by his correspondents around the country and abroad were actually intended for Wallis, even if they were not addressed to him. Such letters are of course included in the present volume together with letters between third parties which throw light on Wallis letters which cannot be gained from elsewhere. In this way, the editors seek to provide as complete a picture as possible of Wallis's intellectual activity during the period concerned.

The volume contains a total of 254 existent or identified letters exchanged between Wallis and thirty-two correspondents in England, Scotland, Ireland, France, Italy, Poland, the Dutch Republic, and the Spanish Netherlands. More than thirty-five of these letters have never appeared in print before, while many of the others have only been published in part or are to be found in publications which are either rare or not widely accessible. Only a very few of the letters have previously appeared in the form of a critical edition.

Much of the work on the volume has been generously funded by the Arts and Humanities Research Council (AHRC), and was carried out in the framework of the Oxford-based project on John Wallis entitled 'Harmony and Controversy in Seventeenth-Century Thought' between 2007 and 2010. The editors would like to thank the AHRC for its generous support throughout the duration of that project. Since September 2010 the edition of the correspondence of John Wallis has become an integral part of the Cultures of Knowledge project—a collaboration between the Bodleian Libraries and the Humanities Division of the University of Oxford funded by The Andrew W. Mellon Foundation. The editors are deeply indebted to the Mellon Foundation and to the Director of the Cultures of Knowledge project, Howard Hotson, for enabling this transfer to take place and for providing necessary financial support for the conclusion of this volume.

The editors would like to express their gratitude also to David Cram for his enthusiasm and assistance in bringing the edition to Oxford, and to the Faculty of Linguistics, Philology, and Phonetics for accommodating it before its move to the Faculty of History.

The excellent facilities of the Oxford Faculty of History, and the congenial and frequently inspirational environment provided by friends and colleagues in the Cultures of Knowledge project have enabled this volume to be brought to a rapid conclusion under near-ideal conditions. For the wonderful support given to them, the editors should like to thank James Brown, Miranda Lewis, Kim McLean-Fiander, Leigh Penman, Anna Marie Roos, and Helen Watts. Howard Hotson has been a tower of strength in all respects, and is to be especially thanked. Without James Brown's steady hand in coordinating and organizing matters and his ready willingness to find solutions to problems at short notice, work on the volume would not have moved so smoothly. The editors are extremely grateful to him for his steadfast efforts.

One of the joys of working on a major correspondence edition is the scholarly collaboration such work engenders. Such assistance can take many forms, ranging from information on previously unknown manuscripts or printed versions of letters to willingness to read edited material and suggest ways in which improvements might be made. The editors are extremely conscious of the help given to them in the course of their work and take great pleasure in acknowledging this. Specifically, they should like to express their gratitude to Michael Hunter, Noel Malcolm, Per Landgren, Kate Bennett, Robert Hatch, Kelsey Jackson Williams, Alexander Häussler, Katerina Oikonomopoulou, Jason Rampelt, Richard Sharpe, William Poole, and Charles Webster.

Editorial assistance during 2008 was provided by Ellie Stedall. The careful way in which she made transcriptions from manuscripts and the energy she brought to the project were sorely missed after her departure for Cambridge.

Invaluable advice on transcriptions has been generously provided by Jutta Heinle. The patience with which she has followed the conclusion of the volume is more deeply appreciated than she can know.

The Vogel Foundation kindly provided financial support for the editors in order for them to obtain digital images of Wallis material formerly contained in the Macclesfield Collection and now housed in Cambridge University Library. The editors are extremely grateful to Menso Folkerts for his help and advice in this regard.

The editors would like to express two special debts of heartfelt gratitude. They are indebted to Jackie Stedall, who has provided invaluable support, advice, and encouragement throughout the course of work on the volume. She has also kindly enabled work on the edition to be presented to colleagues in Oxford on a number of occasions. The editors are also indebted to Pietro Corsi, who has been a constant source of sound advice and moral support since the edition arrived in Oxford. His tireless work as head of the steering committee of the AHRC project has been particularly appreciated.

Initial transcriptions of a substantial number of letters were made by Uwe Mayer. The editors are grateful to him for his lasting and conscientious input into the edition. Siegmund Probst has continued to share generously in his knowledge, and has been constantly willing to provide assistance wherever he can.

The editors' greatest debt is to Kim McLean-Fiander, who has taken time off preparing the *Calendar of the Correspondence of John Wallis* in order to proof-read and check a substantial part of the book manuscript. Not a few of her suggestions have found their way into the commentaries of the letters. At the same time her industry and care in preparing the *Calendar*, for which they are immensely grateful, will ensure that future work on the edition will be considerably facilitated.

Staff at numerous libraries and archives have provided often invaluable help in preparing the present volume. The editors should like to express their particular gratitude to Adam Perkins of Cambridge University Library, Rupert Baker and Keith Moore of the Library of the Royal Society, Frances Harris of the British Library, and Östlund Krister of the Universitetsbibliotek, Uppsala. Once again, the editors are especially indebted to Simon

Bailey, the Keeper of the Archives of the University of Oxford, not only for pointing them in the direction of new material but also for ready support on details of University history.

The editors are grateful to the following persons and institutions for granting permission to publish copyright material held in their possession: The Syndics of Cambridge University Library; the Librarian of the Bibliotheek der Rijksuniversiteit, Leiden; the British Library Board; the Librarian of the Bodleian Library, Oxford; the Keeper of the Archives, University of Oxford; the Director of the Bibliothèque Nationale de France; the Royal Society; the Director of the Bibliothèque de l'Observatoire, Paris; the National Archives, Kew; the Librarian of the Universitetsbibliotek, Uppsala; the Director of the Biblioteca Nazionale, Florence.

Finally, the editors should like to thank editorial staff at Oxford University Press, especially Keith Mansfield, Elizabeth Hannon, and, more recently, Clare Charles, for their continued help in realizing the edition, their care in proof reading, and their patience in the light of often considerable delays in the submission of material.

Philip Beeley
Christoph J. Scriba
Oxford and Hamburg, September 2011

CONTENTS

Introduction — xvii
    Correspondence and conflict with Hobbes xviii
    Wallis as mediator between Gregory and Huygens xix
    Debate on the laws of motion xx
    Debate on the theory of tides xx
    Correspondence with Dulaurens, Huygens, and Hevelius xxi
    Correspondence with Bertet and Sluse xxii
    The heritage of Pierre de Fermat xxiv
    Correspondence with Borelli xxiv
    Leibniz and the Royal Society xxv
    Publication of Wallis's *Mechanica* xxv
    Scientific papers of Horrox and Merry xxvii
    The ceremonial opening of the Sheldonian Theatre xxviii
    Foreign visitors to Oxford xxviii
    Cryptanalysis and the fate of Clarendon xxix

Editorial principles and abbreviations — xxxi

Correspondence — 1
    1. WALLIS to OLDENBURG, 6/[16] October 1668 1
    2. GOTT to WALLIS, 19/[29] October 1668 3
    3. WALLIS: A Conjecture on Mr Gott's Proposal of an
       Artificiall Spring 4
    4. WALLIS to BROUNCKER, 22 October/[1 November] 1668 . . . 6
    5. WALLIS to GREGORY, 22 October/[1 November] 1668 6
    6. COLLINS to PELL, 23 October/[2 November] 1668 7
    7. COLLINS to WALLIS, 26 October/[5 November] 1668 10
    8. GREGORY to WALLIS, 26 October/[5 November] 1668 10
    9. WALLIS to HEVELIUS, 26 October/[5 November] 1668 13
    10. PELL to COLLINS, 28 October/[7 November] 1668 15
    11. DULAURENS to OLDENBURG, October? 1668 18

Contents

12. OLDENBURG to WALLIS, early November 1668 18
13. WALLIS to GREGORY, 2?/[12?] November 1668 19
14. HUYGENS to WALLIS, [3]/13 November 1668 19
15. WALLIS to COLLINS, 3/[13] November 1668 23
16. WALLIS to BROUNCKER, 4/[14] November 1668 25
17. WALLIS to OLDENBURG, 7/[17] November 1668 37
18. WALLIS to HEVELIUS, 12/[22] November 1668 41
19. WALLIS to HUYGENS, 13/[23] November 1668 41
20. COLLINS to PELL, 14/[24] November 1668 44
21. WALLIS to OLDENBURG, 14/[24] November 1668 46
22. WALLIS to OLDENBURG, 15/[25] November 1668 48
23. OLDENBURG to WALLIS, 16/[26] November 1668 52
24. OLDENBURG to WALLIS, 17/[27] November 1668 53
25. WALLIS to OLDENBURG, 19/[29] November 1668 53
26. OLDENBURG to WALLIS, 24 November/[4 December] 1668 . . 56
27. WALLIS to OLDENBURG, 26? November/[6? December] 1668 56
28. WALLIS: Paper on Dulaurens, 27 November/[7 December] 1668 60
29. WALLIS: Paper on Hobbes, 27 November/[7 December] 1668 63
30. WALLIS to OLDENBURG, 30 November/[10 December] 1668 . 66
31. WALLIS to GREGORY, November/early December 1668 67
32. OLDENBURG to WALLIS, 1/[11] December 1668 67
33. OLDENBURG to WALLIS, 2/[12] December 1668 68
34. WALLIS to OLDENBURG, 3/[13] December 1668 68
35. WALLIS to OLDENBURG, 5/[15] December 1668 70
36. OLDENBURG to WALLIS, 8/[18] December 1668 74
37. WALLIS to OLDENBURG, 9/[19] December 1668 74
38. WALLIS to OLDENBURG, 9 and 10/[19 and 20] December 1668 80
39. WALLIS to OLDENBURG, 12/[22] December 1668 82
40. OLDENBURG to WALLIS, 14/[24] December 1668 86
41. OLDENBURG to WALLIS, 15/[25] December 1668 86
42. GREGORY to WALLIS, c.20/[30] December 1668 86
43. WALLIS to OLDENBURG, Oxford, 21/[31] December 1668 . . . 87
44. OLDENBURG to WALLIS, 26 December 1668/[5 January 1669] 93
45. OLDENBURG to WALLIS, 29 December 1668/[8 January 1669] 93

Contents

46. OLDENBURG to WALLIS, 31 December 1668/[10 January 1669] 93
47. WALLIS to OLDENBURG, 31 December 1668/[10 January 1669] 94
48. WALLIS to OLDENBURG, 2/[12] January 1668/9 95
49. WALLIS to GREGORY, 8/[18] January 1668/9 98
50. WALLIS to COLLINS, 8/[18] January 1668/9 99
51. COLLINS to WALLIS, 12/[22] January 1668/9 99
52. WALLIS to OLDENBURG, 12/[22] January 1668/9 100
53. OLDENBURG to WALLIS, 14/[24] January 1668/9 102
54. WALLIS to HEVELIUS, 15/25 January 1668/9 102
55. WALLIS to HEVELIUS, 15/25 January 1668/9, enclosure 104
56. JOHN WALLIS JR to WALLIS, [15]/25 January 1668/9 144
57. OLDENBURG to WALLIS, 16/[26] January 1668/9 144
58. WALLIS to COLLINS, Oxford, 18/[28] January 1668/9 144
59. WALLIS to COLLINS, 19/[29] January 1668/9 147
60. WALLIS to OLDENBURG, 19 and 21/[29 and 31] January 1668/9 150
61. OLDENBURG to WALLIS, 26 January/[5 February] 1668/9 . . 153
62. JOHN WALLIS JR to WALLIS, late January 1668/9 154
63. COLLINS to WALLIS, late January/February 1668/9 154
64. OLDENBURG to WALLIS, 6/[16] February 1668/9 154
65. OLDENBURG to WALLIS, 8/[18] February 1668/9 155
66. WALLIS to OLDENBURG, 11/[21] February 1668/9 155
67. OLDENBURG to WALLIS, 16/[26] February 1668/9 158
68. HEVELIUS to WALLIS, [11]/21 March 1668/9 158
69. COLLINS to WALLIS, 21?/[31?] March 1668/9 160
70. WALLIS to COLLINS, 23 March/[2 April] 1668/9 161
71. WALLIS to OLDENBURG, 15/[25] April 1669 164
72. OLDENBURG to WALLIS, 17/[27] April 1669 166
73. WALLIS to OLDENBURG, 24 April/[4 May] 1669 166
74. OLDENBURG to WALLIS, 1/[11] May 1669 169
75. LAMPLUGH to WILLIAMSON, 5/[15] May 1669 169
76. OLDENBURG to WALLIS, 8/[18] May 1669 171
77. OLDENBURG to WALLIS, 8/[18] May 1669, enclosure 171
78. WALLIS to OLDENBURG, 10/[20] May 1669 178
79. NEILE to OLDENBURG, 13/[23] May 1669 181
80. OLDENBURG to WALLIS, 15?/[25?] May 1669 183
81. WALLIS to OLDENBURG, 17/[27] May 1669 184

Contents

82. NEILE to OLDENBURG, 20/[30] May 1669 186
83. OLDENBURG to WALLIS, 24 May/[3 June] 1669 188
84. WALLIS to OLDENBURG, 29 May/[8 June] 1669 189
85. NEILE to OLDENBURG, 1/[11] June 1669 191
86. OLDENBURG to WALLIS, 4/[14] June 1669 194
87. WALLIS to OLDENBURG, 7/[17] June 1669 194
88. OLDENBURG to WALLIS, 12/[22] June 1669 200
89. NEILE to OLDENBURG, 15/[25] June 1669 201
90. WALLIS to OLDENBURG, 15/[25] June 1669 207
91. COLLINS to WALLIS, 17/[27] June 1669 210
92. WALLIS to OLDENBURG, 19/[29] June 1669 213
93. NEILE to OLDENBURG, 23 June/[3 July] 1669 219
94. WALLIS to COLLINS, 24 June/[4 July] 1669 220
95. WALLIS: Corrections to Newton's *Art of Practical Gauging*, June 1669 223
96. WALLIS: Corrections to *Mechanica* 227
97. OLDENBURG to WALLIS, 26 June/[5 July] 1669 228
98. BOYLE to WALLIS, 3/[13] July 1669 228
99. OLDENBURG to WALLIS, 5/[15] July 1669 228
100. LAMPLUGH to WILLIAMSON, 13/[23] July 1669 229
101. WALLIS to OLDENBURG, 16/[26] July 1669 230
102. WALLIS to BOYLE, 17/[27] July 1669 232
103. OLDENBURG to WALLIS, 23 July/[2 August] 1669 237
104. WALLIS to OLDENBURG, 29 July/[8 August] 1669 237
105. WALLIS to BOYLE, second half of July 1669 240
106. WALLIS to LANGHAM, second half of July–first half August 1669 240
107. OLDENBURG to WALLIS, 4/[14] August 1669 240
108. BOYLE to WALLIS, 7/[17] August 1669 241
109. WALLIS to OLDENBURG, 15/[25] August 1669 241
110. WALLIS to OLDENBURG, 15/[25] August 1669, enclosure . . 244
111. WALLIS to BOYLE, 17/[27] August 1669 244
112. WALLIS to LANGHAM, 24 August/[3 September] 1669 247
113. WALLIS to SLUSE, 10/[20] September 1669 248
114. ELLIS to WALLIS, 7/[17] October 1669 251
115. OLDENBURG to WALLIS, 14?/[24?] October 1669 252
116. WALLIS to OLDENBURG, 16/[26] October 1669 253
117. WALLIS to OLDENBURG, 17/[27] October 1669 256
118. WALLIS to BROUNCKER, 10/[20] November 1669 258

Contents

119. The Reasonableness of the University's Exemption from Charges 259
120. WALLIS to COLLINS, ? November 1669 265
121. OLDENBURG to WALLIS, c.8/[18] December 1669 265
122. OLDENBURG to WALLIS, c.8/[18] December 1669, enclosure 266
123. WALLIS to OLDENBURG, 9/[19] December 1669 269
124. WALLIS to BROUNCKER, 21/[31] December 1669 270
125. WALLIS for OLDENBURG, c.21/[31] December 1669 277
126. OLDENBURG to WALLIS, 24 December 1669/[3 January 1670] 279
127. WALLIS to SCHOOTEN, second half of 1669 279
128. BORELLI to WALLIS, second half of 1669 (i) 280
129. BORELLI to WALLIS, second half of 1669 (ii) 280
130. SCHOOTEN to WALLIS, December 1669 – January 1669/70 . 280
131. OLDENBURG to WALLIS, 6/[16] January 1669/70 281
132. WALIIS to OLDENBURG, 7/[17] January 1669/70 281
133. COLLINS to WALLIS, 8/[18] January 1669/[1670] 287
134. WALLIS to OLDENBURG, 9/[19] January 1669/[1670] 287
135. WALLIS to COLLINS, 11/[21] January 1669/70 289
136. WALLIS to BORELLI, 13/[23] January 1669/70 294
137. WALLIS to GIOVANNI ALFONSO BORELLI, Books for Borelli 296
138. WOOD to WALLIS, 8/[18] February 1669/[1670] 299
139. HYRNE to WALLIS, 28 February/[10 March] 1669/70 299
140. SLUSE to WALLIS, [28 February 1669/70]/10 March 1670 . . 302
141. COLLINS to OLDENBURG, February/March ? 1669/70 303
142. CHILDREY to WARD, 4/[14] March 1669/70 305
143. WALLIS: Reply to Collins's Question on Algebraic Roots . . 313
144. OLDENBURG to WALLIS, 8/[18] March 1669/[1670] 319
145. WALLIS to HYRNE, 9/[19] March 1669/70 320
146. WALLIS to OLDENBURG, 10/[20] March 1669/[1670] 322
147. WALLIS to WOOD, 10/[20] March 1669/[1670] 324
148. OLDENBURG to WALLIS, c.16/[26] March 1669/[1670] 327
149. WALLIS to OLDENBURG, 19/[29] March 1669/[1670] 327
150. CHILDREY to WALLIS, c.20/[30] March 1669/70 334
151. WALLIS to OLDENBURG, 24 March/[3 April] 1669/70 334
152. OLDENBURG to WALLIS, 26 March/[5 April] 1669 337
153. WALLIS to SLUSE, 29 March/[8 April] 1670 339
154. WALLIS to OLDENBURG, 29 March/[8 April] 1670 341

155. WALLIS to CHILDREY, c.29 March/[8 April] 1670 343
156. ? to WALLIS, mid March 1669/70 343
157. HYRNE to WALLIS, 2/[12] April 1670 343
158. WALLIS to HYRNE, 4/[14] April 1670 359
159. OLDENBURG to WALLIS, 9/[19] April 1670 363
160. WALLIS to OLDENBURG, 9/[19] April 1670 363
161. CHILDREY to OLDENBURG, 12 and 15/[22 and 25] April 1670 365
162. WOOD to WALLIS, 15/[25] April 1670 368
163. OLDENBURG to WALLIS, 16/[26] April 1670 368
164. WALLIS to WOOD, 16/[26] May 1670 368
165. WALLIS to WOOD, 18/[28] May 1670 377
166. WALLIS to VERNON, early June 1670 377
167. WALLIS for OLDENBURG, 11/[21] July 1670 377
168. COLLINS to WALLIS, 14/[24] July 1670 378
169. SLUSE to WALLIS, [15]/25 July 1670 378
170. OLDENBURG to WALLIS, 18/[28] July 1670 380
171. WALLIS to OLDENBURG, 22 July/[1 August] 1670 381
172. WALLIS to COLLINS, 23 July/[2 August] 1670 385
173. WOOD to WALLIS, 23 July/[2 August] 1670 387
174. OLDENBURG to WALLIS, 28 July/[7 August] 1670 387
175. COLLINS to WALLIS, end of July 1670 387
176. WALLIS to COLLINS, 4/[14] August 1670 388
177. WALLIS to WOOD, 4/[14] August 1670 389
178. WALLIS to OLDENBURG, 4/[14] August 1670 390
179. OLDENBURG to WALLIS, 6/[16] August 1670 393
180. VERNON to WALLIS, ?[25 August]/4 September 1670 393
181. AUBREY to WALLIS, 27 August/[6 September] 1670 393
182. WALLIS to NEILE, August/September 1670 394
183. ? to WALLIS, August/September 1670 395
184. COSIMO III DE' MEDICI to WALLIS, 3/[13] October 1670 . . 395
185. WALLIS to AUBREY, 24 October/[3 November] 1670 395
186. [WALLIS] to OLDENBURG, 25 October/[4 November] 1670 . 396
187. OLDENBURG to WALLIS, 1/[11] November 1670 398
188. WALLIS to OLDENBURG, 3/[13] November 1670 398
189. OLDENBURG to WALLIS, 8/[18] November 1670 401
190. WALLIS to COSIMO III DE' MEDICI, 9/[19] November 1670 401
191. WALLIS to OLDENBURG, 15/[25] November 1670 405
192. WALLIS to OLDENBURG, 24 November/[4 December] 1670 . 408

193. BORELLI to WALLIS, [26 November]/6 December 1670 . . . 410
194. OLDENBURG to WALLIS, 1/[11] December 1670 413
195. WALLIS to OLDENBURG, 12/[22] December 1670 413
196. WALLIS to BROUNCKER, 12/[22] December 1670 416
197. OLDENBURG to WALLIS, 17/[27] December 1670 417
198. WALLIS: Lecture for the Prince of Orange, 20/[30] Dec. 1670 417
199. WALLIS to BERTET, late 1670 420
200. HUYGENS for WALLIS, late 1670–late 1671 420
201. BERTET to WALLIS, beginning of 1671 426
202. BROUNKER to WALLIS, 15/[25] February 1670/1 427
203. COLLINS to WALLIS, February/early March? 1671 430
204. COLLINS to BERNARD, 16/[26] March 1670/71 431
205. WALLIS to COLLINS, ?March 1670/1 435
206. COLLINS to WALLIS, 21/[31] March 1670/1 435
207. WALLIS to OLDENBURG, 23 March/[2 April] 1670/1 439
208. WALLIS to OLDENBURG, 23 March/[2 April] 1670/1, enclosure 441
209. OLDENBURG to WALLIS, 30 March/[9 April] 1671 442
210. WALLIS to OLDENBURG, 7/[17] April 1671 442
211. OLDENBURG to WALLIS, 9/[19] May 1671 448
212. WALLIS to OLDENBURG, 9/[19] May 1671 448
213. WALLIS to OLDENBURG, 13/[23] May 1671 448
214. OLDENBURG to WALLIS, mid/end May 1671 451
215. BERTET to WALLIS, May/June 1671 452
216. WALLIS to OLDENBURG, 2/[12] June 1671 452
217. WALLIS to OLDENBURG, 10/[20] June 1671 454
218. WALLIS to OLDENBURG, 27 June/[7 July] 1671 455
219. WALLIS to OLDENBURG, 16/[26] July 1671 459
220. HOBBES to the ROYAL SOCIETY, end of July 1671 472
221. OLDENBURG to WALLIS, 1/[11] August 1671 475
222. WALLIS to OLDENBURG, 4/[14] August 1671 475
223. OLDENBURG to WALLIS, 5/[15] August 1671 477
224. WALLIS to OLDENBURG, 5–9/[15–19] August 1671 479
225. WALLIS to OLDENBURG, 10/[20] August 1671 479
226. OLDENBURG to WALLIS, 10/[20] August 1671 482
227. WALLIS to OLDENBURG, 13/[23] August 1671 482
228. HOBBES to the ROYAL SOCIETY, first half of August 1671 . 485
229. OLDENBURG to WALLIS, 15/[25] August 1671 486
230. WALLIS to OLDENBURG, 16/[26] August 1671 488

Contents

231. WALLIS to the ROYAL SOCIETY, August 1671 489
232. HOBBES to the ROYAL SOCIETY, early September 1671 . . . 497
233. POCOCKE to WALLIS, 21 September/[1 October] 1671 . . . 501
234. WALLIS to POCOCKE, 23 September/[3 October] 1671 . . . 501
235. WALLIS to HOBBES/ROYAL SOCIETY,
 September 1671 (i) 503
236. HOBBES to the ROYAL SOCIETY, September 1671 510
237. WALLIS to HOBBES/ROYAL SOCIETY,
 September 1671 (ii) 514
238. OLDENBURG to WALLIS, 4/[14] November 1671 519
239. WALLIS to LAMPLUGH, 6/[16] November 1671 519
240. WALLIS to CREW, 6/[16] November 1671 520
241. WALLIS to OLDENBURG, 6/[16] November 1671 520
242. OLDENBURG to WALLIS, 11/[21] November 1671 522
243. WALLIS to SCARBOROUGH, 16/[26] November 1671 523
244. WALLIS to SCARBOROUGH, 21 November/
 [1 December] 1671 523
245. WALLIS to SCARBOROUGH, 21 Nov./[1 Dec.] 1671,
 enclosure 524
246. BERTET to WALLIS, [21 November]/1 December 1671 529
247. WALLIS to OLDENBURG, 23 November/[3 December] 1671 . 535
248. OLDENBURG to WALLIS, 26 November/[6 December] 1671 . 537
249. WALLIS to OLDENBURG, 27 November/[7 December] 1671 . 538
250. OLDENBURG to WALLIS, 5/[15] December 1671 540
251. WALLIS to BOYLE, 13/[23] December 1671 541
252. COLLINS to VERNON, 14/[24] December 1671 543
253. WALLIS to BERTET, 19/[29] December 1671 548
254. OLDENBURG to WALLIS, 30 December 1671/
 [9 January] 1672 554

Biographies of correspondents   555

List of manuscripts   567

Bibliography   569

List of letters   599

Index: persons and subjects   609

INTRODUCTION

In October 1668, when this volume of letters begins, John Wallis was approaching the age of fifty-two. He had been Savilian professor of geometry at Oxford for almost twenty years, and had seen the Royal Society emerge from inauspicious beginnings in London in the mid-1640s to become the leading scientific organization of its day. Intellectually and emotionally he was at home in the University of Oxford and the Royal Society, and both of these institutions were more indebted to him than they were often prepared to acknowledge. Together with Seth Ward (1617–88), the Savilian professor of astronomy, Wallis had contributed decisively in establishing modern mathematical science at Oxford, and as the keeper of the university's archives he had fought fiercely to defend its rights and privileges from encroachments originating either in Whitehall or more often in the city of Oxford itself. The future would bring many more battles on these fronts. At the same time, Wallis was committed to promoting the interests of the Royal Society, attending meetings whenever he was in London—more often than not on University business—and corresponding regularly with its secretary Henry Oldenburg (1618?–77).

Alongside the publication of books and learned articles, correspondence and the development of communication networks were at the core of the knowledge culture which enabled the stupendous growth in scientific learning in the seventeenth century to take place. But the growth of knowledge and the associated promise of improvements to the human condition are only part of the history of early modern science, albeit a major one. These aspirations collided almost inevitably with the limits of human nature itself. While Oldenburg continued to promote and practice a spirit of scientific collaboration which had its roots deep in the tradition of Jan Amos Comenius (1592–1670) and Samuel Hartlib (c.1600–62), Wallis found himself increasingly confronted by a different reality: a Republic of Letters in which defence of social status, institutional power, and disciplinary authority often made quarreling and dispute the order of the day.

Early experiences had lasting effects. Having at first underestimated the significance of Pierre de Fermat's (1607/8–65) challenges on number theory,

Introduction

and feeling deceived by Blaise Pascal's (1623–62) anonymously posed prize questions on the cycloid, Wallis sought thereafter to avoid being drawn into discourse of a kind he felt unsuitable for the promotion of knowledge. Intellectual commerce was for him to be conducted through the open medium of correspondence—a demand which he was not always able to live up to entirely himself, as his early dispute with Roberval over the rectification of the semi-cubical parabola had displayed. When Oldenburg forwarded him a set of problems posed by the Dutch mathematician and surveyor Jacob van Wassenaer (c.1607–?), he obliged his friend by duly sending solutions, but left no doubt about his dislike of this kind of challenge, speaking of his weariness of 'such Problemes from beyond Sea'. For, as he explained to Oldenburg, the result of such questions would often be dispute or quarreling: 'If they be not solved; they insult. If they be; they cavil, & be angry. At lest, they take up time & give trouble to no purpose' (No. 37). Despite such misgivings he also obliged the antiquary John Aubrey (1626–97), no doubt as a favour to Aubrey himself, after he had forwarded to Wallis a mathematical problem received from one of his Jesuit contacts in Paris (Nos. 181 and 185).

Correspondence and conflict with Hobbes

If there was one contemporary with whom Wallis certainly did not avoid disputes it was the author Thomas Hobbes (1588–1679). By the time at which the present volume begins, many battles in their long drawn out war had already been fought. The publication of Hobbes' *Opera philosophica* in the Netherlands in 1668 set off a new stage of fighting. Almost immediately Wallis drafted a paper for publication in *Philosophical Transactions*, aimed at reminding readers of his fundamental objections to the philosopher's mathematical writings (No. 29). Oldenburg evidently felt that the learned public did not need reminding in this way and decided against publishing Wallis's latest piece—perhaps also because he deemed the ongoing public dispute between James Gregory (1638–75) and Christiaan Huygens (1629–95) to be quite sufficient for his readership. The Savilian professor in the meantime drew up plans of his own for republishing his mathematical works on the continent so as to make them, too, available to a wider audience (No. 35; see also No. 130). When Hobbes printed the second edition of his latest quadrature of the circle in 1669, Wallis promptly responded by writing a review for the *Philosophical Transactions* in which he yet again demolished the philosopher's arguments (No. 125). This brought no relief: Hobbes's publication of his mathematical endeavours was as unrelenting as Wallis's destructive replies. The Royal Society, naturally fearful of being brought into association with a man whose political and religious views were anathema to those in power, was generally glad to support Wallis in the dispute. The appearance of Hobbes's *Rosetum geometricum* in the summer of 1671 started another round of exchanges which continued up to the end of the year, with Hobbes on the one side producing a series of broadsheets and with Wallis on the other side publishing his responses in Oldenburg's journal (Nos. 218–20, 222–3, 225, 227–32, 235–7). A successful prosecution of the war against Hobbes even took on national importance. In a remarkable letter from Oldenburg we learn that the king had instructed William Brouncker (1620–84), president of the Royal Society, that Hobbes's first broadsheet from late summer 1671 (No. 220) should be answered in a manner not just comprehensible to mathematicians. The intention was clear: all should be able to see the ineptitude of his mathematical reasoning which would also throw light on his reasoning in general. Oldenburg duly passed on the request to Wallis (No. 229).

Wallis as mediator between Gregory and Huygens

As a man who could speak with authority on mathematical matters, Wallis was regularly called upon by the Royal Society to deliver his expert opinion on the work of others or on matters of dispute not involving himself. Indeed, in cases where he was not involved personally, Wallis would often display an exceedingly high degree of equanimity. When James Gregory published his *Vera circuli et hyperbolae quadratura* (1667) and in the following year his *Geometriae pars universalis* (1668), they made such a strong impact on mathematical discussion at the time that they led directly to his being elected fellow of the Royal Society. But they also engendered a dispute with another Royal Society fellow, Christiaan Huygens, who claimed priority for some of the results. More importantly, Huygens denied the validity of Gregory's method for proving the transcendence of π: that no finite combination of the four elementary arithmetical processes together with the extraction of roots could serve to express the area of an elliptic, circular, or hyperbolic sector in terms of the inscribed triangle and the circumscribed quadrilateral. Wallis discussed the topic in long letters to Huygens and Brouncker, eventually declaring himself dissatisfied with central points in Gregory's arguments (Nos. 6, 8, 10, 14–17, 19, 21, 27).

Introduction

Debate on the laws of motion

Despite chronically poor finances, the Royal Society flourished as a centre of scientific debate. When, on 22 October 1668, members reconvened after a ten-week break, they decided, at the proposal of Robert Hooke (1635–1703), to take new steps to ascertain the true laws of motion.[1] Huygens and Christopher Wren (1632–1723) were to be asked to develop further their own ideas and to make submissions to the Society on the topic. A contribution from Wallis was also commissioned (No. 22), and in the January 1668/9 issue of *Philosophical Transactions* articles containing Wren's and Wallis's laws were published. Huygens' contribution was at first suppressed, his laws being regarded as largely identical with those of Wren, and the ensuing dispute was resolved only when Oldenburg agreed to publish the Dutch mathematician's article (which had since appeared in French in the *Journal des Sçavans*) in the April issue of *Philosophical Transactions*, together with an account of the reasons why it had not been published earlier.

The printing of Wallis's laws of motion led to a secondary debate between him and William Neile (1637–70), son of the courtier Paul Neile (1613–before 1686). The young mathematician's laws were presented to the public at the meeting of the Royal Society on 29 April 1669,[2] but the exchanges between him and Wallis were conducted almost privately and through the hands of Oldenburg. Neile's approach, which employed the notion of 'springiness' based on the internal motion of particles, found little favour with Wallis. Communication between the two men eventually came to an end with increasing levels of frustration on both sides in view of their failure to overcome fundamental differences of opinion (Nos. 77–9, 81–2, 84–5, 87, 89–90, 92–3).

Debate on the theory of tides

The debate on the laws of motion was soon followed by another protracted debate concerning the theory of tides. The starting point this time was the theory which Wallis had originally formulated in his letter to Robert Boyle (1627–91) of 25 April 1666 and which had been published in the issue of *Philosophical Transactions* for August of that year. In the theory, Wallis argued against Galileo Galilei's (1564–1642) hypothesis that the earth's dual motion—that is to say, its diurnal motion around its axis and its annual

---

[1] *The History of the Royal Society*, ed. T. Birch, II, 315.
[2] *The History of the Royal Society*, ed. T. Birch, II, 361.

Introduction

motion around the sun—are the cause of the acceleration and retardation of the movement of the oceans. By considering the earth and the moon to have a common centre of gravity, Wallis believed he could account more satisfactorily for the variations of the tides between spring and neap. His theory was considered but rejected by Henry (or Harry) Hyrne (c.1626–after 1672), private secretary to Edward Conway (c.1623–83), third viscount Conway and Killultagh. In a long letter to Wallis, Hyrne produced both empirical and theoretical arguments against the foundations of his hypothesis (No. 139). Encouraged by the Savilian professor, Hyrne subsequently drafted an hypothesis of his own employing a third motion of the earth in addition to its diurnal and annual motion (No. 157). However, it failed to impress Wallis who, in reply, not only provided substantial reasons for its untenability, but also claimed, rather damningly, that Hyrne's approach was not at all new (No. 158).

Further opposition to Wallis's hypothesis of tides came from Joshua Childrey (1625–70), who had already made a name for himself by publishing an account of natural rarities of England, Scotland, and Wales under the programmatic title *Britannia Baconica* (1662). In a letter to Wallis's former colleague, Seth Ward, Childrey suggested that Wallis had failed to provide a satisfactory account of the tides because he had ignored empirical data concerning the role of winds in bringing about high tides and inundations. In setting out his animadversions on Wallis's hypothesis, Childrey drew extensively on the historical evidence of chronicles and the reports of pilots and seamen (No. 142). Wallis gave a detailed response to Childrey. In it, he did not question the importance of the winds in relation to flooding, but he emphasized that his business was to provide a sound theoretical model based on regular motions, rather than to consider such 'Accidental Extravagances' as high winds and storms (No. 149).

Correspondence with Dulaurens, Huygens, and Hevelius

Intellectual commerce of the Royal Society with continental Europe, which had been built up principally by Oldenburg, and to a somewhat lesser extent by Robert Moray (1608/9–73), continued in all its facets during the period covered by the present volume. Not all this commerce was constructive. Wallis's dispute with François Dulaurens (d. c.1675) which had begun after he had written a highly critical review of the French mathematician's *Specimina mathematica* (1667) was brought to an end after Oldenburg decided against publishing Wallis's latest reply in the *Philosophical Transactions* (No. 28). Oldenburg had evidently recognized that nothing was to be gained by prolonging the public side of the quarrel in which solely the adversaries had been interested.

Most of Wallis's correspondence with Huygens was taken up with questions relating to Gregory, but the Dutch mathematician did send him his proof of the measurement of the area between the cissoid and the asymptote (No. 200). Huygens had announced his possession of this proof more than ten years earlier. After receiving it, Wallis included the proof as an appendix to the third part of his *Mechanica: sive, de motu, tractatus geometricus*, published in 1671.

Collaboration with the Danzig astronomer Johannes Hevelius (1611–87). which had begun through Wallis's offer of assistance in carrying out observations and calculations in April 1649, continued through the period covered by the present volume. Their correspondence was always one with intellectual commerce at heart—in the best spirit of the Republic of Letters. For his forthcoming major work *Machina coelestis*, Hevelius required data on the fixed stars collected in ancient star-charts. Having already sent Hevelius a transcript of Oxford manuscripts containing the observations of Ulug Beg[3], Wallis now produced an extensive catalogue of stars based on a Greek codex of the *Almagest* contained among the rich treasures of Oxford's Bodleian Library. Not only did the Savilian professor provide the latitude and longitude of the fixed stars according to Ptolemy's (85?–165?) catalogue, but also he collated Ptolemy's data against two Arabic manuscripts—an Arabic version of the *Almagest* itself, and Al Sufi's (903–86) book on the constellations of the fixed stars (Nos. 54 and 55).

Correspondence with Bertet and Sluse

The trading of ideas in the seventeenth century importantly also took the form of the exchange of books and other publications. The London intelligencer John Collins (1625–83) had a network of correspondents in continental Europe from which he would procure books in exchange for English publications. Notoriously, the wares produced by printers in London and elsewhere in the British Isles were often difficult to obtain abroad. Collins's correspondence with the diplomat James Vernon (1637–77) in Paris, and with John Pell (1611–85), Edward Bernard (1638–97), and Wallis in England, is replete with information on books and the book trade at the time (Nos. 6, 10, 15, 20, 70, 91, 94, 135, 176, 204, 206, 252).

---

[3]See WALLIS–HEVELIUS 5/[15].IV.1664; *The Correspondence of John Wallis (1616–1703)*, ed. P. Beeley and C. J. Scriba, II, 103–6.

Introduction

It was through Collins's efforts to obtain scientific books from France that Wallis established contact with the French Jesuit mathematician Jean Bertet (1622–92) (No. 149). Indeed, Wallis was evidently called upon to assure Bertet that he could expect 'candid and upright dealing' from Collins in committing himself to this intellectual trade. Not all of Bertet's letters to Wallis have survived, and those that have are purely mathematical in scope. On noting a fundamental difference between the approach used by Grégoire de Saint-Vincent (1584–1667) to quadratures and that employed in *Arithmetica infinitorum* (1656), Bertet wrote to Wallis in November 1671. His long letter, in which he sought to demonstrate the comparative limits to Wallis's method, provoked an equally prolix but respectful response, in which the Savilian professor sought to provide demonstrations to the contrary (Nos. 246 and 253).

Notable among the new correspondents in the present volume is René François de Sluse (1622–85). Already in 1667 Oldenburg had sought to initiate an exchange of letters with Sluse, having heard of the value placed by contemporary mathematicians on his *Mesolabum* (1659). News of the impending publication of the second edition of this work was met with excitement among the mathematical community in England (see Nos. 6, 10, and 20), and Sluse was persuaded to send fifty copies across the Channel for distribution. The Liège mathematician was certainly gratified by the adulatory review of the second edition of *Mesolabum* in the *Philosophical Transactions*, which may well have been produced by Collins. Wallis was evidently reluctant to write a review, having had bad experiences recently with Dulaurens and Gregory (No. 34). Correspondence between Wallis and Sluse might not have come about at all had Wallis not passed through London in the summer of 1669 and on this occasion been shown by Oldenburg the letter he had recently received from his correspondent in Liège. Unaware of Wallis's contribution to the topic, Sluse praised Christopher Wren's (1632–1723) method for generating a hyperbolic-cylindroidal solid, on which Wren had published an article in the June 1669 issue of *Philosophical Transactions*. This method was seen as being important, as it provided a suitable model for grinding hyperbolic lenses. In order to set matters straight, Wallis wrote to Sluse in September 1669, pointing out that he, and not Wren, had devised the fundamental principle involved. Wallis's letter was sent by Oldenburg as an enclosure to one of his own (Nos. 113 and 140). Intellectual commerce across less common routes such as that between London and Liège was often unreliable. After Oldenburg's letter failed to arrive, he wrote again, in January 1670, sending a slightly modified copy of Wallis's earlier communication. Subsequent letters exchanged between Wallis and Sluse were concerned with mathematical news and principally with Giambattista Riccioli's (1598–1671) argument against the Copernican system based on the accelerated motion of heavy bodies (Nos. 153 and 169).

The heritage of Pierre de Fermat

The anticipated arrival of new publications from abroad could also have effects quite different from those in the case of the second edition of Sluse's *Mesolabum*. Having been so ill-prepared for Fermat's challenges on number theory in 1656/7, Wallis was geared for action by reports of the printing of Jacques de Billy's (1602–79) new edition of Bachet's *Diophantus* with Fermat's commentaries. Above all, he wanted to ensure that his long Latin letter to Brouncker of August 1668 would be forwarded in suitable manner to the mathematical community in Paris before the book arrived. This would provide evidence of the success of his own work on Fermat's negative theorem, i.e. $x^2 - ny^2 = 1$, where n is a non-square integer. Wallis suggested sending a copy with Oldenburg's nephew, the Heidelberg law professor Heinrich von Coccejus (1644–1719), who would soon be returning to the continent. Not wishing to appear as the initiator of this move himself, he suggested to Oldenburg that he should present the letter as something he had had in his hands for some time, but which Wallis had not consented to have sent over to France until then (No. 207).

Correspondence with Borelli

The exchange of books also features centrally in Wallis's correspondence with the mathematician and natural philosopher Giovanni Alfonso Borelli (1608–79), who by this time was living in exile in Rome under the protection of Queen Christina of Sweden (1626–89). Borelli's writings were well known to Wallis, and he discussed them frequently in the letters he exchanged with Oldenburg. After a meeting with Vernon, probably in Rome, Borelli sent Wallis a copy of his *Theoricae mediceorum planetarum a causis physicis deductae* (1666) (No. 48). Somewhat later, during the second half of 1669, he shipped a package of his books to England for distribution, including a number of publications specifically intended for Boyle (No. 129). Collins and Wallis subsequently conferred as to which books should be sent in return (Nos. 133 and 136). Their package, which included a substantial number of works by Boyle together with other publications chosen by Wallis and Collins, was sent by ship to Palermo in February or March 1670. A letter from Wallis was also enclosed (Nos. 136 and 137). When controversy broke out between Borelli and Honoré Fabri (1607–88) over the nature of the path of a descending projectile, Wallis inclined to agree with Borelli, noting that he had earlier suggested that an experimental investigation be carried out on this topic by the Royal Society. Fabri's position was not helped by his reputation for caveling 'very often, without just cause', as Wallis reported to Oldenburg in November 1670 (No. 192). Indeed, Wallis would soon himself become embroiled in a controversy with Fabri over the question of the compounding of forces in dynamics.

Leibniz and the Royal Society

While Wallis was reluctant to review the second edition of Sluse's *Mesolabum*, he readily accepted the task of providing a qualified report for the Royal Society on Gottfried Wilhelm Leibniz's (1646–1716) *Hypothesis physica nova* (1671), when a copy of this new publication arrived in London in the spring of 1671. Leibniz had in fact been motivated to draw up his physical hypothesis after he had seen a copy of the April 1669 issue of *Philosophical Transactions* containing Huygens' laws of motion and Oldenburg's note on the background to the dispute over their publication. Wallis wrote a glowing review of the *Hypothesis physica nova*, noting many points of agreement between that work and propositions contained in his *Mechanica* (Nos. 209 and 210). Oldenburg, keen to promote the interests of his promising young fellow-countryman, published the review in full in *Philosophical Transactions* and quoted from it extensively in a letter he sent to Leibniz. A shorter review of Leibniz's *Theoria motus abstracti* (1671) received the same treatment (No. 216). Leibniz's work having received such approbation, the Royal Society readily agreed to have the two books reprinted in London by their printer, John Martyn (1617/18–80).

Publication of Wallis's *Mechanica*

The publication of *Mechanica* provides in many ways the backdrop to the present volume. Wallis first announced the work, and indeed presented part of it, when he attended the meeting of the Royal Society in London on 30 April 1668.[4] The members present on that occasion took an immediate interest, urging him to complete his book on mechanics as quickly as possible. In view of the nature of the book, the Royal Society's council decided that it should be printed under subscription by the London printer Moses Pitt

---

[4] *The History of the Royal Society*, ed. T. Birch, II, 275.

Introduction

(1639–97), one of only a few printers in the English capital prepared and willing to print mathematical books. Since Wallis's teaching duties demanded that he spend most of his time in Oxford, it was not possible for him to supervise the printing process, and this task was taken on instead by his friend Collins. Despite the good postal connections between Oxford and London, the geographical distance between author and printer inevitably led to delays in the correcting of proofs and in the carrying out of corrections (see No. 96). Pitt repeatedly complained to the Royal Society over the low price which had been set for the book, considering the amount of work involved in typesetting and the relatively small print-run (Nos. 176 and 178).

Oldenburg did his utmost to promote the *Mechanica*. Already in January 1668/9 he informed the French astronomer Adrien Auzout (1622–91) that Wallis's book was being printed.[5] In May he let Huygens know that printing was not taking place as quickly as had been hoped.[6] But by the autumn he was able to notify both Huygens[7] and Sluse[8] that the first part of the *Mechanica* was about to appear. Nonetheless, it was not until December 1669 that Oldenburg felt confident to send Wallis a draft announcement of its publication. After he had received Wallis's approval, Oldenburg published the announcement in the December issue of *Philosophical Transactions*, almost a year after his message to Auzout (No. 122). During the following two years Oldenburg kept his correspondents abreast of developments concerning the other two parts of the book. Ultimately, the greatest problem was not the distance between the author and the printer or the disagreement between Pitt and the Royal Society over the price which had been set, but Wallis's state of health. As a result of a long bout of quartan ague he was nearly incapacitated for almost twelve months between the summers of 1670 and 1671, and at times feared the worst. One senses his desperation when, in November 1670, he urges the printers to make haste and remarks that he would not willingly have part three of his *Mechanica* be a posthumous work (No. 188).

The cost of producing mathematical books, particularly those requiring plates, and the time involved in proof-reading and making corrections made

---

[5] OLDENBURG–AUZOUT 2/[12].I.1668/9; *The Correspondence of Henry Oldenburg*, ed. A. R. Hall and M. B. Hall, V, 298.

[6] OLDENBURG–HUYGENS 31.V/[10.VI].1669; *The Correspondence of Henry Oldenburg*, ed. A. R. Hall and M. B. Hall, V, 581.

[7] OLDENBURG–HUYGENS 7/[17].X.1669; *The Correspondence of Henry Oldenburg*, ed. A. R. Hall and M. B. Hall, VI, 269.

[8] OLDENBURG–SLUSE 10/[20].XI.1669; *The Correspondence of Henry Oldenburg*, ed. A. R. Hall and M. B. Hall, VI, 309.

Introduction

the alternative publication of journal articles increasingly attractive to scientific writers in the seventeenth century. Articles also offered the advantage of allowing a much quicker establishment of priority in discovery and a more rapid response to an ongoing debate. A close relation existed between the form of the letter and the form of the learned article, which permitted the former with relative ease to be transformed into the latter. Indeed, many articles which appeared in *Philosophical Transactions* and similar journals on the continent were originally letters to the editor. In the present volume, thirteen of Wallis's letters to Oldenburg were intended by Wallis for publication. Nine of these were actually published in Oldenburg's journal, including Wallis's reviews of Leibniz and an article providing an example of his continued interest in experimental philosophy: an investigation based on observations with the baroscope and thermoscope (No. 132). Perhaps not surprisingly in view of the importance attached by the Royal Society to rebuffing Hobbes, four of Wallis's published articles concerned the author of *De corpore* (1655) and his mathematical endeavours.

Scientific papers of Horrox and Merry

Wallis's continuing efforts to see England's recent scientific heritage preserved in a suitable manner are evident throughout the present volume. Although Jeremiah Horrox's (1617?–41) account of the transit of Venus had already been published by Hevelius, Wallis was keen to see further papers of his former Cambridge friend appear in print, having been commissioned to carry out this task by the Royal Society. A central problem in this regard was the whereabouts of the most substantial part of Horrox's papers (No. 73). It was rumoured that they had fallen into the hands of his brother, Jonas Horrox, who had fled to Ireland during the Civil Wars. In a letter to Robert Wood, a former associate of Hartlib, who lived near Dublin, Wallis asked the mathematician to make enquiries on his behalf (No. 147).

Wallis was similarly prepared to arrange for the printing of Thomas Merry's (d. 1682) 'Invention and Demonstration of Hudden's Rules for Reducing Equations', an exposition of Jan Hudde's (1628–1704) *De reductione aequationum* (1659). The possible printing of this together with work by John Kersey (1616–77) was seriously considered. With this aim in view, Merry handed the book manuscript to Collins, who in turn gave it to Wallis. However, the projected publication was not realized. Merry's manuscript was instead deposited in the library of the Savilian professors at Oxford (No. 206).

Introduction

The ceremonial opening of the Sheldonian Theatre

At Oxford, the four years in office of vice-chancellor John Fell (1625–86) came to an end with the spectacular opening of the Sheldonian Theatre, designed by Wren and made possible through the generous benefaction of Gilbert Sheldon (1598–1677). Celebrations stretching over several days in July 1669 were overshadowed not only by a scandalous performance of the University jester or Terrae filius, but also by a venomous speech given by the University orator, Robert South (1634–1716). There were verbal attacks on the new philosophy as well as on the Royal Society. In addition, respected individuals were not spared abuse, including Susanna Wallis (1622–87), the Savilian professor's wife. In long letters to Oldenburg and to Boyle, Wallis recounted the unfolding of events (Nos. 101 and 102; see also No. 100).

Foreign visitors to Oxford

In contrast, the ceremonial visit to Oxford by the crown prince of Tuscany two months earlier had been an unqualified success. From the report provided by Thomas Lamplugh (1615–91), we learn that Cosimo de' Medici during his time in Oxford was entertained by a lecture on architecture by Wallis, based on a ceiling construction employing short timbers which he had designed while still a fellow at Queens' College, Cambridge in 1644 (No. 75).

Wallis was rather less prepared for another royal visitor from abroad in the following year. After receiving a request from Oldenburg that he should look after a distinguished but unnamed guest arriving shortly from London, Wallis welcomed a young nobleman bearing the title Count of Schaumburg in Oxford on 22 July 1670. Wallis gave the visitor the statutory tour of the University, introducing him to its officers and heads of houses. Accidentally hearing the count being addressed in much higher terms by one of his entourage, Wallis was filled with embarrassment—which he later communicated in a rather reproachful letter to Oldenburg (No. 171). The nobleman travelling incognito was in fact Karl II (1651–85), the eldest son of the Elector Palatine, Karl I Ludwig (1617–80). He did not make an official visit to Oxford until September 1680, on which occasion he was created Doctor of Physic.[9]

Five months later, Willem (1650–1702), prince of Orange and future king of England, visited Oxford. As part of the planned events during the visit, Wallis prepared a short lecture demonstrating the property of infinite approximation (No. 198). For reasons unknown, the lecture was not given;

---

[9] See *The Life and Times of Anthony Wood*, ed. A. Clark, II, 495.

Introduction

but the very fact that it was prepared tells us much more. Barely fifty years after the founding of the first mathematics lecturership at Oxford, the discipline had become so established that it was deemed suitable for presentation to visiting royalty. The mathematical sciences were now as truly at home at the University of Oxford as they were at the universities of Pisa or Leiden. Undoubtedly, Wallis had played an important part in bringing about this change.

Cryptanalysis and the fate of Clarendon

Politics would be conspicuously absent from the present volume if not for one letter of remarkable significance. The dismissal, impeachment, and subsequent exile of Edward Hyde (1609–74), first earl of Clarendon, continued to have repercussions up to the end of 1669. When the House of Commons met in October 1669, different factions struggled to fill the vacuum caused by his demise.[10] Joseph Williamson (1633–1701), who coordinated the restoration government's intelligence-gathering activities, was evidently concerned to keep Clarendon under surveillance even in his Montpellier exile. Suspicion was no doubt increased because the former lord chancellor used cipher in communications with his friends in England. Evidently, a considerable number of intercepted letters from Clarendon were conveyed to Wallis for deciphering through the hands of John Ellis (c.1646–1738) (No. 114). We do not know what Wallis felt about this task, particularly considering that Clarendon had been chancellor of the University of Oxford, albeit only for a brief time. The Savilian professor would have been aware that he did not have much choice in the matter. He had received his professorship largely on account of his skills as a decipherer, and it was precisely on this account that he was useful to those in power. Throughout his life he would be caught between state and academic duty, and would have cause to reflect on the intimate connections between these two sides of his professional life.

---

[10]See J. Spurr, *England in the 1670s. 'This Masquerading Age'*, Oxford and Malden, Mass., 2000, 10.

EDITORIAL PRINCIPLES AND ABBREVIATIONS

All letters in the volume are preceded by an account (*Transmission*) of the various manuscript and printed forms in which they have been handed down. In the case of those letters whose text has not survived, the reasons for assuming that they did exist at some time are given.

The *Transmission* section also places each letter in context, records, when known, how it was conveyed to the addressee, and supplies additional information such as postmarks, details of notes appended to manuscripts, enclosures, and so on.

Manuscript and printed sources are denoted according to the following scheme:

W original manuscript in Wallis's hand
w copy of Wallis manuscript in scribal (or identified) hand
C original manuscript in correspondent's hand
c copy of correspondent's manuscript in scribal (or identified) hand
E contemporary edition

Where there is more than one source in a particular category, these are numbered successively W^1, W^2, ..., w^1, w^2, ..., and so on.

All letters contained in the volume are dated according to both the old style or Julian calendar employed in England until 1752 and the new style or Gregorian calendar widely used on the Continent, with the form not given in a particular letter placed in square brackets. In the period covered by the present volume the difference between the two calendars was ten days. To accommodate the English year, which began on Annunciation or Lady Day (25 March) and which permitted a new style date such as 16 February 1663 to be expressed in a number of ways old style—6 February 1662, 6 February 1662/3 or even 6 February 1663—the most common form (1662/3) has been used in brackets where a correspondent on the Continent employing new style has not supplied the date old style himself. (For reasons of legibility, only the Gregorian calendar has been used in creating the Index of Letters.)

Editorial principles and abbreviations

Where the place at which a letter was written can only be surmised, this also is set in square brackets.

The spelling, capitalization, and punctuation of manuscript and printed sources has been retained throughout. Contractions have been silently expanded, except where they are still in common use today, and thorn has been altered to 'th'. The use of i/j and u/v in Latin has been modernized. All symbols, the ampersand, and use of superscripts to denote pounds, shillings, and pence have likewise been retained.

All underlining in manuscripts is reproduced as italics. The reproduction of italics from printed sources has been treated diplomatically. In mathematical passages, letters used to indicate points or places in figures, and likewise all algebraic formulae, have been italicized where the writer or printer has not already done this.

Editorial signs

⟨text⟩	uncertain reading
⟨— —⟩	illegible words (the number of dashes indicates the number of illegible words to a maximum of three)
...	words omitted
[paper torn]	Editor's remarks (N.B. upright square brackets contained in text or in variant readings of the critical apparatus are always either employed by the author himself or represent a contemporary addition, as indicated)
‖	new paragraph within a variant reading
add.	added
alt.	contemporary alteration to text by someone other than the author
corr.	corrected
|text *del.*|	word or words deleted
ed.	editor
ins.	inserted
suppl.	supplied

The critical apparatus shows the development of text through its various stages. Each successive stage replaces the preceding one. Thus stage (1) is superseded by stage (2) and this in turn by stage (3). Further subdivisions are indicated by letters: (a) is replaced by (b) and then by (c), (aa) is replaced by (bb), (aaa) by (bbb), and so on.

Editorial principles and abbreviations

As in the case of the critical apparatus, but placed above this, marginal annotations to texts are referenced by means of line numbers. Editorial comments (footnotes) are indicated by numerical superscripts.

Astronomical and mathematical symbols

♈	Aries, vernal equinox	$a.b :: c.d$	geometrical proportion
∞	equality	£	pounds
s	shillings	d	pence
~	similar	♓	Pisces
$a)b(c$	division	▭	rectangle
♊	Gemini	♄	Saturn, Saturday
♂	Mars, Tuesday	□	square
☿	Mercury	☉	Sun
☽	Moon	△	triangle
⌒	segment	♋	Cancer
♍	Virgo	♀	Venus
♃	Jupiter	♉	Taurus
♌	Leo	♎	Libra
♏	Scorpio	♐	Sagittarius
♑	Capricorn	♒	Aquarius

CORRESPONDENCE

1.
WALLIS to HENRY OLDENBURG
Oxford, 6/[16] October 1668

Transmission:

W Letter sent: LONDON *Royal Society* Early Letters W1, No. 64, 4 pp. (p. 3 blank) (our source). On p. 4 beneath address in Oldenburg's hand: 'Rec. oct. 7. 68. concerning Lalovera de Cycloide. Answ.' Postmark on p. 4: 'OC/7'.—printed: OLDENBURG, *Correspondence* V, 80–1.

Reply to: OLDENBURG–WALLIS IX/early X.1668.

Oxford. Octob. 6. 1668.

Sir,

Your severall letters,[11] recommending persons hither, I have received, & endeavoured to observe as there was occasion. The receit of Lalovera's book[12] (for which I thank you) I did not presently give you the trouble of a particular letter to yourself to signify, because I presumed you might understand it from what I presently wrote[13] to my Lord Brouncker & Mr Collins concerning it. To what you ask concerning it, I have this to say. His demonstrations, though somewhat perplex & obscure, (at lest they seem so to mee, who have not read a former book[14] of his to which they refer,) yet I take, for the main, to be sound; and his Methods likewise; though in

---

8 To *add*.
11 his (*1*) likewise; (*2*) Methods

---

[11] letters: i.e. OLDENBURG–WALLIS IX/early X.1668 as one of several now missing letters from Oldenburg.

[12] Lalovera's book: i.e. LALOUBÈRE, *Veterum geometria promota, in septem de cycloide libris*, Toulouse 1660.

[13] what I presently wrote: see WALLIS–COLLINS 8/[18].IX.1668 and WALLIS–COLLINS 10/[20].IX.1668. No letter from Wallis to Brouncker concerning Laloubère's book has survived.

[14] former book: i.e. LALOUBÈRE, *Quadratura circuli et hyperbolae segmentorum*, Toulouse 1651.

1. WALLIS to OLDENBURG, 6/[16] October 1668

the applications of them there be (certainly) some mistakes. I think I shall hardly take the pains strictly to examine all his demonstrations, & particular methods; because I had long before performed the same things by methods of my own (published in my book de Cycloide,[15] 1659, the year before his, published in 1660 as it seems, though I never saw it till now, nor heard of it till this book came over; wondering at a passage in Borellus,[16] which I saw but a little before, seeming to cite some such book:) which seem, at lest to mee, much more clear than those of Lalovera. Yet I take his methods (as I sayd) to bee sound; because his calculations, so far as I have compared them with mine, do for the most part agree: And where they differ, though the mistake be certainly on his side, yet I suppose rather to arise from some mis-calculation, or mistake in the application of his methods, then from any fundamental errour in the methods themselves.

I think I shall not much further examine him then I have done, for the reasons mentioned; & shall| therefore, if you please, send you the book [2] again. But I had rather give you the price of it, if you can spare it, to have it ly by mee. But shall there in be ruled by you.

The Transactions of August, & September, I have not yet read (though, I beleeve, I am concerned in them,) having sought them at our Book-sellers several times in vain; (& therefore can say nothing of them;) but I am now told I shal quickly have them. I am

Sir,
Your friend to serve you
John Wallis.

These [4]
For Mr Henry Oldenburg,
in the Old Palmal, near
St. James's
London.

---

2 & |the *del.*| particular
4 Cycloide, *(1)* a ye *breaks off* *(2)* 1659, *(a)* that *(b)* the
8 his *(1)* as I s *breaks off* *(2)* methods
12 in *(1)* his *(2)* the
19 sought |them *add.*| at

---

[15] my book de Cycloide: i.e. WALLIS, *Tractatus duo*, Oxford 1659.
[16] Borellus: possibly BORELLI, *Euclides restitutus, sive, Prisca geometriae elementa*, Pisa 1658.

2.
Samuel Gott to Wallis
Bristol, 19/[29] October 1668

Transmission:

C Letter sent (fragment): OXFORD *Bodleian Library* MS Add. D. 105, f. 34ʳ–35ᵛ (f. 35ᵛ blank). On f. 34ᵛ Wallis has written a conjecture on Gott's proposal for an artificial spring.

[5ʳ] *[paper torn]* with their own 2 Natural Motions Circular about their Axis & Progressive in their Orbite & then I only desire others according to the other Hypothesis to assigne any such probably continual External Mover of Earth whose own Magnetical Virtue is only to its Poles or Centers of Rest whereas the Planetary Virtue of the Aether & Starrs is Circular & produceth thereby a perpetual Motion which as I have said is Natural to their Forms or Spirits in which that Virtue doth forbere though Violent to their Bodys of Matter which other wise would rest in their own Spherical Station & Position as the diurnal Motion of the Starrs by the Aether is also Violent to their particular Planetary Forms or Spirits though Natural to its own Form or Spirit. So that the Earth which is only Magnetical & Polar hath neither in itself any Principle of Motion (but only to Rest & Fixation in its own Polar Position as any Terrella or Needle dislocated or moved out of its Polar Position) like the Starrs nor any such continual external mover like the Aether & therefore is not moved by any other nor doth move in itself but rest Fixed & Immovable in its own Polar Position. I have repeated these 3 Phaenomena becaus they seem to me to be very material & unles I may be otherwise satisfied I must offer[17] them to the World. I have another Opusculum[18] wherin I much needed Dr Bathersts[19] Supervision. It is a Rhapsody of some Juvenile Poems & so forward which I gave my Son who is at Tunbridge Schole & his Master will now print them. I know him to be a good Critike & hope he will be carefull of them. I pray let me hear whither you shall be

2 desire (*1*) any (*2*) others
13 out of its (*1*) or any ⟨—⟩ (*2*) Polar Position) like

---

[17] offer: i.e. [GOTT], *The Divine History of Genesis of the World explicated and illustrated*, London 1670. This work was published anonymously and was reviewed in *Philosophical Transactions* No. 60 (20 June 1670), 1083–4.

[18] Opusculum: not identified.

[19] Bathersts: i.e. Ralph Bathurst (1619/20–1704), founder member of the Royal Society, fellow and later president of Trinity College, Oxford, *ODNB*.

3. WALLIS: A Conjecture on Mr Gott's Proposal of an Artificiall Spring

at London this next Term & when I hope God willing to meet you [paper torn] & I always wish I were as near to you in place as in affection

Your assured Friend
Samuel Gott.

5   Bristol October 19 1668.

For my honored friend [34
Dr John Wallis at his
house in
Oxford

3.

WALLIS

A Conjecture on Mr Gott's Proposal
of an Artificiall Spring

Transmission:

W Note: OXFORD *Bodleian Library* MS Add. D. 105, f. 34^v (written on reverse of GOTT–WALLIS 19/[29].X.1668).—printed: SCRIBA, *John Wallis: A Conjecture on Mr Gott's Proposal of an Artificiall Spring (1668)*.

Wallis evidently wrote this conjecture soon after receiving Gott's letter of 19 October 1668. It responds to a proposal which was probably contained in the now missing part of that letter.

10   A Conjecture on Mr Gott's proposall, of an
Artificiall Spring, which should Ebbe & Flow
at certain vicissitudes; suppose, every
twelve hours.

1 next *add*.
2 were (*1*) nearer to you (*2*) as near to you in place as in affection
5 Bristol ... 1668. *at 90° in left margin*

3. WALLIS: A Conjecture on Mr Gott's Proposal of an Artificiall Spring

Your artificial Spring (or Clepsydra) which should Ebbe & flow every twelve hours, or the like vicissitudes: I take to be of some such form as this. Suppose the Vessel ABC (unseen) into which by the Pipe DE a supply of water is continually filling it: which yet that it may not exceed a certain hight, as that of AFC, provision is made by the wast pipe FB, which shall be large inough to carry away, as wast, what ever surmounts that hight: And yet DE so large as that it be never lower. Out of this Vessel, at G, suppose a conceled pipe GHI; which (because of the same constant depth of water GC) shall at the same constant rate convey water into the Vessel MNO; filling it in 1, 2, 3, 4, &c hours, to the hight I_1, I_2, I_3, I_4, &c. and, consequently, in 12 hours to the height of HL. Then, by another concealed Pipe $KLOP$, so soon as the water comes to the hight of HL, this vessel will begin to empty it self (but not before,) as by a Siphon. Which Pipe if wee suppose it so large as to empty twice as much as P, as is brought in at H; the vessel will in 12 hours more, be emptied from H, to 11, 10, &c, till it come as low as IK: Then, the Air entering at K, will stop the course of water at P; till in 12 hours more it come to be filled, as before, to HL; & then to empty again. And thus continually, in a constant sucession.

This your project, may, no doubt, be so ordered as, for the general to succeed: But, though (because of the equal depth GC) it will fill equally, or at the same rate, in equal times: Yet it will not empty at the same rate; but faster when its depth is IH, or I_{10}, than when it is but I_6, or I_4, And the

10 rate (*1*) carry (*2*) convey
13 vessel *add*.
15 as to (*1*) carry off twice (*2*) carry away twice (*3*) empty
17 IK: (*1*) Then (*2*) Then

5. WALLIS to GREGORY, 22 October/[1 November] 1668

Pipe KLO is not necessaryly to be the just double of GHI; but more or less according to the different depths GC, and IH. For the deeper the water is above G, or K; the greater will be the quantity of water issuing out at G or P, supposing the pipes unchanged.

4.
WALLIS to WILLIAM BROUNCKER
22 October/[1 November] 1668

Transmission:

Manuscript missing.

Existence and date: Mentioned in WALLIS–COLLINS 3/[13].XI.1668.
Enclosure: WALLIS–GREGORY 22.X/[1.XI].1668.

5.
WALLIS to JAMES GREGORY
22 October/[1 November] 1668

Transmission:

Manuscript missing.

Existence and date: Mentioned in GREGORY–WALLIS 26.X/[5.XI].1668 and WALLIS–COLLINS 3/[13].XI.1668.
Answered by: GREGORY–WALLIS 26.X/[5.XI].1668.
Enclosure to: WALLIS–BROUNCKER 22.X/[1.XI].1668.

Wallis apparently expressed dissatisfaction with proposition 11 of Gregory's *Vera circuli et hyperbolae quadratura*, where it is claimed that the area of a sector of a central conic cannot be expressed analytically in terms of the areas of an inscribed triangle and a circumscribed trapezium. As he points out to Collins in WALLIS–COLLINS 3/[13].XI.1668, he had not at this time seen Gregory's preface to *Exercitationes geometricae*, in which the Scottish mathematician vehemently rejects Wallis's criticism.

3 issuing *(1)* at *(2)* out at
4 supposing the |the *del. ed.*| pipes

6.
JOHN COLLINS to JOHN PELL
London, 23 October/[2 November] 1668

Transmission:

C Letter sent: LONDON *British Library* Add. MS 4278, f. 344ʳ–344ᵛ. Postmark: 'OC/22'.
Answered by: PELL–COLLINS 28.X/[7.XI].1668.

In view of the disagreement between the postmark and the date given, it is possible that Collins was mistaken in dating this letter.

Reverend Sir

I spake to Mr Pitts[20] to send you Mr Gregories[21] Exercitationes Geometricae which he saith he accordingly did. Therein you will find the Quadrature of the Hyperbola further advanced, but possibly his Approaches about the Circle will not be understood till you have his booke[22] to which that[23] is an Appendix, the grand Designe of which booke is to make good this assertion.

> That it is impossible by any Analyticall Operations whatsoever to obtaine, or by any Aequation whatsoever to expresse, the true exact quantitie of any portion of a Circle Ellipsis Hyperbola into equall portions, or Logarithme Curve.

In the first part of his praeface &c you see he intends Hugenius, in the latter part Dr Wallis who[24] never opposed the assertion as untrue, but not as compleatly demonstrated. Now the Hyperbola is so farre advanced that (with Dr Wallis) it may be thought little more can be added, most mens thoughts are busy whether the Area of a Circle and its portions may not be computed without tables by some like Series. Mr Gregorie hath done

---

3 did *add.*
16 without tables *add.*

---

[20]Pitts: i.e. Moses Pitt (1639–97), London printer and bookseller, *ODNB*.
[21]Gregories: i.e. GREGORY, *Exercitationes geometricae*, London 1668.
[22]booke: i.e. GREGORY, *Vera circuli et hyperbolae quadratura*, Padua 1667. The following year it was reprinted and bound together with Gregory's *Geometriae pars universalis*, Padua 1668.
[23]that: i.e. the first section of *Exercitationes geometricae*: 'Appendicula ad veram Circuli & Hyperbolae Quadraturam'.
[24]who: i.e. in WALLIS–COLLINS 8/[18].IX.1668 and WALLIS–COLLINS 26.IX/[6.X].1668.

6. COLLINS to PELL, 23 October/[2 November] 1668

what he can towards it, doth suspect[25] that some such Series for the Circle may likewise lurke, is unwilling it should remaine as it were opprobium Geometrarum and is desirous any Man would fairly enter into the Debate or Consideration of the Assertion above. He intends another Small treatise[26] wherein to exhibit easy methods for dividing an Elliptick line to divide a Semicircle in a given Ratio by a line issuing through any Point in the Diameter. The Logarithme of any number being propounded to give the Number without tables, two of these Propositions he hath already done in his booke[27] de quadratura Circuli &, the latter whereof is in effect in any ranke of Continuall Proportionalls to require any one in the ranke, which he asserts he can now much more easily doe by Approximations relating to his Converging Series, to any exactnesse required. And the worst trouble that will be enjoined, is the Solution of a Quadratick aequation, and by the same methods will render all the rootes of any Aequation.

I sent for some of the 2^d Edition[28] of Slusius, and from Amsterdam they write it is shipd off. When arrived God Willing I intend to send you one. Here is a new Councill of Trade[29] erected and yesterday I spake to the L. Brereton about perusing all such bookes of trade as his Lordship hath, and his Lordship gave me Directions to write to you to send up all such as you can find together with the Seiur d'Taneur[30] printed at Paris 1650 in $4°$ on the 10 of Euclid et des quantitez irrationelles, whereabout I must earnestly desire your answer. Other bookes of Trade his Lordship saith there are, which you cannot come at, but that his Lordship intends there comming up amongst other Goods ere long. His Lordship intimates as if you were willing to fit up somewhat more for the Presse and your owne Letters[31] seeme to hint as much. I should be glad to be further informed about it. Dr Wallis thinking we| goe on but slowly here, with his Tracts[32] de Motu, [34

---

11 now much more easily *add.*
20 printed at Paris 1650 in $4°$ *add.*

---

[25]suspect: i.e. presumably in a now missing letter from Gregory to Collins.
[26]Small treatise: not identified.
[27]booke: i.e. GREGORY, *Vera circuli et hyperbolae quadratura*, Padua 1667.
[28]2^d Edition: i.e. SLUSE, *Mesolabum*, 2nd ed., Liège 1668. Cf. SLUSE–OLDENBURG [24.IX]/4.X.1668; OLDENBURG, *Correspondence* V, 65–6.
[29]Council of Trade: a commission of 20 October 1668 (old style) established a Council of Trade 'for Keeping a control and super-inspection of his Majesty's Trade and Commerce'. See ANDREWS, *British Committees, Commissions, and Councils of Trade and Plantations, 1622–1675*, Baltimore 1908, 92–4.
[30]Taneur: i.e. Jacques-Alexandre le Tenneur. Cf. COLLINS–PELL 18/[28].VII.1668.
[31]Letters: see for example PELL–COLLINS 6/[16].IX.1668.
[32]Tracts: i.e. Wallis's *Mechanica: sive, de motu, tractatus geometricus*, the first part

6. COLLINS to PELL, 23 October/[2 November] 1668

Statica, Mechanica, et Calculo Centri gravitatis, desires that Paper be sent to Oxford, to print off there his Tract de Sectionibus,[33] Angularibus, Algebra,[34] and some Miscellanies.[35]

Mr Barrow hath a Tract[36] of Opticks ready for the Presse which he lately brought up to London with him, wherein are many Solid Problemes solved by new Curves, that are naturally fitt for the buisinesse but not so Geometricall as the Conicks whereby he likewise Solveth those Problemes. Dr Rawlinson[37] as likewise the Earle of Northumberland[38] are lately dead, as also Vincent Wing[39] who hath a folio Booke[40] of Astronomy at the Presse, where the Coppy notwithstanding is left compleate. Wishing your happinesse and arrivall here I remaine

October 23th 1668

Your affectionate Servitor
to command
John Collins

I presume you received Andersons booke[41] with the Transactions though I have not heard from you since I sent it

---

of which was eventually published in 1670. Cf. COLLINS–PELL 21.IV/[1.V].1668; *British Library* MS. Add. 4278, f. 334^{r}: 'Dr Wallis hath brought up as much Manuscript, de Motu, de Statica, Mechanica, et Calculo Centri Gravitatis as will make two quarto Bookes as big as his Workes already extant, and Mr Pitts doth undertake them'.

[33]Sectionibus: i.e. WALLIS, *De sectionibus angularibus tractatus*, first published in English, London 1684. The Latin edition appeared in WALLIS, *Opera mathematica*, vol. II, Oxford 1693.

[34]Algebra: i.e. WALLIS, *Treatise of Algebra*, London 1685. The Latin edition appeared in WALLIS, *Opera mathematica*, vol. II, Oxford 1693.

[35]Miscellanies: presumably WALLIS, *Cono-Cuneus: or, the Shipwright's Circular Wedge*, London 1684; *A Defense of the Treatise of the Angle of Contact*, London 1684, and *A Discourse of Combinations, Alternations, and Aliquot Parts*, London 1685. The Latin editions of all three tracts appeared in WALLIS, *Opera mathematica*, vol. II, Oxford 1693.

[36]Tract: i.e. BARROW, *Lectiones opticae*, London 1669. Cf. BARROW–COLLINS 13/[23]. III.1668/9, RIGAUD, *Correspondence of Scientific Men* II, 68–70.

[37]Rawlinson: i.e. Richard Rawlinson (d. 1668), mathematician and Fellow of the Queen's College, Oxford. Cf. COLLINS–VERNON ?.I.1671/2; RIGAUD, *Correspondence of Scientific Men* I, 151–6; WALLIS–OLDENBURG 30.III/[9.IV].1668.

[38]Northumberland: i.e. Algernon Percy, tenth earl of Northumberland (1602–68), *ODNB*.

[39]Wing: i.e. Vincent Wing (1619–68), astronomer and land surveyor, *ODNB*. See TAYLOR, *The Mathematical Practitioners*, 222–3.

[40]Booke: i.e. WING, *Astronomia Britannica*, London 1669.

[41]booke: i.e. ANDERSON, *Stereometrical Propositions variously applicable*, London 1668.

8. GREGORY to WALLIS, 26 October/[5 November] 1668

To the Reverend Doctor
John Pell at the house
of the right honourable the
Lord Brereton at Brereton
in Cheshire
Stonebagg[42]

7.
JOHN COLLINS to WALLIS
26 October/[5 November] 1668

Transmission:

Manuscript missing.

Existence and date: Referred to in and answered by WALLIS–COLLINS 3/[13].XI.1668.

As emerges from Wallis's reply, Collins sent this letter together with a copy of Gregory's *Exercitationes geometricae*, which had recently come off the press.

8.
JAMES GREGORY to WALLIS
London, 26 October/[5 November] 1668

Transmission:

c^1 Copy of missing letter sent (with corrections in Oldenburg's hand): LONDON *Royal Society* Early Letters G1, No. 21, 4 pp. (pp. 3 and 4 originally blank) (our source). At top of p. 4 in unknown hand: 'A Letter from Mr James Gregorie to Dr Wallis, containing answers to several Objections against his mathematical Writings. Entered LB. Suppl.', and on lower right at 90° in Oldenburg's hand: 'Mr Gregory's letter to Dr Wallis Oct. 26. 1668'.—printed: TURNBULL, *James Gregory*, 51–3.
c^2 Copy of c^1: LONDON *Royal Society* Letter Book Supplement Original 3, pp. 455–8.
c^3 Copy of c^2: LONDON *Royal Society* Letter Book Supplement Copy 3, pp. 382–5.

Reply to: WALLIS–GREGORY 22.X/[1.XI].1668.

---

[42]Stonebagg: i.e. the postbag for Stone, Staffordshire, from where this letter would be conveyed to Brereton. Stone was an important staging point on the arterial postal route to Ireland.

8. GREGORY to WALLIS, 26 October/[5 November] 1668

Reverend Sir

Those objections[43] yee make against my writings I wish yee had formerly mentioned them, that so I might have had occasion to answer them in print. As for that yee say I demonstrat the 11 prop:[44] only in a sector indefinitly considerd: I hope that is sufficient, seing all problems before the resolution are indefinite, otherwayes the analyticks serve to no purpose; That instance yee bring in concerning the trisection of an angle, sayes nothing, because in this particular analysis of the affected cubick equation (viz when a semicircle is trisected) the root is composed of the determinating quantityes by the first analytick operations; but in this my sixt operation all analysis is naturally infinite.

I answerd this same objection (albeit in different terms) in the preface[45] to my *Geom: pars universalis*; only to hint, these cannot produce quantityes non analytical according to my acception, because they can bee brought to an analyticall equation; for I call the resolution of affected equations, also extraction of roots.

As to that yee say, I cannot apply my demonstration to a semicircle or greater Sectors: I hope I neede not tell you, neither unto any Geometer, that the demonstration is the same thing, supposing in place of the triangle & trapezium, yee put any complicat polygons, & so the demonstration is the same in all sectors. I doe presume, that yee doe exceedingly mistake mee when I say that the triangle is analyticall yea commensurable with its sector, for (allthough this be true) yet the sector cannot be exhibited, because (as it is demonstrated in the 11 proposition) the Sector cannot be analytical with both the triangle & trapezium, & therefore being analytical with the triangle, the Triangle is not analyticall with the Trapezium; and therefore the Triangle & the Diameter of the circle are not analyticall among themselves, (because if they be, then the trapezium shall be analytical with them both, as it is easily demonstrated) and therefore having the diameter, this triangle cannot be found analytically, & therefore all your consequences evanish: All this (in my opinion) is sufficiently intimated in those words yee [2] cite, if so be ye had continued 2 lines more. I think the obscurity of my two petitions, as also of the Scholium of the 11 may easilie be passed by, seing

---

26–27 the Triangle is ... therefore the Triangle *add.* Oldenburg
31 evanish: and this *corr.* Oldenburg

---

[43] objections: i.e. in WALLIS–GREGORY 22.X/[1.XI].1668.
[44] 11 prop: i.e. GREGORY, *Vera circuli et hyperbolae quadratura*, prop. 11.
[45] preface: i.e. GREGORY, *Geometriae pars universalis*, Padua 1668, sig. †3^v–†6^r.

8. GREGORY to WALLIS, 26 October/[5 November] 1668

(as I supposed) I shewd unto you in my last postscript,[46] that my intent can be deduced from the ii onely by an axiom tacitly supposed in al Geometrical problems. I doe easily perceive, that yee propose these objections only to be doubted of by the more ignorant, for yee cannot but perceive them to be easilie answered without my reply; it is enough to me to have surpassed the greatest difficultyes viz the 11 prop: & the cons: 10. As for the rest I am ready at all times to give satisfaction to any Geometer, who doubts of them. It is hard to settle the principia of Geometrie (for in this particular there are some wanting) as it is evident from 6 def: 5 Euclid[47] & one[48] of the postulata of the first: neither is it to be done by any particular person, but by several: it is enough to mee to have made the truth evident to intelligent persons, as I have done to severall, before my return[49] to England. If your method of demonstrating the same thing in your Arith: infinit:[50] be any wayes cogent, it is either very obscurely set down, or otherwayes yee are exceedingly wronged by all, whoever I have heard speak of it: however if yee had thought so your selfe, being in a matter altogether Geometricall, I think yee might have affirmed it more confidently; for very modest persons have boldly affirmed that in Geometrie, which afterwards they have recanted. I was allwayes confident enough, that my method was sure, but I kept it up, not knowing how to evite the opprobrie of the ignorant, which neverthelesse of my circumspection I must subject my self to.

No more but expecting that yee will excuse my boldnesse in troubling you. I rest

Sir

your humble servant
J. Gregorie

London
octob: 26 1668

---

21 my self for no more *corr.* Oldenburg

[46] postscript: i.e. to a now missing letter to Wallis, GREGORY–WALLIS IX/X.1668, to which WALLIS–GREGORY 22.X/[1.XI].1668 was the reply.

[47] Euclid: i.e. EUCLID, *Elements* VI, def. 5. Cf. WALLIS, *De Postulato quinto; et definitione quinta lib. 6. Euclidis; disceptatio geometrica*; *Opera mathematica* II, 665–78.

[48] one: probably EUCLID, *Elements* I, post. 5.

[49] return: Gregory returned to England from Italy via Paris around Easter 1668.

[50] Arith. Infinit.: i.e. WALLIS, *Arithmetica infinitorum*, Oxford 1656.

9.
WALLIS to JOHANNES HEVELIUS
Oxford, 26 October/[5 November] 1668

Transmission:

W Letter sent: UPPSALA *Universitetsbibliotek* Waller Ms gb-01783, f. 1^r–2^v (f. 2^r blank) (our source).
w Copy of letter sent: PARIS *Bibliothèque Nationale* Fonds latin 10348, IX, pp. 155–6.
E First edition of part of letter sent: OLHOFF, *Excerpta ex literis*, 115–16 (incorrectly dated 26 October 1666).

Wallis acknowledges receipt of his copy of Hevelius's recently-published *Cometographia*. A total of five copies of *Cometographia* had been sent from Danzig via Amsterdam to Oldenburg in London, the other copies being intended for Ward, Hooke, the Royal Society, and Oldenburg himself.

Oxonii, Octob. 26. 1668 st.vet.

Clarissime et eruditissime Vir,

Cum multis tibi obstrictus sim nominibus; omnino parum est quod valeam retribuere, nudas gratias. Neque hoc tantum pro Exemplaribus quae mihi variis vicibus pro tua humanitate dono misisti, nil tale merito: sed maxime ob Libros ipsos editos, quibus eruditorum gentem ditasti. Non possum utique ego non mirari (nedum laudare) indefessam illam tuam in Caelestibus Phaenomenis observandis industriam, et incredibilem laborem; quasi quidem his unicis nocte dieque intentus esses: Sed interim tantam Calculi diligentiam, qua molestias tantum non insuperabiles vincis; quanta totum hominem (nequid ultra dicam) mereri videatur. Verum et, post devoratos hos labores; scribendi opus, quo omnia in ordinem redigantur in publicum prodita, tertio saltem homini facessat negotium. At interim tu quarti nedum et quinti munus obis, dum Caelatorem agis, totamque Editionem, tuis Typis, tuis Sumptibus, tuis curis, ornas. Quantum itaque Te Virum praedicare debeam, qui unus isthaec ita praestas omnia, ut, si haec singula totidem praestarent, merito laudandi essent.

Evolvi[51] nuper (post priora scripta) Cometographiam tuam; nondum ea quidem quam meretur diligentia, sed summa carpens, utcunque evolvi. Et, quamquam ego is non sum qui de tuis judicium ferre valeam, (ut qui Caelestibus observandis ita parum sum versatus;) non possum tamen quin

---

[51]Evolvi: Wallis probably received his exemplar of Hevelius's *Cometographia* in mid-October. Cf. OLDENBURG–HEVELIUS 28.X/[7.XI].1668; OLDENBURG, *Correspondence* V, 112–14; BIRCH, *History of the Royal Society* II, 313.

9. WALLIS to HEVELIUS, 26 October/[5 November] 1668

accuratam illam in Observando diligentiam, et calculi labores maximos, agnoscere et praedicare. Ipsamque quam tractas Hypothesin, quamquam non ii simus qui de rebus tam procul dissitis quicquam certi statuere valeamus, omnino dignam esse censeo quae perpendatur, nec sua probabilitate carere. Quod quidem prope totum id est quod in obscuro hoc de Cometis negotio hactenus sperare datum est.

Pag. 377. ubi Stellae Novae in Ceto observatores memoras, non incommodum fuisset ea inseruisse quae ego Tibi antehac descripsi,[52] ex D. Johannis Palmeri[53] Nostratis, (nunc Archidiaconi Northamptoniensis,) Planisphaerio Catholico, (Anno 1658 Anglice edito.[54]) Quippe ille Stellam hanc (ex nostratibus saltem) omnium primus observavit, et aliis indicavit, (jam ante Phocylidis[55] librum editum:[56]) atque inde ab anno saltem 1639 vicissitudines apparendi et disparendi notavit.

Video te, pag. 378, nullas ejusdem Observationes memorare, post mensem Octobris 1660. Cum tamen alibi ejusdem alias habeas Observationes, viz. Anno. 1661, 1662. Quod facit ut non miror te| de hyeme jam proxime [1^v praeterita nihil dixisse: qua tamen ego illam saepius conspexi paulo minorem quam est ea quae in Mandibula. (Fallor, annon hac ipsa nocte illam conspexerim, sed admodum exiguam.) Credo utique illa jam Anno 1660 scripta, sinon et impressa fuisse, utut jam modo prodierint.

Reliqua, credo, ad Machinam tuam Caelestem reservas; quam propediem praelo subjectum iri spem facis[57]. Cui si ea quae de Palmero nostro (Horroxii[58] olim familiari) supra diximus inseras; non erit fortassis incommodum, etiam in speratum illud inserere Phaenomenon de ipsius (per aliquot annos) disparitione antequam ad Meridianum pervenerit; (de quo in sequente quadam Epistola mentionem feceram.) Quamquam enim illud jam non

---

[52]descripsi: i.e. WALLIS–HEVELIUS 30.III/[9.IV].1662/3.

[53]Palmeri: i.e. John Palmer (1612–79), mathematician and rector of Eaton and Ecton, Northamptonshire; from 1665 archdeacon of Northampton. See TAYLOR, *The Mathematical Practitioners*, 213.

[54]edito: i.e. PALMER, *The Catholique Planisphaer: Which Mr Belgrave calleth The Mathematical Jewel*, London 1658.

[55]Phocylidis: i.e. Johannes Phocylides Holwarda (1618–51), Frisian astronomer and natural philosopher, professor of philosophy at the University of Franeker from 1639–51.

[56]editum: probably HOLWARDA, Πανσέληνος ἐκλειπτικὴ διαυγάξουσα. *Id est Dissertatio astronomica quae occasione ultimi lunaris anni 1638 deliquii manductio sit ad cognoscendum*, Franeker 1640.

[57]facis: Hevelius's *Machina coelestis* was not in fact published until 1673 (pars prior) and 1679 (pars posterior).

[58]Horroxii: i.e. Jeremiah Horrox (Horrocks) (1618–41), English astronomer, *ODNB*. Wallis was commissioned by the Royal Society to edit his *Opera posthuma*, published in 1672/3.

10. PELL to COLLINS, 28 October/[7 November] 1668

conspiciatur: nullus tamen dubito quin ille haec temere non dixerit; et fieri olim aliquando poterit, hoc ipsum iterum contingere.

Tu interim Vale, Vir Clarissime; et perge, quod facis, rem Caelestem ornare.

Tui Observantissimus
Johannes Wallis.

[?ᵛ] Amplissimo Doctissimoque
Viro, Dn. Johanni Hevelio,
et Consuli Dantiscano;
Amico plurimum honorando.
Gedanum.

10.
JOHN PELL to JOHN COLLINS
[Brereton], 28 October/[7 November] 1668

Transmission:

C Copy of letter sent: LONDON *British Library* Add. MS 4278, f. 127ᵛ–128ʳ.

Reply to: COLLINS–PELL 23.X/[2.XI].1668.

October 28. 1668

Sir,

Yours[59] of Octob. 23 tells me that I have not yet answered yours[60] of Sept. 22. I will therefore now answer them both. I am to thank you for the bookes[61] which you have sent and those which you promise to send, Slusius[62] etc. I have received *two* coppies of Exercitationes Jac. Gregorii. I guess that Mr Pitt intended one of them for Mr Brancker.:[63] Who in his

---

[59]Yours: i.e. COLLINS–PELL 23.X/[2.XI].1668.

[60]yours: i.e. COLLINS–PELL 22.IX/[2.X].1668; *British Library* MS. Add. 4278, f. 343ʳ–343ᵛ.

[61]bookes: i.e. GREGORY, *Exercitationes geometricae*, London 1668 and ANDERSON, *Stereometrical Propositions variously applicable*, London 1668.

[62]Slusius: i.e. SLUSE, *Mesolabium*, 2nd ed., Liège 1668.

[63]Brancker: i.e. Thomas Brancker (1633–76), mathematician and school master, *ODNB*.

10. PELL to COLLINS, 28 October/[7 November] 1668

letter,[64] dated Octob 7, prayed Mr Pitt[65] to binde up the book (which, in May last, the Dean of Ripon[66] left in his hands for me) and then to deliver it to My L. Br...[67] for me. I hope Mr Pitt hath done so before this time.

I doe not perceive any relation between Mr Andersons[68] last Probleme[69] and his Stereometrical propositions.[70] Perhaps, it was put there for a *Specimen* of his comments[71] on *Diophantus*: Which, you say, He is writing. If he doe all the rest at the same rate, it will be a Rare booke. But Memus[72] his comment upon Apollonius Pergaeus, did not hinder Commandinus[73] from writing[74] on the same Author.

I have here Gregor. à Sancto Vincentio[75] and Wallisii.[76] Epistol. Commer. Which are cited[77] by *Mr Ja. Gregory* in his Exercitationes. But he quotes also His owne[78] Geometriae pars Universalis; and His Circuli & Hyperbolae quadratura; and Transactiones[79] philosophicae mensis Julii. I have not yet seen any of those three: but I desire to see them; as also the

---

13 philosophicae *add.*

---

[64]letter: i.e. BRANCKER–PELL 7/[17].X.1668.

[65]Pitt: i.e. Moses Pitt (1639–97), London printer and bookseller, *ODNB*.

[66]Dean of Ripon: i.e. John Wilkins (1614–72), theologian and natural philosopher, *ODNB*. Wilkins was made dean of Ripon in 1660.

[67]Br...: i.e. William Brereton (1631–80), third Baron Brereton of Leighlin; natural philosopher and founder member of the Royal Society, *ODNB*.

[68]Andersons: i.e. Robert Anderson (*fl.*1666–96), silki-weaver and mathematician, *ODNB*.

[69]last Probleme: at the end of *Stereometrical Propositions*, Anderson formulates a problem in number theory: 'To find two such numbers, that their Product being added to the sum of their Squares, the sum shall be a Square, and its Root commensurable.' (105).

[70]Stereometrical propositions: i.e. ANDERSON, *Stereometrical Propositions: variously applicable; but particularly intended for gageing*, London 1668.

[71]comments: it was evidently rumored that Anderson planned to write on Diophantus.

[72]Memus: i.e. Johannes Baptista Memus (15th century). His Latin translation of Apollonius's *Opera* was published in Venice in 1537.

[73]Commandinus: i.e. Federico Commandino (1509–75), Italian humanist and mathematician.

[74]writing: i.e. COMMANDINO (ed.), *Apollonii Pergaei conicorum Libri quattuor*, Bologna 1566.

[75]Vincentio: i.e. SAINT–VINCENT, *Opus geometricum*, Antwerp 1647.

[76]Wallisii: i.e. WALLIS, *Commercium epistolicum*, Oxford 1658.

[77]cited: Saint-Vincent's *Opus geometricum* and Wallis's *Commercium epistolicum* are quoted by Gregory in his *Exercitationes*, p. 23.

[78]owne: i.e. GREGORY, *Geometriae pars universalis*, Padua 1668, and *Vera circuli et hyperbolae quadratura*, Padua 1667.

[79]Transactiones: i.e. GREGORY, *Mr. Gregories Answer to the Animadversions of Mr. Hugenius upon his Book, De vera Circuli & Hyperbolae Quadratura, Philosophical Transactions* No. 37 (13 July 1668), 732–5.

10. PELL to COLLINS, 28 October/[7 November] 1668

small treatise,[80] which you say He hath in hand. In the second page of his Preface he quotes[81] words of a praestantissimus Geometra (I think he meanes Dr Wallis) but he tells us not whether those words[82] be *printed* in some Treatise that I have not seene, or were onely *written* in some letter. In his 16th page I have put out[83] the 20th line, [hoc est ex ducti etc.]

I am glad to heare that so much of Doctor Wallis is under the Press. Dr Rawlinson[84] could write very faire and grave neatly: It is not unlikely that he hath left some exercises ready for the press. ou write for a Book,[85] written by Le Sieur de Taneur,[86] des quantitez irrationelles, printed at Paris in 4to, 1650. I never saw it. Here is a book of about 40 sheets| printed at Paris in 4to, 1640, with a privilege for 5 yeares: Its Title is Traitè des Quantitez incommensurables et le dizieme livre d'Euclide. The Authors name is not expressed. When it first came forth, P. Mersenne sent[87] me a coppy, but could not then learn the Authors name. Perhaps the Author reprinted it,[88] ten yeares after, and adjoined his name. If so; it is likely That second Edition was better than This of 1640; which I will send you, if you desire it. If my Lord name the bookes of Trade, which I can come at, they may be sent to you before his Lordship[89] returns hither. I finde none yet, but Malines[90] and such like, which are every where to be had.

You inquire concerning my fitting of somewhat more for the press. I hope, my freinds doe not think that all my time is spent in Mathematicks; though I profess a desire to polish some of my rough draughts, that they may be thought not unworthy to be preserved by some understanding Reader.

---

11 Quantitez (*1*) Irrationelles (*2*) incommensurables

---

[80]small treatise: cf. COLLINS–PELL 23.X/[2.XI].1668.

[81]quotes: see GREGORY, *Exercitationes geometricae*, sig. A2^v. Wallis was indeed intended. See COLLINS–PELL 14/[24].XI.1668.

[82]words: in fact the words concerned were contained in WALLIS–COLLINS 8/[18].IX.1668.

[83]put out: the 20th line was a printer's error, repeating the line preceding it.

[84]Rawlinson: cf. COLLINS–PELL 23.X/[2.XI].1668.

[85]Book: i.e. LE TENNEUR, *Traité des quantitez incommensurables*, Paris 1640.

[86]Taneur: i.e. Jacques-Alexandre le Tenneur(1610–60), French mathematician and sometime counsellor to the provincial senate of Guyenne.

[87]sent: i.e. in September 1640. See note to CAVENDISH–MERSENNE 22.VIII/[1.IX].1640; MERSENNE, *Correspondence* X, p. 99.

[88]reprinted it: Le Tenneur's *Traité des quantitez incommensurables* did not appear in a second edition.

[89]Lordship: i.e. William Brereton. Brereton Hall in Cheshire was the family home.

[90]Malines: i.e. Gerard Malynes (de Malines) (*fl.* 1585–1641), Antwerp-born merchant and writer on economics, active in London from 1585, ODNB. Pell is no doubt referring to his *Consuetudo, vel lex mercatoria, or The ancient law-merchant Divided into three parts*, first published in London in 1622.

But whether I shall live to see them printed, am not sollicitous. If in the meane time Men, of more leisure and skill, doe print better bookes of the same Arguments, I shall rejoice to see my self so presented.

a DIEU

11.
FRANÇOIS DULAURENS to HENRY OLDENBURG
October? 1668

Transmission:

Manuscript (or printed flysheet) missing.

Existence and date: Mentioned in JUSTEL–OLDENBURG [31.X]/10.XI.1668; OLDENBURG, *Correspondence* V, 121–4.
Answered by: WALLIS: *Paper on Dulaurens* 27.XI/[7.XII].1668.

Justel, in his letter to Oldenburg, notes simply: 'M. du Laurens vous a addressé une lettre contre M. Wallis qui est pleine de fiel et dinjures.' Although Wallis replied, Oldenburg ultimately chose not to continue the controversy in the *Philosophical Transactions*. Cf. WALLIS–OLDENBURG 2/[12].VII.1668 (ii), WALLIS, *Correspondence* II, 471–6, and WALLIS–OLDENBURG 18/[28].VII.1668, WALLIS, *Correspondence* II, 498–524.

12.
HENRY OLDENBURG to WALLIS
early November 1668

Transmission:

Manuscript missing.

Existence and date: Mentioned in and answered by WALLIS–OLDENBURG 7/[17].XI. 1668.

In this letter Oldenburg considered the question of how to continue the publication of WALLIS–OLDENBURG 18/[28].VII.1668 (WALLIS, *Correspondence* II, 499–524) of which the first part had been printed in *Philosophical Transactions* No. 39 (21 September 1668). In this regard he enclosed the two differing versions, which Wallis had sent to him as enclosures to WALLIS–OLDENBURG 20/[30].VII.1668. See WALLIS, *Correspondence* II, 498–9.

13.
Wallis to James Gregory
2?/[12?] November 1668

Transmission:

Manuscript missing.

Existence and date: Mentioned in Wallis–Collins 3/[13].XI.1668.
Enclosure to: Wallis–Collins 3[13].XI.1668.

This letter, in which Wallis again discussed the method employed by Gregory in *Vera circuli et hyperbolae quadratura*, was sent through the hands Collins as an enclosure to Wallis–Collins 3/[13].XI.1668.

14.
Christiaan Huygens to Wallis
Paris, [3]/13 November 1668

Transmission:

C^1 Letter sent (in scribal hand with corrections by Huygens): London Royal Society Early Letters H1, No. 58, 4 pp. (our source). On p. 4. in Oldenburg's hand: 'M. Hugens latin letter to Dr Wallis from Paris Nov. 13. 68.'—printed: Huygens, *Œuvres complètes* VI, 278–81.
C^2 Draft of letter sent: Leiden Bibliotheek der Rijksuniversiteit Hug. 45, No. 1671, 4 pp.

Reply to: Wallis–Huygens 31.VIII/[10.IX].1668.
Answered by: Wallis–Huygens 13/[23].XI.1668.

Huygens sent this letter as an enclosure to Huygens–Oldenburg 3/[13].XI.1668 (Oldenburg, *Correspondence* V, 126–8), which was read at the meeting of the Royal Society on 12 November 1668. See Birch, *History of the Royal Society* II, 320.

Celeberrimo Doctissimoque Viro
Doctori Jo. Wallisio, Chr. Hugenius
S. P.

Literas[91] tuas ultima Augusti mensis datas, ultima demum Octobris ad me pertulit Dr. Rychius,[92] e quo, cum ad diem illis adscriptum non attendissem,

---
[91]Literas: i.e. Wallis–Huygens 31.VIII/[10.IX].1668.
[92]Rychius: i.e. Dirk de Rycke. Cf. Wallis–Huygens 31.VIII/[10.IX].1668; Wallis, *Correspondence* II, 568–73.

14. HUYGENS to WALLIS, [3]/13 November 1668

lectionemque dum ille abscesissit distulissem, tam longae morae causam non rescivi, etsi aliqua excusatione usum memini, nec postea ad me rediit, etsi aliud promiserat. Itaque te rogo (vir amicissime) nequid ob tam serum responsum de me serius suspiceris, qui quamlibet poenam subire dignus
5 essem, si literas tam humaniter scriptas gratissimasque, tum paternae tum mecum initae amicitiae commemorationem habentes usque adeo neglexissem. Verum quidem est jam triennio fere epistolarum commercium[93] non tecum solum, sed et cum caeteris Regiae Societatis Clarissimis Viris, mea, ut videri potest, culpa interruptum mansisse, qua de re si queruntur, com-
10 munis ipsis aliisque omnibus aliarum regionum amicis meis expostulatio est. Sed profecto non nisi cogente necessitate et quantumlibet invitus ita facere coactus sum cum crescenti in dies epistolarum scribendarum oneri me parem diutius esse non posse animadvertissem, nisi si, id unum agere contentus, reliqua studia omitterem. Nolim autem existiment quicquam proinde de pristino
15 affectu decessisse, aut me oblitum, eximiae humanitatis ac benevolentiae, qua advenam me excepistis[94] omnique officiorum genere demeruistis; imo crebro jucundissimum illud tempus memoria repeto, vultusque ac nomina simul insignium virorum, quibus tam innotescere contigit. Quin et praeclaro Regiae Societatis instituto, si quicquam alius, ab eo tempore favi, cujus
20 nuper quoque cum historiam elegantissimo stylo conscriptam[95] perlegerem haud exiguam inde voluptatem cepi. Quod si sors tulisset ut apud vos potius quam hic vitam ducerem, utilior fortasse vobis opera fuisset mea, voluntas & effectus pronior esse nequisset; eoque non nimium dolere debes, haec uti evenerunt contigisse. Ego vero ex quo huc concessi id quod commendas
25 curare pro viribus non destiti; nequid nempe invidiae atque obtrectationis ex mutua aemulatione inter hanc nostram[96] vestramque Academiam suboriatur, atque id imposterum quoque operam dare pergam; ac nemo quidem est e nostrorum numero, qui aliter quam oporteat, erga te animatus sit, quive non maximi te faciat, ut proinde iniquus ipse futurus sis, si Gallos tibi

---

[93]commercium: Huygens's exchanges with correspondents in England had ceased after his departure for Paris on [11]/21 April 1666. A letter to him from Oldenburg of 15/[25] May 1666, as well as one from Moray of 14/[24] May 1666 which it enclosed, remained unanswered, although they were directed to Paris.

[94]excepistis: Wallis first met Christiaan Huygens at Gresham College on 8/[18] April 1661 while he was carrying out business in London for his father, Constantijn Huygens. During the following days they met again a number of times. See HUYGENS, Œuvres complètes XXII, 572–3.

[95]conscriptam: i.e. SPRAT, *The History of the Royal Society of London, For the Improving of Natural Knowledge*, London 1667.

[96]nostram: i.e. the Montmor Academy in Paris.

parum equos ex Leotaudo,[97] Sorberio,[98] aut Laurentio[99] aestimes; quorum unus[100] ut scis ob procaciam poenam sustinuit, omnes vero quatenus immerito te lacescunt bonis indignationem movent. Eum vero, quem ultimo loco nominavi, quod attinet, cum tibi eruditione mathematum scientia longissime concedat, altercandi vero dicteria congerendi peritia nihilo forsan inferior futurus, doleo cum illo litem exortam; ac nunc etiam suadeo, ut tanquam indignum te certamen quam primum omittas, quando ille saniora consilia respuens bili suae temperare nequit. Caeterum uti de illatis tibi injuriis pro jure amicitiae apud me conquereris, liceat mihi invicem de meis apud te dicere.|

Rectene an secus Jacobi Gregorii demonstrationes[101] reprehenderim non jam quaeram, cum ex iis, quae novissime nunc respondi,[102] quo loci sit controversia illa, intellecturus sis. Credebam equidem prima illa discussione mea nihil eum offensum iri, namque et non sine laude de summa operis locutus sum,[103] et quae parum evidenter demonstrata demonstrata erant, examinare concessum putabam. Quod vero accuratiora suis de Dimensione Circuli antea me edidisse subjunxeram, illud et vere et modeste, ut existimo, dixi, illud reticens nempe, quod merito objicere poteram, praecipua illum theoremata quae hoc spectant ex meo libello desumsisse, quippe plane eadem ac tantum alia via demonstrasse, idque nulla facta mei mentione. Sed nec inique eum tulisse primas illas animadversiones meas responso[104] suo satis

---

[97]Leotaudo: i.e. Vincent Léotaud (1595–1672), French Jesuit mathematician.

[98]Sorberio: i.e. Samuel Sorbière (1615–70). French physician and active member of the Montmor Academy. On the latter's decline he encouraged Jean-Baptiste Colbert (1619–83) to found the Académie Royale des Sciences. His *Relation d'un voyage en Angleterre, où sont touchées plusieurs choses, qui regardent lestat des sciences, & de la religion, & autres matieres curieuses*, Paris 1664, published after his return from a visit to England in 1663, during which he was elected fellow of the Royal Society (together with Christiaan Huygens and Balthasar de Monconys), caused such uproar in France that he was temporarily exiled to Nantes. See SORBIÈRE–OLDENBURG [5]/15.XII.1663; OLDENBURG, *Correspondence* II, 133–6.

[99]Laurentio: i.e. François Dulaurens q.v.

[100]unus: i.e. Sorbière. Cf. BIRCH, *History of the Royal Society* I, 317.

[101]demonstrationes: i.e. GREGORY, *Vera circuli et hyperbolae quadratura*.

[102]respondi: i.e. HUYGENS–GALLOIS ?.XI.1668, published as 'Extrait d'une lettre de M. Hugens à l'Auteur du Journal, touchant la Réponse que M. Gregory a faite à l'examen du Livre intitulé Vera Circuli & Hyperboles Quadratura, dont on a parlé dans le V. Iournal de cette année', *Journal des Sçavans* (12 November 1668), 109–12.

[103]locutus sum: cf. Huygens's critical review of J. Gregory's *Vera circuli et hyperbolae quadratura*, *Journal des Sçavans* (2 July 1668), 52–6.

[104]responso: i.e. GREGORY, 'Mr. Gregories Answer to the Animadversions upon his book, De vera circuli & hyperbolae quadratura; as they were publish'd in the Journal des Scavans of July 2. 1668', *Philosophical Transactions* 37 (13 July 1668), 732–5.

14. HUYGENS to WALLIS, [3]/13 November 1668

moderato ac civili mihi persuaserat. Ex eo vero cum penitius argumentum hoc inspexisset, melioraque aliquanto initio proditis reperissit, successu ut videtur ferocior factus neque exspectans quid ei repositurus essem, acerbissimo scripto[105] in me nihil tale metuentem homo invehitur, plagiique sese accusatum praetexens publice me mendacii insimulat; opuscula,[106] quae a multis annis edidi pro viribus vilipendit, cum tamen, ut dixi, praecipuam suorum partem inde mutuatus sit, denique quacunque potest carpit, obtrectatur, insultat. Atque id eo molestius tuli quod Regiae Societati se insertum, titulo libelli praefert, quae cum viris sapientibus abundat, nec male erga me quod sciam affectis, debuisset sane moderari praecipitem illam adversarii mei audaciam. Sed et nunc si quid in eum juris habent ad retractandum compellere deberent, ne mihi necesse sit existimationem meam publice quoque vindicare, sicut publice impetita est. Quam enim invitus ad contentiones ejusmodi descendam, qui me norunt sciunt. Quod vero ad accusationem attinet, qua contendit, me non misisse ad vos Regulam[107] super hyperbolae dimensione fundatam, ad inveniendam Aeris gravitatem in datis a terra elevationibus, Illustrissimus Eques Moraeus, si epistolas meas asservavit, in iis alicubi eam reperiet, atque ego, si opus esset, responsum[108] ejus proferre possum, quo regulam illam non displicuisse Geometris verstris significavit; sed credibile est, Gregorium non admodum curiose rei veritatem exquisivisse. Caeterum cum libellus iste contumeliosus allatus est, responsio[109] mea, quae cum his literis ad vos deportatur typis jam excusa erat, quae licet nunc moderata sit, primoque ac non secundo Gregorii scripto conveniens, tamen quo minus in lucem exiret obsistere nolui, cum tanto majorem mihi factam injuriam omnes agnituri sint, quanto longius ab inferenda abfuisse videbunt. Et de his quidem jam nimium multis, sed veniam dabis spero justae indignationi.

Quod dimensionem hyperbolae per Logarithmos a Barovio[110] vestro eandem mecum repertam fuisse admones, cave existimes id mihi| parum [3]

---

[105]scripto: i.e. GREGORY, *Exercitationes geometricae*, preface.

[106]opuscula: i.e. HUYGENS, *Theoremata de quadratura hyperboles, ellipsis, et circuli, ex dato portionum gravitatis centro*, Leiden 1651, and HUYGENS, *De circuli magnitudine inventa*, Leiden 1654.

[107]Regulam: i.e. the rule Huygens enclosed in HUYGENS–MORAY [8]/18.VIII.1662; HUYGENS, *Œuvres complètes* IV, pp. 200–3 and 205–6.

[108]responsum: i.e. MORAY–HUYGENS 22.VIII/[1.IX].1662; HUYGENS, *Oeuvres complètes* IV, pp. 216–7. Cf. WALLIS–OLDENBURG 19/[29].XI.1668.

[109]responsio: i.e. HUYGENS–GALLOIS ?.XI.1668. Huygens apparently intended to enclose a copy of his reply to Gregory, but according to Oldenburg this did not arrive. See OLDENBURG–HUYGENS 18/[28].XI.1668; OLDENBURG, *Correspondence* V, pp. 176–8.

[110]Barovio: i.e. Isaac Barrow. Cf. WALLIS–HUYGENS 13/[23].XI.1668.

gratum esse, si enim ille me aeque atque ego illum reperisse eum ignoravit, utrique in solidum laus debetur. At si verum fateri volumus, magna ejus pars ad Gregorium a Sancto Vincentio pertinet, qui primus spatiorum hyperbolicorum comparationem instituit; quod ego considerans, regulam illam jam olim inventam venditare non curavi cum et multa potiora hactenus praestiterim. Vale vir Clarissime meque Tui studiosissimum ama.

Dabam Paris. 13 Nov. 1668

15.
WALLIS to JOHN COLLINS
Oxford, 3/[13] November 1668

Transmission:

W Letter sent: CAMBRIDGE *Cambridge University Library* MS Add. 9597/13/6, f. 204^r–204^v (our source). On f. 204^v over deleted address at 180⁰ in Collins' hand: 'Of the Correspondency beteene the Circle & Hyperbola. Postmark on f. 204^v: 'NOV/4'.—printed: RIGAUD, *Correspondence of Scientific Men* II, 506–7.

Enclosure: WALLIS–JAMES GREGORY 2?/[12?].XI.1668 (missing).

As emerges from the text of this letter, Wallis enclosed his latest letter to Gregory (WALLIS–GREGORY 2?/[12?].XI.1668).

Oxford, Novemb. 3, 1668.
Sir

I thank you for yours[111] of Octob. 26, and the book[112] you sent with it, which I received last Saturday toward night. I thought to have sent you the enclosed[113] by the last post; but before I had transcribed it (thinking fit to keep a copy of it) I was otherwise diverted. Mr. Gregory is certainly in the wrong, & therefore I am sorry to see him write at that rate he doth. And I could have wished (but that it is now too late) that his angry preface & the first leaf of his book had been suppressed. I wrote him a large letter[114]

---

15 (but ... late) *add.*

---

[111] yours: i.e. COLLINS–WALLIS 26.X/[5.XI].1668.
[112] book: evidently GREGORY, *Exercitationes geometricae*.
[113] enclosed: i.e. WALLIS–JAMES GREGORY 2?/[12?].XI.1668.
[114] letter: i.e. WALLIS–GREGORY 22.X/[1.XI].1668.

15. WALLIS to COLLINS, 3/[13] November 1668

of Octob. 22 (when I knew not of this preface) enclosed in one[115] to my Lo. Brounker. Whether it came to his hands before he went out of town, I know not. If not, I desire you will send it after him, with this enclosed, which you may if you have opportunity shew my Lord before it goes. I would be content to have his Lordship's sense upon the former, & upon this whole matter.

The series for the Circle, answerable to that of the hyperbola, is the same which I sent[116] you a while since. For that which in the Ellipse, answeres to the Asymptotes of the hyperbola, are the two Equal diameters, which in the circle are any two crosse diameters intersecting at right angles, answering to Asymptotes so crossing. And the ordinates to one of these diameters, answer to the ordinates to the Asymptotes, terminated in the curve.

I was not against printing a comment[117] on the Clavis Math:[118] if any think fit to do it; but onely that I thought it not necessary, & would swell a manual into a volume. What peece I shall else print I have not yet determined. But an Introduction to Algebra[119] I have not yet ready. That[120] at London gets on so slowly, that, if I had been aware of it I would never have given way to print it there: And I doubt I must yet be forced to have it finished here.

The post is going & I can adde no more but that I am

Yours, &c.
J. W.

---

1 (when ... preface) *add.*
10 to (*1*) these (*2*) one of these
15 I (*1*) do not (*2*) have not

---

[115]one: i.e. WALLIS–BROUNCKER 22.[1.XI].1668.

[116]sent: i.e. the appendix to WALLIS–COLLINS 26.IX/[6.X].1668.

[117]comment: cf. COLLINS–WALLIS 2/[12].II.1666/7 and WALLIS–COLLINS 5/[15].II.1666/7.

[118]Clavis Math.: i.e. OUGHTRED, *Clavis mathematicae*. Wallis who had already been involved in producing the third Latin edition, published in Oxford in 1652, proposed the printing of the fourth edition, published in Oxford in 1667. See STEDALL, *Ariadne's Thread: The life and times of Oughtred's Clavis*, 46–50.

[119]Algebra: cf. COLLINS–PELL 23.X/[2.XI].1668.

[120]That: i.e. WALLIS, *Mechanica: sive, de motu, tractatus geometricus*, 3 parts, London 1670–1.

4ᵛ] For Mr John Collins at
the three Crowns in
Blooms-bury market
at
London.

16.
WALLIS to WILLIAM BROUNCKER
Oxford, 4/[14] November 1668

Transmission:

W Draft letter: LONDON *Royal Society* Early Letters W1, No. 65, 6 pp. (our source). At top right of p. 1 endorsement in Oldenburg's hand: 'Read Dec: 3: 68 Entered LB. 2. 303'.—printed: HUYGENS, *Œuvres complètes* VI, 282–9; OLDENBURG, *Correspondence* V, 138–9 (partly).
w Copy of letter sent: LONDON *Royal Society* Letter Book Original 2, pp. 303–13. On margin of p. 303 in Oldenburg's hand: 'The Original of the Letter Dr Wallis had again.'

As indicated in this letter, and also by remarks in WALLIS–OLDENBURG 26?.XI/[6?.XII]. 1668, Wallis had intended to send this letter to Brouncker, but changed his mind on receipt of Gregory's latest attack on Huygens. He eventually sent the draft to Oldenburg in late November 1668, indicating that it should be shown to Brouncker. The letter was read at the meeting of the Royal Society on 3/[13] December 1668, though on account of its length not in its entirety. See BIRCH, *History of the Royal Society* II, 332.

For my Lord Brouncker.

Oxford Novemb. 4. 1668.

My Lord

This paper concerns a controversy wherein this Royall Society may to some seem concerned; but indeed is not so. It is concerning a Book lately published at Padua by Mr James Gregory now¹²¹ a Member of this Society, entituled,

---

1 (*1*) These for Mr James Gregory at (*2*) For Mr John Collins
1 at Padua *add.*
1 Gregory (*1*) now a (*2*) recently (*3*) now a

---

¹²¹ now: Gregory was elected and admitted to the Royal Society on 11/[21] June 1668, having been proposed as candidate by Collins a week earlier. See BIRCH, *History of the Royal Society* II, 291.

16. WALLIS to BROUNCKER, 4/[14] November 1668

Vera Circuli et Hyperbolae Quadratura in propria sua proportionis specie. This Book soon after it came over into England, was by another Member[122] of this Society sent[123] to mee, desiring (in general terms) my opinion of it. And after a slight perusal of the whole, to see what matter it conteined, and
5 examining the Demonstration of the more leading propositions, the account I gave him in a familiar letter[124] (of which I kept no Coppy) was, as I remember, to this purpose; that *it seemed to mee to contain divers things (so far as I could judge upon a slight perusall) ingeniously demonstrated though obscurely; amongst which was a new methode of approximation for*
10 *the squaring of the Circle, which was allso equally applicable to the Ellipsis and Hyperbola.*

And more than to this purpose I do not remember that I did write. Nor do I yet see reason to retract what was then sayd in favour of it. It seemes your Lordships opinion was also asked,[125] which what it was your
15 Lordship doth best know. And so of some others.[126] Out of all which the Publisher[127] of the *Philosophicall Transactions* collected that Character[128] thereof which is inserted in those of *March* last; in such words as he thought fittest to express what he did apprehend to be their concurrent sense. Nor do I see any necessity of receding from them save that where he speakes
20 of the *Termination of Converging Series,* in stead of *the Termination,* by a mistake[129] it is sayd *the Summe.* Nor is the Royal Assembly[130] (that I know of) further concerned in it. Soon after; those in France, in their *Journal des*

7-8 divers (*1*) ingenious things whi⟨ch⟩ (*2*) things (*a*) well (*b*) ingeniously demonstrated
12 than (*1*) this (*2*) to this purpose
17 last; (*1*) Not in their own words, but in such (*2*) in such words as
18 did *add.*
19 any (*1*) reason (*2*) necessity

---

[122] Member: almost certainly Sir Robert Moray. See WALLIS–MORAY? ?.II.1667/8; WALLIS, *Correspondence* II, 432–3.

[123] sent: i.e. MORAY?–WALLIS ?.II.1667/8; WALLIS, *Correspondence* II, 432.

[124] letter: i.e. WALLIS–MORAY? ?.II.1667/8; WALLIS, *Correspondence* II, 432–3.

[125] asked: Brouncker gave a short account of Gregory's *Vera circuli et hyperbolae quadratura* at the meeting of the Royal Society on 27 February 1668 (old style). See BIRCH, *History of the Royal Society* II, 253.

[126] others: most notably John Collins.

[127] Publisher: i.e. Henry Oldenburg.

[128] Character: i.e. the review of J. Gregory's *Vera circuli et hyperbolae quadratura* in *Philosophical Transactions* No. 33 (13 March 1667/8), 640–4.

[129] mistake: This mistake was explicitly corrected in *Philosophical Transactions* No. 40 (19 October 1668), 812.

[130] Assembly: cf. HUYGENS–WALLIS [3]/13.XI.1668.

16. WALLIS to BROUNCKER, 4/[14] November 1668

Sçavans, published an opinion[131] of Monsieur *Huygens*, to whom, it seems, it had been referred[132] to consider & deliver an opinion of it. Hee there takes notice that those of England had in the generall given it a favourable character, (to which he addeth a like of his own;) but that they had sayd nothing as to that particular *whether it were therein demonstrated that it is impossible Analytically to square the Circle & Hyperbola*, (that is, by Addition, Subduction, Multiplication, Division, & Extraction of Roots, which he calls[133] Analytical operations.) And delivers his opinion in the *negative*; and that *supposing all to bee true which is demonstrated in his 11th proposition (where that demonstration is supposed to ly) it proves no more but that it cannot be performed by his methode; not that it cannot be done at all: unlesse it be supposed that the termination of a converging series can be no other way found but by his methode; or at lest, that if it may be found any other way, it may be found this way allso; which is not* (he sayd) *demonstrated.* To which Mr Gregory published an Answere,[134] inserted in the *Transactions* of *Julie* last. Which were both made publike before I had seen either of them. Which were both sent[135] to mee and my opinion desired concerning them. My answere[136] was; That, (beside some other particulars of lesse| moment) the proposition mentioned, did not so much as affirm, that the circle could not analytically be squared; & therefore it was not be expected it should bee there demonstrated; nor was the Demonstration to be blamed

---

1 *Huygens,* (*1*) whom it seems was (*2*) to (*3*) to whom, it seems,
5 as to *add.*
6 *Analytically* (*1*) *(as he speaks* (*2*) *to square*
8 and *add.*
10-11 *but* (*1*) the termination (*2*) *that it cannot be performed by his methode;* (*a*) not, that there can be no other methode of (*aa*) the (*bb*) terminating a Converging Series, as that, if there were, (*b*) not, that it (*aaa*) could not (*bbb*) cannot
12 *unlesse* |*it add.*| be
21 the (*1*) Propositi⟨on⟩ (*2*) Demonstration

---

[131]opinion: i.e. Huygens's critical review of J. Gregory's *Vera circuli et hyperbolae quadratura, Journal des Sçavans* (2 July 1668), 52–6.
[132]referred: probably at the suggestion of Sir Robert Moray.
[133]calls: i.e. in the introduction to Huygens's critical review of J. Gregory's *Vera circuli et hyperbolae quadratura.*
[134]Answere: i.e. GREGORY, 'Mr. Gregories Answer to the Animadversions of Mr. Hugenius upon his Book, De vera Circuli & Hyperbolae Quadratura', printed in *Philosophical Transactions* No. 37 (13 July 1668), 732–5.
[135]sent: i.e. probably with OLDENBURG–WALLIS ?.VII/VIII.1668.
[136]answere: probably WALLIS–OLDENBURG 3/[13].VIII.1668.

16. WALLIS to BROUNCKER, 4/[14] November 1668

for not proving what the Proposition did not assert. And to the same purpose I wrote[137] myself to Monsieur Huygens.

I confess, My Lord, I had not all this while observed (nor was I told it) that Mr Gregory did pretend, therein, to have *demonstrated* that the Circle could not be analytically squared; which made mee give such answere. And though, in his Preface, pag. 5. hee intimate something of it; yet when hee presently adds, *Verum certe est me hanc demonstrationem integram ad phrasem geometricam non reduxisse; nam ut hoc perficiatur, opus est non parvo volumine, &c.* I did not think that I was to expect a formed demonstration of it; but onely an intimation of some principles from whence such a demonstration (hee supposed) might be formed.

But presently after, hee sent[138] mee word that he did not onely *Affirm* it, but *did thinke hee had Demonstrated*, that it was not possible *Analytically to square* the Circle *in any methode whatsoever*, pressing that I would positively *assent to it*, or *give my reasons* why I did not. By which I perceived that Mr Huygens (who, it seems, knew this before, though I did not,) had more reason to make that Exception to the Demonstration, than I was aware of, who knew not of that pretense.

I have severall times[139] since signified to him, that although I was satisfyed, as to my own judgement, that it could not be done; yet I was not satisfied, that is was by him demonstrated. And have given him severall reasons (though, it seems, not such as satisfy him,) why I was so unsatisfyed. As, That there be many Lemmata or Suppositions which he doth either postulate or silently take for granted, which, though they may be true, yet are not so clear but that there is reason they should be proved,

---

5 give *add.*
6 pag. 5. *add.*
10 of (*1*) such (*2*) some
12 presently (*1*) upon that (*2*) after
13 had |there *add.* and *del.*| Demonstrated
14-15 Circle (*1*) what (*2*) *in any methode whatsoever* (*a*) . And did press that I would either positively declare my (*b*) , pressing that I would positively *assent to it*,
19 times (*1*) sent him my reasons (though it seems not such as have satisfied him,) why I was not satisfyed to declare that hee ha⟨th⟩ (*2*) signified to him (*3*) since signified to him
24 postulate or (*1*) take (*2*) silently take

---

[137] wrote: i.e. WALLIS–HUYGENS 31.VIII/[10.IX].1668.
[138] sent: in a now missing letter, possibly GREGORY–WALLIS ?.IX.1668.
[139] times: i.e. in WALLIS–GREGORY 22.X/[1.XI].1668 and a number of other letters which are now missing.

16. WALLIS to BROUNCKER, 4/[14] November 1668

before his Demonstration can bee judged full & perfect. That the exception made by Mr Huygens, is not yet removed: which is as much as to say, that, although his 10th proposition be demonstrated, the converse of it is not: And, consequently, it proves onely that the converging series cannot his way be terminated, not that it can no way be terminated analytically. In summo, That hee hath no where proved this Consequence, That, if the *Sector* be not in *his way* analytically composed of its Triangle & Trapezium (which is the whole of his 11th proposition,) then the *Circle* can *no way* be analytically squared. And have desired[140] him to give a cleare demonstration of this consequence; Presuming that, in applying himself so to do, hee would either shew mee more light then yet I see, or else meet with such insuperable difficulty as to discover to himself that I had reason not to bee hasty in affirming his Demonstration to be full & perfect.

And I have the more reason to insist upon the proof of that Consequence; because, though other less objections could be all removed, this one

---

1 can bee |there (?) *del.*| judged
1 perfect. (*1*) That (*a*) the except⟨ion⟩ (*b*) chief exception (*c*) as to the chief exception which Mr Huygens makes, above mentioned, his Reply takes no notice of it: and though (*aa*) it to himself (*bb*) to himself that exception seem groundless, yet hee (*2*) That the (*aaa*) excepti⟨on⟩ (*bbb*) chief exception (*ccc*) exception
3 that (*1*) alt⟨hough⟩ (*2*), although
4 his |way *add.*| be
5 can *add.*
7 composed (*1*), of the (*2*) of its
7-8 (which is the whole of his 11th proposition,) *add.*
8 *Circle* (*1*) cannot *any way* (*2*) can *no way*
9 desired him |first *del.*| to give
10 to |do *del.*| do
11 insuperable (*1*) doubts (*2*) difficulty
13 himself that (*1*) he (*2*) I had (*a*) no reason (*b*) reason not to (*aa*) give such Attestation (*bb*) bee hasty in affirming
13-14 perfect. || |Hee hath (*1*) given some (*2*) attempted some (*3*) attempted Answere (first in writing, since in print,) to some (*a*) Reasons (*b*) of these Reasons; which he calls *Objections against his Doctrine*. Hee should rather have called them *Objections against his Demonstration*; (for I did not object against his *Doctrine* at all, as having many years since Demonstrated the same myself, though he take no notice of it, in my *Arithmetica Infinitorum*, prop. 190. with the *Scholium* annexed:) But such as do no more satisfy mee than my Reasons did him. *del.*| || And I
14-15 Consequence; |because *del.*| because

---

[140] desired: i.e. apparently in WALLIS–GREGORY 22.X/[1.XI].1668.

16. WALLIS to BROUNCKER, 4/[14] November 1668

great one seemes to mee insuperable; That his 11th proposition, though ever so well demonstrated, shews onely, that the Sector *indefinitely considered* can not be so com|pounded as is there sayd: Or, (which is equivalent,) not *every* Sector. Notwithstanding which, it might well inough be possible, that some Sector (if not all) might be Analyticall to its Triangle or Trapezium: (And I think he doth allow it so to bee, or even Commensurable, *pag.* 29. *Respondeo hoc esse verissimum, &c.*) Like as, in this Equation, for the Trisection of an Arch $3r^2a - a^3 = r^2c$ indefinitely taken; the Root a, is not Analyticall with r and c, (that is, the Proportion of the Chord of the Single Arch, to that of the Triple arch & the Radius, cannot be universally designed by those he calls Analyticall operations; or the value of a analytically compounded of r and c, as he speakes; that is, it cannot be designed by commensurable numbers & surd Roots:) as Chartes,[141] Schoten[142] & others agree. Yet in some cases (though not universally) it may happen to be not onely Analytical, but even commensurable, (as, for instance, if $c = 2r$ be the Subtense of a Semicircle, a may be equal to r or to $-2r$.) Now if but some one sector (though not all, or the Sector indefinitely taken,) be found Analytical with his Triangle, or Trapezium; there be many ways (as, by its proportion to the whole, by its &c,) by the help of this One, to square the whole Circle.

I might adde allso (though this may more safely be avoided by altering his construction,) That his whole processe concerns onely such Sectors as are lesse than a Semicircle. For if it be a Semicircle (to say nothing of those that be greater) the two Tangents will never meet in F (as his figure supposeth) to make his Trapezium. And therefore it proves nothing directly as to the Semicircle, much lesse as to the circle itself. Now though we cannot, Analytically, trisect some lesser Archs; yet the Semicircle wee can. And though we cannot, Analytically, assign the Center of Gravity of a Sector; yet we can of a Circle. To prove therefore, that lesser Sectors (at lest some of them) are not Analyticall to their Triangles, or to the Square of the

---

8 indefinitely taken; *add.*
12 is, |supposing *del.*| it
18 ways |by the help of this one *add. and del.*| (as
21 be *add.*
28 of (*1*) the whole (*2*) a Sector
29 that |in *del.*| lesser

---

[141] Chartes: i.e. Descartes.
[142] Schoten: c.f. SCHOOTEN, *Appendix de cubicarum aequationum resolutione*, in DESCARTES, *Geometria*, ed. Schooten, 2 vols, Amsterdam 1659–61, I, 345–68.

16. WALLIS to BROUNCKER, 4/[14] November 1668

Diameter; doth not presently prove, that the Semicircle, or Circle, are not so. But on this Objection I lay lesser weight. For though it shew a fault in the Demonstration; yet it is onely such a fault as may bee amended: which, I doubt, the former cannot bee.

Hee hath attempted Answere, (first in Writing,[143] & since in Print[144]) to some of those Reasons above mentioned: Which he calls *Objections against his Doctrine*: (Hee should rather have called them *Objections against his Demonstration*; For I did not object against his *Doctrine* at all; as having many years since demonstrated the same myself, though he take no notice of it, in my *Arithmetica Infinitorum, prop.* 190. with the *Scholium* annexed:) But such as do no more satisfy mee, than my Reasons did him.

Hee is yet very earnest to have mee satisfied that his Demonstration is good; & to have a like approbation from the Royal Society: thinking that he hath hard measure, that having (as he is confident) the truth on his side, not onely those of France have declared against him; but those of this Royal Society seem at lest tacitely so to do. Which hee complains of with some regrett, both in Letters[145] to mee, & in his printed Preface to his *Exercitations*. Professing that he desires no more but a fair Character granting what he hath done, shewing in what he hath failed, and proposing what yet rests to do the work, that so if he or any man else can adde what is wanting it may be supplyd.

I shall therefore represent to your Lordship (as to a very competent Judge) how the state of his Demonstration stands, as to my apprehension. As well that he may be satisfied that I have considered, & do understand (in some measure) the strength of his Demonstration, and not reject it unconsidered; As that your Lordship allso may judge, whether it be Obstinacy or Reason that holds mee yet unsatisfied.

---

2 on *add.*
8 I (*1*) never objected (*2*) did not object
14 that |he *add.*| hath
17 Preface *add.*
18-21 Professing that ... supplyd. *add.*
26 and not reject it unconsidered; *add.*
27 be (*1*) reason or obstinacy (*2*) ⟨—⟩ that holds (*3*) Obstinacy or Reason that holds

---

[143]Writing: i.e. in a now missing letter to Wallis.
[144]Print: i.e. in GREGORY, *Exercitationes geometricae*.
[145]Letters: i.e. in now missing letters to Wallis.

16. WALLIS to BROUNCKER, 4/[14] November 1668

First, therefore, I take it as evident & confessed; that there is not in his Book any such Proposition formally layd down to be demonstrated, as *That the Circle cannot Analytically be squared*: Nor any Demonstration which doth in terms| conclude any such Proposition. All therefore that wee are [4] to inquire after is but, whether from what he hath demonstrated, such a proposition may bee directly inferred.

I could have wished therefore, that he would himself have drawn up his demonstration into form, & not left us onely to seeke materialls for it as they ly scattered up & down: That there might be no occasion for him to complain that I have not represented the strength of his demonstration to the best advantage. But since he hath not done it; and yet would have it thought that the thing is fully demonstrated, though the demonstration be not put into form: I shall lay down the severall branches of that demonstration in the best order I can, & shew which of them I judge to be proved, & which not.

1. A Sector (lesse than a Semicircle) indefinitely taken, is the Termination of a Converging Series infinitely continued; beginning with the Inscribed Triangle & Circumscribed Trapezium; (& so onward, by continuall Bisection, with Inscriptions & Circumscriptions respectively;) whose respective Converging terms are continually in the same manner analytically compounded of those next foregoing; and so continually approaching as that at length they become coincident each with other; & with the Sector.

2. And, in particular; the two first (or any two respective antecedents) being a, b; the two next consequently will be $\sqrt{ab}$, $\frac{2ab}{a+\sqrt{ab}}$. Both which are proved in Schol. prop. 5. from the Antecedent propositions, with that which next follows.

3. These component terms a, b, (the Triangle & Trapezium,) supposing the chord Analytical with the Radius, are Analyticall each to other. Which is necessary to this business, & may be easyly proved.

4. And, consequently, what is Analytical to either of them, is Analyticall to both. Which may be proved from Def. 6. 7. and Petit. 1.

---

2 in his Book *add.*
2 demonstrated, |as *del.*| as
5 wee (*1*) can hope to find, is (*2*) are to inquire after |of *del.*| is but
7 would (*1*) have himself (*2*) himself have
9 as they ly *add.*
11 it; and *add.*
26-27 Trapezium,) (*1*) are Analyticall each to other (*2*) supposing either of them Analyticall to the square of the Radius, are Analyticall each to other (*3*) supposing the (*a*) ⟨—⟩ (*b*) Chord Analytical with the Radius, are Analyticall each to other.

16. WALLIS to BROUNCKER, 4/[14] November 1668

5. Now in any such converging series, if there can be found a quantity which may in the same manner be analytically compounded (without introducing any extrinsick quantity) of the two first, & of the two second, converging terms; by help of this quantity, that Series may be Analytically terminated: That is, the Termination thereof may by Analytical operations be compounded of the two first converging Terms. This is proved by Prop. 10. (as it is now reformed & explained in his Reply[146] to Mr Huygens;) But not the converse of it; because the converse of that proposition is not proved. Which is the exception of Monsieur Huygens.

6. And, consequently, if the two first Converging terms be Analytical each to other; the Termination wil be Analyticall to both of them. Which may be proved from Petit. 1.

7. If therefore any quantity can be found analytically composed of a, b (the two first terms of the converging series proposed,) and, in the very same manner, of (the two next terms $\sqrt{ab}$, $\frac{2ab}{a+\sqrt{ab}}$, (without the intermistion of any other quantity, in either of the composition then may this series be analytically terminated. This (but not the converse of it) follows from §5.

8. And consequently (when as a, b, be analyticall each to other, viz. §3,) the Termination thereof, (that is, the *Sector indefinitely taken*,) will be analytical with both of them. Which follows from §7.

9. And consequently, *every such Sector* with its respective Triangle & Trapezium. For such Sector indefinitely taken, is any such Sector whatsoever.

10. Now of *Some* Sectors the chord is analytical with the Radius, as, for instance, of the Quadrantal Sector; and consequently the Triangle & Trapezium are Analyticall with the Square of the Radius, or of the Diameter. As

[5] is easyly proved.| 11. And therefore their Sectors will then be so; & there-

---

8-9 it *(1)* . *(2)* ; because ... Huygens.
14 converging *add.*
18 consequently (*(1)* since *(2)* when as
21 *such add.*
22 any |such *add.*| Sector
23-25 Now *(1)* the Triangle & Trapezium of Some *(a)* Sectors *(b)* such Sectors *(c)* Sectors *(d)* such Sectors (though not of all, onely *(aa)* such *(bb)* of such indefinitely taken,) as, for instance, of the *(aaa)* Quad⟨rantal⟩ *(bbb)* Quadrantal Sector; are Analyticall with the square of the Radius *(2)* of *Some* Sectors *(aaaa)* (as those whose chord is analytical with the Radius,) *(bbbb)* the chord is analytical withe the Radius, as, for instance, of the Quadrantal Sector; *(aaaaa)* not ⟨—⟩ *(bbbbb)* and consequently the Triangle & Trapezium are Analyticall with the Square of the Radius

---

[146]Reply: i.e. GREGORY, 'Mr. Gregories Answer to the Animadversions of Mr. Huge-

16. WALLIS to BROUNCKER, 4/[14] November 1668

fore may be analytically squared: That is, their proportion to the Square of the Radius or the Diameter, may be designed by Analyticall operations, or by commensurable quantities & surd Roots. Which may be proved from Def. 6, 7, and Petit. 1.

12. Now to some at lest of those Sectors, the Circle is Analyticall; and particularly to the Quadrantall; As being the quadruple thereof.

13. And will therefore be Analyticall to their Triangles, Trapezia, and Squares of the Radius and diameter; (and so may be analytically squared.) Which may be proved from Def. 6. 7. and Petit. 1.

14. But, on the contrary, if in such a converging series no such quantity can be found, (as is mentioned §5) which can, in the same manner, be analytically compounded of the two first, & of the two second, converging terms: then cannot this series, be *by that process* analytically terminated. For that process supposeth such a quantity; at §5 &c.

15. And if *not by that processe*, then not at all. Which is yet to be proved.

16. If therefore no such quantity can in the same manner be analytically compounded of a, b, and of $\sqrt{ab}$, $\frac{2ab}{a+\sqrt{ab}}$, then cannot this series, *by that process*, be analytically terminated. For that processe supposeth such a quantity, at §7. &c. Or it may be proved from §14.

17. And, consequently, the Circle cannot *by that process* be analytically squared. For that process supposeth such a termination, at §8. &c.

18. And if *not by this process*, then not at all. Which is to be proved.

19. Now if the first terms were $a^3 + a^2b$, $ab^2 + b^3$, and the second terms (in like manner compounded of the first, as in the series proposed,) $ba^2 + b^2a$, $2b^2a$: No such quantity could in the same manner be analytically compounded of the two first, & of the two second, converging terms. Which is proved at Prop. 11.

5-6 some |at lest *add.*| of those Sectors, (*1*) and particularly to the Quadrantall, the circle is Analyticall, (*2*) the Circle is Analyticall; and particularly to the Quadrantall; As being the quadruple thereof.
7 Analyticall to (*1*) those Sectors, (*2*) |to *del. ed.*| their Triangles,
8 so *add.*
20 Or it ... §14. *add.*
23 is |yet *del.*| to
26 analytically *add.*

---

nius upon his Book, De Vera Circuli & Hyperbolae Quadratura', printed in *Philosophical Transactions* No. 37 (13 July 1668), 732–5.

16. WALLIS to BROUNCKER, 4/[14] November 1668

20. And therefore, not if the two first be a, b, & the two second $\sqrt{ab}$, $\frac{2ab}{a+\sqrt{ab}}$. Which connexion deserves to be cleared.

21. Therefore this series cannot be *by that process* analytically terminated. By §14.

22. And therefore can *not at all* be analytically terminated. To be proved by §15.

23. Therefore the *Sector indefinitely taken* is not analyticall with its Triangle & Trapezium. For the *Sector indefinitely taken* is this termination.

24. Therefore not *every sector*. For if every sector, then any sector whatsoever; that is, the sector indefinitely taken.

25. Therefore *no* sector: At lest no sector which is analyticall with the whole Circle, and whose Triangle & Trapezium are analyticall with the square of the Radius or of the Diameter. Which consequence, I doubt, will hardly be made good.

26. Therefore the Circle cannot be *by that process* analytically squared. For that process supposeth some such Sector which shall be analyticall with the Circle, & with its own Triangle & Trapezium, and these with the Square of the Radius or Diameter, at §10, 12.

27. Therefore the Circle can in noe manner be Analytically squared. Which is to be proved from §18.

This, my Lord, I take to be the true Anatomy of that demonstration, which from his principles should prove that the Circle cannot at all be analytically squared. Which to mee, I confesse, seemes somewhat lame at §15, 18, 25, and those which depend on these. Especially at §15, & 25. The former of which is that which Mr Huygens excepts against as not proved: The latter seems to mee as much or more considerable. Till this be supplyed, his argument seemes to mee to be⟨are⟩ such a forme as this; If the Sector be so compounded, the circle may be analytically squared; But the Sector is not so compounded; Therefore the circle cannot be analytica⟨lly⟩ [6] square⟨d.⟩| Which Syllogism is peccant in form, though the propositions be

---

1 the |two *add.*| first
5 therefore (*1*) *not at all.* (*2*) *not at all* terminated analytically. (*3*) can *not at all* be analytically terminated.
11 sector (*1*) with (*2*) which
13 I doubt, *add.*
17 analyticall (*1*) with the Circle its Triangle (*2*) with the Circle,
24 18, (*1*) 20, 25. (*2*) 25, and those which depend on these.
29 Therefore (*1*) it cannot be square⟨d.⟩ (*2*) the circle cannot be analytica⟨lly⟩ square⟨d.⟩

16. WALLIS to BROUNCKER, 4/[14] November 1668

true; & therefore the conclusion follows not. Unlesse we suppose not onely the Consequence of the Major Proposition to be demonstrated; but the Converse of that consequence; which is not done.

I shall onely adde two things; The first is that however I am not satisfyed that this demonstration is full & perfect, yet this hinders not but that divers other things in that Book, may be very ingenious & well demonstrated. For this proposition, be it true or false, demonstrated or not demonstrated, doth not at all influence the rest of the Book, or enervate the other demonstrations.

The other is, that this Author need not be very solicitous for the supplying of what is defective in this demonstration; because the work is done allready, the thing itself being proved long since in my *Arithmetica Infinitorum*; prop. 190. with the *Scholium* annexed to it. Where it is proved, that what was before demonstrated to be the true proportion between the Circle & the Square of its Diameter or Radius, or between the Diameter & the Perimeter; cannot be expressed either by Rational Numbers or Surd Rootes: (or, as this Author speakes, is not Analyticall;) without supposing an odde number to be equally divided into two integers; and a forming of Equations between the Laterall & the Quadratick, between the Quadratick & the Cubick, &c.; that is, which shall have more then one Root but fewer than two, and more then two but fewer then three &c. which are impossible.

Notwithstanding all which, My Lord, if Mr Gregory shall supply these defects; or otherwise make it evident to the Royall Society that these consequences are allready proved though I have not been so quick-sighted as yet to see it: I shall very willingly consent that the Royall Society shall give as full an attestation thereof as hee can desire.[147]

---

1 not onely *add.*
6 things (*1*) there (*2*) in that Book
11-12 the work is done allready, the *add.*
12 since *add.*
14 it is (*1*) demonstrated that to designate what before the true (*2*) proved, that what was before demonstrated
17 (or, as this Author speakes, is not Analyticall;) *add.*
24 are (*1*) all good (*2*) allready proved

---

[147] desire: the Royal Society had no interest in becoming involved in proceeding with the dispute. See the introductory remarks to GREGORY, 'An Extract of a Letter of Mr. James Gregory to the Publisher, containing some Considerations of his, upon Mr. Hugens his Letter, printed in Vindication of his Examen of the Book, entitled Vera Circuli & Hyperbolae Quadratura', printed in *Philosophical Transactions* No. 44 (15 February 1668/9), 882–6, and WALLIS–OLDENBURG 14/[24].XI.1668.

Having had no other design in all this but to satisfy his importunity, which I could hardly avoyd without being uncivill. I am

My Lord
Your Honours very humble servant
John Wallis.

17.
WALLIS to HENRY OLDENBURG
Oxford, 7/[17] November 1668

Transmission:

W Letter sent: LONDON *Royal Society* Early Letters W1, No. 66, 2 pp. (our source). Postmark on p. 2: 'NO/9'.—printed: OLDENBURG, *Correspondence* V, 135–7.

Reply to: OLDENBURG–WALLIS early XI.1668.

Oxford Novemb. 7. 1668.

Sir,

As for Mr Gregory, I do not mean to trouble myself farther with him. I am onely sorry that I have, upon his importunity, taken so much pains, to displease him. Yet, after all his ranting, he is certainly in an Error; For what he doth pretend to, neither is by him demonstrated, nor can it (his way) bee done. But he is not capable of being advised; & therefore must take his course. I would be content to know, what My Lord Brounker thinks of it; especially since my last[148] to him. I wish I had known Hevelius's his resolution[149] as to the Library, a little sooner. For one of the sets I sent for, was for it. They formerly had, by his gift, his Selenography, but no more.

6 Oxford (*1*) Octob. (*2*) Novemb.
11 it (*1*) (that way (*2*) (his
15 to *add.*
16 more. (*1*) So that, if I am to present the (*2*) And

---

[148] my last: i.e. WALLIS–BROUNCKER 22.X/[1.XI].1668, in which WALLIS–GREGORY 22.X/[1.XI].1668 was enclosed. Wallis decided against sending WALLIS–BROUNCKER 4/[14].XI.1668 immediately. At the end of November, he sent it by carrier to Oldenburg with the request that it be shown to Brouncker.

[149] Hevelius's his resolution: i.e. Hevelius's request that one of the copies of his *Cometographia* (Danzig 1668), which he had sent in September 1668, be presented to the Bodleian Library. See HEVELIUS–OLDENBURG [19]/29.IX.1668 (OLDENBURG, *Correspondence* V, 50–1).

17. WALLIS to OLDENBURG, 7/[17] November 1668

And these which I last had were for them, but are not yet delivered to the Curators.

If I am to present the Cometography gratis; then there will be so much mony to be refunded to them. And that which you now send, must either be returned to you, or ly here till I meet with a Chapman for it. If Hee shall think good to present a copy of all, to the University of Cambridge allso, (to whom, I think he hath not yet sent any;) I beleeve it would be very well taken.

As to the Tydes; I do rather expect that these at the Full-moon should be very high than those last at the New; (though I advised that both should be observed;) For it was the New-Moon that made the high Tydes at Candlemas (if I do not mis-remember) & therefore I expect the Full-moon should do it now at Allhollantide. And there appearing (as you say) nothing extraordinary at the New, confirms me in expectation of it at the Full. I had written thus far in answer to that by the Post, before I had the pacquet by the Carrier.

The Observations of the Bristoll Tydes[150] I shal return you suddenly; but will not charge this with it. I have acquainted the Vice-Chancellor[151] (in the morning) with Hevelius's present, (& have this evening sent it him;) hee tells mee that he intends to send him a letter of thanks in the Name of the University.

If you write before that come to your hands, you may intimate so much. In the paper of the Bristol Tydes; I do a little wonder that he fixeth the Annuall high-tydes on those (be it at New or Full) which happen nearest

---

1 to the Curators *add.*
4 to be *(1)* returned *(2)* refunded
4 must *add.*
15 Tydes *(1)* Ob *breaks off* *(2)* because you call it a Copy, I do not return, (unlesse you send for them.) But I should else have thought it the Original. *(3)* I shal return you suddenly; but will not charge this with it.
17 (in the morning) *add.*
20 If you write ... so much. *add.*
22 those *(1)* next *(2)* (be

---

[150] Observations of the Bristoll Tydes: probably the observations made in Hongroad near Bristol by Samuel Sturmy (1633–69), writer on seamanship, *ODNB*. Oldenburg received Sturmy's observations about mid-October by way of BEALE–OLDENBURG 12/[22].X.1668 (OLDENBURG, *Correspondence* V, 83). They were read at the meeting of the Royal Society on 19 November 1668 (BIRCH, *History of the Royal Society* II, 326–7), and printed under the title 'An Account of some Observations, made this present year by Capt. Samuel Sturmy in Hong-Road within four miles of Bristol, in Answer to some of the Queries concerning the Tydes, in No. 17 & No. 18' in *Philosophical Transactions* No. 41 (16 November 1668), 813–17.

[151] Vice-Chancellor: i.e. John Fell.

17. WALLIS to OLDENBURG, 7/[17] November 1668

the Aequinox, be it before or after it. For this doth not agree with what account[152] I once heard given to the R. Society by another, of the Tydes about Chepstow bridge; that their highest tydes there, are at St Davids & Michaelmas Stream. I should rather think, that it is allways before the vernall Equinox, & allways after the Autumnal: (But because both are pretty near it, perhaps it was thought to refer to the Equinox.) And perhaps, allways at the New Moon before the one; & the Full moon after the other. And I think it were not amisse if he were desired hereafter to take a little notice of this in particular, whether it be not so.|

I do not know by what accident the single sheet,[153] which you now send,[154] came into your hands. For I knew not that it was ever sent nor was it intended to bee: & I was a little surprised to see it. It was that I first wrote, but finding it to swell too much; I did it over again[155] to contract it; & you will see the whole of this sheet (for substance) to be not so much as the first leaf of the two sheets. I suppose that I did by a meer mistake in folding up the two sheets, unwittingly clap in that (being at hand) thinking it had been a part of the other. Which if I did, it was pure heedlesseness without any intention: nor did I know it was done. So that if you had not sent them up I should never have understood what you meant by the single sheet. However, since it is so; you may now omitt what is in the first leaf of the two sheets, till you come to Porro (ut minutiora quaedam praeteream) &c. And print on, the next (in this months transactions) as being a part of the former letter of July. 18. but omitted in the last either by a mistake, or

---

1 agree (*1*) of (*2*) with what account I |once *add.*| heard
4 it (*1*) alt *breaks off* (*2*) is
6 Equinox.) (*1*) Allways, (*2*) And perhaps, allways (*a*) before the (*b*) at
10 single |sheet, *add.*| which you now (*1*) sent (*2*) send
12 it (*1*) ind *breaks off* (*2*) intended
16 hand) (*1*) as a (*2*) thinking
18 intention: (*1*) However (*2*) nor
20 in the (*1*) beginning (*2*) first

---

[152]account: i.e. the observations presented at the meeting of the Royal Society on 12 December 1666 by Henry Powle (1630–92), ODNB. See WALLIS–OLDENBURG 7/[17].III.1667/8 (ii); WALLIS, *Correspondence* II, 435–9.

[153]single sheet: i.e. the earlier, incomplete draft of WALLIS–OLDENBURG 18/[28].VII.1668 (there edited as W^1), which already had been printed in *Philosophical Transactions* No. 39 (21 September 1668); WALLIS, *Correspondence* II, 499–506.

[154]send: enclosed in OLDENBURG–WALLIS early XI.1668.

[155]did it over again: i.e. the revised version of WALLIS–OLDENBURG 18/[28].VII.1668 (there edited as W^2); WALLIS, *Correspondence* II, 507–24.

17. Wallis to Oldenburg, 7/[17] November 1668

for want of room. (whether the Rayling Print come or not.) For my meaning was that the whole letter of July 18. should have come together. Only (as it now happens) the first part was intended to be a little more contracted then as it is now printed.

If the Print of his new Railing come over timely inough; you shall have some small reply to it; perhaps in 8 or 10 lines; and no more. Unless possibly I may withall shew that his whole second chapter[156] (for substance) is taken out of Oughtred; & cite paragraph by paragraph whence it is taken. Onely this I observe for the present: that there is not any one thing for which I blame him; which he doth either deny, or defend. Onely rails at mee for taking notice of it. I think I shall send you all back on Tuesday next by Dr Alestree[157] who then comes up. Who allso will pay you the 3^l 10^s (for so much I think it should be, though your figures be blotted so as not to be read) for those bookes I receive to day; which were for him; & I have allready caused them to be delivered to him. I add no more at present, but that I am

Yours &c
J. W.

These
For Mr Henry Oldenburg
in the Palmal near
St James's
London.

1 (whether (*1*) P *breaks off* (*2*) the Rayling Print come or not.) *add.*
5 If (*1*) he f *breaks off* (*2*) the
6 lines; (*1*) unless (*2*) and
10 which |he *add.*| doth either (*1*) defe *breaks off* (*2*) deny,
11 back (*1*) by (*2*) on

---

[156]Chapter: i.e. of GREGORY, *Exercitationes geometricae*, chapter two: 'Analogia inter lineam meridianam planispherii nautici: & tangentes artificiales geometrice demonstrata, &c'.

[157]Alestree: i.e. Richard Allestree (1621/2–81), regius professor of divinity in the University of Oxford and provost of Eton College, *ODNB*.

18.
WALLIS to JOHANNES HEVELIUS
12/[22] November 1668

Transmission:

Manuscript missing.

Existence and date: Mentioned in WALLIS–OLDENBURG 14/[24].XI.1668.

Possibly this letter or a copy of WALLIS–HEVELIUS 26.X/[5.XI].1668 thanking Hevelius for the gift of his recently-published *Cometographia* was sent as an enclosure to OLDENBURG–HEVELIUS 11/[21].XII.1668; OLDENBURG, *Correspondence* V, 237–9.

19.
WALLIS to CHRISTIAAN HUYGENS
Oxford, 13/[23] November 1668

Transmission:

W Letter sent: LEIDEN *Bibliotheek der Rijksuniversiteit* Hug. 45, No. 1676, 4 pp. (p. 3 blank) (our source).—printed: HUYGENS, *Œuvres complètes* VI, 296–8.

w Part copy of *W* (in Oldenburg's hand): LONDON *Royal Society* Early Letters W1, No. 67, 4 pp. (p. 3 blank). On p. 4 in Oldenburg's hand: 'Dr Wallis's letter to Hugens Nov. 13. 68.'

Reply to: HUYGENS–WALLIS [3]/13.XI.1668.

This letter was sent via Oldenburg, who transmitted it to Huygens as an enclosure to a letter of his: OLDENBURG–HUYGENS 18/[28].XI.1668; OLDENBURG, *Correspondence* V, 176–8.

Oxoniae Novemb. 13. 1668.

Nobilissime Vir,

Humanissimas tuas literas,[158] et mihi gratissimas, datas Parisiis 13° Novemb. stilo vestro; hic accepi Novemb. 12°. stilo nostro. Cur autem meae[159] Tibi tam sero traderentur, nescio. Scripseram utique eodem die qui adscriptus erat: Quo (si memini) D. Ryckius,[160] hinc descessit; cui paulo post Londinum (uti jusserat) transmittebantur: unde intra octiduum se abiturum

---

[158] literas: i.e. HUYGENS–WALLIS [3]/13.XI.1668.

[159] meae: i.e. WALLIS–HUYGENS 31.VIII/[10.IX].1668.

[160] Ryckius: i.e. Dirk de Ryke Cf. WALLIS–HUYGENS 31.VIII/[10.IX].1668; WALLIS, *Correspondence* II, 568–73.

19. WALLIS to HUYGENS, 13/[23] November 1668

dixerat, recta in Galliam traiecturum, et saltem ante finitum Septembrem tibi affuturum Parisiis. Sed omnino fieri potest (quod peregrinantibus usu venire solet) ut minus expedito itinere usus fuerit quam speraverat. At non multum interest; quum nihil inibi contineretur quod magnam postulabat festinationem.

Quod Jac: Gregorium spectat: haereo quid dicam. Ut qui tam male usus est favore praeterito, ut nesciam annon poenitere oporteret eorum quae in ipsius gratiam dixeram. De libello[161] ejus, tum nuper edito,[162] sententiam rogatus,[163] responderam generalia quaedam: prout ex levi inspectione licuit, nondum singulis examinatis: neque eram ultra solicitus. Quibus alii[164] (ut videtur) alia addebant, unde quem edidit characterem[165] formavit D. Oldenburgius. Post editas tuas animadversiones,[166] ejusque responsum;[167] nesciebam adhuc quod prae se ferret, legitime demonstratum ibidem esse, non posse Circulum Analytice (ut loquitur) quadrari. Quamquam enim hujusmodi quid videatur in Praefatione sua insinuare; falsus tamen ibidem videbatur, quod *Demonstratio integra ad Phrasem Geometricam reducta nondum esset*. Quae quidem ego sic interpretatus eram, ac si insinuasset, Rem veram sibi videri, (uti et videtur mihi,) sed nondum a se legitime demonstratam esse secundum Geometricum rigorem. Quod fecit ut ea scriberem[168] quae proximis literis legisti. Statim autem atque illas miseram; ex insperato interpellatus sum, se illud non tantum ut Verum asserere, sed ut a se legitime Demonstratum; et postulare ut ego vel idem statim pronunciare vellem, vel rationes in contrarium proferre quibus ipse respondeat. Quorum cum prius non potuerim, posterius feci; ad hunc fere sensum: Nempe,

3 usus *add.*

---

[161] libello: i.e. GREGORY, *Vera circuli et hyperbolae quadratura*.

[162] edito: J. Gregory's *Vera circuli et hyperbolae quadratura* had been reprinted and annexed to his *Geometriae pars universalis*, Padua 1668.

[163] rogatus: i.e. the request which was almost certainly made by Sir Robert Moray and to which Wallis replied in WALLIS–MORAY? ?.II.1667/8; WALLIS, *Correspondence* II, 432.

[164] alii: i.e. Brouncker and Collins. See WALLIS–BROUNCKER 4/[14].XI.1668.

[165] characterem: i.e. the review of Gregory's *Vera circuli et hyperbolae quadratura* published by Oldenburg in *Philosophical Transactions* No. 33 (16 March 1668), 640-4, the beginning of which reads: 'This tract perused by some very able and judicious mathematicians, and particularly by the lord Viscount Brouncker and the reverend Dr. John Wallis, receiveth the character of being very ingeniously and very mathematically written, etc.'

[166] animadversiones: i.e. Huygens' critical review of J. Gregory's *Vera circuli et hyperbolae quadratura*, printed in *Journal des Sçavans* (2 July 1668), 52–6.

[167] responsum: i.e. GREGORY, 'Mr. Gregories Answer to the Animadversions of Mr. Hugenius upon his Book, De Vera Circuli & Hyperbolae Quadratura', printed in *Philosophical Transactions* No. 37 (13 July 1668), 732–5.

[168] scriberem: i.e. WALLIS–HUYGENS 31.VIII/[10. IX].1668.

19. WALLIS to HUYGENS, 13/[23] November 1668

(praeterquam quod multa Lemmata praesumpserat, quae utut vera forent, non tamen statim postulanda erant, sed Demonstranda potius, priusquam Demonstratio perfecta dici debeat:) haec duo saltem impediebant, quo minus illud pronunciare possem; Nimirum, Quod Conversa propositionis Decimae non fuerit demonstrata, (quod ipsum fere cum tu objeceras, nihil reposuerat:) Quodque, utut ea vel maxime demonstrata fuisset; adeoque (quod vult Propositio 11^a) *Sectorem indefinite sumptum* non esse Analyticum &c; [2] adeoque non *Omnem*: non inde tamen sequatur, *Nullum* esse;| adeoque nec, Circulum non posse sic quadrari; cum ad hoc non opus sit ut *Omnes* sed ut *Certi quidam* Sectore essent (ut loquitur) *Analytici* cum Quadrato Radii vel Diametri. (Quod non temere obiectum esse, exemplis aliquot ostendi). Quae tantum abest ut ipsi satisfaciant, ut mihi non multo mitius quam Tibi irascatur: tum quod ego non statim pro ipso pronunciaverim: tum quod per me stetisse autumet, quod a Societate Regia nondum obtinuerit (quod maxime vellet) ejusmodi Testimonium. Habes itaque (Vir Clarissime) quid ego hac in re senserim. Quid sentiant alii, nondum audio: Sed necdum scio quid sit quod tu reponis,[169] Utut enim D. Oldenburgium id accepisse putem; nondum tamen hic accepimus. Reliqua quod spectat; Quid illud sit quod ad Equitem Moraeum te olim scripsisse[170] dicis, non memini me vidisse, ut nihil ea de re sit quod dicam: non dubito tamen quin ille futurus sit tibi satis aequus. Quod autem tibi non displicuisse dicas, eadem Barovio nostro fuisse cognita quae tu indicas: quodque (summam rei quod spectat) Spaciorum Hyperbolicorum cum Logarithmis comparationem, Gregorio de Sancto Vincentio primitus attribuendam agnoscas: Omnino illud facis quod Virum ingenuum decet. Quae autem in te petulantius dicta sunt, cave credas mihi placuisse: etiam paria quasi passo, non tantum in privatis ad me literis,[171] sed in posterioris libelli[172] praefatione; quam Tu

---

25 in te *add.*

[169]reponis: i.e. HUYGENS–GALLOIS ?.XI.1668. This letter was published as 'Extrait d'une lettre de M. Hugens à l'Auteur du Journal, touchant la Réponse que M. Gregory a faite à l'examen du Livre intitulé Vera Circuli & Hyperboles Quadratura, dont on a parlé dans le V. Iournal de cette année', *Journal des Sçavans* (12 November 1668), 109–12 (HUYGENS, *Œuvres complètes* VI, 272–6). Huygens had apparently intended to send a copy of this reply with HUYGENS–WALLIS [3]/13.XI.1668, but according to Oldenburg, through whose hands the letter was transmitted, no such enclosure arrived. See OLDENBURG–HUYGENS 18/[28].XI.1668; OLDENBURG, *Correspondence* V, 176–8.

[170]scripsisse: i.e. HUYGENS–MORAY [8]/18.VIII.1662 and appendix, HUYGENS, *Œuvres complètes* VI, 200–2, 205–6. Cf. MORAY–HUYGENS 22.VIII/[1.IX].1662, HUYGENS, *Œuvres complètes* VI, 216–17.

[171]literis: i.e. GREGORY–WALLIS 26.X/[5.XI].1668 and other now missing letters.

[172]libelli: i.e. GREGORY, *Exercitationes geometricae*, London 1668.

(credo) prius quam ego videras. Ubi postquam multa acerbe satis questus est Gregorius; ἀποσπασμάτιον quoddam ex epistola[173] quadam mea ad alium scripta decerptum, imperfecte recitat (omissis quae erant praecipua) nec satis fideliter. Non enim ego (quod ait) contra *Doctrinam* ejus quicquam objeceram (utpote quam ipse, ni fallor, olim demonstraveram, *Arithm. Infin. Schol.* prop. 190.) sed contra *Demonstrationem*. (Et quidem diserte dixeram, *rem ipsam me neutiquam repudiare*, sed non vidisse me, ubinam *Demonstraretur*.) Sed neque *Admittebam* (quod affirmat ille) *demonstratum esse, quod Sector ABIP non sit analytice compositus,* &c; sed, quod non sit, *sic ut dicebatur,* analytice compositus; hoc est, non eo modo quem ad Prop. 10. indicaverat. Quae duo (missis reliquis) monenda duxi, ne sensum meum ex ejus verbis secus quam est concipias. Denique, nolim ut existimes, Reg. Societatem de Te secus quam par est sentire. Vale, Vir Nobilissime, et amare pergas,

Tui studiosissimum,
Johannem Wallis.

Nobilissimo Eruditissimoque Viro, [4]
D. Christiano Hugenio a Zulychem,
Parisiis.

20.
JOHN COLLINS to JOHN PELL
London, 14/[24] November 1668

Transmission:

C Letter sent: LONDON *British Library* Add. MS 4278, f. 345^r–345^v. On f. 345^r endorsement in Pell's hand: 'Received 16th November'. Postmark: 'NO/14'.

Reverend Sir

I have yours[174] of the 28th of October. In answer whereto say as to Anderson that I thinke I signifyed before that he had high conceits of himselfe, and

---

2 questus est *(1)* , *(2)* Gregorius;

[173] epistola: i.e. WALLIS–BROUNCKER 4/[14].XI.1668.
[174] yours: i.e. PELL–COLLINS 28.X/[7.XI].1668.

20. COLLINS to PELL, 14/[24] November 1668

perschaunce you have met with something of that nature. The Prop at the end of his Booke,[175] is irrelative to the rest, and not so full as Bachet.[176]

As to Mr Gregorie he is gone home with the Archbishop[177] of St Andrewes into Scotland who hath praeferred him to a Lectureship at St Andrewes. The rest of the said Mr Gregories Bookes published in Italy are not yet arrived. He doth indeed cheifely write against Hugenius, though Dr Wallis be intended in the latter part of his praeface, and what the said Mr Gregorie there answers was wrote in a private Letter[178] to my selfe. I have not yet received Slusius[179] though I thinke I may praesume, it is aboard of a ship in the Thames. As to the note inclosed about Tenn pounds, I have this to say, that no part of the former that I lent by your request is yet repaid, and the which in it selfe is not one Moity of what the L. B.[180] oweth mee, from whome I have never as yet received any thing, though frequently promised, and as soone as accomplished I shall willingly performe your request,[181] if I continue in the Employment now under his Lordship. But I find it will be a laborious Drudgery, and that I can better spend my time in other affaires and therefore have thoughts of relinquishing the same. Not else but that I am

London 14th November
1668

Your humble Servitor
John Collins

[5v] To the Reverend Doctor John Pell
at the house of the right honourable
the Lord Brereton
at Brereton
In Cheshire
Stonebagg[182]

---

[175] Booke: i.e. ANDERSON, *Stereometrical Propositions variously applicable*, London 1668.
[176] Bachet: i.e. Claude-Gaspard Bachet, Sieur de Méziriac (1580–1639). His edition of Diophantus's *Arithmetica* was published in Paris in 1621.
[177] Archbishop: i.e. James Sharp (1618–79), *ODNB*.
[178] Letter: i.e. WALLIS–COLLINS 8/[18].IX.1668.
[179] Slusius: i.e. SLUSE, *Mesolabium*, 2nd ed., Liège 1668.
[180] L.B.: i.e. Lord Brereton.
[181] request: i.e. Pell's request for money.
[182] Stonebagg: i.e. the postbag for Stone, Staffordshire, on the arterial postal route to Ireland. Cf. COLLINS–PELL 23.X/[2.XI].1668.

21.
WALLIS to HENRY OLDENBURG
Oxford, 14/[24] November 1668

Transmission:

W Letter sent: LONDON *Royal Society* Early letters W1, No. 68, 2 pp. (our source). On p. 2 beneath address in Oldenburg's hand: 'Rec. Nov. 16. Answ. Nov. 17.' Postmark on p. 2: 'NO/16'.—printed: OLDENBURG *Correspondence* V, 161–3.

Enclosures: HUYGENS–WALLIS [3]/13.XI.1668, WALLIS–HUYGENS 13/[23].XI.1668, and WALLIS–HUYGENS 31.VIII/[10. IX].1668.

This letter and its enclosures served to inform members of the Royal Society of the current state of the dispute between Huygens and Gregory.

Oxford. Novemb. 14. 1668.

Sir,

I would not have sent you so large a pacquet by the Post, but that I thought it was convenient that you should see as soon as may be both what Hugenius says to me[183] & what answer[184] I make him; (& that you may the better understand both, I send you allso the copy of what I wrote to him before:[185]) The rather, because, there beeing so much in his which concerns the R. Society, it may possibly be proper for my Lo. Brouncker, Sir Robert Moray, & whom else you think fit, to see both his & mine. I have told you formerly, that Mr Gr:[186] in my opinion is certainly in the wrong. And therefore, since this is that which must at last be sayd, I thought it as proper for me to say it at first, before I see what Mr H. now sends by way of reply to him: And, withall, to represent the R. Society as uningaged (as indeed they are;) because I take that to be more honourable for them; than to appear & retreat. You may possibly think also that (especially things being so) Mr Gr. may have demeaned himself a little too intemperately towards M. H. considering him as a person of quality, as one well skilled in these things, and as in the right. And therefore I could not well say less then I have done; Especially having before appeared to him in the behalf of Mr Gregory; so long as he would permit himself to be looked upon as in a capacity to be defended. The summe of his Demonstration (since he

---

[183] says to me: i.e. HUYGENS–WALLIS [3]/13.XI.1668.
[184] answer: i.e. WALLIS–HUYGENS 13/[23].XI.1668.
[185] wrote to him before: i.e. WALLIS–HUYGENS 31.VIII/[10.IX].1668.
[186] Mr Gr:: i.e. James Gregory.

21. WALLIS to OLDENBURG, 14/[24] November 1668

will needs defend what I would have waved for him) depends upon these two Syllogisms. If the Sector, indefinitely taken, *can be*, in such manner as he speakes, Analytically compounded; Then the Circle *can be* Analytically squared: But the Sector *cannot bee* so compounded: Therefore the Circle *cannot be* so squared. Which Syllogism is manifestly peccant in form. The Minor of that, hee proves by another in the same form. If there *can be* a quantity in the same manner compounded of the two first, & of the two second Terms; Then the Sector *can be* Analytically compounded: But there *cannot be* any such quantity: Therefore the Sector *cannot be* so compound. Both are false Syllogisms; and can conclude nothing, be the premises never so true, unless not onely the consequences of the Majors, but allso the *Converse* of those consequences be demonstrated. (Which any Logician, though no Mathematician, will easyly discern.) Otherwise, he might as well argue: If Virgil *be* the same with Homer; Then *both are* Poets: But Virgil is *not* the same with Homer: Therefore *Neither of them* is a Poet. This I say to yourself; But you need not trouble Mr Gr. with it: For I have allready sayd the same in effect to him, more than once. But he having once persuaded himself that the Demonstration is good; and (as he tells mee expressely) *loathing to revise what hath once passed him*; Hee cannot but be angry with any that would undeceive him. I have yet been so carefull all along in what I say, as to speak onely my own opinion; without declaring for others who's opinions I know not: lest while I go about to right myself, I should wrong them. If it be thought necessary that more be sayd or done in reference to the R. Society: yourselves upon the place, are more competent judges of that than I. When you have done with the inclosed, you may please to return them to mee, all but that which I now send to Mr H. which I shall desire you to seal & send away; (if, at lest, my Lo. Br. &c do not see reason to countermand it:) and, withall, to supply mee with the Reply which Mr H. now sends, to which his Letter refers. Which is all I shall say at present to that affaire.

The great winds which all this week wee have had here, make mee not much to doubt that your Spring-tydes (as well as the Winds) have been high at London. For I take them both to proceed from the same cause; & sutable to my hypothesis. And I expect your next will tell mee, that it is so come to pass as I expected it would bee. But I expect allso, that your seamen who tell you the Tydes are high; will tell you allso (because they know no other)

8 Sector *(1)* may *(2)* can
19 cannot *(1)* be *(2)* but
23 thought *add.*

that the Winds are the cause of it. For so they have used to say in other years. I adde onely, that I am

Yours &c.
J. Wallis.

Pray send me one other whole set of Hevelius's workes; and a second of all except the Cometography,[187] That is, one Cometography, & two of all the rest. I suppose you have the mony from D. Alestree;[188] & the pacquet[189] I sent by him on Tuesday,[190] & one of Thursday[191] with one enclosed[192] for Hevelius.

For Mr Henry Oldenburg, in [2]
the Old Palmal near
St James's,
London.

22.
WALLIS to HENRY OLDENBURG
Oxford, 15/[25] November 1668

Transmission:

W Letter sent: LONDON *Royal Society* Early Letters W1, No. 69, 4 pp. (our source). At top right of p. 1 in Oldenburg's hand: 'Dr Wallis's (*1*) Theory of (*2*) Account of the Laws of Motion, sent to Mr Oldenburg, and produced at the Society Nov. 26. 1668.' At top left of p. 1 Oldenburg has noted and deleted: '(*1*) What time (*2*) This produced at the Society. ⟨N⟩ov. 26. and orderd to be enterd.' On p. 4 beneath address in Oldenburg's hand: 'Rec. Nov. 18. 68.', and to the left of address, again in Oldenburg's hand: 'Dr Wallis of motion. of Nov. 15. 68. Rec. Nov. 18. 68.' Beneath this in unknown hand: 'Entred.'—printed: OLDENBURG, *Correspondence* V, 164–7 (Latin original); 167–70 (English translation).

---

[187]Cometography: i.e. HEVELIUS, *Cometographia: totam naturam cometarum... exhibens*, Danzig 1668.

[188]Alestree: i.e. Richard Allestree. Cf. WALLIS–OLDENBURG 7/[17].XI.1668.

[189]pacquet: i.e. the packet in which Wallis returned the two versions of WALLIS–OLDENBURG 18/[28].VII.1668. Wallis had already announced the dispatch of this packet in WALLIS–OLDENBURG 7/[17].XI.1668.

[190]Tuesday: i.e. 10 November 1668.

[191]Thursday: i.e. 12 November 1668.

[192]one enclosed: i.e. WALLIS–HEVELIUS 12/[22].XI.1668.

22. WALLIS to OLDENBURG, 15/[25] November 1668

w Copy of letter sent in Oldenburg's hand (with additions and emendations given in WALLIS–OLDENBURG 26?.XI/[6?.XII].1668): LONDON Royal Society Classified Papers III (1), No. 42, 2 pp.
E First edition (based on *w*): *Philosophical Transactions* No. 43 (11 January 1668/9), 864–6 ('A Summary Account given by Dr. John Wallis, of the General Laws of Motion, by way of Letter written by him to the Publisher, and communicated to the R. Society, Novemb. 26. 1668.').

Oxoniae Novemb. 15. 1668.

Petis, Vir Clarissime, (si mentem tuam satis assequor,) ut, quae mea sunt de Motibus aestimandis Principia, paucis aperire velim. Id autem, si meministi, jam olim factum est; non modo in illo Opere[193] quod ante octo menses Regiae Societati exhibitum,[194] eorum jussu Praelo subjectum est: sed et jamdudum in duobus Scriptis Regiae Societati ante plures Annos exhibitis, quae et te penes sunt. Quorum alterum,[195] ex generalibus Motus Principiis, rationem reddit, qui fieri possit, ut Homo flatu suo (Vesicam inflando) saltem Centipondium elevare potis sit: (quod Experimentum, ante sexdecim vel octodecim annos Oxoniae exhibitum, coram ipsis aliquoties repetitum fuit:) Alterum[196], varia de Experimento, *Torricelliano* dicto, Phaenomena, ex Principiis Hydrostaticis exponit.

Summa rei huc redit.[197]

1. Si Agens ut A, efficit ut E; Agens ut $2A$, efficiet ut $2E$; $3A$, ut $3E$, &c. caeteris paribus: Et, universaliter, mA ut mE; cujuscunque Rationis Exponens sit m.

---

2 Petis, V. C. ut quae *w* E

[193] in illo Opere: i.e. WALLIS, *Mechanica: sive, de motu, tractatus geometricus*, 3 parts, London 1670–71.
[194] exhibitum: At the meeting of the Royal Society on 30 April 1668, Wallis had shown members part of his *Mechanica*, whereupon he 'was desired by the Society to hasten its publication'. See BIRCH, *History of the Royal Society* II, 275.
[195] alterum: see WALLIS, *Mechanica* III, 15, prop. 3 ('Inflata Vesica pondus elevare'), 759–67; WALLIS, *Opera mathematica* I, 1056–60. At the meeting of the Royal Society on 4 March 1662/3, Wallis had given an account of this experiment, in which a weight is raised by the blowing of a bladder. See BIRCH, *History of the Royal Society* I, 206.
[196] Alterum: i.e. Wallis's account of Goddard's barometric experiments, which he had presented to the Royal Society on 20 August 1662 and which was printed in his *Mechanica* III, 14, prop. 7 and 8, 714–9; WALLIS, *Opera mathematica* I, 1035–6. See BIRCH, *History of the Royal Society* I, 104–6.
[197] Summa rei huc redit: for the following §§ 1–8 cf. WALLIS, *Mechanica* I, 1, prop. 1–30, 9–31; (WALLIS, *Opera mathematica* I, 580–94), and for §§ 9–14 cf. WALLIS, *Mechanica*, III, 11, prop. 1–14, 660–82; (WALLIS, *Opera mathematica* I, 1002–12).

22. Wallis to Oldenburg, 15/[25] November 1668

2. Ergo, Si Vis ut V moveat Pondus P; Vis ut mV movebit mP, caeteris paribus: puta, per eandem Longitudinem eodem Tempore; hoc est, eadem Celeritate.

3. Item, Si Tempore T moveat illud per Longitudinem L; Tempore nT, movebit per Longitudinem nL.

4. Adeoque, Si Vis V Tempore T, moveat Pondus P, per Longitudinem L; Vis mV, Tempore nT, movebit mP, per Longitudinem nL. Et propterea, ut VT (factum ex viribus et tempore) ad PL (factum ex pondere et longitudine:) sic $mnVT$, ad $mnPL$.

5. Quoniam Celeritatis gradus sunt Longitudinibus eodem Tempore transactis Proportionales; seu, quod eodem recidit, Reciproce Proportionales Temporibus eidem Longitudini transigendae impensis: erit $\frac{L}{T} \cdot C :: \frac{mL}{nT} \cdot \frac{m}{n} C$. Hoc est, Gradus Celeritatum, in ratione Composita ex Directa Longitudinum et Reciproca Temporum.

6. Ergo, propter $VT \cdot PL :: mnVT \cdot mnPL$: Erit $V \cdot \frac{PL}{T} :: mV \cdot \frac{mnPL}{nT}$: Hoc est, $V \cdot PC :: mV \cdot mPC = mP \times C = P \times mC$.

7. Hoc est, Si Vis V movere potis sit Pondus P, Celeritate C: Vis mV movebit vel idem Pondus P, Celeritate mC; vel eadem Celeritate, Pondus mP; vel denique quodvis Pondus ea Celeritate, ut factum ex Pondere et Celeritate sit mPC.

8. Atque hinc dependet omnium Machinarum (pro facilitandis motibus) construendarum ratio. Nempe, ut qua ratione Augetur Pondus, eadem Minuatur Celeritas; quo fiat, ut Factum ex Celeritate et Pondere, eadem Vi movendo, idem sit. Puta $V \cdot PC :: V \cdot mP \times \frac{1}{m} C = PC.$| [2]

9. Si Pondus P, Vi V, Celeritate C latum, in Pondus quiescens (non impeditum) mP directe impingat: ferentur utraque Celeritate $\frac{1}{1+m} C$. Nam, propter eandem Vim majori Ponderi movendo adhibitam, eadem ratione minuetur aucti Celeritas. Nempe $V \cdot PC :: V \cdot \frac{1+m}{1} P \times \frac{1}{1+m} C = PC$. Adeoque Alterius Impetus (intellige Factum ex Pondere et Celeritate) fiet $\frac{1}{1+m} PC$; Reliqui $\frac{1}{1+m} mPC$.

10. Si in Pondus P (Vi V) Celeritate C latum; directe impingat aliud, eadem via, majori Celeritate insequens; puta Pondus mP, Celeritate nC, (adeoque Vi mnV latum:) ferentur ambo Celeritate $\frac{1+mn}{1+m} C$. Nam $V \cdot PC ::$ $mnV \cdot mnPC :: V + mnV = \frac{1+mn}{1} V \cdot \frac{1+mn}{1} PC = \frac{1+mn}{1} P \times \frac{1+mn}{1+m} C$. Adeoque praecedentis Impetus fiet $\frac{1+mn}{1+m} PC$; subsequentis, $\frac{1+mn}{1+m} mPC$.

11. Si Pondera contrariis Viis lata, sibi directe occurrant sive impingant mutuo; puta, Pondus P (vi V) Celeritate C, Dextrorsum; et Pondus

---

1 mV (1) moveat (2) movebit W

22. Wallis to Oldenburg, 15/[25] November 1668

mP, Celeritate nC, (adeoque Vi mnV,) Sinistrorsum: Utriusque Celeritas, Impetus, et Directio, sic colliguntur. Pondus dextrorsum latum, reliquo si quiesceret, inferret Celeritatem $\frac{1}{1+m}C$, adeoque Impetum $\frac{1}{1+m}mPC$, dextrorsum; sibique retineret hanc eandem Celeritatem, adeoque Impetum $\frac{1}{1+m}PC$ dextrorsum; (per § 9.) Pondusque sinistrorsum latum (simili ratione) reliquo si quiesceret, inferret Celeritatem $\frac{mn}{1+m}C$, adeoque Impetum $\frac{mn}{1+m}PC$ sinistrorsum; sibique retineret hanc eandem Celeritatem, adeoque Impetum $\frac{mn}{1+m}mPC$ sinistrorsum. Cum itaque Motus utrinque fiat; Impetus dextrorsum prius lati, jam aggregatus erit ex $\frac{1}{1+m}PC$ dextrorsum, et $\frac{mn}{1+m}PC$ sinistrorsum; adeoque reapse vel dextrorsum vel sinistrorsum, prout ille vel hic major fuerit, eo impetu qui est duorum differentia. Hoc est, (posito + signo *Dextrorsum*, et − *Sinistrorsum* significante,) Impetus erit $+\frac{1}{1+m}PC - \frac{mn}{1+m}PC = \frac{1-mn}{1+m}P$; Celeritas $\frac{1-mn}{1+m}C$; (adeoque Dextrorsum vel Sinistrorsum, prout 1 vel mn major fuerit.) Et similiter, Impetus Sinistrorsum prius lati, erit $+\frac{1}{1+m}mPC - \frac{mn}{1+m}mPC = \frac{1-mn}{1+m}mPC$; Celeritas $\frac{1-mn}{1+m}C$: Adeoque Dextrorsum vel Sinistrorsum, prout 1 vel mn major fuerit.

12. Si vero Pondera nec eadem directe via procedant, nec directe contraria, sed oblique sibi mutuo impingant: moderandus erit praecedens calculus pro Obliquitatis mensura. Impetus autem Oblique impingentis, ad ejusdem Impetum qui esset si directe impingeret (caeteris paribus,) est in ea ratione qua Radius ad Secantem anguli Obliquitatis: Quae quidem Consideratio, cum Calculo priore debite adhibita, determinabit quaenam futura sint sic Oblique impingentium Celeritas, Impetus, et Directio: Hoc est, Quo Impetu, qua Celeritate, et in quas partes ab invicem resilient, quae sic impingunt. Eademque est ratio Gravitationis gravium oblique descendentium, ad eorundem perpendiculariter descendentium Gravitationem. Quod [3] alibi Demonstramus.|

13. Si, quae sic impingunt Corpora, intelligantur non absolute dura (prout hactenus supposuimus) sed ita ictui cedentia ut elastica tamen vi se valeant restituere; hinc fieri poterit ut a se mutuo resilient ea corpora

---

3 Impetum, (*1*) $\frac{1}{1+m}PC$ dextrorsum; (per $ 9.) Pondusque sinistr *breaks off* (*2*) $\frac{1}{1+m}mPC$ dextrorsum; sibique W

22 Obliquitatis: |(Quod etiam intelligendum est, ubi perpendiculariter, sed oblique cadit in percussi superficiem, non minus quam ubi viae motuum se mutuo oblique decussant.) *add*. Oldenburg (on Wallis's instruction)| Quae W Obliquitatis: (Quod...decussant.) Quae w E

23. OLDENBURG to WALLIS, 16/[26] November 1668

quae secus essent simul processura; (et quidem plus minusve prout haec vis restitutiva major minorve fuerit:) nempe si Impetus ex vi restitutiva sit progressivo major.

14. In motibus acceleratis et retardatis, Impetus pro singulis momentis is reputandus est qui gradui Celeritatis tum acquisito convenit. Ubi autem per Curvam fit motus, ea reputanda est, in singulis punctis, motus Directio, quae est Rectae ibidem Tangentis.

Et siquando motus tum acceleratus vel retardatus sit, tum et per Curvam fiat, (ut in Vibrationibus Penduli;) Impetus aestimandus erit, pro singulis punctis, secundum tum gradum accelerationis, tum Obliquitatem ibidem Tangentis.

Atque hae sunt (quantum ego judico) Generales Motuum leges; quae, ad casus particulares, Calculo sunt accommodandae. Quos tamen si sigillatim persequi vellem, Epistolae limites transilirem. Vale.

Tuus
Joh: Wallis.

These [4
For Mr Henry Oldenburg
at his house in the Palmal
near
St James's
London.

23.
HENRY OLDENBURG to WALLIS
16/[26] November 1668

Transmission:

Manuscript missing.

Existence and date: Mentioned in WALLIS–OLDENBURG 19/[29].XI.1668.
Answered by: WALLIS–OLDENBURG 26?.XI/[6?.XII].1668.

8 sit *add.* W
14 transilirem. |Neque commode fieri potest sine Schematum apparatu, quibus hic abstinendum putavi. Vale. *add. Oldenburg (on Wallis's instruction)*| Vale. W transilirem: Neque ... putavi. Vale. *w E*

25. WALLIS to OLDENBURG, 19/[29] November 1668

At the meeting of the Royal Society on 12/[22] November 1668, Oldenburg was instructed to request that Wallis participate in the task of examining and comparing recent writings on the laws of motion. This was probably the letter in which he passed on this request. See BIRCH, *History of the Royal Society* II, 320.

24.
HENRY OLDENBURG to WALLIS
17/[27] November 1668

Transmission:

Manuscript missing.

Existence and date: Mentioned in and answered by WALLIS–OLDENBURG 19/[29].XI. 1668; noted on WALLIS–OLDENBURG 14/[24].XI.1668.
Reply to: WALLIS–OLDENBURG 14/[24].XI.1668.

25.
WALLIS to HENRY OLDENBURG
Oxford, 19/[29] November 1668

Transmission:

W Letter sent: LONDON *Royal Society* Early Letters W1, No. 70, 4 pp. (our source). On p. 4 beneath address in Oldenburg's hand: 'Rec. Nov. 20. 68. Answ. Nov. 24.' Postmark: 'NO/20'—printed: OLDENBURG, *Correspondence* V, 192–4.

Reply to: OLDENBURG–WALLIS 17/[27].XI.1668.
Answered by: OLDENBURG–WALLIS 24.XI/[4.XII].1668.

Oxford Nov. 19. 1668.

Sir,

I have, this morning, yours[198] of Nov. 17. but not that,[199] which it refers to, of the day before: nor did I hear of the books till this letter: While I am writing this (lest the Post bee gone) I send to inquire after them. I had rather you had not sent those Opuscula,[200] but onely notice of them; for

6 onely (*1*) word (*2*) notice

---

[198] yours: i.e. OLDENBURG–WALLIS 17/[27].XI.1668.
[199] that: i.e. OLDENBURG–WALLIS 16/[26].XI.1668. This letter reached Wallis not later than 26 November; see WALLIS–OLDENBURG 26?.XI/[6?.XII].1668.
[200] opuscula: cf. WALLIS–HEVELIUS 14/[24].XI.1668.

25. WALLIS to OLDENBURG, 19/[29] November 1668

I had before told you wee had bought them for the Library, & so it had been but discounting so much mony, & the thing had been the same. The Cometography[201] (which, by the like means, wee had double) served to make up one of that double sett which I last sent for; & had I known of those,
5 I should amongst them have sent for so much less. And therefore, if they be not come away before this come at you, pray send mee those opuscula but once; that is (in all) One whole sett, & (beside it) a Selenography[202] distinct. If they be sooner come away, it is no great matter: onely then they must ly by till I meet with some other opportunity of another Chapman
10 which possibly may not be long. There is one or two more who I expect may be willing to take off a sett of them; & these will serve to make up one of them. I have sent you by mine[203] of Nov. 15. my hypothesis of motion. It should have come with my last pacquet, but that it was not transcribed time inough; & (through the neglect of myself or my man or both) it was forgot to
15 be sent the next day as I intended. But I suppose, before this time you have it being sent on Tuesday. Dr Wren (to whom I pressed the like upon your Letter) told mee he meant to send you his[204] on Tuesday last if he could finish it so soon. (I have not since seen him).| I have (now) those opuscula; [2] but no letter. The Vice-chancellors[205] letter[206] last sent (being supposed to
20 be sent after the receit of these bookes) will contain his thanks for these as well as the other. So that it will not be necessary for him (I think) to write a second: (at lest if for *libro* you could (if not yet sent away) make it *libris*. They shall be putt into the Library with the Inscriptions you have given them; (those formerly bought for the Library, being otherwise disposed of.)
25 Mr Gregory's impudence will reflect on himself, not on mee. And I am sorry for it, for his sake. I gave him private notice[207] of his mistake, more then once; & told him what did remain to be proved. But did withall, endeavour

8 then (*1*) so (*2*) they
15 But I ... on Tuesday. *add.*
17 last *add.*
21 it *add.*

---

[201]Cometography: i.e. HEVELIUS, *Cometographia: totam naturam cometarum... exhibens*, Danzig 1668.
[202]Selenography: i.e. HEVELIUS, *Selenographia: sive, lunae descriptio*, Danzig 1647.
[203]mine: i.e. WALLIS–OLDENBURG 15/[25].XI.1668.
[204]his: i.e. Wren's theory of the collision of bodies, which he presented to the Royal Society in the meeting on 17 December 1668; BIRCH, *History of the Royal Society* II, 335.
[205]Vice-chancellors: i.e. John Fell.
[206]letter: cf. WALLIS–OLDENBURG 7/[17].XI.1668.
[207]notice: see WALLIS–GREGORY 22.X/[1.XI].1668.

25. WALLIS to OLDENBURG, 19/[29] November 1668

(in my former[208] to Mr Hugens) to fetch him off as well as the nature of the business would bear. Hee must now fetch off himself as well as hee can. And I shal hereafter take heed of giving him advise. For if what private advise is given him must be answered in Print, & there mis-represented; hee cutts off the opportunity of friendly advise. What you say of M. Hugens his account[209] he gave of some experiments of motion, is true inough; (I was then present:) but it is true allso that he gave us no account of any principles by which he did calculate but onely of the Result; (and, I well remember, that when hee had so done it, he was carefull to blott out all his writing, that it might not from thence appear how he did calculate:) But hee is to [3] remember allso; that the thing had before been done by Mr Rook[210] &| Dr Wren; & that his calculation at that time was but to try whether his & theirs did agree: & it was found so to do: His answers proving to bee the same which theirs before had done. But this, I suppose, is onely mentioned by him in reference to what he [is] now printing; bout the Laws of motion. Not, to that of the Hyperbole (with which I do not know that this hath any affinity.) But for what concerns that, he refers to some letter[211] written to Sir Rob. Moray.

What concerns the Testimony[212] in the Transactions, is not intended to cast any blame on you; but only to say; that it was intended to give onely an approbation in general, without undertaking for every punctilio in the

---

1 (in (*1*) mine (*2*) my former to Mr Hugens) (*a*) en *breaks off* (*b*) to
3 advise (*1*) , if (*2*) . For
5 of (*1*) being (*2*) friendly
12 & (*1*) were (*2*) that
13 proving (*1*) the sa *breaks off* (*2*) to
15 he his now *corr ed.*
17 But |for *add.*| what (*1*) he con *breaks off* (*2*) concerns

---

[208]former: i.e. WALLIS–HUYGENS 13/[23].XI.1668.
[209]account: i.e. the account in HUYGENS–OLDENBURG [3]/13.XI.1668 (OLDENBURG, *Correspondence* V, 126–7) on the experiments which had been performed on 13 April 1661 during Huygens's visit to London. Cf. MORAY–OLDENBURG 10/[20].X.1665 (OLDENBURG, *Correspondence* II, 559–62).
[210]Rook: i.e. Lawrence Rooke (1619/20–62), from 1652 professor of astronomy at Gresham College; from 1657 professor of geometry at Gresham College, *ODNB*.
[211]letter: i.e. HUYGENS–MORAY [8]/18.VIII.1662 (HUYGENS, *Œuvres complètes* IV, 200–3 and 205–6). See HUYGENS–WALLIS [3]/13.XI.1668.
[212]Testimony: i.e. the short remarks by Wallis and Brouncker at the beginning of the review of J. Gregory's *Vera circuli et hyperbolae quadratura*, printed in *Philosophical Transactions* No. 33 (16 March 1667/8), 640–4; the detailed part of the review was provided by Collins 'upon a more particular examination of this Book'.

book. I have them not at hand, & therefore cannot speak particularly: But by my Mr Gregories to mee,[213] it seems that instead of the word *Summe*, it should have been *Termination* (of a converging series;) which, if need be, may be mentioned[214] amongst some Errata to be amended: Numb. 40, I think, I have not seen. I know not what it is that made the Tydes not answere expectation, unless the wind.

The post is going: I am

Yours &c.

For Mr Henry Oldenburgh, in [4]
the Palmal near St James's
London.

26.
HENRY OLDENBURG to WALLIS
24 November/[4 December] 1668

Transmission:

Manuscript missing.

Existence and date: Mentioned in and answered by WALLIS–OLDENBURG 26?.XI/ [6?.XII]. 1668; noted on WALLIS–OLDENBURG 19/[29].XI.1668.

This letter evidently concerned the publication of Wallis's letter to Oldenburg on the laws of motion, i.e. WALLIS–OLDENBURG 15/[25].XI.1668.

27.
WALLIS to HENRY OLDENBURG
[Oxford], 26? November/[6? December] 1668

Transmission:

W Letter sent: LONDON *Royal Society* Early Letters W1, No. 71, 2 pp. (our source). Manuscript badly damaged at the edges. On p. 2 beneath address in Oldenburg's hand:

2 by *add*.
6 expectation, |I know not, *del.*| unless

---

[213] my Mr Gregories to mee: probably GREGORY–WALLIS 26.X/[5.XI].1668.

[214] mentioned: This error had in fact already been corrected in *Philosophical Transactions* No. 40 (19 October 1668), 812.

27. WALLIS to OLDENBURG, 26? November/[6? December] 1668

'Rec. Nov. 27. 68. Answ.' Postmark: 'NO/27'.—printed: OLDENBURG, *Correspondence* V, 203–5.

Reply to: OLDENBURG–WALLIS 16/[26].XI.1668 and OLDENBURG–WALLIS 17/[27].XI.1668.

Through damage to the manuscript, the date and place of this letter are missing. The date is conjectured from the postmark.

Sir

I have all your letters of last week, & that[215] in particular which was at first want⟨ing.⟩ The Tydes being higher then ordinary a day sooner then they should have been, seems ⟨to show⟩ that they would have been much higher the following days, had not the violence of the wind kept them out. But I do not much expect that the following spring tydes should make any great amends for it. I rather think the water was driven up into the other Channell, & that the Tydes on the Severn were thereby made much higher. But all this is but conjecture. The last parcell of books are not yet come to hand, by reason (as they tell mee) that the wagon was come away before the books came: but I expect them this week which will be time inough. I think it were not amisse that Mr Hevelius should be moved to send some copies over of his *Saturni facies*[216] for those that are willing to have them; for I perceive those here are willing to have his works compleat, notwithstanding that (which I suppose is his reason) Mr Huygens seemes in that point to have guessed[217] the more luckyly. I do not find that wee have, in the publike library, the book you mention of Tychonis Historia Coelestis;[218] & therefore if you please to send us one copy, it will do well. I have, as you advertised, looked somewhat into Borellus;[219] & mean to do it further. I do

8 that *(1)* all *(2)* the Tydes *(a)* all *(b)* on
12 should *(1)* should send *(2)* be moved to send
13 for *(1)* the *(2)* those that are willing to have them; *(a)* & *(b)* for

---

[215]that: i.e. OLDENBURG–WALLIS 16/[26].XI.1668. The other letter of the previous week was OLDENBURG–WALLIS 17/[27].XI.1668.
[216]*Saturni facies*: i.e. HEVELIUS, *Dissertatio de nativa Saturni facie ejusque variis phasibus, certa periodo redeuntibus*, Danzig 1656.
[217]guessed: i.e. in HUYGENS, *Systema Saturnium, sive de causis mirandorum Saturni phaenomenon et comite ejus planeta novo*, The Hague 1659.
[218]Tychonis Historia Coelestis: i.e. TYCHO BRAHE, *Historia coelestis*, ed. Albert Curtz, Augsburg 1666.
[219]Borellus: i.e. BORELLI, *De vi percussionis*, Bologna 1667.

27. WALLIS to OLDENBURG, 26? November/[6? December] 1668

not find but that for the main his result agrees well inough with that of my methode; though he goes other ways to come at it, and (as at lest it seemes to mee) more perplex. I am well inough contented that you print my last latine letter[220] (as yours[221] of Novemb. 24 intimates, which I have just now received:) But you may then add at Numb. 12. after those words *ad secantem anguli obliquitatis*, this Parenthesis, *(Quod etiam intelligendum est ubi percussio, non perpendiculariter, sed oblique cadit in percussi superficiem, non minus quam ubi viae motuum se mutuo oblique decussant.)* which was intended to be comprehended in what was before sayd in general words, but perhaps the Reader would not so easily apprehend it. And then at the end, next before *Vale*, you may adde, *Neque fieri commode poterit sine schematum apparatu, quibus hic abstinendum putavi.* And I like well inough that you send over those pieces[222] of mine which I did put into Latine, to those who are putting the transactions into Latine in Holland:[223] For I doubt they will not do it better for the expressing of my sense: but they must allso have notice to make those little alterations in the schemes as I formerly mentioned[224] on that occasion. I am not sorry that my last[225] to Hugenius was gone before wee had his printed reply;[226] (& I hope yours[227] to him told him that his was not then come to hand:) For hee will thereby see, that I was so far from approving Mr Gregorie's mistake, that I had there made the very same exception which he now makes but had omitted in his

1 for the main *add.*
3 inough *(1)* contend *breaks off (2)* contented
9 to be comprehended *add.*
18 printed *add.*

---

[220] latine letter: i.e. WALLIS–OLDENBURG 15/[25].XI.1668.
[221] yours: i.e. OLDENBURG–WALLIS 24.XI/[4.XII].1668.
[222] pieces: i.e. WALLIS–BOYLE 25.IV/[5.V].1666 and WALLIS–OLDENBURG 18/[28].VII.1666. These two letters, containing Wallis's hypothesis of the tides, had been printed in *Philosophical Transactions* No. 16 (6 August 1666). Wallis translated them at Oldenburg's request in early 1668. See WALLIS–OLDENBURG 1/[11].II.1667/8.
[223] Holland: A Latin translation of the volumes of *Philosophical Transactions* which appeared between 1665 and 1670 was published under the title *Acta Philosophica Societatis Regiae in Anglia* by Hendrick Boom in Amsterdam, 1671–81.
[224] mentioned: see WALLIS–OLDENBURG 18/[28].VIII.1668; WALLIS, *Correspondence* II, 286–7.
[225] my last: i.e. WALLIS–HUYGENS 13/[23].XI.1668.
[226] printed reply: i.e. HUYGENS–GALLOIS ?.XI.1668, published in *Journal des Sçavans* (12 November 1668), 109–12 ('Extrait d'une lettre de M. Hugens...touchant la Réponse que M. Gregory a faite à l'examen du livre intitulé Vera Circuli & Hyperboles Quadratura').
[227] yours: i.e. OLDENBURG–HUYGENS 18/[28].XI.1668 (OLDENBURG, *Correspondence* VI, 176–7, 177), which enclosed WALLIS–HUYGENS 13/[23].XI.1668.

27. WALLIS to OLDENBURG, 26? November/[6? December] 1668

first papers. I know not what Mr Gregory hath to do further but either to say nothing, or to confesse his mistake. But if hee shall still (as I doubt hee will, since this sayth nothing more then what I had written to him before, nor even that so clearly,) ins⟨ist⟩ to talk at the former rate, hee will but betray his ignorance, & unskillfullness in the nature of a demonstration. I shall send back the printed paper by the next oppor⟨tunity⟩ because I intend first to transcribe it. I send you now onely a letter[228] which I had ⟨for⟩merly written to be sent to my Lo. Brouncker (upon the sight of Mr Grey: printed ⟨paper[229]⟩) and had begun to transcribe it; but upon the receit of Mr Gregories last ranting letter ⟨I did⟩ not think fit to bestow further pains upon him. What I send is my first foul dr⟨aft⟩ and I have not time so much as to read it over ⟨before⟩ it goes; it being *[paper torn]* And, when my Lo. hath

[2] read it| *[paper torn]* not taking particular notice of receit of the booke, I can say nothing further than that the letter seems to take it for gran⟨ted⟩ by returning thanks for them. But it is not now to be helped there, since hee knows not that I sent it open. But I shall supply it shortly upon another occas⟨ion.⟩ I am

Yours &c.
J. Wallis.

These
For Mr Henry Oldenburg, in
the Palmal near
St James's
London.

7 *[In left margin in Wallis's hand:]* ⟨I h⟩ave taken ⟨the⟩ letter out again, to send by the carrier; not to swell the postage of this.

9 begun *(1)* trans *breaks off (2)* to
9 of *(1)* his *(2)* Mr Gregories
11 time *(1)* ⟨...⟩ *(2)* so
14 it *add.*

---

[228] letter: i.e. WALLIS–BROUNCKER 4/[14].XI.1668. According to Wallis's marginal note this letter was sent separately to Oldenburg by the Oxford carrier.

[229] printed paper: presumably GREGORY, 'Mr. Gregories Answer to the animadversions of Mr. Hugenius upon his book, De vera circuli et hyperbolae quadratura; as they were publish'd in the Journal des Scavans of July 2. 1668', printed in *Philosophical Transactions* No. 37 (13 July 1668), 732–5.

28.
WALLIS:
Paper on Dulaurens
Oxford, 27 November/[7 December] 1668

Transmission:

W Paper sent: LONDON *Royal Society* Early Letters W1, No. 72, 3 pp.

This paper, written in the form of a letter addressed to Oldenburg, was sent, together with that of the same date on Hobbes, as an enclosure to WALLIS–OLDENBURG 30.XI/[10.XII]. 1668. In the paper Wallis replies to Dulaurens's latest attacks, contained in a now missing letter or flysheet. Justel refers to Dulaurens's communication which provoked the reply in JUSTEL–OLDENBURG [31.X]/10.XI.1668 (OLDENBURG, *Corespondence* V, 121–2), and at the same time recommends that it not be published in the *Philosophical Transactions*. No doubt for this reason Oldenburg also decided against publishing Wallis's paper. Cf. WALLIS–OLDENBURG 30.XI/[10.XII].1668.

Mr Oldenburg. Oxoniae Novemb. 27. 1668.

Vidi ego (Vir Clarissime) *Dulaurentii* rescriptum[230] alterum, contra ea quae ego (speciminis loco) ex multis pauca (quo ipsius importunitati satis facerem) in libro suo animadvertenda notaveram.[231] Quod quidem, praeter
5 opprobria, convitia, aliaque impotentis animi indicia, (quae mihi negligenda duxi;) nihil est quod responsione indigeat. Nam ex iis omnibus quae ego ut perperam posita notaveram, (sive ut parum sana, sive ut minus accurate tradita,) ne unum quidem est quod ille vel negat, vel defendit, vel

3 quo (*1*) ipsi importunus (*2*) ipsius importunitati
4 in libro suo *add.*
4 notaveram. (*1*) In (*2*) Quod
6 responsione (*1*) dign⟨um⟩ (*2*) indigeat

---

[230] rescriptum: i.e. Dulaurens's letter or flysheet atacking Wallis, which was addressed to Oldenburg. Cf. JUSTEL–OLDENBURG [31.X]/10.XI.1668; OLDENBURG, *Correspondence* V, 121–2.

[231] notaveram: i.e. WALLIS, 'Another Letter ... concerning ... specimina mathematica Francisci Du Laurens', *Philosophical Transactions* No. 34 (13 April 1668), 654–5; WALLIS, 'Some Animadversions ... on ... Responsio Francisci Du Laurens', *Philosophical Transactions* No. 38 (17 August 1668), 744–50; WALLIS, 'A second Letter ... on the same printed Paper of Franciscus Du Laurens', *Philosophical Transactions* No. 39 (21 September 1668), 775–9; WALLIS, 'A Continuation of Dr. Wallis his second Letter, to the printed Paper of Mr. Du Laurens', *Philosophical Transactions* No. 41 (16 November 1668), 825–32.

28. WALLIS: Paper on Dulaurens, 27 November/[7 December] 1668

diffitetur: Quod autem reponit unicum; nempe, quod haec *Incuriae*, non *Inscitiae*, imputanda velit; (quod fatentis est:) modo conviciis abstenuisset, ferri posset. An autem tam multiplex Incuria, non et Inscitiae multum vel arguat vel conjunctum habeat Lectori judicandum permitto. Cum autem, post editum de primo primae partis Capitis specimen satis amplum, non et reliquum librum similiter percurram, ratio est, quoniam non tanti rem esse judico, et bonas horas melius collocari posse non dubito. Conviciis autem si responsum velit exquirat oportet petulantem aliquem juvenem et conviciandi gnarum, cujus aetati et moribus melius conveniet, convitianti et muliebriter rixanti paria reponere. Qui, modo matheseos peritia saltem [2] leviter tinctus fuerit, (ut enim penitus imbutus sit, non erit opus;)| si ex *Dulaurentio* triumphato aliquid sibi gloriae accessurum putaverit: materiam mihi satis amplam inveniet qua se exerceat. Quoniam vero tam importunus urget, ut velim paginas et lineas indicare ubi extant ea quae ex Oughtredi *Clavi* celate nomine desumpta dixeram: erat mihi in animo *Dulaurentii* secundum caput ad examen vocare (speciminis loco) ut qua pagina quo versu ipsius singula membra apud Oughtredum (vel ipsis verbis vel tantundem significantibus) reperiantur; nisi quod Lectorem hujusmodi tricis delirendum non putaverim. Id saltem insinuasse sufficiat; si Lectori vocaverit, caput illud integrum, satis quidem verbose traditum, et quod plus quam paginas viginti[232] (*in quarto* ut loquuntur) complet; cum primoribus quindecim articulis Capitis Sexti *Clavis*[233] Oughtredi, (qui quasi

---

5 permitto. (*1*) Post autem editum (*2*) Cum autem, post editum
7 posse (*1*) judico (*2*) non dubito.
10 matheseos (*1*) per saltem (*2*) peritia saltem
11-12 si |otio abundet otio, et *del.*| ex *Dulaurentio*
12 putaverit: (*1*) ampl⟨am⟩ (*2*) materiam
14 et (*1*) indicare unde ea desumpserit (*2*) lineas indicare ubi extant ea
16 mihi (*1*) semel in animo, ut ex ejus ipsius (*2*) in animo *Dulaurentii*
17 ut (*1*) ubi ip⟨sius⟩ (*2*) qua pagina
19-22 putaverim. (*1*) Id (*2*) Id saltem |speciminis loco *add. and del.*| insinuasse sufficiat; (*a*) ex iis omnibus quae caput illud integrum constituunt, quod (*b*) si Lectori vocaverit, caput illud integrum satis quidem |satis *del.*| verbose traditum, et quod plus quam paginas (*aa*) quadraginta (*bb*) viginti (*in quarto* ut loquuntur) complet; (*aaa*) vix quicquam est quod non in Oughtredi Clavis (*bbb*) cum primoribus quindecim articulis Capitis Sexti *Clavis* Oughtredi

---

[232] viginti: cf. WALLIS–OLDENBURG 3/[13].XII.1668. The correction of quadraginta to viginti was clearly carried out by Oldenburg on Wallis's instruction.
[233] *Clavis*: i.e. OUGHTRED, *Clavis mathematicae*.

28. WALLIS: Paper on Dulaurens, 27 November/[7 December] 1668

tribus paginis, *in Octavo*, continentur,) conferre; (saltem si adjungat Cap. XV. artic. 1, et 4:) tam manifesto indicio comperiet, tum *multa* verbatim inde transcripta esse, tum et *omnia* ibidem ab *Oughtredo* succincte tradita, (quae *Dulaurentius*, quo pompam ornet, fusius protrahit, sed non melius:) ut non sit opus mihi ad singulos para⟨gra⟩phos digitem intendere. Quod cum viderit Lector; ipsius judicio permitto, an Author, ex quo tot in uno Capite desumpta sunt, non debeat uspiam in toto libro nominari. Quod non ideo dictum est quasi haec sola, fuit ea omnia quae ex Oughtredo desumpta putanda sunt, (nam et ipse plura fatetur;) sed ut haec speciminis loco considerentur. Et quidem illud quod inter sua *nova inventa* numerat, et de quo| non parum gloriatur, nempe *solutio eversae aequationis* [3] $+aa - pa + q = 0$, *absque ullius ad rectam p parallelae ope*; *quam tamen parallelam*, *omnes* (inquit) *alii, ut idem praestent ducere coguntur*, (sic utique est in sua *Replicatione*; et, ad eundem sensum, in opere ipso, pag. 174.) Hoc ipsum, inquam, (praeterquam quod res levicula sit, et quam quilibet in Analyticis Tyro praestabit, si imperetur;) apud Oughtredum habetur Cap. XIX. in constructione Problematis 18. Sed mihi non est animus ad singula excurrere, vel examen particulare ultra caput primum continuare, aut serram porro reciprocare. Sed neque solicitus ero, utrum Incuriae fuerint an Ignorantiae vitia.

Quod ad Hobbii opera spectat, &c. vide # #[234]

---

3 verbatim (*1*) descripta (*2*) inde transcripta
3 omnia (*1*) ibidem (*2*) ibidem
6 Lector; *add.*
7 tot (*1*) desu⟨mpta⟩ (*2*) in uno Capite desumpta
5-7 Quod cum ... nominari. *add. on following page*
8 est *add.*
16-19 Cap. XIX. (*1*) in solutione Problema (*2*) in constructione Problematis (*a*) . Quoque id sibi facessere volet negotii, etiam alibi, non dico spicilegium, sed *Messem* satis amplam reperiet. Mihi (*b*) . 18. Sed Mihi non est animus ad singula excurrere, (*aa*) vel serram ultra reciprocandi (*bb*) vel (*aaa*) examen (*bbb*) particulare examen ultra caput primum continu⟨are⟩ (*cc*) vel examen particulare ultra caput primum continuare, aut serram porro reciprocare.

---

[234] vide # #: i.e. the accompanying paper on Hobbes.

29.
WALLIS:
Paper on Hobbes
Oxford, 27 November/[7 December] 1668

Transmission:

W Paper sent: LONDON *Royal Society* Early Letters W1, No. 73, 3 pp.

This paper, written in the form of a letter addressed to Oldenburg, was sent, together with that of the same date on Dulaurens, as an enclosure to WALLIS–OLDENBURG 30.XI/[10.XII].1668.

Wallis's evident aim that Oldenburg should print his latest paper on Hobbes in the *Philosophical Transactions* was not realized. Cf. WALLIS–OLDENBURG 30.XI/[10.XII].1668.

For Mr. Oldenburg.   Oxoniae Nov. 27. 1668.

##[235]

Quod ad *Hobbii* Opera[236] spectat, quae iterata vice dicis jam juncta prodire quae per partes prius prodierunt; atque, cum reliquis, ipsius Scripta Mathematica: Certe si singula fuerint jam olim satis refutata; non expectandum   5
est ut eo meliora jam sint quod juncta prodeant. Qua in libro *De Corpore*[237] admiserat, in re Mathematica, Paralogismos et Pseudographemata; refellit meus *Elenchus*[238] *Geometriae Hobbianae*. Ejusdem Replicationibus aliquot,[239] oppositae sunt a me *Debita Correctio*,[240] et *Hobbiani Puncti Dispunctio*,[241] Ejusdem *Sex Dialogis*; de *Mathematica Hodierna*;[242] (ubi et   10

---

3 jam (*1*) pro⟨dire⟩ (*2*) juncta prodire
4 prodierunt; (*1*) aut (*2*) atque
5 Mathematica: (*1*) jam dudum refutata; (*2*) Certe si singula
6 ut (*1*) juncta (*2*) eo meliora
9 a me *add.*

---

[235] # #: cf. the end of the accompanying paper on Dulaurens.

[236] Opera: i.e. HOBBES, *Opera philosophica, quae Latine scripsit, omnia. Ante quidem per partes, nunc autem, post cognitas omnium objectiones, conjunctim & accuratius edita*, Amsterdam 1668.

[237] *De Corpore*: i.e. HOBBES, *Elementorum philosophiae sectio prima, De Corpore*, London 1655.

[238] *Elenchus*: i.e. WALLIS, *Elenchus geometriae Hobbianae*, Oxford 1655.

[239] aliquot: i.e. HOBBES, *Six lessons to the professors of the mathematiques*, London 1656.

[240] *Correctio*: i.e. WALLIS, *Due Correction*, Oxford 1656.

[241] *Dispunctio*: i.e. WALLIS, *Hobbiani puncti dispunctio*, Oxford 1657.

[242] *Hodierna*: i.e. HOBBES, *Examinatio et emendatio mathematicae hodiernae qualis explicatur in Libris Johannis Wallisii ... Distributa in sex Dialogos*, London 1660.

29. WALLIS: Paper on Hobbes, 27 November/[7 December] 1668

de suis *Circuli Quadraturis* agitur;) cum Septimo; qui est *Dialogus Physicus*; *de Natura Aeris*; (ubi et de *Duplicatione Cubi* agitur;[243]) respondit meus *Hobbius Heautontimorumenos*.[244] Ipsius denique Tractatui *De Principiis et Ratiocinatione Geometrarum*; (ubi contra *Euclidem* potissimum agitur; sed et de suis *Pseudo-tetragonismis*, de *Divisione Arcus in ratione data*, et *Mediis quotlibet Proportionalibus*;) Respondent *Animadversiones*[245] meae *Transactionibus Philosophicis* insertae, mensis *Augusti*, Ann. 1666. Sed et ab aliis aliquot, (qua Nostratibus, qua Transmarinis,) refutata fuerunt de suis Pseudo-graphematis non pauca, tum in scriptis editis, tum et manuscriptis, (quod etiam in Aula Regia notius est quam ut dicta opus sit.)| Quod et [2 ea successu ab omnibus factum est; ut jamdudum destiterint Mathematici ad unum omnes aliud (in Mathematicis saltem) ab *Hobbio* sperare, quam Paralogismos et Pseudographemata. Sed et alia ipsius Scripta (Physica, Ethica, Politica, Theologica,) non absimilem sortem apud alios Scriptores experta sunt; utut id eadem evidentia fieri, quae in rebus Mathematicis haberi solet, Subjecti ratio non ferat. Si itaque (quod non existimo) *post auditas* (ut loquitur) *omnium Objectiones*,[246] haec emendaverit omnia, in quibus foedissime errasse certum erat: Esto. Si vero, (quod magis credo, novi siquidem genium viri,) *post auditas omnium Objectiones* eadem vel

---

1 qui (*1*) dicitur (*2*) est
8 aliquot, ((*1*) tum et ex (*2*) qua
9-10 pauca, (*1*) (quod etiam in Aula (*2*) tum in Scriptis editis, tum et manuscriptis, (quod |etiam *add.*| in Aula
10 opus sit.) |Verum et ipsius reliqua Opera, (quamquam ea me mi *breaks off* (*1*) Phil⟨osophica⟩ (*2*) Physica, Ethica, (*a*) Theologi⟨a⟩ (*b*) Politica, Theo⟨logia⟩ *del.*| Quod et
11 jamdudum (*1*) destiterit Orbis (*2*) destiterint Mathematici
13 ipsius (*1*) Opera (*2*) Scripta
15 utut (*1*) non eadem (*2*) id eadem evidentia |fieri *add.*|
18 quibus (*1*) errasse es (*2*) foedissime errasse
19 siquidem (*1*) hominis genium (*2*) genium viri,)

---

[243] agitur: i.e. HOBBES, *Dialogus Physicus, sive de Natura Aeris ... Item de Duplicatione Cubi*, London 1661.

[244] *Heautontimorumenos*: i.e. WALLIS, *Hobbius heauton-timorumenos. Or a Consideration of Mr. Hobbes his Dialogues*, Oxford 1662.

[245] *Animadversiones*: i.e. WALLIS, 'Animadversions ... upon Mr. Hobs's late Book, De Principiis et Ratiocinatione Geometrarum', *Philosophical Transactions* No. 16 (6 August 1666), 289–94.

[246] *post ... Objectiones*: cf. the full title of Hobbes's *Opera philosophica*.

29. WALLIS: Paper on Hobbes, 27 November/[7 December] 1668

non mutata, vel non in melius mutata, (quippe hoc solet,) iterato emittat: Sperare forsan poterit, quod *nondum imbuta Posteritas* sera, juxta cum eo sensura sit; verum illud ab hujus seculi viris factum iri, jamdudum se *desperasse* dixit aliquoties. Quippe qui *a Geometris fere omnibus se dissentire*, Recte pronunciat (nisi quod vox *fere* delenda sit,) et quod ipse *vel solus non insaniat, vel insaniat solus*,[247] Verum non necesse erit, ut, quoties ille sua sensa iterato profert, toties responsa iterentur. Quippe non tam in proclivi est Veritas praesertim *Geometrica*, quin ut a Paralogismis et Pseudographematis, saltem ubi semel judicantur, se potius sit defendere. Neque tam stupida est *tota Geometrarum Natio*, ut, saltem monitis sibi paralogismis impetu patientur: Quod et ipse hactenus expertus est. Dum vero, [3] *post auditas omnium objectiones,*| (Hoc est, postquam Paralogismi sui jam toti passim Orbi luce clarius detecti fuerint,) sat frontis habet eadem iterato ut vera propinare (de Mathematicis suis praesertim loquor, in quibus se omnium maxime ludibrio exposuit;) audacior saltem est (non dico, prudentior) quam *Dulaurentius* ipse: qui, utut vix mitius irascatur quod errata sua (lacessitus) detexeri ni, non tamen ea (sic detecta) festinet defendere, sed excusare potius; non diffitendo

Incuriam, ne Inscitiam porro proderet.

Tuus &c.
J W

19-1 eadem (*1*) iterato emittat, (*2*) vel non mutata, vel non in (*a*) mutata (*b*) melius mutata
2 quod (*1*) sera (*2*) sera posteritas (*3*) sera Posteritas et nondum imbuta (*4*) *nondum imbuta Posteritas* sera
3 sensura (*1*) fuerit; (*2*) sit; verum illud ab hujus seculi viris |factum iri, *add.*| jamdudum
4 Quippe (*1*) quem (*2*) qui
5-6 delenda (*1*) fere (*2*) sit,) et |quod *add.*| ipse (*a*) vel insaniet solum, vel solus non insaniat (*b*) *vel solus non insaniat, vel insaniat solus.*
6-7 non |ideo *del.*| necesse |erit, *add.*| ut, ... profert, toties (*1*) et responsa (*2*) responsa iterentur.
10 sibi *add.*
12 objectiones, (*1*) sat (*2*) (Hoc est, |fronte *del.*| postquam
12-14 jam (*1*) toti Orbi (*2*) toti passim Orbi luce clarius (*a*) innotuerint) (*b*) detecti fuerint,) (*aa*) eadem (*bb*) sat frontis habet eadem (*aaa*) iterato p *breaks off* (*bbb*) iterum proponendo, se porro ostentum ludib⟨rium⟩ (*ccc*) iterato propinare (*ddd*) iterato (*aaaa*) propinare (*bbbb*) ut vera propinare
17 (lacessitus) *add.*
18-19 excusare (*1*) satagit, (*2*) potius; (*a*) Incuriam (*b*) non diffitendo Incuriam

---

[247] *vel solus ... insaniat solus*: cf. the dedicatory letter to Hobbes's *De principiis et ratiocinatione geometrarum*.

30.
WALLIS to HENRY OLDENBURG
Oxford, 30 November/[10 December] 1668

Transmission:

W Letter sent: LONDON *Royal Society* Early Letters W1, No. 74, 2 pp. (our source).—printed: OLDENBURG, *Correspondence* V, 210.

Enclosures: Wallis's papers on Dulaurens and Hobbes, dated 27.XI/[7.XII].1668.

Oxford. Nov. 30. 1668.

Sir

You have here, what I have to say to my two Antagonists; which I have purposely written in two papers[248] that you may use one or both as you see cause; (*a*) You will supply the coach (*b*) They or neither. They should have come in my last packet,[249] of Friday last, (which I hope you received on Saturday, by Bartlets coach,) but that I durst not stay it till these were finished, lest the Coach should be gone without them. I have since received the Bookes[250] you sent; for which I suppose it will not be long before I receive the mony for you, as also for that which this week I expect. I presume by that time this comes at you, it will be seasonable to gratulate My Lord Brounker & yourself your enterance[251] on your Office for another year. When you see Mr Boyle, pray do mee the favour to present my service to him, with my humble thanks, for his Book[252] delivered to mee from Mr

---

5 cause; (*1*) & (*2*) or neither. (*a*) You will supply the coach (*b*) They
7 till (*1*) this was (*2*) these were

---

[248] two papers: i.e. Wallis's papers on Dulaurens and Hobbes, dated 27.XI/[7.XII].1668.
[249] last packet: presumably the packet, which contained WALLIS–BROUNCKER 4/[14].XI.1668, as reported in WALLIS–OLDENBURG 26?.XI/[6?.XII].1668. Wallis sent this packet by the Oxford carrier on Friday, 27 November.
[250] the Bookes: i.e. the copies of Hevelius's books, requested by Wallis in WALLIS–OLDENBURG 14/[24].XI.1668 and WALLIS–OLDENBURG 19/[29].XI.1668.
[251] enterance: Brouncker and Oldenburg were re-elected president and secretary of the Royal Society respectively on 30 November 1668; see BIRCH, *History of the Royal Society* II, 331.
[252] his Book: probably BOYLE, *A Continuation of New Experiments Physico-Mechanical touching the Spring and Weight of the Air and their Effects. The I. part ... Whereto is annext a short discourse of the Atmospheres of Consistent Bodies*, Oxford 1669. This book was presented to the Royal Society at the meeting on 30 November 1668; see BIRCH, *History of the Royal Society* II, 330.

Davis.²⁵³ And give my service allso to such others of my friends as you see occasion from

Your friend & servant
John Wallis.

[2] For Mr Henry Oldenburg
at his house in the Palmal
near St James's
London.

31.
WALLIS to JAMES GREGORY
November/early December 1668

Transmission:

Manuscript missing.

Existence and date: Referred to in WALLIS–OLDENBURG 21/[31].XII.1668.
Answered by: GREGORY–WALLIS *c.*20/[30].XII.1668.

This letter, which Wallis sent to Gregory in Scotland, evidently concerned the topic of the continuing dispute with Huygens.

32.
HENRY OLDENBURG to WALLIS
1/[11] December 1668

Transmission:

Manuscript missing.

Existence and date: Mentioned in and answered by WALLIS–OLDENBURG 3/[13].XII. 1668 and WALLIS–OLDENBURG 5/[15].XII.1668.

In this letter, Oldenburg evidently gave a report on the reception of Wallis's hypothesis on motion, which had been read at the meeting of the Royal Society on 26 November, and passed on four questions which had been raised by William Neile. He also informed Wallis of a discussion in the Royal Society on the nature of elasticity, but without mentioning

---

1 mee (*1*) by Mr (*2*) from

---

²⁵³Davis: i.e. Richard Davis (1617/18–*c.*95), Oxford bookseller, *ODNB*.

that the ideas he cited had originated from Hooke. Oldenburg also asked for more precise instructions from Wallis on a number of book commissions.

33.
Henry Oldenburg to Wallis
2/[12] December 1668

Transmission:

Manuscript missing.

Existence and date: Referred to in and answered by Wallis–Oldenburg 5/[15].XII.1668.

This letter apparently accompanied a copy of Tycho Brahe's *Historia coelestis*, Augsburg 1666, and enclosed a letter to a German visitor, probably Benedict Wasmer of Bremen, then present at Oxford. See *Bodleian Library* MS e. 533, f. 180^v.

34.
Wallis to Henry Oldenburg
Oxford, 3/[13] December 1668

Transmission:

W Letter sent: London *Royal Society* Early Letters W1, No. 75, 4 pp. (our source). At top of p. 1 in Oldenburg's hand: 'Read Dec 10: 68. Entered L.B. 2. 343'. On p. 4, also in Oldenburg's hand: 'Extract of a letter of Dr Wallis to M. Old. (*1*) Concerning Motion, and of all (*2*) Of Springenes as the cause of all rebounding; and together with an Answer to some suggestions about Motions.' Beneath address Oldenburg has noted: 'Rec. Dec. 5. 68.' Postmark: 'DE/4'.—printed: Oldenburg, *Correspondence* V, 218–19.

w^1 Part copy of letter sent: London *Royal Society* Letter Book Original 2, pp. 343–4.

w^1 Copy of w^1: London *Royal Society* Letter Book Copy 2, pp. 394–6.

Reply to: Oldenburg–Wallis 1/[11].XII.1668.

Dec. 3. 1668. Oxford

Sir,

Yours[254] of Dec. 1. I receive just now, when the Post is upon going. My order for Historia Caelestis[255] I intended to be positive (unlesse you have

---

[254]Yours: i.e. Oldenburg–Wallis 1/[11].XII.1668.

[255]Historia Caelestis: i.e. Tycho Brahe, *Historia Coelestis*, ed. Albert Curtz, Augsburg 1666.

34. WALLIS to OLDENBURG, 3/[13] December 1668

reason for the contrary) I know not well in what words it was written. Of that[256] to my Lo. Br: there is no hast; but of your convenience is. In that[257] of mine which concerns Du-Laurens you must needs mend one mistake; I have sayd *quadraginta* paginas when I should have sayd but *viginti*. I did it out of my memory having remembred some what of 20, & thought it had been 20 leaves, whereas it is but 20 pages (his second chapter.) Pray amend it before it be forgotten. Of Slusius[258] I hear nothing yet, but from you. But I shall be afraid of giving characters, having had so bad success in the two last.[259]

That all rebounding comes from Springynesse, is my opinion. & therefore you see mee express that as the onely reason in my short Hypothesis[260] which you lately had.

That quiescent matter hath no resistance to motion (save what it may have from circumstantiall incumberances, or, if there be any innate propensity to the contrary motion, as in gravity is sup|posed,) I take for granted amongst most of the moderns; & I see nothing to the contrary why I should not be of that opinion.

Whether motion passe out of one subject into another, must be first explained; for in a sense it doth, in a sense it doth not. You will see by my hypothesis what I think of it. In summe, in all percussion the body striking looseth of its swiftness, & the other gains if before at rest. If both before were in opposite motion, both loose of their motion & both going from the other in such proportion as is there expressed.

Whether any motion perish, I am not yet resolved what to say. I have much thought of it; & see somewhat *pro* & somewhat *contra*. But if wee say none perisheth; I doubt wee must say, that none begins. (Which seems hard as to Voluntary motions especially.) Else the world must needs grow more

---

10 Springynesse, (*1*) I am apt to beleeve; (*2*) is my opinion.
11 in *add.*
18 into (*1*) one (*2*) another,
19 for (*1*) I do no (*2*) in
22 loose |of their *add.*| |motion *add. ed.*| & both gain |from the other *add.*| in
27 especially.) (*1*) Els *breaks off* (*2*) Else

[256]that: i.e. WALLIS–BROUNCKER 4/[14].XI.1668.
[257]i.e. Wallis's paper on Dulaurens, dated 27.XI/[7.XII].1668.
[258]Of Slusius: i.e. of SLUSE, *Mesolabum*, 2nd ed., Liège 1668.
[259]having had ... two last: presumably an allusion to Wallis's reviews of Dulaurens and James Gregory.
[260]short Hypothesis: i.e. WALLIS–OLDENBURG 15/[25].XI.1668.

35. WALLIS to OLDENBURG, 5/[15] December 1668

& more in commotion, till it become infinitely so. But (as I sayd at first) I affirm nothing positively in that point.

Whether different motions destroy one another. I think, onely so far as to make up a compound motion or of both: Which how it is to bee computed, my paper of motion shews. There being all the principles from whence the result is onely to be a calculation. I onely adde; that (whose so ever they are) none of these suggestions are strange or New; but have been all dis|cussed [3] in such as have written of these *subjects*. This short account in answere to all your quere's is all I have time to say, but the post be gone: in as great hast as I can write.

I am
Your &c.
J. Wallis.

For Mr Henry Oldenburg [4]
in the Palmal near
St James's
London.

35.
WALLIS to HENRY OLDENBURG
Oxford, 5/[15] December 1668

Transmission:

W Letter sent: LONDON *Royal Society* Early Letters W1, No. 76, 4 pp. (our source). At top of p. 1 in Oldenburg's hand: 'Read December 10: 68. Entered L.B. 2. 344'. On p. 4, also in Oldenburg's hand: 'Dr Wallis's Letter to M. Old. enlarging upon the subject of Motion, discoursed of in (*1*) his former letter of Decemb. 3: 1669. (*2*) the next foregoing letter'. Beneath address Oldenburg has noted: 'Rec. Dec. 7. 68. Read at the Society Dec. 10. 68. order'd to be entred in the Letter-book'.—printed: OLDENBURG, *Correspondence* V, 220–2.
w^1 Copy of letter sent: LONDON *Royal Society* Letter book Original 2, pp. 344–6.
w^2 Copy of w^1: LONDON *Royal Society* Letter book Copy 2, pp. 396–9.

4 Which (*1*) what (*2*) how it is to bee |computed, *add.*| my
7 strange or *add.*
8 account (*1*) is all (*2*) in
9 to (*1*) ad *breaks off* (*2*) say,
9 gone (*1*) . I am (*2*) : in as great hast as

35. WALLIS to OLDENBURG, 5/[15] December 1668

Reply to: OLDENBURG–WALLIS 1/[11].XII.1668 and OLDENBURG–WALLIS 2/[12]. XII. 1668.

This letter is an enlarged version of WALLIS–OLDENBURG 3/[13].XII.1668 in reply to Oldenburg's report on the reception of Wallis's account on the laws of motion (WALLIS–OLDENBURG 15/[25].XI.1668) at the meeting of the Royal Society on 27 November 1668. The present letter was read at the meeting of the Royal Society on 10 December 1668; see BIRCH, *History of the Royal Society* II, 333.

Oxford. Dec. 5. 1668.

Sir,

To the Quaere's in yours[261] of Dec. 1. because it came to hand not an hour before the post was going, I could by mine[262] of Dec. 3. give you but a very summary account. And therefore I adde this, to the same purpose, but a little more expresse.

What you say was started in the Society (but not, by whom,[263]) *That the springyness of Bodies is the* Onely *cause of their rebounding*. My opinion is, (& hath been a good while, & oft declared,) that (beside *Repercussion* which I suppose was not intended to be excluded, being one manifest cause; as when a Racket returns the Ball;) there is no other cause (that I know of) of Rebounding, but Springyness. And therefore, you see, in my Hypothesis sent you, I assign no other. And I think all Phaenomena may from thence (& from Repercussion) be salved. But that which is added, that *if therefore there were a body perfectly hard, it would not rebound at all*, doth not follow, unlesse that against which it lights be so too, & without any opposite motion; For else it may be made to rebound either by that opposite repercussion, or by the springyness of that other Body.

To the other 4 Quaere's (from allso I know not whom;[264])

---

3 Quaere's (*1*) of (*2*) in
11 Ball;) (*1*) I kno *breaks off* (*2*) there
12 therefore, (*1*) I (*2*) you
14 salved. (*1*) To the other 4 Quere's, from allso I know not whom (*2*) But that which (*a*) doth (*2*) is
16 too (*1*) ; for it may (*2*) , &

---

[261] yours: i.e. OLDENBURG–WALLIS 1/[11].XII.1668.
[262] mine: i.e. WALLIS–OLDENBURG 3/[13].XII.1668.
[263] by whom: in fact, it was Hooke who performed the experiments at the meeting of the Royal Society on 27 November (and later on 10 December); see BIRCH, *History of the Royal Society* II, 328 and 333.
[264] whom: in fact, these questions were raised by William Neile; see BIRCH, *History of the Royal Society* II, 333.

35. WALLIS to OLDENBURG, 5/[15] December 1668

1. *Whether Quiescent Matter have any resistence to motion*: I look upon it as taken for granted by most of our moderns; that (supposing it uningaged from circumstantiall incumberances, & all innate propensions to a contrary motion, such as that of Gravity is wont to be accounted,) that it is indifferent as to rest or motion, without any aversenesse to either; as allso indifferent as to any direction of motion, this way or that way. And accordingly doth remain as it is, either in rest or motion, & this with the same direction & celerity, till some positive cause alter it. Which I think I have demonstrated in those sheets of my book[265] allready printed.

2. *Whether Motion may pass out of one Subiect into another.* Must be distinguished, that we may know what is meant by it. But, (without disputing the School question, De migratione Accidentium;) if the meaning be no more but this, That the Force or impetus whereby one body is moved, may, by percussion, cause another body, against which it strikes, to be put into motion; &, withall, loose somewhat| of its own strength or swiftnesse; [2 experience tells us clearly inough that it is so: And, in what proportion this is, my late Hypothesis teacheth. But whether this be that which the Schools call Migratio accidentis, is onely to dispute of words.

3. *Whether no Motion in the World perish, nor new motion be generated.* Needs distinguishing allso. And therefore I know not what to say positively in it, till I know what is meant by it. There being therein somewhat to be sayd *pro* and *con*, which makes mee not so forward to speak positively, as perhaps some who have lesse considered of it, & do therefore take notice but of what may be sayd for the one side. But, that a Body once in motion, may (as to that motion) cease to bee so; I do not doubt. Nor, that

---

3 to (*1*) contrary motions, (*2*) a contrary motion,
4 indifferent |either *del.*| as
5 either; |or *del.*| as
7 remain (*1*) eith *breaks off* (*2*) as
12 the (*1*) Peripatet *breaks off* (*2*) School
15 strength or *add.*
18 accedentis *corr.*
21 , till I know what is meant by it *add.*
22 somewhat (*1*) of wh *breaks off* (*2*) to
23 positively, (*1*) then (*2*) as
24 notice (*1*) of (*2*) but

---

[265] book: i.e. printed sheets of Wallis's *Mechanica: sive, de motu, tractatus geometricus*, the first two parts of which were published in 1670.

35. WALLIS to OLDENBURG, 5/[15] December 1668

a body (if any such be) perfectly at rest, may be putt into motion. What further may be intended in that quaere, must be more clearly explained before I know what I am to say to it.

4. *Whether different motions meeting, destroy one another.* I think they do it no otherwise then by making a compound motion of both. But this compound motion, as the case may happen, may chance to be a Rest. As for instance, a Motion directly Upward, compounded (in the same Body) with another directly Downward, equally swift, is equivalent with Rest.

This is the clear account of my thoughts as to those Quaere's: (Which perhaps I might have expressed more appositely, if I had known who's they are:) consonant to that short synopsis[266] of my Doctrine of Motion which I lately sent you. Of which I have this to adde in reference to one of your letters in pursuance of it; where you tell mee that *the Society in their present disquisitions have rather an Eye to the Physical causes of Motion, & the Principles thereof, than the Mathematical Rules of it.* It is this. That the Hypothesis I sent, is indeed of the *Physical* Laws of Motion, but *Mathematically* demonstrated. For I do not take the Physical & Mathematical Hypothesis to contradict one another at all. But what is Physically performed, is Mathematically measured. And there is no other way to determine the Physical Laws of Motion exactly, but by applying the Mathematical measures & proportions to them.

[3] Which is all I have at present to say to this subject.| I have this day received your other letter[267] of Dec. 2. with Tycho's Historia Caelestis. And the Letter inclosed[268] was delivered here to your Countryman.[269] I have not yet seen Mr Hobs's ⟨ne⟩w Book;[270] but thought it might be easyly answered without book, since I know his way so well. I do not know whether (since what I have published is allmost all out of print) these in Holland would not be willing inough to reprint those things there: which would be better done,

---

1 body |supposed *del.*| (if
7 Motion |directly *add.*| Upward, compounded |(in the same Body) *add.*| with another |directly *add.*| Downward,
13 the (*1*) Society's (*2*) Society in their
23 with (*1*) the (*2*) Tycho's
25 but (*1*) you (*2*) thought

---

[266]short synopsis: i.e. WALLIS–OLDENBURG 15/[25].XI.1668.
[267]other letter: i.e. OLDENBURG–WALLIS 2/[12].XII.1668.
[268]Letter inclosed: probably a now missing letter from Oldenburg to Benedict Wasmer.
[269]Countryman: probably Benedict Wasmer of Bremen, then present at Oxford. See *Bodleian Library* MS e. 533, f. 180^v.
[270]Book: i.e. HOBBES, *Opera philosophica, quae latine scripsit, omnia*, Amsterdam 1668.

37. WALLIS to OLDENBURG, 9/[19] December 1668

& with more opportunity of dispersing, then here. And, if so, I would fit up some corrected copies to send them. You may please by your correspondency there to inquire. For I think they may be in themselves (what ever they may be thought in the Vogue of men) as considerable as those of Mr Hobs. I have
5 not received here the money due for the bookes you have sent in the two last parcells: but will take care it shall not be long before you have it. I am

                                                                    Yours &c.
                                                                    John Wallis.

These   [4]
10 For Mr Henry Oldenburg in the
Palmal near St James's
London.

36.
HENRY OLDENBURG to WALLIS
8/[18] December 1668

Transmission:

Manuscript missing.

Existence and date: Mentioned in postscript to WALLIS–OLDENBURG 9/[19].XII.1668. Date given in introduction to excerpt of WALLIS–OLDENBURG 12/[22].XII.1668;

In this letter, which was received by Wallis on the morning of 10 December, Oldenburg gave further details of the Royal Society's view on the difference between mathematical and physical theories of motion. He apparently also asked permission to publish WALLIS–OLDENBURG 5/[15].XII.1668. Oldenburg also enclosed three mathematical problems which had been sent to the Royal Society by the Dutch mathematician and surveyor Jacob van Wassenaer (*c.*1607–?). See WALLIS–OLDENBURG 12/[22].XII.1668.

37.
WALLIS to HENRY OLDENBURG
Oxford, 9/[19] December 1668

Transmission:

W Letter sent: LONDON *Royal Society* Early Letters W1, No. 77, 4 pp. (our source). On p. 4 in Oldenburg's hand: 'Accepi d. 11. Decemb. A Copy of this was sent to Hevelius

28 be (*1*) mo *breaks off* (*2*) better

37. WALLIS to OLDENBURG, 9/[19] December 1668

jan. 25. 1669.'—printed: OLDENBURG, *Correspondence* V, 232–4 (Latin original), 234–7 (English translation).
w^1 Copy of letter sent (with additional note by Oldenburg): PARIS *Bibliothèque Nationale* Nouv. acq. lat. 1641, f. 112^r–113^v.
w^2 Copy of w^1: PARIS *Bibliothèque Nationale* Fonds latin 10348, IX, pp. 160–3.

Enclosure to: WALLIS–OLDENBURG 9 and 10/[19 and 20].XII.1668.

Although this letter was addressed to Oldenburg, it was also intended for Hevelius; Wallis explicitly suggests its transmission to him in WALLIS–OLDENBURG 9 and 10/[19 and 20].XII.1668. Oldenburg sent the copy w^1 to Hevelius as enclosure to OLDENBURG–HEVELIUS 23.I/[2.II].1668/9 (OLDENBURG, *Correspondence* V, 352–4).

Oxoniae, Decemb. 9. 1668.

Quos nuper acceperam (Vir Clarissime) tua cura transmissos, Authoris munificientia donatos, Celeberrimi Hevelii libros,[271] Bibliothecae Bodleianae destinatos; in D. Vice-cancellarii[272] manus (uti par erat) tradendos curavi. Qui et literas gratulatorias[273] propediem misit, quas te dudum accepisse non diffido, tua cura ad Authorem Celeberrimum Doctissimumque transmittendas. Eosdem ego (cum id meae curae demandaverit D. Vice-Cancellarius) curavi in duo Volumina redigendos, nitide compacta et deaurata. Quorum alterum, Epistolas duas[274] (de Motu Lunae Libratorio, et Eclipsibus aliquot;) Mercurium in Sole visum,[275] (una cum Venere in Sole visa, et mirabilis in Ceto Stellae Historiola;) Prodromum Cometicum,[276] (ubi de nuperorum

---

[271] libros: cf. WALLIS–OLDENBURG 30.XI/[10.XII].1668.
[272] Vice-cancellarii: i.e. John Fell.
[273] literas gratulatorias: on this letter of thanks to Hevelius, written on behalf of the University Of Oxford by John Fell, cf. WALLIS–OLDENBURG 7/[17].XI.1668.
[274] Epistolas duas: i.e. HEVELIUS, *Epistolae II. Prior: De motu lunae libratorio, in certas tabulas redacto. Ad ... Johannem Bapt. Ricciolum ... Posterior: De utriusque luminaris defectu anni 1654. Ad ... Petrum Nucerium*, Danzig 1654.
[275] Mercurium in Sole visum: i.e. HEVELIUS, *Mercurius in sole visus Gedani, Anno Christiano MDCLXI, d. III Maji, st. n. ... Cui annexa est, Venus in sole pariter visa, anno 1639, d. 24 Nov. st. v. Liverpoliae, a Jeremia Horroxio: nunc primum edita, notisque illustrata. Quibus accedit succincta Historiola, novae illius, ac mirae stellae in collo Ceti, certis anni temporibus clare admodum affulgentis, sursu omnino evanescentis ...*, Danzig 1662.
[276] Prodromum Cometicum: i.e. HEVELIUS, *Prodromus cometicus, quo historia, cometae Anno 1664 exorti cursum, faciesque, diversas capitis ac caudae accurate delineatas complectens*, Danzig 1665.

37. WALLIS to OLDENBURG, 9/[19] December 1668

Cometarum priore agitur;) et Prodromi Mantissam,[277] (ubi etiam de posteriore agitur,) continet: Alterum, ejusdem Cometographiam.[278] Eadem denique in Bibliothecam Bodleianam reponenda curavi, juxta ejusdem Volumen primum, quo Selenographia, cum Appendice,[279] continetur: quod ab ipsius
5 Authoris munificentia jam olim acceperat[280] Bibliotheca Bodleiana. Sed et adscripta sunt Lemmatia, quae et Authoris munificentiam, et tuam curam testentur. Optassem etiam ut Tractatum, de Nativa Saturni facie;[281] aliumque de Eclipsi quadam Solari[282] exiguum, adjunxisset; (quos etiam memini me ipsius olim dono accepisse:) vel etiam, si hic venales fuissent,
10 adjunxissem ipse: Quo ipsius edita omnia in Bodleiana Bibliotheca simul prostarent. Et quidem, nisi habeat ille causas cur secus mallet, non incommodum forsan erit ut illorum etiam aliquot exemplaria in Angliam transmittat. Qui enim ipsius reliqua habent (quod saltem ab Academicis nostris hic intelligo) nollent et haec deesse. Verum hoc ipsius arbitrio permittendum
15 erit. Praestolamur autem ejusdem magni Viri, quam pollicitus est, Machinam Coelestem;[283] opus Hevelio dignum, quod bonis avibus mature proditurum spero. In eum finem, memini me antehac Fixarum Catalogum ex Observationibus Uleg-Beigi, ex| Bibliothecis nostris transcribendum, atque [2] ad Hevelium mittendum.[284] Eundem, cum notis D. Hyde[285] (Bibliotecae

---

10 adjunxissem |ego del.| ipse W
18 in Bibliothecis nostris w^1

---

[277]Prodromi Mantissam: i.e. HEVELIUS, *Descriptio cometae anno Aerae Christ. MDCLXV exorti, cum genuinis observationibus, tam nudis, quam enodatis, mense Aprili habitis Gedani. Cui addita est Mantissa Prodromi cometici, observationes omnes prioris cometae MDCLXIV, ex iisque genuinum motum accurate deductum, cum notis, & animadversionibus exhibens*, Danzig 1666.

[278]Cometographiam: i.e. HEVELIUS, *Cometographia*, Danzig 1668.

[279]Selenographia, cum Appendice: i.e. HEVELIUS, *Selenographia: sive, Lunae descriptio ... Addita est, lentes expoliendi nova ratio*, Danzig 1647.

[280]jam olim acceperat: in early 1651. See WALLIS–HEVELIUS 21/[31].I.1650/1.

[281]de Nativa Saturni facie: i.e. HEVELIUS, *Dissertatio de nativa Saturni facie, ejusque variis phasibus, certa periodo redeuntibus*, Danzig 1656.

[282]de Eclipsi quadam Solari: possibly HEVELIUS, *Eclipsis solis observata Gedani anno ...1649, die 4 Novembris st. Greg.*, Danzig 1650.

[283]Machinam Coelestem: Hevelius's *Machina coelestis* appeared in Danzig 1673, followed by a second volume in 1679.

[284]me antehac Fixarum Catalogum ...ad Hevelio mittendum: Wallis had sent his transcript of Ulug Beg's catalogue of fixed stars together with WALLIS–HEVELIUS 5/[15].IV.1664; WALLIS, *Correspondence* II, 103–6.

[285]D. Hyde: i.e. Thomas Hyde. (1636–1703), oriental scholar, appointed Bodley's Librarian in 1665; Laudian professor of Arabic, 1691, regius professor of Hebrew and canon of Christ Church 1697, *ODNB*.

37. WALLIS to OLDENBURG, 9/[19] December 1668

Bodleianae Protobibliothecarii,²⁸⁶) ex eo tempore Typis editum,²⁸⁷ jam antehac misissem; nisi quod id suae curae permittendum petiverit D. Robertus Moray,²⁸⁸ Eques, (quem et dudum id praestitisse non diffido.) Habemus autem, uti audio, alicubi in Bibliothecis nostris, istiusmodi Tabulas alias, sive Persicas, sive Arabicas; quas, si commode potero, curabo, ab istarum linguarum peritis transcribi, et Latinitate donari; quo earundem etiam copia fiat. Quod et eo potius faciendum autumo, quoniam Tabulas Persicas invenio ab Editore²⁸⁹ Observationum Tychonicarum (in Historia Coelesti²⁹⁰ nuper editarum) laudatas, et earundem ἀποσπασμάτια quaedam digna quae edantur habita; unde et Tabulas integras non ingratas futuras conjicio. Verum hoc, cum ab aliis dependere necesse sit, (quippe me istarum linguarum non ita peritum profiteor ut id in me solum suscipiam,) non aliter spondere possum, quam si ab illorum (qui apud nos non ita multo sunt) otio hoc obtinere potero. Praesertim cum D. Pocockius²⁹¹ (linguarum Orientalium peritia nulli secundus) jam a longo tempore morbo detineatur; D. Hyde jam per aliquot annos in novo Catalogo librorum omnium in Bodleiana Bibliotheca ornando fuerit occupatus;²⁹²

---

2 petiverit D. Robertus Moray, Eques, |et D. Henr. Oldenburg *add. Oldenburg*| (quem et *W* petiverint D. Robertus Moray Eques, et |(*1*) D. Henri⟨cus⟩ Oldenburg, (*2*) Tu ipse *alt. Oldenburg*| (quos et w^1
11 necesse est, w^1
12 me *add. W*
15 Pocockius (|harum *del.*| linguarum *W*
17 Bibliotheca (*1*) ex *breaks off* (*2*) ornando *W*

---

²⁸⁶Protobibliothecarii: Hyde had been appointed sub-librarian of the Bodleian Library in 1659. Contrary to what Wallis writes, this was no longer his position.

²⁸⁷eundem ... editum: i.e. Hyde's edition of Ulug Beg's catalogue, *Jadâwil-i mawâdi'-i thawâbit dar tûl u'ard kih bi-rasad yâftah ast Ulugh Baik Sive tabulae long. ac lat. stellarum fixarum ex observatione Ulugh Beighi, Tamerlani Magni nepotis ... Ex tribus ... MSS. Persicis ... luce ac Latio donavit, et commentariis illustravit T. Hyde ...*, Oxford 1665. Oldenburg had announced the dispatch of this work in OLDENBURG–HEVELIUS 24.VIII/[3.IX].1666 (OLDENBURG, *Correspondence* III, 215–8, 216).

²⁸⁸Moray: i.e. Robert Moray (Murray) (1608–73), army officer and politician, founder member of the Royal Society, *ODNB*.

²⁸⁹Editore: i.e. Albert, Graf von Curtz, S.J. (1600–71), professor of philosophy and mathematics at various Jesuit colleges, finally rector of Neuburg.

²⁹⁰Historia Coelesti: i.e. TYCHO BRAHE, *Historia Coelestis. Ex libris commentariis manuscriptis observationum vicennalium ... Tichonis Brahe Dani*, ed. Lucius Barrettus (i.e. Albert von Curtz), 2 parts, Augsburg 1666.

²⁹¹D. Pocockius: i.e. Edward Pococke, q.v.

²⁹²in novo Catalogo ... ornando fuerit occupatus: Hyde's *Catalogus impressorum librorum bibliothecae Bodlejanae in academia Oxoniensi* was published in Oxford in 1674.

37. WALLIS to OLDENBURG, 9/[19] December 1668

et D. Bernard[293] jam statim in Bataviam abiturus est quo Apollonii Pergaei Conicorum libros septem quos hic Arabice habemus cum eorundem codice uno aut altero ibidem conferat, ut deinceps emendatius edat;[294] et D. Clark,[295] tum in praeparando Volumine altero Bibliis Polyglottis addendo, tum in Abulfedae Geographia Arabica an Persica ex Manuscriptis aliquot quos hic habemus recensenda, (quam et aliquando editurum speramus,) jam aliquandiu fuerit occupatus: Atque ex paucis qui porro sunt harum linguarum peritis nescio an illud obtinuero. Interea temporis (nisi hoc jam ante factum fuerit) non incommodum existimo, ut Hypothesin meam, de Marinis Aestibus,[296] ad Hevelium mittas; (qui, nisi fallor, linguam nostram aliquatenus saltem intelligit, ut qui in Anglia nonnihil temporis olim posuerit:) quoniam ab illo, quem ibidem suppono Telluris Epicyclum, salvari forsan inveniat exiguas, praesertim in Lunari Motu, inaequalitates, quae Aequationes Menstruas spectant: fortassis etiam in Planetarum aliis, praesertim ubi sunt in situ Telluri proximo: Sed et (ob similem causam, quae ex Satellite Saturni provenire possit) istiusmodi exiguas Saturni inaequalitates, quae Horroxium| nostrum[297] (quod ex illius literis Manu-scriptis reperio) [3] satis suspensum olim tenuerunt. (Forsan, et in Jove, ob similem Satellitum

18 *[Note in margin in Oldenburg's hand w^1:]* Te, Clarissime Heveli, indigitat Author, ut facile intelligis. Fac igitur, sciam, quaeso, (*1*) ut (*2*) num rationibus tuis conveniat, libellum hic commemoratum, hactenus ineditum, 20. vel 25. philyris forte constantem, praelo vestro una cum tuis, dehinc edendis, committere.

4 in (*1*) parando (*2*) praeparando *W*
6 (quam et aliquando editurum speramus,) *add. W*
14 in (*1*) aliis (*2*) Planetarum *W*

---

[293] D. Bernard: i.e. Edward Bernard, q.v.

[294] ut deinceps emendatius edat: This edition was eventually prepared by Edmond Halley and published under the title *Apollonii Pergaei conicorum libri octo: libri quatuor Graece et Lat. cum Pappi Alexandrini lemmatis et Eutocii Ascalonitae commentariis*, Oxford 1710.

[295] D. Clark: i.e. Samuel Clarke (1624–69), oriental scholar; architypographus and superior bedel of law in the University of Oxford, *ODNB*. Clarke had assisted Brian Walton in his edition of *Biblia sacra polyglotta*, 6 vols., London 1655–57, and was preparing a seventh volume at the time of his death. His transcript of the 'Geography' of the Arab historian and geographer Abul Feda (1273–1331) was left in manuscript. See WALLIS–OLDENBURG 23.XI/[3.XII].1671.

[296] Hypothesin meam, de Marinis Aestibus: i.e. WALLIS–BOYLE 25.IV/[5.V].1666 (WALLIS, *Correspondence* II, 200–22), published in *Philosophical Transactions* No. 16 (6 August 1666), which contained Wallis's hypothesis of the tides.

[297] Horroxium nostrum: i.e. Jeremiah Horrox. On Wallis's engagement in digesting Jeremiah Horrox's astronomical papers and preparing an edition of his *Opera posthuma*, see WALLIS–OLDENBURG 6/[16].IV.1664 and WALLIS–OLDENBURG 21.IX/[1.X].1664; WALLIS, *Correspondence* II, 106–10, 160–2.

37. WALLIS to OLDENBURG, 9/[19] December 1668

suorum causam.) Sed et dum Horroxii mentionem facio, subit animum interrogare, annon ea quae de ipsius Observatis adhuc supersunt (quae ex sparsis chartis, a Societate Regia rogatus, utcunque collegi,) velit Hic, (qui jamjam illius Venerem in Sole visam,[298] publici juris fecit,) cum suis edere. Utut enim ea nondum compleverat Author (saltem nisi, quod suspicor, quae ipse in meliorem ordinem redegerat perierint,) digniora tamen sunt quam ut pereant. Et quidem, si in eodem Volumine cum Doctissimi Viri Observatis prodeant, spes est, ut tutius in posterorum usum conserventur, quam si seorsum in exiguo libello emittantur.

Atque haec fere sunt quae jam dicenda habeo. Nisi et hoc additum velis; Stellam illam in Ceto miram, mensibus praeteritis Octobri, et Novembri, et hoc Decembri, serenis noctibus me saepius conspexisse; sed obscuriorem illis quae tertiae magnitudinis, aut etiam quartae, haberi solent.

Historiam Coelestem (ex Tychonis et aliorum Observationibus congestam) quam nuper accepi, aliquatenus evolvi. Reperio autem, quod doleo, menda Typographica, saltem in Literis, saepius admissa; quod facit ut suspicor idem in Numeris contigisse; (et istiusmodi nonnulla jam comperi;) quod in hujusmodi rebus magni est momenti, cum ea ex sensu restitui non possint. Si tu aliquid commercii habes cum iis quibus haec curae fuerunt, aut esse possunt; rem facies non malam si author esse possis ut eorum aliqui curent impressos codices, cum Manu-scriptis Autographis, summa cum diligentia conferri, atque omnia menda numeralia accurate notari, atque tum demum erratorum Catalogum imprimi et in bonum publicum emitti: unde suos quisque libros corrigere possint. Quod eo potius moneo, quoniam nullum in libro edito erratorum Catalogum reperio; cum tamen res quasi impossibilis sit, in tot passim numeris, errata preli aliquammulta non admissa esse. Quod et ab iis omnibus, qui hujusmodi tractant negotia, summopere factum vellem. Tu interim Vale.

Tuus,
Johannes Wallis.

[4] Clarissimo Viro D. Henrico, Oldenburg, Societatis Regiae Secretario, Londini.

---

14 Tychonicis w^1

---

[298] Venerem in Sole visam: i.e. HORROX, *Venus in sole visa*. Hevelius printed this work as an annex to his *Mercurius in sole visa*, Danzig 1662.

38.
WALLIS to HENRY OLDENBURG
Oxford, 9 and 10/[19 and 20] December 1668

Transmission:

W Letter sent: LONDON *Royal Society* Early Letters W1, No. 78, 2 pp. (our source). Postmark on p. 2: 'DE/11'.—printed: OLDENBURG, *Correspondence* V, 230–1.

Reply to: OLDENBURG–WALLIS 8/[18].XII.1668.
Enclosure: WALLIS–OLDENBURG 9/[19].XII.1668 and paper with van Wassenaer's three mathematical problems sent with OLDENBURG–WALLIS 8/[18].XII.1668.

Wallis wrote this letter on the evening of Wednesday, 9 December 1668. After receiving Oldenburg's letter of 8 December the following morning, he replied to that in a separately dated postscript.

Oxford Dec. 9. 1668.
Sir

I give you the inclosed[299] in Latine rather than in English, because though directed to yourself yet it doth as well concern Mr Hevelius, to whom perhaps
5 you may think fit to transmit it. To what I have there sayd, of the Historia Celestis[300] (which I do not repeat,) I adde that the Book I have, (beside that it hath suffered some dammage by the wett, which I suppose may be common with it & the rest that came over,) in the Liber Prolegomenos hath two leaves cutt (at pag. 105 and pag. 107.) as if they were to be taken out
10 & two others substituted in their room; but those Substitutes are wanting, & the Table of Errata if any such were. And pag. 110. it is sayd that there are Icones of Uraniburg,[301] Wandesburg,[302] & some other places: Whereas I have amongst the plates of Cuts (which are in all but three) onely one of Places, which have indeed Letters A. B. C. D. E. but what they signify

---

5 of the Historia Celestis *add.*
9 leaves |cutt *add.*| (at pag. 105 and pag. 107.) (*1*) ha *breaks off* (*2*) as

[299] inclosed: i.e. WALLIS–OLDENBURG 9/[19].XII.1668.
[300] Historia Celestis: i.e. TYCHO BRAHE, *Historia coelestis*, ed. Albert von Curtz, 2 parts, Augsburg 1666.
[301] Uraniburg: i.e. Uraniborg, Tycho Brahe's observatory on the Danish island of Hven in the Øresund, which was built around 1576–80 and destroyed in 1601.
[302] Wandesburg: palace in Danish Holstein built by Heinrich Rantzau (1526–99) around 1564. The central tower contained an observatory, from which Tycho Brahe conducted astronomical investigations in the years 1597–98.

38. WALLIS to OLDENBURG, 9 and 10/[19 and 20] December 1668

is no where expressed. If you have any supply of these; you may please to let mee know it. I remember allso you sayd you had Two of these Bookes to sell. If the other be not disposed of; pray send mee that allso. And, if you had one or two more I could tell how to dispose of them. I have little else to adde at present, unless possibly to morrow morning before the post goes I may meet with one by this post which may call for an answere; from

Yours to serve you
John Wallis.

Dec. 10. 1668.

What is above, I had written last night. Since which I have received yours[303] this morning. As to what you say of the Physical cause of motion: If it be onely, Why this way? & thus fast? & with thus much force? the Mathematick hypothesis satisfyes. But if it be, Whence it comes to pass that there [is] any Motion in the World? I doubt wee must make that for a Postulatum; That there is Motion, as well as, That there is Matter. And refer both to the same original cause. And, if we allow motion to begin, we must postulate That there is a Vis motrix even in resting Bodies. At lest I know not at present what to say more to it.

As to mine[304] of Dec. 5. you may do with it as you see cause. But you have not told mee, who's these quaere's are. The Paper of Problemes,[305] I have transcribed & do here return you. You do not tell mee, who sends it, nor to whom, nor by whom my Answer is desired. (whether sent to the Society at large; or, as a challenge to mee by name.) The truth is, I am allmost weary of such Problemes from beyond Sea, for the result is commonly but some quarelling. If they be not solved; they insult. If they be; they cavil,

---

14 be, *add.*
14 that there any *corr. ed.*
15 for *add.*
23–24 (whether sent ... by name.) *add.*

---

[303] yours: i.e. OLDENBURG–WALLIS 8/[18].XII.1668.
[304] mine: i.e. WALLIS–OLDENBURG 5/[15].XII.1668.
[305] Paper of Problemes: i.e. the three mathematical problems proposed by Jacob van Wassenaer, which had been enclosed in OLDENBURG–WALLIS 8/[18].XII.1668. Wallis replied to these problems with WALLIS–OLDENBURG 12/[22].XII.1668.

39. WALLIS to OLDENBURG, 12/[22] December 1668

& be angry. At lest, they take up time & give trouble to no purpose. If he that proposeth them can solve them, let him tell the world how, if he think fit: Which, in my judgement, tends more to the advancement of publike knowledge, than chalenging others to find out what they have found allready. But no more of this at present, the Post being now going.

The occasion of the inclosed, is cheefly to take occasion to supply what (if I understood you aright) you thought wanting in Mr Vice-Chancelours letter,[306] viz. to signify to Hevelius your care in transmitting the Bookes to the Library.

These
For Mr Henry Oldenburg,
in the Old Palmal, near
St James's
London.

39.
WALLIS to HENRY OLDENBURG
12/[22] December 1668

Transmission:

E First edition of part of missing letter sent: WALLIS, *Opera mathematica* II, 599–601.

Jacob van Wassenaer sent three geometrical problems from the Low Countries to the Royal Society, whence they were forwarded (without naming the author) by Oldenburg to Wallis with OLDENBURG–WALLIS 8/[18].XII.1668. Wallis quickly made a transcription and returned the problems with WALLIS-OLDENBURG 9/[19].XII.1668. His solution was part of an otherwise now missing letter to Oldenburg. See Wallis's comments in WALLIS–OLDENBURG 21/[31].XII.1668.

1 & (*1*) troub *breaks off* (*2*) give
3 fit: (*1*) Why (*2*) Which
3 to (*1*) advance (*2*) the advancement
6 The (*1*) occas *breaks off* (*2*) occasion
8 to Hevelius *add.*

---

[306] Mr Vice-Chancelours letter: presumably the letter of John Fell, which Oldenburg passed on to Hevelius with OLDENBURG–HEVELIUS 11/[21].XII.1668 (OLDENBURG, *Correspondence* V, 237–9).

39. WALLIS to OLDENBURG, 12/[22] December 1668

Quod attinet ad Problemata tria Jacobi a Wassenaer, quae abhinc biduo a te recepi: hoc habeas Responsum.

Primum hoc est. *In* A *Conchoidis Gibbositas seu Convexitas est. Quaeritur altera ejus convexitas in* D; *ubinam illa sit ut Tangens sit in* D. *viz. quaenam longitudo* DE.

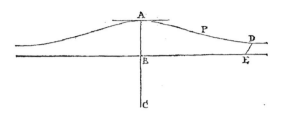

Dico. Notum est, in Conchoide, utrinque a Vertice A, dari punctum quoddam, ut P, in quo incipit Conchoides convexitatem suam mutare. In hoc Puncto, nulla est Tangens, (sed, hujus loco, Secans; quae ex Tangente extrinsecus partis Superioris, & Tangente intrinsecus partis Inferioris curvae componitur:) sed, extra hoc punctum P, ubivis sumi potest punctum D, (quippe nullum-non Convexitatem suam habet, quam tangat recta.) Et quidem si sumatur propius ad A, tangens erit supra curvam; si remotius, infra curvam. Illud autem punctum P quomodo reperiatur, jamdudum docuit D. Hugenius, in libro[307] De magnitudine Circuli; & D. Schootenius in Notis[308] ad Cartesii Geometriam, pag. 258 (Editionis Anni 1659.) Et quidem, Pro dato Puncto Tangentem ducere; vel quae huic sit ad angulos rectos; ibidem docetur, pag. 249. Sin illud velit Proponens, ut Tangens in D ita se habeat, ut, quomodocunque producta, nusquam occurrat Conchoidi eamque secet, (quod captiosum esset:) Dico, istiusmodi punctum D nusquam (extra punctum A) reperiri. Saltem nisi quis velit Regulam BE tangentem dici Conchoidis in infinitum continuatae.|

Tertium hoc est. *Sit* DG *hyperbola;* AE *parallela ipsi* BH [*Asymptotae:*] *Invenire punctum* E, *& longitudinem ipsius* EF, [centrum scilicet & radium circuli per $F G$ transeuntis,] *ita ut* FK *sit ad* GH, *ut 2 ad 1*.

---

[307] libro: i.e. HUYGENS, *De circuli magnitudine inventa*, Leiden 1654.
[308] Notis: i.e. DESCARTES, *Geometria a Renato DesCartes*, ed. F. van Schooten, 2 vols., Amsterdam 1659.

39. WALLIS to OLDENBURG, 12/[22] December 1668

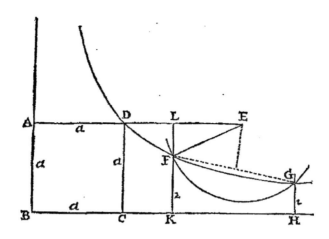

Dico. Si (in Asymptota BC, utcunque producta) sumpto ubivis puncto K, sumatur ipsi BK aequalis KH; & ducantur KF HG (parallelae alteri Asymptotae BA) hyperbolae occurrentes in F G; & jungatur F G recta: Quae hujus puncto medio insistit perpendicularis, secabit rectam AE in puncto E. Junctaque E F, habetur recta quaesita. Demonstratio facilior est quam ut debeat apponi.

Videatur, ex adscripto Schemate, solam spectari rectam AE quae inscripto Quadrato $ABCD$ adjacet. Sed constructio mea pariter procedit ubicunque in Asymptota BA, sumatur punctum A.

Secundum est paulo adhuc abstrusius; (quod itaque in postremum locum rejeci.) Sic utique se habet. *Sit* AB *(parabolae) latus rectum. Quaeritur* BF [*circuli radius*] *ita ut* [*arcus*] BC *sit aequalis ipsi* CD [*arcui circulari.*]

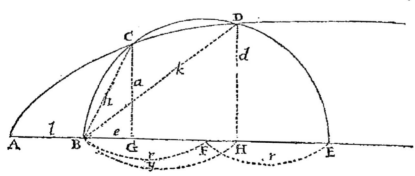

Intelligantur demitti perpendiculares CG DH; jungique BC BD rectae. Ponantur autem $AB = l$, $BF = r$, $BC = h$, $BD = k$, $BH = y$, $DH = d$, $CG = a$; adeoque (sinus duplici arcus) $d = \frac{2a}{r}\sqrt{\ :\ r^2 - a^2\ :}$ & $d^2 = \frac{4r^2a^2 - 4a^4}{r^2}$.

39. WALLIS to OLDENBURG, 12/[22] December 1668

Ponatur item $BG = e$; adeoque (propter $BG, GC, GE, \because$, hoc est $e, a,$ $2r-e, \because$) $a^2 = 2re - e^2$; & $h^2(= a^2 + e^2) = 2re$. Ergo (quadratum subtensae dupli arcus) $k^2 = (\frac{4r^2h^2 - h^4}{r^2} = \frac{8r^3e - 4r^2e^2}{r^2} =)8re - 4e^2$.

Item (propter $BE, BD, BH \because$, hoc est $2r, k, y \because$,) $y = (\frac{k^2}{2r} =) 4e - \frac{2e^2}{r}$.

Tum (propter Parabolam) $AG \cdot AH :: CGq \cdot DHq$. Hoc est

$$l + c \cdot (l + y =)l + 4e - \frac{2e^2}{r} ::$$
$$(a^2 \cdot d^2 = \frac{4r^2a^2 - 4a^4}{r^2} :: r^2 \cdot 4r^2 - 4a^2 ::) r^2 \cdot 4r^2 - 8re + 4e^2.|$$

Ergo factum ab extremis facto a mediis.)
$4lr^2 - 8lre + 4le^2 + 4r^2e - 8re^2 + 4e^3 = lr^2 + 4r^2e - 2re^2$.
Adeoque $3lr^2 - 8lre - 6re^2 + 4le^2 + 4e^3 = 0$.
Et (ductis omnibus in $3l$)
$9l^2r^2 - 24l^2re - 18lre^2 + 12l^2e^2 + 12le^3 = 0$.
Seu $9l^2r^2 - 24l^2re - 18lre^2 = -12l^2e^2 - 12le^3$.
Et (additis utrinque $16l^2e^2 + 24le^3 + 9e^4$)
$9l^2r^2 - 24l^2re - 18lre^2 + 16l^2e^2 + 24le^3 + 9e^4 = 4l^2e^2 + 12le^3 + 9e^4$.
Et (sumptis radicibus) $3lr - 4le - 3e^2 = 2le + 3e^2$.
Hoc est, $3lr - 6le - 6e^2 = 0$. Seu $\frac{1}{2}lr - le - e^2 = 0$.
Et (resolvendo aequationem)
$\frac{\sqrt{:2lr - l^2:} - l}{2} = e$. Et $\frac{lr + l^2, -l\sqrt{:2lr - l^2:}}{2} = e^2$.

Est autem (ut supra ostensum) $a^2 = 2re - e^2$.
Sed & (propter Parabolam) $a^2 = l^2 + le$.
Ergo $2re - e^2 = l^2 + le$. Hoc est $2re = e^2 + le + l^2$.
Seu (propter valorem e modo inventum)

$$r\sqrt{:2lr + l^2:} - lr(= \frac{lr + l^2, -l\sqrt{:2lr + l^2:}}{2}$$
$$+ \frac{l\sqrt{:2lr + l^2:} - l^2}{2} + l^2 =) \frac{lr + 2l^2}{2}.$$

Hoc est, $2r\sqrt{:2lr + l^2:} - 2lr. = lr + 2l^2$.
Seu $3lr + 2l^2 = (2r\sqrt{:2lr + l^2:} =) \sqrt{:8lr^3 + 4l^2r^2}$.
Et (sumptis quadratis) $9l^2r^2 + 12l^3r + 4l^4 = 8lr^3 + 4l^2r^2$.
Adeoque $9lr^2 + 12l^2r + 4l^3 = 8r^3 + 4lr^2$.
Hoc est, $8r^3 - 5lr^2 - 12l^2r - 4l^3 = 0$.
Quae quidem Aequatio Cubica, ordinata & resoluta, exhibet
$r = \frac{\sqrt{C:5741 + 96\sqrt{249}.} + \sqrt{C:5741 - 96\sqrt{249}.}}{24} l$. Quod erat quaesitum.

Et quidem, si ponatur latus rectum $l = 1$; erit circuli radius $r = 1.688637$ proxime.

Fieri forte potest ut ad hanc (vel similem) aequationem, paulo expeditius perveniatur. Sed haec sufficiant raptim scribenti. Vale.

40.
HENRY OLDENBURG to WALLIS
14/[24] December 1668

Transmission:

Manuscript missing.

Existence and date: Mentioned in WALLIS–OLDENBURG 21/[31].XII.1668.

In this letter Oldenburg evidently continued the discussion on the nature of motion on Neile's behalf and once again took up the topic of Hevelius's books.

41.
HENRY OLDENBURG to WALLIS
15/[25] December 1668

Transmission:

Manuscript missing.

Existence and date: Mentioned in and answered by WALLIS–OLDENBURG 21/[31]. XII.1668.
Reply to: WALLIS–OLDENBURG 9/[19].XII.1668 (ii).

In this letter Oldenburg informed Wallis that the author of the three geometrical problems he had forwarded to him was Jacob van Wassenaer.

42.
JAMES GREGORY to WALLIS
c.20/[30] December 1668

Transmission:

Manuscript missing.

43. WALLIS to OLDENBURG, Oxford, 21/[31] December 1668

Existence and date: Mentioned in WALLIS–OLDENBURG 2/[12].I.1668/9 and WALLIS–OLDENBURG 12/[22].I.1668/9.

As emerges from WALLIS–OLDENBURG 21/[31].XII.1668, Wallis had written to Gregory on the controversy with Huygens in November or early December. Oldenburg apparently informed Wallis that a reply should have reached him before his own letter of 26 December 1668 (old style), indicating that it was probably sent around 20 December—see WALLIS–OLDENBURG 2/[12].I.1668/9. By the time he wrote WALLIS–OLDENBURG 12/[22].I.1668/9 the letter had arrived, 'having been long in the warehouse at London by reason of the wagons not going in Christmas week'.

43.
WALLIS to HENRY OLDENBURG
Oxford, 21/[31] December 1668

Transmission:

W Letter sent: LONDON *Royal Society* Early Letters W1, No. 79, 4 pp. (p. 3 blank) (our source). At top of p. 1 in Oldenburg's hand: 'Enter'd LB. 2. 354'. On p. 4 beneath address in Oldenburg's hand: 'Rec. Dec. 21. 68.', and: 'An Extract of a letter of Dr Wallis to M. Old. containing a further explanation of the former Queries touching Motion'. Postmark on p. 4: 'DE/21'.—printed: OLDENBURG, *Correspondence* V, 272–5.
w^1 Copy of letter sent: LONDON *Royal Society* Letter book Original 2, pp. 354–6.
w^2 Copy of w^1 LONDON *Royal Society* Letter Book Copy 2, pp. 409–11.

Reply to: OLDENBURG–WALLIS 14/[24].XII.1668 and OLDENBURG–WALLIS 15/[25]. XII. 1668.

After receiving this letter, Oldenburg apparently sent it to William Neile for inspection. See NEILE–OLDENBURG 28.XII.1668/[7.I.1669] (OLDENBURG, *Correspondence* V, 286–7).

Oxford. decemb. 21. 68.

Sir

Yours[309] of Dec. 15. came to me so late on Thursday that I could not answere it by that Post. Wassenaer's Problemes,[310] especially the first & third, do not argue any great profoundness of Skill; & none of them are very accurately proposed; (but must be helped out, by the sight of the schemes, to be understood;) but well inough for one that is (as you say)

---

1 |Oxford *add.* Oldenburg| decemb.

[309]Yours: i.e. OLDENBURG–WALLIS 15/[25].XII.1668.
[310]Wassenaer's Problemes: see WALLIS–OLDENBURG 12/[22].XII.1668.

43. WALLIS to OLDENBURG, Oxford, 21/[31] December 1668

illiterate. In my solution,[311] (in the last line save one or two) where you have $r = 1,688635$ proxime; pray make it $r = 1,688637$ proxime, (changing the 5 for 7:) for though, as it is, it be near inough, yet 7 will be more accurate. I hope you have amended that allready (putting 20 for 40) in
5 my answere[312] to Dulaurens. You may adde[313] to the end of that answere (before I speak of Mr Hobs,) —: *sed Lectori permitto judicandum. Cur autem quae ego pridem miseram, et junctim; tu serius edideris, et per partes: non meum est respondere. Sed nec responso opus est.* And then, I hope, I have done with him. I suppose you have heard nothing of what Mr Gre-
10 gory says to my last letter[314] (which was sent to him into Scotland) or to Mr Hugens's Reply:[315] because you say nothing of it. (When you do, pray let mee hear of it.) Nor whether Dulaurens have yet printed his written paper.[316] I am very desirous (& so are some others) that the *Historia Coelestis* should be exactly compared with the Originals; & an accurate
15 Table of the Numerall Errata's that have scaped the Presse, be printed. or it is of great moment, in such a work; & impossible to be supplied by conjecture: And therefore you shall do well to importune it by your correspondency if you have any in those parts. Is *Cassini*'s Ephemerides,[317] drawn from

    6 *[In left margin in Wallis's hand:]* I think those words wil make sense with [what] is there preceding: if not, you must help it; for have no copy of what I sent you.

4 that *add*.
4 in *(1)* the *(2)* my
10 Scotland) *(1)* or of *(2)* or to
11 it. *(1)* I ⟨—⟩ will *(2)* (When
18 parts. *(1)* Doth *Cassini*'s Ephemerides agr *breaks off* *(2)* Is *Cassini*'s Ephemerides,

---

[311] my solution: i.e. WALLIS–OLDENBURG 12/[22].XII.1668.

[312] answere: i.e. Wallis's paper on Dulaurens, dated 27.XI/[7.XII].1668. Wallis's request that Oldenburg make the said correction is contained in WALLIS–OLDENBURG 3/[13].XII.1668.

[313] You may adde: Oldenburg decided not to publish Wallis's paper on Dulaurens in the *Philosophical Transactions*; nor did he add this passage to Wallis's manuscript.

[314] letter: i.e. the now missing letter WALLIS–GREGORY XI/early XII.1668. Gregory's reply, JAMES GREGORY–WALLIS c.20/[30].XII.1668, did not reach Wallis before early January 1668/9.

[315] Mr Hugens's Reply: i.e. HUYGENS–GALLOIS ?.XI.1668, published as 'Extrait d'une lettre de M. Hugens à l'Auteur du Journal, touchant la Réponse que M. Gregory a faite à l'examen du Livre intitulé Vera Circuli & Hyperboles Quadratura, dont on a parlé dans le V. Iournal de cette année', *Journal des Sçavans* (12 November 1668), 109–12.

[316] written paper: i.e. Dulaurens's letter, addressed to Oldenburg but written against Wallis, to which Wallis replied with his paper of 27.XI/[7.XII].1668. Dulaurens seems not to have published his letter.

[317] *Cassini*'s Ephemerides: i.e. CASSINI, *Ephemerides Bononienses mediceorum syderum*

43. WALLIS to OLDENBURG, Oxford, 21/[31] December 1668

Borelli's hypothesis,[318] or some other? He shall do well hereafter to publish his Ephemerides sooner; that they may be dispersed before the times be past; as that for 1668 now is, within a few days. I have received some of your mony, but not all; (by reason that he who was to have a considerable part of the Books, hath been for some time out of town, & is not yet returned:) when I have the rest, I shall give you account of the whole together. I have no copy of the Latine pieces I sent you concerning the Tydes:[319] (as indeed, I seldome transcribe what I send you, but send you the first & onely draught of it;) & therefore if you send those into Holland, it will be convenient that either you send or keep a Transcript; that wee may at lest have a copy here *in omnem eventum*. Since I had written thus far, I have received yours[320] of Dec. 14, & (this evening) the 2^d *Historia Coelestis*, (brought by Moor's Wagon) & have collated it this evening, & find it perfect; But find by it, that the former wants two Leaves; which are at Pag. CV &c. of the liber Prolegomenos, to be substituted for two other that are there to be cut out. (If you can get a supply of them it will be a curtesy.) & this latter is not so much dammaged by the wet as the former was. I make no question but if I had two or three more here, they would be disposed of; so that you need not doubt sending for some of them. I guesse by your last letter that you had these from Hevelius, which (before) I did not know. In one of the Copies of Hevelius his Prodromus, I have the sheet B twice; so that if you want that sheet in any other copy, I can supply you. If you send to Hevelius

drawn
4 all; (*1*) because (*2*) (by reason that he |who *add.*| was
7 Tydes: (*1*) & ther *breaks off* (*2*) (as indeed, I seldome (*a*) write ⟨—⟩ (*b*) transcribe
11–13 I (*1*) have (*2*) have received |yours of Dec. 14, & (this evening) *add.*| the 2^d *Historia Coelestis*, |(brought by Moor's Wagon) *add.*| &
14 Pag. (*1*) CXXV (*2*) CV |&c. *add.*| of
19 that (*1*) h *breaks off* (*2*) you

---

ex hypothesibus et tabulis Jo. Dominici Cassini . . . ad observationum opportunitates praemonstrandas deductae, Bologna 1668. Oldenburg had received this work in early December 1668; see OLDENBURG–HEVELIUS 11/[21].XII.1668 (OLDENBURG, *Correspondence* V, 237–9).

[318]*Borelli*'s hypothesis: this is contained in BORELLI, *Theoricae mediceorum planetarum ex causis physicis deductae*, Florence 1666.

[319]Latine pieces . . . concerning the Tydes: i.e. Wallis's Latin translation of his hypothesis of the tides WALLIS–BOYLE 25.IV/[5.V].1666 (WALLIS, *Correspondence* II, 200–22) and the appendix WALLIS–OLDENBURG 18/[28].VII.1666 (WALLIS, *Correspondence* II, 251–63), both of which had been published in *Philosophical Transactions*. Cf. also WALLIS–OLDENBURG 26.XI/[6.XII].1668.

[320]yours: i.e. OLDENBURG–WALLIS 14/[24].XII.1668.

43. WALLIS to OLDENBURG, Oxford, 21/[31] December 1668

for any books; you may desire him allso to send the Scheme RRR in his Selenography, supernumerary; for I want it in my own book. This Jacobus a Waessenaer, I suppose may be the same whom Schoten mentions in his Additament[321] to his Comment on Des Chartes' Geometry: But then he must be pretty ancient; for Schoten mentions a book[322] of his published in the year 1640.

The third of the late Quaere's,[323] to which you desire a further answere. *Whether no motion in the World perish*, &c; that is, (as you now explain it,) *whether any of that motion, that was first* (or at any time since) *impressed in matter be lost, or* (onely) *communicated from one parcell of the matter to another; so that though this or that body do cease to be moved, yet the motion itself ceaseth or perisheth not*. Is a question, which I find Mathematicians, as well as Naturalists, sparing to determine positively;| and, you knowe, I am sparing and wary in asserting Universall Negatives. Yet you have, to this, my answere, full inough, (if it be observed,) in my answere[324] to the fourth. For I there intimate my judgement, that *motion may be extinguished*, & I shew you *how*; that is, a Motion compounded of two contrary forces, may be extinguished by each other, & become equivalent with Rest. If it be objected, that the motion of two bodies so meeting, is not lost, but both in effect preserved by checking the contrary motion in its opposite: this will bee but to cavil at words. For if this be agreed, that two bodies in motion do mutually stop each other, so that the motion of both be thereby extinguished

[2]

---

7 (*1*) To the (*2*) The third of the |late *add.*| Quaere's
12 *ceaseth* (*1*) not. (*2*) or
15 the (*1*) ⟨th⟩ *breaks off* (*2*) fourth.
16 intimate |my judgement *add.*| , that *motion may be* (*1*) *lost*, (*2*) *extinguished*,
18 may (*1*) extinguish (*2*) be extinguished by
18 with (*1*) rest in both of them (*2*) Rest. If (*a*) you (*b*) it
21 be (*1*) granted, (*2*) agreed,
22 other *add*.
22 be |thereby *add.*| extinguished & (*1*) they remain ⟨—⟩ in rest (*2*) at rest remain in rest (*3*) both

---

[321] mentions in his Additament: i.e. DESCARTES, *Geometria a Renato DesCartes*, ed. F. van Schooten, 2 vols., Amsterdam 1659, I, 369.

[322] book: i.e. WASSENAER, *Den on-wissen wis-konstenaer J. Stampioenius ontdeckt door zyne ongegronde weddinge ende mis-lucte solutien van syne eygene questien*, Leiden 1640.

[323] Quaere's: i.e. William Neile's four queries on Wallis's hypothesis of motion, WALLIS–OLDENBURG 15/[25].XI.1668. See WALLIS–OLDENBURG 3/[13].XII.1668 and WALLIS–OLDENBURG 5/[15].XII.1668.

[324] my answere: i.e. WALLIS–OLDENBURG 5/[15].XII.1668.

43. WALLIS to OLDENBURG, Oxford, 21/[31] December 1668

& both remain at rest; Whether this motion shal be sayd to be lost, or to bee onely virtually preserved, is as men shall please to call it. The other suggestion of Mr H;[325] *that springyness is the cause of rebounding*; is (as I sayd before) not new to mee; & I think you are my witness that you had it in my hypothesis of motion, before he started it[326] in the society; I am sure, before you signified any such thing to mee: so that I suppose he doth not think mee to have robbed him of his notion. How hee came by it (whether by a conjecture or a certainty) hee knows best; I was forced to it by the necessity of a consequence from those principles which I layd down in my first & second chapter[327] (allready printed) where is couched the foundations of those demonstrations (or calculations rather) which are to bee deduced in some following chapters, De Percussione, et Motuum Acceleratione, &c. in their due places. The reason which forced mee to bee of that opinion, you have breefly, in the Hypothesis[328] I sent: For if the force *prorsum* (supposing bodies perfectly hard) do (as is there argued) require that both it, & the body it directly strikes, should move the same way (that way which the greater force determines) and both at the same swiftness; I found, that, upon this account, there could be no rebounding: and therefore was necessitated (as

---

1 be *(1)* lost, *(2)* lost, or to bee |onely *add.*| virtually
3 the *(1)* caus *breaks off* *(2)* cause
5 society; *(1)* at lest *(2)* I am sure,
7 notion. *(1)* Whe *breaks off* *(2)* How
9 a consequence from *add.*
9 in *add.*
10 is *(1)* couched *(2)* couched
11–12 demonstrations *(1)* which are to follow *(2)* (or calculations rather) which are *(a)* to follow in those chapters which *(b)* to bee deduced in *(aa)* the following chapter *(bb)* some following chapters
14 the *(1)* P *breaks off* *(2)* Hypothesis
16 directly *add.*
16 the *(1)* mor *breaks off* *(2)* greater *(a)* d *breaks off* *(b)* force
17 both *add.*
18 there *(1)* was *(2)* could

---

[325] other suggestion of Mr H: i.e. Hooke's opinion, that elasticity is the cause of reflection; see WALLIS–OLDENBURG 5/[15].XII.1668.
[326] started it: Hooke performed experiments to evince this idea for the first time in the meeting of the Royal Society on 26 November 1668; BIRCH, *History of the Royal Society* II, 328.
[327] my first & second chapter: i.e. of Wallis's *Mechanica: sive, de motu, tractatus geometricus*, the first part of which appeared in 1670.
[328] Hypothesis: i.e. WALLIS–OLDENBURG 15/[25].XI.1668.

43. WALLIS to OLDENBURG, Oxford, 21/[31] December 1668

you see in my hypothesis) to have recourse, to that of Elasticity, in one or both of the bodies. And, if the rest of the hypothesis bee admitted; this, seemes to mee, to be unavoidable. Whether I had before heard or read of this notion, before I was by this argument forced to it; I do not remember: But, that I have (either before or since) mett with it in print, I am very certain: & (as I remember) in more authors than one; though I cannot perfectly tell their names. But I do not take the notion to be very common: (and I am not sorry if Mr H. bee of my mind.) But it is a notion which, I think, I have well digested; & am very confident it must hold. Yet I do not so lay it upon Elasticity, but that I do acknowledge that a repercussion is (oft) a concomitant cause. And from one, or both, of these causes (as the case may happen,) I deduce the Equality of the Angle of Reflexion to that of Incidence: of which I do not know that I have yet met with any satisfactory account in other Authors.

I thank you for the Transactions[329] you sent. Which I have cursorily read over: & find no considerable errata in that which concerns mee. These few I took notice of, (beside what you have amended) pag. 852. l. ult. *what* is put for *which*. pag. 826. *et fiunt* is put for *ut fiant*. pag. 827. l. 23. *Telluri* for *Tellure*. *lin. ult. quod* for *quid*. You may mend these with a pen, at lest in the copy you send to Du-laurens. I think I shall have some papers for M. Hevelius within a few days; part of what I intimated in my latine letter[330] to you. My intimation of sending that to him; was but to supply what I apprehended you thought wanting in the Vice-chancellours letter;[331] in not taking express notice that you had sent the Books. If it were; that hee did not by a particular letter to your self thank you for your care; that was it which hee left to mee to signify to you; which (being but a complement) I might possibly forget it: & desire your pardon for it. I am

Yours &c
J. Wallis.

---

8 sorry (*1* that (*2*) if)
11 of (*1*) the cause (*2*) these causes
17 (beside what you have amended) *add*.
18 l. 23. *add*.
21 my (*1*) latest (*2*) latine

---

[329]Transactions: i.e. *Philosophical Transactions* No. 41 (16 November 1668). Wallis's errata are noted in No. 42 (14 December 1668), 852.

[330]latine letter: i.e. WALLIS–OLDENBURG 9/[19].XII.1668.

[331]Vice-chancellours letter: i.e. Fell's letter to Hevelius, which was enclosed to OLDENBURG–HEVELIUS 11/[21].XII.1668 (OLDENBURG, *Correspondence* V, 237–9).

These
For Mr Henry Oldenburg,
in the Palmal near
St. James's
London.

44.
HENRY OLDENBURG to WALLIS
26 December 1668/[5 January 1669]

Transmission:

Manuscript missing.

Existence and date: Mentioned in WALLIS–OLDENBURG 31.XII.1668/[10.I.1669] and WALLIS–OLDENBURG 2/[12].I.1668/9.

45.
HENRY OLDENBURG to WALLIS
29 December 1668/[8 January 1669]

Transmission:

Manuscript missing.

Existence and date: Mentioned in and answered by WALLIS–OLDENBURG 31.XII.1668/[10.I.1669].

As emerges from Wallis's reply, Oldenburg touched on Neile's hypothesis of motion.

46.
HENRY OLDENBURG to WALLIS
31 December 1668/[10 January 1669]

Transmission:

Manuscript missing.

Existence and date: Mentioned in and answered by WALLIS–OLDENBURG 2/[12].I.1668/9.

47.
WALLIS to HENRY OLDENBURG
Oxford, 31 December 1668/[10 January 1669]

Transmission:

W Letter sent: LONDON *Royal Society* Early Letters W1, No. 80, 4 pp. (p. 2 and p. 4 blank) (our source). At top right of p. 1 in Oldenburg's hand: 'Enter'd LB. 2. 362.' On p. 2 beneath address: 'Rec. jan. 1. 68/9.', and above address, again in Oldenburg's hand: 'An Extract of a letter from D. Wallis to M. Old. concerning Motion, and in particular, compound motion.' Postmark on p. 4: 'JA/1'.—printed: OLDENBURG, *Correspondence* V, 287–8.
w^1 Part copy of letter sent: LONDON *Royal Society* Letter Book Original 2, pp. 362.
w^2 Copy of w^1: LONDON *Royal Society* Letter Book Copy 2, pp. 418–19.

Reply to: OLDENBURG–WALLIS 29.XII.1668/[8.I.1669].

Oxford Dec. 31. 1668.

Sir

This is, onely to tell you that I have received yours[332] of Dec. 29. since eleven of the clock & the Post goes by 12. I thank you for it. But that[333] of Saturday,
5 which you mention, is not yet come to hand. As to the Hypothesis of Mr Neile,[334] which you mention, I can say nothing to it, (but leave it to himself to manage,) further than as my own hypothesis expresseth allready. What that is which wee call springynesse; & what, Gravity: I do not determine: but from those things, what ever they are, & from what ever cause they
10 do proceed, I am to give account of the effects, I ascribe to them. The one (from what ever cause) is the principle of the motion of restitution; & the other, of tendency downward. I know Des-Cartes & others do attempt to assign causes[335] of both; but I have not yet seen any hypothesis, that doth fully satisfy my apprehensions; & therefore I do not, as to that, determine
15 any thing. What I mean by a Compound motion from two movents, is not hard to apprehend.

---

7 than *add.*

---

[332] yours: i.e. OLDENBURG–WALLIS 29.XII.1668/[8.I.1669].
[333] that: i.e. OLDENBURG–WALLIS 26.XII.1668/[5.I.1669].
[334] Neile: i.e. William Neile, q.v.
[335] causes: see for example DESCARTES, *Principia philosophiae*, Amsterdam 1644, IV, §§20–7, 199–204.

48. WALLIS to OLDENBURG, 2/[12] January 1668/9

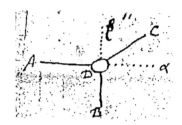

If the Body D receive a stroke from A, & another from B, at the same time; it will by this compound force be carried neither to α nor to β, but to some intermediate point as to C. so that I call DC a motion compounded of $D\beta$ & $D\alpha$, & in such proportion near to β or to α, as the two strengths shall require; & in a streight or crooked line as there shall be occasion. But I can adde no more but that I am

Yours &c.
Jo. Wallis.

[3] For Mr Henry Oldenburg
in the Palmal near St. James's.
London.

48.
WALLIS to HENRY OLDENBURG
Oxford, 2/[12] January 1668/9

Transmission:

W Letter sent: LONDON *Royal Society* Early Letters W1, No. 81, 2 pp. (our source). Endorsed by Oldenburg beneath address on p. 2: 'Rec. Jan. 4. 1668/9'. Postmark on p. 2: 'IA/4'.—printed: OLDENBURG, *Correspondence* V, 302–4.

Reply to: OLDENBURG–WALLIS 29.XII.1668/[8.I.1669].

Oxford. Jan. 2. 1668./9.

Sir

To yours[336] of Dec. 29. I answered[337] by the last Post, as the time (which was

---

4 DC (*1*) to be (*2*) a motion compounded

[336] yours: i.e. OLDENBURG–WALLIS 29.XII.1668/[8.I.1669].
[337] answered: i.e. WALLIS–OLDENBURG 31.XII. 1668/[10.I.1669].

48. WALLIS to OLDENBURG, 2/[12] January 1668/9

scarce a quarter of an hour) would permit. The summe (as I remember) was to this purpose: That as to the businesse of Springyness & Gravity, what they are, & from what Physical principles they proceed, I cannot undertake to satisfy another till I am better satisfied my self. I onely take for granted what sense & experience shews: viz. that there is in Springy bodyes an inclination to restore themselves to their former position; & in Heavy bodies, to move downward: Without determining what that inclination is, or from what cause. If Mr Neil[338] can make out an hypothesis to clear that, I shall not be his adversary.

Your two letters[339] of Dec. 26. & Dec. 31. came both to my hands this morning. But what[340] you say should have come, before the first of these, from Mr Gregory to mee, I have not received. And, so many Posts having passed since, makes mee suppose it may be lost. The inclosed papers[341] which I received in those this morning, (of Mr Gregory to Mr Hugens) I have read over, & that is all; & herewith return them. I have formerly told him my opinion, & my reasons; which were not kindly taken. I did defend him (in my first letter to Mr Hugens) as far as the matter would bear; without his asking: but that he did not admit of. I have advised what I thought was to bee done: but hee is of another mind. All that I have to desire further of him, is, that hee will give mee leave to hold my peace; And (since I can do him no service) to be unconcerned.

In my paper that concerns Waessaners Problems;[342] in that foul draught which I kept it is *Si in Asymptota BC (utcunque producta)* &c. If in yours

2 Gravity, (*1*) from (*2*) what
11 the first of *add.*
14 in those *add.*
16 I (*1*) would have defended (*2*) did defend him
17 Hugens) (*1*) |as far as the matter wo *add. and del.*| but that defense he doth not (*2*) as
20 of him, *add.*
20 peace; (*1*) without either to be concerned (*2*) And
22 that (*1*) paper (*2*) foul draught
23 I (*1*) keep (*2*) kept

---

[338] Neil: i.e. William Neile, q.v.
[339] two letters: i.e. OLDENBURG–WALLIS 26.XII.1668/[5.I.1669] and OLDENBURG–WALLIS 31.XII.1668/[10.I.1669].
[340] what: i.e. JAMES GREGORY–WALLIS c.20/[30].XII.1668.
[341] inclosed papers: probably JAMES GREGORY–OLDENBURG 15/[25].XII.1668. (OLDENBURG, *Correspondence* V, 250–5). Gregory's response to Christiaan Huygen's critique of his *Geometriae pars universalis* was published by Oldenburg in *Philosophical Transactions* No. 44 (15 February 1668/9), 882–6.
[342] Waessaners Problems: i.e. the mathematical problems posed by Jacob van Wassenaer,

48. WALLIS to OLDENBURG, 2/[12] January 1668/9

it bee BE, it is a mistake in transcribing, which you may please to amend. But rather by BC than by BH.

As to Borelli's Hypothesis;³⁴³ I understand from Mr Vernon³⁴⁴ that Borelli sent mee one of them a good while since: But it never came to my hands. (I suspect it may be that which (falling into other hands) was bought for my Lo. Brouncker:) So that I do not yet know his Hypothesis. That design³⁴⁵ had been performed amongst ourselves long since, had Mr Rook lived.

The notion of Resilition upon the account of Springynesse, is in Borelli³⁴⁶ de vi percussionis, pag. 47. (though he have something else where that seemes not so well consistent with it.) and (as I remember) in a French writer which Mr Collins lent me when I was last in London; but I have forgotten the name of it. I think it was that of which hee had three volumes in a broad quarto.

That which made me the more particularly observe it there, was, that in such occurrence hee makes the Resilition to be allways in the lighter body. But certainly the notion is in divers other writers though I cannot name them: Else I should have the more looked on it as new, when I mett it there. Which I did not, but as a thing familiar to mee; and, had it not been for the circumstance I mention, I should scarce have so much taken notice of it as to remember it. I cannot say whether they make this the onely cause: But

2 BH | because the point H is del.|
3 to (1) Cassini's (2) Borelli's
3 that (1) hee (2) Borelli
12 Collins (1) left me (2) lent me
18 there. (1) Whereas I did (2) Which
19 mee; (1) onely (2) and
21 they (1) say this is (2) make this

---

solutions to which Wallis presented in WALLIS–OLDENBURG 12/[22].XII.1668.

³⁴³Borelli's Hypothesis: i.e. Borelli's attempt to explain how the elliptical orbits of planetary bodies could be understood by means of the effects of three kinds of motion. See BORELLI, *Theoricae mediceorum planetarum a causis physicis deductae*, Florence 1666.

³⁴⁴Vernon: i.e. Francis Vernon, q.v. It is probable that Vernon met Borelli when he was in Rome in March 1668. Later the same year he returned to Oxford and no doubt informed Wallis personally that Borelli had sent him a copy of his hypothesis.

³⁴⁵design: when Lawrence Rooke died in June 1662 he was finishing work on his 'Ephemerides', which contained accurate observations of the satellites of Jupiter. See BIRCH, *History of the Royal Society* I, 97–8; WARD, *Lives of the Professors of Gresham College*, London 1740, 92.

³⁴⁶BORELLI, *De vi percussionis*, Bologna 1667, 47.

49. WALLIS to GREGORY, 8/[18] January 1668/9

I must take it to bee till I know of some other: My hypothesis affords mee no other, but that & Repercussion, (or what is equivalent.) What I sayd in mine to the 3^d Query,[347] (that two contrary equal forces must extinguish each other;) I sayd it, because I do not see how to avoyd it. And I think Borelli (by his principles) must say it too. For though he seem to bee (though not peremptory) of a contrary opinion, (that no motion is lost,) yet when hee salves it by being virtually preserved in checking a contrary motion; (as pag. 128, 135, 171.) I take it to be (for substance) what I say, though in other words. But the truth is, hee is a little imbrangled by some notions that do seem to clash with one another; but which a little help would reconcile, & expedite with much ease & clearnesse. I have nothing yet ready for Hevelius, but shall have shortly. Mean while I am

your friend & servant
Joh. Wallis.

These For Mr Henry Oldenburg
at his house in the Palmal
near St James's
London.

49.
WALLIS to JAMES GREGORY
8/[18] January 1668/9

Transmission:

Manuscript missing.

Existence and date: Referred to in WALLIS–COLLINS 20/[30].I.1668/9 and WALLIS–OLDENBURG 12/[22].I.1668/9.

1 must *add.*
1 bee (*1*) sin *breaks off* (*2*) till (*a*) I can tell (*b*) I know
3 equal (*1*) motions (*2*) forces
5 (though not peremptory) *add.*
7 (as pag. 128, 135, 171.) *add.*
8 be (*1*) for substance | and *add. and del.* | (*2*) (for substance)
9-11 But ... clearnesse. *add.*

---

[347] 3^d Query: evidently a mistake for the fourth query considered in WALLIS–OLDENBURG 5/[15].XII.1668.

51. COLLINS to WALLIS, 12/[22] January 1668/9

Reply to: GREGORY–WALLIS *c.*20/[30].XII.1668.

Wallis sent this letter together with that to which it replied as an enclosure to WALLIS–OLDENBURG 12/[22].I.1668/9, so that members of the Royal Society could see both. He requested that Oldenburg either send it to Gregory at his leisure 'or not send as shal be thought fit', while Gregory's letter to him was to be returned. Cf. GREGORY–COLLINS 15/[25].II.1668/9; RIGAUD, *Correspondence of Scientific Men* II, 184–7.

50.
WALLIS to JOHN COLLINS
8/[18] January 1668/9

Transmission:

Manuscript missing.

Existence and date: Referred to in WALLIS–COLLINS 18/[28].I.1668/9.

According to WALLIS–COLLINS 18/[28].I.1668/9, Wallis sent this letter in a parcel directed to the printer Moses Pitt. Evidently it concerned Gregory's dispute with Huygens, on which Collins shortly afterwards produced a report: 'The State of the Controversy between Mr. Hugenius and Mr. James Gregory'. See HUYGENS, *Œuvres complètes* VI, 372–6 and HUYGENS–OLDENBURG [20]/30.III.1668/9; OLDENBURG, *Correspondence* V, 450–2.

51.
JOHN COLLINS to WALLIS
12/[22] January 1668/9

Transmission:

Manuscript missing.

Existence and date: Referred to in WALLIS–COLLINS 18/[28].I.1668/9.

52.
WALLIS to HENRY OLDENBURG
Oxford, 12/[22] January 1668/9

Transmission:

W Letter sent: LONDON *Royal Society* Early Letters W1, No. 82, 2 pp. (our source). Endorsement by Oldenburg beneath address on p. 2: 'Rec. Jan. 14. 1668/9 Answ. d. ead.' Oldenburg has also noted twelve memoranda with no relation to the content of the letter.—printed: OLDENBURG, *Correspondence* V, 336–7.

Enclosures: GREGORY–WALLIS *c.*20/[30].XII.1668 and WALLIS–GREGORY 8/[18].I.1668/9.

As emerges from Oldenburg's endorsement, he replied this letter on the same day as he received it, i.e. by OLDENBURG–WALLIS 14/[24].I.1668/9.

Oxford, Jan. 12. 1668/9

Sir,

Mr Gregory's letter[348] is at last[349] come to hand, having been long in the ware-house at London by reason of the wagons not going in Christmas week. I send it you inclosed, with my Answer[350] to it: which you may shew our friends as you see occason, & then send at your leisure or not send as shal be thought fit. And when you have done with his letter send it me again; & that[351] of mine (or a copy of it) written to my Lord Brouncker of Novemb. 8. (or thereabouts) which you say is Registred concerning the same subject; of which I kept no copy.

Mr Gregories to mee (beside some unhandsome language in the first the page) is but a transcript of part of what he answer[352] to M. Hugens. That

7 And *(1)* if *(2)* when
8 it) *(1)* to my *(2)* written
9 which you say is Registred *add.*
12 part of *add.*

---

[348]letter: i.e. GREGORY–WALLIS *c.*20/[30].XII.1668.

[349]at last: Wallis inquired after Gregory's response to the letter he had sent to him in Scotland in Wallis–Oldenburg 21/[31].XII.1668. Having later heard from Oldenburg that a response was on its way, he suspected at the time of writing WALLIS–OLDENBURG 2/[12].I.1668/9 that it might have gone missing.

[350]Answer: i.e. WALLIS–GREGORY 8/[18].I.1668/9.

[351]that: i.e. WALLIS–BROUNCKER 4/[14].XI.1668.

[352]answer: i.e. GREGORY–OLDENBURG 15/[25].XII.1668; OLDENBURG, *Correspondence* V, 250–5. Cf. GREGORY–COLLINS 15/[25].II.1668/9 (TURNBULL, *James Gregory,*

52. WALLIS to OLDENBURG, 12/[22] January 1668/9

answer of his, I think may be well inough printed,[353] without doing any hurt, (with some preface declaring yourself not interested, nor the society, in the expressions of it, but onely & as done to satisfy the desire of Mr Gregory;) Because I beleeve M. Hugens doth expect it; & means at once to say what what he hath to say both to it, & to his Appendicula,[354] of which in the last he had not taken notice. And it will rather shorten then lengthen the contest. For if M. Hugens first answere that, & then this be otherwise published it will make a double work, & more papers. But you may be allso advised by others to who's judgement I submit mine. But you may then advertise Mr Gregory: that his friends do not approve of his way of language: & therefore he ought for the future to take notice of M. Hugens with more of respect (as he deserves) or that it will not be thought proper for the Transactions. Mr Hugens Hypotheses & Propositions of Motion; I have scarce had time more then to read[355] over. But they seem to mee very ingenious, & his way of demonstration allso. Hee proceeds from other principles; but, if I mistake not, his & mine may be well inough accommodated. Taking his *duram* in Hyp. 2. as opposed to *molle* in such a sense as this admits a transposition of parts (as in Lead & other soft bodies which upon percussion change their figure,) and supposing his *dura* there to be allso *aequaliter Elastica*, from which I am to *derive* that Resilition which he *postulates*.

10 that (*1*) such (*2*) his
11 to *add*.
13 Hugens (*1*) by the (*2*) Hypothesis
14 very *add*.
15 if (*1*) thi *breaks off* (*2*) I mistake not
19 and (*1*) his *dura* adding (*2*) supposing his *dura* there to be allso

---

68): 'I perceive by Dr Wallis his last to me that he is not satisfied with my answer to Hugenius, if so he hath seen it'.

[353] printed: Gregory's reply to Huygens was published, with a preface from Oldenburg, under the title 'An Extract of a Letter of Mr. James Gregory to the Publisher, containing some Considerations of his, upon M. Hugens his Letter, printed in Vindication of his Examen of the Book, entitled *Vera Circuli & Hyperbola Quadratura*' in *Philosophical Transactions* No. 44 (15 February 1668/9), 882–6.

[354] Appendicula: Gregory published his 'Appendicula ad veram Circuli & Hyperbolae Quadraturam' as part of his *Exercitationes Geometricae*, London 1668; HUYGENS, *Œuvres complètes* VI, 315–21.

[355] read: at the meeting of the Royal Society on 7 January 1668/9, a letter from Huygens dated 9 January 1668 (new style) together with his theory of motion was read. It was ordered that the papers on the theory of motion be copied and sent to those members of the Society who were concerned with the topic. Wallis is among those named. See BIRCH, *History of the Royal Society* II, 337.

Mr Neile's hypothesis[356] I cannot say nothing further to, till I see it drawn up in a methodicall body, that I may see where to find a foundation of a Mathematicall Calculation, by which he is to infer somewhat equivalent to what wee call a Spring.

The messenger is going: I can onely adde that I am

Yours &c.
J Wallis.

These
For Mr Henry Oldenburg
in the Palmal near
St James's
London.

[2]

53.
HENRY OLDENBURG to WALLIS
14/[24] January 1668/9

Transmission:

Manuscript missing.

Existence and date: Oldenburg noted on the cover to WALLIS–OLDENBURG 12/[22].I.1668/9 that he had received that letter on 14 January and answered it on the same day.

54.
WALLIS to JOHANNES HEVELIUS
Oxford, 15/25 January 1668/9

Transmission:

W Letter sent: PARIS *Bibliothèque Nationale* Nouv. acq. latines 1641, f. 114^r. (our source).
w Copy of letter sent: PARIS *Bibliothèque Nationale* Fonds Latin 10348, pp. 159–60.

Enclosure: Ptolemy's Catalogue of Stars.

---

[356] hypothesis: Neile did not send Oldenburg his hypothesis of motion until May 1669. Oldenburg sent Wallis a copy as enclosure to OLDENBURG–WALLIS 8/[18].V.1669.

54. WALLIS to HEVELIUS, 15/25 January 1668/9

This letter to Hevelius was sent through the hands of Oldenburg, who added a letter of his own (OLDENBURG–HEVELIUS 23.I/[2.II].1668/9; OLDENBURG, *Correspondence* V, 352–4) and a copy of WALLIS–OLDENBURG 9/[19].XII.1668. Wallis enclosed a catalogue of stars based on a Greek codex of the *Almagest*, giving the latitude and longitude of the fixed stars according to Ptolemy's catalogue. These data are collated against two Arabic manuscripts: the one an Arabic version of the *Almagest*, the other Al Sufi's book on the constellations of the fixed stars.

Amplissimo Celeberrimoque Viro. D. Johanni Hevelio,
Consuli Dantiscano. Joh: Wallis Geometriae
Professor Savilianus Oxoniae.
S.

Habes hic (Celeberrime Vir) ex Manuscriptis nostris Oxoniensibus Excerpta quaedam; quae tibi usui fore posse existimabam in Opere[357] quod sub manibus habes, De Fixarum locis restituendis. Nempe Fixarum Longitudines et Latitudines secundum Almagesti[358] exemplar MS Graecum; et Versionem Arabicam MS. et denique secundum Astronomi Celeberrimi Al-Suphi[359] Tabulas. Unde restitui possunt numeri Ptolemaici in vulgatis Editionibus luxati; quod quidem per Librariorum errores toties transcribentium, per tot quot a Ptolemaeo hactenus secula transierunt, non potuit non multis in locis accidere. Numeros Graecos, Graecis notis, uti invenimus, expressos habes; Arabicos vero, notis numerariis quibus nunc passim utimur. Eos cum Uleg-beigi[360] numeris, ex suis observationibus depromptis, (quos antehac ex Tribus MSS. Arabico-Persicis inter se collatis accepisti,) conferre poteris.

Accedunt etiam ex Chrysococcae[361] Codicibus MSS. excerpta quaedam.

7 restituendis *add.*
9 Astronomi *add.*
10 Ptolemaici *add.*
12 tot (*1*) quod (*2*) quot
14 utimur. (*1*) Eosdem (*2*) Eos

---

[357] Opere: i.e. HEVELIUS, *Machina coelestis*, two parts, Danzig 1673–79.
[358] Almagesti: i.e. PTOLEMY, *Almagest*.
[359] Al-Suphi: i.e. Al Sufi (Abd al Rahman) (903–86), Persian astronomer, lived at the court of Emir Adud ad-Daula in Isfahan.
[360] Uleg-beigi: i.e. Ulug Beg (1394–1449), Usbek astronomer. His tables of longitudes and latitudes of the fixed stars, edited and translated into Latin by Thomas Hyde, were published in Oxford in 1665. Cf. OLDENBURG–HEVELIUS 24.VIII/[3.IX].1666; OLDENBURG, *Correspondence* III, 215–21.
[361] Chrysococcae: i.e. Georgios Chrysococces (d. 1336), Greek physician and astronomer active in Constantinople. Excerpts of his *Syntaxis Persarum*, based on Persian astronomical tracts and containing a catalogue of fixed stars, were published by Boulliau in 1645.

55. WALLIS to JOHANNES HEVELIUS, 15/25 January 1668/9, enclosure

Usus sum, in omnibus, opera praesertim Egregii viri D. Edwardi Bernardi,[362] qui (ut in aliis, sic) in Lingua speciatim Arabica, et in rebus Mathematicis peritus est: Sed et Celeberrimi Edwardi Pocockii[363] SS. Th. D, et linguarum tum Hebraicae tum Arabicae Professoris; cujus etiam Codicibus MSS. usus sum. Extant quidem hic Tabularum Arabicarum, Persicarum, aliarumque MSS. plura volumina, digna quidem quae in lucem edantur; ex quibus haec pauca, utpote rem praesentem propius spectantia, excerpenda curavi. Tu illis utere, et Vale.

Scribebam raptim, Oxoniae
die 15 Januarii 1668. Stilo Angliae.
25 Januarii 1669. Stilo Gregoriano.

55.
WALLIS to JOHANNES HEVELIUS
Oxford, 15/25 January 1668/9,
enclosure:
Ptolemy's Catalogue of Stars

Transmission:

W Catalogue sent: PARIS *Bibliothèque Nationale* Nouv. acq. latines 1641, f. 115^r–118^v. In two places on f. 115^r Oldenburg has corrected 'Min. 12'.42''' to 'grad. 12.42''' in accordance with Wallis's instructions in WALLIS–OLDENBURG 21/[31].I.1668/9.

Enclosure to: WALLIS–HEVELIUS 15/25.I.1668/9.

Cl. Ptolemaei Canon Constellationum Stellarum Fixarum secundum Longitudinem et Latitudinem, ex Almagesti Codice MS. Graeco elegantissime exarato, in Bibliotheca Bodleiana Oxoniae, descriptus. Collatis etiam cum duobus Codicibus MSS. Arabicis Clarissimi Pocockii: Altero, versionis Arabica Almagesti; Altero, Libri Figurarum Coelestium Celeberrimi Astronomi Al-Suphi; qui ante 700 (circiter) annos Constellationum Imagines descripsit, et commentariis illustravit, Canonemque Ptolemaicum (additis Longitudini

6 volumina, (*1*) qua (*2*) digna
13 Graeco *add.*

---

[362]Edwardi Bernardi: i.e. Edward Bernard, q.v.

[363]Edwardi Pocockii: i.e. Edward Pocock (1604–91), q.v. Cf. WALLIS–OLDENBURG 9/[19].XII.1668.

55. WALLIS to JOHANNES HEVELIUS, 15/25 January 1668/9, enclosure

Ptolemaicae, grad. 12.42′.) ad suum tempus rectificavit. Numeros autem in Codd. MSS. Arabicis, quoties ab illo Graeco MS. differebant, ex adverso posui, (deducto prius Longitudinum numeris lapsionis additamento illo grad. 12.42′.) quos ab invicem ita distinxi, ut numeri Al-suphiani lineam rectam subscriptam habeant; numeri vero ex Almagesto Arabico, non habeant. Significant autem (praeter notos graecarum literarum valores) in Canone graeco, $\overset{''}{=} \frac{1}{2}$. $\Gamma' \frac{1}{3}$. $\delta' \frac{1}{4}$. $\epsilon' \frac{1}{5}$. $\varsigma' \frac{1}{6}$. $\omega''' \frac{2}{3}$. $\bar{o}\, o$. Interim omisimus (brevitatis gratia) Signorum characteres (ut qui ex impressis libris suppleri satis poterunt,) et Latitudinum plagam, et stellarum item magnitudines, sed et stellarum nomina: utpote ipsos Longitudinum et Latitudinum Numeros praecipui momenti existimantes qui comparentur. Reliqua, si opus fuerit, suppleri poterunt; sed longioris temporis erit quam jam suppetebat, omnia describere. Interim facile est animadvertere numerorum discrimina in his inter se collatis, et cum edito Graeco codice; saepissime ab Accentus figuram vel omissam vel perperam positam evenisse; putam $\lambda\Gamma$ 33, et $\lambda\Gamma'$ $30\frac{1}{3}$, et similibus: Item ob characterum similitudinem $\overset{''}{=} \frac{1}{2}$. $\varsigma' \frac{1}{6}$. Item quod in Graeco edito promiscue fere semper ponitur (exceptis paucis ex posteribus constellationibus) Γ', tum pro manuscripto Γ' tum ω''' (quod forte fuerat $\Gamma\Gamma'$ vel ıГ vel Γ'' vel tale quid) hoc est, pro $\frac{1}{3}$ et $\frac{2}{3}$. Item literae Graecae α, δ, ob similitudinem in veteri scriptione saepe confunduntur. Idem observavi aliquoties in $\epsilon\theta\zeta\bar{o}$ (hoc est $\epsilon\theta\zeta\bar{o}$ in quibus si leviculus absit lineola $^\Gamma$ ductus, non ita facile determinabitur quid ponendum sit. Item in Arabicis MSS. praesertim illo Almagesti, numerorum characteres Arabicis literis alphabeticis expressis, ob magnam inter se similitudinem, haud facile erat nonnumquam determinare (aliter quam ex fide reliquerum Codicum) ex duobus similibus uter ponendus erat. Ordo stellarum, est secundum Almagestum Graecum[364] editum Basilae 1538 per Simonem Grinaeam: Ubi ordo in reliquis locus est, id ex adverso notatur.

---

1 Ptolemaicae, min. 12′.42″. *corr. Oldenburg*
3 Longitudinum numeris lapsionis *add.*
4 illo min. 12′.42″. *corr. Oldenburg*
7 Interim (*1*) omissi sunt (*2*) omisimus
12 quam jam suppetebat, *add.*
24 Arabicis ... expressis *add.*
25 nonumquam *add.*

---

[364] Almagestum Graecum: The editio princeps of the Greek text of the *Almagest*, produced by the German scholar Simon Grynaeus (Gryner) was published in Basel in 1538.

55. Wallis to Johannes Hevelius, 15/25 January 1668/9, enclosure

	Ursae minoris Ἄρκτου μικρᾶς Longit.	Lat.	Almag. Arab. et Al-suphi.	
1	ο̄ς′	ξςο̄		
2	β″	ο		
3	ις	οδΓ′	…… {	74 · 0′ 74 · 0
4	κω″	οεω″		
5	Γω″	οζω″		
6	ιζς′	οβ″Γ′		
7	κςς′	οδ″Γ′		
	Informis			
1	ιΓ	οας′		

	Ursae majoris	ἄρκτοῦ μεγάλης.		
1	κεΓ′	λο″Γ′		
2	κε″Γ′	μΓ′		
3	κςΓ′	μΓ	26 · 30′	
4	κςς′	μξς′		
5	κςω″	μξ		
6	κης	ν″		
7	ο̄″	μΓ″Γ′		
8	β″	μδΓ′		
9	θ	μβ		
10	ια	μδ	…… {	45 · 0′ 45 · 0
11	ιω″	λε		
12	ε″	κθΓ′		
13	ςΓ′	κηΓ′		
14	εω″	λς′	…… {	36 · 0 36 · 0
15	ε″Γ′	λΓ′		
16	ιζω″	μθ		
17	κβς′	μδ″	45 · 30	
18	Γς′	να		
19	Γ′	νς″	… {	3 · 0 46 · 30 3 · 0 46 · 30

55. WALLIS to JOHANNES HEVELIUS, 15/25 January 1668/9, enclosure

20	$\kappa\beta\omega''$	$\kappa\theta\Gamma'$	
21	$\kappa\delta\varsigma'$	$\kappa\eta\delta'$	
22	$\alpha\omega''$	$\lambda\epsilon\delta'$	
23	$o\overset{''}{-}\Gamma'$	$\kappa\epsilon\overset{''}{-}\Gamma'$	
24	$\iota\Gamma'$	$\kappa\epsilon$	
25	$\iota\beta\varsigma'$	$\nu\Gamma\overset{''}{-}$	
26	$\iota\eta$	$\nu\epsilon\omega''$	
27	$\kappa\theta\overset{''}{-}\Gamma'$	$\nu\delta$	

Informes

1	$\kappa\zeta\overset{''}{-}\Gamma'$	$\lambda\theta\overset{''}{-}\delta'$	
2	$\kappa\Gamma'$	$\mu\alpha\Gamma'$	
3	$\iota\epsilon$	$\iota\zeta\delta'$	
4	$\iota\Gamma\Gamma'$	$\iota\theta\varsigma'$	
5	$\iota\varsigma\varsigma'$	κ	
6	$\iota\beta\varsigma'$	$\kappa\beta\overset{''}{-}\omega''$	...... $\begin{cases} 22\cdot 45 \\ \underline{22\cdot 45} \end{cases}$
7	$\iota\alpha\varsigma'$	$\kappa\Gamma$	.... $\underline{22\cdot 20}$
8	$\bar{o}\,\bar{o}$	$\kappa\beta\delta'$	

Draconis. $\delta\varrho\acute{\alpha}\kappa o\nu\tau o\varsigma$.

1	$\kappa\varsigma\omega''$	$o\varsigma\overset{''}{-}$	
2	$\iota\alpha\overset{''}{-}\Gamma'$	$o\eta\overset{''}{-}$	
3	$\iota\Gamma\Gamma'$	$o\epsilon\omega''$	$13\cdot 10$ / $\underline{13\cdot 10}$
4	$\kappa\zeta\Gamma'$	$\pi\Gamma'$	
5	$\kappa\theta\omega''$	$\pi\epsilon\overset{''}{-}$	...... $75\cdot 30$ / $\underline{75\cdot 30}$
6	$\kappa\delta\omega''$	$\pi\beta\Gamma'$	
7	$\beta\Gamma'$	$o\eta\delta'$	
8	$\kappa\eta\overset{''}{-}\Gamma'$	$\pi\Gamma'$	
9	$\iota\theta\overset{''}{-}$	$\pi\alpha\Gamma'$	
10	η	$\pi\alpha\omega''$	
11	$\kappa\overset{''}{-}$	$\pi\Gamma$	.... $\underline{73\cdot 0}$
12	$\zeta\omega''$	$o\eta\overset{''}{-}\Gamma'$	
13	$\kappa\beta\overset{''}{-}\Gamma'$	$o\zeta\overset{''}{-}\Gamma'$	
14	$\iota\omega''$	$\pi\overset{''}{-}$	

55. Wallis to Johannes Hevelius, 15/25 January 1668/9, enclosure

15	καω″	παΓ′	22 · 20	{ 81 · 40 81 · 40
16	κϛϛ′	πδ′	......	{ 82 · 15 82 · 15
17	ιΓΓ′	πδ″		
18	κΓ′	πΓ″		
19	ια″Γ′	πδ″Γ′		
20	κηω″	πζ″		
21	καω″	πϛ″Γ′		
22	θ	παδ′		
23	θΓ′	πΓ′	....	{ 83 · 0 83 · 0
24	ηΓ′	πδ″Γ′		
25	ι	οη		
26	ιΓ′	οδω″		{ 13 · 0 13 · 0
27	ιβω″	ο		
28	ζΓ′	ξδω″	.... 74 · 40	
29	ιαϛ′	ξε″		
30	ιθϛ′	ξαδ′		
31	ιΓϛ′	νϛδ′		

Κηφέως. Cephei.

1	θ	οεω″	{ 5 · 0 5 · 0
2	Γ	ξδδ′	
3	ζΓ′	οαϛ′	8 · 0
4	ιϛω	ξθ	
5	θΓ′	οβ	
6	ι	οδ	
7	κη″	ξε″	
8	ζ″	ξβ″	
9	ιϛΓ′	ξδ′	
10	ιζΓ′	ξαδ′	
11	ιθ	ξαΓ′	

Informes

1	ιΓω″	ξδ
2	καΓ	νθ″

55. WALLIS to JOHANNES HEVELIUS, 15/25 January 1668/9, enclosure

	$Ba\acute{\omega}\tau o\upsilon$	Boötae		
1	$\beta\Gamma'$	$\nu\eta\omega''$		
2	$\delta\varsigma'$	$\nu\eta\Gamma'$		
3	$\theta\omega''$	$\xi\varsigma'$	5 · 40	
			5 · 40	
4	$\theta\omega''$	$\nu\delta\omega''$		
5	$\iota\theta\omega''$	$\mu\theta$		
6	$\kappa\varsigma\omega''$	$\nu\Gamma''\Gamma'$		
7	$\epsilon\omega''$	$\mu\eta\omega''$		
8	$\epsilon\omega''$	$\nu\Gamma\delta'$		
9	ϵ	$\nu\zeta''$		
10	$\zeta\omega$	$\mu\varsigma''$	········ {	46 · 10
				46 · 10
11	η''	$\mu\epsilon''$		
12	$\eta\varsigma'$	$\mu\alpha\omega''$	{ 8 · 30 41 · 20 }	
			{ 8 · 30 41 · 20 }	
13	$\varsigma\omega''$	$\mu\alpha\omega''$		
14	ζ	$\mu\beta''$		
15	$\zeta\omega''$	$\mu\Gamma$	········ {	40 · 20
				40 · 20
16	$\bar{o}\bar{o}$	$\mu\delta$	········ {	40 · 15
				40 · 15
17	$\kappa\epsilon\omega''$	$\mu\alpha\omega''$		
18	$\kappa\iota$	$\mu\beta\varsigma'$		
19	$\epsilon\Gamma'$	$\kappa\eta$		
20	$\kappa\alpha\Gamma'$	$\kappa\eta$	········ 23 · 0	
21	κ''	$\kappa\varsigma''$		
22	$\kappa\alpha\Gamma$	$\kappa\epsilon$	········ 25 · 30	
	Informis			
1	$\langle\kappa\zeta\rangle$	$\lambda\alpha''$	········ 23 · 50	
	$\Sigma\tau\epsilon\varphi\acute{a}\nu o\upsilon\ \beta o\varrho\epsilon\acute{\iota}o\upsilon.$ Coronae Boreae.			
1	$\iota\delta\omega''$	$\mu\delta''$		
2	$\iota\alpha\omega''$	$\mu\varsigma''$	········ {	46 · 10
				46 · 10
3	$\iota\alpha''\Gamma'$	$\mu\eta$		
4	$\iota\varsigma\omega''$	ν''	13 · 40	
			13 · 40	

55. WALLIS to JOHANNES HEVELIUS, 15/25 January 1668/9, enclosure

5	$\iota\zeta'$	$\mu\delta''\delta'$		
6	$\iota\theta\varsigma'$	$\mu\delta''\Gamma'$		
7	$\kappa\alpha\Gamma'$	$\mu\varsigma\varsigma'$		
8	$\kappa\alpha\omega''$	$\mu\theta\Gamma'$		

$\tau o\tilde{\upsilon}\ \dot{\epsilon}\nu\ \gamma\acute{o}\nu\alpha\sigma\iota\nu$

1	$\iota\zeta\omega''$	$\lambda\zeta''$		
2	$\Gamma'\omega''$	$\mu\Gamma$		
3	$\alpha\omega''$	$\mu\varsigma'$		
4	$\kappa\eta$	$\lambda\zeta\varsigma'$		
5	$\iota\varsigma\omega''$	$\mu\eta$	$\cdots\cdots\ 43\cdot 0$	
6	$\kappa\beta$	$\mu\theta''$		
7	$\kappa\zeta\omega''$	$\nu\beta$		
8	ϵ''	$\nu\beta''\Gamma'$		
9	$\alpha\omega''$	$\nu\delta$		
10	α''	$\nu\Gamma$		
11	$\varsigma''\Gamma$	$\nu\omega''$	$3\cdot 50$	$53\cdot 10$
			$\underline{3\cdot 50}$	$\underline{53\cdot 10}$
12	$\iota\varsigma$	$\nu\Gamma''$	$9\cdot 30$	
13	ι	$\nu\varsigma''$	$\cdots\cdots$	$56\cdot 10$
				$\underline{56\cdot 10}$
14	$\iota\alpha\varsigma'$	$\nu\eta''$		
15	$\iota\delta$	$\nu\theta''\Gamma'$		
16	$\iota\epsilon\Gamma'$	$\xi\Gamma$	$\cdots\cdots$	$60\cdot 20$
				$\underline{60\cdot 20}$
17	$\iota\varsigma\Gamma'$	$\xi\alpha\delta'$		
18	$''\Gamma'$	$\xi\alpha$		
19	$\kappa\beta\varsigma'$	$\xi\theta\Gamma'$	$25\cdot 10$	
20	$\iota\epsilon\Gamma'$	$o\delta'$		
21	$\iota\varsigma''\Gamma'$	$o\alpha\delta'$		
22	$\iota\theta\omega''$	$o\beta\delta$	$\cdots\cdots$	$72\cdot 0$
				$\underline{72\cdot 0}$
23	ω''	$\xi\delta$	$\cdots\cdots$	$60\cdot 15$
				$\underline{60\cdot 15}$
24	$\kappa\epsilon\Gamma'$	$\xi\Gamma$	$\cdots\cdots\ 53\cdot 0$	
25	$\iota\epsilon\omega''$	$\xi\epsilon''$		
26	$\iota\Gamma\omega''$	$\xi\Gamma\omega''$	$13\cdot 20$	
			$\underline{13\cdot 20}$	

55. Wallis to Johannes Hevelius, 15/25 January 1668/9, enclosure

27	$ις'$	$ξδδ'$		
28	$ιας'$	$ξ$		
	Informis			
1	$βω''$	$λης'$		

	$λύρας.$			
1	$ιζΓ'$	$ξβ$	$7 \cdot 20.$	
2	$κΓ'$	$ξβω''$		
3	$κΓ'$	$ξα$		
4	$κ'Γω''$	$ξ$		
5	$β$	$ξαΓ'$		
6	$αω''$	$ξΓ'$		
7	$κα$	$νες'$	$\cdots\cdots\cdots$	$\begin{cases} 56 \cdot 10 \\ \underline{56 \cdot 10} \end{cases}$
8	$κ\underset{}{''}Γ'$	$νε$		
9	$κδς'$	$νεΓ'$		
10	$κα$	$νδ\underset{}{''}δ'$		$\begin{cases} 24 \cdot 0 \\ 24 \cdot 0 \end{cases}$

	$ὄρνιθος$			
1	$δ\underset{}{''}$	$μθ$	$\cdots\cdots\cdots$	$\begin{cases} 49 \cdot 20 \\ \underline{49 \cdot 20} \end{cases}$
2	$θ$	$ν\underset{}{''}$	$\cdots\cdots\cdots\cdots$	$\underline{50 \cdot 20}$
3	$ιςΓ'$	$νδ\underset{}{''}$		
4	$κη\underset{}{''}$	$νζΓ'$		
5	$θς'$	$ξ$		
6	$ιθΓ'$	$ξδω''$		
7	$κβ\underset{}{''}$	$ξθω''$		
8	$κας'$	$οα\underset{}{''}$		$41 \cdot 30$
9	$ιβω''$	$οδ$	$\begin{cases} 16 \cdot 0 \\ \underline{17 \cdot 10} \end{cases}$	$\underline{44 \cdot 0}$
10	$\underset{}{''}Γ'$	$μθ\underset{}{''}$		
11	$Γ\underset{}{''}Γ'$	$νβς'$		
12	$ςω''$	$μδ$		
13	$ι$	$νες'$		
14	$ιδ\underset{}{''}$	$νζ$		
15	$ας'$	$ξδ$		
16	$βω''$	$ξδ\underset{}{''}$		
17	$ιβς'$	$ξΓ\underset{}{''}δ'$	$2 \cdot 10$	

55. WALLIS to JOHANNES HEVELIUS, 15/25 January 1668/9, enclosure

	Informes			
1	$\iota\omega''$	$\mu\theta\omega''$	$13\cdot 40$	$49\cdot 0$
2	$\iota\Gamma''\Gamma'$	$\nu\alpha\omega''$		

	$K\alpha\sigma\sigma\iota\epsilon\pi\epsilon\acute{\iota}\alpha\varsigma$			
1	$\zeta''\Gamma'$	$\mu\epsilon\Gamma'$		
2	$\iota''\Gamma'$	$\mu\varsigma''\delta'$		
3	$\iota\Gamma'$	$\mu\zeta''\Gamma'$	$\underline{13\cdot 0}$	
4	$\iota\varsigma\omega''$	$\mu\theta$		
5	$\kappa\omega''$	$\mu\epsilon''$		
6	$\kappa\zeta$	$\mu\zeta''\delta'$		
7	$\alpha\omega''$	$\mu\zeta\Gamma'$		
8	$\iota\delta\omega''$	$\mu\delta\Gamma'$		
9	$\iota\zeta\omega''$	$\mu\epsilon$		
10	$\beta\Gamma''$	ν		
11	$\iota\epsilon$	$\nu\beta\omega''$		
12	$\zeta''\Gamma'$	$\nu\alpha\omega''$	$7\cdot 30$	
13	$\Gamma\Gamma'$	$\nu\alpha\omega''$		$\underline{21\cdot 40}$

	$\Pi\epsilon\varrho\sigma\acute{\epsilon}\omega\varsigma.$			
1	$\kappa\Gamma'\omega''$	μ''		
2	$\alpha\varsigma'$	$\lambda\zeta''$	$\dots\dots\dots\dots$	$\underline{36\cdot 30}$
3	$\beta\omega''$	$\lambda\delta''$		
4	$\kappa\zeta''$	$\lambda\beta\Gamma'$		
5	ω''	$\lambda\delta''$		
6	α''	$\lambda\alpha\varsigma'$		
7	$\delta''\Gamma'$	λ		
8	$\epsilon\Gamma'$	$\kappa\zeta''\Gamma'$		
9	ζ	$\kappa\zeta\omega''$		
10	$\zeta\omega''$	$\kappa\zeta\Gamma'$		
11	$''$	$\kappa\zeta$	$\dots\dots\dots$	$\left\{\begin{array}{l}26\cdot 0\\ \underline{26\cdot 0}\end{array}\right.$
12	$\kappa\theta\omega$	$\kappa\Gamma$		
13	$\kappa\theta\varsigma'$	$\kappa\alpha$		
14	$\kappa\zeta$	$\omega''\kappa\alpha$		
15	$\kappa\varsigma''\Gamma'$	$\kappa\beta\delta'$	$\dots\dots\dots\dots$	$22\cdot 55.$
16	$\iota\delta''\Gamma'$	$\kappa\eta$		
17	$\iota\Gamma'$	$\kappa\eta\varsigma'$		
18	$\iota\beta\Gamma'$	$\kappa\epsilon$		

55. Wallis to Johannes Hevelius, 15/25 January 1668/9, enclosure

[6ʳ]

19	ιδ	κϛδ′		
20	ιδϛ′	κδ″	15·10	
21	ιϛ′Γ′	⟨ιη⟩″δ′		
22	ϛ″Γ′	κδ″Γ′	········	$\begin{cases} 21 \cdot 50 \\ \underline{21 \cdot 50} \end{cases}$
23	ηω″	ιθδ′		
24	ηΓ′	ιδ″δ′		
25	δϛ′	ιβ		
26	ϛΓ′	ια		

Informes

1	ια″Γ′	ιη		
2	ιε	λα		
3	κδω″	κω″	25·40	$\underline{31 \cdot 40}$

Ἡνιόχου

1	β″	λ	
2	βΓ′	λα″Γ′	············ 30·50
3	κε	κβ″	
4	β″Γ′	κ	
5	αϛ′	ιεδ′	
6	β″Γ′	ιΓΓ′	
7	κβ	κω″	
8	κβϛ′	ιη	
9	κβ	ιη	
10	ιθ″Γ′	ιϛ′	
11	κεω″	ε	
12	κϛ	ν″	········ $\begin{cases} 8 \cdot 30 \\ \underline{8 \cdot 30} \end{cases}$
13	κϛΓ′	ιβϛ′	············ $\underline{12 \cdot 20}$
14	κω	ιϛ′	············ 13·0

Ὀφιούχου.

1	κδ″Γ′	λϛ	············ 36·15
2	κη	κζδ′	
3	κθ	κϛ″	········ $\begin{cases} 26 \cdot 45 \\ \underline{26 \cdot 45} \end{cases}$
4	ιΓΓ′	λΓ	
5	ιδω″	λα″Γ′	

55. WALLIS to JOHANNES HEVELIUS, 15/25 January 1668/9, enclosure

6	$\iota\eta\Gamma'$	$\lambda\Gamma''\Gamma'$	$\begin{cases} 8\cdot 20 & 28\cdot 45 \\ \underline{8\cdot 20} & \underline{24\cdot 30} \end{cases}$
7	ϵ	$\iota\zeta$	$\cdots\cdots\cdots\cdots 17\cdot 20$
8	ς	$\iota\varsigma''$	
9	$\kappa\varsigma\omega''$	$\iota\epsilon$	
10	$\beta\Gamma'$	$\iota\Gamma\omega''$	
11	$\Gamma\Gamma'$	$\iota\delta\Gamma'$	$\begin{cases} 4\cdot 20 \\ \underline{4\cdot 10} \end{cases}$
12	$\kappa\alpha\varsigma'$	ζ''	
13	$\kappa\varsigma\omega''$	$\beta\delta'$	$\begin{cases} 23\cdot 20 \\ \underline{24\cdot 40} \end{cases}$
14	$\kappa\Gamma$	$\beta\delta'$	
15	$\kappa\delta\Gamma'$	α''	
16	$\kappa\epsilon$	$\Gamma\omega''$	$25\cdot 50\quad 0\cdot 15$ $\hspace{4em} \underline{0\cdot 20}$
17	$\kappa\epsilon''\Gamma'$	$\bar{o}\delta'$	$\underline{25\cdot 30}$
18	$\kappa\zeta\varsigma'$	α	
19	$\iota\beta\varsigma'$	$\iota\alpha''\Gamma'$	
20	$\iota\alpha\omega''$	$\epsilon\Gamma'$	$\underline{10\cdot 40}$
21	$\iota\omega''$	$\Gamma\varsigma'$	$9\cdot 40$
22	$\theta''\Gamma'$	$\alpha\Gamma'$	$\cdots\cdots\cdots \begin{cases} 1\cdot 40 \\ \underline{1\cdot 40} \end{cases}$
23	$\iota\beta\Gamma'$	ω''	
24	$\iota\omega''$	$''\delta'$	
	Informes		
1	β	$\kappa\eta\varsigma'$	
2	$\beta\omega''$	$\kappa\varsigma\Gamma'$	
3	Γ	$\kappa\epsilon$	$\cdots\cdots\cdots \begin{cases} 0\cdot 20 \\ \underline{0\cdot 20} \end{cases}$
4	$\omega''\Gamma'$	$\kappa\zeta$	$\cdots\cdots\cdots \begin{cases} 3\cdot 20 \\ \underline{3\cdot 20} \end{cases}$
5	$\delta\omega''$	$\lambda\Gamma$	
	Ὄφεω ὀφιούχου.		
1	$\iota\eta''\Gamma'$	$\lambda\eta$	
2	$\lambda\alpha$	$\omega''\mu$	$\begin{cases} 21\cdot 40 \\ \underline{21\cdot 40} \end{cases}$

55. Wallis to Johannes Hevelius, 15/25 January 1668/9, enclosure

3	$\kappa\alpha\Gamma'$	$\lambda\varsigma$	$\underline{24\cdot 20}$	
4	$\kappa\beta$	$\lambda\alpha\delta'$	$\cdots\cdots\cdots$	$\left\{\begin{array}{l}34\cdot 15\\ \underline{34\cdot 15}\end{array}\right.$
5	$\kappa\alpha\Gamma'$	$\lambda\zeta\delta'$		
6	$\kappa\Gamma\varsigma'$	$\kappa\beta''$	$\cdots\cdots\cdots$	$\left\{\begin{array}{l}42\cdot 30\\ \underline{42\cdot 30}\end{array}\right.$
7	$\kappa\alpha\omega''$	$\kappa\theta\delta'$		
8	$\kappa\delta''\Gamma'$	$\kappa\varsigma''$		
9	$\kappa\delta\Gamma'$	$\kappa\epsilon\Gamma'$		
10	$\kappa\varsigma\Gamma'$	$\kappa\delta$		
11	$\kappa\eta''\Gamma'$	$\iota\varsigma''$	$\underline{28\cdot 40}$	
12	$\eta\varsigma'$	$\iota\varsigma\delta'$		
13	$\kappa\Gamma\omega''$	ι''		
14	$\kappa\zeta$	η''		
15	$\kappa\zeta\omega''\Gamma'$	$\iota''\Gamma'$	$\cdots\cdots\cdots\cdots$	$\underline{10\cdot 30}$
16	$\Gamma\omega''$	κ		
17	$\eta\omega''$	$\kappa\alpha\varsigma'$	$\cdots\cdots\cdots$	$\left\{\begin{array}{l}21\cdot 20\\ \underline{21\cdot 15}\end{array}\right.$
18	$\iota\eta\Gamma'$	$\kappa\zeta$		

Ὀϊσου.

1	$\iota\varsigma$	$\lambda\epsilon\omega''$	$\underline{10\cdot 10}$	$\left.\begin{array}{l}35\cdot 20\\ 39\cdot 20\end{array}\right\}$
2	$\varsigma\omega''$	$\lambda\theta\varsigma'$		
3	$\epsilon''\Gamma'$	$\lambda\theta''\Gamma'$	$\cdots\cdots\cdots\cdots\;39\cdot 7$	
4	$\delta\omega''$	$\lambda\theta$		
5	$\Gamma\Gamma'$	$\lambda\zeta\omega''$	$\cdots\cdots\cdots$	$\left\{\begin{array}{l}18\cdot 40\\ 38\cdot 40\end{array}\right.$

Ἀετοῦ.

1	$\zeta\varsigma'$	$\kappa\varsigma''\Gamma'$	
2	$\delta''\Gamma'$	$\kappa\zeta\Gamma'$	
3	$\Gamma''\Gamma'$	$\kappa\theta\varsigma'$	$\cdots\cdots\cdots\cdots\;29\cdot 0$
4	$\delta\omega''$	λ	
5	$\varsigma\varsigma'$	$\lambda\alpha''$	
6	ς	$\lambda\alpha''$	$5\cdot 0$
7	$\kappa\theta\omega''$	$\kappa\eta\omega''$	

55. WALLIS to JOHANNES HEVELIUS, 15/25 January 1668/9, enclosure

8	$\alpha\varsigma'$	$\kappa\varsigma\Gamma'$	.... $\Big\{$	$26\cdots\cdots$
				$26\cdot 40$
9	$\kappa\beta\varsigma'$	$\lambda\varsigma\Gamma'$		
	Informes			
1	$\Gamma\omega''$	$\kappa\alpha\omega''$		
2	$\nu\underline{''}\omega''$	$\iota\theta\varsigma'$	$\Big\{$	$8\cdot 50$
				$\underline{8\cdot 40}$
3	$\kappa\varsigma$	$\kappa\epsilon$		
4	$\kappa\eta\varsigma'$	κ		
5	$\kappa\theta\omega''$	$\iota\epsilon\underline{''}$		
6	$\kappa\alpha\varsigma'$	$\iota\eta\varsigma$	$21\cdot 40.$	

$\Delta\epsilon\lambda\phi\tilde{\iota}\nu o\varsigma$

1	$\iota\zeta\omega''$	$\kappa\theta\varsigma'$		
2	$\iota\eta\omega''$	$\kappa\theta$		
3	$\iota\eta\omega''$	$\kappa\zeta\underline{''}\delta'$		
4	$\eta\Gamma'$	$\lambda\beta$	$\Big\{$	$18\cdot 30$
				$\underline{18\cdot 30}$
5	$\kappa\varsigma$	$\lambda\Gamma\underline{''}\Gamma'$	$20\cdot 10$	
6	$\kappa\alpha\Gamma'$	$\lambda\beta$		
7	$\kappa\Gamma\varsigma'$	$\lambda\Gamma\varsigma'$	$\Big\{$	$20\cdot 30$
				$\underline{23\cdot 20}$
8	$\iota\zeta\underline{''}$	$\lambda\delta'\underline{''}$	$\underline{17\cdot 20}$	$\underline{34\cdot 0}$
9	$\iota\zeta\Gamma'$	$\lambda\alpha\underline{''}\Gamma'$	$\Big\{$ $17\cdot 40$	
			$\underline{17\cdot 30}$	
10	$\iota\theta$	$\lambda\alpha\underline{''}$		

$\H{I}\pi\pi o v\ \pi\varrho o\tau o\mu\tilde{\eta}\varsigma.$

1	$\kappa\varsigma\underline{''}\Gamma'$	$\kappa\underline{''}$	$\Big\{$	$26\cdot 20$
				$\underline{26\cdot 20}$
2	$\kappa\eta$	$\kappa\omega''$		
3	$\kappa\varsigma\Gamma'$	$\kappa\epsilon\underline{''}$	....	$\underline{25\cdot 20}$
4	$\kappa\zeta\omega''$	$\kappa\epsilon$		

$\H{I}\pi\pi o v.$

1	$\iota\zeta\underline{''}\Gamma'$	$\kappa\varsigma$		
2	$\iota\beta\varsigma'$	$\iota\beta\underline{''}$	....	$12\cdot 10$
3	$\beta\varsigma'$	$\lambda\alpha$	$\underline{1\cdot 50}$	
4	$\kappa\varsigma\omega''$	$\iota\theta\omega''$		

55. Wallis to Johannes Hevelius, 15/25 January 1668/9, enclosure

5	δ''	$\kappa\epsilon''$	
6	ϵ	$\kappa\langle\delta\rangle$	
7	$\kappa\theta$	$\lambda\epsilon$	
8	$\kappa\eta''$	$\lambda\delta''$	
9	$\kappa\varsigma\varsigma'$	$\kappa\theta$	
10	$\kappa\zeta$	$\kappa\theta''$	
11	$\iota\eta''\Gamma'$	$\iota\eta$	$19\cdot 0$
12	κ''	$\iota\theta$	$20\cdot 50$
13	$\kappa\alpha\Gamma'$	$\iota\epsilon$	
14	κ''	$\iota\varsigma$	$21\cdot 10$
15	$\theta\varsigma'$	$\iota\varsigma''\Gamma'$	$9\cdot 20$
16	η	$\iota\varsigma$	
17	$\epsilon\Gamma'$	β''	⋯⋯⋯ $\begin{cases} 22\cdot 30 \\ \underline{22\cdot 30} \end{cases}$
18	$\kappa\Gamma\omega''$	$\mu\alpha\varsigma'$	$\begin{cases} 28\cdot 40 \\ \underline{21\cdot 40} \end{cases}$
19	$\iota\zeta\omega''$	$\lambda\delta\delta'$	
20	$\iota\beta\Gamma'$	$\lambda\varsigma''\Gamma'$	

	$\mathring{A}\nu\delta\rho o\mu\epsilon\delta\alpha\varsigma.$		In Al-suphi MS.
1	$\kappa\epsilon\Gamma'$	$\kappa\delta''$	deest folium hu-
2	$\kappa\varsigma\Gamma'$	$\kappa\zeta$	jus constellationis.
3	$\kappa\delta\varsigma$	$\kappa\Gamma$	
4	$\kappa\Gamma\omega''$	$\lambda\beta$	
5	$\kappa\delta\omega''$	$\lambda\varsigma''$	
6	$\kappa\epsilon$	$\lambda\beta\Gamma'$	
7	$\epsilon\omega''$	$\mu\alpha$	$19\cdot 40.$
8	$\kappa\omega''$	$\mu\beta$	
9	$\kappa\beta\varsigma'$	$\mu\delta$	
10	$\kappa\delta\varsigma'$	$\iota\zeta''$	
11	$\kappa\epsilon\omega''$	$\iota\epsilon''\Gamma'$	
12	$\Gamma''\Gamma'$	$\kappa\epsilon\Gamma'$	⋯⋯⋯ $26\cdot 20.$
13	$\alpha''\Gamma'$	λ	
14	β	$\lambda\beta''$	
15	$\iota\varsigma''\Gamma'$	$\kappa\eta$	
16	$\iota\zeta\varsigma'$	$\lambda\zeta\Gamma'$	
17	$\iota\epsilon\varsigma'$	$\lambda\epsilon\omega''$	⋯⋯⋯ $35\cdot 20$
18	$\iota\beta\Gamma'$	$\kappa\theta$	
19	$\iota\beta$	$\kappa\eta$	

55. Wallis to Johannes Hevelius, 15/25 January 1668/9, enclosure

20	$ι ς'$	$λ ε''$		
21	$ι β ω$	$λ δ''$		
22	$ι δ ς'$	$λ β''$	········	30 · 30
23	$ι α ω''$	$μ α$	········	44 · 40

Τριγώνου.

1	$ι α$	$ι ς''$
2	$ι ς$	$κ ω''$
3	$ι ς Γ'$	$ι θ ω''$
4	$ι ς'' Γ'$	$ι θ$

Κριοῦ.

1	$ς ω''$	$ζ Γ'$		
2	$ζ ω''$	$η Γ'$		
3	$ι α$	$ζ ω''$		
4	$ι α''$	$ς$	········	$\begin{cases} 6 \cdot 40 \\ \underline{6 \cdot 40} \end{cases}$
5	$ς''$	$ε''$		
6	$ι ζ ω''$	$ς$		
7	$κ α Γ'$	$δ'' Γ'$		
8	$κ'' Γ'$	$α ω''$		$\begin{cases} 23 \cdot 50 \\ \underline{23 \cdot 50} \end{cases}$
9	$κ ε Γ'$	$β''$		
10	$ι ς$	$α'' Γ'$		$\begin{cases} 27 \cdot 0 \\ \underline{27 \cdot 0} \end{cases}$
11	$ι θ ω$	$α''$	········	$\begin{cases} 1 \cdot 10 \\ \underline{1 \cdot 10} \end{cases}$
12	$ι η$	$α''$		
13	$ι ε$	$ε δ'$		

Informes

1	$ι ω''$	$ι$			
2	$κ δ ω''$	$ι''$	···	21 · 40 10 · 10	$\Big\}$
				$\underline{21 \cdot 40}$ $\underline{10 \cdot 10}$	
3	$κ α Γ'$	$ι β ω''$			
4	$ι θ ω''$	$ι α ς'$			
5	$ι θ ς'$	$ι ω''$			

55. Wallis to Johannes Hevelius, 15/25 January 1668/9, enclosure

|Ταύρου.

1	κςς'	ς	
2	κς	ζδ'	
3	κδ'Γ'	η″	$\begin{cases} 24\cdot 40 \\ \underline{24\cdot 40} \end{cases}$
4	καΓ'	θδ'	
5	κθω″	θ″	$\begin{cases} 24\cdot 20 \\ \underline{24\cdot 20} \end{cases}$
6	Γω″	η	
7	ςω″	ιβω″	
8	Γ	ιδ″Γ'	
9	ιβς'	ι	
10	ιΓ	ιΓ	············ 10·20
11	θ	ε″Γ'	
12	ιΓ'	δδ'	
13	ι″Γ'	ε″Γ'	
14	ιβω″	ες'	
15	ιβ″Γ'	Γ	$\underline{11\cdot 50}$
16	ιζ″	δ	$\begin{cases} 17\cdot 10 \\ \underline{16\cdot 40} \end{cases}$
17	κΓ'	ε	23·0
18	κ	Γ″	
19	κζω″	β″	
20	ιεω″	δ	$\underline{10\cdot 40}$
21	κεω″	ε	deest in Arabico et Al-suphi, ut communis cum Heniocko.
22	ιβ	″⟨δ⟩	4·0
23	ιαω″	δ	············ 0·15.
24	ζ	ω	
25	θ	α	
26	η	ε	
27	η″	ζΓ'	········ $\begin{cases} 7\cdot 10 \\ \underline{4\cdot 10} \end{cases}$
28	ιβ	Γ	
29	ιαω″	ε	
30	βς'	δ″	

55. WALLIS to JOHANNES HEVELIUS, 15/25 January 1668/9, enclosure

31	$\beta\Gamma'$	$\Gamma\omega''$	$\begin{cases} 12\cdot 30 \\ \underline{12\cdot 30 \quad 3\cdot 20} \end{cases}$
32	$\Gamma\omega''$	$\Gamma\Gamma'$	$5\cdot 0$
33	$\Gamma\omega''$	ϵ	$\underline{3\cdot 20}$

Informes

1	$\kappa\epsilon$	$\iota\zeta''$	$\underline{20\cdot 0}$
2	κ	β	
3	$\kappa\alpha$	$\alpha''\delta'$	$\begin{cases} 24\cdot 0 \\ \underline{24\cdot 0} \end{cases}$
4	$\kappa\varsigma$	β	
5	$\kappa\theta$	$\varsigma\Gamma'$	
6	$\kappa\theta$	$\zeta\omega''$	
7	$\kappa\zeta$	ω''	$\cdots\cdots\cdots\underline{2\cdot 40}$
8	$\kappa\theta$	α	
9	α	$\alpha\Gamma'$	
10	$\beta\Gamma'$	$\Gamma\Gamma'$	
11	$\Gamma\Gamma'$	$\alpha\delta'$	

$\Delta\iota\delta\acute{\upsilon}\mu\omega\nu.$

1	$\kappa\Gamma\Gamma'$	θ''	$\cdots\cdots\begin{cases} 9\cdot 40 \\ \underline{9\cdot 40} \end{cases}$
2	$\kappa\varsigma\omega''$	$\varsigma\delta'$	
3	$\iota\varsigma\omega''$	ι	
4	$\iota\eta\omega''$	$\zeta\Gamma'$	
5	$\kappa\beta$	ϵ''	
6	$\kappa\delta$	$\alpha''\Gamma'$	$\cdots\cdots\begin{cases} 4\cdot 50 \\ \underline{4\cdot 50} \end{cases}$
7	$\kappa\varsigma\omega''$	$\beta\omega''$	
8	$\kappa\alpha\omega''$	$\beta\omega''$	
9	$\kappa\varsigma\varsigma'$	Γ'	$\begin{cases} 23\cdot 10 \quad 3\cdot 0 \\ \underline{23\cdot 10 \quad 3\cdot 0} \end{cases}$
10	$\iota\Gamma$	α''	
11	$\iota\eta\delta'$	β''	⟩ transponuntur in Arabico.
12	$\kappa\alpha\omega''$	$''$	
13	$\kappa\alpha\omega''$	$''$	$\cdots\cdots\begin{cases} 6\cdot 0 \\ \underline{6\cdot 0} \end{cases}$
14	ς''	α''	
15	η''	$\alpha\delta'$	

55. WALLIS to JOHANNES HEVELIUS, 15/25 January 1668/9, enclosure

16	$\iota\varsigma$	Γ''	$\begin{cases} 10\cdot 10 \\ \underline{10\cdot 10} \end{cases}$
17	$\iota\beta$	ζ''	
18	$\iota\delta\omega''$	$\langle\iota\rangle''$	

Informes

1	$\delta\varsigma'$	ω''
2	ς''	$\epsilon''\Gamma'$
3	$\iota\epsilon\varsigma'$	$\beta\delta'$
4	$\kappa\eta\Gamma'$	$\alpha\Gamma'$
5	$\kappa\varsigma\Gamma'$	$\Gamma\Gamma'$
6	$\kappa\varsigma$	δ''
7	ω''	$\beta\omega''$

Καρκίνου.

1	$\iota\Gamma$	Γ'	$\begin{cases} 10\cdot 20 \quad 0\cdot 40 \\ \underline{10\cdot 20} \quad \underline{0\cdot 40} \end{cases}$
2	$\zeta\omega''$	$\alpha\delta'$	
3	η	$\alpha\varsigma'$	
4	$\iota\Gamma$	$\beta\omega''$	$\begin{cases} 10\cdot 20 \\ \underline{10\cdot 20} \end{cases}$
5	$\iota\alpha\Gamma'$	$\bar{o}\varsigma'$	
6	$\iota\varsigma''$	ϵ''	
7	$\eta\Gamma'$	$\iota\alpha''\Gamma'$	
8	$\beta\omega''$	α	
9	$\zeta\varsigma'$	ζ''	$\begin{cases} 7\cdot 30 \\ \underline{7\cdot 30} \end{cases}$

Informes

1	$\iota\theta\varsigma''$	$\beta\Gamma'$	$\begin{cases} 15\cdot 10 \\ \underline{19\cdot 40} \end{cases}$
2	$\kappa\alpha\varsigma'$	$\epsilon\omega''$	$\begin{cases} 21\cdot 40 \\ \underline{21\cdot 40} \end{cases}$
3	$\iota\delta$	$\delta''\Gamma'$	
4	$\iota\varsigma$	$\zeta\delta'$	.... $5\cdot 15$

55. Wallis to Johannes Hevelius, 15/25 January 1668/9, enclosure

    Λέοντος.

1	$ιηΓ'$	$ι$		
2	$κας'$	$ζς'$	$\left\{\begin{array}{ll} 24 \cdot 10 & 7 \cdot 30 \\ & \underline{7 \cdot 30} \end{array}\right.$	
3	$κδΓ'$	$ιβ$		
4	$κδς'$	$θ''$		
5	$\bar{ο}ς'$	$ια$		
6	$βς'$	$ια$		
7	$ω$	$δ''$	$0 \cdot 40$	$4 \cdot 30$
8	$β''$	$\bar{ο}ς'$		
9	$Γ''$	$α''Γ'$		
10	$\bar{ο}\bar{ο}$	$δ$	$\cdots\cdots\cdots \left\{\begin{array}{l} 0 \cdot 15 \\ \underline{0 \cdot 15} \end{array}\right.$	
11	$κζΓ'$	$\bar{ο}$		
12	$κδς'$	$Γω''$		
13	$κζΓ'$	$δς'$	$\cdots\cdots\cdots\cdots 2 \cdot 10$	
14	$β''$	$δδ'$	$\underline{2 \cdot 15}$	
15	$θς'$	$Γ$	$\cdots\cdots\cdots \left\{\begin{array}{l} 0 \cdot 10 \\ \underline{0 \cdot 10} \end{array}\right.$	
16	$ζ$	$δ$		
17	$ιΓ'$	$εΓ'$	$\underline{13 \cdot 0}$	
18	$ιβΓ'$	$βΓ'$	$\underline{12 \cdot 20}$	
19	$ιαΓ'$	$ιβδ'$		
20	$ιδς'$	$ιΓω''$		
21	$ιδΓ'$	$ιας'$	$\cdots\cdots\cdots \left\{\begin{array}{l} 11 \cdot 20 \\ \underline{11 \cdot 20} \end{array}\right.$	
22	$ιςΓ'$	$θω''$	$\cdots\cdots\cdots\cdots 11 \cdot 20$	
23	$κΓ'$	$ε''Γ'$	$\cdots\cdots\cdots\cdots 9 \cdot 40$	
24	$καω''$	$αδ'$		
25	$κδ$	$ω''{''}Γ'$	$\underline{21 \cdot 40}$	$\underline{5 \cdot 50}$
26	$κζ''$	$Γε'$	$\cdots\cdots\cdots \left\{\begin{array}{l} 0 \cdot 20 \\ \underline{3 \cdot 0} \end{array}\right.$	
27	$κδ''$	$ια''Γ'$		

55. Wallis to Johannes Hevelius, 15/25 January 1668/9, enclosure

	Informes		
1	ς	$\iota\Gamma\Gamma'$	$0\cdot 10$
2	$\eta\varsigma'$	$\iota\epsilon''$	
3	$\iota\zeta''$	$\alpha\varsigma'$	
4	$\iota\varsigma\varsigma'$	$''$	
5	$\iota\eta$	$\beta\omega''$	
6	$\kappa\delta''\Gamma'$	λ	
7	$\kappa\delta\Gamma'$	$\kappa\epsilon$	$\underline{21\cdot 20}$
8	$\kappa\eta''$	$\kappa\epsilon''$	

	$\Pi\alpha\varrho\vartheta\acute{\epsilon}\nu o\upsilon.$		
1	$\kappa\epsilon\Gamma'$	$\delta\delta'$	$\begin{cases} 26\cdot 20 \\ \underline{26\cdot 20} \end{cases}$
2	$\kappa\zeta$	$\epsilon\omega''$	
3	ω''	η	
4	$\bar{o}\varsigma'$	ϵ''	
5	$\kappa\theta$	ς	$\cdots\cdots\cdots\begin{cases} 0\cdot 10 \\ \underline{0\cdot 10} \end{cases}$
6	$\eta\delta'$	$\alpha\varsigma'$	
7	$\iota\varsigma\varsigma'$	$\beta''\Gamma'$	
8	$\iota\zeta\varsigma'$	$\beta''\Gamma'$	$\cdots\cdots\cdots\cdots 2\cdot 10$
9	$\kappa\alpha$	$\alpha\omega''$	
10	$\iota\delta\Gamma'$	η''	
11	$\eta\varsigma'$	$\iota\Gamma''\Gamma'$	
12	$\iota\varsigma$	$\iota\alpha\omega''$	$\underline{10\cdot 10}$
13	$\iota\beta\varsigma'$	$\kappa\varsigma'$	$\cdots\cdots\cdots\begin{cases} 15\cdot 10 \\ 15\cdot 10 \end{cases}$
14	$\kappa\varsigma\omega''$	β	
15	$\kappa\delta''\Gamma'$	$\eta\omega''$	
16	$\kappa\varsigma\Gamma'$	$\Gamma\Gamma'$	
17	$\kappa\zeta\delta'$	ς	$\begin{cases} 27\cdot 0 & 0\cdot 10 \\ \underline{27\cdot 0} & \underline{0\cdot 10} \end{cases}$
18	$\bar{o}\bar{o}$	α''	

55. Wallis to Johannes Hevelius, 15/25 January 1668/9, enclosure

19	$\kappa\eta$	Γ	.... $\underline{0 \cdot 20}$
20	ω''	α''	
21	$\kappa\eta$	η''	
22	$\varsigma\Gamma'$	ζ''	$\begin{cases} 6 \cdot 40 \\ \underline{6 \cdot 40} \end{cases}$
23	$\zeta\Gamma'$	$\beta\omega''$	
24	$\eta\Gamma'$	$\iota\alpha\omega''$	
25	ι	$''$	
26	$\iota\beta\omega''$	$\bar{o}''\Gamma'$	.... $\begin{cases} 9 \cdot 50 \\ \underline{9 \cdot 50} \end{cases}$

Informes

1	$\iota\delta\omega$	Γ''	
2	$\iota\theta$	Γ''	
3	$\kappa\beta\delta'$	$\Gamma\Gamma'$	
4	$\kappa\zeta\Gamma'$	$\zeta\Gamma'$	.... $\underline{7 \cdot 20}$
5	$\kappa\eta\varsigma'$	$\eta\Gamma'$	
6	ϵ	$\zeta''\Gamma'$	

$X\eta\lambda\tilde{\omega}\nu.$

1	$\iota\eta$	ω''	
2	$\iota\zeta$	β''	
3	$\kappa\beta\varsigma'$	$\eta''\Gamma'$	
4	$\iota\zeta\omega''$	η''	
5	$\kappa\delta$	$\alpha\omega''$	
6	$\kappa\alpha\Gamma'$	$\alpha\delta'$	$24 \cdot 20$
7	$\kappa\zeta''\Gamma'$	$\delta''\delta'$	
8	Γ	Γ''	.... $3 \cdot 15$

Informes

1	$\kappa\varsigma\varsigma'$	θ	
2	$\Gamma\omega''$	$\varsigma\omega''$	
3	$\delta\Gamma'$	$\theta\delta'$	
4	Γ''	$''$	
5	$\bar{o}\Gamma'$	$\bar{o}\Gamma'$	.... $\underline{3 \cdot 0}$
6	$\alpha\varsigma'$	α''	
7	$\kappa\Gamma$	ζ''	
8	$\alpha\varsigma'$	η''	.... $\begin{cases} 8 \cdot 10 \\ \underline{8 \cdot 10} \end{cases}$
9	β	$\alpha\omega''$	.... $\begin{cases} 9 \cdot 40 \\ \underline{9 \cdot 40} \end{cases}$

55. WALLIS to JOHANNES HEVELIUS, 15/25 January 1668/9, enclosure

$\Sigma\kappa o\rho\pi\acute{\iota}ov.$

1	$\varsigma\Gamma'$	$\alpha\Gamma'$	....	$\left\{\begin{array}{l} 1\cdot 0 \\ \underline{0\cdot 20} \end{array}\right.$
2	$\epsilon\omega''$	$\alpha\omega''$		
3	$\epsilon\omega''$	ϵ	....	$5\cdot 20$
4	ς	$\zeta\overset{''}{\underline{}}\Gamma'$		
5	ζ	$\alpha\omega''$	....	$1\cdot 30$
6	$\varsigma\Gamma'$	$\overset{''}{\underline{}}$		
7	$\iota\omega''$	$\Gamma\overset{''}{\underline{}}\delta'$		
8	$\iota\beta\omega''$	δ		
9	$\iota\delta\overset{''}{\underline{}}$	$\epsilon\overset{''}{\underline{}}$		
10	$\theta\Gamma'$	$\varsigma\overset{''}{\underline{}}$	....	$\left\{\begin{array}{l} 6\cdot 30 \\ \underline{6\cdot 30} \end{array}\right.$
11	$\iota\omega''$	$\varsigma\omega''$	....	$3\cdot 40$
12	$\iota\eta\overset{''}{\underline{}}$	$\iota\alpha$		
|13	$\iota\eta\overset{''}{\underline{}}\Gamma'$	$\iota\epsilon$		
14	κ	$\iota\eta\omega''$		
15	$\kappa\varsigma'$	$\iota\eta$	....	$19\cdot 30$
16	$\kappa\Gamma\varsigma'$	$\iota\theta\overset{''}{\underline{}}$		
17	$\kappa\eta\varsigma'$	$\iota\eta\overset{''}{\underline{}}\Gamma'$		
18	$\bar{o}\overset{''}{\underline{}}$	$\iota\varsigma\omega''$		
19	$\kappa\rho$	$\iota\epsilon\varsigma'$	....	$\left\{\begin{array}{l} 15\cdot 20 \\ \underline{15\cdot 20} \end{array}\right.$
20	$\kappa\zeta\overset{''}{\underline{}}$	$\iota\Gamma\Gamma'$		
21	$\kappa\zeta$	$\iota\Gamma\overset{''}{\underline{}}$		

Informes

1	$\alpha\overset{''}{\underline{}}$	$\iota\Gamma\delta'$	$\left\{\begin{array}{l} 1\cdot 10 \\ \underline{1\cdot 10} \end{array}\right.$	
2	$\kappa\epsilon\overset{''}{\underline{}}$	$\varsigma\varsigma'$	....	$14\cdot 10$
3	$\kappa\epsilon\overset{''}{\underline{}}$	$\alpha\varsigma'$	$\underline{29\cdot 30}$	$\left.\begin{array}{l} 4\cdot 10 \\ \underline{4\cdot 10} \end{array}\right\}$

$T o\xi\acute{o}\tau ov.$

1	$\theta\overset{''}{\underline{}}$	$\varsigma\Gamma'$	$\left\{\begin{array}{l} 4\cdot 30 \\ \underline{4\cdot 30} \end{array}\right.$	
2	$\zeta\omega''$	$\varsigma\overset{''}{\underline{}}$		
3	η	$\kappa\Gamma'$	....	$\left\{\begin{array}{l} 10\cdot 50 \\ \underline{10\cdot 50} \end{array}\right.$

55. WALLIS to JOHANNES HEVELIUS, 15/25 January 1668/9, enclosure

4	θ	$\iota\zeta''\underline{''}\Gamma'$	....	$\begin{cases} 1\cdot 30 \\ \underline{1\cdot 30} \end{cases}$
5	$\varsigma\omega''$	$\beta\zeta''\Gamma'$		
6	$\iota\xi\Gamma'$	$\Gamma\varsigma'$		
7	$\iota\Gamma$	$\Gamma\underline{''}$		$\begin{cases} 3\cdot 45 \\ 3\cdot 50 \end{cases}$
8	$\iota\epsilon\Gamma'$	$\underline{''}\delta'$		
9	$\iota\epsilon\omega''$	$\beta\varsigma'$		
10	$\iota\zeta\omega''$	$\alpha\underline{''}$		
11	$\iota\theta\varsigma'$	β		
12	$\kappa\alpha\Gamma'$	$\beta\underline{''}\Gamma'$		
13	$\kappa\beta\Gamma'$	$\delta\underline{''}$		
14	$\kappa\beta\varsigma''\underline{''}\Gamma'$	$\varsigma\underline{''}$		
15	$\kappa\epsilon\Gamma'$	$\epsilon\underline{''}$		$\begin{cases} 25\cdot 40 \\ \underline{25\cdot 40} \end{cases}$
16	$\kappa\theta\underline{''}$	$\epsilon\underline{''}\Gamma'$		
17	$\kappa\zeta\omega''$	β		
18	$\kappa\beta\omega''$	$\alpha\underline{''}\Gamma'$		$\begin{cases} 22\cdot 20 \\ \underline{22\cdot 20} \end{cases}$
19	$\kappa\delta\underline{''}\Gamma'$	$\beta\varsigma''\underline{''}\Gamma'$		
20	κ	$\beta\underline{''}$		
21	$\iota\zeta\omega''$	$\delta\underline{''}$		
22	$\iota\varsigma\Gamma'$	$\varsigma\underline{''}\delta'$		
23	$\iota\zeta\omega''$	$\kappa\Gamma$		
24	$\iota\varsigma$	$\iota\eta$		
25	$\varsigma\omega''$	$\iota\Gamma$		
26	$\kappa\zeta\Gamma'$	$\iota\Gamma\underline{''}$		
27	$\kappa\Gamma\underline{''}$	$\kappa\varsigma$		$\begin{cases} 26\cdot 50 \\ \underline{26\cdot 50} \quad 22\cdot 10 \end{cases}$
28	$\kappa\zeta\Gamma'$	$\delta\underline{''}\Gamma'$		$\begin{cases} 27\cdot 40 \\ \underline{24\cdot 40} \end{cases}$
29	$\kappa\eta\underline{''}\Gamma'$	$\delta\underline{''}\Gamma'$		
30	$\kappa\eta\underline{''}\Gamma'$	$\epsilon\underline{''}\Gamma'$		
31	$\kappa\theta\omega''$	$\varsigma\underline{''}$		

Αἰγόκερω.

1	$\zeta\Gamma'$	$\zeta\Gamma'$	
2	$\zeta\omega''$	$\varsigma\omega''$	$4\cdot 40$
3	$\zeta\Gamma'$	ϵ	

55. Wallis to Johannes Hevelius, 15/25 January 1668/9, enclosure

		4	θ	η	$\begin{cases} 5 \cdot 40 \\ \underline{5 \cdot 40} \end{cases}$
		5	θ	$\overset{\prime\prime}{-}\delta'$	
		6	$\eta\omega''$	$\alpha\overset{\prime\prime}{-}\delta'$	$9 \cdot 0$
		7	$\eta\overset{\prime\prime}{-}\Gamma'$	$\alpha\overset{\prime\prime}{-}$	
		8	$\varsigma\varsigma'$	ω''	
		9	$\iota\alpha\omega''$	$\Gamma\overset{\prime\prime}{-}\Gamma'$	
Arab.	Graec.	10	$\iota\alpha\overset{\prime\prime}{-}\Gamma'$	$\overset{\prime\prime}{-}\varsigma'$	.... $\begin{matrix} 0 \cdot 50. \\ \underline{0 \cdot 50} \end{matrix}$
	MS.				
13	13	⎧ 11	$\iota\varsigma\omega''$	$\zeta\omega''$	⎫ transponun-
12	11	⎨ 12	$\iota\alpha\omega''$	$\eta\omega''$	⎬ tur hae Stellae
11	12	⎩ 13	$\iota\overset{\prime\prime}{-}\Gamma'$	$\varsigma\overset{\prime\prime}{-}$	⎭ in MSS.
		14	$\kappa\varsigma$	$\varsigma\overset{\prime\prime}{-}\Gamma'$	$20 \cdot 10.$
		15	$\kappa\Gamma'$	ς	$23 \cdot 0.$
		16	$\iota\eta\omega''$	$\delta\delta'$	
		17	$\iota\varsigma\omega''$	δ	
		18	$\iota\varsigma\omega''$	$\beta\overset{\prime\prime}{-}\Gamma'$	
		19	$\iota\varsigma\omega''$	$\bar{o}\bar{o}$	
		20	κ	$\overset{\prime\prime}{-}\Gamma'$	$\begin{cases} 21 \cdot 0 \\ \underline{21 \cdot 0} \end{cases}$
		21	$\kappa\overset{\prime\prime}{-}\Gamma'$	$\delta\overset{\prime\prime}{-}\Gamma'$	$\begin{cases} 23 \cdot 20 \\ \underline{23 \cdot 20} \end{cases}$
		22	$\kappa\epsilon$	$\delta\overset{\prime\prime}{-}$	$\underline{23 \cdot 20}$
		23	$\kappa\alpha\overset{\prime\prime}{-}\Gamma'$	$\beta\varsigma'$	$\begin{cases} 24 \cdot 50 \\ \underline{24 \cdot 50} \end{cases}$
		24	$\kappa\varsigma\Gamma'$	$\beta\bar{o}$	
		25	$\kappa\varsigma\overset{\prime\prime}{-}\Gamma'$	Γ'	
		26	$\kappa\omega''$	$\bar{o}\bar{o}$	$\begin{cases} 28 \cdot 40 \\ \underline{28 \cdot 40} \end{cases}$
		27	$\kappa\zeta\omega''$	$\beta\overset{\prime\prime}{-}\Gamma'$	
		28	$\kappa\eta\omega''$	$\delta\Gamma'$	

$\overset{\frown}{\Upsilon}\delta\rho o\chi\acute{o}ov.$

1	$\bar{o}\Gamma'$	$\iota\epsilon\overset{\prime\prime}{-}\delta'$	
2	$\epsilon\Gamma'$	$\iota\alpha$	$\begin{cases} 6 \cdot 20 \\ \underline{6 \cdot 20} \end{cases}$
3	$\epsilon\varsigma'$	$\theta\omega''$	
4	$\kappa\varsigma\overset{\prime\prime}{-}$	$\eta\overset{\prime\prime}{-}\Gamma'$	

55. Wallis to Johannes Hevelius, 15/25 January 1668/9, enclosure

5	$\kappa\zeta\Gamma'$	$\varsigma\delta'$	
6	$\iota\zeta\omega''$	ϵ''	
7	$\iota\varsigma\varsigma'$	η	
8	$\iota\delta\omega''$	$\eta\omega''$	
9	θ''	$\eta''\delta'$	
10	$\iota\alpha\omega''$	$\iota''\delta'$	
11	$\iota\beta$	θ	
12	$\iota\Gamma\Gamma'$	η''	
13	$\varsigma\varsigma'$	σ	
14	ζ	$\Gamma\varsigma'$	
15	$\eta\omega'$	$''\Gamma'$	
16	$\alpha\omega''$	$\alpha\omega''$	
17	$\Gamma\varsigma'$	δ'	.... 4 · 0
18	$\iota\alpha\omega''$	ζ''	
19	$\iota\alpha\Gamma'$	ϵ	
20	$\delta\omega''$	$\epsilon\omega''$	
21	$\eta\Gamma'$	ι	
22	$\zeta''\Gamma'$	θ	$\underline{6 \cdot 50}$
23	$\iota\epsilon$	β	
24	$\iota\delta''\Gamma'$	$\bar{o}\varsigma'$	
25	$\iota\zeta\omega''$	$\alpha\varsigma'$	
26	κ	$''$	
27	κ''	$\alpha\omega''$	
28	$\iota\theta$	Γ''	
29	$\iota\theta''\Gamma'$	$\delta\varsigma'$	
30	$\kappa''\Gamma'$	$\eta\delta'$	
31	$\kappa\beta\Gamma'$	$\iota\alpha$	$\left\{\begin{array}{l} 22 \cdot 40 \\ \underline{22 \cdot 40} \end{array}\right.$ 12 · 0
32	$\kappa\Gamma\varsigma'$	$\iota''\Gamma'$	26 · 10
33	$\kappa\alpha\omega''$	$\iota\delta$	
34	$\kappa\beta\varsigma'$	$\iota\delta''\delta'$	
35	$\kappa\Gamma\varsigma'$	$\iota\epsilon\omega''$	23 · 15
36	$\iota\varsigma$	$\iota\delta\varsigma'$	
37	$\iota\eta\Gamma'$	$\iota\epsilon''\delta'$	
38	$\iota\zeta''$	$\iota\epsilon$	.... 15 · 15
39	$\iota\alpha''\Gamma'$	$\iota\delta''\delta'$	.... $\left\{\begin{array}{l} 14 \cdot 50 \\ \underline{14 \cdot 50} \end{array}\right.$

55. Wallis to Johannes Hevelius, 15/25 January 1668/9, enclosure

40	$\iota\beta\Gamma'$	$\iota\epsilon\Gamma'$	$\begin{cases} 12\cdot 40 \\ \underline{12\cdot 40} \end{cases}$	$14\cdot 20$
41	$\iota\Gamma\varsigma'$	$\iota\delta$		
42	ζ	$\kappa\Gamma$		

Informes

1	$\kappa\varsigma\omega''$	$\iota\epsilon''$		
2	$\kappa\theta\omega''$	$\iota\delta\omega''$	$\cdots\cdots$	$\begin{cases} 14\cdot 20 \\ \underline{14\cdot 20} \end{cases}$
3	$\kappa\theta$	$\iota\eta\delta'$		

$\dot{I}\chi\vartheta\acute{v}\omega\nu.$

1	$\kappa\alpha\omega''$	$\theta\delta'$		
2	$\kappa\alpha\varsigma'$	ζ''	$\begin{cases} 24\cdot 10 \\ \underline{24\cdot 10} \end{cases}$	$9\cdot 15$
3	$\kappa\varsigma$	$\theta\Gamma'$		
4	$\kappa\eta\varsigma'$	θ''		
5	$\bar{o}\omega''$	ζ''		
6	$\kappa\varsigma$	δ''		
7	$\kappa\theta\omega''$	Γ''		
8	ς	$\varsigma\Gamma'$		
9	$\iota\alpha$	$\epsilon''\delta'$		
10	$\iota\Gamma$	$\Gamma''\delta'$		
11	$\iota\zeta\varsigma'$	$\beta\delta'$		
12	$\kappa\varsigma'$	$\alpha\varsigma'$	$\begin{cases} 20\cdot 30 \\ \underline{20\cdot 30} \end{cases}$	
13	$\kappa\Gamma$	ς		
14	$\kappa\beta''$	β	$\cdots\cdots$	$\begin{cases} 22\cdot 20 \\ \underline{22\cdot 20} \end{cases}$
15	$\kappa\Gamma\Gamma'$	ϵ	$23\cdot 0$	
16	$\kappa\varsigma''$	$\beta\Gamma'$		
17	$\kappa\eta\Gamma'$	$\delta\omega''$	$\begin{cases} 28\cdot 40 \\ \underline{28\cdot 40} \end{cases}$	
18	$\bar{o}\omega''$	$\zeta''\delta'$		
19	β''	η''		
20	$\bar{o}''$	$\alpha\omega''$	$\cdots\cdots$	$\begin{cases} 1\cdot 20 \\ \underline{1\cdot 20} \end{cases}$
21	$\bar{o}\varsigma'$	$\alpha''\Gamma'$		

55. WALLIS to JOHANNES HEVELIUS, 15/25 January 1668/9, enclosure

22	$\bar{o}\omega''$	$\epsilon\Gamma'$		$\begin{cases} 0\cdot 20 \\ \underline{0\cdot 20} \end{cases}$
23	$\bar{o}{''\!\!\!\!-}$	θ		
24	β	$\kappa\alpha{''\!\!\!\!-}\delta'$		
25	$\alpha\omega''$	$\kappa\alpha\omega''$		
26	$\kappa\eta\omega''$	κ		
27	$\kappa\zeta\omega''$	$\iota\theta{''\!\!\!\!-}\Gamma'$		
28	$\kappa\zeta$	$\kappa\Gamma$	$\cdots\cdots\cdots$	$\begin{cases} 20\cdot 20 \\ \underline{20\cdot 20} \end{cases}$
29	$\kappa\epsilon\omega''$	$\iota\delta\Gamma'$	$\cdots\cdots\cdots\cdots 13\cdot 0$	
30	$\kappa\varsigma\omega''$	$\iota\Gamma\delta'$	$\underline{26\cdot 20}$	$\begin{cases} 13\cdot 0 \\ \underline{13\cdot 0} \end{cases}$
31	$\kappa\zeta\omega''$	$\iota\beta$		
32	$\beta{''\!\!\!\!-}\varsigma'$	$\iota\varsigma$		$\begin{cases} 2\cdot 10 \\ \underline{2\cdot 10} \end{cases}$
33	$\kappa\theta{''\!\!\!\!-}\Gamma'$	$\iota\epsilon\Gamma'$		
34	$\bar{o}\bar{o}$	$\iota\alpha{''\!\!\!\!-}\delta'$		
	Informes			
1	$\alpha\varsigma'$	$\beta\omega''$		
2	$\beta\delta'$	$\beta{''\!\!\!\!-}$		
3	$\bar{o}\omega''$	$\epsilon{''\!\!\!\!-}$		
4	$\beta\Gamma'$	$\epsilon{''\!\!\!\!-}$		

	$K\acute{\eta}\tau o \upsilon\varsigma.$			
1	$\iota\zeta\omega''$	$\zeta{''\!\!\!\!-}\delta'$	$\cdots\cdots\cdots\cdots 7\cdot 15$	
2	$\iota\zeta\omega''$	$\iota\beta\Gamma'$		
3	$\iota\beta\omega''$	$\iota\alpha{''\!\!\!\!-}$		
4	$\iota{''\!\!\!\!-}$	$\iota\delta'$	$\cdots\cdots\cdots$	$\begin{cases} 14\cdot 0 \\ \underline{14\cdot 0} \end{cases}$
5	$\iota\varsigma$	$\eta\varsigma'$	$\begin{cases} 10\cdot 10 \\ \underline{9\cdot 10} \end{cases}$	
6	$\iota\beta\omega''$	$\varsigma\Gamma'$		
7	$\zeta\Gamma'$	$\delta\varsigma'$		
8	Γ	$\kappa\delta{''\!\!\!\!-}$		
9	$\Gamma\Gamma'$	$\kappa\eta$		
10	$\varsigma\omega''$	$\kappa\epsilon\varsigma'$		
11	ζ	$\kappa\zeta{''\!\!\!\!-}$	$7\cdot 40$	
12	$\kappa\beta$	$\kappa\epsilon\Gamma'$		

55. Wallis to Johannes Hevelius, 15/25 January 1668/9, enclosure

13	$κΓ$	$λ\overset{''}{-}Γ'$		
14	$κε$	$κ$		
15	$ιθω''$	$ιεω''$	$3·40·$	$\left.\begin{array}{l}15·20\\\underline{15·20}\end{array}\right\}$
16	$ιε$	$ιεω''$	$\underline{14·0·}$	
17	$ια$	$ιΓω''$		$\underline{33·40}$
18	$ιω$	$ιδω''$	$\underline{8·40}$	
19	$θΓ'$	$ιΓ$	$9·40$	
20	$θ$	$ιδ$		
21	$δω''$	$θω''$	$\left\{\begin{array}{l}9·20\\\underline{4·20}\end{array}\right.$	
22	$εω''$	$κ\,ΓΓ'$		$\left\{\begin{array}{l}20·20\\\underline{20·20}\end{array}\right.$

| Ὠρίωνος. |

1	$κζ$	$ιϛ\overset{''}{-}$	....	$\left\{\begin{array}{l}13·50\\\underline{13·50}\end{array}\right.$
2	$β$	$ιζ$		
3	$κδ$	$ιζ\overset{''}{-}$		
4	$κε$	$ιη$		
5	$δΓ'$	$ιδ\overset{''}{-}$		
6	$ϛΓ'$	$ια\overset{''}{-}Γ'$	$20·20.$	
7	$ϛ\overset{''}{-}$	$ι$		
8	$ϛ$	$θ\overset{''}{-}δ'$		
9	$ζΓ'$	$ηδ'$		
10	$ϛω''$	$ηδ'$		
11	$αω''$	$Γ\overset{''}{-}δ'$		
12	$δω''$	$δδ'$	$\left\{\begin{array}{l}7·20\\\underline{4·20}\end{array}\right.$	
13	$κζ\overset{''}{-}Γ'$	$ιβω''$	$\left\{\begin{array}{l}27·30\\\underline{27·30}\end{array}\right.$	$\begin{array}{l}19·40\\\underline{19·40}\end{array}$
14	$κϛΓ'$	$κ$		
15	$κεΓ'$	$κΓ'$		
16	$καϛ'$	$κω''$	$\left\{\begin{array}{l}24·10\\\underline{24·10}\end{array}\right.$	$20·0$
17	$κ\overset{''}{-}$	$η$	············	$8·40$
18	$ιθΓ'$	$ηϛ'$		
19	$ιη$	$ιδ'$	···········	$\underline{15·15}$

55. WALLIS to JOHANNES HEVELIUS, 15/25 January 1668/9, enclosure

20	$\iota\varsigma\,\Gamma'$	$\iota\beta\stackrel{\prime\prime}{-}\Gamma'$		
21	$\iota\epsilon\varsigma'$	$\iota\delta\delta'$		
22	$\iota\delta\stackrel{\prime\prime}{-}\Gamma'$	$\iota\epsilon\stackrel{\prime\prime}{-}\Gamma'$		
23	$\iota\delta\stackrel{\prime\prime}{-}\Gamma'$	$\iota\zeta\varsigma'$		
24	$\iota\epsilon\,\Gamma'$	$\kappa\,\Gamma'$		
25	$\iota\varsigma\,\Gamma'$	$\kappa\alpha\stackrel{\prime\prime}{-}$		
26	$\kappa\epsilon\,\Gamma'$	$\kappa\delta\varsigma'$		
27	$\kappa\zeta\,\Gamma'$	$\kappa\delta\stackrel{\prime\prime}{-}\Gamma'$		
28	$\kappa\eta\varsigma'$	$\kappa\epsilon\omega''$		
29	$\kappa\Gamma\stackrel{\prime\prime}{-}\Gamma'$	$\kappa\epsilon\stackrel{\prime\prime}{-}\Gamma'$		
30	$\kappa\varsigma\stackrel{\prime\prime}{-}\Gamma'$	$\kappa\eta\,\Gamma'$	$\begin{cases} 26\cdot 30 \\ \underline{26\cdot 30} \end{cases}$	$\begin{matrix} 28\cdot 40 \\ \underline{28\cdot 40} \end{matrix}$
31	$\kappa\varsigma\,\Gamma'$	$\kappa\theta\varsigma'$	$\begin{cases} 26\cdot 40 \\ \underline{26\cdot 40} \end{cases}$	
32	$\kappa\zeta$	$\kappa\theta\stackrel{\prime\prime}{-}\Gamma'$		
33	$\kappa\zeta\omega''$	$\lambda\omega''$		
34	$\kappa\varsigma\stackrel{\prime\prime}{-}$	$\lambda\stackrel{\prime\prime}{-}\Gamma'$	$\begin{cases} 26\cdot 10 \\ \underline{26\cdot 10} \end{cases}$	
35	$\kappa\stackrel{\prime\prime}{-}\Gamma'$	$\lambda\alpha\stackrel{\prime\prime}{-}$	$\begin{cases} 19\cdot 50 \\ \underline{19\cdot 50} \end{cases}$	$31\cdot 10$
36	$\kappa\alpha$	$\lambda\delta'$		
37	$\kappa\Gamma\,\Gamma'$	$\lambda\alpha\varsigma'$		
38	$\bar{o}\varsigma'$	$\lambda\Gamma\stackrel{\prime\prime}{-}$		

Ποταμοῦ

1	$\iota\eta\,\Gamma'$	$\lambda\alpha\stackrel{\prime\prime}{-}$	....	$\begin{cases} 31\cdot 50 \\ \underline{31\cdot 50} \end{cases}$
2	$\iota\eta\stackrel{\prime\prime}{-}\Gamma'$	$\kappa\eta\delta'$		
3	$\iota\eta$	$\kappa\theta\stackrel{\prime\prime}{-}\Gamma'$		
4	$\iota\alpha\omega''$	$\kappa\eta\delta'$	$\begin{cases} 14\cdot 40 \\ \underline{14\cdot 40} \end{cases}$	
5	$\iota\Gamma\varsigma'$	$\kappa\epsilon\stackrel{\prime\prime}{-}\Gamma'$	$13\cdot 40$	
6	$\iota\varsigma'$	$\kappa\epsilon\,\Gamma'$		
7	$\varsigma\,\Gamma'$	$\kappa\varsigma$		
8	$\epsilon\stackrel{\prime\prime}{-}$	$\kappa\zeta$		
9	$\beta\stackrel{\prime\prime}{-}\Gamma'$	$\kappa\zeta\stackrel{\prime\prime}{-}\Gamma'$		
10	$\kappa\zeta$	$\lambda\beta\stackrel{\prime\prime}{-}\Gamma'$		
11	$\kappa\delta\,\Gamma'$	$\lambda\alpha$	$\begin{cases} 24\cdot 40 \\ \underline{24\cdot 40} \end{cases}$	

55. WALLIS to JOHANNES HEVELIUS, 15/25 January 1668/9, enclosure

12	κδϛ′	κη∥Γ′		
13	κβ	κη		
14	ιζϛ′	κε∥		
15	ιδ∥ϛ′	κΓ∥Γ′		
16	ιβϛ′	κΓ∥	⋯⋯⋯⋯⋯ 23·50	
17	ι∥	κΓδ′		
18	εϛ′	λβϛ′		
19	ε∥Γ′	λδ∥Γ′		
20	η∥Γ′	λη∥		
21	ιΓ∥Γ′	ληϛ′		
22	ιζ∥	λθ		
23	καΓ	μαΓ′		
24	κα∥	μβ∥		
25	κβϛ′	μΓδ′		
26	κδω″	μΓΓ′	24·20.	
27	δϛ′	νΓΓ′	⋯⋯⋯⋯ { 50·20 / <u>50·20</u> }	
28	ε	να∥δ′		
29	κηϛ′	νΓ∥Γ′		
30	κε∥Γ′	νΓϛ′		
31	ιζ∥Γ′	νΓ		
32	ιδ∥Γ′	νΓ∥		
33	ια∥Γ′	νβ	⋯⋯⋯⋯⋯ 52·30	
34	ζ∥	νΓ∥	{ 0·10 / <u>0·10</u> }	

Λαγωοῦ.

1	ιθ	λε	{ 19·40 / <u>19·40</u> }	
2	ιθ∥Γ′	λϛ∥		
3	καΓ′	λεω″		
4	καΓ′	λϛω″		
5	ιθϛ′	λθδ′		
6	ιϛϛ′	μεδ′		
7	κε∥Γ′	μα∥		
8	κδ∥Γ′	μαΓ′	{ 24·20 44·20 / <u>24·20</u> <u>44·20</u> }	
9	α	μδ	⋯⋯⋯⋯ { 45·15 / 44·15 }	

55. WALLIS to JOHANNES HEVELIUS, 15/25 January 1668/9, enclosure

10	$\kappa\theta$	$\mu\epsilon''\Gamma'$	
11	$\bar{o}\bar{o}$	$\lambda\eta\Gamma'$	
12	$\beta\omega''$	$\lambda\eta\varsigma'$	
	Κυνός		
1	$\iota\zeta\omega''$	$\lambda\theta\varsigma'$	
2	$\iota\theta\omega''$	$\lambda\epsilon$	$9 \cdot 40$
3	$\kappa\alpha\Gamma'$	$\lambda\varsigma''$	
4	$\kappa\Gamma\Gamma'$	$\lambda\zeta''\delta'$	
5	$\kappa\epsilon\Gamma'$	μ	
6	κ''	$\mu\beta\omega''$	
7	$\iota\varsigma\varsigma'$	$\mu\alpha\delta'$	
8	$\iota\varsigma$	$\mu\beta''$	$16 \cdot 10.$
9	$\iota\alpha$	$\mu\alpha\Gamma'$	
10	$\iota\delta\omega''$	$\mu\varsigma''$	
11	$\iota\varsigma\varsigma'$	$\mu\epsilon''\Gamma'$	
12	$\kappa\delta\omega''$	$\mu\varsigma\varsigma'$	
13	$\kappa\alpha\omega''$	$\mu\zeta$	
14	$\kappa\varsigma\omega''$	$\mu\alpha''\delta'$	
15	$\kappa\Gamma\omega''$	$\nu\alpha''$	
16	$\kappa\Gamma$	$\nu\epsilon\varsigma'$	
17	$\theta\omega''$	$\nu\vartheta''\delta'$	.... $\begin{cases} 53\cdot 45 \\ \underline{53\cdot 45} \end{cases}$
18	$\beta\varsigma'$	$\nu\omega''$	
	Informes		
1	$\iota\theta''$	$\kappa\epsilon\delta'$	.... $25\cdot 10.$
2	ϵ	$\xi\alpha''$	
3	$\iota\alpha\Gamma'$	$\nu\eta''\delta'$	$1\cdot 30.$
4	$\iota\Gamma$	$\nu\zeta$	
5	$\iota\delta\varsigma$	$\nu\varsigma$	
6	$\kappa\eta$	$\nu\epsilon''$	
7	$\bar{o}\Gamma'$	$\nu\zeta\omega''$	
8	$\beta\Gamma'$	$\nu\theta''\Gamma'$	
9	$\kappa\theta$	$\nu\theta\omega''$	
10	$\kappa\varsigma$	$\nu\zeta\omega''$	
11	$\kappa\beta\varsigma'$	$''$	
	Προκυνός.		
1	$\kappa\epsilon$	$\iota\delta$	

55. Wallis to Johannes Hevelius, 15/25 January 1668/9, enclosure

2	$\kappa\theta''$	$\iota\varsigma\varsigma'$	$\begin{cases} 29\cdot 10 \\ \underline{29\cdot 10} \end{cases}$	

Ἀργους.

1	$\iota\Gamma$	$\mu\beta''$	$10\cdot 20$	
2	$\iota\delta\Gamma'$	$\mu\Gamma\Gamma'$	$11\cdot 20$	
3	$\eta''\Gamma'$	$\mu\epsilon$		
4	$\eta\omega''$	$\mu\varsigma$		
5	$\epsilon\Gamma'$	$\mu\epsilon''$	$\ldots\ldots\ldots\ldots 40\cdot 30$	
6	$\varsigma\Gamma'$	$\mu\zeta\delta'$		
7	$\epsilon\Gamma'$	$\mu\theta''\delta'$	$\ldots\ldots\begin{cases} 49\cdot 30 \\ \underline{49\cdot 30} \end{cases}$	
8	$\theta\Gamma'$	$\mu\theta''\Gamma'$	$\ldots\ldots\begin{cases} 49\cdot 30 \\ \underline{49\cdot 30} \end{cases}$	
9	η''	$\mu\theta\delta'$		
10	$\iota\delta$	$\mu\theta''\Gamma'$		
11	δ	$\nu\Gamma$		
12	δ	$\nu\eta\omega''$	$4\cdot 10$	
13	$\iota\varsigma\varsigma'$	$\nu\epsilon''$	$\begin{cases} 16\cdot 0 \\ 10\cdot 10 \end{cases}$	
14	$\iota\beta\varsigma'$	$\nu\eta\omega''$		
15	$\iota\Gamma\omega''$	$\nu\zeta\delta'$	$\underline{14\cdot 10}$	
16	$\iota\varsigma''$	$\nu\zeta''\delta'$	$\underline{17\cdot 0}$	
17	$\kappa\alpha\varsigma'$	$\nu\eta\omega''$	$\ldots\ldots\begin{cases} 58\cdot 20 \\ \underline{58\cdot 20} \end{cases}$	
18	$\iota\eta\varsigma'$	ξ		
19	$\kappa\alpha$	$\nu\theta\Gamma'$		
20	$\kappa\Gamma$	$\varsigma'\nu\varsigma\Gamma'$	$\begin{cases} 23\cdot 0 \\ \underline{23\cdot 0} \end{cases}$	$\begin{matrix} 56\cdot 40 \\ \underline{56\cdot 40} \end{matrix}$
21	$\kappa\delta\Gamma'$	$\nu\zeta\omega''$	$\ldots\ldots\begin{cases} 57\cdot 0 \\ \underline{57\cdot 0} \end{cases}$	
22	$\epsilon\omega''$	$\nu\alpha''$		
23	$\varsigma\varsigma'$	$\nu\epsilon\omega''$		
24	δ	$\nu\zeta\varsigma'$		
25	$\theta\varsigma'$	ξ		
26	θ	$\xi\alpha\delta'$		
27	$\bar{o}\varsigma'$	$\nu\alpha''\varsigma'$	$\ldots\ldots\begin{cases} 51\cdot 30 \\ \underline{51\cdot 30} \end{cases}$	

55. Wallis to Johannes Hevelius, 15/25 January 1668/9, enclosure

28	$\kappa\theta\Gamma'$	$\mu\theta$		
29	$\kappa\eta$	$\mu\Gamma\Gamma'$	....	$\begin{cases} 43\cdot 30 \\ \underline{43\cdot 30} \end{cases}$
30	$\kappa\theta$	$\mu\Gamma''$		
31	$\iota\delta\varsigma'$	$\nu\alpha''$	....	$\begin{cases} 54\cdot 30 \\ \underline{54\cdot 30} \end{cases}$
32	$\iota\zeta''$	$\nu\alpha\delta'$	$18\cdot 10.$	
33	$\iota\alpha\varsigma'$	$\xi\Gamma$		
34	$\iota\theta$	$\xi\delta''$		
35	$\bar{o}\bar{o}$	$\xi\Gamma''\Gamma'$		
36	η''	$\xi\theta\omega''$		
37	$\iota\epsilon\varsigma'$	$\xi\epsilon\omega''$		
38	$\kappa\alpha\Gamma'$	$\xi\epsilon''\Gamma'$		
39	$\kappa\varsigma$	$\xi\zeta\Gamma'$		
40	α	$\xi\beta''\Gamma'$		
41	η	$\xi\beta\delta'$		
42	δ	$\xi\epsilon''\Gamma'$		
43	$\kappa\varsigma'$	$\xi\epsilon\omega''$		
44	$\iota\zeta\varsigma'$	$o\epsilon$		
45	$\kappa\theta$	$o\alpha''\delta'$		

$\overset{\text{\'}}{\Upsilon}\delta\rho o\upsilon.$

1	$\iota\delta$	$\iota\epsilon$		
2	$\iota\Gamma\Gamma'$	$\iota\Gamma''\varsigma'$	....	$\begin{cases} 13\cdot 10 \\ \underline{13\cdot 10} \end{cases}$
3	$\iota\epsilon\Gamma'$	$\iota\alpha''$		
4	$\iota\epsilon''$	$\iota\delta\delta'$	....	$\begin{cases} 14\cdot 45 \\ \underline{14\cdot 45} \end{cases}$
5	$\iota\zeta''\Gamma'$	$\iota\beta\delta'$	$\underline{23\cdot 50}$	$\left.\begin{array}{l} 12\cdot 0 \\ \underline{12\cdot 0} \end{array}\right\}$
6	$\kappa\Gamma$	$\iota\alpha''\Gamma'$	$\begin{cases} 20\cdot 20 \\ \underline{20\cdot 20} \end{cases}$	$14\cdot 40$
7	$\kappa\Gamma\Gamma'$	$\iota\beta\omega''$	....	$\begin{cases} 19\cdot 20 \\ \underline{19\cdot 20} \end{cases}$
8	$\kappa\eta''\Gamma'$	$\iota\epsilon\Gamma'$		
9	$\bar{o}\omega''$	$\iota\delta''\Gamma'$		
10	$\kappa\eta''$	$\iota\zeta\varsigma'$		

55. Wallis to Johannes Hevelius, 15/25 January 1668/9, enclosure

Ordo in MS. Graeco	Ordo in Arab:				
		11	$\kappa\theta\varsigma'$	$\iota\theta''\delta'$	
		12	$\bar{o}\bar{o}$	κ''	
		13	ς	$\kappa\varsigma''$	
19	16	14	$\iota\eta$	$\kappa\epsilon''$	$\cdots\cdots\cdots \begin{cases} 24\cdot 40 \\ \underline{24\cdot 40} \end{cases}$
20	17	15	κ	$\kappa\varsigma$	
18	15	16	κ	$\varsigma'\kappa\varsigma\delta'$	$\cdots\cdots\cdots \begin{cases} 23\cdot \dfrac{15}{45} \\ \underline{23\cdot 15} \end{cases}$
14	14	17	κ	$\kappa\varsigma\delta'$	$\cdot\begin{cases} 8\cdot 40 \\ \underline{8\cdot 40} \end{cases} \cdot \begin{cases} 26\cdot 0 \\ \underline{26\cdot 0} \end{cases}$
15	et	18	$\kappa\Gamma$	$\kappa\beta\varsigma'$	
16	Al-	19	α''	$\kappa\epsilon''\delta'$	
17	suphi	20	$\beta\Gamma'$	$\lambda\varsigma$	$\cdots\cdots\cdots \begin{cases} 30\cdot 10 \\ \underline{30\cdot 10} \end{cases}$
		21	$\iota\beta\varsigma'$	$\lambda\alpha\Gamma'$	
		22	$\iota\alpha''$	$\lambda\Gamma\varsigma'$	$\begin{cases} 14\cdot 30 \\ 14\cdot 30 \end{cases}$
		23	$\iota\varsigma\varsigma'$	$\lambda\alpha\Gamma'$	
		24	$\bar{o}\bar{o}$	$\lambda\Gamma\omega''$	$\cdots\cdots\cdots \begin{cases} 13\cdot 40 \\ \underline{13\cdot 40} \end{cases}$
		25	$\iota\Gamma''$	$\iota\zeta\Gamma'$	$\begin{cases} 17\cdot 40 \\ \underline{17\cdot 40} \end{cases}$

Informes.

1	$\iota\beta''$	$\kappa\Gamma\delta'$
2	$\iota\alpha$	$\iota\Gamma$

|$K\rho\alpha\tau\tilde{\eta}\rho o\varsigma$.

1	$\kappa\varsigma\Gamma'$	$\kappa\Gamma$	
2	β''	$\iota\theta''$	
3	$\bar{o}\bar{o}$	$\iota\eta$	
4	ζ	$\iota\eta''$	
5	$\kappa\theta\Gamma'$	$\iota\Gamma\omega''$	
6	$\theta\varsigma'$	$\iota\varsigma\varsigma'$	
7	$\iota\alpha\Gamma'$	$\iota\alpha''\Gamma'$	$\begin{cases} 1\cdot 40 \\ \underline{1\cdot 40} \end{cases}$

$K\acute{o}\rho\alpha\kappa o\varsigma$.

1	$\iota\epsilon\Gamma'$	$\kappa\alpha\omega''$

55. Wallis to Johannes Hevelius, 15/25 January 1668/9, enclosure

2	$\iota\delta\Gamma'$	$\iota\theta\omega''$	
3	$\iota\varsigma\omega''$	$\iota\eta\varsigma'$	$\cdots\cdots\cdots\cdots 18\cdot 20.$
4	$\iota\Gamma''$	$\iota\delta''\Gamma'$	
5	$\iota\varsigma\omega''$	$\iota\beta''$	
6	$\iota\zeta$	$\iota\alpha''\delta'$	
7	κ''	$\iota\eta\varsigma'$	

$K\epsilon\nu\tau\alpha\upsilon\rho\sigma\upsilon.$

1	ι''	$\kappa\alpha\omega''$	
2	ι	$\iota\eta''\Gamma'$	$\cdots\cdots\cdots\cdots 8\cdot 50$
3	$\theta\varsigma'$	κ''	
4	ι	κ	
5	$\varsigma\varsigma'$	$\kappa\epsilon\omega''$	
6	$\iota\epsilon\omega''$	κ''	$\cdots\cdots\cdots\; 22\cdot 30$ / $\underline{22\cdot 30}$
7	$\theta\varsigma'$	$\kappa\zeta''$	
8	$\iota\eta\varsigma'$	$\kappa\beta\Gamma'$	
9	$\iota\theta\varsigma'$	$\kappa\Gamma''\delta'$	
10	$\kappa\beta$	$\iota\eta\delta'$	
11	$\kappa\beta''$	$\kappa''\Gamma'$	
12	$\iota\Gamma\Gamma'$	$\kappa\eta\Gamma'$	
13	$\iota\delta$	$\kappa\theta\Gamma'$	
14	$\iota\epsilon\varsigma'$	$\kappa\eta$	
15	$\iota\varsigma\Gamma'$	$\kappa\varsigma''$	
16	$\kappa\beta''\Gamma'$	$\kappa\epsilon\delta'$	$\underline{17\cdot 50}$
17	$\kappa\zeta''$	$\kappa\delta$	
18	$\iota\eta$	$\lambda\Gamma''$	
19	$\iota\zeta\omega''$	$\lambda\alpha$	
20	$\iota\varsigma''\Gamma'$	$\lambda\Gamma$	$\cdots\cdots\cdots \begin{cases} 30\cdot 20 \\ \underline{30\cdot 20} \end{cases}$
21	$\iota\beta\varsigma'$	$\lambda\alpha''\Gamma'$	$\underline{9\cdot 10.} \begin{cases} 34\cdot 50 \\ \underline{34\cdot 50} \end{cases}$
22	θ	$\lambda\zeta\omega''$	
23	$\epsilon''\Gamma'$	μ	
24	ϵ	$\mu\Gamma$	$\cdots\cdots\cdots \begin{cases} 40\cdot 20 \\ \underline{40\cdot 20} \end{cases}$
25	$\beta\omega''$	$\mu\alpha$	
26	$\beta\omega''$	$\mu\varsigma\varsigma'$	
27	Γ''	$\mu\varsigma''\delta'$	

55. Wallis to Johannes Hevelius, 15/25 January 1668/9, enclosure

28	$\iota\eta\Gamma'$	$\mu\beta\overset{''}{\_}\delta'$	········ $\begin{cases} 40\cdot 45 \\ 40\cdot 45 \end{cases}$
29	$\iota\varsigma\Gamma'$	$\mu\Gamma$	
30	$\iota\zeta\omega''$	$\mu\Gamma\overset{''}{\_}\delta'$	Al-suphi dicit non conspici.
31	ι	$\nu\alpha\varsigma'$	············ $\underline{51\cdot 20}$
32	$\iota\epsilon\Gamma'$	$\nu\alpha\omega''$	
33	$\varsigma\Gamma'$	$\nu\epsilon\varsigma'$	
34	$\iota\alpha\varsigma'$	$\nu\epsilon\Gamma'$	
35	$\eta\Gamma'$	$\mu\delta\varsigma'$	········ $\begin{cases} 41\cdot 10 \\ \underline{41\cdot 10} \end{cases}$
36	$\kappa\delta\varsigma'$	$\mu\epsilon\Gamma'$	
37	$\iota\delta\omega''$	$\mu\theta\varsigma'$	

$\Theta\eta\varrho\iota'ov.$

1	$\kappa\eta$	$\kappa\delta\overset{''}{\_}\delta'$	············ $24\cdot 50$
2	$\kappa\epsilon\overset{''}{\_}$	$\kappa\theta\varsigma'$	$25\cdot 50$
3	α	$\kappa\alpha\delta'$	
4	$\delta\varsigma'$	$\kappa\alpha$	
5	$\Gamma\varsigma'$	$\kappa\epsilon\varsigma'$	$\begin{cases} 3\cdot 0 \\ \underline{3\cdot 0} \end{cases}$
6	$\bar{o}\varsigma'$	$\kappa\zeta$	
7	$\bar{o}\overset{''}{\_}\varsigma'$	$\kappa\theta$	$\begin{cases} 0\cdot 30 \\ \underline{0\cdot 30} \end{cases}$
8	$\delta\omega''$	$\kappa\eta\overset{''}{\_}$	
9	$\Gamma\omega''$	$\lambda\varsigma$	············ $30\cdot 10.$
10	$\epsilon\omega''$	$\lambda\Gamma\varsigma'$	
11	$\kappa\beta$	$\lambda\alpha\Gamma'$	$20\cdot 0.$ Al-suphi dicit non conspici.
12	$\kappa\alpha\overset{''}{\_}\Gamma'$	$\lambda\overset{''}{\_}$	
13	$\kappa\Gamma$	$\kappa\theta\Gamma'$	
14	$\eta\overset{''}{\_}\Gamma'$	$\iota\zeta$	
15	$\theta\Gamma'$	$\iota\epsilon\Gamma'$	
16	$\epsilon\omega''$	$\iota\Gamma\Gamma'$	
17	$\varsigma\omega''$	$\iota\alpha\overset{''}{\_}\Gamma'$	

55. Wallis to Johannes Hevelius, 15/25 January 1668/9, enclosure

18	κζϛ'	ια$\stackrel{''}{-}$Γ'	27 · 20	11 · 30 ⎫
19	κϛ$\stackrel{''}{-}$	ι		<u>11 · 30</u> ⎭

	Θυμιατηρίου.			
1	κζω''	κβω''		
2	Γ	κε$\stackrel{''}{-}$δ'	⎰	0 · 20
			⎱	<u>0 · 20</u>
3	κϛϛ'	κϛ$\stackrel{''}{-}$		
4	κω''	αΓ'	········· ⎰	30 · 20
			⎱	<u>30 · 20</u>
5	κεϛ'	λδϛ'		
6	κε	λΓΓ'		
7	κε$\stackrel{''}{-}$Γ'	λδδ'	⎰ 20 · 50	34 · 0
			⎱ <u>20 · 50</u>	<u>34 · 0</u>

	Στεφάνου νοτίου.			
1	ōϛ'	κδ$\stackrel{''}{-}$	⎰ 9 · 10	21 · 30
			⎱ <u>9 · 10</u>	<u>21 · 30</u>
2	ιαω''	κα		
3	ιΓϛ'	κΓ	········· ⎰	20 · 20
			⎱	<u>20 · 20</u>
4	ιδ$\stackrel{''}{-}$Γ'	κ		
5	ιϛϛ'	ιη$\stackrel{''}{-}$		
6	ιξ	ιζϛ'		
7	ιϛΓ'	ιϛ	⎰	16 · 50
			⎱	<u>16 · 50</u>
8	ιϛ$\stackrel{''}{-}$	ιεϛ'		
9	ιεϛ'	ιεΓ'		
10	ιδω''	ιδ$\stackrel{''}{-}$Γ'		
11	ια$\stackrel{''}{-}$Γ'	ιδω''		
12	θω''	ιε$\stackrel{''}{-}$Γ'		
13	θϛ'	ιη$\stackrel{''}{-}$		

55. Wallis to Johannes Hevelius, 15/25 January 1668/9, enclosure

$\overset{\text{'}}{I}\chi\vartheta\acute{v}o\varsigma\ \nu o\tau\acute{\iota}ov.$

1	ζ	κΓ	
2	ōω″	κΓ	$\begin{cases} 0\cdot 10 & 20\cdot 20 \\ \underline{0\cdot 30} & 20\cdot 20 \end{cases}$
3	δς′	κβδ′	Pro duabus primoribus, Arabs et Al-suphi unicam habent.
4	εΓ′	κβ″⸺	.... $\begin{cases} 5\cdot 40 \\ \underline{5\cdot 30} \end{cases}$
5	δΓ′	ιςδ′	
6	κες′	ιθ″⸺	
7	ας′	ιες′	
8	κη″⸺Γ′	ιδω″.	
9	κεΓς′	ιε	············ 16·0
10	κα″⸺Γ′	ις″⸺	
11	κα	ιης′	
12	κς′	κβδ′	$\begin{cases} 26\cdot 0 \\ \underline{26\cdot 0} \end{cases}$

Informes. Desunt in Al-suphi.

1	η	κβΓ′	
2	ιας′	κβς′	
3	ιδ	κας′	············ 21·0
4	ιβ	κ″⸺Γ′	
5	ιΓ″⸺Γ′	ιζ	
6	ιΓ″⸺Γ′	ιδ″⸺Γ′	············ 27·50.

Finis Canonis Constellationum Ptolemaici.

[3ᵛ] | Canonion Stellarum aliquot illustrium (de quibus fuse Ptolemaeus lib. Syntax. ex veterum suisque observationibus:) Ex Tabulis Persicis depromptum; Collatis duobus MSS. Codicibus (Baroccianis) in Bibliotheca Bodleiana Oxoniae.

55. WALLIS to JOHANNES HEVELIUS, 15/25 January 1668/9, enclosure

Observationes (atque inde Canonion hoc) factae sunt in Persia imperante Melixa, die primo primi mensis sive Pharuartae hora diei sexta jam exeunte atque instante septima, Imperii Melixani anno primo, et die 13 mensis Martii apud Romanos, quando sol in principio ipso Arietis constitit: Inde ad nostrum tempus (inquit Author) fluxere anni persici 338. A nato autem Christo 1346. Quare Anni Domini, minus $\varrho\eta$ = anno Persico.

Urbes ubi factae sunt Observationes Persicae pleraeque (ex MSS. Barocc. 58 et 166)

55. WALLIS to JOHANNES HEVELIUS, 15/25 January 1668/9, enclosure

	μῆχος	πλάτος
χαζαρία	οε κ	λς λ
τυβήνη	οβ ō	λη ō

Ex Chrysiciae Tr. de Epochis. MS Barocc. $\boxed{58}$ fol. 182. (Quoad sensum sed non verbatim.)

$$\text{Anni O. C.} - 6116 = \text{Anno Arabico. ut } \overline{,\zeta\lambda\zeta} - \overline{,\varsigma\varrho\iota\varsigma} = $$
$$ \text{☿} \kappa\alpha = \tau\tilde{\omega} \ \tau\tilde{\omega}\nu \ \mathring{\alpha}\dot{\varrho}\varrho\alpha\beta\omega\nu \ \overset{\prime}{\epsilon}\tau\epsilon\iota. \qquad \text{aequal.}$$

$$\text{Anni O. C.} - 5197 = \text{Anno } \tau\tilde{\omega}\nu \ P\mathring{\omega}\mu\alpha\iota\omega\nu. \text{ ut } \overline{,\zeta\lambda\zeta}-\overline{\varsigma\varrho G\zeta}=$$
$$ ^\dagger\overline{\omega\bar{\mu}} = \tau\tilde{\omega} \ \tau\tilde{\omega}\nu \ P\mathring{\omega}\mu\alpha\iota\omega\nu \ \overset{\prime}{\epsilon}\tau\epsilon\iota.$$

$$\left.\begin{array}{l}\text{Anni O. C.} - 6139 \\ \text{An. Nat. } \chi^{\text{ti}} - 631\end{array}\right\} = \left\{\begin{array}{l}\text{Anno Persarum}\\ \text{seu ab Rege}\\ \text{Jasingerda}\end{array}\right\} \text{ ut } \overline{,\zeta\lambda\zeta} - {}_\varsigma\varrho\lambda\vartheta = $$
$$\overline{\omega G \eta} = \tau\tilde{\omega} \ \tau\tilde{\omega}\nu \ \pi\varrho\omega\tilde{\omega}\nu \ \overset{\prime}{\epsilon}\tau\epsilon\iota.$$

$749 - 378 = \frac{371}{4} = 92\frac{3}{4}$

Anni Persarum — 378: — 13 Martii ὀπιδορμήτως, (ut 13, 12,) = 1° dici Pharuartae, seu primi Mensis Persarum = initio anni Persici, ut $\overline{\psi\mu\vartheta} - \overline{\tau o\eta} = \tau\overline{o\alpha}$. Tum $\frac{\tau\overline{o\alpha}}{\delta}=\overline{G\beta}$. (τὰ γὰρ Γ΄ εἰάθησαν ὡς ἀργά.) ὑφείλομεν ἀπὸ τῶν $\overline{\text{IΓ}}$ τοῦ μαρτίου ὀπισθορμήτως μῆχρις οὗ ἔληξεν ὁ ἀριθμός. ἔληξε δὲ εἰς τὴν $\overline{\kappa}$ τοῦ δεκεμβρίου. ἥτις ἐστὶ $\overline{\iota\beta}$ τοῖς βαδίζουσιν ὀπισθορμήτως = 1° dici Pharuartae. ἰστέον ὅτι ἀεί ποτε λάμβανε τον φενρουαρίων μετὰ $\overline{\kappa\eta}$ ἐπὶ τῶν ὑφειλομένων — &c.

13	
371	(92
44	

Mart.	13	
Febr.	28	
Jan.	31	
Dec.	-20	retrarsum
	92.	ad diem 12

5

Bulliald. Astron. Philolaic. p. 218. de Apogaeo Martis anno 1° Persarum subdubitat; et $\overline{\theta}$ mavult quam $\overline{\epsilon}$. At vero $\overline{\epsilon}$ legitur in tribus Codd. MSS^{is} Bibliothecae Bodleianae Oxoniae.

---

†forte $\overline{\alpha\omega\mu}$

56.
JOHN WALLIS JR to WALLIS
[15]/25 January [1668]/1669

Transmission:

Manuscript missing.

Existence and date: Mentioned in WALLIS–OLDENBURG 11/[21].II.1668/9 (the circumstances indicate that the date given for the present letter is new style).

Wallis's son evidently reported that Edward Bernard and he had arrived in Leiden shortly before Franz de le Boë (1614–72), for whom he had brought a letter from Oldenburg, was elected to the office of Rector Magnificus of the University. In a postscript to a letter to the orientalist Samuel Clarke (1624–69), Bernard had written before leaving London: 'Pray commend my respects to Dr Wallis & acquaint him that I have sent his Letters & the Bundle & will by Gods permission returne him an answer att large, when I am in Holland.' (*British Library*, MS Birch 4275, f. 40^r–41^v, f. 41^r). There is no evidence that Bernard wrote to Wallis from Leiden. Cf. WOOD, *Athenae Oxoniensis* IV, 704.

57.
HENRY OLDENBURG to WALLIS
16/[26] January 1668/9

Transmission:

Manuscript missing.

Existence and date: Mentioned in and answered by WALLIS–OLDENBURG 19/[29].I.1668/9.

58.
WALLIS to JOHN COLLINS
Oxford, 18/[28] January 1668/9

Transmission:

W Letter sent: CAMBRIDGE *Cambridge University Library* MS Add. 9597/13/6, f. 205ra–205^{vd} (our source). At top of f. 205^{vd} in Collins's hand: 'About Areas of Segments'. Postmark 'IA/18'.—printed: RIGAUD, *Correspondence of Scientific Men* II, 510–12.

Reply to: COLLINS–WALLIS 12/[22].I.1668/9.

58. WALLIS to COLLINS, Oxford, 18/[28] January 1668/9

The date of the present letter emerges from its postmark and from the postscript to WALLIS–COLLINS 19/[29].I.1668/9. As that letter was sent on a Tuesday, the last post, to which Wallis there refers, would have been the day before.

Jan. 1668./9. Oxon.

Sir,

To yours[365] of Jan. 12 I had before answered, as to so much as concerns Mr Gregorie's letter[366] in mine[367] of Jan. 8. sent in a parcell directed to Mr. Pits. And I have since sent his letter up to Mr. Oldenburg enclosed in one[368] of Jan. 12. with an answere.[369] As to Dr. Newtons[370] design of calculating a table for Segments of Circles: I do not know any more convenient or expedite way than according to a Specimen which I have heretofore sent[371] you. What may best serve the Gaugers turn I shall not determine: But otherwise I should think it much better to make the Radius equal to 1. (with as many places as you please for Decimal fractions:) & consequently the Square of Radius (rather then the Semicircle) equal to 1. And then, for every number in your Table, you will need but one subduction, one Addition, & one Extraction of the Square Root (to a very few places) which, to those versed in it, is done presently; & sooner then so much of Division. And when hee is a little onward in his work, he will presently see how great leaps he may safely take without danger of missing one unite in what decimall place he please. So that I do not question but hee may dispatch a great many places in a very short time: Without any Rectification by the 3^d, 4th or 5th differences; which, in so quick a work as this is, I think will be of very little advantage. Yet if hee would go the other way to work; hee must, at convenient distances, make essays, whether the 1st, 2^d, 3^d, 4th, or 5th differences will serve the turn; & then it will be easy to apply them in such manner as Torporly[372] (if I remember the name aright) in the manuscript you shewed

3 answered, (*1*) (to (*2*), as to
5 sent (*1*) it (*2*) his letter
8 you. (*1*) But (*2*) What
10 much *add.*
18 dispatch (*1*) ⟨—⟩ places in an afternoon (*2*) a great

---

[365]yours: i.e. COLLINS–WALLIS 12/[22].I.1668/9.
[366]letter: i.e. GREGORY–WALLIS c.20/[30].XII.1668.
[367]mine: i.e. WALLIS–COLLINS 8/[18].I.1668/9.
[368]one: i.e. WALLIS–OLDENBURG 12/[22].I.1668/9.
[369]answer: i.e. WALLIS–GREGORY 8/[18].I.1668/9.
[370]Newtons: i.e. John Newton (1622–78), D.D., astronomer and mathematician, *ODNB*.
[371]sent: possibly in WALLIS–COLLINS 8/[18].I.1668/9.
[372]Torporly: i.e. Nathaniel Torporley (1564–1631), English mathematician and divine,

58. WALLIS to COLLINS, Oxford, 18/[28] January 1668/9

mee,| or as Mercator[373] in his Logarithmotechnia[374] prop. 12. directs. A
Table thus computed, is easyly applyed to the other hypothesis (of making
the Semicircle equal to 1.) by dviding any place you have occasion to use;
(or all, if you would so reduce the whole table,) by 1.57080 – (the proportion
of the Semicircle to the Square of Radius, being as 1.57080 (proxime) to 1.)

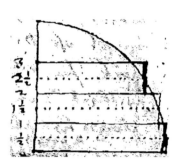

I should onely direct this alteration in the former directions (of September last) that were as I take the Ordinates in the Quadrant $= \sqrt{\ }: R^2 - a^2 ::$
and take a successively $= 1, 2, 3$, &c, or $= 0, 1, 2, 3$, &c; (the one answering to the circumscribed figure, the other to the inscribed;) it will be more
accurate to take it $= \frac{1}{2}, 1\frac{1}{2}, 2\frac{1}{2}, 3\frac{1}{2}$, &c; which will be the midle lines of the
parallelograms partly inscribed & partly circumscribed; As for instance; if
$R = 1$. and this divided into 10000 parts; then is $a = 0.00005$, $a = 0.00015$,
$a = 0.00025$, &c. And then each of these $\sqrt{\ }: R^2 - a^2$: multiplied in the
altitude of one such part, that is, into 0.0001, gives you the parallelogram
(partly inscribed, partly circumscribed,) answering to the 1st, 2^d, 3^d, 4th, &c
parts of the erect radius; reconed from the transverse Radius, or the Base.
And when you have finished the Table for all the $\sqrt{\ }: R^2 - a^2$: you may
begin to collect the aggregates, either from the top, or from the bottom,
which you please; & the Aggregates (because to be multiplied by 0.0001,)
are to be depressed four places downwards in the table. I have given here
a short specimen, which is easy to continue; at pleasure. I adde onely that
I am

---

15 &c. (*1*) each of which respectively multiplied into the hight of one (*2*) And

---

ODNB. See COLLINS–BAKER 19/[29].VIII.1676; RIGAUD, *Correspondence of Scientific Men* II, 4–10.

[373] Mercator: i.e. Nicolaus Mercator (1620–87), German mathematician, born in Danish Holstein; from 1657 until 1682 mathematics teacher in England, mainly in London. Elected fellow of the Royal Society 1666, *ODNB*.

[374] Logarithmotechnia: i.e. MERCATOR, *Logarithmotechnia: sive methodus construendi Logarithmos nova, accurata, & facilis*, London 1668, 23–4.

Your friend to serve you,
John Wallis.|

$R^2 - a^2$	$\sqrt{\,}: R^2 - a^2 :$
0.9999999975	0.99999999875
0.9999999775	0.99999998875
0.9999999375	0.99999996875
0.9999998775	0.99999993875
0.9999997975	0.99999989875
0.9999996975	0.99999984875
0.9999995775	0.99999978875
&c.	&c.

Summa in 0.0001	Numeri tabulares	Alt.
0.99999999875	0.000099999999875	0.0001
1.99999998750	0.000199999998750	0.0002
2.99999995625	0.000299999995625	0.0003
3.99999989500	0.000399999989500	0.0004
4.99999979375	0.000499999979375	0.0005
5.99999964250	0.000599999964250	0.0006
6.99999943125	0.000699999943125	0.0007
&c.	&c.	&c.

These
For Mr John Collins
next door to the three
Crowns in
Blooms-bury Market
London.

59.
WALLIS to JOHN COLLINS
[Oxford] 19/[29] January 1668/9

Transmission:

W Letter sent: CAMBRIDGE *Cambridge University Library* MS Add. 9597/13/6, f. 206^r–207^v (our source). At top of f. 207^v in Collins's hand: 'About Laloveras Series. Postmark: 'IA/20.—printed: RIGAUD, *Correspondence of Scientific Men* II, 507–10.

59. WALLIS to COLLINS, 19/[29] January 1668/9

Jan. 19. 1668./9.

Sir,

what, I suppose Mr Oldenburg intends in the next Transactions,[375] though it contain divers of the principles on which I proceed in my Hypothesis of Motion, is not intended as any summary of my book[376] now printing; nor is it at all in that method, however upon the same principles. To the other Question, The chapter de Centro Gravitatis must stand where it doth, & is not to be removed. The next, De Calculo Centri Gravitatis, though I once thought of taking it out there, & putting it by itself; yet considering that cannot be done, without altering the Numbers of the figures; &, in pursuance of that, going over the whole work anew, in which the figures are cited over & over again, many times in the same page; it would make so much work to make that alteration all along, & would be subject to so many mistakes, that I think there will be a necessity of letting it stand where it doth; & proceeding in some following chapters as they were at first designed; without dividing it into two parts. But what you speak of putting out these 3. or 4. chapters alone, cannot at all be; they being necessaryly connected with what is to follow, de Vecte, Cochlea, Trochea, Tympano, &c; & that de Motuum acceleratione et percussione, which must all go together, because they frequently cite propositions out of the precedent chapters. If that de Calculo Centri gravitatis be taken out; it is all can be done, & that not without much trouble for the reasons mentioned. The Series of which you inquire[377] in pag. 398 of Lalovera,[378] de Cycloide I have looked upon; but it's complicated so with other things; that I see not how to give an account of it without reading over most of the book; nor can it well be otherwise understood; which| at present, having many things on my hands at once, I [20

6 it (*1*) any thing (*2*) at all
13 along, (*1*) that (*2*) &
17 be; (*1*) because (*2*) they
22 much *add.*
22 mentioned. (*1*) What you aks of the method (*2*) The Series
23 Lalovera, (*1*) I have looked upon it; (*2*) de Cycloide I . . . upon;

---

[375]Transactions: i.e. *Philosophical Transactions* No. 43 (11 January 1668/9), which contained Wallis's and Wren's accounts of the laws of motion.
[376]book: i.e. Wallis's *Mechanica: sive, de motu, tractatus geometricus*, which Collins was seeing through the press. Chapters IV (De centro gravitatis) and V (De calculo centri gravitatis) constitute the second part of the work.
[377]inquire: presumably in COLLINS–WALLIS 12/[22].I.1668/9.
[378]Lalovera: i.e. LALOUBÈRE, *De cycloide Galilaei et Torricellii propositiones viginti*, Toulouse 1658.

59. WALLIS to COLLINS, 19/[29] January 1668/9

am not in a capacity to do. And I should rather give an account of the things from my own principles then study to be perfect in his: His whole methode all along being somewhat perplexed. Though (because I find the results for the most part agree with mine) I take it to be sound though dark. But the general design there, I take to bee, to shew, how by having the summe of lines making the plain of those figures, (the circle & hyperbole,) he proceeds to a summe of squares to find the Solid Ungule, or the Moment of that plain; & so to the summes of Cubes, to find the Moment of that Ungle, & so on. Or, which is equivalent, from the squaring of a plain whose lines are as the lines in an Hyperbola or Circle; to the squaring of a 2^d 3^d 4th &c plain, whose lines are in the Duplicate, Triplicate, quadruplicate, &c proportion of those lines. Which is the methode he prosecutes through out his Book: But I have not taken the pains to make myself master of those methodes, because those of my own are (to me at lest) lesse perplexed & more simple than those of his; & bring about the same effects with more ease & clearness. If to others it shall seem otherwise; they will have their choise to use which they please.

But his design there is not at all like that series which wee have about the Hyperbola; which I suppose you aim at. You may, with my humble service to him, signify so much to my Lo. Brounker; from

Your friend to serve you,
Joh: Wallis.

I wrote[379] to you by the last post in answer to a former letter.[380] The series in that letter designed, is the series for the Circle answering to that we have for the Hyperbola.|posing The Centers in each being C. the Asymptotes in the Hyperbola answering to the two Conjugate equal diameters in the Ellipsis or Circle; & the points a, b, c, d, e, &c in one, corresponding to those in the

---

5 there, (*1*) seemes (*2*) I take
6 (the...hyperbole,) *add.*
7 to (*1*) summe (*2*) a summe
7 squares (*1*) making a (*2*) to find the
9 the (*1*) summe (*2*) squaring
10 lines (*1*) of (*2*) in
13 to (*1*) study (*2*) make
18 there *add.*
18 series *add.*

---

[379] wrote: i.e. WALLIS–COLLINSS 18/[28].I.1668/9.
[380] letter: presumably COLLINS–WALLIS 12/[22].I.1668/9.

other: The rectangles $a\alpha$, $b\beta$, $c\gamma$, $d\delta$, $e\epsilon$, &c in the one, being equall, suppose B^2, B^2, B^2 &c, or $\frac{B^2}{a}a$, $\frac{B^2}{b}b$, $\frac{B^2}{c}c$, &c: in the other $a\sqrt{\,}: B^2-a^2$. $b\sqrt{\,}: B^2-b^2$. $c\sqrt{\,}: B^2-c^2$. &c. (supposing the angles at C to be right angles).

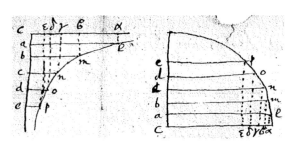

And the lines al, bm, cn, do, ep, &c: in the one are $\frac{B^2}{a}$, $\frac{B^2}{b}$, $\frac{B^2}{c}$, $\frac{B^2}{d}$, $\frac{B^2}{e}$, &c: in the other, $\sqrt{\,}: B^2-a^2.$, $\sqrt{\,}: B^2-b^2.$, $\sqrt{\,}: B^2-c^2.$, $\sqrt{\,}: B^2-d^2$: &c. And for the calcule of the consequent parallelograms cl, am, bn, co, &c: you are in the one, to Divide; in the other, to extract a square root: And then, to multiply the result into one of those parts $C\alpha$, or ab.

For Mr John Collins
next to the three Crowns
in Blooms-bury Market
London.

60.
Wallis to Henry Oldenburg
Oxford, 19 and 21/[29 and 31] January 1668/9

Transmission:

W Letter sent: LONDON *Royal Society* Early Letters W1, No. 83, 4 pp. (our source). Endorsement by Oldenburg beneath address on p. 2: 'Rec. jan. 22. 1669. Answ. jan. 26. Postmark on p.2: 'IA/22'.—printed: OLDENBURG, *Correspondence* V, 342–4.

Reply to: OLDENBURG–WALLIS 16/[26].I.1668/9.
Answered by: OLDENBURG–WALLIS 26.I/[5.II].1668/9.

2 other (*1*) B^2-a^2, B^2-b^2, B^2-c^2 &c (*2*) $a\sqrt{\,}: B^2-a^2$.
3 (supposing...angles) *add.*
7 calcule (*1*) in the one sum the (*2*) of

60. WALLIS to OLDENBURG, 19 and 21/[29 and 31] January 1668/9

Oxford Jan. 19. 1668./9.

Sir,

I must first thank you for your civility to my Son,[381] & Mr Bernard[382] before they went to Holland. Next I am glad of what you tell me of Bartholinus.[383] For it is but too true that the Book[384] is full of Typographical Errors. But if Bartholinus do in good earnest intend a more correct Edition;[385] I think you may then spare sending for more Copies of this.

But I should advise allso that Bartolin would not omit those other Additions besides Ticho's Observations; which are in this Historia Coelestis. As to Dulaurens, I do not know whether you mean that he is printing his last[386] or a third paper. If onely the other, you have an answere,[387] which I think you have not yet printed. If it be a third; you may adde to the Close of the last these words. [Quae autem de Scripto ejus secundo dicta sunt; eadem et Tertio vel etiam Quarto, (si plura scripturus sit,) accommodanda erunt. Non enim opus erit ea sigillatim retexere, magis quam Librum ipsum: Neque ut ego de illo porro sim solicitus.]

I am next to tell you that I sent yesterday, by Bartlet, (in a paquet directed to Mr Clenden[388] at Mr Stag's[389] house in Drury Lane, where I use to lodge,) a paquet inclosed for you when you| there call for it; (because it was to big to come by the Post;) containing Papers[390] to be sent to Hevelius.

17 in a *(1)* seal *breaks off* *(2)* parcell *(3)* paquet

---

[381]son: i.e. John Wallis jr, q.v.

[382]Mr. Bernard: i.e. Edward Bernard, q.v. The main purpose of his visit to Leiden in December 1668 was to complete a commissioned transcription of books V–VII of the textit*Conics* of Apollonius of Perga; the transcript is now *Bodleian Library* MS Thurston 1.

[383]Bartholinus: i.e. Rasmus Bartholin (1625–98), professor of geometry, and later professor of medicine, at the University of Copenhagen.

[384]Book: i.e. TYCHO BRAHE, *Historia coelestis*, ed. Albert von Curtz, 2 parts, Augsburg 1666.

[385]Edition: in *Specimen recognitionis nuper editarum observationum astronomicarum ... Tychonis Brahe*, Copenhagen 1668, Bartholin had severely criticized the Curtz edition of Tycho Brahe's *Historia Coelestis* on account of the many errors it contained. After Frederick III of Denmark (1609–70) had purchased Tycho Brahe's manuscript collection of astronomical observations from Ludwig Kepler (1607–63) in 1662, he commissioned Bartholin to direct their publication. The new edition was, however, never completed.

[386]last: Dulaurens wrote to Oldenburg, in reply to Wallis, in October 1668.

[387]answere: i.e. WALLIS–DULAURENS 27.XI./[7.XII].1668. Contrary to Wallis's wish, Oldenburg chose not to protract the dispute with Dulaurens in the *Philosophical Transactions*, and therefore did not print Wallis's latest response.

[388]Clenden: i.e. John Clendon (or Clendan) (*fl.*1653–68), nephew ('cousin') of John Wallis; a chandler or lamp maker in London.

[389]Stag's: not identified.

[390]Papers: i.e. WALLIS–HEVELIUS 15/[25].I.1668/9.

60. WALLIS to OLDENBURG, 19 and 21/[29 and 31] January 1668/9

The contents whereof you will see when you open them. Of M. Hugens's Papers I can adde no more than is in my last,[391] having had no leisure to examine them further:

I am

Your friend to serve you,
John Wallis.

In the Papers for Hevelius, page. 2. (in the Title of the whole) there is in two places *Min. 12.′42.″*
which should be *grad. 12.42 ′*.
Pray mend it.[392]

Jan. 21. 1668.

Sir

Having lost the last post (which was gone before my letter[393] came[)]; I have opened it to adde this in answere to yours[394] of Jan. 16. which is but just now come to hand, when this Post is almost ready to go.

The two experiments you there mention, do not contradict, or directly concern what is in my paper[395] at §7. & therefore do not require any alteration in it. For that is of *Motus Æquabilis*; whereas the two Experiments are about *Motus acceleratus & retardatus*, of [which] I do not speak till §14. and there but breefly.

I will inquire as you direct for such a student; & give you further account hereafter. What you speak of accommodating mine & M. Hugens's hypothesis, may be done hereafter; but at present my hands are full. I still think

---

13 came; I have *corr. ed.*
16 mention, (*1*) I do not (*a*) take (*b*) judge (*2*) do not
16 or (*1*) concern (*2*) directly concern
19 of I do not *corr. ed.*

---

[391] last: i.e. WALLIS–OLDENBURG 12/[22].I.1668/9.

[392] Pray mend it: As the original manuscript of WALLIS–HEVELIUS 15/[25].I.1668/9 reveals, Oldenburg carried out this request.

[393] letter: i.e. the first part of the present letter to Oldenburg.

[394] yours: i.e. OLDENBURG–WALLIS 16/[26].I.1668/9.

[395] paper: i.e. Wallis's paper on the laws of motion (Wallis–Oldenburg 15/[25].XI.1668.), which was published in the *Philosophical Transactions* No. 43 (11 January 1668/9), 864–6.

that Mr. Gregories' answere may well inough be published for the reasons I gave before; which is confirmed by what you adde.³⁹⁶|

I should have told you in one of my former letters, that I have all your mony for the books you sent hither: & that so soon as I hear of mony payd for mee in London, which I expect dayly, I will direct so much to be payd out of it to you.

You may please in your publishing: to mention the date of my paper de motu; & the time when that & the rest came to hand, & were first published in the Society.

Yours &c.
J. W.

For Mr Henry Oldenburg
in the Palmal near St
James's
London.

61.
Henry Oldenburg to Wallis
26 January/[5 February] 1668/9

Transmission:

Manuscript missing.

Existence and date: Referred to in Oldenburg's endorsement on the cover of WALLIS–OLDENBURG 19 and 21/[29 and 31].I.1668/9.
Reply to: WALLIS–OLDENBURG 19 and 21/[29] and 31.I.1668/9.

21-2 & give you further account ... what you adde. *add.*

---

³⁹⁶ adde: i.e. the addendum which Gregory sent to Oldenburg which the latter inserted in the text published in *Philosophical Transactions*. Cf. GREGORY–COLLINS 20/[30].I.1668/9; TURNBULL, *James Gregory*, 63.

62.
JOHN WALLIS JR to WALLIS
late January 1668/9

Transmission:

Manuscript missing.

Existence and date: Referred to in WALLIS–OLDENBURG 11/[21].II. 1668/9.

63.
JOHN COLLINS to WALLIS
late January/February 1668/9

Transmission:

Manuscript missing.

Existence and date: Referred to in WALLIS–COLLINS 23.III/[2.IV].1668/9; probably a reply to WALLIS–COLLINS 19/[29].I.1668/9.

64.
HENRY OLDENBURG to WALLIS
6/[16] February 1668/9

Transmission:

Manuscript missing.

Existence and date: Mentioned in and answered by WALLIS–OLDENBURG 11/[21]. II.1668/9.

In this letter Oldenburg sent Wallis a copy of the cipher which had recently been sent by Huygens, and an account of spring tides.

65.
HENRY OLDENBURG to WALLIS
8/[18] February 1668/9

Transmission:

Manuscript missing.

Existence and date: Mentioned in and answered by WALLIS–OLDENBURG 11/[21].II. 1668/9. This letter was sent with a copy of *Philosphical Transactions* No. 43 (11 January 1668/9).

66.
WALLIS to HENRY OLDENBURG
Oxford, 11/[21] February 1668/9

Transmission:

W Letter sent: LONDON *Royal Society* Early Letters W1, No. 84, 2 pp. (our source). Endorsement by Oldenburg beneath address on p. 2: 'Rec. Febr. 12. 1669. Answ. febr. 16. of Hyperbolic try of Wren'. There is also a calculation by Oldenburg with no relation to the content of the letter.—printed: OLDENBURG, *Correspondence* V, 390–1.

Reply to: OLDENBURG–WALLIS 6/[16].II.1668/9 and OLDENBURG–WALLIS 8/[18].II.1668/9.

Oxford Febr. 11. 1668./9.

Sir,

I have yours[397] of Febr. 6, and Febr. 8. And, in the first, Mr Hugens his Ænigma[398] in Cypher; in the second, the Transactions. That the Cypher is not an absolute security, I have heretofore, on another occasion, sufficiently

---

[397] yours: i.e. OLDENBURG–WALLIS 6/[16].II.1668/9 and OLDENBURG–WALLIS 8/[18].II. 1668/9.

[398] Ænigma: as a means to protecting discoveries in advance of publication, Huygens had proposed in HUYGENS–OLDENBURG [27.I]/6.II.1668/9; OLDENBURG, *Correspondence* V, 360–1, that members of the Royal Society submit them for registration in the Register-Book concealed behind anagrams or ciphers. In that letter Huygens concealed in this way his discovery of a new theorem on hyperbolic lenses. Oldenburg subsequently communicated the cipher to Wallis in his letter of 6/[16] February 1668/9. See BIRCH, *History of the Royal Society* II, 344–5.

66. WALLIS to OLDENBURG, 11/[21] February 1668/9

discovered to M. Hugens: But it may be a competent security, amongst persons tolerably ingenious, & inough to serve such a turn. I do not know what it is that his Cipher involves. But, if it should chance to, a Methode of Turning an Hyperbolical Solid, by ordinary Instruments, by the Application of Streight Line: It is that which Dr Wren[399] & myself have, divers years since, discoursed of: &, particularly, of late, in the years past. And, that the figure so to be made, is an Hyperbole: is by mee demonstrated in Print,[400] nine or ten years agoe. Of which Dr Wren, is able to give you an account. The Suggestion of that operation being his: & the demonstration that the body is Hyperbolicall, being mine. But whether hee think fit yet to divulge it, I know not. If not; let not this be to his prejudice.

I expected, yours[401] of Feb. 8. would have given a further account of the spring-tydes, which that[402] of Feb. 6. did begin.

Amongst the *Errata*[403] in the Transactions, not onely that of p. 865, l. 24; but that of p. 866. l. 3. should have been omitted, for *Sinistrasum* is right: If any danger of a mistake, it were better helped by a comma before it, or by putting it after *prius*. But, instead of these two, you may put pag. 865. lin. antepen. *reapse* (not *re adse*:) & p. 866. l. 1. pro *mm* lege *mn*; et pro P, lege PC. But neither of them are very considerable.

My son,[404] by a Letter[405] from Leyden of Jan. 25. tells mee that they came thither just before the Election of their Rector Magnificus; who is, for the year following, Dr Sylvius,[406] (to whom you vouchsafed him a letter:) which hee willed mee to signify to your self & Mr Hake;[407] but, his letter coming to hand but this morning; (I know not by what means; for I had last

---

3 of (*1*) Forming (*2*) Turning
9 Suggestion (*1*) being his own (*2*) of that operation
10 yet *add.*

---

[399] Wren: Wren's paper, entitled 'Generatio Corporis Cylindroidis Hyperbolici, elaborandis Lentibus Hypberbolicis accommodati', and apparently based on his collaboration with Wallis, was published in *Philosophical Transactions* No. 48 (21 June 1669), 961–2.
[400] Print: i.e. WALLIS, *Tractatus duo*, 97–8; *Opera mathematica* I, 554–5.
[401] yours: i.e. OLDENBURG–WALLIS 8/[18].II.1668/9.
[402] that: i.e. OLDENBURG–WALLIS 6/[16].II.1668/9.
[403] *Errata*: i.e. *Philosophical Transactions* No. 43 (11 January 1668/9), 876.
[404] son: i.e. John Wallis jr.
[405] Letter: i.e. JOHN WALLIS JR–WALLIS [15]/25.I.1668/9.
[406] Sylvius: i.e. Franz de le Boë (Franciscus Sylvius) (1614–72), Dutch chemist and physician, founder of the iatrochemical school of medicine.
[407] Hake: i.e. Theodore Haak (1605–90), translator and natural philosopher, *ODNB*.

66. WALLIS to OLDENBURG, 11/[21] February 1668/9

week one[408] of a later date:) I beleeve it is now no news to you. Pray, let mee hear by your next, How Esquire Boyle doth. I am

Your friend & servant
J. Wallis.

My Hypothesis[409] of motion, which (you say) My Lord Brounker observes to be different from that of Dr Wren[410] & Mr Hugens;[411] may (I suppose) be made Equivalent by taking in my §13. where the whole business of Resilitionis dedu[ced] from the Spring.

See whether this be the true cipher of Hugenius[412]; for it's possible in transcribing, a number may be mistaken

a	b	c	d	e	h	i	l	m	n	o	p	r	s	t	u	y
5	2	2	1	4	1	2	3	3	3	1	3	2	2	3	2	4	1

[2] For Mr Henry Oldenburg,
in the Old Palmal, near
St James's
London

---

9 Resilition is dedu *corr. ed.*

---

[408]one: i.e. JOHN WALLIS JR–WALLIS late January/[early February] 1668/9.

[409]Hypothesis: i.e. 'A Summary Account given by Dr. John Wallis, of the General Laws of Motion, by way of Letter written by him to the Publisher', printed in *Philosophical Transactions* No. 43 (11 January 1668/9), 864–6.

[410]Wren: i.e. 'Dr. Christopher Wrens Theory concerning the same Subject; imparted to the R. Society Decemb. 17. last', printed in *Philosophical Transactions* No. 43 (11 January 1668/9), 867–8.

[411]Hugens: Huygens sent Oldenburg his laws of motion in his letter of [26 December 1668]/5 January 1669 and they were subsequently presented to the Royal Society on 7/[17] January—see BIRCH, *History of the Royal Society* II, 337. Only after they had appeared in French in the *Journal des Sçavans* were Hugens' laws of motion published in Latin in England: 'A Summary Account of the Laws of Motion, communicated by Mr. Christian Hugens in a Letter to the R. Society, and since printed in French in the Iournal des Scavans of March 18, 1669. st. n.', printed in *Philosophical Transactions* No. 46 (12 April 1669), 925–8.

[412]cipher of Hugenius: the transcription Wallis received from Oldenburg was in fact correct; each letter appears in the concealed sentence as often as the number beneath it indicates. In plain text Huygens' theorem is 'Lens e duabus composita hyperbolicam aemulatur'.

67.
HENRY OLDENBURG to WALLIS
16/[26] February 1668/9

Transmission:

Manuscript missing.

Existence and date: Referred to in Oldenburg's endorsement on the cover of WALLIS–OLDENBURG 11/[21].II.1668/9.

Evidently this letter particularly concerned Wren's method of constructing a hyperbolic cylindroid.

68.
JOHANNES HEVELIUS to WALLIS
Danzig, [11]/21 March 1668/9

Transmission:

C Copy (in Hevelius's hand) of (missing) letter sent: PARIS *Observatoire de Paris* C1–9, No. 1357, 3 pp. (our source).
c Copy (in scribal hand) of *C*: PARIS *Bibliothèque Nationale* Fonds Latin 10348, IX, pp. 164–5.

Reply to: WALLIS–HEVELIUS 15/[25].I.1668/9.

Domino
Wallisio
Oxonii

Clarissime ac Doctissime Vir

Laudes illas eximias, quas in me abunde confers, sic equidem interpretor, quasi a mero tuo erga me amore atque sincero affectu profluxerint; gratissimae tamen extiterunt ex eo, quod sentiam me haud parum esse inflammatum, (dum nostrae quales quales opellae Tibi Laudatissimo Viro, aliisque Tui similibus Viris Eruditissimis non usque adeo displicent) ad alia praestantiora ac sublimiora, cum Deo et die elaboranda ac edenda. Quare mecum divinam implorabitis opem, quo Gratiam Suam, fixamque valetudinem benignissime nobis concedat, ad ea omnia feliciter in Nominis sui Gloriam peragenda. Sperque etiam haud levis affulget conatus nostros eo felicius successuros,

8 Viro, (*1*) Tu⟨i⟩ (*2*) aliisque Tui *C*

68. HEVELIUS to WALLIS, [11]/21 March 1668/9

cum re ipsa experiar, quanto ardore et affectu eos voveatis, ac promoveatis, dum praeter illum jam olim Ulug Beigi transmissum Fixarum catalogum,[413] denuo alium ex plurimis praestantissimis manuscriptis, summa diligentia inter se collatis[414] superaddere haud fucritis gravati. Quod sincerissimi amoris erga me vestri, nec non Benevolentiae erga Philosophiam sideralem, singulare munus, ut [ambabus] manibus amplector, sic operam daturus sum, si non paribus, saltem aliis quibusdam gratissimis officiis, officiosam hanc vestram voluntatem rursus demereri non nequeam; totus quoque Orbis Eruditus suo tempore studia illa vestra, sine dubio, magnopere depraedicaturus, si porro allaboretis, quo pariter reliqua| Manuscripta illa egregia, de quibus in literis[415] Tuis loquutus es, in lucem quantocquo prodeant. Horrexis observata,[416] cum Tu digniora existimes, quam ut pereant, non intermittam, ut cum meis simul in publicum proferantur:[417] exspecto igitur illa, una cum hypothesi[418] Tua de Marinis estibus avidissime. De caetero noli mirari, quod libellum[419] de Saturno, una cum reliquis meis opusculis non transmiserim: quippe ne unicum quidem exemplar tum temporis supererat, alias lubentissime et illum concessissem; memini tamen me Illustriss. Vestris Academiis duo Exemplaria etiam hujus opusculi per Cl. Dominum Oldenburg obtulisse;[420] an illud autem Laudatissima Universitas Oxoniensis

6 ambabibus *C corr. ed.*
11 locutus es *c*

---

[413]catalogum: i.e. the transcription of Ulug Beg's catalogue of stars which Wallis sent Hevelius with WALLIS–HEVELIUS 5/[15].IV.1664.

[414]plurimis ... collatis: i.e. Wallis's transcription of Ptolemy's catalogue of stars, sent as enclosure to WALLIS–HEVELIUS 15/[25].I.1668/9.

[415]literis: see WALLIS–HEVELIUS 15/[25].I.1668/9.

[416]observata: at the request of the Royal Society, Wallis had carried out the task of ordering Horrox's astronomical papers for publication. See BIRCH, *History of the Royal Society* I, 395, 413–14, and WALLIS–OLDENBURG 21.IX/[1.X].1664; WALLIS, *Correspondence* II, 160–2. In his letter to Oldenburg of 8/[18] May 1665 (WALLIS, *Correspondence* II, 184), Wallis had made recommendations on the printing of Horrox's papers, having already in his letter of 21/[31] January 1664/5 (WALLIS, *Correspondence* II, 174–9) suggested that several of Horrox's results anticipate those of Hevelius.

[417]proferantur: Hevelius also expresses his willingness to publish Horrox's papers as they had been digested by Wallis in HEVELIUS–OLDENBURG 11/[21].III.1668/9; OLDENBURG, *Correspondence* V, 440–3. See BIRCH, *History of the Royal Society* II, 359.

[418]hypothesi: i.e. Wallis's hypothesis on the tides, published in *Philosophical Transactions* No. 16 (6 August 1666). See WALLIS–BOYLE 25.IV/[5.V].1666; WALLIS, *Correspondence* II, 200–22.

[419]libellum: i.e. HEVELIUS, *Dissertatio de nativa Saturni facie, ejusque variis phasibus, certa periodo redeuntibus*, Danzig 1656.

[420]obtulisse: Wallis held Hevelius to be mistaken in thinking that he had sent two presentation copies of his *Dissertatio de nativa Saturni facie* to Eng-

acceperit? percontari poteris. Nuper quoque Astrosophorum gratia aliquot exemplaria ejusdem libelli, quae mihi rursus optato ad manus venere, Londinum destinavi, cum observationibus quibusdam Eclipsium. Quod vero Johannis Palmerii[421] Vestratis ejusque observationum quas mihi communicare[422] anno 1663 dignatus es in Cometographia nostra haud fecerim mentionem, ex eo evenit, quod illo tempore, ea ipsa Cometographia jam penitus fere erat conscripta, imo jam eousque typis exscripta;[423] postmodum vero nec occasio data est, ea de re disserendi, nec amplius hucusque in mentem venit; quodsi autem forte iis de rebus nobis sermo erit, omni laude Viri illius Egregii ejusque observationum mentionem faciam. Tantum enim abest, ut quisquam cogitet me per studiose ea silentio praeteriisse, vel cujuscunque Sua obliterare voluisse, ut potius omnibus modis ea ab omni oblivione nunquam non conservem, propugnem:| ut quemlibet ingenuum Virum decet. [3] Deus Te tanquam salutare Sidus, diu nostro Orbi jubeat lucere. Dabam Gedani Anno 1669 die 21 Martii

T. Cl.
Studiosissimus
J. Hevelius

69.
JOHN COLLINS to WALLIS
21?/[31?] March 1668/9

Transmission:

Manuscript missing.

Existence and date: Mentioned in and answered by WALLIS–COLLINS 23.III/[2.IV]. 1668/9.

According to Wallis, he received this letter from Collins on 22 March. It is therefore probable that it had been sent from London the previous day.

4-5 ejusque ... dignatus es *add.* C

---

land for the universities of Oxford and Cambridge. See WALLIS–OLDENBURG 25.IV/[4.V].1669.

[421]Palmerii: i.e. John Palmer (1612–79), archdeacon of Ecton, Northamptonshire, and friend and associate of the astronomers Samuel Foster (c.1600–52) and John Twysden (1607–88). See TAYLOR, *The Mathematical Practitioners*, 213.

[422]communicare: i.e. WALLIS–HEVELIUS 30.III/[9.IV].1663; WALLIS, *Correspondence* II, 82–6.

[423]exscripta: as Hevelius explains, the corresponding parts of his *Cometographia* (Danzig 1668) were already typeset before Wallis's letter with details of Palmer's observations arrived. Cf. WALLIS–OLDENBURG 24.IV/[4.V].1669.

70.
WALLIS to JOHN COLLINS
Oxford, 23 March/[2 April] 1668/9

Transmission:

W Letter sent: CAMBRIDGE *Cambridge University Library* MS Add. 9597/13/6, f. 208^r–208^v (our source). On f. 208^v in Collins's hand: 'Dr Wallis about Mr Pitts' and diagrams of an ellipse and a hyperbola. Postmark: 'MR/24'.—printed: RIGAUD, *Correspondence of Scientific Men* II, 513–4.

Reply to: COLLINS–WALLIS 21?/[31?].III.1668/9.

Oxford March. 23. 1668./9.

Sir,

I gave no Answere to your former letter,[424] because I have yet had no time to prosecute that Inquiry; And for the same reason I can say little to that[425] I received yesterday. Your notion therein, as to the particular smal parts, is true. But it will not hold as to the Aggregate of such parts. For though it be true, that as *Each of* the (Sines or their smal Parallelograms) GH, to GK: so GK, to *the respective* (Tangent or its little Parallelogram) GF: Yet it will not follow, that therefore *All the* GH's, to *as many* GK's; so those GK's, to *All the* GF's: (the proportions of GH to GK being every where various.)

It is	1	.	100	::	100	.	10000.
	2	.	100	::	100	.	5000.
	3	.	100	::	100	.	$3333\frac{1}{3}$.
	4	.	100	::	100	.	2500.
But not	10	.	400	::	400	.	$20833\frac{1}{3}$.

So that I do not yet see how to make that advantage which seemes to bee designed hereby. As to Mr Hobs's undertakings;[426] it wil be time to speak to them, when I see the performance.

---

[424]letter: i.e. COLLINS–WALLIS I/II?.1668/9.

[425]that: i.e. COLLINS–WALLIS 21?/[31?].III.1668/9.

[426]undertakings: cf. COLLINS–GREGORY 15/[25].III.1668/9; TURNBULL, *James Gregory*, 70–3, 71: 'Mr. Hobs hath 3 sheetes in the Presse intituled Quadratura Circuli, Cubatio Sphaerae, et Duplicatio Cubi breviter demonstrata.' Hobbes's *Quadratura circuli, cubatio sphaerae, duplicatio cubi, breviter demonstrata* was published by Andrew Crooke in London in 1669.

70. WALLIS to COLLINS, 23 March/[2 April] 1668/9

I have sent this morning, by More's Wagon, a Bundle of Books, directed to Mr Pits;[427] in which is contained, as is below expressed. You may please to consider of them with Mr Pits, what price to put upon them; or of a like value in other Bookes.

I find (since I came home) that I had some of my Cycloides[428] left; so that you need not be solicitous of returning me that you had. But I wish I had the two Papers of Cuts, & the printed sheet's[429] after N, (which is the last I have.) I am

Sir

Your friend to serve you
John Wallis.

5. Opera Mathematica. pars prima,
 { Oratio inauguralis.
   Opus Arithmeticum.   } each ... 69 sheets
   Contra Meibomium. - -

9. ...... Pars altera,
 { De Angulo Contactus.
   De Sectionibus Conicis.   } each, 52 sheets, & 1 Plate.
   Arithmetica Infinitorum.
   Observatio Ecclipseos.

 } Pot paper, 4°. Pica. With cuts &c.

12. Commercium Epistolicum .... each $24\frac{1}{2}$ sheets.
4. De Cycloide &c .... each 17 sheets.
3 large Plates

8. Elenchus Geometriae Hobbianae   each 9 sh: 2 Plates.
6. Due Correction &c .... each 9 sh: 1 Pl:
3. Hobbiani Puncti dispunctio .... each 3 sh: ......

} Crown paper, 8°. long primer

---

[427] Pitts: i.e. Moses Pitt (1639–97), London printer and bookseller, *ODNB*.

[428] Cycloides: i.e. WALLIS, *Tractatus duo*, Oxford 1659.

[429] sheets: i.e. proof sheets of his *Mechanica: sive, de motu, tractatus geometricus*, then at the press.

70. WALLIS to COLLINS, 23 March/[2 April] 1668/9

ᵛ] For Mr John Collins
next to the three Crowns
in Blooms-bury Market
London.

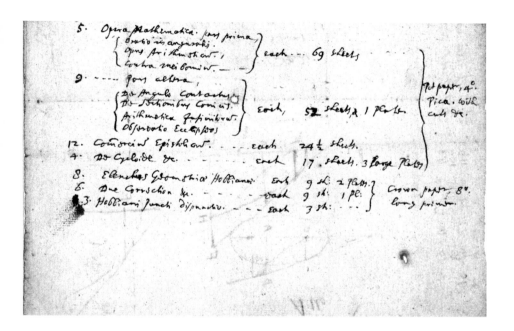

71.
WALLIS to HENRY OLDENBURG
Oxford, 15/[25] April 1669

Transmission:

W Letter sent: LONDON *Royal Society* Early Letters W1, No. 85, 2 pp. (our source). Endorsed by Oldenburg beneath address on p. 2: 'Rec. Apr. 16. 1669. Answ. Apr. 17. 1669'.—printed: OLDENBURG, *Correspondence* V, 493–4.

Answered by: OLDENBURG–WALLIS 17/[27].IV.1669.

Oxford Apr. 15. 1669

Sir

I have not seen Mr Childries[430] Britannia[431] Baconica, nor know what the Hypothesis is, or the Almanack, to which he refers.[432] What mine is, you may acquaint him: where hee will see the Moons Apogaeum & Perigaeum concerned: But mine will rather expect high Tides at both of them, caeteris paribus: This being but one of the considerations to be taken notice of amongst many. But, whether it be mentioned in my first[433] or last letter about that business, I do not well remember[.] You may do well to send him both of them. Mr. St.'s[434] design will do the Society no great hurt. Onely hee doth, it seems, thus requite the Societies making a Dr of him,[435] & speaking

---

4 or the Almanack *add.*
9 You ... of them. *add.*

---

[430] Childries: i.e. Joshua Childrey (1625–70), astrologer and antiquary, *ODNB*.

[431] Britannia Baconica: i.e. CHILDREY, *Britannia Baconica: or, the natural rarities of England, Scotland, & Wales. According as they are to be found in every shire*, London 1660.

[432] refers: i.e. in CHILDREY–OLDENBURG 22.III/[1 April] 1668/9; OLDENBURG, *Correspondence* V, 454–6. Oldenburg had evidently conveyed all or part of this letter to Wallis. The Almanack referred to which Childrey refers is his *Syzygiasticon Instauratum. Or, an Ephemeris of the places and aspects of the planets, as they respect the ☉ as center of their orbes, calculated for the year of the incarnation of God, 1653*, London 1653.

[433] first or last letter: i.e. WALLIS–BOYLE, 25 April/[5 May] 1666 and WALLIS–OLDENBURG 18/[28] July 1666; WALLIS, *Correspondence* II, 200–22 and 251–62.

[434] Mr. St.'s: i.e. Henry Stubbe (1632–76), who, in 1669, began a massive attack, at first privately, and then through publications, on the aims and ideals of the Royal Society as these had been expressed in Sprat's *History of the Royal Society* (1667) and Glanvill's *Plus ultra* (1668).

[435] making ... him: in the title to Stubbe's second article containing observations made on his voyage to Jamaica in 1665, 'An Enlargement of the Observations, Formerly Publisht

71. WALLIS to OLDENBURG, 15/[25] April 1669

of him so oft as a person more considerable than those who know him have reason to think him.

I forgot, when I was last in London, to speak to you about a note[436] I had from you in January last concerning an Anatomical Operator; For though I do not know one presently to recommend; yet when I know the terms more particularly I may possibly give you a further account. You mention 20^l & diet; the person to have a good hand at it, diligent, & humble. I suppose by a good hand, you mean onely one dextrous & intelligent; and who by experience may come to be more knowing: But if you mean one allready experienced & perfect at it; such a one will either hardly be found, or else will think of putting himself into some standing way of livelyhood: By humble; if it be meant meerly in the condition of a servant: it will be the lesse inviting to ingenious persons & scholars of any standing, who will scarce be willing to go from the University to such a condition. You do not particularly say whether it be an Operator to the Society (though I suppose that be your meaning) or to some eminent person of it; nor with whom he is to have his diet. When I know more particularly what the terms intended are, & what to be expected I shall the better know how to direct my inquiries. If it be intended of one to Goe in an ingenious way, as a gentleman, but of no proud or peevish nature: it will be proper to propose it to some Master of Arts or Batchelour of Arts, who have studied Anatomy & have some competent skill in it. If to be in a meaner servile imployment; it must be some servitour (for others will hardly condescend to it) & wee shall hardly find one, in that condition, very fit for it; but possibly such as may bee in a capacity of being taught. In either capacity, as I shall be directed, I shall be very willing to do the Society what Service I can in finding out one for their turn. Resting

Your friend to serve you
John Wallis.

---

6 I (*1*) will (*2*) may
7 to (*1*) be (*2*) have
8 onely *add.*

Numb. 27', printed in *Philosophical Transactions* No. 36 (15 June 1668), 699–709, he is called 'Dr. Stubbes'.

[436]note: this note from Oldenburg to Wallis on the topic of an anatomical operator is now missing. Numerous physiological experiments were carried out at meetings of the Royal Society in December and January 1668/9, and at a meeting of Council on 18 March 1668/9 it was declared 'that it was necessary to have another curator'. See BIRCH, *History of the Royal Society* II, 355.

You may please to send me your
thoughts of it by the first opportunity.

These
For Mr Henry Oldenburg,
5 in the Palmal near
St James's
London.

[2]

72.
HENRY OLDENBURG to WALLIS
17/[27] April 1669

Transmission:

Manuscript missing.

Existence and date: Indicated by Oldenburg's endorsement on the cover of WALLIS–OLDENBURG 15/[25].IV.1669.
Reply to: WALLIS–OLDENBURG 15/[25].IV.1669.

This letter possibly enclosed HEVELIUS–WALLIS [11]/21.III.1668/9, which had been sent by Hevelius with a letter to Oldenburg of the same date. Cf. WALLIS–OLDENBURG 24.IV/[4.V].1669.

73.
WALLIS to HENRY OLDENBURG
Oxford, 24 April/[4 May] 1669

Transmission:

W Letter sent: LONDON *Royal Society* Early Letters W1, No. 86, 2 pp. (our source). Endorsement by Oldenburg beneath address on p. 2: 'Rec. April 26. 69. Answ. May 1. 69'. Postmark on p. 2: 'AP/26'.—printed: OLDENBURG, *Correspondence* V, 499–500.

Reply to: OLDENBURG–WALLIS 17/[27].III.1668/9.
Answered by: OLDENBURG–WALLIS 1/[10].IV.1669.

73. WALLIS to OLDENBURG, 24 April/[4 May] 1669

Oxford Apr. 24. 1669.

Sir,

I received yours[437] with that[438] from Hevelius inclosed; for which when you write you may please to return him my thanks. The sume of it (beside thanks to mee for the papers I sent[439] him;) is that hee desires Mr Horrox's observations, which hee is willing to print with his own. And desires also a sight of my Hypothesis de Marinis aestibus. It should seem that hee hath not our Transactions:[440] & therefore (since, I suppose, he sends his Books to the Society) it might be convenient that the Society send him a sett of those Transactions from the first: Which hee would take as a respect from them. Hee says, the Reason hee did not send (with the rest) his Hypothesis[441] of Saturn, was, because hee had none left; but addeth, that hee did however send of them to each of the Universities one by you, & inquires whether it were received. (Which I suppose was a mistake of his; for there came none of these hither.) He says allso, that he hath since recovered some of them, & designes to send them. And that the reason he mentioned not Mr Palmer,[442], was, because what was sayd of that business was written before mine came; & he afterward forgot to insert it; but will do it with the next occasion hee hath to speak of it. I am glad Hevelius will publish[443] those observations of Horrocks; but am sorry they are missing (as Mr Collins wrote[444] mee word.) I find (by the perusing your letters to mee) that they were sent to yourself, & communicated to the Society, from whom you sent[445] mee their thanks. I had first sent part of it; which was returned to mee, to adde more, & to draw up a narrative of his life: which I did, & sent all back again; & had their finall thanks[446] with their resolution to have them printed. And, I

---

19 the (*1*) first (*2*) next

---

[437]yours: i.e. OLDENBURG–WALLIS 17/[27].IV.1669.

[438]that: i.e. HEVELIUS–WALLIS [11]/21.III.1668/9.

[439]sent: i.e. the transcription of Ptolemy's catalogue of stars which Wallis enclosed with WALLIS–HEVELIUS 15/[25].I.1668/9.

[440]Transactions: Wallis's hypothesis on the flux and reflux of the sea was published in *Philosophical Transactions* No. 16 (6 August 1666). See WALLIS–BOYLE 25 April/[5 May] 1666; WALLIS, *Correspondence* II, 200–22.

[441]Hypothesis of Saturn: i.e. HEVELIUS, *Dissertatio de nativa Saturni facie*, Danzig 1656.

[442]Palmer: i.e. John Palmer (1612–79), rector of Ecton in Northamptonshire.

[443]publish: see HEVELIUS–OLDENBURG [11]/21.III.1669; OLDENBURG, *Correspondence* V, 440–2, 441.

[444]wrote: possibly in the now missing letter COLLINS–WALLIS 21?/[31?].III.1669.

[445]sent mee their thanks: i.e. OLDENBURG–WALLIS 29.IX/[9.X].1664; WALLIS, *Correspondence* II, 162–3. See BIRCH, *History of the Royal Society* I, 470–1.

[446]finall thanks: there is no record of this, nor of the Royal Society's resolution to have

73. WALLIS to OLDENBURG, 24 April/[4 May] 1669

remember, your Printer[447] had it communicated to him, but first deferred, and after declined the printing of it. I mention these particulars that the consideration of them may possibly be some direction where to inquire after them if they bee not yet come to hand.

I do not hear that St.[448] hath any incouragement from hence for his design; nor do I know that this university is disaffected to the Society. If any single persons be so; I suppose they are very few; & will not openly aver it. Nor do I know that our designes do crosse one another. But I would not have you insist on that argument, (which you seem to lay some store upon,) that the University doth not meddle with Experimentall Phylosophy. For it is a great mistake, (Experimentall Philosophy being as properly appertaining to the constitution as any other; though, perhaps, in former times it have not been so much in fashion, as it now is here as well as with you:) You should rather say: It is no disparagement to the Universityes, for others to pursue philosophicall studies allso. The other insinuation will do hurt. When the Florentine Prince &c. comes[449] I hope they will find all due respect here. I am

Yours &c

John Wallis.

For Mr Henry Oldenburg, [2]
at his house in the
Palmal near
St James's
London.

---

2 after (*1*) delayed (*2*) declined
13 as it...with you: *add.*

Horrox's papers printed. See WALLIS–OLDENBURG 8/[18].V.1665 (WALLIS, *Correspondence* II, 184) and WALLIS–OLDENBURG 7/[17].V.1666 (WALLIS, *Correspondence* II, 223–5). Cf. BIRCH, *History of the Royal Society* II, 473.

[447]Printer: i.e. John Martyn (1617/18–80), printer to the Royal Society and bookseller at The Bell in St Paul's Churchyard, London, *ODNB*.

[448]St. Stubbe: see WALLIS–OLDENBURG 15/[25].IV.1669.

[449]comes: on Cosimo de' Medici's visit to Oxford see LAMPLUGH–WILLIAMSON 5/[15].V.1669.

74.
HENRY OLDENBURG to WALLIS
1/[11] May 1669

Transmission:

Manuscript missing.

Existence and date: Indicated by Oldenburg's endorsement on the cover of WALLIS–OLDENBURG 24.IV/[4.V].1669.
Reply to: WALLIS–OLDENBURG 24.IV/[4.V].1669.

75.
THOMAS LAMPLUGH to JOSEPH WILLIAMSON
Oxford, 5/[15] May 1669

Transmission:

C Letter sent: KEW *The National Archives* SP 29/259, No. 183, 2 pp. (p. 2 blank). On p. 2 in unknown hand: 'Pr. Toscanys Entertainment'.

Lamplugh reports on Cosimo de' Medici's ceremonial visit to Oxford at the beginning of May 1669, as part of which Wallis gave a lecture on architecture. See also WOOD, *Life and Times* II, 156–61, and CRINO, *Un principe di Toscana*, 97–103.

Dearest Sir

Yesterday the Prince[450] of Tuscany was solemnly received here; a Degree was preferrd to him, but he thankfully declined it. His own Physitian,[451] who was Dr. in Bononia[452] was admitted ad eundem here; & that was all that was done in the Convocation, besides a Speech by the publique Orator[453]

---

3 who *add.*

[450]Prince: i.e. Cosimo III de' Medici (1642–1723), who had set sail from Portugal for England on 9 March 1668/9 and remained in the country until 4 June 1669 (old style).
[451]Physitian: i.e. Giovanni Battista Gornia (17th Century), professor of medicine at Pisa and personal physician to Cosimo de' Medici. On his incorporation as doctor of physic at Oxford see WOOD, *Fasti* II, 310. See also GORNIA–OLDENBURG [30.VI]/10.VII.1669; OLDENBURG, *Correspondence* VI, 86–8, and BIRCH, *History of the Royal Society* II, 374–5.
[452]Bononia: i.e. Bologna.
[453]Orator: i.e. Robert South (1634–1716), Church of England clergyman and theologian, University orator and canon of Christ Church, Oxford, *ODNB*.

75. LAMPLUGH to WILLIAMSON, 5/[15] May 1669

& some ordinary Dispensations proposed to show the Prince the manner of proceedings in that House. We had no Creation; 'twas not thought fitt, for many obscure persons would have crept out in the crowd to the great dishonour of Degrees. The Vicechan:[454] & Dr.s in their Scarlet, upon the ringing of S. Marys bell met in the Vestry, & went from thence to wait upon the Prince at his Inne, which was the Angell. Where the Vicech: made a short Speech to thank him for honouring the University with his Presence, & to know his command. After that the Vicech. & several DDrs. waited upon him to Madgdalen, where at the entrance a short speech was made by Mr Bayly.[455] From thence he went to All Soules; from thence to S. John's where he was received by a Speech[456] in the Library. From thence he rid in his Coach to Ch: Church where a speech[457] was made in the Hall; & at his comeing down the staires, severall of the students, who stood in files prelected him with short Odes, Sapphiques &c. like so many pop. gunns, which took well.

After dinner, the Vicech: & all the DDrs. in their formalities, went from the Convocation to the Angell & attended the Prince in the Street, (the Scholars standing orderly on both sides, while they passed thorow). In the Convocation House there was a Seat reard up on high as the Vicech: Seat, & a chair of State for the Prince. The House was very full, but the Masters sat very orderly in their seates, no crowding at all. From thence he went to the library, where he was receivd with a speech,[458] & visited there, with which he was well pleased. From thence he went to hear Dr. Wallis read, who pitched upon a peice[459] of Architecture, which took well. From there he went to the morall philosophy School, where 2[460] of Qu. Coll: Tabarders[461] disputed on that new philos. & did it exceeding well. At last

---

[454] Vicechan: i.e. John Fell (1625–86), vice-chancellor of the University of Oxford from 1666 to 1669, *ODNB*.

[455] Bayly: i.e. Thomas Baylie (Bayly) (?–1699), Fellow of Magdalen College.

[456] Speech: i.e. by an unnamed gentleman commoner. See WOOD, *Life and Times* II, 160.

[457] speech: i.e. by William Wigan (?–1700), Student of Christ Church.

[458] speech: i.e. by Thomas Hyde (1636–1703), oriental scholar and Bodley's Librarian, *ODNB*.

[459] Architecture: cf. WALLIS, *Mechanica: sive, de motu, tractatus geometricus* III, 589–91; *Opera mathematica* I, 953–5. See NEWMAN, *The Architectural Setting*, 174; WREN, *Parentalia: or, memoirs of the family of the Wrens*, London 1750, 338–9.

[460] 2: i.e. John Mill (?–1707) and Henry Smith (?–1673). According to Wood, the questions disputed were 'An stellae Medicaeae sunt lunae Jovis? Aff.' and 'An detur vacuum? Neg.'. See WOOD, *Life and Times* II, 161.

[461] Tabarders: i.e. bachelors. The name evidently derives from the short gowns ('tabards') which they wore.

77. HENRY OLDENBURG to WALLIS, 8/[18] May 1669, enclosure

he heard a lesson[462] in the Musick School, & so returnd to his lodging. I doubt not but you will have the relation from far better hand; so pardon this hasty scribble, I am just near setting my Wife[463] & children towards Dorsett, this to see if they can recover their healthes; so I write this with almost one foot in the stirrup. I am

Dear Sir

Alban Hall Sincerely yours
May. 5. 69. T. L.

76.
HENRY OLDENBURG to WALLIS
8/[18] May 1669

Transmission:

Manuscript missing.

Existence and date: Mentioned in and answered by WALLIS–OLDENBURG 10/[20].V. 1669.
Enclosure: William Neile's Hypothesis of Motion.

77.
HENRY OLDENBURG to WALLIS
8/[18] May 1669, enclosure:
William Neile's Hypothesis of Motion

Transmission:

C Original paper (subsequently revised): LONDON *Royal Society* Classified Papers III (i), No. 53, f. 1ʳ–9ᵛ (our source). At foot of f. 1ʳ in Oldenburg's hand: 'Mr William Neile Hypoth. of Motion. Read April 29. 1669. Entered R.B. 4. 59'. Together with f. 9ʳ all versos blank, except f. 6ᵛ, which contains addition to f. 7ʳ. On f. 7ᵛ in unknown hand: 'Mr William Neiles Theory of Motion read at the Society'; on f. 9ᵛ in unknown hand: 'Mr Neile's Theory of Motion read before the Society Apr. 29. 1669. Ent[e]red.'—printed: OLDENBURG, *Correspondence* V, 519–24 (Latin original); 524–8 (English translation).

---

[462]lesson: i.e. musical performance. First Mr Chrispine, an undergraduate of Christ Church, presented a song; then Mr Withie performed a division on the bass-viol. See WOOD, *Life and Times* II, 158.

[463]Wife: i.e. Katherine Lamplugh (1633–71), daughter of the mathematician Edward Davenant of Gillingham, Dorset. Davenant instructed both of his daughters, and John Aubrey, in algebra. See AUBREY, *Brief Lives* I, 201.

77. HENRY OLDENBURG to WALLIS, 8/[18] May 1669, enclosure

c^1 Copy of original paper (sent to Wallis) (manuscript missing).
c^2 Part copy of revised paper: LONDON *Royal Society* Register Book IV, pp. 59–67.

Enclosure to: OLDENBURG–WALLIS 8/[18].V.1669.

William Neile's theory of motion was read at the meeting of the Royal Society on 29 April 1669, after which it was ordered that the theory be entered in the Register Book. Neile was requested to complete the theory and 'to consider how to verify his principles by experiments, and to accommodate them to the rules of Dr. Wren and Monsr. Huygens'. At the meeting the following week, Neile was reminded of this and in reply indicated that the work had already been started; see BIRCH, *History of the Royal Society* II, 361, 362. The next day, Neile wrote to Oldenburg that he would revise and stylistically polish the theory sometime in the future (see NEILE–OLDENBURG 7/[17].V. 1669; OLDENBURG, *Correspondence* V, 517–18). This evidently prompted Oldenburg to send Wallis a copy of the theory he already had at his disposal rather than waiting for the promised revised version. The content of the original paper cannot be determined with certainty. After the sentence ending 'relinquenda censeo' in the middle of f. 5^r the ink is visibly different and corresponds to that used for making stylistic improvements to earlier parts of the text. This suggests that these passages and corrections were added sometime afterward and were not included in the copy which Oldenburg sent to Wallis on 8 May. The Register Book copy incorporates Neile's corrections, but leaves out all of the text in the revised paper from the beginning of f. 6^r to the end of f. 7^r. The final passages contained on f. 8^r (with the exception of 'debet ergo S. reverti velocitate $SE + 2AE$') is designated in the Register Book as an 'Appendix', no doubt reflecting their being a later addition.

Sint duo cubi aequales A. et B. (quos assumo solum melioris explicationis gratia) cubi autem sint istiusmodi ut nulla habeant inania spatiola interspersa inter eorum particulas neque ullum motum intestinum. Jam supponantur duo cubi aequali velocitate moveri, donec concurrant ita, ut tota
5 quadrata superficies unius, integrae quadratae superficiei alterius eodem tempore applicetur; tunc sequetur, ut in instanti concursus ambo moveri desinant.

Fac cubum A. advenisse a manu dextra ac B. a sinistra jam dico, vel uterque moveri desinet vel ambo reflectentur uti satis liquet. Nunc autem si
10 A. motum impertierit cubo B, necesse est, ut cubus A. moveatur simul cum B. eodem tempore pauxillum versus sinistram, id est, aliquantulum comitetur cubum B. illorsum; nam si supponamus, cubum A. protinus sisti eodem tempore, quo contingit cubum B, neque omnino progredi, nulla prorsus erit ratio, quare ullum motum cubo B. communicaret vel adferret. Cur enim B.
15 suo loco cederet, quum nihil urgeret neque eum impelleret? Sin vero cubus A. tempore concursus admittatur moveri paululum versus sinistram, sequetur ob paritatem, quae omni respectu intervenit inter duos cubos, cubum

2 sint istiusmodi ut *add.*
17 cubos, (*1*) quod cubus B eodem tempore movebit (*2*) cubum ... iri

77. HENRY OLDENBURG to WALLIS, 8/[18] May 1669, enclosure

B eodem tempore motum iri aliquantulum versus dextram eadem de causa, quae cubum *A*. tempore concursus versus sinistram advexit. *B*. autem prius supponebatur illo tempore moveri versus sinistram deduxit. *B*. autem prius supponebatur illo tempore moveri versus sinistram; Quae duae res simul consistere nequeunt. Ergo amborum motus desinet tempore concursus.

Nunc supponamus cubum *A* moveri velocitate majori et cubum *B*. minori; nihilo secius amborum motus desinet tempore concursus: quoniam *B*. vel aliquem sortietur effectum versus inhibendum cubum *A* aut nulla omnino velocitas ei inhibendo sufficiet, quia quantalibet velocitas major erit solummodo velocitas cubi *B*. multiplicata vel adaucta sed si cubus *B*. desinat moveri versus dextram tempore quo cubum *A* contingat, nulla erit ratio quare vel minimum effectum supra cubum *A*. sortiretur. Sequetur igitur quod cubus *B* tempore concursus aliquantulum moveri debeat versus dextram sed ob eadem rationem cubus *A* pellet cubum *B* pauxillum versus sinistram et proinde *B* habebit duos motus contrarios simul quod est impossibile (scilicet ut eadem materiae pars duabus diversis viis eodem tempore simul insistat) ergo amborum motus desinet tempore concursus.|

Ob parem rationem consequetur quosvis motus differentes qui simul occurrunt se mutuo inhibituros licet non directe opponantur quod consideranti satis forte patebit. Hic igitur solum insinuandum duxi. Quia opus esset quamplurium forsan verborum dispendio rem totam penitus elucidare.|

Nunc supponatur cubus *A*. ferri contra *B*. quiescentem sequetur ut post appulsum ambo cubi procedant versus sinistram eadem velocitate qua prius cubus *A* movebatur (supposito interim nullum aliud corpus quidquam in eos agere.) Nam si *A*. non impertiatur aliquantillum motus cubo *B*, nulla plane velocitas ei movendo sufficiet ob rationem praedictam: sin autem *B*. prorsus inhabilis sit ad impediendum motum cubi *A*, motus cubi *A* eodem prorsus tenore continuabitur: *B* autem quum sit quiescens, nullam quidem habere potest virtutem agendi in cubum *A*: si enim ullam habere possit; illa sane virtus sive potentia non poterit esse motus, quoniam *B*. est in statu quietis. Si igitur fieri possit, sit propensitas quaedam ad motum; quum autem *B* penitus quiescat, quaenam ratio jubeat illum ad unam potius quam

2 concursus versus sinistram (*1*) deduxit (*2*) advexit
20 satis (*1*) forsan (*2*) forte
21 forsan *add.*
28 prorsus *add.*
30 potentia (*1*) non possit (*2*) non poterit
31 igitur *add.*
32-1 illum (*1*) una potius quam altera semita motus (*2*) ad unam ... semitam motus

77. HENRY OLDENBURG to WALLIS, 8/[18] May 1669, enclosure

alteram semitam motus propendere? An hac via atque illa atque omnia via quaquaversum simul propendere potis erit? Par igitur est, concludi, B nullam habere potentiam agendi in cubum A, et consequenter cubi A motus eodem tenore perseverabit; nequit enim sibi ipsi esse impedimento. Oportet igitur cubum B propulsari a cubo A eadem plane velocitate, qua antea cubus A adventabat, id est, cubus A motum suum cubo B communicabit sive similem in eo generabit. Atque si A attingat B, eadem plane via jamdudum moventem, nihil erit rationis, quare cubus A minorem haberet potentiam propellendi cubum B, quam in priore casu; majorem vero illi velocitatem impertire non potest, quia cubus A suo ipsius motui nullo mode adiumento esse valeat, aut eum ullatenus augere quod etiam de cubo B dici potest.

Hanc doctrinae partem me libris[464] domini Hobs acceptam referre lubens agnosco.| [4^r]

Quum ex praecedentibus colligatur ubicunque sit resistentia sive reactio corporis, ibi motum inesse, solam utpote causam quae talem effectum naturaliter producere valeat, ob ineptitudinem materiae quiescentis ad motum ullo modo impediendum; quae utique continuo cedat cuicunque corpori impellenti sine ulla morula aut repugnantia; et quoniam hanc resistendi facultatem omnibus apud nos corporibus plus minusve inesse plerumque constat, idcirco ut motum illis attribuamus prorsus conveniet. Quae quidem resistentia cum undiquaque opponatur cuilibet impetui externo, necesse est ut innumerabili pene varietate instruatur: motus iste intestinus minutissimarum particularum in quibuslibet corporibus, quae scilicet renitatur cuicunque impulsui externo, tanta quantam saepe conspicimus promptitudine, ut ictu oculi longe citius corpora ab invicem reflectantur; et tamen in firmis corporibus ea est particularum coarctatio et constipatio, ut sese ab invicem expedire nequeant, sed mutuis ac mira varietate alternis motibus: quasi irretitae ac illaqueatae desineantur; neque interea totum corpus seipsum transferre valeat sine impulsu externo, cum motus particularum intestinus in nullam partem potius quam in aliam, ob varietatem praedictam, propendeat. Jam si suponamus, corpus A. esse congeriem minutissimarum particularum, summa varietate moventium invicem ac motarum cum intervallis quietis, secundum regulas praedictas, quae quidem particulae modo uniantur modo dividantur ab invicem per motum mox inhibentem mox

2 propendere potis (*1*) est (*2*) erit?
6 cubo B (*1*) propagabit (*2*) communicabit
19 apud nos *add.*
22 varietate (*1*) donetur (*2*) instruatur

---

[464]libris: presumably HOBBES, *De corpore*.

77. Henry Oldenburg to Wallis, 8/[18] May 1669, enclosure

discutientem; sequetur ut unaquaeque particula in parte temporis exilissima moveatur hac illac sursum deorsum, prorsum retrorsum, ac uno verbo, pene undiquaque versum neque tamen multum dimoveatur respectu circumvicinarum particularum.| Nihilominus aliquae particulae exteriores possunt interim avolare ac aliae vicissim a corporibus vicinis subingredi corpus A. Nunc si concedamus, minutissimas istas particulas una aliqua via saepius quam ulla alia promoveri, idque forsan majori velocitate ac paucioribus aut brevioribus morulis, quam antea solebant, consequetur ut totum corpus A brevissimo temporis spatio interposito transferri possit secundam regulas praedictas cum intervallis quietis. Plura hic adiiesse hac de re liceret, at mihi animus non est, omnibus objectionibus occurrere, sed ea tantum indicare, quae ad sensa mea enucleanda necessaria indicabam, quae si non omnibus satisfaciant, fortasse in causa erit quod mentem meam (uti in re nova facile solet evenire) prima facie haud recte assequantur, vel quod illi id possibile judicent, quod mihi omnem captum superare videtur; ut si quis dicere velit, corpus posse agere in distans nulla intercedente materia; corpus praeditum esse anima, sensu, ratione, aut similia diceret ego nihil haberem, quod regererem nisi hoc tantum, mihi videri non posse concipi haec tanquam possibilia. Unicuique igitur in tali casu sensa sua libera esse relinquenda censeo. Ego non possum praevidere quas objectiones alius in animo habeat; ea solum in praesenti congerere potui quae si ab aliis promerentur (me hujus negotii penitus inscio) eos satis, hanc esse (qualis hic traditur) naturam motus, evicisse putarem: Mihi autem valde conscius sum, quam pronum sit labi ac errare in argumento adeo involuto ac recondito, ac tam parum adhuc a doctis ingeniis trito, quale nempe est minutissimarum particularum motus, quem solum oculo rationis perspicere valeamus. Ei igitur gratias habebo qui

3 versum |(si ita dicam) *del.*| neque
10 at mihi (*1*) non in animo (*2*) animus non
16 corpus (*1*) habere animam, sensum, rationem, ac similia (*2*) praeditum esse anima, sensu, ratione, aut similia
18 regererem (*1*) , sed no *breaks off* (*2*) nisi hoc
21 congerere (*1*) satis habui (*2*) potui
21 ab aliis (*1*) exhiberentur eos satis (*2*) promerentur ... eos satis
23 pronum (*1*) est (*2*) sit
24 errare in (*1*) mat *breaks off* (*2*) subjecto (*3*) argumento adeo involuto ac (*a*) inconspicuo (*b*) recondito
25 trito, (*1*) qualis (*2*) quale nempe

77. Henry Oldenburg to Wallis, 8/[18] May 1669, enclosure

mihi errorem meum (si talis fuerit) methodo ac ratiocinio, quibus illis hic insisto clarioribus monstraverit. Mei non est instituti omnia quae ad motum spectant explicare quod immensi operis esset sed principia solum motus siquidem possim eruere. Caetera iis qui peritiam meipso longe majorem habent tractanda relinquens.| [6ʳ]

In primo casu[465] corporum aequalium, necesse est ut ambo corpora R. et S. inhibeantur (ambo enim procedere nequeunt) et cum hoc fiat utrinque potentia aequali ei quae illa ad concursum advexit oportet ut particularum motus intestinus in corpore S. cum utrinque aequali potentia urgeatur ac proinde progredi non potest revertatur versus dextram eadem velocitate qua prius properabat ad sinistram rivulus enim iste particularum (ut dicere liceat) cum derepente a priore cursu inhibeatur necesse est ut retrorsum resiliat pari velocitate quoniam repentina ista ablatio motus versus sinistram propulsatione versus dextram eodem tempore perseverante (quippe quae minime impediatur) idem motus praedominium relictura est versus dextram qualis fuit antea versus sinistram referretur igitur corpus S. versus dextram pari velocitate qua prius adveniebat versus sinistram et ob eandem causam corpus R simili modo revertetur versus sinistram.

In 2.° casu potentia corporis R augeatur velocitate AE. et potentia corporis S. minuitur velocitate AE. quod tantumdem est ac si velocitas corporis R. augeretur $2AE$ vel OE. S. igitur debet reverti velocitate $SE + 2AE$ vel OE et ob subtractionem potentiae OE a corpore S. Corpus R revertetur velocitate $RE - 2AE$ vel OE.

In 3.° casu potentia corporis R augeatur velocitate AE ergo corpus S. revertetur velocitate $SE = 0 + OE$ et potentia corporis S. minuitur velocitate AE proinde corpus R revertetur velocitate $RE - OE = O$.| [7ʳ]

In 4.° casu Potentia corporis S. versus sinistram minuitur velocitate OE quia velocitate AE minuitur ob additionem velocitatis AE, factam ad corpus R, et minuitur velocitate AS ob subtractionem velocitatis AS. factam

---

1-2 methodo (*1*) clariori (*2*) illa quam hic insisto clariori (*3*) ac ratiocinio ... clarioribus
7 (ambo ... nequeunt) *add.*
10 progredi non (*1*) possit (*2*) potest
16 versus dextram (*1*) eadem celeritate (*2*) pari velocitate
24 augeatur velocitate OE *corr. ed.*
25 minuitur velocitate OE *corr. ed.*
27 versus sinistram *add.*
28-3 quia velocitate ... versus sinistram |$2AE$. vel OE. *corr. ed.*| *add.*

---

[465] In primo casu: in this and the following cases Neile refers to Wren's theory of motion, published in *Philosophical Transactions* No. 43 (11 January 1668/9), 867–8.

77. HENRY OLDENBURG to WALLIS, 8/[18] May 1669, enclosure

a corpore S; et minuitur velocitate SE ob additionem velocitatis SE. factam ad corpus S versus dextram; minuitur ergo Potentia corporis S. versus sinistram $2AS + AE = RE$. R igitur post apulsum habebit velocitatem $OE - RE$. id est corpus S. deficiet a potentia sufficienti ad inhibendum corpus R. velocitate OR. Igitur Corpus R progredietur versus dextram velocitate OR. atque etiam potentia corporis R augetur velocitate AE et augetur subtractione velocitatis AS a corpore S. et rursus augetur velocitate OR quia potentia corporis S. sinistram versus minuitur velocitate $SE = OR$ igitur Potentia corporis R augetur AE debet ergo corpus S. promoveri versus dextram velocitate SE (quam prius habuit) $+RS$.

Si quis error in hoc calculo reperiri contigerit is nihil est necesse ut veritati praedictorum principiorum ullo modo officiat agnita enim (quaenam sit) calculi veritate, causa forsan haud difficulter eruetur.

In 1.° casu inaequalium velocitas ac magnitudo utrinque se mutuo compensabunt proinde potentia R et S aequatur. Caetera priore methodo absolvi possunt.|

Ubicunque corpora R et S. moventur velocitatibus RE. et SE ita ut potentia sint aequali, ibi revertentur iisdem velocitatibus: quoniam necesse est ut se mutuo inhibeant a priori progressu ob paritatem virium; motus autem intestinus particularum cum subito impediatur in uno cursu continuo, proruet in contrariam partem pari velocitate: quantum enim aufertur motui particularem intestino versus unam partem, tantum additur motui intestino in contrariam partem ob ablationem impedimentorum; tanto enim plures particulae eodem tempore simul ferentur in contrariam partem.

In caeteris casibus sic argumentari licet. Siquidem corpus R moveretur tali velocitate ut potentiam, aequalem potentiae corporis S. obtineret. Corpus S. reverteretur priori sua velocitate, id est velocitate SE. sed potentia corporis R augetur velocitate AE sibi addita, atque iterum augetur velocitate AE ob ablationem velocitatis AE. a corpore S: tantum enim potentiae additur corpori R quantum aufertur a corpore S: igitur potentia corporis R.

3 $RE - OE$ *corr. ed.*
4 a potentia (*1*) idonea (*2*) sufficienti
5 atque etiam *add.*
9 augetur OE *corr. ed.*
10 velocitate SE |(quam prius habuit) *add.*| $+OE$ *corr. ed.*
11 reperiri (*1*) contigat (*2*) contigerit
17 R et S. (*1*) aequali velocitate concurrunt (*2*) moventur
20 continuo *add.*
26 corporis S. (*1*) consequeretur (*2*) obtineret
30 corporis R. (*1*) excedit (*2*) superat

superat potentiam corporis S velocitate $2AE$: debet ergo S. reverti velocitate $SE + 2AE$ tantum enim additur motui reversivo intestino (ut ita dicam:) Cum autem potentia corporis S deficit a potentia aequali potentiae corporis R. velocitate $2AE$. debet proinde R. reverti velocitate $RE - 2AE$. Si $REerit = 2AE$ tunc corpus R non omnino revertetur ob defectum potentiae in corpore S. ac si RE velocitas minus sit quam $2AE$, quantum deficit RE a $2AE$, tantum deficiet corpus S. a potentia sufficienti ad inhibendum corpus R. R igitur progredietur ea velocitate qua deficit S.

78.
WALLIS to HENRY OLDENBURG
Oxford, 10/[20] May] 1669

Transmission:

W Letter sent: LONDON *Royal Society* Early Letters W1, No. 87, 4 pp. (p. 3 blank) (our source). Endorsement by Oldenburg beneath address on p. 3: 'Rec. May 12th 69.' At 90° in unknown hand: 'Wallis about Neile. 69'. Postmark on p. 3: 'MA/12'.—printed: OLDENBURG, *Correspondence* V, 540–2.

Reply to: OLDENBURG–WALLIS 8/[18].V.1669.

This letter constitutes Wallis's reply to William Neile's hypothesis or theory of motion, a copy of which Oldenburg had enclosed in his letter of 8 May.

Oxford May. 10. 1669.

Sir,

I thank you for yours[466] of the 8th instant; which I received this day, with the Paper inclosed,[467] of Mr Neiles Hypothesis. His principles & mine be not the same; as will easyly appear by comparing both. What the differences are, I had rather hee take out of my hypothesis, (from which I do not yet see

---

2 enim additur (*1*) (si ita dicam) motui reversivo, cum (*2*) motui reversivo ... dicam:) Cum
3-4 potentia corporis S (*1*) movuntur (*2*) deficit ... corporis R
4 debet (*1*) ergo (*2*) proinde
5 Si RE |erit *add.*| $= 2AE$
11 which ... this day, *add.*

---

[466] yours: i.e. OLDENBURG–WALLIS 8/[18].V.1669.
[467] inclosed: i.e. a copy of William Neile's hypothesis of motion.

78. WALLIS to OLDENBURG, 10/[20] May 1669

reason to recede,) than by way of Reply to his Paper. Both because that is the shorter way; & because I find men commonly lesse displeased with what is sayd (different from their opinions) at large (without any particular respect to them) then when it is expressely (eo intuitu) directed to them.

    I did think upon what passed before (finding it by your letters so uneasyly assented to, that any Motion can be extinguished) that Mr Neile's opinion had allso been the same with theirs who thought that no motion was lost or extinguished: But I find him in this paper so much of the contrary opinion, that it is but a chance that all is not lost; (&, were it once so, none could begin:) for whenever motions do in the lest cross one another, (by his principles) both are extinguished: and none continued but what meets with either nothing or with what is perfectly at rest or moving precisely the same way, which would in a very short time extinguish all motion, were it not releeved by another principle, That the lest imaginable body in motion, would give motion (& the same celerity which it self had) to a whole world of quiescent bodies if in its way. Which will indeed put quiescent Bodies into a new motion. But, withall, will by degrees, make well nigh all motions to be the same way. For since, by his principles, no Body can be turned out of its way; (but must either move directly onward, carrying all before it with its own swiftnesse, or else be perfectly stopped:) what ever oblique motions come to be stopped (as innumerably many must needs be in every moment,) those bodies so stopped can never be moved again but by falling into the way of some of those yet in motion; so that the *Ways* of Motion (the Plaga toward which they go) must still bee fewer and fewer: till at length they come to be either few; or none; according as some of the last surviving ways of Motion do chance either to hit or misse one another. Nor can this be at all avoided. For so long as he allows nothing either of Resilition, or Obliquation (but that every thing must either go straight on, or quite stop;) it is not possible that any new Way, can be begun which before was not; nor any be restored which is once extinguished.

5 so (*1*) difficult to (*2*) uneasyly
10 (by his principles) *add.*
14 releeved (*1*) with (*2*) by
17 degrees, (*1*) put all bodies into motion (*2*) make ... to be
19 onward, (*1*) or be perfect *breaks off* (*2*) carrying
20 ever (*1*) moti *breaks off* (*2*) oblique motions
21 stopped (*1*) (as will be (*2*) (as innumerably many
25 either *add.*
26 to (*1*) misse one anot *breaks off* (*2*) hit or
28 must *add.*

78. WALLIS to OLDENBURG, 10/[20] May 1669

But I find myself fallen into Argument, before I was aware. I onely adde what follows, to rectify a mistake, wherein I suspect (by what I now find Mr Neile's opinion to be) that hee did formerly mis-apprehend a concession of mine. For when I did assent formerly, that Matter of itself is indifferent as to Rest or Motion; my meaning was no more but this, (and I think it was clearly inough expressed,) That, as Matter at Rest will (without any new cause to keep it in that posture) of itself continue to Rest, till some positive cause put it into Motion; so, being in Motion, will (without any new supervenient cause) of itself continue so to move, till by some positive cause it be stopped, or diverted. A positive cause being| equally necessary, [2] & in the same degree of strength to stop a Motion, as to begin it. Nor will a Body (once in motion) more Rest of itself, than (being at Rest) Move of itself. But if (when he sayd he was glad that I was of his opinion in that point) hee did take my opinion to be (as I perceive his is) that not onely matter as matter, but matter as at Rest, is indifferent as to Rest or Motion, (and so indifferent as that a Body in motion, though never so little, and never so swift, shall carry before all quiescent matter in its way though never so much, with the same swiftness as itself would have moved if it had found nothing in its way;) hee did very much mistake my meaning. For when I sayd that Matter of itself was so indifferent as to Rest or Motion, that being at Rest it would of itself continue to Rest, & being in Motion it would of itself continue to Move, (without any supervenient cause;) till some contrary positive cause did alter its condition: It is manifest that my meaning must needs be, that Matter considered as at Rest, is prone to Rest; like as Matter considered as in Motion, is prone to that Motion; though matter of itself be indifferent either to Rest or to Motion, according as it shall happen to be in this or that condition. Which is so far different from his opinion, that (according to this) when a Body in Motion lights on a Body at Rest, that is no more prone to give Motion to this, than this to give Rest to that. Nor doth this receive any more of Motion from that, than (in its proportion) that receives Rest from this. The abatement of motion (by retardation) in

4 indifferent (*1*) either as (*2*) as to Rest or | to *del.*| Motion
5 (and I ... expressed,) *add.*
10 or diverted. *add.*
13 if (*1*) he did take my opinion (*2*) (when
16 so (*1*) small, (*2*) little,
24 to Rest; *add.*
30 proportion) (*1*) it gives (*2*) that receives

that, being equivalent to the Motion (or Increase of Motion) produced in this. Which is wholly inconsistent with his hypothesis. But, whether his or mine be to take place, it is not my meaning here to dispute.

I am

Sir

Yours &c.
John Wallis.

[4] These
For Mr Henry Oldenburg,
in the Palmal near
St James's
London.

79.
WILLIAM NEILE to HENRY OLDENBURG
London, 13/[23] May 1669

Transmission:

C Letter sent: LONDON *Royal Society* Early Letters, N1, No. 21, 4 pp. (p. 4 blank) (our source).—printed: OLDENBURG, *Correspondence* V, 542–4.

Reply to: WALLIS–OLDENBURG 10/[20].V.1669.
Answered by: WALLIS–OLDENBURG 17/[27].V.1669.
Enclosure to: OLDENBURG–WALLIS 15?/[25?].V.1669.

This letter is William Neile's response to Wallis's criticism of his hypothesis or theory of motion. Although addressed to Oldenburg, it was clearly intended for Wallis. Oldenburg duly sent a now missing copy as an enclosure to his letter to Wallis of 15? May 1669.

London May the 13th. 69.

Sir

I am much obliged to Dr. Wallis that he has been pleased so farre to consider[468] and weigh those principles concerning motion you know my desire is if I could only to find some firmenesse in the foundations the Superstructures I confesse it will passe my Skill to carry much further than the very beginnings but the foundations themselves I thinke are to be grounded upon

---

[468] consider: i.e. WALLIS–OLDENBURG 10/[20].V.1669.

79. NEILE to OLDENBURG, 13/[23] May 1669

reason for I doubt experiment will hardly ever cleare the nature of motion in minute particles for very probably there is no quantity of matter liable to sense but does farre exceed the magnitude of those divisions and subdivisions which are made by motion in minute particles sense may tell us that
a whole considerable quantity of matter is moved out of it's place but sense will not tell us after what fashion the motion is performed in the minute particles of that matter or whether it were with intervalls of rest or no. I am now in hopes to come to some satisfaction about my opinions concerning motion for before I could not well relie upon my owne opinion though it seemed cleere enough to mee and now I hope it will either be confuted or aproved. I confesse Dr. Wallis says right that it is but a chance that all motion is not extinguished. but it is such a chance as I think can never unlesse it were very difficultly (that is to say by an extraordinary disposition of things ordered by the same power that made the matter and motions) bring things totally to a cessation of motion for I conceive it is easie for the motion to be continued upon the principles I have alledged if you will supose the variety of the motion to be great enough in the same plaga or line of motion which you may supose as great as you please and I beleive can never imagine it so great as it is in nature and then I thinke it will be more difficult to imagine how motion can come to stoppe totally in a lumpe of matter then how it can be continued I confesse if the lumpe of matter were quiescent it would be very difficult to introduce so much motion into the minute particles,| as would not soon be extinguish'd because there was [2] not a preceding variety of motion sufficient to keepe it alive but where that innumerable variety of motion is praeexistent I thinke nothing but a power that could bring a chaos to order can putt a totall stoppe to the motions for the least particle of matter no sooner rests but there may be suposed ten thousand thousand other particles ready (as it were) to make warre upon it and to thrust it out of its place or at least to cutte of something from it or if it be at once surrounded on every side it may soon be delivered from that constipation either in whole or in part I can easily imagine particles to stoppe and to continue quiescent if their neighbours would not trouble them but where I presuppose nothing but confusion how to make a durable quiet I think is very difficult. I grant what Dr Wallis says that if a particle stop it cannot be moved on again in the same waie (or Plaga of motion) but by some other particle which was in a preceding motion in that Plaga

17 in the ...of motion *add*.
20 totally *add*.
26 totall *add*.

that is to say if the motion were all at once extinguished in that line or Plaga in infinitum it could never naturally be renewed in that Plaga again, but in any Plaga or line of motion particles are continually moving some a litle forward some a litle backward some stopping at the same time some moving out of the Plaga others coming into it so that if men could no better scape bullets in a battle then particles can continue quiescent all the men on both sides would soon be killed or fewe left alive: I thinke so many accidents must contribute to extinguish motion in any one Plaga that it might last many myryiads of years without stoppe and perchance never stoppe unlesse it please almighty God supernaturally to make it doe so. but I think here is enough said of this matter for one time. I shall only adde that I did not before mistake Dr Wallises opinion because I concluded nothing about it. but I thought indeed (I remember) that he had not so much dissented from the opinion that quiescent matter receives motion without repugnancy as now I find| but I hope this businesse of motion will not alway's lye hidde in obscurity but that wee shall either find quiescent matter to have a power of resistance or none and find motion to have a repugnancy to be stopped or none but I doubt it is not to be determined by sense but by reason yet of late people are growne so distrustful of reason that they take her to bee like a lyer who so often say's that which is false that though he may speake true yet no body ought to relye upon it barely for his word. I am afraid to growe too tedious so I rest

your humble servant
W. Neile.

80.
HENRY OLDENBURG to WALLIS
15?/[25?] May 1669

Transmission:

Manuscript missing.

Existence and date: Mentioned in WALLIS–OLDENBURG 17/[27].V.1669.
Enclosure: NEILE–OLDENBURG 13/[23].V.1669.

2 in infinitum *add.*
7 or ... alive: *add.*
20 though (*1*) they (*2*) he

Oldenburg enclosed a copy of Neile's latest letter which was clearly intended for Wallis. Since Wallis received Oldenburg's letter on 17 May, it is probable that Oldenburg sent it on 15 May, the 16 May being a Sunday.

81.
Wallis to Henry Oldenburg
Oxford, 17/[27] May 1669

Transmission:

W Letter sent: LONDON *Royal Society* Early Letters W1, No. 88, 2 pp. (our source). Endorsed by Oldenburg beneath address on p. 2: 'Rec. May 19. 69.' In unknown hand on p. 2: 'Wallis about Neile. 69'.—printed: OLDENBURG, *Correspondence* V, 550–1.

Reply to: NEILE–OLDENBURG 13/[23].V.1669.

Oxford. May. 17. 1669.

Sir,

As to Mr Neils letter,[469] the copy of which I do just now receive from you:[470] I shal not adde much to what I have sayd of that Hypothesis in my last. The main of which was this; That I know not how to grant his great *Postulatum*, That the smallest particle of Matter in motion, meeting with another though never so great, shall carry it before it (supposing no obstacle by a contrary or decussant motion) at the same rate at which it self did before move. Which *Postulatum* needs to be proved. I could grant, that it might carry it before it; but not, at the same rate. For the same *Vis* which carried A two spaces in a certain time; when it finds another quiescent (suppose equall to it) will carry both, that is $2A$, but not above half so fast. For to carry $2A$ requires double the strength of what carries A at that rate. And therefore the same strength (in the same time) will carry it but half so far.

Which seems to mee agreeable not onely to sense, but to Reason allso, which hee appeals to: As grounded on that generall notion; that *Effecta (caeteris paribus) sunt causis suis proportionalia*.

16-17 As grounded ... *proportionalia*. add.

---

[469] letter: NEILE–OLDENBURG 13/[23].V.1669.

[470] from you: Oldenburg sent Wallis a copy of Neile's letter as an enclosure to OLDENBURG–WALLIS 15?/[25?].V.1669.

81. WALLIS to OLDENBURG, 17/[27] May 1669

The objection which I did occasionally move, was but ex abundanti; for though that & a thousand more were answered, yet the *Postulatum* denyed, remains unproved; & which will not easyly be granted gratis. Yet the Objection itself is of more weight than to bee easyly blown away. For notwithstanding his Salvo, from the great variety in the same line of motion; (which my objection did provide for;) I say this Variety, will lessen so fast (& never be recruited) that (to use his similitude) where Bullets fly so thick, and Bodies be so thronged, all would be soon killed, or few left alive. For instance, suppose in the same line of motion, there were found, a, b, c, d, e, f, &c, moving with so many severall varieties, of swiftness, or direction, or both, (which variety of direction can be but two, that is forward or backward; for any other will leave the line.) Now if wee suppose c to overtake d, then d is killed, (for though both continue to move the same way, yet onely at the rate of c, the faster; so that though d move, yet its varietie is destroyd; c and d going now but for one;) but if c meet d, both are killed; and must there ly dead till either a collateral motion strike them out of the line (& so farwell) or till (suppose) b finding them at rest presse them to serve under him; But, if thus; it will be at the rate b; (not that of c or of d;) So that the varieties of c and d remain defunct without revival, onely that of b surviving; which, when by and by it shal meet with e, will be killed allso. 'Tis true that f finding b, c, d, e, ly dead in the way, may (without reviving those dead varieties) take all along with him under his conduct; (but at his own rate:) but so soon as a meets them, they are all dead. It is true allso, that as c, d, might (as was sayd before) be struck out of the line; so is it possible that g, h, crossing that road, may kill each other there; but if it be while f with his convoy be just passing, they wil kill the convoy too, but if it be just at such a time as none is passing (which will be a great chance, for b c d e f will by this time fill the way for some continued length together) then g, h, or one of them, (and all the convoy that they have there killed,) may be taken into the next convoy that either comes along, or

18 rest (*1*) take them along with (*2*) presse them to serve under
23 him (*1*) at (*2*) under his conduct; (but at
26 if (*1*) wh *breaks off* (*2*) it be
27 (*1*) those (*2*) the convoy
30 them, (*1*) may be taken in the (*2*) (and

crosses, that road; whereby the Party may indeed be increased, but none of those varieties revived which were before destroyed. So that here is nothing but slaughter still. And this happening in every line of motion; though wee should suppose that there is in each at present a very great variety; yet, so many thousands being extinguished every moment, (without any recruits,) there will not need either Miracles, or Myriads of years, to reduce them to a much smaller number. And whether those few that shall at length be left, shall or shall not extinguish one another, will be but (as I sayd the last time) as they shall chance to hit or misse. And This inconvenience (which you had for substance the last time, & now a little more explicitely,) is not to be avoided, so long as hee allows not any thing of Resilition, or of Compound Motion. These are the present thoughts of

Sir

Your friend & servant
John Wallis.

For Mr Henry Oldenburg [2]
in the Palmal near St
James's
London.

82.
WILLIAM NEILE to HENRY OLDENBURG
London, 20/[30] May 1669

Transmission:

C Letter sent: LONDON *Royal Society* Early Letters, N1, No. 22, 2 pp. (our source).— printed: OLDENBURG, *Correspondence* V, 558–9.

Reply to: WALLIS–OLDENBURG 17/[27].V.1669.
Answered by: WALLIS–OLDENBURG 29.V/[8.VI].1669.
Enclosure to: OLDENBURG–WALLIS 24.V/[3.VI].1669.

1 the (*1*) Convoy wil in (*2*) Party
1 increased, (*1*) to a greater bulk (*2*) but
4 at present *add.*
7 few *add.*
8 (as ... time) *add.*
9 chance *add.*
11 allows (*1*) nothing (*2*) not any thing

82. NEILE to OLDENBURG, 20/[30] May 1669

This letter constitutes William Neile's response to Wallis's latest criticism of his hypothesis or theory of motion. Although addressed to Oldenburg, the letter was clearly intended for Wallis. Oldenburg duly sent a now missing copy as an enclosure to his letter to Wallis of 24 May 1669.

May 20. 69.

Sir

I confesse it is a very difficult nice matter to make out the manner of motion in small particles, but I think reason will not leave one altogether to incertaintie in it, if it doe, I shall not be much discontented with that uncertaintie, but in case the rules I conceive doe not prove to be right, I shall be very glad to be convinced and to be showed other rules more firmly proved concerning the motion of small particles. or if there can be none firmly established, I should be very glad that there might be an end putte to disquisition in that matter. I should be very glad to have it determined whether in a lumpe of matter there be any intestine motion in the particles or no: if there be in what manner performed: to have the nature of motion and quiet cleerly explained, that is, what effect they will have upon one another, and to have those principles reconciled to the apearances of nature. I am not now inquiring any thing concerning other principles but only endeavoring to defend my owne. Dr Wallis is pleased to say that motion by those rules of mine will soon come to be generally killed or stopped. but I aplyed the similitude to show that quiet will soon be killed, or motion revived. I grant that motion is killed too, but I endeavored to keep both sides pretty neare in equal vigor whereas Dr Wallis condemnes motion (in that waie I goe) to distruction giving the victory to quiet. Dr Wallises obiection as I conceive comes to this, that particles by coalition doe indeed increase their party but diminish the variety of motion. which I grant they doe for the time. but this coalition is soon broken, by oblique or decussant motions. Supose $b.$ $c.$ $d.$ and $e.$ ioyned into one. whilst they continue united, $b.$ and $e.$ cannot move to come nearer one another, but if $c.$ and $d.$ be strucke out the line, then $b.$ and $e.$ may come together and $c.$ and $d.$ or something equivalent to them| may come again into the line in some other place, and come to be moved again with newe velocitys. So that I conceive any particle may come into any place, and come to be moved, at one time or other, by any velocity imaginable. which

7 convinced (*1*) by (*2*) and to be showed
11 if ... performed: *add.*
19 pretty neare *add.*
22 indeed *add.*

perpetual circulation and shifting of places amongst the particles, will for ought that I see keepe motion sufficiently alive. I confesse that if any particular degree of velocity were quite extinguished in any one line of motion, that degree could never be revived again in the same line of motion, and if the disiunction of the particles did not goe on as fast as the coalition does, all would soone come to a coalition. but I doe not conceive that there ever is a coalition of any considerable number of particles which continues for any considerable time, but there is a continual comminution of particles made by decussant motions, and no particle any sooner possesses it selfe of any place. but it is soon iustled out of it again. but I leave it to the determination of learned men, who I hope will not be displeased though my opinions should chance not to satisfie them. I don't find that Dr. Wallis obiects any thing particularly, against what I alledged in behalfe of that opinion that quiescent matter has no resistance to motion. I am afraid to be too tedious to him or to your selfe so I rest,

Your humble Servant
W. Neile

83.
Henry Oldenburg to Wallis
24 May/[3 June] 1669

Transmission:

Manuscript missing.

Existence and date: Mentioned in Oldenburg's endorsement on Neile–Oldenburg 20/[30].V.1669 and in Wallis–Oldenburg 29.V/[8.VI].1669.

This letter enclosed a copy of Neile–Oldenburg 20/[30].V.1669, which was clearly intended for Wallis.

7 of particles *add.*
10 it *add.*
12 should (*1*) not (*2*) chance not

84.
WALLIS to HENRY OLDENBURG
Oxford, 29 May/[8 June] 1669

Transmission:

W Letter sent: LONDON *Royal Society* Early Letters W1, No. 89, 2 pp. (our source). Endorsed by Oldenburg beneath address on p. 2: 'Rec. May 31. 1669.' At 90° in unknown hand: 'Wallis to Neile. 69'.—printed: OLDENBURG, *Correspondence* V, 573–4.

Reply to: NEILE–OLDENBURG 20/[30].V.1669 and OLDENBURG–WALLIS 24.V/[3.VI]. 1669.

Wallis began the present letter on 27 May, but postponed sending it on receiving Oldenburg's letter of 24 May. After he had taken account of that letter, Wallis supplied it with the new date of 29 May.

Oxford May 29. 1669.

Sir,

To Mr Neil's Postulatum; my Objection was; that I thought it should be Proved, not Postulated; which still remains. Till then, he will give mee leave to suspend my assent. To what I added *ex abundanti*; That by his Principles Motion must needs diminish: and its innumerable varieties be reduced to none or much fewer. Hee grants, that I prove, 1. That no new Way or Line of Motion can begin, which is not allready. 2. That, if all motion in any one line happen to cease; no motion can, in that line, begin again. 3. That if any particular degree of velocity, do therein cease; that degree can never, in that line, be restored. 4. That Particles by coalition (as when b, c, d, e, though at first so many severall varieties, do by coalition, which I proved unavoidable, come to be one,) though they increase their party, do diminish the variety of motion, *for the time*. To this I must adde; that, if for the time, then for ever. What hee suggests, to avoid it, That by decussant motions c, d may bee struck out of the line, whereby the coalition is broken: doth not help it (unless he had shewed also that the variety is thereby restored.) For, while b, c, d, e, move as one; (suppose forwards;) the decussant motion which strikes d, doth stop it, and (with it) b, c, which were close behind it,

---

1 May (*1*) 27. (*2*) 29.
8 if (*1*) in any line of (*2*) all motion (*a*) happe *breaks off* (*b*) in
17 (unless...restored.) *add.*
18 (suppose forwards;) *add.*
19 close *add.*

84. Wallis to Oldenburg, 29 May/[8 June] 1669

(all waiting there till another remove them,) so that e (or what else goes forward with it or shall afterward be carried before it,) will still bee but as one. And though c, d, (or some other equivalent,) may be cast into that line again; yet they will never bring with them a new variety of motion in that line; but must (before they can move in it) be there at rest, & become one by coalition with the next that comes that way. So that, decussant motions, serve indeed to destroy varieties, but never to begin or to restore any.

......a..........b..........c..........d..........e..........f......

These coalitions therefore being, every moment, in every line of motion, innumerably frequent; must needs, every moment, destroy innumerable varieties.

Since therefore he grants 5. That if disjunction of particles did not go on as fast as coalition doth, all would soon come to a coalition: 'Tis manifest what the consequence must needes be, when as those disjunctions do not at all help the matter. Nay, they do the contrary. For the decussant motion which disjoins d from e, doth but stop not onely b, c, d, but itself also all his own party; (without giving motion to anything:) & over this by a coalition. Nor can any part of what is thus stopped, be moved again; but, by a new coalition with some praeexistent motion; not by beginning or restoring any lost variety of motion.

I might easyly adde a great deal more: But this, if well considered, is inough to do the work. And, the more distinctly he considers it, the more evident hee will see the consequence.

For all his decussant motions, (whether by striking out or by thrusting in,) though they may destroy, can never give motion, in the line $a\ f$. And whatever variety, therein, is once stopped; is lost forever. And whatever particle is once stopped, can never after move but onely under the conduct of some of those varieties which had never been stopped. I am

Yours &c
John Wallis.

6 by coalition *add*.
6 that (*1*) finds the ⟨—⟩ (*2*) comes
18 doth (*1*) onely (*2*) but
19 (without...anything:) *add*.
21 restoring (*1*) another (*2*) any lost
26 motions (*1*) thou *breaks off* (*2*) (whether
30 stopped. (*1*) I onely add an (*2*) I am

85. Neile to Oldenburg, 1/[11] June 1669

This was intended you by the last post: But your letter[471] coming so late to hand, that I had hardly half an hours time to write; the Post was gone before I had finished my letter, which made mee change the date of it.

As to the note concerning Mr. Hugens;[472] I do not much doubt but that hee will in time be satisfied, that my Rules of Motion are true: And, therefore, if his be so too, they must needs agree.

[2] These
For Mr Henry Oldenburg, in
the Palmal, near
St James's,
London.

85.
William Neile to Henry Oldenburg
[London], 1/[11] June 1669

Transmission:

C Letter sent (including appendix): London *Royal Society Early Letters*, N1, No. 24, 4 pp. (our source). On p. 4 in unknown hand: 'Neile to Wallis'.—printed: Oldenburg, *Correspondence* VI, 3–5.

Reply to: Wallis–Oldenburg 29.V/[8.VI].1669.
Answered by: Wallis–Oldenburg 7/[17].VI.1669.
Enclosure to: Oldenburg–Wallis 4/[14].VI.1669.

This letter represents Neile's response to Wallis's most recent criticism of his theory of motion. Although the letter is addressed to Oldenburg it was clearly intended for Wallis himself. Oldenburg duly sent a now missing copy as an enclosure to his letter to Wallis of 4/[14] June 1669.

June 1./69.

---

[471] letter: i.e. Oldenburg–Wallis 24.V/[3 VI].1669.
[472] Hugens: evidently, Odenburg had reported on Huygens's recent letter: Huygens–Oldenburg [19]/29.V.1669; Oldenburg, *Correspondence* V, 554–6. There, in respect of his and Wallis's laws of motion, Huygens wrote: 'Je ne scay si Mr. Wallis aura pu reduire ses regles au mesme sens des nostres; car je n'y vois pas beaucoup de rapport' (554).

85. Neile to Oldenburg, 1/[11] June 1669

Sir

I must crave leave to dissent from Dr. Wallises opinion, for I doe not see as yet any cause for the final destruction or diminution of motion upon my principles. neither doe I thinke that Dr Wallises opinion gains any advantage by what I have granted. I desire you will please to remember what I have often told you that I supose almost an infinite multiplicity of motions at the same time existing, and an extreme minutenesse of particles, and that these particles are not invariably the same, but are comminuted by motion into various other particles. here, it may be, a peice is paired off by motion from a particle; there, may be, a peice is added and severall may make a coalition; but what perchance may the summe of the coalition amount to, afore any seperation is made? it may bee, to the quantity of a sphaere of the diameter of the hundredth thousandth part an inche, or as much lesse as you please. and if there be any disiunction or seperation made afterwards, that disiunction is made by motion, and is motion in the particle impelled for the time it lasts. if $b.$ $c.$ $d.$ $e.$ be united and be all in motion together in the line $a.f.$ no particle of those, can be struck out of the line, whilst they are in motion; but suposing them to be quiescent, then $c.$ and $d.$ may be strucke out of the line. consequently the particular motions of $c.$ and $d.$, which were lost before, come to be renewed; and if not with the same velocities, yet with others as good for the purpose of keeping motion alive: so that I see no necessity, that if $c.$ and $d.$ have once lost their varieties of motion they should never be recovered; for that they doe recover by being strucke out of the line, and when they are once gone out| of the line they are left to take [2] their fortunes and to undergoe in like manner as many millions of varieties as you please. I confesse to make $b.$ and $e.$ come nearer together in the line $a.$ $f.$ $c.$ and $d.$ or some particle by going out of the line, must leave some vacuitie between $b.$ and $e.$ which may be easily done, if that which should have succeeded to fill up the roome be suposed to be stopped from coming to fill it up. or if nothing were at that time about to succeed: therefore there may be some vacuitie made between $b.$ and $e.$ and consequently the variety of

3 or (*1*) determination (*2*) diminution
8 by motion *add.*
16 in ... *a.f. add.*
17 be (*1*) pared off (*2*) struck out of the line
22 once *add.*
25 in like manner *add.*
30 Or ... succeed: *add.*

85. NEILE to OLDENBURG, 1/[11] June 1669

motion between the points or distances *b.* and *e.* may be again revived, and in like manner *b.* and *e.* may come to ioyne, and afterward's be strucke out of the line, and then there will be roome for *a.* and *f.* to aproach nearer and in like manner *a.* and *f.* after they are ioyned may again be disunited, and the varieties of motion revived. the decussant motions make new variety's by striking quiescent particles out of the line, which having vacuities make room for new aproximations. and the decussant motions by thrusting in newe particles into the line, helpe to preserve motion in the line: those newe particles receiving, after they have been first quiescent, the motion of other neighboring particles, so that though many particles moving in the line *a. f.* be stopped by decussant motions, yet they may have first communicated their motion to some of those new interposed quiescent particles, which may supplye their offices toward's the propagation of motion. I confesse the whole businesse is very obscure and difficult, but I thinke it is as obscure and difficult to bringe things generally to quiet by these principles as to continue [3] motion.| but I leave it to the determination of learned men, amongst whom I knowe Dr Wallis is as likely to sifte it to the bottome as any body. I know not what he may, but I confesse that if I should endeavor to prove a decrease of motion from these principles I should no sooner make a motion stoppe but I should think I had not garded it enough from being putte in motion again, and that, without keeping united to the particles with which it is at present ioyned; for, what is there to hinder it from being moved and stopped alternately any ways as long as you please? either in the whole together, or the parts of that particle severally; because, you know, I make them not invariable or indivisible: the like may be said of every parcel or particle of matter. no litle particle of space is any sooner filled up with quiescent matter, but it is either in whole or in part perchance made empty again, and leaves roome for new particles to succeed into it. which may again give place to others, and for ought I see make a perpetual shufling and circulation of motion. if those particles that goe out of a line, may have motion; kept alive in them with vicisitudes of quiet, and that the particles, that come into the line, may have their motion again renewed, and the motion in the line be perpetually continued by the interposition or interception of new

1 or distances *add.*
8 line: (*1*) by (*2*) those newe particles
17 I (*1*) knowe (*2*) supose (*3*) knowe
23 any ways *add.*
24 severally; *add.*
33 or interception *add.*

vacuities, I know not what obstruction there will be to the perpetuation of motion.| but as I said before I must leave it to better iudgments. For that opinion that quiescent matter has no resistance to motion. I thought I had alledged proof enough of it and Dr Wallis showed no insufficiency in what I alledged.

[4]

I remain Sir
your humble Servant.
W. Neile.

86.
HENRY OLDENBURG to WALLIS
4/[14] June 1669

Transmission:

Manuscript missing.

Existence and date: Mentioned in WALLIS–OLDENBURG 7/[17].VI.1669.
Enclosure: NEILE–OLDENBURG 1/[11].VI.1669.

As emerges from WALLIS–OLDENBURG 7/[17].VI.1669, Oldenburg enclosed with this letter a copy of NEILE–OLDENBURG 1/[11].VI.1669.

87.
WALLIS to HENRY OLDENBURG
Oxford, 7/[17] June] 1669

Transmission:

W Letter sent: LONDON *Royal Society* Early Letters W1, No. 90, 4 pp. (our source).—printed: OLDENBURG, *Correspondence* VI, 14–19.

Reply to: NEILE–OLDENBURG 1/[11].VI.1669 and OLDENBURG–WALLIS 4/[14].VI.1669.

This letter to Oldenburg represents Wallis's reply to Neile's latest defence of his hypothesis of motion. The deleted postscript indicates that Wallis intended to write a letter to Oldenburg himself two days later, on Wednesday, 9 June. There is no evidence that a letter of that date was written. The ordinal numbers in the fourth paragraph are evidently a later addition.

Oxford June. 7. 1669.

87. WALLIS to OLDENBURG, 7/[17] June 1669

Sir,

Mr Neil's paper[473] of Jun. 1. in yours[474] of Jun. 4. (which I received this morning,) needs no reply; as having nothing in it, which was not sayd before; & answered.

The onely argument on foot (& I do not mean to stir a new one) is, from the necessary decay of Motion according to his principles.

That hee supposeth very great variety; was sayd long since: (& hee says no more now.)

But it hath been shewed (& it is the onely thing I had to prove) That these Varieties will presently diminish as to their number, (in every line of motion:) Because they are, every moment, either destroyed by 1 contrariant, or 2 decussant motions; or 3 swallowed up by swifter which overtake them; and 4 can never be restored, or 5 multiplied, or 6 recruited by new ones. All which hath been before so clearly & so strongly demonstrated; that nothing but his not attending it, could leave him unsatisfied.

I will onely endeavour to say the same again, if it be possible, a little plainer (for, stronger, it cannot be sayd;) & if this be not understood, I shal despair of speaking intelligibly.

Suppose, in any one line of Motion (& there is the same reason of all) so many of the nearest movers to be a, b, c, d, e, f, (or, if these be not inough, as many hundred thousand more as you please,) all between them either being at rest, or onely crossing the line: And, these moving severally, at so many several rates of swiftness, (suppose, in the proportion of the numbers 6, 3, 5, 4, 2, 1; or what other he please to assigne:) some forward, some backward: Making six varieties of motion.

Let us first consider d; which we will suppose to move (for instance) at the rate of 4.

If now we suppose d to be hitt by any collateral motion; it is manifest that, by his principles, d is stopped (as well as that which hits it; of which here wee take no notice, having onely the line af in consideration;) and so the variety of 4, is lost: Unlesse possibly d have onely a piece broken off, (the rest escaping;) or have first given its motion to h, which may continue to move at the rate of 4, when the other is stopped.

But (because this is not impossible) let us next suppose this motion not to be yet destroyed; but that d (scaping the pikes,) or h (his substitute,) or dh jointly, do continue to move at the rate of 4: and, with all, c at the rate of 5.

---

[473] paper: i.e. NEILE–OLDENBURG 1/[11].VI.1669.
[474] yours: i.e. OLDENBURG–WALLIS 4/[14].VI.1669.

87. WALLIS to OLDENBURG, 7/[17] June 1669

Now if c be so hitt as was before sayd of d; then is the rate 5 destroyed, unless (as was sayd of d) some substitute of his as g, do continue that motion.

But, granting both motions, for the present to continue: If c, (or cg, or g,) move forward; and d (or dh, or h,) backward: So soon as these meet, both are destroyed; & the rates of 4 and 5, perish. And so, of six varieties, there will be but 4 left.

Or (that wee may be most favourable,) let us suppose c and d (or their substitutes) not to meet, but both to move forward. Yet, at last, c the swifter, overtaking d the slower, will swallow up that of 4, (all moving at the rate of 5:) and so our six varieties at lest reduced to 5, without any new one in the lieu of it.

If it be sayd, that, g or h, or as many more as you please may be thrown into the line which before were not in it. It is true they may; but not so as to bring with them any new variety of motion: For the decussant motions which cast them in, can onely give them Rest (not motion) in the line af;| [2] till c or d pick them up. All the advantage therefore will be but this; that, in stead of cd, you may now have $cgdh$ all moving as one, at the rate of 5. That of 4 being extinguished, & no new variety come in the room of it.

If it be sayd; that though $cgdh$ do thus by coalition become one; yet gd may (by a decussant motion) be struck out; and c, h separated. It's true they may; but not so as to give us a new variety.

If he say, Yes; for $c\ h$ which were before moved jointly as one, shal now move as Two. Here is a double mistake.

For, if we should grant them to move as Two; yet both at the rate of 5, (4 is not again restored;) so that wee have yet but Five varieties; not Six, as at first.

But, indeed wee cannot have so much. For c, h, will not (as is supposed) continue to move as two, (not so much as at the same rate of 5;) for gd cannot be struck out, but that they must first be stopped; and that which stops d the formost, stops allso cg that follow. So that nothing goes forward but h.

If it be yet sayd; that they may by the next which comes this way, suppose b or a, be put into motion again: It is true they may; but not at the rate of 4 (which is irrecoverably lost,) but at that of 6 or 3, (which are the varieties of a, b). So that still wee have but 5 varieties; and but five moving parties, instead of our six.

2 as g *add.*
10 at lest reduced *add.*
19 though *add.*
22 which *add.*
27 For *(1)* d *(2)* c

87. Wallis to Oldenburg, 7/[17] June 1669

If it be again replyed, That still more may be thrown in between. It must still be answered; that, between *a* and *f*, though new *Particles* may be cast in, yet no new *Varieties of Motion*; because nothing of Motion can there be had, till some of those yet surviving do pick them up. So that by what ever shuffling *c*, *g*, *d*, may be cutt off from *h*, and joined to *b*; or, thrown out of the line, & others brought in; yet never shall the lost variety of 4, be brought in again; nor any new one in stead of it, to make up the former number of varieties. Nor can ever those six varieties (now reduced to Five) become Six again.

$$\begin{matrix} & & . & & & & & & \\ & & . & & & & & & \\ a....b....c....g....d....h....k\ .\ e....f.... & & p\ q \\ & . & & . & & & & & \\ & . & & . & & & & & \end{matrix}$$

If (which is the last shift) it shall be sayd; that, though not within *af*, yet somewhere in the same line a great way off (suppose at *p* or *q*) there may some particle yet be found to move at the rate of 4; & so this variety not quite lost. This helps not the matter at all. For, if so; then there were, before, Two distinct parties moving at the rate of 4, now there is but One. Or, if never so many; yet at lest there are fewer by One, than before that of *d* was extinguished. Nor can there ever be so many again.

And what is thus sayd of the rate of 4, extinguished in *d*; will hold of all others which ever come to be stopped, or to be swallowed up. That is, when they are once stopped or swallowed up; they are lost for ever. Nor can any new one come to make up the number.

And here it is that Mr Neil is to shew his skill; if, by any art, he can restore us this lost variety, of motion at the rate of 4, (at lest, so many parties moving at that rate, as before wee had;) or can give us any new Varieties in stead of it, that our number of varieties bee not lessened.

Now when *cgdh* (or the remains of them, or their substitutes,) thus move as One, at the rate of 5: If any collateral motion strike the formost of them; they are all stopped. And so the variety of 4 and 5 are both lost.

If it be sayd, that it is a great chance if the decussant motion light just upon the leader of the party; but may somewhere fall upon the flank; & so, the rear may be cutt off, and stopped; yet the foreman may keep on his way (so as to preserve that variety;) and presse more company which hee finds
[3] loitering in the way.|

19 will (*1*) be say *breaks off* (*2*) hold

87. WALLIS to OLDENBURG, 7/[17] June 1669

I answere; it is not at all strange that h (the leader of the party $cgdh$) should hit, or be hitten. For, it is very possible, that just as h with his followers come one way, k with his followers may come to crosse the line: and if the two captains do so meet; the whole of both parties are cutt, and those two motions lost. (And this, in contrariant motions, happens allways.)

But suppose it do not so happen, that the whole of both: Yet the lest that can happen, at every decussation; is, to stop the whole of one, and part of the other. For since the ways of dh, & k, do crosse in a point; if k be got past the line (but not all his company) before h come; h falling on the flank of k, cuts off his rear; but withall, stops himself, and all his own party. But indeed if h come first, hee may scape himself with the losse of his rear; but k, with all his, are cutt off. So that, upon every decussation, one of the two parties are wholly destroyed; & at lest part of the other, if not the whole of it allso.

But let us allow h, for the present to scape the skirmish, and to move forward at the rate of 5; having left cgd (or so many of them as shall not first be struck out of the line; and what ever else shall be cast into the way & not removed,) to be gathered up by b, (which follows slowly at the rate of 3,) as by some substitute of a fore-runner of his, to which b hath given motion. (For between that of b, & that of h, there can be no other motion in the line af:)

Tis manifest in this case, that neither b, nor any of his substitutes, at the rate of 3, can ever overtake h, that is gone before at the rate of 5: but do perpetually loose ground of him, & follow at a greater distance. And what ever in the mean time is cast into the way, must either be struck out again; or ly there at rest, till either b (or some of his) do come upon them, or else h (or some beyond him) come back to them.

Suppose wee then that h turn back: so soon as this (or some of his) do meet with b (or some of his) both parties, if not before, are at lest now destroyed: And, as 4 before; so, now, 3 and 5, are quite lost. (And that also in the bargain, which brought h back.)

If hee say, Noe; but h, before he turn back, may have given motion to k, which moving forward may preserve the rate of 5.

I say: this cannot bee. For h can never turn back till hee be first stopped, and, bee afterwards, (by e, f, or some beyond these,) beaten back. Now e

9 (but ... company) *add.*
20-21 (For ... line af:) *add.*
24 & *(1)* grow *(2)* follow

or f or any beyond them, can never come back upon h, so long as k (or what ever else have received that motion from h) moves forward. For there is no room for k, e, to passe by each other, without meeting to their own destruction. So that, not onely h but all his substitutes; that is, what ever was sent before him at the rate of 5, (for none of these can passe e or f,) must be first stopped, and then either beat out of the way, or turned back allso. So that before h, or any for him, can turn back, the whole effect of its former motion must cease; and the rate of 5 be quite extinguished. And that also of e or f which brings him back.

But let us reprieve h a while longer. Yet since hee moves faster before, than b can follow after; the gap must needs grow wider and wider: (And much more, if b should be supposed to move backwards toward a:) since that nothing can come in between, to give any motion in the line af.

[4] Nor can this be helped, till either h (as was sayd before) or some beyond him, come back; or else till a (or some behind him) over take b.|

Now if we should suppose a to move slower then b, this cannot bee done. And if it should move faster than b, but slower than h; it will not serve the turn: For though b will be thus swallowed up of a, (and so the variety of 3 be lost,) yet the inconvenience remains as before: the gap between ab and h, will stil grow wider.

Let us therefore allow to a, a swifter motion (suppose at the rate of 6:) which having gathered up, in the way, $bcgd$ (or their remains, or substitutes,) will (if itself be not first destroyed,) overtake h: & so, at length, not onely the rates of 4. 3, but 5 allso be swallowed up of a. So that of our first six varieties, wee have but 3 left.

Now if a with his party moving forward at the rate of 6, meet with e moving backward at the rate of 2; Both parties are quite stopped. And of all the six varieties which at first wee had, there is none left but that of f; (moving back at the rate of 1:) which may gather up $ehdcba$ (or so much of them as is left, or is come into their room:) But onely at the rate of 1, and but in one company. So that in stead of six varieties, and six parties; there is but one variety left; and that but in one party: Nor can all his art, by shuffling, or cutting, or striking out, or casting in, give us two for this one; or ever restore any of the rates which are lost, or give any new one to supply the number.

1 or any beyond them, *add.*
15 (or ... him) *add.*
24 the rates of *add.*

And what is thus shewed of these six reduced to one; will in the same way come to passe, if for those six, there had been six hundred thousand. For so soone as every variety hath been once stopped, it is for ever destroyed.

And what is shewed in the line af; must, by the same reason, happen in every Line. Whereby instead of that great variety which wee had at first, wee have as great a scarcenesse; (for none live longer than till hee hath been once hit;) All these being reduced, either to few or none, as the case may happen. Which was to be demonstrated.

Excuse the length of this Demonstration; which though it say no more than what in substance was sayd before: yet, to meet with all cases, would require some length.

If this do not satisfy; It will be incumbent on him (not in the general, to talk of varieties, of shuffling, of dissolution, at large, but) to find out some art whereby the lost variaties of 6, 3, 5, 4, 2, may be restored, or at lest as many new varieties (which had not been before) created to supply them. That so the number of varieties & the number of parties, may be as many as before. Which can never be done by any of the means yet assigned. For the shifting of particles from one party to another (since they cannot come of themselves to make a new party) will not increase the number of parties; & the decussant motions can give no motion in this line; nor is there any means to restore a variety once stopped. Nor can his principles (which allow no resilition nor compounded motion) afford any expedient for any of these mischiefs. I am

yours &c.
John Wallis.

For Mr Henry Oldenburg

88.
HENRY OLDENBURG to WALLIS
12/[22] June 1669

Transmission:

Manuscript missing.

Existence and date: Mentioned in and answered by WALLIS–OLDENBURG 15/[25].VI. 1669.

3 stopped, (*1*) (which will (*2*) it
6 scarcenesse; (*1*) (all th *breaks off* (*2*) (for
25 Wallis. |I write | again *add.* | by Wednesdays coach, which will bee at the Greyhound in Holborn Wednesday night. *del.*|

In this letter, Oldenburg apparently requested from Wallis a copy of his recently published tract *Thomae Hobbes quadratura circuli, cubatio sphaerae, duplicatio cubi; confutata* for the library of the Royal Society. He also raised questions concerning mistakes made by Hobbes in his treatment of quadratures and by Neile in his theory of motion.

89.
WILLIAM NEILE to HENRY OLDENBURG
[London] 15/[25] June 1669

Transmission:

C Letter sent (including appendix): LONDON *Royal Society* Early Letters, N1, No. 23, 12 pp. (p. 11 and p. 12 blank) (our source).—printed: OLDENBURG, *Correspondence* VI, 37–42.

Reply to: WALLIS–OLDENBURG 7/[17].VI.1669.
Answered by: WALLIS–OLDENBURG 19/[29].VI.1669.
Enclosure to: OLDENBURG–WALLIS 4/[14].VI.1669.

In this letter, William Neile responds to Wallis's latest criticism of his theory of motion. Although addressed to Oldenburg, the letter was clearly intended for Wallis himself. Oldenburg probably sent a copy as an enclosure to a now missing letter to Wallis.

Sir

My last letters[475] I confesse were much the same as those a litle before; I told you as much. I had not the other's by mee. nor I know not well what other replye to give to the same obiection then a litle to insist upon the same kind of answere. I did not well understand the force of that obiection nor doe not yet, it may be that is my dulnesse I hope Dr Wallis will pardon it. I shall make my answere as breif as I can, for I supose a litle hint will serve if I am in the right and if I am in the wrong; I ought so much the more to spare the Doctor's patience. I am beholding to him that he would give so strict and large a scrutinye to my principles, I can't pretend to any art of legerdemaine, but I confesse if I may speake my foolish opinion in this businesse I thinke motion may be as easily continued by this hypothesis as a certain brewer suposed it easie for him (as the storie goes) to goe to Lawe as long as there was water in the thames, because mingling of that with his drinke would still suplye him with money. the comparison may be something extravagant. But when a particle is at rest if one motion will not light upon it to move

---

[475]letters: i.e. NEILE–OLDENBURG 1/[11].VI.1669. Cf. WALLIS–OLDENBURG 7/[17].VI.1669.

89. Neile to Oldenburg, 15/[25] June 1669

it, another may, it is but making so many more motions like taking so many more barrels of water out of the thames. and indeed unlesse there be some such easie helpe, I supose Dr Wallis's lawe will be much too hard for mine in this matter. but this may be a litle extraneous to the businesse, I hope the vain excursion may be pardoned.| But if there be no one particle at any [2] time quiescent but is liable to assault's on every side, I desire Dr Wallis would showe what there is to hinder it from receiving motion. and if there be enemies (as I may call them) on every side to impugne it, which sure may easily be, if there be multiplicity of motions enough, I know not what is wanting to give it a cause of motion. if the particle d. could only move in the line $a.f.$ it would never recover its varietie of motion when once it has lost it. (suposing no other motions but in the line $a.f.$) because it can never move again in that line but by coalition with another particle. but by the helpe of a decussant motion that which stopped d. may be strucke out of the line, and d. may be moved on again the same waie as before; and that which moved it may be stopped by a decussant motion and yet d. may move on alone by it selfe, and consequently recover the varietye of it's motion in the line $a.f.$: or it may recover it by being strucke out of the line. The like may be said of any other particle, unlesse wee should supose, that all the motion's extinguish together, and that, when a particle is quiescent, it's neighbors are all so too; and if one particle can not be brought to a durable rest, how shall more then one be brought to it, or what decaye of motion can ensue. When different motions meet they stoppe, but there is no fence to secure those particles from being moved again some waie or other, so that if no particle can remain in quiet for any considerable time, how shall wee want motion.| if when 2. particles are stopped there is so [3] much lesse quantitye of moving matter in the world at that time, and when those 2. particles are put in motion again there is so much more quantity of moving matter in the world at that time, then there is as much moving matter restored as was taken away before. and if there may be continually as much moving matter restored, as was lost before; where is the decaye of motion? the velocities that are restored, may serve as well for the purpose as those that were taken awaie. and though in everie moment of time there may be a variation of motion as to the whole quantity of matter, yet that variation may in every moment be inconsiderable as to the whole, and what

5 at ...quiescent *add.*
8 call (*1*) it (*2*) them
33 awaie. *add.*
34 of motion *add.*

89. Neile to Oldenburg, 15/[25] June 1669

is gotte in one moment as to the whole, may be lost again in another and in re versa the like. if the same particle of matter may come successively into any place and be moved successively by any motion whatsoever, (and if you please) to move alone by it selfe, how comes it that that particle, if it have once lost its variety of motion can never recover it again? it matters not to say that if that particle recover it's motion, some other loseth it's motion, for so the lost motion be still suplyed it is no matter how many are lost. I am not to make motion increase in the worlde, but to keep it from impairing by these principles if I can. I confesse when a quiescent particle is moved by another in motion there is not two motions made but only one. But it is the motion of two particles. you will say those two particles may be both stopped at once, verye true. and they may be both putt in motion again at once, one being as easie as the other.| nor is there any feare that particles will unite together into too long a line of motion, as long as there is decussant motion's to disioyne them. If there were but one line of motion in the world it would be so. but allowing innumerable, for my part I won't undertake to make sixe particles unite together into one. though if they did, perchance such small ones, as they may be, there were no great harme in it as to this businesse. I don't aprehend enough what it is Dr Wallis would have mee prove. Which I hope he will excuse in this nice matter. let us supose the sixe particles united and quiescent, any of them may neverthelesse be putt into a motion independent from the rest, by being strucke out of the line, but that makes not more motion in the line $a.f$: Verye true, but though collateral or decussant motions give no motion in the line $a.f.$ yet why may not motion be sufficiently preserved in the line $a.f.$ by particles moving in the line from time to time, which continually may propagate motion to quiescent particles in that line; for why may there not be alwayes so many moving particles in the line $a.f.$ that though many of them continually stop, yet they may have first propagated their motion to other quiescent particles? I take the case to be thus, that there is so great a variety of motion in the line $a.f.$ that wherever there is a vacuity, it is soon filled up and a union made of some particles which leave vacuities in

[4]

1 and ... like. *add.*
7 for (*1*) the lost still (*2*) so the lost motion
8 in the worlde *add.*
10 only *add.*
20 Which ... matter. *add.*
31 vacuity, (*1*) between two moving particles they soon fill it up and make (*2*) it is ... particles which

89. NEILE to OLDENBURG, 15/[25] June 1669

the place they deserted, which may again be filled up with other particles, and decussant motions may continually make new vacuities in the line to prevent too great a coalescence of particles. supose $a.f.$ to be a full line of matter quiescent, by emptying intermediate spaces by decussant motions, $a.$ and $f.$ may come to meet:| supose in the point $g.$ and other particles may continually succeed in the roome of $a.$ and $f.$ and again come to meet in $g.$ so that the space $a.f.$ shall continually have motion kept alive in it, and the same case it may be in any other space: therefore motion will not extinguish; And of what length will it need to supose the space $a.f.$ to be? You may supose it as short as you please, so that I thinke there will be no great danger if one should allowe that to be once a full line of matter. The motions in the line doe endeavor to make a coalescency as if they were weaving something like Penelopes webbe, for they no sooner knitte up holes but decussant motions bore them again. when particles in motion meet either only with vacuitie or with quiescent matter, for so longe they are not stopped; and when they are stopped, how can there want others in motion to communicate newe motion to them? if a tennis ball be strucke one waie and then suposed to stoppe, (it matter's not though the racket which strikes it accompanie it) and if then another racket strike it perchance another waie, and then it stoppe, and then another strike it, and so continually, the ball sure will have exercise enough, and be bandied about sufficiently. and I can't see how those rackets can faile to lye in ambush for it, when I can supose as many racket's as I please. and I thinke can never supose more then there may possibly bee in nature. The Dr. makes a constipation of motion to choake it selfe, which I thinke| were verie hard to doe amongst such a multitude of motion's, and where the matter is so fluid, that is so easily mouvable. I doubt if any body would make a model of a great company of particles moving and resting alternately according to those principles; they would have much adoe to calculate the method of it. for I thinke it is hard to calculate the motion of sixe particles without calculating the motion of sixe hundred thousand into into which they may be divided and subdivided: for I dont supose the moving particles to be indivisible invariable atomes. I supose no particle to be nowe

[5]

[6]

2 new *add.*
8 And (*1*) how long (*2*) of what length
12 something like *add.*
17 tennis *add.*
29 sixe (*1*) without calculate the motions of sixe particles (*2*) particles into which they may be divided and subdivided (*3*) particles without
30 into ... subdivided *add.*

89. Neile to Oldenburg, 15/[25] June 1669

actually in motion in the worlde, but what is a parralelepipide prism prisme or cilindrical bodie. for though there might be other figures in motion at first yet they will soon be reduced to them afterwards, because no motion has any power but only straight forward against what it meets directly in it's waye. but these paralelipipedes etc. are not invariable atomes but theye may have still lesse and lesse particles minced and carved out of them. if that in the line $a.f.$ the particles may continually move and rest alternately going sometimes to the right hand and sometimes to the left hand motion may be sufficiently preserved. and the interstices which particles going out of the line may leave will for ought I see necesçarily prevent too great a coalescency of particles: Supose no other motions but in the line $a.f.$ and supose the space to be full with quiescent matter, it will not be denied but that $a.f.$ may be tossed about sometimes to the right hand and sometimes to the left, as long as there is varietie of motion continuing in the line $a.b.$ by making [7] a coalescency still with more and more particles.| but if wee supose decussant motions continually to bore holes in the line howe shall the coalescence growe to be large? if wee allowe six particles to make a coalescence it is not necesçary that therefore a dosen shall make a coalescence afore there be any holes bored betwixt them. yet I supose the decussant particles to stoppe up holes as well as to make them but it is no matter how many are stopped so there be enough still bored. let one man have a libertye to stoppe holes in a sive, and another have as much libertye of making holes with a bodkin and the sive may never come to want holes. if a man be making a snowe ball, if as it is rowled newe snowe still stick on and accumulate, it will soon growe to be a great one, but if wee should supose the snowe to droppe off as fast as it comes on, where will be the encrease. so if coalescent particles have continually intermediate particles dropping of from them, the coalescency may never come to have any great extension. if the particles that are in motion did meet with nothing but oposition, motion would soon be extinguished. but they doe as well meet with quiescent as with moving particles. as long as no particle in the line $a.f.$ need's to be a perpetual inhabitant of it, but may continually wander about from place to place resting by turnes, I can't see how there shall come to be any great coalescent accumulation of particles

1 prism ... bodie. *add.*
5 etc *add.*
32-5 I suppose no ... invariable atomes *text enclosed in box*
5 theye *add.*
6 and carved *add.*

89. NEILE to OLDENBURG, 15/[25] June 1669

in the line $a.f.$ and if there be no such accumulation I don't see why the variety of motion once existing in the line $a.f.$ (there being also the like variety in other decussant lines) shall not| continue variety of motion, by [8] making vibrations of moving matter sometimes on one hand and sometimes on another. supose $a.f.$ a full line of matter moving towards the right hand. For the like reason I may supose such another decussant line passing through $g.$ or or as manie as you please: moving all one waie either upward or downward, $g.$ is therefore to be considered as a common inhabitant of the two lines but $g.$ can not move in both of them at once therefore the place $g.$ must be emptye if if there were such motion's: therefore there is not such an entire coalesency, nor can not be for any considerable length. Dr. Wallis has putt mee upon the right waie of considering this business which is by lines of motion. and my lord Brouncker[476] said a thing to mee which gives mee a great deale of Light. which I confesse I did not at first aprehend. for he allowes (as I conceive) that there shall be as many particles as I please continually strucke out of the line, and consequently that there shall be no general coalescence of particles or of any large extension, but he alledges that the swiftest motions will by often associating themselves to other particles, and communicating their motion to them, bring it to passe, that at last there will be no moving particles lefte in the line but such as are very swift, and the slowest of those will by degrees come to be strucke out out of the line, and consequently the moving particles that are left will be of an equal velocity. which I confesse I knowe not well how to denye at present. but I think it will serve my turne though I should admitt it.| for wee will [9] supose if you please all the motions in the line $a.f.$ to come to be of one and the same velocity, (but supose that to be as great as you please) for I can't see anie inconvenience that will followe, so there be always innumerable motions kept up in the line, $a.f.$ some moving to the right hand, and some to the lefte hand. since I writt this I had the good fortune to meet with my Lord Brouncker again, and I find I was not mistaken in his opinion. but that he thinkes it may be no detriment or inconvenience if one should supose but one degree of velocity in the world. as long as that may be as swift as you please, or as is necescarye. so that nowe by this waie I need not endeavor to keepe up various degrees of motion in the line $a.f.$ it will be

7 or ... please *add*.
10 if ... motion's *add*.
22 of the line, *add*.
33 I (*1*) thinke (*2*) need

---

[476]Brouncker: i.e. William Brouncker (1620?–84), q.v.

sufficient if there be alway's innumerable motion's kept up in it, some to the right hand, and some to the left hand, indeed I thought that the varietye of motion needed not to be destroyed in any line but I think that won't be necessary to the businesse. and now I confesse I am somewhat a weary of this long letter, and I supose Dr Wallis may be more a weary of it if you bring him to participate wholly of this trouble, but I thinke the best waie perchance would be only to sende the heades of it to him, and by that wee may see his iudgment of it, for that is the inconvenience of letters that one is fain| to write more then one would doe, because one knowes not what they will or may perchance except against and one may write lesse because one knowes not what they doe except against. this which my Lord Brouncker sayes will I thinke amount to that which Dr. Wallis sayes that the varietie of motions will be destroyed and but one motion left in the line (nor perchance in the world.) which I think will consist well enough with my hypothesis, but if this waie Dr Wallis doe not allowe of it neither, I can say no more at present but leave it to further consideration. I am sensible here is trouble enough for you for one time, if not too much so I rest

your humble servant
W. Neile

June. 15. 69.

90.
WALLIS to HENRY OLDENBURG
Oxford, 15/[25] June 1669

Transmission:

W Letter sent: LONDON *Royal Society* Early Letters W1, No. 91, 2 pp. (our source). Minor alterations in preparation for publication and supplementation of missing words (on account of torn paper) in Oldenburg's hand. Postmark on p. 2: 'IV/16'.—printed: OLDENBURG, *Correspondence* VI, 33–5.

E First edition of part of letter sent: *Philosophical Transactions* No. 48 (21 June 1669), 971–2.

Reply to: OLDENBURG–WALLIS 12/[22].VI.1669.

The first part of this letter is an account, by Wallis, of his latest tract against Hobbes, entitled *Thomae Hobbes quadratura circuli, cubatio sphaerae, duplicatio cubi; confutata.* Oldenburg published this account, with minor changes, as an anonymous review of that tract in *Philosophical Transactions.*

7 perchance *add.*
10 or may perchance *add.*

90. WALLIS to OLDENBURG, 15/[25] June 1669

Oxford June 15. 1669.

Sir,

In answere to yours[477] of June 12. I hope that by the next meeting of the Society on Thursday,[478] you will receive not onely a book for the Society, but some allso to distribute to other friends; My Lord Brounker,[479] Mr Boyle,[480] Sir Robert Moray,[481] Dr Wren,[482] Mr Collins,[483] & others[484] of our friends as far as they will go. If any thing hinder, it will be the want of Cuts; which wee expected from London last week, but have not received them.

As to your next question, what mistake &c. Mr Hobb's first grand mistake, is, in the demonstration of the first proposition; where these words, *Aut ergo in Triangulo ACG, triangulum rectangulum, cujus vertex sit A, aequale sectori ACL sumi nullum potest; aut PQL, CYP sunt aequalia*; are not at all proved; nor are they true. Hee had onely proved, That *If PQL, CYP, be equall; then such a triangle may bee*; But not the converse, *If those be not equal, then such a triangle cannot bee*. For if PQL be not equal, but a little bigger than CYP; &, consequently, the right-angled Triangle AYQ so much bigger than the sector ACL: it is manifest that a line drawn parallel to the base QY, a little nearer to the vertex A, may cut off a like Right-angled triangle (a little lesse than AYQ) which may be equal to the sector ACL.

Besides this (which overthrows all) his next great mistake, is in the demonstration of the second proposition: Where (supposing, by the first

---

16 than (1) the $\langle-\rangle$ (2) CYP

---

[477] yours: i.e. OLDENBURG–WALLIS 12/[22].VI.1669.

[478] Thursday: i.e. 17 June 1669 (old style). Oldenburg in fact distributed twelve copies of Wallis's *Thomae Hobbes quadratura circuli, cubatio sphaerae, duplicatio cubi; confutata*, including one for the Royal Society, during the meeting which took place on that day. See BIRCH, *History of the Royal Society* II, 382.

[479] Brouncker: i.e. William Brouncker, q.v.

[480] Boyle: i.e. Robert Boyle, q.v.

[481] Moray: i.e. Robert Moray (1608–73), army officer and politician, founder member of the Royal Society, *ODNB*.

[482] Wren: i.e. Christopher Wren (1632–1723), architect, mathematician, and astronomer, founder member of the Royal Society, *ODNB*.

[483] Collins: i.e. John Collins, q.v.

[484] others: apart from those named, the following members received copies: Daniel Colwall (?–1690), merchant and philanthropist, treasurer of the Royal Society, *ODNB*; John Wilkins (1614–72), theologian and natural philosopher, founder member of the Royal Society, *ODNB*; John Lowther (1642–1706), politician and industrialist, active member of the Royal Society, *ODNB*; Robert Hooke (1635–1703), natural philosopher and curator of experiments of the Royal Society, *ODNB*; and John Hoskins (1634–1705), lawyer and natural philosopher, founder member of the Royal Society *ODNB*.

90. WALLIS to OLDENBURG, 15/[25] June 1669

proposition, a square found equal to a circle,) he argues, that because the square takes in, as much of what is left out by the Circle, as [the Circle] takes in, of what is left out by the Square: therefore a cube [answering] to that Square, compared with a Sphere answering to that [Circle] will do so too. (Which would have been well argued of a Cylinder [on] that Circle, of equal hight with a Cube on that Square: but [not] so of a Sphere.) So that he seems here to have mis-taken a Cylinder for a Sphere.

Besides these two (which do influence all that follows,) hee allso (in his second figure) supposeth (untruly) without proof; that (on the common Center A) the Archs drawn by Y, O, h, will cutt the line AG, in the same points b, c, l, where the streight lines eb, zc, kl, (parallel to CG) do cut the sayd line AG: (which do influence all those propositions which do depend on these suppositions.) And other particular mis-takes, too many to enumerate.

As to Mr Neil;[485] he must either change his Hypothesis, or relinquish it. For, as it now stands; let his imaginable varieties of Motion be never so many, it is fully demonstrated, that not one of them can subsist longer than till it do either meet a contrary, or be overtaken by a swifter, or do run against another which is crossing his way. (For so soon as any of these three do happen, that variety is for ever extinguished.) If therefore hee will but allow, that every of these varieties (of moving parties) which now are, may be reasonably supposed once within half an hour, either to be overtaken by a swifter, or to meet a contrary, or at lest to runne against one that is crossing his way: Hee must then grant, that it may| as reasonably be supposed, that Once within half an hour, all motion will bee at end.

And he will find no evasion to avoid this consequence, wherein I have not allready prevented him.

All that can be hoped for is but, that two or three (or some small number) of the last, may escape a brush; & move onward so many severall ways without meeting, crossing, or overtaking one another. But (unless he

---

2 the Circle *suppl. Oldenburg*
3 answering *suppl. Oldenburg*
4 Circle *suppl. Oldenburg*
5 on *suppl. Oldenburg*
6 not *suppl. Oldenburg*
30 But (*1*) (without (*2*) (unless

---

[485] Neil: cf. OLDENBURG–WALLIS 8/[18].V.1669, enclosure; NEILE–OLDENBURG 13/[23].V.1669; NEILE–OLDENBURG 20/[30].V.1669; NEILE–OLDENBURG 1/[11].VI. 1669; NEILE–OLDENBURG 15/[25].VI.1669.

makes the world infinitely extense) even their journey will soon be at an end. But, of this, inough. I am,

Yours &c.
John Wallis.

For Mr Henry Oldenburg,
at his house in the Palmal,
near St James's
London.

91.
JOHN COLLINS to WALLIS
[London], 17/[27] June 1669

Transmission:

C Draft of letter sent CAMBRIDGE *Cambridge University Library* MS Add. 9597/13/6, f. 209^r–209^v (our source). On f. 209^v in Collins' hand: 'To Doctor Wallis June 17 1669'.— printed: RIGAUD, *Correspondence of Scientific Men* II, 514–16.

Reverend Sir

I yesterday left with Mr Pitts[486] to be sent to you, and he saith it was delivered to the Oxford Waggoner, that went on this Morning, to wit Doctor Newtons booke[487] of Guageing, and Andersons 3 sheetes,[488] which be pleased to accept. Dr Newton is said to have made 600 Numbers more in readinesse for the further enlargement of his Table, When you Write to Mr Peter Schooten[489] I should be glad you would vouchsafe to put a few quaeries[490] to

---

11 to the Oxford Waggoner ... went on *add.*

[486] Pitts: i.e. Moses Pitt (1639–97), London printer and bookseller *ODNB*.

[487] booke: i.e. NEWTON, *The Art of Practical Gauging*, London 1669.

[488] sheetes: i.e. ANDERSON, *Gaging Promoted. An Appendix to Stereometrical Propositions*, London 1669, which Collins describes to Gregory as being 'about 3 sheets'. See COLLINS–GREGORY 25.XI/[5.XII].1669 (TURNBULL, *James Gregory*, 73–4).

[489] Schooten: i.e. Pieter van Schooten, q.v.

[490] queries: Wallis posed these queries in a now missing letter to Pieter van Schooten: WALLIS–SCHOOTEN VI–XII.1669. Cf. WALLIS–COLLINS 24.VI/[4.VII].1669. He reports on Schooten's reply (SCHOOTEN–WALLIS XII.1669–I.1670) in WALLIS–COLLINS 11/[21].I.1669/70.

91. COLLINS to WALLIS, 17/[27] June 1669

him to witt, what became of Golius[491] his Manuscripts particularly of Vietas Harmonicon Coeleste,[492] Ad harmonicon coeleste libri quinque priores and some Remaines of Anderson[493] the Scot which were sent by Sir Alexander Hume[494] over thither to be printed What mathematicall booke[495] it is that Hudden hath now in the Presse at Amsterdam, whether Vossius[496] intends to enlarge his fathers Booke[497] de Scriptoribus Mathematicis (which the Preface sayth is the most crude peice that ever Gerrard Vossius[498] published) how he approoves of those Bookes here unknowne and not to be had viz one[499] intituled Sclot en Sclutel (the Lock and Key) and of Martin Wilkins[500] his Officina Algebrae belgice which hath above 300 Problemes solved in it and whether he the said Schooten will take care to procure and leave with such persons as he shall be directed to some few such bookes as

8 here ... had *add.*
10 belgice *add.*
11 to *(1)* |buy *inadvertently not del.*| and send *(2)* procure ... some few

---

[491] Golius: i.e. Jakob van Gool (1596–1667), Dutch orientalist and mathematician, professor of mathematics and Arabic in the University of Leiden. A catalogue of his library, including his celebrated collection of mathematical and oriental manuscripts, was published already in 1668. See HUYGENS–DOUBLET [19]/29.VI.1668; HUYGENS, *Œuvres complètes* VI, 227–8. On account of dispute among the heirs, the collection was not auctioned until 1696.

[492] Coeleste: i.e. VIÈTE, *Ad harmonicon coeleste libri quinque priores* (unpublished). Viète was working on this, a complete theory of cosmology, at the time of his death. Six manuscripts, two of them autograph, representing different levels of completion survive. See VAN EGMOND, *A Catalog of François Viète's Printed and Manuscript Works*, 386–7.

[493] Anderson: i.e. Alexander Anderson (1581/2–after 1621), Scottish mathematician, who became a pupil and friend of Viète in Paris. Collins is possibly referring to two now missing works on stereometry, known to have been in the possession of Alexander Hume. See COLLINS–GREGORY ?.II.1667/8 (RIGAUD, *Correspondence of Scientific Men* II, 174–9, 178).

[494] Hume: apparently Hume had taken possession of Anderson's papers after his death.

[495] booke: Collins had evidently heard from Oldenburg that a book by Hudde on optics was soon to be published in Amsterdam. From Hudde's correspondence with Spinoza it appears that he had assembled a small work on dioptrics. However, no book by Hudde on this topic was ever published. See COLLINS–GREGORY 15/[25].III.1668/9 (TURNBULL, *James Gregory*, 70–73, 71).

[496] Vossius: i.e. Isaac Vossius (1618–1689), Dutch philologist, *ODNB*.

[497] Booke: i.e. VOSSIUS, *De quatuor artibus popularibus, de philologia et scientiis mathematicis, cui opera subjungitur: chronologia mathematicorum libri III, editio nova*, Amsterdam 1660.

[498] Vossius: i.e. Gerardus Joannes Vossius (1577–1649), Dutch humanist scholar, father of Isaac Vossius, founder of Athenaeum Illustre in Amsterdam, *ODNB*.

[499] one: probably MARTINI, *Slot en sleutel van de navigation, ofte groote zeevart*, Amsterdam 1659. Martini also wrote *Oprecht, grondlich en rechtsinnigh school-boeck van de wyn-royeryen enz*, Amsterdam 1663.

[500] Wilkins: i.e. WILKENS, *Officina algebrae*, Groningen 1636.

91. COLLINS to WALLIS, 17/[27] June 1669

I (possibly for the use of the Societie) may send to him for, his money shall be advanced or reimbursed to him either at Amsterdam or Middleburgh, one John Jacob Ferguson hath lately in 1667 at the Hague published in low Dutch a booke[501] intituled Labyrinthus Algebra wherein he solves Cubick and biquadratick Æquations by such new Methods as render the rootes in their proper Species when it may be done to wit in whole or mixt Numbers, Fractions, or Surds either simple compound or Universall, and likewise improoves the generall Method, thereby accomplishing as much as Hudden in annexis[502] Geometriae Cartesianae, seemed to promise about it, this part is translated by Mr Old[503] and by me almost transcribd which annexed to Kinkhuysens Introduction[504] (with your helpe Advice or Assistance which the Lord Brounckner may possibly crave, the bookes being translated at his Lordships desire) will render the said Introduction very acceptable. I hope ere long to send you both to peruse in the *[breaks off]*

June 17th 1669

---

1 Societie) (*1*) shall (*2*) may
2 advanced or *add.*
2 to *add.*
3 in 1667 ... Hague *add.*
4 Cubick and biquadratick *add.*
5 by (*1*) new Methods wherein he accomplishes (*2*) such new Methods as render the rootes (*a*) , when (*aa*) |are *add.*| mixt (*bb*) they are either mixt (*b*) in their proper Species when it may be done to wit in whole or mixt Numbers, Fractions, or Surds either simple compound or Universall, and |likewise *add.*| improoves the generall Method, thereby accomplishing
9 Cartesianae (*1*) promised (*2*) seemed to promise
9 it, (*1*) which is (*2*) this part
10 Mr Old *add.*
13 the (*1*) same very acceptable (*2*) said

---

[501] booke: i.e. FERGUSON, *Labyrinthus algebrae*, The Hague 1667.

[502] annexis: i.e. HUDDE, *Epistolae duae quarum altera de aequationum reductione, altera de maximis et minimis agit*, in: *Geometria a Renato Des Cartes*, ed. Schooten, Amsterdam 1659.

[503] Old: i.e. Henry Oldenburg.

[504] Introduction: i.e. KINCKHUYSEN, *Algebra ofte stel-konst*, Haarlem 1661. This work was translated by Nicolaus Mercator into Latin; sheets containing the translation in Mercator's hand are interleaved in the printed Dutch text. The translation bears the title *Algebra sive (1) Ars Analytica (2) Logistica Speciosa, Tyronum usui conscripta* and has been augmented and annotated by Isaac Newton. This interleaved copy is now *Bodleian Library* Savile G. 20 (4). See COLLINS–GREGORY 14/[24].III.1671/2; TURNBULL, *James Gregory*, 224–6, 225; HOFMANN, *Nicolaus Mercator*, 72–3; SCRIBA, *Mercator's Kinckhuysen-Translation*.

92.
WALLIS to HENRY OLDENBURG
Oxford, 19/[29] June] 1669

Transmission:

W Letter sent: LONDON *Royal Society* Early Letters W1, No. 15, 4 pp. (our source). On p. 4 beneath address in Oldenburg's hand: 'Rec. june 21. 69.' Wallis dated the letter falsely '1665'. Postmark on p. 4: 'IV/21'.—printed: OLDENBURG, *Correspondence* VI, 55–9.

Reply to: NEILE–OLDENBURG 15/[25].VI.1669.

This letter represents Wallis's reply to NEILE–OLDENBURG 15/[25].VI.1669, a copy of which Oldenburg probably forwarded to Wallis in a now missing letter.

Oxford, June. 19. 166[9].

Sir

There is nothing in Mr Neile's letter[505] of June 15. which is not distinctly answered in mine[506] of June. 7. And, if I were by him, I could shew him where; but, because I am not, I can onely desire him to consider distinctly of every objection he makes, whether it be not there provided for: &, I am very sure, that he must needs find it is so. That which confounds him in this businesse, is, that he onely pleaseth himself, with the confused apprehension of innumerable varieties, & mutuall interfeers; without considering particularly what the influence of each will bee. For had hee considered this; hee must needs see; 1. that all slower motions (if not otherwise destroyed) will be swallowed up of the swifter: & so the number of varieties will decrease, in every line. 2. that all contrariant motions, do mutually destroy each other; which doth doubly decrease the number of varieties, (by killing two at once;) & this also in every line. 3. that in every decussation, the same happens in one of the two lines, if not in both: For, if the foremost of both the crossing parties do meet, the whole of both parties are stopped (& so the number of varieties is abated in both:) If not the foremost of both meet; yet at lest the foremost of one party will run against the side of the other; in which

---

1 1665 *corr. ed.*
3 not *(1)* abundantl *breaks off (2)* distinctly
9 considering *(1)* the *(2)* particularly
15 same |same *del. ed.*| happens

---

[505]letter: i.e. NEILE–OLDENBURG 15/[25].VI.1669.
[506]mine: i.e. WALLIS–OLDENBURG 7/[17].VI.1669.

92. Wallis to Oldenburg, 19/[29] June 1669

case one of the parties is wholly stopped (I mean, that whose leader runs against the others flank) & so the number of varieties is in his line abated; the other party is (not bored through, as he speakes, but) curtailed; all that is behind the knock being stopped. 4. that there being (in every line) all
5 these ways of decreasing varieties; there is no way of restoring any of them. Or, if he thinks there bee; hee should (without talking, in general, of innumerable varieties) have shewed, by what particular chance this can possibly be. Till then; I must say, as I have sayd before, that if it may be reasonably supposed that once within halfe an hour each party shal be either overtaken
10 by a swifter, or meet a contrariant, or run against another which is crossing his way (for then it is manifest his leader must be stopped, & therefore the whole party:) then it must as well be supposed that once within half an hour all motion shall bee at an end. All that can bee hoped, is but that two or three (or some small number of) varieties may at last persist to move
15 several ways, without enterfering: & even these few (unlesse he will suppose the world infinite) will soon be at their journies end. 5. that what is sayd of the number of varieties; is proved allso as to the number of *moving-parties*; that so soon as ever (for instance) six parties are by any accident reduced to four or five; it is not possible that for these four or five, wee can ever have
20 six (moving-parties) again.

When therefore he says, hee doth not know what it is I would have him prove, to make good his hypothesis: It is his own fault, not mine: For I had sayd very distinctly; That here it is hee is to shew his skill; if when six varieties of celerity, or six *parties-in-motion*, are reduced to the number of
25 five or four, hee can by any art bring them to the number of six again. And this is to be answered, not by asking *Why not?* but by shewing *How so.*| [2]

4 stopped (*1*) by stopping that which is (*2*) . 4.
6 hee (*1*) should (*a*) shew (*b*) have showed, by what (*aa*) possible (*bb*) particular means they (*aaa*) coul *breaks off* (*bbb*) can be observed (*2* should
11 stopped (*1*) with (*2*), &
12 as *add.*
15 few *add.*
18 *moving-parties*; (*1*) so soon as (*2*) For (*3*) that
19 or five; *add.*
19 or five *add.*
20 (moving-parties) *add.*
23 if (*1*) he can by any art (*2*) when
24 of celerity, *add.*
25 five or *add.*

92. WALLIS to OLDENBURG, 19/[29] June 1669

I have shewed him, why not; often inough all ready: when he tries to shew mee, How so; hee wil find himself at a losse.

But because he doth not yet think fit to attempt it; I wil attempt it for him, that he may see how far it will go. And I will do it with the greatest advantage for his hypothesis that I can.

1. Suppose a, b, c, d, e, f, moving in the same line, as six several particles, or parties; (all between them being either void, or at rest, or onely crossing the line, not moving in it.) By *parties*, I mean, Aggregates of contiguous particles.

2. Suppose c, (not to be stopped, or struck out; for then it is all one as if at first hee had not been there, or but at rest: nor to meet d, for then both are stopped, and are as if from the first they had been both at rest: but) to overtake d. Whereby of six marching parties, wee have but five; cd being but one. I say, it is impossible that for these five (by his principles) wee can ever have six again. For if so; it must be by reason either of its meeting, or overtaking, or being overtaken by, some other marching party in the same line; or by reason of some other cast in, or there found at rest; or of somewhat struck out: or at lest by a combination of some two or more of these cases. But neither any, nor all of these suppositions, will ever give us six for those five. Which I shew by parts.

3. Not by meeting. For if cd meet e; both are stopped, & instead of making five to be six, they are made but three. (And the like, if any other two should be supposed to meet, as ef or ab.)

4. Not by overtaking. For if cd overtake e; then our five (instead of being made six) are made but four. (And the like if a overtake b, or e overtake f, &c.)

5. Not by being overtaken. For if cd be overtaken by b: tis as before; our five are made but four. (And the like if any other be overtaken.)

6. Not by any there at rest, either before or behind cd; as g, h. For so long as they remain at rest, they are indeed more parties, but not (which is the thing in question) more *moving-parties*.

---

6 (*1*) 1. Suppose a, b, c, d, e, f, moving in the same line as six several parties, (*a*) and 2. let us then suppose (*b*) either resting between them all (*c*) at rest, between (*2*) 1. suppose
12 from the first *add.*
14 one. (*1*) For, i *breaks off* (*2*) I say,
14 (by his principles) *add.*
19 ever (*1*) make (*2*) give
20 parts. | Therefore *del.* |
27 tis as before; *add.*
29 rest, (*1*) ⟨—⟩ in (*2*) which is the same case, (for whe *breaks off* (*3*) either

92. Wallis to Oldenburg, 19/[29] June 1669

7. Not by these being put into motion. For g moves not till b come at him; nor h till cd come there: and then, though the parties may be greater, they will yet bee but five (as before) not six.

8. Not by any cast in between. For suppose g, h, (not to have been there at first, but) to be thrown in: the case is the same as before; for what ever is so thrown in, must there be first at rest (as at number 6.) and then can but be taken in to the partie next coming that way, (as at number 7;) So that we have still but five moving parties; not six.

9. Not by stopping first, & then striking out a whole party (suppose e) by collateral motions. For this doth still diminish not increase the number of marching parties, in that line.

10. Not by stopping first & then striking out, the hinder peece of a moving party or particle; (suppose b, from bg.) For though this may increase the number of *parties*, yet not of *moving parties*, in the line proposed. For the piece stopped is (if not struck out) at best at rest in the line af.| [3]

11. Not by restoring motion to the piece thus stopped: Suppose when overtaken by a. For still the number of parties in motion is but five (as before) not six. All the difference is but that in stead of a, bg, wee have now ab, g.

12. Nor is it recompensed by creating a new motion in another line. For (beside that this is nothing to the purpose, for wee are now inquiring, not of the motions in another line, but of restoring that lost in af,) the particle b or e struck out of this line, will no more increase the number of moving parties in that other line, than did (at numb. 8.) the particles g h (thrown in) increase the number of moving parties in this line: for the case is just the same.

13. Not by boaring a hole (as he speakes) or striking out a middle piece: suppose d between c and h (whereby in stead of one cdh, wee might have two parties c, h:) For this cannot bee. For d cannot be struck out, without being first stopped: & what stopps d doth stop c allso; and h goes on alone (as g did, at numb. 10.) So that still wee have but five moving parties in the line af: (cd being at rest, like b was, at number 10.)

1 by (*1*) any others are (*2*) these
5 before; (*1*) as were (*2*) for
7 to (*1*) another (*2*) the
8 but *add*.
21 now (*1*) speaking of (*2*) inquiring
28 one *add*.
32 (*cd* ... number 10.) *add*.

92. WALLIS to OLDENBURG, 19/[29] June 1669

Nor can he assign any other imaginable way, according to his principles, whereby he can again restore us six *moving-parties* (as at first we had) in stead of these five to which number (by the coalition of *cd*) they are reduced.

And what is sayd of this one, lost by the coalition of *cd*, holds as well of any other, so lost. And what is sayd of the line *af*; holds as well of every other line. That is, upon every coalition made by meeting, crossing, or overtaking, any moving particle or party; the number of moving-parties is fewer at lest by one (if not by two or more) than before it was: & that never to be repaired. And what the issue of this will be, if in half an hour (or half a minute) every moving party should either be overtaken by a swifter, or meet a con[trari]ant one, or run against a decussant moving-party: I leave to him to collect.

I have now answered his *Why not*, (just as I had done before, for all this was sayd in mine of June 7: and, the substance of it, from the first.) If hee be not satisfyed; let him try to answere my *How so*. Which is the task I sett him the last time; (to which hee replies nothing, but a generall noise of *innumerable varieties*, and a *Why not*, which signify nothing.) I say, let him try to marshall his moving parties so, as that when the number six is once by a coalition reduced to five, (whether of varieties of celeritie, of which hee seems allready to despair; or of moving-parties;) they may be restored to the number of six again; and tell mee by what incounters this may bee effected. Till that be done; I think I need say no more but that I am,

Yours &c.
J. Wallis

1. a.......b.......c.......d..............e.........f...............
2. ...a........b..............cd...........e..............f............
3. ......a........b..............¢ɟ¢.......................f.........
4. ....a.......b....................cde..............f........
5. ....a.....................bcd........e............f.......
6.8.12. .....a.....b....ɟ.......cd......ℎ.....e.........f.....
7.8.12. ....a........bg.............cdh........e.........f...

2 whereby (*1*) for these five movin *breaks off* (*2*) he |can *add.*| again
3 number *add.*
8 at lest *add.*

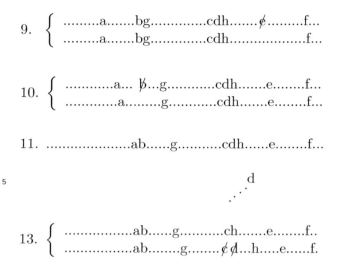

|As to what is sayd of my Lord Brounecker;[507] I do not know how far hee [4] is concerned in it. But, to him, a few words will be a demonstration. That the ways of decreasing the number of *moving-parties* (I do not say, *moving particles*,) are evident: (by overtaking, by meeting, & by crossing of motions.)

That the ways of increasing or repairing that number, are impossible. For 1. Nothing by coming-in from another line, can increase the number of moving-parties in this line: Because it must here be first at rest before it can in this line be moved. 2. Nothing here at rest, by being put into motion can do it: for it cannot receive motion but by being taken-in to a party allready in motion, & therefore increaseth not the number of parties in motion. 3. No party-in-motion, can by separation be made two moving-parties (Suppose by striking out a piece in the middle:) Because nothing can strike out a middle piece of a party in motion, without first stopping it; & what stops that, stops all behind it: & therefore doth not, of this one, make two *moving* parties. 4. Nor is there any other way (by his principles) to do it.

Therefore the number of parties in motion, will still grow fewer & fewer. Quod erat demonstrandum.

For Mr. Henry Oldenburg,
in the Palmal, near St James's
London.

---

[507]Brounecker: i.e. William Brouncker.

93.
WILLIAM NEILE to HENRY OLDENBURG
London, 23 June/[3 July] 1669

Transmission:

C Letter sent (including appendix): LONDON *Royal Society* Early Letters, N1, No. 8, 2 pp. (p. 2 blank) (our source).—printed: OLDENBURG, *Correspondence* VI, 65–6.

Reply to: WALLIS–OLDENBURG 19/[29].VI.1669.
Enclosure to (?): OLDENBURG–WALLIS 26.VI./[5.VII].1669.

This letter represents Neile's response to Wallis's detailed reply on his theory of motion (WALLIS–OLDENBURG 19/[29].VI.1669). It concludes the discussion between the two men on this topic. Although addressed to Oldenburg, Neile's letter was clearly intended for Wallis himself. Oldenburg probably sent a copy as an enclosure to OLDENBURG–WALLIS 26.VI./[5.VII].1669.

June 23.

Sir

I shall not goe about to deny Dr Wallises demonstration[508] I thanke him for it very much. if I have troubled him unsuccessfully I shall be sorry and I am sorry I did not apprehend him sooner. but I hope he will excuse it. I very much wishe that this businesse may be cleared and setled some waie or other. I desire not nature to adapt it selfe to my fancye, all that I desire is to comprehend the lawes and method of nature in in this businesse (if I could) and the causes for it. I hope this businesse will come to be better setled then yet it is, but perchance wee may come to find that there is too much obscurity in any waie we can goe. if not I shall be glad of it.

if wee shall supose that the motions tend all one waie in a line and with equal velocity moving with interstices of vacuity, in like manner supose an infinite number of these lines meeting in a centre, these motions (I conceive) will be no hindrance to one another by decussation but only iust in the centre, therefore I supose the motion of those lines may continue because there will be no coalescency nor no decussation but only iust in the centre. but this is a speculation I doe but only suggest at present, and shall not take

---

4 and...sooner. *add.*
8 in this businesse *add.*
11 we (*1*) shall (*2*) can

---

[508]demonstration: i.e. WALLIS–OLDENBURG 19/[29].VI.1669.

time to prosecute. all that I can say at present is, that I can not find what the flawe is in my principles, I confesse I am at a stand at present how to make aplication of them. I shall offer another thing that is, this, when I first fell upon the consideration of this matter I conceived: that two particles as the two cubes *A*. and *B*. which I proposed meeting together would make an exchange of their motions, as suposing that neither of them would have anye repugnancy to receive the motion of the other. but afterwards I conceived that neither of them could give a motion to the other without moving a litle the same waie at the same time which made mee reiect that hypothesis whether it ought to be again resumed I know not. but being at a losse at present in the consideration of these matters I should be glad to knowe Dr. Wallises iudgement about them. so with my thankes to him and your selfe for your consideration of this businesse I remain

your humble servant
W. Neile

there may perchance be something altered also concerning decussant motions, but I know not how to find a cause for compound or mixt motion.

94.
WALLIS to JOHN COLLINS
Oxford, 24 June/[4 July] 1669

Transmission:

W^1 Letter sent CAMBRIDGE *Cambridge University Library* MS Add. 9597/13/6, f. 210^r–210^v (f. 210^v originally blank) (our source). On f. 210^v in Collins's hand: 'Dr Wallis about Doctor Newton'. On f. 210* corrections to WALLIS, *Mechanica*.—printed: RIGAUD, *Correspondence of Scientific Men* II, 516–17.
W^2 Copy of letter sent: OXFORD *Bodleian Library* MS Add. D. 105, f. 22^r. On f. 22^v part of Wallis's calculation on which he bases his corrections to Newton's tables.

Reply to: COLLINS–WALLIS 17/[27].VI.1669.
Enclosures (?): Corrections to John Newton's *Art of Practical Gauging* and to Wallis's *Mechanica*.

It is probable that Wallis used the opportunity of this letter to send Collins his latest corrections to the proof sheets of his *Mechanica* which was being typeset by William

12-1 If wee shall ... time to prosecute. *text enclosed in box*
4 I conceived: *add.*
5 proposed (*1*) meeting (*2*) : I conceived that in meeting together these would (*3*) meeting

94. WALLIS to COLLINS, 24 June/[4 July] 1669

Godbid and printed by Moses Pitt. He possibly also enclosed corrections to Newton's tables, having read the latter's *Art of Practical Gauging* the previous day.

Oxford. June 24. 1669.

Sir,

I received, & thank you for, the two Bookes[509] you sent; & did yesterday look them (slightly) over. And find divers things proper inough for the subjects they handle. In that of Dr N. I find, the Presse was not well corrected, by reason of which many Errata have scaped; as well as some mistakes in the calculation: divers of which are amended in the table of Errata. But I find (as I did suspect, by Dr N.'s discourse with mee, & told him of it,) that, in his Table[510] of Segments, hee did not take his Data precisely inough, to bear a calculation of so many places: For in his table of 10 places, the three last figures are not accurate: As I found by examining some few, by accurate computation;[511] As for instance

	At number	Instead of	The accurate numbers are	which
vers. sine.	250.	0.19550,11110,	0.19550,11094,78−	agree but
	750.	0.80449,88890,	0.80449,88905,22+	to seven
	490.	0.48726,84781,	0.48726,84529,00+	places of
	510.	0.51273,15219,	0.51273,15471,00−	Decimals.

supposing the whole Diameter = 1000 | and the circle = 1. | which are his suppositions.

But however; 7 places, are much as the usuall Tables of Sines &c reach unto; & is abundantly sufficient for the business it is designed for.

I find a smal mistake in his praeparatory Table (in pag. 2, 3, of the sheet K next before that of Segments;) where, his numbers should have been

---

5 Dr Newton (his Art of Guaging) the Presse W^2
6 which (*1*) divers (*2*) many W^1
10 in (*1*) the (*2*) his W^1

---

[509]Bookes: i.e. John Newton's *The Art of Practical Gauging*, London 1669, and Robert Anderson's *Gaging Promoted. An appendix to Stereometrical propositions*, London 1669.
[510]Table: i.e. Newton's 'Table shewing the Area of the Segments of a Circle'.
[511]computation: for the corrections to the versed sines of 250 and 490 see Wallis's *Corrections to Newton's Art of Practical Guaging* (enclosure).

94. WALLIS to COLLINS, 24 June/[4 July] 1669

| First number, | 0.017,4532,9252− | not | 0.017,4532,9259 | & all the inter- |
| Last number, | 1.745,3292,5199+ | | 1.745,3292,5900 | mediate numbers proportionable. |

It's possible this smal (undiscovered) mistake, (in taking the first number of this table;) might influence the whole Table of Segments which follows; if (as I suppose he did) he calculated that Table upon the credit of this. It were not amisse, if he were advertised of it, before he proceed with the rest of his numbers.

When I write next to Van-Schooten,
I will propose the Quere's you mention.[512]

Mr Wing's new Book[513] I have not yet seen; & therefore can give you no opinion upon it. But Dr Wren[514] (who hath more considered that subject then I have done;) can much better satisfy you in that question than I.

I sent yesterday, by Mrs Margaret Lichfield;[515] another parcell of Copy (directed to Mr Pits;[516]) She lodgeth at Mr Moxon's[517] in Russel Street. If you or Mr. Pits go that way (before shee send it,) you may there have it. There remain about two such parcells, of the chapter we are now about.

I suppose, before this time, Mr Oldenburg hath given you one of my bookes[518] in Answere to that[519] of Mr Hobbs.

No more at present, but that I am,

Your friend to serve you
John Wallis.

---

3 taking *add.* W^1
6-7 It were ... his numbers. *add.* W^1
13 Mrs M. L. W^2
15 (before she send it,) *add.* W^1
17 you (*1*) a Copy (*2*) one W^1

---

[512]mention: cf. COLLINS–WALLIS 17/[27].VI.1669.
[513]Book: i.e. WING, *Astronomia Britannica*, London 1669.
[514]Wren: i.e. Christopher Wren.
[515]Lichfield: cf. WALLIS–COLLINS 5/[15].II.1666/7.
[516]Pits: i.e. Moses Pitt. (1639–97), London printer and bookseller, *ODNB*.
[517]Moxon's: i.e. Joseph Moxon (1627–91) printer, engraver, and instrument maker, elected Fellow of the Royal Society 1678, *ODNB*.
[518]bookes: i.e. WALLIS, *Thomae Hobbes quadratura circuli, cubatio sphaerae, duplicatio cubi; confutata*, Oxford 1669.
[519]that: i.e. HOBBES, *Quadratura circuli, cubatio sphaerae, duplicatio cubi, breviter demonstrata*, London 1669. Cf. BIRCH, *History of the Royal Society* II, 382.

95.
WALLIS to JOHN COLLINS
Oxford, 24 June/[4 July] 1669, enclosure (i):
Corrections to Newton's *Art of Practical Gauging*

Transmission:

W Working notes: OXFORD *Bodleian Library* MS Add. D. 105, f. 22^v (written on reverse of Wallis's copy of WALLIS–COLLINS 24.VI/[4.VII].1669).

Enclosure to (?): WALLIS–COLLINS 24.VI/[4.VII].1669.

Wallis probably received a copy of John Newton's *The Art of Practical Gauging* and Robert Anderson's *Gaging Promoted. An appendix to Stereometrical Propositions* from Collins in early June 1669. After cursorily reading the former on 23 June (old style), he noted the numerous errors contained in the 'Table of Segments' and probably sent these as an enclosure to WALLIS–COLLINS 24.VI/[4.VII].1669.

Suppos: Rad. = 1.000, 0		Quadrati ordinatim-applicatorum	Ordinatim-applicatae
Dist. a Centro,	5	0.999999, 75	0.999999, 875−
	1, 5	7, 75	8, 875−
	2, 5	3, 75	6, 875−
	3, 5	0.999987, 75	3, 875−
	4, 5	79, 75	0.999989, 875−
	5, 5	69, 75	84, 875−
	6, 5	57, 75	78, 875−
	7, 5	43, 75	71, 874+
	8, 5	25, 75	63, 874+
	9, 5	09, 75	54, 874−
	10, 5	0.999889, 75	44, 873+
	11, 5	67, 75	33, 873−
	12, 5	43, 75	21, 872−
	13, 5	17, 75	08, 871−
	14, 5	0.999789, 75	0.999894, 869+
	15, 5	59, 75	879, 868−
	16, 5	27, 75	863, 866−
	17, 5	0.999693, 75	846, 863+
	18, 5	59, 75	828, 860+
	0.019, 5	0.999619, 75	0.999809, 857−
		summa,	19.998667, 419

95. WALLIS: Corrections to Newton's *Art of Practical Gauging*, June 1669

Summa ducta in 0.001 } truncus quadrantis cujus altitudo $\frac{20}{1000}$
est 0.01999,86674,19 } Radii vel $\frac{10}{1000}$ Diametri. Subductus

ex Quad-
ratis } $0.78539,81633,97\frac{1}{2}-$, Relinquit

semi-
segm: } $0.76539,94959,77\frac{1}{2}$; cujus duplum est

segm: $9.53079,89919,57-$ } cujus sin. vers. $\frac{980}{1000}$ Radii vel $\frac{490}{1000}$ Diametri.

posito Radio = 1. adeoque

circ: $= 3.14159,26535,898-$. Ergo, posito

circ = 1.

segm: $= \frac{1.53079,89919,57-}{3.14159,26535,90-} = 0.48726,84529,00+$

Sed Newtoni tabula habet $= 0.48726,84781$,

quae itaque est erronea.

$3.14159,26535,90)\quad 1.53079,89919,57\quad (0.48726,84529,00$
$\overline{12.5663.70614,36}(4$
$27416,19305,21$
$\overline{25.132,74122,87}(8$
$2283,45182,34$
$\overline{21.99,11485,75}(7$
$84,33696,59$
$\overline{6.2,83185,31}(2$
$21,50511,28$
$\overline{18.84955,59}(6$
$265555,69$
$\overline{25.1327,41}(8$
$14228,28$
$\overline{12.566,37}(4$
$1661,91$
$\overline{15.70,80}(5$
$91,11$
$\overline{6.2,83}(2$
$28,28$
$\overline{28.27}(9$
1

1 (*1*) Summa ducta in 0.001 est } truncus quadrantis cujus altitudo a Diametro seu
 0.01998,66741,9 } Centro, sit $\frac{20}{1000}$ Radii vel $\frac{10}{1000}$ Diametri.

Subducti
ex quadr. } (*a*) 0.7859,866 (*b*) 0.78539,816

(*2*) Summa ducta ... $0.78539,81633,97\frac{1}{2}$.

95. WALLIS: Corrections to Newton's *Art of Practical Gauging*, June 1669

Iterum, posito Radio=1.

adeoque semiperipheria; vel planum circuli .... $\left.\right\} = 3.14159, 26535, 89793\frac{1}{4} -$

erit Arcus grad. 60. vel sector grad. 120 .... $\left.\right\} = 1.04719, 75511, 96597\frac{3}{4} -$

sinus versus grad. 60. et ejusdem Cosinus; .... $\left.\right\} = 0.5$

$\frac{1}{4}$ subtensae arcus grd. 120. $= \frac{1}{4}\sqrt{3}$; vel Triang: inscript. $\left.\right\} = 0.43301, 27018, 92219\frac{1}{3} -$

magnitudo segmenti, $= 0.61418, 48493, 04378\frac{1}{3} +$

Ergo, posito Circulo, $= 1$.

magnitudo segmenti, erit $= \dfrac{0.61418,48493,04370\frac{1}{2}}{3.14159,26535,89793\frac{1}{4}}$

hoc est (ut dividendo patet.) $= 0.19550, 11094, 77889\frac{1}{4} +$

Sed, Newtoni tabula habet, $0.19550,11110$.

$3 = (1)\ 0.51359, 87755, 90299 - (2)\ 1.04719, 75511, 96597\frac{3}{4} -$
$6 = (1)\ 0.09058, 60737, 06079\frac{1}{2}\ (2)\ 0.61418, 48493, 04378\frac{1}{3} +$
$8 = (1)\ \dfrac{0.09050,60737,06079\frac{1}{2}}{3.14159,26535,89793\frac{1}{4}}\ (2)\ \dfrac{0.61418,48493,04370\frac{1}{2}}{3.14159,26535,89793\frac{1}{4}}$
$9 = (1)\ 0.02883, 49428, 09265\ (2)\ 0.19550, 11094, 77885\frac{1}{4} +$

95. WALLIS: Corrections to Newton's *Art of Practical Gauging*, June 1669

$3.14159, 26535, 89793,)$ $\quad \overline{0.61418, 48493, 04378, 3}$ $\quad (0.19550, 11094, 77885 \tfrac{1}{4}$

$$\begin{array}{r} \overline{3.1415, 92653, 58979, 3(1} \\ 30002, 55839, 45399, 0 \\ \overline{28.274, 33388, 23081, 4(9} \\ 1728, 22451, 22317, 6 \\ \overline{15.70, 79632, 67949, 0(5} \\ 157, 42818, 54368, 6 \\ \overline{15.7, 07963, 26794, 9(5} \\ 34855, 27573, 7 \\ \overline{3.1415, 92653, 6(01} \\ 3439, 34920, 1 \\ \overline{3.141, 59265, 4(1} \\ 297, 75654, 7 \\ \overline{28.2, 74333, 9(09} \\ 15, 01320, 8 \\ \overline{12.56637, 1(4} \\ 2, 44683, 7 \\ \overline{21.9911, 5(7} \\ 24772, 2 \\ \overline{21.991, 1(7} \\ 2781, 1 \\ \overline{25.13, 3(8} \\ 267, 8 \\ \overline{251, 3(8} \\ 16, 5 \\ \overline{15, 7(5} \\ 18, \\ \overline{6.(2} \\ .2 \end{array}$$

96.
WALLIS to JOHN COLLINS
Oxford, 24 June/[4 July] 1669, enclosure (ii):
Corrections to Wallis's *Mechanica*

Transmission:

W Note sent: CAMBRIDGE *Cambridge University Library* MS Add. 9597/13/6, f. 210*.

Enclosure to (?): WALLIS–COLLINS 24.VI/[4.VII].1669.

This note contains corrections to the first part of Wallis's *Mechanica: sive, de motu, tractatus geometricus*, which was being seen through the press by Collins in London. Since there is no record of further correspondence between Wallis and Collins before completion of the printing of this part at the end of 1669, it is probable that these corrections accompanied WALLIS–COLLINS 24.VI/[4.VII].1669. None of the corrections were carried out until the publication of the second edition in volume 1 of Wallis's *Opera mathematica* in 1695.

pag. 9. l. pen. *l* ad *r. pag. 10. l. 24* 6 El.

l. 28 quam quod.

p. 14. l. 10. designatorum, indiciis . p. 23.
Marg. Dele. 19
p. 24 marg. Dele. 17. p. 25. marg. Fig. 20, 21.
p. 35. l. 27. dele C. p. 38. l. 7. dele &c.
l. 22. judicium:
p. 39. l. 29. (propter. p. 57. l. 32. impediatur.
p. 61. l. 1. esset
p. 70. l. 11 Ascendat. p. 72. tit. CAP. III.
p. 75. l. 9. per 31.
p. 78. l. pen. vel. p. 86. l. 10. accommodanda
p. 96. l. 16. sastinentur.
p. 98. l. 27. d4P. p. 99. l. 23. 20P, illic suspensa. p. 101. l. 18. Perpendiculare.
p. 103. l. 19 succedaneo. p. 105. l. 20. Succedaneo. p. 108. l. 26. Quas.

---

15 p. 103 l. 19 succedaneo *add.*

97.
HENRY OLDENBURG to WALLIS
26 June/[5 July] 1669

Transmission:

Manuscript missing.

Existence and date: Mentioned in WALLIS–OLDENBURG 16/[26].VII.1669 and answered by WALLIS–OLDENBURG 29.VII/[8.VIII].1669.

In this letter, Oldenburg evidently reported on a request by the Navy Commissioners, which Brouncker had presented at the meeting of the council of the Royal Society on 24 June, that the Society 'undertake the weighing of the wrecks in the Thames at Woolwich'. The president was instructed to reply that the Society did not have the means at its disposal for carrying out such work, but that nevertheless it was willing to offer advice. See BIRCH, *History of the Royal Society* II, 385.

Oldenburg also discussed the solution to Alhazen's Problem which Huygens had recently sent (see HUYGENS–OLDENBURG [16]/26.VI.1669; OLDENBURG, *Correspondence* VI, 42–4) and probably conveyed a copy of Neile's latest letter (NEILE–OLDENBURG 23.VI/[3.VII]. 1669).

98.
ROBERT BOYLE to WALLIS
3/[13] July 1669

Transmission:

Manuscript missing.

Existence and date: Mentioned in and answered by WALLIS–BOYLE 17/[27].VII.1669.

99.
HENRY OLDENBURG to WALLIS
5/[15] July 1669

Transmission:

Manuscript missing.

Existence and date: Mentioned in WALLIS–OLDENBURG 16/[26].VII.1669 and answered by WALLIS–OLDENBURG 29.VII/[8.VIII].1669.

Oldenburg evidently introduced a certain Mr Verinus—possibly a pseudonym for the German philologist Johann Ludwig Prasch (1637–90)—who wished to be granted a degree by the University.

100.
Thomas Lamplugh to Joseph Williamson
Oxford, 13/[23] July 1669

Transmission:

C Letter sent: KEW *The National Archives* SP 29/262, No. 174, 2 pp. On p. 2 in unknown hand: 'Alban hall Oxon. Jul. 13. 69.'

Dearest Sir

I had intended a larger accompt of the businesse of our Hick,[520] but that I am unexpectedly calld away by Sir Edward Norrys[521] to pay the last office to his deceased father[522] who departed this life last sunday night, & is to be buried this afternoon.

Your letter[523] of the 7th instant came not to my hands untill yesterday by the Post. I saw not the person you mention in it, to whom I should be kind for his brother's sake: you need have namd no other but your self to incite me to be kind to any person that comes from you.

I have spoken to Dr Wallis who I hope will provide you a perfect narrative[524] by the next Post. The Terrae filius yesterday[525] was very abusive,[526] but not witty, for which it is likely he will suffer. Dr Compton[527] performed

---

[520] Hick: probably William Hicks (*c*.1630–82), author of the miscellany *Oxford Jests*, evidently first published in 1669, ODNB. See WOOD, *Life and Times* II, 176, and WOOD, *Athenae Oxoniensis* III, 490.

[521] Norrys: i.e. Sir Edward Norris (*fl*.1675–1708), politician, sometime MP for Oxfordshire and for Oxford. Like his father, he lived at the manor in Weston on the Green, Oxfordshire.

[522] father: i.e. Sir Francis Norris (1609–1669), politician, sometime MP for Oxfordshire, ODNB.

[523] letter: i.e. WILLIAMSON–LAMPLUGH 7/[17].VII.1669.

[524] narrative: Wallis in fact decided against writing to Williamson. See WALLIS–OLDENBURG 16/[26].VII.1669.

[525] yesterday: i.e. Monday, 12 July 1669. On that day, as already two days earlier, on Saturday, 10 July, the Act and Vespers, which previously had been held in the University Chruch of St Mary the Virgin, took place for the first time instead in the newly-opened Sheldonian Theatre. Abusive satires of the Terrae filii, presented during the ceremonies, created a scandal. See WALLIS–OLDENBURG 16/[26].VII.1669. The dedication and official opening of the Sheldonian Theatre had taken place on 9 July. See DE BEER, *Diary of John Evelyn* III, 530–4.

[526] abusive: see Wallis's reports in WALLIS–OLDENBURG 16/[26].VII.1669 and WALLIS–BOYLE 17/[27].VII.1669.

[527] Compton: i.e. Henry Compton (1631/2–1713), canon of Christ Church (installed 24 May 1669). Thereafter appointed bishop of Oxford, 1674, and bishop of London, 1675, ODNB. On 10 July 1669, Compton was inceptor in theology at the first commemoration held in the Sheldonian theatre.

his part so well, that he came off with great applause, gained reputation to himself, & did the University a great deal of honour. I am now in haste

Alban Hall
July. 13. 69

(Dearest Sir,)
Yours sincerely
T. Lamplugh

For his ever honourd
friend Joseph Williamson
Esquier att his
lodgings in
Whitehall.

101.
WALLIS to HENRY OLDENBURG
Oxford, 16/[26] July] 1669

Transmission:

W Letter sent: LONDON *Royal Society* Early Letters W1, No. 92, 2 pp. (our source). On p. 2 beneath address in Oldenburg's hand: 'Rec. july 17. 69. Answ. by Count Zani july 22. 69.' Postmark on p. 2: 'IY/19'. — printed: OLDENBURG, *Correspondence* VI, 129–30.

Reply to: OLDENBURG–WALLIS 26.VI/[6.VII].1669 and OLDENBURG–WALLIS 5/[15].VII. 1669.

July 16. 1669. Oxon.

Sir,

I have yours[528] of June 26, & of July 5. but have had no time to answer either of them. Friday, July 9. was the Dedication of our New Theater.[529] In the morning was held a convocation in it: Wherein was read, first, the Archbishops[530] Instrument of Donation:[531] then, a Letter[532] of his, intimat-

---

[528] yours: i.e. OLDENBURG–WALLIS 26.VI/[6.VII].1669 and OLDENBURG–WALLIS 5/[15].VII.1669.

[529] New Theater: on the consecration of the Sheldonian Theatre and the events pertaining to the Act and Vespers over the following days see WOOD, *Life and Times* II, 165, and DE BEER, *Diary of John Evelyn* III, 530–4.

[530] Archbishops: i.e. Gilbert Sheldon (1598–1677), Archbishop of Canterbury (from 1663), *ODNB*.

[531] Instrument of Donation: i.e. Sheldon's deed of gift, dated 25 May 1669; *Oxford University Archives* SEP/X/1, and NEP/supra/Reg Ta, pp. 264–5 (copy).

[532] Letter: i.e. Sheldon's letter of 28 May 1669; *Oxford University Archives* NEP/supra/Reg Ta, p. 266 (copy).

101. WALLIS to OLDENBURG, 16/[26] July 1669

ing that he had designed two thousand pounds to buy a purchase for the endowment of it: then, a letter of thanks,[533] to be sent him in the name of the University. After this, Dr South,[534] as University Orator, made a long Oration; which consisted, first, of Satyrical Invectives, against Cromwell, Fanaticks, the Royal Society, & New Philosophy: next of Encomiasticks, in prayse of the Archbishop, the Theater, the Vice-chancellor,[535] the Architect,[536] & the Painter;[537] And, lastly, of Execrations, against Fanaticks, Comprehension, & New Philosophy. The Oration ended, some honorary degrees were conferred, & the convocation dissolved. The Afternoon, was spent in Panegyrick Orations; & in reciting Poems of severall sorts of Verse, composed in prayse of the Archbishop, the Theater, &c, & crying down fanaticks. The whole action began & ended with a Noyse of Trumpets, & twise was interposed variety of other Musick, Vocall & Instrumentall, purposely composed for this occasion. On Saturday & Munday those exercises appertaining to the Acte & Vespers, which were wont to be performed in St Maries church, were now had in the Theater: where the Terrae filius,[538] for both days, were so abhominably scurrilous, (& permitted so to proceed, without the lest check or interruption, from any of those who were to govern the Exercise;) and gave so generall offense to all honest spectators; that (I beleeve) the University hath thereby lost more reputation, than they have gained by all the rest. All or most of the heads of houses, & eminent persons in the University, with their relations, being represented, in the foulest language, as

---

10 of Verse, *add.*
11 Theater, &c, (*1*) & ⟨—⟩ (*2*) & crying
17 so (*1*) intolerably (*2*) abhominably
20 thereby *add.*
22 with their relations, *add.*
22 represented, (*1*) as ⟨—⟩ but (*2*) in ... language,

---

[533]letter of thanks: i.e. Letter of thanks to the Archishop of Canterbury, dated 9 July 1669; *Oxford University Archives* NEP/supra/Reg Ta, pp. 267–8 (copy).

[534]South: i.e. Robert South (1634–1716), public orator of the University from 1660, *ODNB*.

[535]Vice-chancellor: i.e. John Fell (1625–86) dean of Christ Church, vice-chancellor of the University of Oxford 1666–9, *ODNB*.

[536]Architect: i.e. Christopher Wren (1632–1723), *ODNB*.

[537]Painter: i.e. Robert Streater (1621–79), serjeant painter to the crown (1660–79), *ODNB*.

[538]Terrae filius: a 'native son' or member of the University of Oxford chosen to speak at the public Act. The terrae filius on 10 July 1669 was Henry Gerard of Wadham College. Thomas Hayes of Brasenose College also acted as terrae filius during the festivities. See HENDERSON, *Putting the Dons in Their Place*, 32–8 (the Latin text of Gerard's speech is 41–8, and the English translation is 49–57).

a company of whore-masters, whores, & Dunces. During this solemnity, (& some dayes before & after it,) the Duke of Yorks Players,[539] have usually acted two Plays a day, (at their Theater behind the town-Hall;) which (for ought I hear) hath been much the more innocent Theater of the two. The Vice-chancellor hath since imprisoned the two Terrae filios, & ('tis sayd) means to expell them.[540]

Mr Williamson[541] (I am told) desired I would write him an account of these affairs: which, if the proceedings had been such as I could have commended, I should readyly have done: but, being as they are, I desire you will present him my service, & excuse mee that I have not thought fit to write him any thing:[542] from

Sir

Yours &c

These   [2]
For Mr Henry Oldenburg,
in the Palmal, near
St James's
London.

102.
WALLIS to ROBERT BOYLE
Oxford, 17/[27] July 1669

Transmission:

W Letter sent: LONDON *Royal Society* Boyle Letters 5, f. 174^r–175^v (our source). The names 'Marya Hastings' and 'Huntingdon' on f. 175^r have been crossed out. Postmark on f. 175^v damaged.—printed: BOYLE, *Correspondence* IV, 141–4.
w Part copy of *W*: LONDON *British Library* Add. MS 6193, f. 69^r–70^r.

4 Theater *add.*
9 readyly *add.*

---

[539] Duke of Yorks Players: i.e. the theatre company established by the poet and playwright William Davenant (1606–68), *ODNB*.
[540] Gerard was expelled together with Thomas Hayes on 22 July 1669: see WOOD, *Life and Times* II, 166.
[541] Williamson: i.e. Joseph Williamson (1633–1701), *ODNB*.
[542] any thing: Williamson had in fact already been informed of the proceedings by Thomas Lamplugh. See LAMPLUGH–WILLIAMSON 13/[23].VII.1669.

102. WALLIS to BOYLE, 17/[27] July 1669

E^1 BOYLE, *Works* (1744) V, 514–5.
E^2 BOYLE, *Works* (1772) VI, 458–60.
Reply to: BOYLE–WALLIS 3/[13].VII.1669.

Oxford July. 17. 1669.

Sir

After my humble thanks for the honour or yours[543] of July 3. I thought it not unfit to give you some account[544] of our late proceedings here. Friday July 9. was the Dedication[545] of our New Theater. In the morning; was held a Convocation in it, for entering upon the possession of it. Wherein was read, first, the Arch Bishop's[546] Instrument of Donation[547] (sealed with his Archiepiscopal Seal) of the Theater, with all its furniture, to the end that St Maries Church may not be further profaned by holding the Act in it: Next, a Letter[548] of his, declaring his intention to lay out 2000$^£$ for a purchase to endow it; Then, a Letter[549] of thanks to be sent from the University to him; wherein he is acknowledged to be both our Creator, & Redeemer; for, having not onely built a Theater for the Act, but, which is more, delivered the Blessed Virgine from being so profaned for the future; he doth (as the words of the letter are) non tantum Condere, hoc est Creare, sed etiam Redimere. These words (I confess) stopped my mouth from giving a Placet to that letter when it was put to the Vote. I have since desired Mr Vice-chancellor[550] to consider whether they were not liable to a just exception: Hee did, at first, excuse it; but, upon further thoughts, I suppose hee will think fit to alter them before the letter be sent & registred. After the Voting of this

---

12 both *add.*

[543] yours: i.e. BOYLE–WALLIS 3/[13].VII.1669.
[544] account: cf. WALLIS–OLDENBURG 16/[26].VII.1669 and DE BEER, *The Diary of John Evelyn* III, 530–4.
[545] Dedication: the dedication of the Sheldonian Theatre on 9 July 1669 (old style).
[546] arch Bishop's: i.e. Gilbert Sheldon (1598–1677), Archbishop of Canterbury (from 1663), *ODNB*.
[547] Instrument of Donation: i.e. Sheldon's deed of gift, dated 25 May 1669; *Oxford University Archives* SEP/X/1, and NEP/supra/Reg Ta, pp. 264–5 (copy). See TYACKE, *History of the University of Oxford* IV, 173–7.
[548] Letter: i.e. Sheldon's letter of 28 May 1669; *Oxford University Archives* NEP/supra/Reg Ta, p. 266 (copy). See WOOD, *Life and Times* II, 165.
[549] Letter of thanks: i.e. Letter of thanks to the Archishop of Canterbury, dated 9 July 1669; *Oxford University Archives* NEP/supra/Reg Ta, pp. 267–8 (copy).
[550] Vice-chancellor: i.e. John Fell (1625–1686), dean of Christ Church, vice-chancellor of the University of Oxford 1666–9, *ODNB*.

102. WALLIS to BOYLE, 17/[27] July 1669

Letter, Dr South[551] (as University Orator) made a long Oration: The first part of which, consisted of Satyricall Invectives; against Cromwel, Fanaticks, the Royal Society, & new Philosophy: The next, of Encomiasticks in praise of the Archbishop, the Theater, the Vice-chancellor, the Architect,[552] & the Painter:[553] The last, of Execrations; against Fanaticks, Conventicles, Comprehension, & New Philosophy; damning them, ad Inferos, ad Gehennam,[554] The Oration being ended, some honorary degrees were conferred,[555] & the convocation dissolved. The Afternoon was spent in Panegyrick Orations, & reciting of Poems, in several sorts of Verse, composed in praise of the Archbishop, the Theater &c; & crying down Fanaticks. The whole Action, began & ended with a Noyse of Trumpets: and twice was interposed Variety of Musick, Vocal & Instrumentall; purposely composed for this occasion.|

On Saturday & Munday, those Exercises appertaining to the Act & Vespers, which were wont to be performed in St Maries Church, were had in the Theater. In which, beside the number of proceeding Doctors (9 in Divinity, 4 in Law, 5 in Physick, & 1 in Musick,) there was little extraordinary; but onely that the Terrae Filius,[556] for both days, were abhominably scurrilous; & so suffered to proceed without the lest check or interruption, from Vice-chancellor, Provicechancellors, Proctors,[557] Curators, or any of those who were to govern the Exercises; which gave so general offense to all honest Spectators; that, I beleeve the University hath thereby lost more reputation, than they have gained by all the rest. All or most of the heads of houses, & eminent persons in the University, with their relations, being represented as a company of Whore-masters, Whores, & Dunces. And, among the rest, the

---

[551]South: i.e. Robert South (1634–1716), public orator of the University from 1660, *ODNB*.

[552]Architect: i.e. Christopher Wren. He took his inspiration for the building from the Theatre of Marcellus in Rome; in constructing the roof Wren employed the geometrical flat floor devised by Wallis. See WREN, *Parentalia: or, memoirs of the family of the Wrens*, London 1750, 338–9, and WALLIS, *Mechanica: sive, de motu, tractatus geometricus* III, 589–91; *Opera mathematica* I, 953–5; NEWMAN, *The Architectural Setting*, 174.

[553]Painter: i.e. Robert Streater (1624–79), serjeant painter to the crown (1660–79), *ODNB*.

[554]ad Inferos ... Gehennan: i.e. TERTULLIAN, *Apologetic* XLVII, 12.

[555]conferred: see WOOD, *Fasti oxoniensis* II, 315–6.

[556]Terrae Filius: i.e. Henry Gerard of Wadham College and Thomas Haynes of Brasenose College. Cf. WALLIS–OLDENBURG 16/[26].VII.1669.

[557]Proctors: i.e. Nathaniel Alsop (17th century) of Brasenose College and James Davenant (d. 1717), Fellow of Oriel College.

102. WALLIS to BOYLE, 17/[27] July 1669

Excellent Lady,⁵⁵⁸ which your letter mentions, was, in the broadest language, represented as guilty of those crimes, of which (if there were occasion) you would not stick to be her compurgator: &, (if it had been so,) shee might (yet) have been called Whore in much more civill language. During this solemnity (& for fo⟨ur⟩ days before & since) have been constantly acted (by t⟨he⟩ Vice-chancellors allowance) two Stage-playes in a day, ⟨besides⟩ those of the Duke of York's⁵⁵⁹ house,) at a Theater erected f⟨or⟩ that purpose at the Town-hall: Which (for ought I hear) was much the more innocent Theater of the Two. It hath been here a common fame for divers weeks (before, at, & since the Act,) that the Vice-chancellor had given $300^£$ bond (some say $500^£$ bond) to the Terrae filius, to save them harmless, what ever they should say, provided it were neither Blasphemy, nor Treason. But this I take to be a slander. A lesse incouragement would serve the turn with such persons. Since the Act (to satisfy the common clamour) the Vice-chancellor hath imprisoned both of them: &, 'tis sayd, he means to expell them. But inough of this.|

[5ʳ] I am next to acquaint you with a discourse of another nature; and which, I trust, will bee, if not more acceptable, yet lesse ungratefull then the former. Since the Act, Sir James Langham⁵⁶⁰ (with his Lady,⁵⁶¹ & some other persons of quality) did mee the honour to dine with mee. Hee is a person of whom I have sometime heard you speak with very good respect: & whom (I think) I may safely represent as one who honours you very much: & who hath divers times told mee, how ambitious hee should bee, of obtaining a nearer acquaintance with you. Hee was telling mee of an excellent Lady, of whom he hath a very great esteem: and I have the less reason to think him mis-taken in his judgement, because he hath had all the opportunities of knowing her very well; having married a Sister⁵⁶² of hers, &, upon that occasion, beeing throughly acquainted with the state of that family. It is the Lady Marya⁵⁶³ Hastings, a daughter of the Countess of

---

29 Marya Hastings *later crossed out*

⁵⁵⁸*Excellent Lady*: i.e. Susanna Wallis (1622–87). The conduct of Wallis's wife had been impugned in Gerard's speech.
⁵⁵⁹Duke of York's: i.e. the theatre company established by the poet and playwright William Davenant (1606–68), *ODNB*.
⁵⁶⁰Langham: i.e. James Langham , q.v.
⁵⁶¹Lady: i.e. Penelope Langham (née Holles) (d. 1684), third wife of Sir James Langham.
⁵⁶²Sister: i.e. Elizabeth Langham (née Hastings) (1635–64), daughter of Ferdinando Hastings (1609–56), sixth earl of Huntingdon.
⁵⁶³Lady Marya: i.e. Mary Hastings (d. 1679), daughter of Ferdinando Hastings (1609–56), sixth earl of Huntingdon.

102. WALLIS to BOYLE, 17/[27] July 1669

Huntingdon,[564] & sister to the present Earle,[565] & to the Lady Elizabeth in whom Sir James did think himself very happy while he enjoyed her as his wife. Of this Lady he gives so high a Commendation; for her temper, her parts, her worth, her vertues, her piety, & every thing else; as makes him extremely solicitous to see her happy in a sutable consort: & doth profess himself so serious in it, that, were it a thing lawfull, there is no Lady hee knows, whom he should sooner have made choise of, to have succeeded her sister in that capacity. It's true, I have not my self been so happy as to know that Excellent Lady; but he that hath had the opportunity of knowing her so well, doth represent her, not onely so accomplished as to make an excellent wife, but particularly as an excellent wife[566] for Esquire Boyle. And doth undertake to answere all objections, not onely which you might make as to the Lady, but (which perhaps may be the greater difficulty) as to the condition allso: And hopes hee shall be able to show you reasons, not onely, to marry her; but, to merry. I should have added (though that be a consideration of less moment than some of the rest) that her Portion will be at lest 4000$^£$, (& so much he will see made good:) her age, about five or six & twenty: And as to her person, vertues, & other per|fections, you will [17 have so many ways of informing yourself, that what I might say would bee superfluous. If I might be a happy Instrument in making two so excellent persons happy in each other, as hee perswades mee I might if you think fitt: I do not know in what else I could more approve my selfe,

Sir,
Your Honours very humble &
affectionate Servant.
John Wallis.

You will oblige mee, to present my service to the very good Lady[567] with whom you are: & to preserve mee a place in her good opinion. And you will doubly oblige my wife, at this time, to afford her the like in yours.

---

1 *(1)* Shrewsbury *(2)* Huntingdon *later crossed out*
2 Sir James *later crossed out*
4 temper, |her *add.*| parts,

---

[564]Countess of Huntingdon: i.e. Lucy Hastings (née Davies) (1613–79), countess of Huntingdon, Irish noblewoman, *ODNB*.

[565]Earle: i.e. Theophilus Hastings (1650–1701), seventh earl of Huntingdon, *ODNB*.

[566]wife: Mary Hastings eventually married William Jolliffe, a baronet, in 1674; this was effectively beneath her status as the daughter of an earl.

[567]Lady: i.e. Katharine, Viscountess Ranelagh (née Boyle) (1615–91), influential Irish noblewoman and sister of Robert Boyle, closely associated with the Hartlib circle, *ODNB*.

104. WALLIS to OLDENBURG, 29 July/[8 August] 1669

These
For the Honourable
Robert Boyle Esquire, at
the Lady Ranelagh's house
in the Palmall, near
St James's.
London.

103.
HENRY OLDENBURG to WALLIS
23 July/[2 August] 1669

Transmission:

Manuscript missing.

Existence and date: Mentioned in and answered by WALLIS–OLDENBURG 29.VII/[8. VIII]. 1669. (Oldenburg's endorsement on the back of WALLIS–OLDENBURG 16/[26].VII.1669 gives the date as 22 July.)
Reply to: WALLIS–OLDENBURG 16/[26].VII.1669.

In this letter, Oldenburg apparently gave his first reaction to events which had taken place in Oxford during the opening of the Sheldonian Theatre. It was evidently carried to Oxford by the Bolognese nobleman and philologist Ercole Zani (1634–84), and served to introduce him to Wallis (cf. Oldenburg's endorsement to WALLIS–OLDENBURG 16/[26].VII. 1669). See BIRCH, *History of the Royal Society* II, 393, and WALLIS–OLDENBURG15/[25]. VIII.1669.

104.
WALLIS to HENRY OLDENBURG
Oxford, 29 July/[8 August] 1669

Transmission:

W Letter sent: LONDON *Royal Society* Early Letters W1, No. 93, 2 pp. (our source). At top of p. 1 in Oldenburg's hand: 'Enter'd LB. 3. 160.' On p. 2 beneath address in Oldenburg's hand: 'Rec. july 30. 69.' and the endorsement 'Dr Wallis's Letter to Mr Oldenb. concerning his thoughts of Mr Neil's hypothesis about motion.' Oldenburg has also noted 'july. 29.' Postmark: 'IY/30'.—printed: OLDENBURG, *Correspondence* VI, 159–61.
w^1 Copy of letter sent: LONDON *Royal Society* Letter Book Original 3, pp. 160–1.
w^2 Copy of w^1: LONDON *Royal Society* Letter Book Copy 3, pp. 196–7.

Reply to: OLDENBURG–WALLIS 26.VI/[6.VII].1669, OLDENBURG–WALLIS 5/[15].VII.1669, and OLDENBURG–WALLIS 23.VII/[2.VIII].1669.

104. WALLIS to OLDENBURG, 29 July/[8 August] 1669

Oxford. July 29. 1669.

Sir,

Our busy time of late, hath cast mee so far behind hand in answering letters, that I have not yet recovered it; for I have divers yet lying on my hands unanswered. To yours[568] of June 26. I can say little as to the weighing of the ships sunk; because I am so much a stranger to the common practise in those affairs: but if I were present with others versed in it, I should not refuse to suggest my thoughts in discourse.

Mr Hugen's optical Probleme[569] I have not had time yet to consider of; but it doth not seem, at first view, to be a matter of very great difficulty. However; if hee have done it already: there will be the less need of doing it again, at lest till I have a little more leisure.

As to what concerns Mr Neil's hypothesis[570] (new, & old,) where the mistake lyes: I have told you my thoughts of it, in my first letter upon that subject. That, by his principles, he makes equal causes to produce unequal effects. As for instance; the force or Vis motrix, which now moves one Pound, so soon as it incounters a body at rest of Ninety nine pounds, must carry both, that is a hundred pound, with the same speed. Which is not to be granted him.

Next, because Body, as such, is indifferent to rest or motion; hee postulates, that Body at rest, as at rest, is so to; & hath no repugnance to motion. Which will not be allowed him neither. For Body at rest hath a repugnance to motion; and body in motion, hath repugnance to rest: though body, as Body, be indifferent to either; & will therefore continue as it is (whether in rest or motion) till some positive alter its condition. And, when such positive

---

8 thoughts (*1*) on (*2*) there about (*3*) in
9 optical *add.*
11 already: *add.*
11 need (*1*) for another (*2*) of
16 Pound, (*1*) must (*2*) so
17 at rest *add.*
22 neither *add.*
22 rest (*1*) is not (*2*) hath

---

[568] yours: i.e. OLDENBURG–WALLIS 26.VI/[6.VII].1669.
[569] Probleme: i.e. Alhazen's problem of finding the point on a spherical mirror where light is reflected between two points on the same plane. See HUYGENS–OLDENBURG [16]/26.VI.1669; OLDENBURG, *Correspondence* VI, 42–44.
[570] hypothesis: see NEILE–OLDENBURG 23.VI/[3.VII].1669.

104. Wallis to Oldenburg, 29 July/[8 August] 1669

cause comes, it acts proportionably to its strength. The lesse the strength is which moves, & the heavier the body to be moved, the slower will be the motion, of which proportion, he takes no notice at all.

Both these faults you will find to bee suggested in my first letter upon his hypothesis. And what hath been since sayd, is but the natural result thereof.

In answere to yours[571] of July 5. I did direct M. Verinus[572] the best I could. I told him the business would be difficult, (because, in granting degrees, wee respect time of study;) but was not impossible; (because wee do sometimes dispense with it upon just occasion:) that it was, almost, wholly in the Vice-chancellors breast as things now stand, who I doubted would not be forward to favor it: That the most advantageous way of presenting it to him, would bee by a testimonial from abroad, of so many years spent [in an] other University, which might (as is oft done) be admitted as if spent here. Whether hee have accordingly applyed himself to the Vice-chancellor,[573] & with what success, I have not yet heard.

What you say in yours[574] of July 23. is much the opinion of some others. The persons you have in your several letters recommended, I have endeavoured to serve as you desired.

I am

Yours &c
John Wallis.

The books[575] sent from Hevelius, are received. Mr Vice-chancellor desires you to return the thanks of the University for the one; & I, mine for the other.

[2] These
For Mr Henry Oldenburg
in the Palmal near

---

1 strength (*1*) & the bigger (*2*) is
3 of ... at all. *add.*
4 you (*1*) find observed (*2*) will find to bee suggested
13 spent |in an *del.*| other *corr. ed.*

---

[571] yours: i.e. Oldenburg–Wallis 5/[15].VII.1669.
[572] Verinus: possibly the German philologist Johann Ludwig Prasch (1637–90), who is known to have used this pseudonym.
[573] Vice-chancellor: i.e. John Fell.
[574] yours: i.e. Oldenburg–Wallis 23.VII/[2.VIII].1669.
[575] books: presumably two copies of Hevelius, *Cometographia*, Danzig 1668.

107. OLDENBURG to WALLIS, 4/[14] August 1669

St James's
London

105.
WALLIS to ROBERT BOYLE
second half of July 1669

Transmission:

Manuscript missing.

Existence and date: Mentioned in WALLIS–BOYLE 17/[27].VIII.1669 as one of two letters which reached Boyle's hands with a delay but before 7 August 1669 (old style). It is therefore probable that the present letter was written sometime between 17 July and the end of that month.

As emerges from WALLIS–BOYLE 17/[27].VIII.1669, this letter, like that preceding it, concerned the events which had taken place in Oxford during the opening of the Sheldonian Theatre.

106.
WALLIS to JAMES LANGHAM
second half of July–first half of August 1669

Transmission:

Manuscript missing.

Existence and date: Mentioned in postscript to WALLIS–LANGHAM 24.VIII/[3.IX].1669 . In WALLIS–BOYLE 17/[27].VII.1669 Wallis describes his meeting with Langham after the Act. It is evident that the letter to which he refers in WALLIS–LANGHAM 24.VIII/[3.IX].1669 would have been written sometime after that meeting, which necessarily must have taken place near the middle of July, and before the letter in late August.

107.
HENRY OLDENBURG to WALLIS
4/[14] August 1669

Transmission:

Manuscript missing.

Existence and date: Referred to in WALLIS–OLDENBURG 15/[25].VIII.1669.

109. WALLIS to OLDENBURG, 15/[25] August 1669

As emerges from WALLIS–OLDENBURG 15/[25].VIII.1669, this letter was probably conveyed by hand by the unidentified person for whom it contained a recommendation. It also enclosed a paragraph from a recently received letter from Huygens on the theory of motion: HUYGENS–OLDENBURG [31.VII]/10.VIII.1669; OLDENBURG, *Correspondence* VI, 161–3, 162.

108.
ROBERT BOYLE to WALLIS
7/[17] August 1669

Transmission:

Manuscript missing.

Existence and date: Mentioned in WALLIS–BOYLE 17/[27].VIII.1669 and partly quoted in WALLIS–LANGHAM 24.VIII/[3.IX].1669.
Reply to: WALLIS–BOYLE 17/[27].VII.1669.

After Boyle had received Wallis's account of his meeting with Sir James Langham in WALLIS–BOYLE 17/[27].VII.1669, he indicated in this letter his willingness to discuss the proposal of a suitable wife when Langham was back in London. Cf. BOYLE, *Correspondence* IV, 144.

[...] As for that part of your letter which relates a discourse &c when I hear of his return to London [...]

109.
WALLIS to HENRY OLDENBURG
Oxford, 15/[25] August 1669

Transmission:

W Letter sent: LONDON *Royal Society* Early Letters W1, No. 94, 2 pp. (our source). Postmark: 'AU/16'.—printed: OLDENBURG, *Correspondence* VI, 189–90.

Reply to: OLDENBURG–WALLIS 23.VII/[2.VIII].1669, and OLDENBURG–WALLIS 4/[14].VIII. 1669.

Enclosure (?): Note on the forthcoming visit of the Vice-chancellor.

Aug. 15. 1669. Oxford.
Sir,

'Tis very little I have to say in answere to yours[576] of Aug. 4. more than that I have endeavoured to be civil to the person recommended; who is returning

---
[576] yours: i.e. OLDENBURG–WALLIS 4/[14].VIII.1669.

109. WALLIS to OLDENBURG, 15/[25] August 1669

(with more company) to morrow toward London. Nor to the note[577] inclosed concerning M. Hugens, with whom I do for the most part concur, though not in all. For I am not yet satisfyed in that notion of Descartes, which hee seems to imbrace, that Motion is onely relative; &, of the two bodies separated, it is indifferent whether of the two be sayd to move. To Mr Neil's question,[578] What is Vis motrix: is not, as to the hypothesis, necessary to answere. For that there is a Vis somewhere to give motion, is as clear as that there is motion. (Else wee must have an effect without a cause.) But, what Force is, may perhaps be better Felt, then expressed in Words. (As well as, what is Motion, Heat, Light &c.) Nor is it allways of one kind; but sometimes as an Impetus within; sometime an Impulsus from without. As for yours[579] of July 23. It need not seem strange that either the Orator[580] or the Terrae filius[581] should fall foul upon a Society of which the King is One, &c. For even the King[582] himself was not spared. Of whom (amongst other things) it was sayd: that he was, in this like, an Angel, that he could, not propagate sobolim;[583] & that when he should obtain his dominions in the world of the moon hee would make there an Archbishop of Cuckolds (archiepiscopum corniculati.) Which whether it were comely or no to be sayd of the King; I leave others to judge. (But I think it very imprudently done of them, thereby to renew the memory of the late story. For it was presently replyed, that seat was not vacant; there was a Bishop there allready;[584] called the

---

7 Vis (*1*) when (*2*) somewhere
8 cause.) (*1*) And (*2*) But,
10 &c.) (*1*) And yet is sometime from within (*a*) with *breaks off* (*b*) from within sometime from without (*2*) Nor
13 King |himself *del.*| is
19 of them, thereby *add.*
20 story (*1*) of a friend concerning a Bishop ⟨—⟩ in the ⟨—⟩, or a friend of theirs (*2*). For

---

[577]note: i.e. the passage on the theory of motion in HUYGENS–OLDENBURG [30.VII]/10.VIII.1669) (OLDENBURG, *Correspondence* VI, 161–3, 162), which Oldenburg had quoted in OLDENBURG–WALLIS 4/[14].VIII.1669.

[578]question: this question was evidently raised in a now missing letter from Neile.

[579]yours: i.e. OLDENBURG–WALLIS 23.VII/[2.VIII].1669.

[580]Orator: i.e. Robert South (1634–1716), public orator of the University from 1660, *ODNB*.

[581]Terrae filius: i.e. Henry Gerard (17th century) of Wadham College, Oxford.

[582]King: i.e. Charles II (1630–85).

[583]Sobolim: i.e. a loanword for 'descendants', derived from the Latin 'soboles' with the Hebrew ending 'im'; evidently intended as a humorous play on 'Angel'

[584]allready: probably a reference to WILKINS, *The discovery of a world in the moone, or, A discourse tending to prove, that 'tis probable there may be another habitable world in*

109. WALLIS to OLDENBURG, 15/[25] August 1669

Man in the Moon.) But, the truth is, I think wee may without breach of charity beleeve, that, as to the ribauldry in the Act, it was by Mr Vice-chancellors[585] particular direction. For if it be true, that the Terrae filius's had commission to say what they would, provided it were not blasphemy or treason; or (as it is most widely expressed) that hee bid them Bite, but not fetch too much bloud; or, that they might take liberty inough, without Killing of men. And if (as is proposed) he did give intimation for such & such particular persons to be abused. (Though he did not put the particular words in their mouthes.) It may well inough be supposed to rest at his door. But himself did not wholly scape; for (beside his other scotches) that which was supposed most to come home, was, that hee was represented as a hater of Women, a prosecutor of whores, but had a man for his bedfellow: (that hee, & another whom I shall not name, had but one hat, one heart, & one Bed.) But I shall trouble you with no more of this ribauldry. Adding onely that I am

Yours &c.

About 10 days hence the Vicechancellor & others will bee at London to admit their new Chancellor, the D. of Ormond; who was chosen[586] Aug. 4. the thing being carried very privately, without any notice given of it to the university till the day before; that there might be no making of parties for any body else. Hee was highly recommended by the Archbishop's letter. In which, was this, amongst other things; that it would be very acceptable to the King. I was out of town at the time; & heard not of it, till the day after.

[2] These
For Mr Henry Oldenburg,
in the Palmal, near St James's
London.

---

3 that *(1)* hee gave leave to the Terrae filius *(2)* the Terrae filius's had commission
6 or *(1)* ⟨sub–⟩ *(2)* that
22 was this, *add.*

---

that planet, London 1638. John Wilkins (1614–72) had been consecrated Bishop of Chester in November 1668, following the fall of the Earl of Clarendon.

[585]Vice-chancellor: i.e. John Fell.
[586]chosen: by letter of 31 July 1669 Sheldon resigned his office of chancellor, recommending James Butler (1610–88), first Duke of Ormond, as his successor. Butler was elected on 4 August 1669. See WOOD, *Life and Times* II, 166.

110.
Wallis to Henry Oldenburg
Oxford, 15/[25] August 1669, enclosure:
Note on the Forthcoming Visit
of the Vice-Chancellor

Transmission:

W Note sent: LONDON *Royal Society* Early Letters W1, No. 127 (our source).—printed: OLDENBURG, *Correspondence* VI, 190.

Enclosure to (?): WALLIS–OLDENBURG 15/[25].VIII.1669.

This otherwise unidentifiable slip of paper, which partly agrees with the wording in WALLIS–BOYLE 17/[27].VIII.1669, was probably a late addition to the earlier letter to Oldenburg, on the paper of which no further space remained. As the postscript to WALLIS–OLDENBURG 15/[25].VIII.1669 reveals, Wallis only discovered the vice-chancellor's plan to journey to London the day after completing that letter.

'Twere not amisse, when the Vice-Chancellor[587] comes up next week, if the King[588] gave him a sharp rebuke for those miscarriages: particularly for what concerns the royal Society & himself. And if he excuse himself by saying he hath expelled the persons:[589] it may be replied, he should not first have set them on work.

111.
Wallis to Robert Boyle
Oxford, 17/[27] August 1669

Transmission:

W Draft of (missing) letter sent: OXFORD *Bodleian Library* MS Add. D. 105, f. 37ʳ (our source).—printed: BOYLE, *Correspondence* IV, 144–6.

Reply to: BOYLE–WALLIS 7/[17].VIII.1669.

---

[587] Vice-Chancellor: i.e. John Fell.
[588] King: i.e. Charles II (1630–85).
[589] persons: i.e. Henry Gerard of Wadham College who together with Thomas Hayes of Brasenose College was expelled on 22 July 1669. See WOOD, *Life and Times* II, 166. Cf. WALLIS–OLDENBURG 16/[26].VII.1669.

111. WALLIS to BOYLE, 17/[27] August 1669

To Esquire Boyle.

Oxford Aug. 17. 1669.

Sir,

I received by the last post the honour of that[590] from you of Aug. 7. But do not perfectly know what did occasion the hysteron proteron of my two last,[591] unless it were either your absence out of Town (which I knew not of) or of his to whom that paquet was directed in which yours (with others[592]) was inclosed, which I suspect allso; (for yours is the first return to any thing in that pacquet:) I might have added (to what was in those letters) that his Majestie himself did not escape the reproches of that day. Of whom (amongst other things) it was sayd, That hee was, in this, like an Angell, that he could not propagare sobolem: And that, when hee shall obtain his dominions in the Moon, hee will make an Archbishop of Cuckoos (Archiepiscopum Corniculati:) with some other things which I do not remember. Which was sayd of him with so little reverence in the face of so great an Assembly, that I think it were not amisse, (when the vice-Chancellor comes up next week, to admit our new Chancellor the D. of Ormond,[593]) if the K. would take so far notice of it, as to give him a check for it. I might adde to it, what Brasier[594] of St John's gave us but last Sunday out the pulpit at St Maries (from Pro.[595] 10. 12. hatred stirs up strife, but love covers all sins;) where (amongst other Insinuations against Fanaticks), he told us, that 'twas no charity to mitigate the penalties of the law, for this was not in the power of any man to do but of the supreme magistrat; and if the Prince himself do so far make himself a party with the offenders, as not to execute the laws

---

10 the reproches of that day *add.*
15 in the face of so great an Assembly, *add.*
17 the K. (*1*) should (*2*) would
18 for it (*1*), & let him (*2*)
20-21 where (*1*) (speaking of Charity) (*2*) (amongst other Insinuations against Fanaticks)
22 law (*1*), against (speaking of fanaticks,) (*2*), for this
23-24 Prince himself (*1*) should |so *add.*| bee of a party (*2*) do so far make himself a party

---

[590] that: i.e. BOYLE–WALLIS 7/[17].VIII.1669.

[591] two last: presumably WALLIS–BOYLE 17/[27].VII.1669 and a second letter written shortly after this: WALLIS–BOYLE second half of VII.1669.

[592] others: there is no record of which letters these were.

[593] i.e. James Butler (1610–88), first Duke of Ormonde, *ODNB*.

[594] Brasier: i.e. John Brazier (d. 1679), Fellow of St John's College, Oxford, who in October 1668 was nominated vicar of St Giles' and later became rector of Aldington, Kent. See WOOD, *Life and Times* II, 145.

[595] Pro.: i.e. *Proverbs* 10,12.

111. WALLIS to BOYLE, 17/[27] August 1669

upon them, this is no charity, but a weakening the strength of the law, till at length he will be so weak himself, as not to stand before the no-charity of those to whom he shews this charity. But such harangues as these (& much worse) are so frequent with us, that it is no news: for 'tis the common theme of a great part of our sermons. But perhaps it will be news, that (to balanse the Duke our Chancellor) the City of Oxford did yesterday choose the Duke of Buckingham[596] to be their High Steward.

For the latter part of your letter: I must refer you for an answere to that worthy person[597] who did to mee undertake to satisfy all the objections that are there made (for I had before represented them as likely to be your objections) for hee told me, the Lady[598] concerned is so much a Philosopher, & so much a Virtuosa, & so well able to judge of worth, that the person of Esquire[599] B. would much more than supply what could be pretended otherwise to be defective. And if she should judge otherwise, she would not be of the opinion of

Sir
Your H.'s very humble servant
J. W.

The Election of our chancellor (Aug. 4) was carried by the vice-chancellor,[600] with so much privacy, that the University had no notice of it till were night (that there might be no opportunity of making a contrary party.) Hee was in the Archbishops letter of resignation, highly recommended; &, particularly, that it would be very acceptable to the King.[601] What orders he had to signify so much from the thing, he knows best.

---

1 them, this (*1*) was (*2*) is
1 but (*1*) weakens (*2*) a weakening
2 he (*1*) would (*2*) will
8 you *add.*
10-11 before (*1*) made them in your behalf,) (*2*) represented them as likely to be your Objections)
11 the (*1*) person (*2*) Lady
12 to judge of |the *del.*| worth

---

[596] Buckingham: i.e. George Villiers (1628–87), second Duke of Buckingham, *ODNB*. Villiers was an old adversary of the Duke of Ormonde, in particular on account of his politics in Ireland.
[597] person: i.e. Sir James Langham. See WALLIS–LANGHAM 24.VIII/[3.IX].1669.
[598] Lady: i.e. Mary Hastings (d. 1679). Cf. WALLIS–BOYLE 17/[27].VII.1669.
[599] Esquire B.: i.e. Robert Boyle.
[600] vice-chancellor: i.e. John Fell.
[601] King: i.e. Charles II (1630–85).

112.
WALLIS to JAMES LANGHAM
Oxford, 24 August/[3 September] 1669

Transmission:

W Draft of (missing) letter sent: OXFORD *Bodleian Library* MS Add. D. 105, f. 37ʳ.

To Sir James Langham.

Oxford. Aug. 24. 1669.

Sir

I received from Mr Boyle, in answere to that[602] I sent him, a Letter[603] which so far as concerns your affair, was in these words.

As for that part of your letter which relates a discourse &c when I hear of his return to London.

Which I thought fit to give you in his own words, because I know not how better to express his sense. I have, in my reply[604] to him, referred him to you to answere those objections hee makes; And that you had beforehand undertaken so to do: the Lady[605] being so much a Philosopher, so much a Virtuosa, & so well able to judge of worth, that the person of Esquire[606] B. would much more then supply what could be pretended to be otherwise defective. And I know you can do it much better, both to his satisfaction & your own, than can

Sir

your very humble servant.

J.W.

3-4 Sir (*1*) In answere to that (*2*) I received
9 sense. (*1*) What is further to be done (*2*) I have, in my reply
13 pretended to be *del. ed.*
10-14 makes; (*1*) which I (*a*) trust (*b*) hope you may do to his (*aa*) satis (*bb*) satisfact⟨ion⟩ (if I do not very much mis-remember) you have undertaken to do & can do (*2*) And that ... you can do

---

[602]that: i.e. WALLIS–BOYLE 17/[27].VII.1669.
[603]Letter: i.e. BOYLE–WALLIS 7/[17].VIII.1669.
[604]reply: i.e. WALLIS–BOYLE 17/[27].VIII.1669.
[605]Lady: i.e. Mary Hastings (d. 1679). Cf. WALLIS–BOYLE 17/[27].VII.1669.
[606]Esquire B.: i.e. Robert Boyle.

113. WALLIS to SLUSE, 10/[20] September 1669

Sir, I was very serious in what I desired of you in the close of my letter.[607] Wherein, if you can well accomodate mee, I shall release you of another promise you made mee.

113.
WALLIS to RENÉ FRANÇOIS DE SLUSE
London, 10/[20] September 1669

Transmission:

w^1 Copy of missing letter sent (with corrections entered in Oldenburg's hand): LONDON *Royal Society* Early Letters W1, No. 95, 4 pp. (our source). At top of p. 1 in Oldenburg's hand: 'This was sent the 2nd time to Slusius, he |not *add.*| having received the first, as he intimates in his letter (*1*) sen⟨t⟩ (*2*) of jan. 2. 1670.' and 'Enter'd LB 3.179'. On p. 4, again in Oldenburg's hand: 'Dr Wallis's Letter |(mentiond in the precedent) *add.*| to Mr Slusius concerning the Occasion of Dr Wren's Invention of the Hyperbolick Cylindroid, and his conference about it with Dr Wallis.'—partly printed in OLDENBURG, *Correspondence* VI, 237–8 (Latin original); 238–9 (English translation).
w^2 Copy of letter sent: LONDON *Royal Society* Letter Book Original 3, pp. 179–82.
w^3 Copy of w^2: LONDON *Royal Society* Letter Book Copy 3, pp. 281–2.

Answered by: SLUSE–WALLIS [28.II]/10.III.1669/70.
Enclosure to: OLDENBURG–SLUSE 14/[24].IX.1669.

On his recent journey through London, Wallis had seen Sluse's letter to Oldenburg of 16 August 1669 (new style) (OLDENBURG, *Correspondence* VI, 178–83), in which Sluse comments favourably on Wren's method for generating a hyperbolic-cylindroidal solid suitable for grinding hyperbolic lenses. In this letter Wallis informs Sluse that he had discovered the fundamental principle employed by Wren. To this end he gives an account of a conversation which he had had with Wren on the topic. The present letter was sent by Oldenburg as an enclosure to his own letter to Sluse of 14/[24] September 1669 (OLDENBURG, *Correspondence* VI, 232–5). Since neither letter arrived, Sluse asked Oldenburg to send him the most substantial parts again. Oldenburg duly complied and sent a copy of Wallis's letter as an enclosure to OLDENBURG–SLUSE 26.I/[5.II].1669/70 (OLDENBURG, *Correspondence* VI, 447–8). From the absence of lines proposed for correction in WALLIS–OLDENBURG 17/[27].X.1669, it is apparent that this copy is not identical with the original letter sent. See WALLIS–OLDENBURG 16/[26].X.1669.

Clarissimo Eruditissimoque Viro Domino
5   Renato Francisco Slusio, Soc. Leod.
Joh. Wallis Geo. Professor Savil. Oxon. salutem.

———

2 shall (*1*) release (*2*) release

———

[607] letter: i.e. the now missing letter WALLIS–LANGHAM, second half of July–first half of August 1669.

113. WALLIS to SLUSE, 10/[20] September 1669

Vidi ego nuperrime apud D. Oldenburg (quem in transitu invisebam) tuas ad eum literas, quae ingeniosum Wrennii nostri, de Cylindroide Hyperbolica, inventum[608] spectant. Quas tamen neque per tempus tum licuit totas perlegere, neque hactenus datum est eas iterum inspicere. Praemium tantum, quo quid agas ostendis, legi: Nempe haec dico observas, primo, rem pariter succedere, si Dolabrae acies obliquo situ ponatur, non modo ad Axem Cylindroidis, sed et ad Horizontem (quod omnino verum est; quippe positionis ad Horizontem ratio ut habeatur non est necesse;) deinde exquiris quo pacto possit, ea methodo, Hyperbola quavis assignata confici.

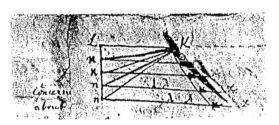

Quaenam sit tua constructio (ut fateor, quod res est) ne legere quidem vacabat (nedum examinare) rem autem ipsam quam simplicissime exponam. Tibi forte non ingratum erit: quo et rei gestae historia nonnihil poterit conducere.

Aderat mihi non ita pridem Wrennius noster, atque inter alia, de quibus colloquebamur, interrogabat me in hunc sensum (nec enim ipsissima verba memini, neque erant scripto consignata:) si ad rectae alicujus puncta quotlibet aequalibus intervallis sumpta, ordinatim-applicentur rectae, quarum quadrata sint, ut numerorum continue consequentium 1, 2, 3, 4, &c. quadrata, eodem aliquo, vel aequalibus quadratis aucta, qualisnam sit ea curva, quae per earum extrema reliqua transeat? Respondi protinus (absque omni cunctatione,) Hyperbolam esse. Petebat, num certus essem, rem ita esse? Respondi certissimum, utpote quod jam olim demonstraveram inter ea, quae [2] meo de Cycloide| tractatui[609] subjunxeram. Et deprompto statim libro ostendebam (ad fig. 27 ibidem) rectas nx tales esse, quales ille ordinatim-applicatas vellet; quae si in situ $[nx]$ ordinatim-applicentur, Hyperbolam

---

1 in situ nk w^1 corr. ed.

[608] inventum: see WREN, 'Generatio corporis cylindroidis hyperbolici, elaborantis lentibus hyperbolicis accomodati', printed in *Philosophical Transactions* No. 48 (21 June 1669), 961–2. Cf. BENNETT, *Christopher Wren*, 34–8.
[609] tractatui: i.e. in WALLIS, *Tractatus duo*, 97–9; WALLIS, *Opera mathematica* I, 555.

113. WALLIS to SLUSE, 10/[20] September 1669

[Kxx] designarent. Quod si manentibus eisdem LK, [nx], ordinatim-applicatis intervalla Ln, nn minora fuerint vel majora, quam [nunc] sunt, prodibit utcunque Kxx Hyperbola, sed cujus aliud sit latus rectum, aliusque ad Asymptoton $L\lambda$ angulus $nL\lambda$, quam nunc est. Quo facto, causam, cur illud interrogaverat hanc indicabat; nempe, quum viderit aliquando in officina, inter alia, venalem corbem quendam vimineum rotundum, ex viminibus tantum rectis contextum, situ obliquo positis (credo, et decussatis) cujus superficies lateralis, Cylindrum extrinsecus excavatum, exhiberet (ea forma, qua salina solent apud nostrates confici, vel trochlearum orbiculi;) animadvertit, uno ex viminibus illis circa Cylindri axem circumducto, manente situ illo ad axem obliquo, descriptum iri superficiem illam concava-convexam, adeoque Torno posse confici cylindroides ejusmodi, per aciem Dolabrae rectam, obliquo ad cylindri axem situ positam, cujus sectio per axem, foret ea linea curva. Sed qualis sit ea linea curva, se nondum satis examinasse dixit, sed Hyperbolam esse non dubitavit.

Respondebam ego, rem omnino ita esse, curvamque illam procul omni dubio meram esse Hyperbolam; sed et, speculationem hanc dignam esse quae melius excoleretur. Urgebam itaque ne pateretur, ut tam elegans speculatio neglecta periret: sed ut in commodam formam redigeret, et typis vulgaret: Quod fecit ille paulo post, ea quo vides forma, subjuncta demonstratione lineari ad morem veterum, qualibus, prae literalibus, ad Recentiorum Analyticorum methodum magis assuescit.

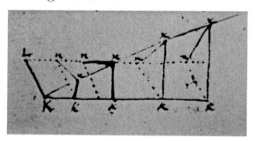

Quod autem problema, mihi ab ipso propositam, negotio suo accomodum fuerit; sic more meo ostendetur. Si intelligatur acies Dolabrae recta [Kxx] in quacunque| ab axe Cylindroidis Ln distantia (quocunque ad Horizontem situ) quocunque ad axem Cylindri situ obliquo posita, manifestum est, per rectam illam [Kxx] transiturum planum aliquod, ut xKk, cui   [3]

1 Hyperbolam Kkk w^1 corr. ed.
1 LK, nk w^1 corr. ed.
2 quam non sunt w^1 corr. ed.
26 recta Kkk w^1 corr. ed.
28 illam Kkk w^1 corr. ed.

parallelus sit axis Cylindroidis Lnn, rectamque aliquam, in illo plano, axi parallelam, ut Kkk (nempe ex parallelis, eam quae sit [axis] proxima) lineam contactus esse, qua planum illud tangat Cylindrum (Cylindroidi inscriptum) axe Lnn radioque LK descriptum: sumptisque in axe Cylindri Ln, partibus continue aequalibus, Ln, nn, &c. atque ad eum perpendicularibus LK, nk. &c. erectisque itidem perpendicularibus $[kx]$, kx &c. Manifestum est, rectas Kk esse, ut 1. 2. 3. &c. numeros continue consequentes; earumque quadrata, ut quadrata horum: et propterea (propter nkx angulum rectum, rectasque nk invicem aequales) junctis xn, quadrata harum esse, ut numerorum quadrata illa, aequalibus quadratis aucta; si itaque Torni ope circa axem Ln, describi intelligantur circuli, quorum radii sint ipsae $[nx]$ rectae, sectio per axem exhibebit ipsissimum nL, Kx (prioris figurae) planum.

Patet hinc; tum, situs ad Horizontem, nullam hic rationem habitam esse (sed situs tantum ad Axem Cylindri,) tum etiam (quod tu hic inquiris,) quo pacto possimus datam Hyperbolam exhibere. Quippe, si exponatur, in figura priore, Hyperbola Kxx, cujus centrum L, semi-axis transversus LK, Axis conjugatus Ln, cui similem torno exhibere imperatum sit; hoc tantum curandum erit, nempe, ut in figura posteriore, sumatur LK (minima aciei Dolabrae ab axe Cylindri, seu Cylindroidis, distantia) quanta fuerit LK prioris, eoque situ ponatur Kxx acies, ut, sumptis in utraque figura aequalibus Ln, aequales itidem sint nx respectivae. Quod obtinebitur, si sumatur, in figura [posteriore] angulus kKx, [aequalis] angulo figurae prioris $nL\lambda$, quem cum Asymptota facit conjugatus axis. Atque haec sunt, quae raptim scribenti dicenda videbantur de hoc negotio: Tibi, spero, futura non ingrata. Vale Lond. Sept. 10. 1669.

114.
JOHN ELLIS to WALLIS
7/[17] October 1669

Transmission:

C Letter sent: LONDON *British Library* Add. MS 32499, f. 32^r–32^v (verso blank).

2 sit axi w^1 *corr. ed.*
6 perpendicularibus Kk, kx, &c. w^1 *corr. ed.*
11 ipsae nk w^1 *corr. ed.*
22 posteriori w^1 *corr. ed.*
22 aqualis w^1 *corr. ed.*

115. OLDENBURG to WALLIS, 14?/[24?] October 1669

As this letter reveals, Wallis was being employed at this time to decipher encrypted letters to and from Edward Hyde, first earl of Clarendon, who since July 1668 had been living in exile in Montpelier. Ellis stood under the patronage of Sir Joseph Williamson, who since the early 1660s played a central role in the restoration government's intelligence-gathering activities. Where appropriate, Williamson would convey the intelligence to his patron, Henry Bennet, secretary of state for the south. The present letter suggests that in respect of the Clarendon letters Bennet was more directly involved.

7: octob: 1669

Sir

What I have received from you gives verry great satisfaction; but I pray for the future put the Number of the pacquet which you send mee in three or fouwer words; that so I may know & be sure I have all you send.

You will finde hereinclosed ten severall peeces[610] in English whereof some are of a new but verry easie Cijpher. My Lords[611] desire you would make what speed you with them. And for the frenche,[612] I shall the Next post sende you a Considerable number.

And so I rest

Sir,

Your most affectionate servant
J. Ellis

115.
HENRY OLDENBURG to WALLIS
14?/[24?] October 1669

Transmission:

Manuscript missing.

---

[610] ten peeces: i.e. the attached ciphers, now *British Library* Add. MS 32499, f. 18^r–29^v. On f. 24^r Wallis has written: 'Letters of the E. of Clarendon, (& to him) in France 1669' and below this 'Of these Letters I had not the Originals, but onely Copies, very faultily transcribed; many of which faults I have corrected in the Margin; but |could *add.*| not |correct *add.*| all. And even the Titles of some of them are wrong'.

[611] Lords: probably Henry Bennet (1618–85), secretary of state for the south, created first Baron Arlington in 1665, *ODNB*. Bennet had promoted Edward Hyde's removal from office in 1667.

[612] the frenche: probably the copies of ciphers in Wallis's hand, now *British Library* Add. MS 32499, f. 35^r–40^v.

116. WALLIS to OLDENBURG, 16/[26] October 1669

Existence and date: Referred to in postcript to WALLIS–OLDENBURG 17/[27].X.1669.
Answered by: WALLIS–OLDENBURG 17/[27].X.1669.

In this letter, Oldenburg apparently informed Wallis that Wren was unhappy about the way their discourse on the hyperbolic cylindroid had been presented in WALLIS–SLUSE 10/[20].IX.1669.

116.
WALLIS to HENRY OLDENBURG
Oxford, 16/[26] October 1669

Transmission:

W Letter sent: LONDON *Royal Society* Early Letters W1, No. 96, 4 pp. (p. 3 blank) (our source). On p. 4 beneath address in Oldenburg's hand: 'Acc. Octob. 18. 69.' Postmark on p. 4: 'OC/18'.—printed: OLDENBURG, *Correspondence* VI, 282–3 (Latin original); 284–5 (English translation).

Enclosure: WALLIS–OLDENBURG 17/[27].X.1669.

As emerges from WALLIS–OLDENBURG 17/[27].X.1669, Wallis's original version of his letter to Sluse of 10/[20] September 1669 had provoked criticism from Wren over the presentation it contained of his discussion with Wallis on the topic of the hyperbolic cylindroid. At the same time Oldenburg wished to publish Wallis's letter to Sluse in the *Philosophical Transactions*. In WALLIS–OLDENBURG 17/[27].X.1669 Wallis proposes alterations for Oldenburg presumably to enter in the Royal Society's copy, while WALLIS–OLDENBURG 16/[26].X.1669 is a rewritten version for possible publication.

Clarissimo Viro, D. Henrico Oldenburg,
Johannes Wallis S.

                   Oxoniae: Octob. 16. 1669.
Clarissime Vir,

Eorum quae ad Cl. *Slusium* scripsi[613] (occasione ipsius ad te Epistolae[614]) haec fere est summa.
 Quod observat,[615] ingeniosissimum *Wrennii* nostri inventum (*de Cylindroide Hyperbolico, Torni ope, per aciem Dolabrae rectam, situ ad Axem obliquo positam, efficiendo,*) non minus succedere si Dolabrae acies obliquo situ ponatur, non modo ad Cylindroidis Axem, sed et ad Horizontem:

---

[613]scripsi: i.e. WALLIS–SLUSE 10/[20].IX.1669.
[614]Epistolae: i.e. SLUSE–OLDENBURG [6]/16.VIII.1669 (OLDENBURG, *Correspondence* VI, 178–83.)
[615]observat: see SLUSE–OLDENBURG[6]/16.VIII.1669 (OLDENBURG, *Correspondence* VI, 178–83, 178).

116. WALLIS to OLDENBURG, 16/[26] October 1669

Omnino verum est. Quippe positionis ad Horizontem ratio ut habeatur, non est necesse.

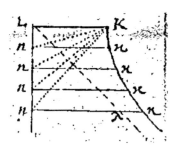

Quod autem, porro, perplexo calculo inquirit; Quo pacto possit, ea methodo, Hyperbola quaevis assignata confici: Id ego breviter et simplicissime expono, (rem totam ab origine repetens, mea methodo;) praemisso primum hoc Lemmate, quod jam olim demonstraveram inter ea quae meo de Cycloide Tractatus[616] subjunxeram; *Si ad Rectae alicujus Puncta [quaelibet], aequalibus intervallis sumpta, ordinatim-applicentur Rectae, quorum Quadrata sint, ut Numerorum continue consequentium 1, 2, 3, 4, &c, Quadrata eodem aliquo vel aequalibus quadratis aucta: quae per harum extrema reliqua transit Curva, est Hyperbola.*

Quippe (in Fig. 27 ibidem, quam hic repeto in figura priore,) Manifestum est, rectas nK, tales esse quales ille ordinatim-applicatas vellet: Quae si in situ nx ordinatimapplicentur, Hyperbolam Kxx designabunt.

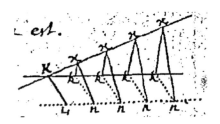

Quod si, manentibus LK, nx, ordinatim-applicatis, Intervalla Ln, nn, Minora fuerint vel Majora quam nunc sunt: prodibit utcunque Kxx Hyperbola; sed cujus aliud erit Latus-rectum, aliusque ad Asymptoton $L\lambda$ angulus $nL\lambda$ quam nunc est.

---

9 quotlibet *corr. ed.*

---

[616]Tractatus: i.e. WALLIS, *Tractatus duo*, Oxford 1659.

116. WALLIS to OLDENBURG, 16/[26] October 1669

Hoc Lemmate praemisso; rem sic expono meo more:

Si intelligatur, in figura posteriore, Acies Dolabrae recta Kxx, in quaecunque ab Axe Cylindroidis Ln distantia, situ quocunque obliquo posita, (quocunque ad Horizontem situ:) Manifestum est, per Rectam illam Kxx, transiturum esse Planum aliquod, ut xKk, cui parellelus sit Cylindroidis Axis Ln; Rectamque aliquam in illo plano, Axi parallelam, ut Kkk, (nempe ex parallelis eam quae sit Axis proxima,) lineam contactus esse qua Planum illud tangat Cylindrum (Cylindroidi inscriptum) cujus Axis Ln, et basis

[2] Radius LK. Sumptisque in Axe Ln,| partibus continue aequalibus Ln, nn, &c; atque ad eum perpendicularibus LK, nk, &c; erectisque itidem ad Planum LKk perpendiculariibus kn, kn, &c: Manifestum est, rectas Kk, esse ut 1, 2, 3, 4, &c, numeros continue consequentes; earumque quadrata, ut quadrata horum; Et propterea (propter angulum xkn rectam, restasque kn invicem aequales,) junctis omnibus xn, Quadrata harum esse, ut Quadrata Numerorum illa aequalibus quadratis aucta.

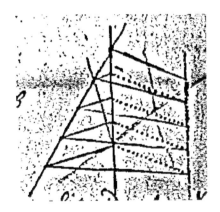

Si itaque, Torni ope, circa Axem Ln, describi intelligantur in Cylindroide Circuli, quorum Radii sint ipsae nx rectae: Sectio per Axem exhibebit ipsissimum $nLKn$ (prioris figurae) Planum.

Patet hinc; Tum situs ad Horizontem nullam hic rationem habitam esse, (sed situs tantum ad Axem Cylindri sive Cylindroidis;) Tum etiam, Quo pacto possimus datam Hyperbolam Torno exhibere.

Quippe si exponatur (in figura priore) Hyperbola Kxx, cujus Centrum L, semi-axis transversus LK, Axis conjugatus Ln; cui similem Torno exhibere imperatum sit: Hoc tantum curandum erit; nempe ut, in figura posteriore, sumatur LK (minima aciei Dolabrae ab Axe Cylindri sive Cylindroidis distantia,) quanta fuerit LK figurae prioris; Eoque situ ponatur Kx acies, ut, sumptis in utraque figura aequalibus Ln respectivis, aequales itidem sint respective nx. Quod obtinebitur, si sumatur in figura posteriore Angulus klx, aequalis angulo figurae prioris $nL\lambda$, quem cum Asymptota facit conjugatus Axis. Vale.

For Mr Henry Oldenburg, [4]
at his house in the Palmal
Near St James's
London.

117.
Wallis to Henry Oldenburg
Oxford, 17/[27] October 1669

Transmission:

W Letter sent: London *Royal Society* Early Letters W1, No. 97, 2 pp. (p. 2 blank) (our source).—printed: Oldenburg, *Correspondence* VI, 280–1.

Reply to: Oldenburg–Wallis 14?/[24?].X.1669.
Enclosure to: Wallis–Oldenburg 16/[26].X.1669.

As emerges from the postscript, Wallis began this reply on 15 October, after he had received Oldenburg's latest letter, informing him of Wren's unhappiness over the presentation of the content of his discussions with Wallis in Wallis–Sluse 10/[20].IX.1669. Wallis's subsequent decision to send it with a summary of his letter to Sluse necessitated its being re-dated to 17 October.

Oxford. Octob. 17. 1669.
Sir

I had no designe of doing Dr Wren the lest prejudice in the letter[617] to Slusius; or to derogate from that his ingenious invention.[618] And therefore if he think any of my words do seem to insinuate any such thing: you may put out what you please of them. As for instance; those words *dum res*

---
8 Octob. (*1*) 15. (*2*) 17.

[617]letter: Wallis–Sluse 10/[20].IX.1669.
[618]invention: i.e. Wren's method of generating hyperbolical cylindroids and the machine for grinding hyperbolical glasses based on this. See Wren, 'Generatio Corporis Cylindroidis Hyperbolici, elaborandis Lentibus Hyperbolicis accommodati', printed in *Philosophical Transactions* No. 48 (21 June 1669), 961–2, and Wren, 'A Description of Dr. Christopher Wren's Engin, designed for grinding Hyperbolical Glasses', printed in *Philosophical Transactions* No. 53 (15 November 1669), 1059–60.

117. WALLIS to OLDENBURG, 17/[27] October 1669

adhuc immatura fuit may be left out. And, a little after, instead of *petebat num certus essem rem ita esse: Respondi, certissimum; utpote quod jam olim demonstraveram,* you may put onely this, *Idque a me jam olim demonstratum esse.* Then for *Quo facto, causam cur illud interrogabat, hanc indicabat* it may be *Quo facto, rem totam indicabat.* Then these words *sed qualis sit ea linea se nondum satis examinasse dixit, sed Hyperbolam esse suspicatum,* may be left quite out. And the words following, thus suppressed, *Annuebam ego, rem omnino ita esse: sed et speculationem hanc suam, dignam esse quae excoleretur. Urgebam itaque, ne pateretur, ut, tam elegans speculatio, neglecta periret, sed ut in commodam formam redigeret et typis vulgaret. Quod fecit ille, paulo post, sua methodo, (nec eo minus sua, quod mecum interea temporis communicaverat,) ea quo vides forma.* And so on, as before. And then I think the whole is not onely truly expressed, but even so innocently as not to have the lest appearance of reflexion upon him. Or, if there bee any thing hee doth yet mis-like, that may be amended too. For I had no other designe, but, from that Character of the Hyperbola of which wee had been speaking, to shew how easy that was which to Slusius seemed so difficult a Probleme.

But I am not at all fond to have it printed; nor was it any motion of mine, but your own. And you may use your own discretion, either to alter it as may be to his content, or not to print[619] it at all.

Or else, if you print Slusius's letter, & think it convenient to adde any thing of mine: I will send you the substance of it so drawn up as not so much as to mention any thing of that discourse that passed between Dr Wren & mee. Which though it were sincerely & candidly related, yet if he do not think it advantageous for him to have it mentioned, I am not at all concerned for it. What concerns the Barometer, I shall send you after upon some other occasion.[620] Resting

Yours &c.
J. Wallis.

I thought to have sent you this yesterday by Monsr Gash,[621] & the French Gentlemen that came in company with him (by whom I received yours the night before) & had written it for that purpose the same night; but upon

19 it (*1*) at all (*2*) any
31 yesterday *add.*

---

[619] print: Oldenburg eventually decided against printing Wallis's letter to Sluse.
[620] occasion: i.e. WALLIS–OLDENBURG 7/[17].I.1669/70.
[621] Gash: neither Mr Gash nor the French gentleman who accompanied him are identified.

second thoughts, I thought fit to stay it one day longer, that what is annexed might come with it; & both as soon (being by the Post) as that one would have come by him.

118.
WALLIS to WILLIAM BROUNCKER
[London], 10/[20] November 1669

Transmission:

E^1 First edition of (missing) letter sent: WALLIS, *Mechanica: sive, de motu, tractatus geometricus*. Pars prima, London 1670, Epistle dedicatory.
E^2 Second edition: WALLIS, *Opera mathematica* I, 573–4 (our source).

The date of the this letter coincides with the completion of the first part of Wallis's *Mechanica* in November 1670. See OLDENBURG–SLUSE 10/[20].XI.1669 (OLDENBURG, *Correspondence* VI, 309) and WALLIS–COLLINS ?.XI.1669. Wallis was at this time in London, staying in lodgings in Little Drury Lane. The finance committee of the Royal Society, of which Wallis was a member, met on 12 November 1669 (old style). See BIRCH, *History of the Royal Society* II, 401.

Honoratissimo Domino
D. Gulielmo Brouncker,
Equiti Aurato;

Baroni Brouncker de Newcastle, Vicecomiti Brouncker de Lions; Serenissimae Regiae Majestatis, pro Re Nautica Commissario; Serenissimae Reginae Cancellario: Regalis Societatis Londini, pro Scientia Naturali promovenda, Praesidi Dignissimo.

En habes tandem (Honoratissime Domine,) eorum partem, quae, Tuis Regiaeque Societatis (cui Tu jam a pluribus annis summo cum honore Praesides) mandatis obsequens, anno ab hinc secundo, prelo commiseram. Speraram equidem Opus integrum breviore tempore absolvendum fore: Sed, partim ob eam,[622] quam causantur Typothetae, rei difficultatem, & insolentiorem typos ponendi modum; partim ob meam Oxonio agentis perpetuam fere a prelo absentiam;[623] partim quod aliis subinde occurrentibus negotiis impediti Typothetae non huic potuerint semper intenti esse; factum est, ut in longius protractum sit negotium quam speraverim. Hinc est, quod, quem totum una

---
[622] eam: cf. WALLIS–COLLINS 11/[21].I.1669/70.
[623] meam ... absentiam: cf. WALLIS–COLLINS 3/[13].XI.1668.

vice proditurum Tractatum destinaveram, jam particulatim prodeat. Cujus jam habes Partem Primam, quae totius Fundamenta continet; &, speciatim, De Libra doctrinam. Quam propediem sequetur Secunda, (quae De Centro Gravitatis erit; ejusque Calculo, in figuris quam plurimis curvilineis, atque ex his oriundis solidis, & superficiebus curvis, satis intricato:) Utpote cujus partem maximam jam absolverunt operae. Et, post illam, Tertia; quamprimum per Preli difficultates licebit. Hoc autem, quicquid est, Tuo potissimum nomini dicandum duxi; non tantum ob eam quam Tibi debemus observantiam, (quae tamen maxima est,) eamque quam a pluribus annis expertus sum tuam in me propensam| Amicitiam: Sed & ob eam, quam cum summa Nobilitate conjunctam habes, summam hujusmodi rerum Intelligentiam. Quanquam enim gravissimis negotiis alias occupatus; tum quae Rem Nauticam spectant, quibus Serenissimo Regi inservis; tum quae Serenissimam Reginam, cujus tu curas negotia; tum quae, cui Tu Praesides, Societatem Regiam: Ea tamen in rebus Mathematicis perspicacia, summoque ingenii acumine fretus es, quasi tu huic tantum negotio intentus esses, & nulli secundus. Verum quidem est, tum ea quae hic habes, tum eorum quae mox insecutura sunt partem maximam, ante plures annos scripta fuisse, (quod norunt saltem ii in quorum privatos usus exscripta fuerunt exemplaria; & quibus jam urgentibus prodeunt:) Sed eo audacius in lucem jam emitto: quod Tibi antehac perlustrata non displicuerint, & Te jubente prodeant. Tu porro perge, quod facis, rem literariam & ornare & promovere: Nec averseris interim,

Tui Observantissimum
Joh. Wallis.

Novemb. 10. 1669.

119.
WALLIS: The Reasonableness of the University's Exemption from Charges
27 November/[7 December] 1669

Transmission:

W Draft paper: OXFORD *University Archives* SP/D/5/37, 6 pp. (p. 6 blank). Endorsement by Wallis on cover: 'Exemption from taxes for the Militia'. Further endorsements in unknown hands: 'Concerning University Priviledg' and '1669 Reasons for University's exemption from Contributions with the City to Musters'.

This paper was provoked by moves on the part of the city of Oxford, in late 1669, to force

119. The Reasonableness of the University's Exemption from Charges

the University to contribute to the local militia. This was seen by the University as a new attempt by the city authorities to remove one of its privileges.

Novemb. 27. 1669.

The Reasonableness of the Universities being exempted
from Contributions with the City of Oxford to Musters
& other Charges.

First this exemption hath been very ancient beyound the time of Memory; & grounded (as it seemes) originally on the consideration had of incouraging Learning & Studies in the University; & on Consideration of the great advantage that comes to the City & Citizens by the Repair of soe many persons from all parts on that account, & yearly expending so much among them.

Which is so much the more manifest, both from history & later experience, by the great complaints of the Inhabitants as being ruined & undone when the University either upon discontents have removed hence to Stanford Northampton & other places; & when of later times by reason of War Pestilence or the like they have been forced for some time to withdraw.

Which equitable consideration, of the Universities being so beneficiall to the City by their presence here, & receiving nothing of advantage from them; may well deserve their being exempted from other charges & contributions.

Secondly; this hath been allways judged so reasonable that upon all differences between the University & town in former times, this hath been allways allowed & judged due to them.

As in the Composition made in Parliament, in the 18th year of Edward the first; with consent of both bodies. And, after the great Conflict, by the Charter of the 29th of Edward the third; upon the Submission of both Bodies.| [2]

And in the Composition between both bodies & confirmed by the King, in the 37th of Henry the 6th.

And in the Charter of the 14th of Henry the 8th confirmed by Act of Parliament, 13° Elizabetha.

And in the Charter of 11° Caroli 1. and many others.

And in the Solemne hearing by consent before the Lords of the Counsell, 17° Elizabetha, & the Orders made thereupon, exemplified under the great Seal.

29 the 14th of *add.*
31 And ...others. *add.*
32 by consent *add.*

119. The Reasonableness of the University's Exemption from Charges

And in the like hearing before the Lords of the Counsell, 10° Jacobi, & the Orders thereupon in like manner exemplified.

At both which hearing the Orders after frequent hearing of Counsel on both side, were drawn up by divers of the Judges & other Commissioners for that purpose.

In all which; Decisions, it is always taken for granted & allowed that Scholars (as well in as out of Colleges) are exempt from publicke charges with the Town; & (except in some particular cases) their servants allso: And that even in such cases as the servants are taxable, (viz. for those merchandizes in which they trade,) they are to be taxed, not by the Mayor or townsmen, but by the Chancellor or his substitutes.

Thirdly, 'tis also manifest (by divers Inquisitions taken on that account, & otherwise,) that not onely their persons, but the houses in which they inhabited, (whoever were the owners of them,) were always exempt from taxes or contributions; as well before as since Colleges were erected.

And the present Colleges have no other ground of being exempted from City charges; but onely as houses inhabited by Scholars. In confidence of which, it was that some of the Deputie Lieutenants did not stick to tell the late vice-chancellor,[624] that they might if they pleased command him to bear a musket himself.

But when anciently a question did arise, whether when houses did belong to Scholars or colleges as the owners, but were inhabited by others, they should in such cases bee taxed as other houses: To prevent differences in that case, the University payd to the Town a considerable summe of mony, in consideration of which the Town agreed, (as by several authentick instru-
[3] ments yet extant appeares,)| that not onely such houses as they then had, but such allso as at any time after they should purchase, should be exempt from such charges. Providing onely, that in case such houses were inhabited by townsmen, not by scholars; those townsmen should be taxable for their

---

3 At *add.*
4 side, *(1)* the or *breaks off (2)* were
6 that *(1)* all Scholars & their *(2)* Scholars
8 & *(1)* their servants allso *(2)* allso their servants *(3)* (except
9 taxable, *(1)* it is to b *breaks off (2)* |(viz. for those merchandizes in which they trade,) *add.*| they
17 onely, *(1)* that *(2)* as
23 prevent |such *del.*| differences

---

[624]late vice-chancellor: i.e. John Fell. Since 25 November 1669, Peter Mews (1619–1706), president of St John's College, was vice-chancellor.

119. The Reasonableness of the University's Exemption from Charges

goods in those houses, (as if they were in any other houses,) but not for the houses in which they were.

Fourthly; beside all the equitable considerations, from the benefit that the citizens have from the Universities residence amongst them: The University & City (though inhabiting one amongst another) are to be considered as two distinct Corporations, as much as if theyr residence were in two several places.

And it is notoriously known, that the City doth not at all contribute to the charges of the University; nor do the University injoy the immunities which the City claim as peculiar to them; As for instance, the share of any gifts or legacies given to the City; Or the liberty of intercommuning in Port-mead, which the City claime as their own; And the like. And therefore it is as little reasonable that the University should bear the burdens with them since they do not share in their benefits; Or that they should contribute to the City charges, since the city doth not at all contribute towards theirs.

Fiftly; beside the former considerations, & the summe of mony payd by the University & accepted by the Town for such immunities, upon solemn compositions heretofore made:

The City do injoy severall advantages & immunities & upon the accomt of the University, which otherwise they had not injoyed.

As for instance; the very foundation of the 4 Aldermen & 8 Assistants (which, with the Mayor, making up the Thirteen,) is not upon any grant to the City or Town of Oxford; but upon a Charter granted by Henry the 3^d, to the University of Oxford. And upon that Charter of the University, it is, that they doe yet stand.

And the Act of Parliament, for the Exemption from the Kings Purveyors within five miles of Oxford, (which was formerly| a very considerable [4] advantage, what ever it be now;) was passed onely upon consideration of the University; without which, there was no more reason why the City of Oxford (with its confines) should be more exempted than other Cities.

And the Act for repayring the High-ways & Bridges within a mile of Oxford, by all the Inhabitants within five miles of it; Was obtained onely

---

5 (though inhabiting one amongst another) *add.*
6 two (*1*) Corporations, as much distinct (*2*) distinct Corporations, as much as
7 two *add.*
11 the (*1*) tow *breaks off* (*2*) City
21 they (*1*) should (*2*) had
22 the (*1*) Alderman & Assistants (*2*) 4 Aldermen & 8 Assistants

119. The Reasonableness of the University's Exemption from Charges

upon the consideration of the University; though the City have by much the greater benefite by it.

And the like may be sayd of the Act for making the River of Thames navigable from Oxford; The Benefit of which doth much more redownd to the Inhabitants of the City, who make much greater use of it than those of the University.

Sixthly; The City of Oxford, have, upon this very account, that the University doth injoy these exemptions, been otherwise considerably eased.

As in the abating their Fee-farm rent more then once. And the advantageous exchange with Oriel College petitioned by the Towne, and granted by the King, expressely; upon the consideration that Scholars houses did not pay contributions with the Town.

And, in the very thing now in question, who can imagine that, otherwise, so great a city as that of Oxford, should not be charged higher to the Militia, (either now, or in former times,) than onely at 35 men, (or, with Binsy,[625] at 36.) When as many market towns, much lesse than Oxford, in other parts of the nation, raise much greater proportions.

How when as it is onely, upon consideration of the Universities not being charged with them, that they are set at so low a proportion as 35 men; it is very unreasonable that they should now goe (which never was done before) to cast part of this final number upon the University.

Seventhly; It doth not seeme at all agreeable to the intentions of the Acts of Parliament on which they would now ground it. For though it be true, that there be not in those Acts any Particular Salvo, for the Universities Privilege of Exemption; (by reason of our Burgesses neglect to have it inserted:) Yet when those Acts do provide that| Corporations should find as formerly, unlesse the Deputy Lieutenants shall see cause to abate the Number: It must be reasonably understood, that the Corporation of the City

---

10 petitioned by the Towne, and *add.*
14 otherwise, *add.*
15 onely *(1)* to raise the *(2)* at
16 many *(1)* small *(2)* market
16 than *(1)* that *(2)* Oxford
18 is *(1)* upon *(2)* onely, upon *(a)* this *(b)* consideration *(aa)* that the *(bb)* of
25 Universities *(1)* exemption *(2)* Privilege

[625] Binsy: i.e. the ancient village of Binsey, to the north-west of Oxford, whose inhabitants were not freemen of the city by incorporation, but who contributed to it through taxes. See WOOD, *City of Oxford* I, 321–3.

119. The Reasonableness of the University's Exemption from Charges

of Oxford, which did, without the Corporation of the University, formerly find 35 men, should, without the University, do so still; (unlesse the Deputy Lieutenants shall think fit to abate them as judging that too great a proportion; which it is not likely that they will:) and that the Corporation of the University, which was allways heretofore exempt from finding any, should so remain exempt. Nor is there any intimation in Act, that the Lieutenants of the County should join two corporations together, to bear that proportion which was formerly born by one of them.

Eightly: The controversy, by those of the City, is purposely mis-stated, to put a colourable pretense upon the business. For the question is not properly, whether University men are to be charged to the Militia: But, whether the Corporation of the University of Oxford, be a part of the Corporation of the City of Oxford. For, it is notorious that the Corporation of the City of Oxford were formerly charged with 35 men. And they are therefore so to be charged by this Act: Now what ever became of the question, whether the University by the Act ought to be charged: certain it is, that there is nothing in the Act to make the Corporation of the University of Oxford to be a part of the Corporation of the City of Oxford, more then it was before. And therefore nothing in it, to make the Corporation of the University chargeable to what is imposed on the Corporation of the City of Oxford.

Ninethly; The service of the King or Kingdome is not at all advanced by this burden indeavoured to bee brought upon the University: For neither the City of Oxford, nor the Deputy Lieutenants of the County, do soe much as pretend (by bringing this new burden upon the University) to raise any greater number of men for the service, than before: for the Corporation of the City of Oxford did before find 35 men (&, with the liberty of Binsy, 36:) and they are not now charged with one more than that number. So that, it is onely to bring a Vexation upon the University; not at all to advantage his Majesties service; that this is now attempted.

1 the Corporation of the *add.*
3 fit to abate them as judging |that *del.*| *add.*
9 The |present *del.*| controversy
14 Oxford (*1*) did some *breaks off* (*2*) were
14 therefore *add.*
15 Now (*1*) though the (*2*) what
19 Corporation of the *add.*
20 to (*1*) those (*2*) what
25 Corporation of the *add.*
26 before (*1*) (with the (*2*) find

120.
Wallis to John Collins
? November 1669

Transmission:

Manuscript missing.

Existence and date: Referred to in Wallis–Oldenburg 9/[19].XII.1669.)

In this letter, Wallis requested of Collins that he present one copy of the first part of his *Mechanica* to William Brouncker, and a further number of copies to Oldenburg, once the book came from the press.

121.
Henry Oldenburg to Wallis
[London], c.8/[18] December 1669

Transmission:

C Letter sent: London Royal Society Early Letters W1, No. 98, 6 pp. (on p. 3) (our source). On pp. 3, 4, and 5 Oldenburg's account of the first part of Wallis's *Mechanica*, p. 2 and p. 6 blank, on pp. 1 and 5 Wallis–Oldenburg 9/[19].XII.1669.—printed: Oldenburg, *Correspondence* VI, 357.

Answered by: Wallis–Oldenburg 9/[19].XII.1669.
Enclosure: Oldenburg's short account of the first part of Wallis's *Mechanica*.

With this letter Oldenburg sent Wallis for his perusal and correction the short account he had written himself of the first part of the *Mechanica* and which he intended to publish in the *Philosophical Transactions*.

Sir,

All I desire in this matter[626] is, that you would alter what may be amisse in this; and suggest some few particulars, considerable above the rest, which may *movere salivam* to the Reader, and doe the Stationer a kindness. And not omit sending it so, that I may have it on Friday night.[627] This very sheet you may please to return with your alterations. and additions to it, which I

---

4 doe the (*1*) Printer (*2*) Stationer
5 This (*1*) part (*2*) very

---

[626]matter: i.e. Oldenburg's short account of the first part of Wallis's *Mechanica*.
[627]night: i.e. 10 December 1669. On that night Oldenburg did in fact receive Wallis's reply.

suppose may be contained in the remaining blank; which together with your letter, will make but a single letter, as this is, I send.

122.
HENRY OLDENBURG to WALLIS
c.8/[18] December 1669, enclosure:
Oldenburg's Short Account of the First Part of Wallis's *Mechanica*

Transmission:

C Draft account (with corrections and additions in Wallis's hand): LONDON *Royal Society Early Letters* W1, No. 98, 6pp. (on pp. 1, 3, 4, and 5) (our source). On p. 1 address to WALLIS–OLDENBURG 9/[19].XII.1669, p. 2 and p. 6 blank, on p. 5 WALLIS–OLDENBURG 9/[19].XII.1669, and on p. 3 OLDENBURG–WALLIS *c.* 8/[18].XII.1669.

E First edition of corrected account: *Philosophical Transactions* No. 54 (13 December 1669), 1086–9.

Enclosure to: OLDENBURG–WALLIS, *c.* 8/[18] December 1669.

For the Presse.
MECHANICA, sive de MOTU Tractatus Geometricus. Auth. Joh. Wallis SS. Th: & Geom. Prof. Saviliano etc.

This Excellent Mathematician having composed a learned Treatise concerning the Doctrine of *Motion* and what thereon depends, managing it in a manner altogether Geometrical, was pleased, upon the importunity of his Friends, to permit the *First part* thereof to come abroad,[628] whilst the others are still in the Printers hands: Which is about one third part of what hee did near two years since impart to the Royal Society and was by them desired to make it publicke. In which First part he delivers

---

6 Mathematician, *(1)* as well *(2)* having *C*
7 depends, *(1)* handli⟨ng⟩ *(2)* managing *C*
8-9 was pleased |, upon the importunity of his Friends, *add.*| to permit ... whilst the *(1)* other two *(2)* others *C*
10-12 Which ... publicke *C add.* W

---

[628]abroad: cf. WALLIS–COLLINS 11/[21].I.1669/70, where Wallis complains that the Printer, William Godbid, 'hath made the year in the title page 1670, whereas I directed it to be 1669'.

122. OLDENBURG to WALLIS, c.8/[18] December 1669, enclosure

1. The *General Rules of motion*, premising thereto, as becomes a strict Reasoner and good Geometrician, the *Definitions* belonging to that Subject; and that done comprising the Rules themselves in 30. Propositions: In which he takes occasion, among many other weighty particulars, to intimate, that 'tis principally the business of a Mechanician, to excogitate and make practicable such Engins, to be interposed between the *Strength* and the *Weight*, as may so moderate the *Celerity* of Motion, as to compensate the *Greatness* of the *Weight*, by the *Slowness* of the *Motion*, or the *Want of Strength*, by the *Length of Time*.

2. Of the *Descent of Heavy Bodies*, and the *Declivity of Motion*, 34 Propositions. In which he doth not think fit to explicate the Physical *Cause* of *Gravity*, whether it proceed from an Innate quality in the Heavy Body itself; or an Universal Tendency of circum-ambient Bodies to the Center, or a Magnetical Power in the Earth, or the like; but contents himself (his purpose requiring no more) to understand by the Word *Gravity*, that Sensible Force of moving downwards or to a certain Point, both the Heavy Body itself, and the lesse powerfull impediments: And in what proportion the different Declivities of Oblique or Sloping Plains, in which a heavy Bodie is supposed to be moved, doth operate to the Helping or Hindering of such motion. All which he delivers as generally applicable to motions produced by any other Force as well as that of Gravity: & directed any other way as well as downward.|

1 thereto, (*1*) like (*2*) as *C*
3 belonging (*1*) thereto (*2*) to that Subject *C*
3 and then comprising *E*
5-8 particulars, to (*1*) mind the *Mechanician* that he ought principally to intend this, to devise such and make (*2*) intimate, that (*a*) the (*b*) 'tis principally the business of a |to del. ed.| Mechanician to excogitate and make practicable such Engins, (*1*) wherewith, by the interposition of (*2*) to be interposed between the ... *Weight*, (*a*) they (*b*) as may *C*
14 Physical *C add. W*
17 himself (*1*) (this being (*2*) (his purpose *C*
20-25 And in what proportion the different ... downward. *C add. W*
21 Impediments: Shewing in this Part also, in what proportion *E*

122. OLDENBURG to WALLIS, c.8/[18] December 1669, enclosure

3. The Doctrine of the *Libra*, or Balance: (containing the fundamentall principles of all *Staticks*;) in 25. *Propositions*; among which he explains the Geometrical Considerations requisite in making both exact *Common Scales*, and the Roman *Statera*.

He illustrates the whole with 80. Figures, all contained in two compendiously contrived Plates. And doth, from their proper principles, demonstrate many of the things which writers commonly Postulate, or take for granted; but which (to make a sure foundation) ought to have demonstrated.

On this of the *Libra*, depends (that which is the subject of the next Part,[629] allmost finished at the Presse) the whole doctrine of the Center of Gravity: and the Calculation thereof. Which (Center of Gravity) hee doth not onely demonstrate to bee; (which others have hitherto Postulated, but not any, that I know of, demonstrated:) but doth from General principles shew how by Calculation to assigne the same, in infinite sorts of Lines, Surfaces, Solides, as well such as are bounded or take their rise from crooked lines; as those which are bounded onely with Streight lines & Plains.

And from the Generall Principles of Motions here layd down; hee doth, in his Third Part,[630] which is to follow; derive the Doctrine of the *Vectis* or Leaver; the *Trochlea*, or Pulley; the *Cochlea*, or Screw; the *Axis in Peritrochio*, or several sorts of Wheel-work; & other such Mechanical Engines derived from these. As likewise the Doctrine of *Percussion* (on which depends that of the *Cuneus* or Wedge; with many other Speculations of a like kind:) And that of *Resilition* or Rebounding; which, (as appears by a short specimen formerly printed in some former Transactions, Number.[631]) he derives

---

1 (*1*) 3. He very distinctly and clearly tr⟨eats⟩ (*2*) 3. The Doctrine of the *Libra*, or Weighing Scales *C corr.* W
1-2 (containing ... Staticks;) *C add.* W
3-4 explaines both the ways of making *C corr.* W
5-9 He illustrates ... demonstrated. *omitted in E*
5 illustrates (*1*) all (*2*) the whole *C*
6-9 And doth ... demonstrated. *C add.* W
10-end On this ... shew how (*1*) to Calculate (*2*) by Calculation to assigne the same, ... Body |in motion which *add.*| it meets ... off these bodies (*a*) both w⟨ayes ⟩ (*b*) one or both wayes ... opposite new motion. *C add.* W
25 in Numb. 43. of these Tracts) *E*

---

[629] Part: i.e. the second part of Wallis's *Mechanica*, which was published later in 1670.
[630] Part: i.e. the third part of Wallis's *Mechanica*, which was not published until 1671.
[631] Number: Oldenburg has left a space for entering the number of the issue of *Philosoph-*

from a Repercussion either of some other Body in motion which it meets with, or from the Elastick force or Spring in one or both of the meeting bodies; which, being compressed by the Collision, doth indeavour to restore it self by casting off these bodies one or both wayes.| Consonant to his Principles here laid down in the 10th, 11th, & 12th Propositions of his First Chapter: where the *Impedimentum* or Obstacle is made sufficient to Retard or Stop a motion; but a *Vis Contraria* or contrary Fore, necessary to give an opposite new motion.

123.
WALLIS to HENRY OLDENBURG
Oxford, 9/[19] December 1669

Transmission:

W Letter sent: LONDON *Royal Society* Early Letters W1, No. 98, 6 pp. (on pp. 1 and 5) (our source). On p. 3 OLDENBURG–WALLIS *c.* 8/[18].XII.1669, p. 2 and p. 6 blank, on pp. 3, 4, and 5 Oldenburg's account of the first part of Wallis's *Mechanica*. On p. 1 beneath address in Oldenburg's hand: 'Rec. dec. 10. 69'.—printed OLDENBURG, *Correspondence* VI, 360.

Reply to: OLDENBURG–WALLIS *c.* 8/[18].XII.1669.
Enclosure: Oldenburg's short account of the first part of Wallis's *Mechanica*.

With this letter, Wallis returned Oldenburg's short account of part one of his *Mechanica*, duly corrected and amended.

This,[632] I think is sufficient in answere to what you desire;[633] at lest as much as I can now write lest I lose the Post.

---

8 motion. In this first Book, he illustrates all with Eighty Figures, contained in two compendiously contrived Plates: And doth, from their proper Principles, demonstrate may of these things, which Writers commonly Postulate, or take for granted; but which (to make a sure Foundation) ought to have been demonstrated. E

10 write (*1*) lesse (*2*) lest

---

ical *Transactions* for 11 January 1668/9, i.e. No. 43, which contained Wallis's 'General Account of the Laws of Motion'.

[632]This: i.e. the corrected and amended version of Oldenburg's short account of his *Mechanica*.

[633]desire: see OLDENBURG–WALLIS *c.* 8/[18].XII.1669.

124. Wallis to Brouncker, 21/[31] December 1669

I did not know the book was abroad till I had your letter: But had before writt[634] to Mr Collins, to present my Lord[635] with one, & furnish you with some for other friends.

The persistance of motion begun, is in prop. 11. & its Scholium.

Dec. 9. 1669. Oxford.

Yours,
J. W.

For Mr Henry Oldenburg, [1]
at his house in the Palmal
near
St James's
London.

124.
Wallis to William Brouncker
21/[31] December 1669

Transmission:

E Second edition: WALLIS, *Thomae Hobbes quadratura circuli, cubatio sphaerae, duplicatio cubi; (secundo edita) denuo refutata*, 8 pp. (p. 2 blank), Oxford 1669.

Written in the form of a letter addressed to Brouncker, this is the second edition of Wallis's refutation of Hobbes's *Quadratura circuli* in which he replies to a later edition of that work, probably published in August 1669. Hobbes had dedicated the second edition of his *Quadratura circuli* to Cosimo III de' Medici after meeting the prince a number of times at the end of May and beginning of June 1669 (see WOOD, *Athenae Oxonienses* III, 1208). Hobbes had also augmented his demonstrations with a critical response to Wallis's first refutation. A review of the second edition of Wallis's refutation appeared in the *Philosophical Transactions* No. 55 (17 January 1670), 1121–[1122]. The tract itself was presented together with the first part of *Mechanica: sive, de motu, tractatus geometricus* at the meeting of the Royal Society on 23 January 1669/70. See BIRCH, *History of the Royal Society* II, 415.

Thomae Hobbes,
Quadratura Circuli,
Cubatio Sphaerae, Duplicatio Cubi;
(secundo edita,) denuo Refutata.

---
[634] writt: i.e. in the now missing letter WALLIS–COLLINS ?.XI.1669.
[635] Lord: i.e. William Brouncker, q.v.

124. WALLIS to BROUNCKER, 21/[31] December 1669

Ad Honoratissimum Dominum, D. Vicecomitem Brouncker.

Nec peregre proficisci (quo Patronum advocem,) nec longo sermone opus erit. (Nobilissime Vir,) quo *Hobbii* nugas Cyclometricas, iterato editas, & Serenissimo Cosimo *Etruriae* Principi dicatas, refellam. Quippe, convulsa Propositione prima, (ne ipso quidem diffidente) reliqua simul ruent. Eam itaque verbatim repeto, & refello.

Prop. I. *Circulo dato Quadratum invenire aequale.* Sit (in Figura prima) Circulus datus $BCDE$, cujus centrum A, divisus quadrifariam a diametris BD, CH. Circulo huic Circumscribatur quadratum $FGHT$, (lege $FGHI$,) quod tangit circulum in punctis B, C, D, E. Ducantur Diagonales GI, HF, secantes circulum in punctis K, L, M, N. Secetur semilatus CG bifariam in O, ducaturque AO secans circulum in P. Per punctum P ducatur recta QR parallela GH, secans AG, AH in Q & R, & AC in Y, compleaturque quadratum $QRST$. Dico quadratum $QRST$ aequale esse Circulo $BCDE$ dato.

Hanc ego Propositionem, jam ante (ad Editionem priorem) falsam esse demonstravi. Eøquod faciat rationem Perimetri ad Diametrum Circuli (ipso non diffitente) majorem quam 22 ad 7, (Contra quam Archimedes, & post illum innumeri, demonstrarunt.) Adeoque Circuli Perimetrum majorem quam est Perimeter Figurae rectilineae circumscriptae. Quae quidem (utut *Hobbio* forsan non ita videantur) satis sunt absurda.

Sed vidamus quam praetendit demnstrationem.

Quoniam enim recta CG secta est bifariam in O, & triangulorum ACG, AYQ bases BG, YQ sunt parallelae, etiam basis YQ secta est bifariam in P, & proinde triangula AYP, APQ sunt aequalia.

In arcu LC sumatur arcus LV aequalis arcui CP, ducaturque AV, secans YP in X. Jam $APL + PQL + CYP = AVL = ACP$ (quia $APL + PQL = AYP$.) Nam (lege Item) $ACV + AVP = ACP = AVL$.

Quare $APL + PQL + CYP = ACV + AVP$.

Ablatis igitur utrinque aequalibus APL, ACV, restant $PQL + CYP = AVP$.|

[4]

Quoniam ergo AVP Sector additus Sectoribus duobus ACV, APL facit integrum Sectorem ACL; etiam duo [trilinea] PQL, CYP addita Sectoribus iisdem ACV, APL facient quantitatem aequalem Sectori integro ACL.

Jam trilineum PQL additum Sectori ALP facit triangulum APQ. Et (quia ALP, ACV Sectores sunt aequales, & triangula AYP, APQ aequalia) trilineum idem PQL additum Sectori ACV facit (lege aequat) triangulum AYP.

33 trilinia *corr. ed.*

124. WALLIS to BROUNCKER, 21/[31] December 1669

Est autem, non falsa quidem, sed superflua, tota hactenus Demonstratio: (Neque alii inservit usui, quam ut se primum, & deinde Lectorem turbet:) Quippe sequentia, per se magis, quam ex hoc apparatu, patent.

Si ergo PQL, CYP sunt aequalia, totum trangulum AYQ aequale erit Sectori integro ACL. Sin PQL sit majus vel minus quam CYP, triangulum AYQ erit majus vel minus Sectore ACL.

Nempe; Propter $ALPY$ quadrilineum, utrique commune. (Quorsum igitur totus ille qui praecessit apparatus quo hoc probetur; quippe nihil notius est, quam, Aequalia aequalibus addita facere aequalia; inaequalibus, inaequalia.)

Sed addo; idem contingere, ubi cunque in CL sumatur P. Puta in p. Quippe similiter dicendum erit, *si pqL, Cyp sunt aequalia, totum triangulum Ayq aequale est sectori integro ACL: sin pqL sit majus vel minus quam Cyp, triangulum Ayq erit majus vel minus sectore ACL.* Nempe, propter quadrilineum $AypL$ utrique commune.

Aut ergo, in triangulo ACG, trinagulum rectangulum, cujus vertex sit A, aequale Sectori ACL sumi nullum potest, aut PQL, CYP sunt aequalia.

Hoc ego negavi (quod & etiamnum pernego) tum ut gratis dictum (nam in ejus probationem ne hilum obtendebatur,) tum ut manifesto falsum. Si enim PQL sit (verbi gratia) majus quam CYP, adeoque Triangulum AYQ majus sectore ACL; Quid impedit quin (ducta paulo subtus qpy,) Triangulum Ayq rectangulum, sit sectori ACL aequale?

Regerit (quae sua solet esse probatio) *non potuisse se credere quod hoc demonstratione indigeret*: sed &, *neque sibi incumbere ut probet verum esse* (sed *accusatori* ut probet esse falsum, quod & factum est) quippe quia *suum non est docere Professorem Publicum*; addo, nec suas ipsius demonstrare propositiones. Sed *id nunc ita factum* spondet, *ut ego ipsius demonstrationes melius intelligam quam vellem*; (quippe, melius quam Ipse vellet, jam intellexeram:) Nempe sic;

Nam, si ACV, ALP, aequalibus addatur dimidium Sectoris (*intellige PAV*) utrinque, fient duo trangula (*non, sed duo Sectores*) Sectori ACL aequalia. Itaque quntum tranguli alterius, erit intra circulum, tantum alterius erit extra.| [5]

Quin dicat velim (saltem apud se cogitet, si nolet me docere,) qua fieri possit, ut utrivis sectorum (peripheria terminatorum) puta ipsi ACV aliquid *addendo*, fieri potest triangulum quod sit totum *intra* circulum. Sed pergentem audiamus:

Quod fieri impossibile est praeterquam in concursu rectae AO cum RQ,& CL, ad P. Alioqui enim aut triangulum aut quantitas AVP non dividetur bifariam.

124. WALLIS to BROUNCKER, 21/[31] December 1669

Imo, simpliciter impossibile est, ut utrivis sectorum ACV, ALP, aliquid addendo, fiat triangulum quod sit totum intra circulum.

Sed omnino fieri potest, ut partim addendo, partim auferendo, aequales quantitates (non quidem PQL, CYP, sed) pqL, Cyp, fiat triangulum Ayq, aequale sectori ACL.

Dum vero excipit ille, *triangulum Ayq, recta Ap, sic non dividi bifariam*; omnino ostendit se rem ipsam non intelligere. Quippe id omnino non est opus. Si enim aequalia sint pqL, Cyp, (sive triangulum Ayq sit ad punctum p aequaliter bisectum, sive secus,) triangulum Ayq tum *rectangulum* erit, *verticem habens A*, atque in triangulo AGC; tum *aequale sectori ACL*, propter reliquum $AypL$ quadrilineum utrique commune. Neque aliunde provenit haec exceptio, quam quod ille (inutili quem diximus apparatu longo, atque ad rem neutiquam faciente) imaginationem suam turbaverat.

Omnino itaque falsum est, quod ille primo gratis affirmabat, atque se nunc demonstrasse affirmat, *non posse sumi in triangulo ACG triangulum rectangulum, cujus vertex sit A*, nisi PQL, CYP sint aequalia: sufficit utique si sint aequalia pqL, cyP.

Videamus demonstrationem secundam.

Aliter, Directe. Sector ACP superat Sectorem ACV quantitate AVP. Ergo ACP superat trangulum AYP quantitate $AVP - CYP$.

Non sequitur. Et quidem eodem jure dicerem ego, (sumpto $Cu = Lp$,) *Sector ACp superat sectorem ACu, quantitate Aup: Ergo ACp superat triangulum Ayp, quantitate Aup − Cyp*; (sumpto ubivis in CL; puncto p:) Quod *Hobbius* non concederet. Adeoque, quae huic subjungit, nullius sunt momenti; Nempe

Superat autem quantitate ipsa CYP. Sunt ergo $AVP - CYP$ & CYP aequalia. Addito ergo utrinque CYP, erunt AVP & $2CYP$ aequalia. Et quia AVP aequalis est ambobus spatiis PQL, & CYP, erunt PQL & CYP aequalia.

Atque ego similiter concluderem, pqL & Cyp aequalia.

Videamus tertiam demonstrationem.

Aliter, Directe. Trilineo CVP (lege XYP) ablato a Sectore AVP, restat triangulum AXP. Ergo trilineo toto CYP ablato ab eodem Sectore AVP (*non quidem, sed a Sectore ACP,*) restabit trangulum AYP.

[6]   Ergo Sector ACP superat triangulum AYP quantitate $AVP - CYP.|$
Non quidem: sed, quantitate CYP.

Illationis error, qui & praecedenti Demonstrationi communis est, & utramque subvertit, hinc ortus videtur, quod modo dixerit *ab eodem sectore AVP*, cum dicendum erat, *a sectore ACP*: Quem excusabit credo, uti solet, ut non *ab ignorantia* sed *ab indiligentia* profectum; & *quem facile erat* (utut

124. WALLIS to BROUNCKER, 21/[31] December 1669

Hobbius id non animadvertit) *cuilibet mediocri Geometra, qui animum ad Diagramma applicaret, cognoscere.* Sec, undecunque sit, Demonstrationem subvertit; & *Hobbii*, quem speraverae, *demonstratae Circuli quadraturae* &c. triumphum; quamque *duraturam se velle* dixerat, & *per me perire nollet*, gloriam evertit penitus.

Adeoque in cassum sunt quae sequuntur; Nempe,

Sed $AVP - CYP$ aequale PQL.

Itaque Sector ACP superat AYP quantitate PQL. Ergo CYP & PQL sunt aequalia. Addito ergo utrinque CYP, erunt AVP & $2CYP$ aequalia. Et est ergo Sector AVP duplus trilinei CYP. Cum igitur idem AVP aequalis sit ambobus trilineis PQL & CYP, erunt ipsa PQL & CYP inter se aequalia. Quorum alterum PQL totum prominet extra Sectorem ACL, alterum nempe CYP totum in eodem Sectore ACL est immersum.

Quare triangula AYP, APQ simul sumpta, id est octava pars totius quadrati $QRST$, aequalia sunt duobus Sectoribus ACP, APL simul sumptis, id est octavae parti totius circuli $BCDE$ dati; & totum quadratum $QRST$ aequale circulo integro $BCDE$.

Atque eadem omnino, eodem tenore, patiter dici possent de Cyp, pqL, atque de CYP, PQL: substitutis ubique, pro Q, P, X, Y, V, minusculis q, p, x, y, u.

Sed superest demonstratio Quarta (ut saltem dici possit, *nos numerus sumus;*) ejusdem commatis,

Aliter. Si triangulum rectangulum AYQ Sectori ACL aequale non sit; supponatur triangulum aliud (primo) minus quam AYQ sed simile, habens verticem in A; latus aq, & basim yq; aequale esse Sectori ACL. Basis autem yq secet arcum CL in p, & rectas AO, AG in r & q.

Quoniam igitur trangulum Ayq aequale est (ut supponitur) Sectori ACL, erunt trilinea qLp, Cyp aequalia.

Recte.

Et quia supponitur qLp dimidium esse Sectoris AVP,—

Nullo modo: Quamquam enim *Hobbius* supponat, in Pseudographemate suo, PQL *dimidium esse Sectoris AVP*; (quod ipse ab initio fraudi fuit; adeoque apparatus ille quem *inutilem* prius insinuavi, etiam ipsi Noxius fuisse deprehenditur:) ut tamen illud de pqL supponatur, non est necesse. Sufficit enim, ad hoc ut triangulum Ayp aequale sit sectori ACL, si saltem sint inter se aequalia pqL, Cyp, (propter commune $AypL$,) utut neutrum sit aequale dimidio ipsius AVP.

Frustra igitur sunt quae sequuntur; Nempe| [7]

Erit Sector ACV una cum trilineo Cyp aequale Sectori ALP una cum trilineo qLp; idemque aequale triangulo Aqr. Rursus quia Triangulum Ayq

124. WALLIS to BROUNCKER, 21/[31] December 1669

aequale est Sectori ACL, erunt trilinea qLp & Cyp aequalia, & ambo simul aequalia Sectori AVP. Et proinde $ACV + Cyp$ aequale dimidio Sectoris ACL, id est triangulo Ayr. Totum parti; quod est absurdum. Similiter Sector ALP una cum trilineo qLp aequale erit triangulo Aqr id est pars toti. Quod est absurdum.

Si Ayq sumeretur supra triangulum AYQ, idem sequeretur absurdum.

Quippe haec Absurda sequuntur, non ex natura rei, seu justa suppositione: sed tantum ex suis falso suppositis.

In cassum item sunt quae hinc colligit;

Est ergo triangulum ipsum AYQ aequale Sectori ACL. Id est octava pars quadrati $QRST$ duobus Sectoribus ACP, APL simul sumptis, id est octavae parti totius circuli $BCDE$ dati; & totum quadratum $QRST$ aequale circulo integro $BCDE$. Inventum est ergo Circulo dato quadratum aequale.

Adeoque non est inventum Circulo dato Quadratum aequale,

Caetera fere ita manent uti prius erant omnia: Hoc est, refutata manent ut prius. Solam propositionem secundam, (quam fassus est fuisse falsam,) quadantenus immutavit; sublato ex multis quae indicaveram uno mendo, pluribus pro eo substitutis. Sed ita foede titubat in re facili, (quippe, data Circuli quadratura, & mediarum quotlibet proportionalium inventione, quae se exhibuisse autumat, quotusquisque est qui nesciat, *Sphaerae Cubum aequalem* dare?) & calculo misere depravato procedit; ut, (si ego reprehenderem & indicarem.) priori locus esset excusationi, quod *de his non erubesceret*, quod *pronunciaverit securius qua* opportuit, quod *non ex ignorantia problematum eminentissimorum orta fuerint*, quod *cuilibet mediocri Geometrae qui animum ad Diagramma applicaret facile erat cognoscere*, quod *sola diligentia opus erat*, utut ipse interim *non omnino fuerit indiligens*. Permittam itaque sua hic utatur diligentia, quo ea detegat (quae quilibet mediocris Geometra modo animum applicet non potest non videre) &, si possit, emendet.

Conatur tandem, sub calcem operis, nonnulla eorum stabilire, quae ego concusseram; seu potius, quae everteram, restituere. Sed irrito conatu omnia. Quippe (inter alia falsa) ubique praesumit gratis (circino deceptus) rectas Ab, AY, (fig. 2.) aequales esse; seu peripheriam centro A per Y ductam, transituram per b punctum, in rectarum eb, AG concursu prius determinatum; atque similiter, aequales esse rectas AO, Ac, seu arcum centro A per O ductum transiturum per c, in rectarum Ab & Zd concursu prius designatum:

124. WALLIS to BROUNCKER, 21/[31] December 1669

(quae secus esse demonstraveram: est utique Quadratum Ab, $[= \frac{25}{32}]^{636}$ Quad. GC, & Quadratum AY, $= \frac{4}{5}$ Quad. CG: Item, Quadr. AO, $= \frac{5}{4}$ Quad. CG, & Quad. AC,$= \frac{32}{25}$ Quad. CG.) Nempe, non calculo, sed circino rem explorans, cum circino not potuerit (in exiguo schemate) rectarum longi- [8] tudines distin|guere, pro aequalibus habuit. Cumque re ad calculos redacta, non invenit voto respondere; invehitur in *Numeros*, in *Arithmeticam*, in *Regulam Auream*, in *Extractiones Radicum*, in *Tabulas Sinuum*, in *Calculi Arithmetici & Geometrici dissensum*, in *Scabiem quam Geometria affricuit Arithmetica*, in *Symbolographiam*, in *Puncta non divisibilia*, in *Lineas non latas*, in *Sectam Mathematicam, Algebristas, Geometriam edoctos, hodiernos Geometras*; quos *universos* sibi queritur *adversos* esse. Adeoque provocat ad *homines liberali ingenio, Geometriam nondum doctos*, ad *Exteros*, ad *Posteros*; nempe ut, quam numquam, nusquam, sibi propitiam sperare possit sententiam, saltem quam possit procul removere, quam possit in longum protelare satagat.

Verum cum utrinque conveniat, ruente prima Propositione, reliqua quae hac nituntur omnia simul corruere; nec ille alia confidentia sperare sustineat, caetera in tuto esse, quam quod eam se putaverit (jam saltem) statuminasse; missis, quas habet innumeras, reliquis nugis (quas mihi singulatim persequi non est animus, nec est necesse,) primum illud, quo caetera dependent, prostrasse sufficiat, (jam secunda vice,) & succenturiatas quas jam attulit Pseudapodeixes.

Caetera qui tanti esse putet, ut sigillatim refutata velit; id ita factum videat, in Refutatione prius edita, ut cordatus nemo, (qui res Mathematicas vel mediocriter intelligit, velitque animum eo applicare,) de eo haesitet. Ut autem Hobbio satisfactum sit, non spondeo; quippe cui (ut secum loquar) *neque intellectum, neque patientiam praestare debeo*, vel etiam sum solicitus.

Decemb. 21. 1669

---

1 Quadratum Ab, $\frac{25}{16}$ *corr.* W

---

[636] Quadratum Ab, $\frac{25}{32}$: cf. WALLIS–OLDENBURG 9/[19].I.1669/70, where Wallis points out the mistake in the printed text and asks Oldenburg to correct the copies which he and Collins had already received.

125.
WALLIS for HENRY OLDENBURG
$c.21/[31]$ December 1669
A Brief Account of Mr. Hobbes's Fundamental Mistake

Transmission:

W Paper sent: LONDON *Royal Society* Early Letters W1, No. 127, 1 p. (original figure inserted in w^1) (our source).
w^1 Copy of paper sent with original figure of *W* inserted: LONDON *Royal Society* Letter Book Original 3, pp. 223–4 (figure inserted between p. 222 and p. 223) (our source of the figure).
w^2 Copy of w^1: LONDON *Royal Society* Letter Book Copy 3, pp. 268–70.
E First edition of paper sent: *Philosophical Transactions* No. 55 (17 January 1670), 1121–2 ('An Accompt of a small Tract, entituled, Thomae Hobbes Quadratura Circuli, Cubatio Sphaerae, Duplicatio Cubi, secundo edita,) Denuo Refutata, Auth. Joh. Wallis. S.T.D. Geom. Prof. Saviliano. Oxoniae, 1669').

This paper, which summarizes Wallis's main argument against Hobbes' latest quadrature of the circle, was published in *Philosophical Transactions* as a review of the work from which the argument is taken: Wallis's printed reply to the second edition of Hobbes' *Quadratura circuli, cubatio sphaerae, duplicatio cubi,* London 1669. Wallis evidently sent this single-page refutation of Hobbes' method to Oldenburg around the same time as he wrote the reply to Hobbes on which it is based (WALLIS–BROUNCKER 21/[31].XII.1669). Oldenburg mentions the single-page refutation in his letter to Huygens a few days later (OLDENBURG–HUYGENS 27.XII.1669/[6.I.1670]; OLDENBURG, *Correspondence* VI, 398). Due to printing delays, Oldenburg's copy, like that sent at the same time to Collins, was without the intended diagram. Wallis rectified this omission when he sent further copies of the single-page refutation with WALLIS–OLDENBURG 9/[19].I.1669/70.

A Brief account of Mr Hobbes's fundamental mistake
in his late quadrature of the circle; by Dr. Wallis.

Mr. Hobbes, Considering, That, *in case it should happen so luckyly* (which was not necessary) *that* QY (the Case of a rightangled triangle QYA equal to the sector LCA, & consequently the square $QRST$ equal to the Circle $BCDE$,) *should, by the arch* CL, *be cut just in the midst at* P: *then would, not onely* (which to his purpose was necessary) QPL, CPY, *be equal each to other* (because of $ALPY$ common both to the triangle and the sector;) *but moreover* (which was not necessary) *each of them equal to the half of* PAV,   5

3 should *(1) fall out (2) happen* W
7 to his *(1) design (2) purpose* W

125. WALLIS for OLDENBURG, c.21/[31] December 1669

(supposing CAV taken equal by construction to LAP:) All which is true, in case of such a lucky hap:

And, finding then (which is true allso) that *this could not All happen unless that intersection at P, were in the line AO* (drawn from the center A to the middle of CG,) *because this must needs passe through the middle of QY*:

Concluded, That *it must needs so happen, or else it was impossible for Any rightangled Triangle, as QYA,* (like to, & part of GCA,) *to be equal to the sector LCA*: Because, *in any other, as qyA, the intersection of CL and qy at p, would not be just in the midst of qy*; *and therefore* (which he supposed necessary, but was not,) $[qpA]$ *not just the half of* $[qyA]$.

Not considering (which is his fundamental mistake) that, *if qpL and Cpy be equal each to other* (though neither of them be equal to the half of PAV, or of pAu; nor yet qp equal to the half of qy, nor $[qpA]$ to the half of $[qyA]$;) *the triangle qyA wil be equal to the sector* $[LCA]$; (because $ALpy$ is common to both;) *and like to the triangle GCA, & a part of it*; which hee thought to have been impossible.

7 That *(1) all this (2) it* W
10 *at p, add.* W
11 not,) *qAp corr.* W
11 half of *qAy corr.* W
14 nor *qAp corr.* W
15 half of *qAy corr.* W
15 sector *LAC corr.* W

127. WALLIS to SCHOOTEN, second half of 1669

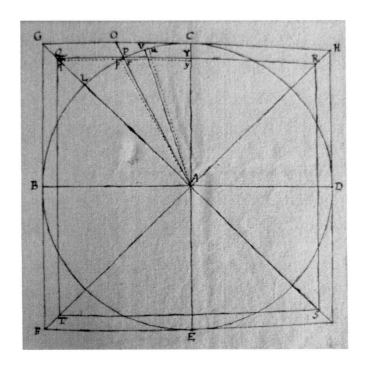

126.
HENRY OLDENBURG to WALLIS
24 December 1669/[3 January 1670]

Transmission:

Manuscript missing.

Existence and date: Mentioned in WALLIS–OLDENBURG 9/[19].I.1669/70.

127.
WALLIS to PIETER VAN SCHOOTEN
second half of 1669

Transmission:

Manuscript missing.

Existence and date: Discussed in advance in COLLINS–WALLIS 17/[27].VI.1669 and WALLIS–COLLINS 24.VI/[4.VII].1669; Schooten's reply referred to in WALLIS–COLLINS 11/[21].I.1669/70.

In this letter, Wallis apparently raised various questions suggested by Collins in COLLINS–WALLIS 17/[27].VI.1669. It appears he also discussed the idea of having his anti-Hobbesian tracts reprinted in the Netherlands, where Hobbes had recently published his *Opera philosophica, quae latine scripsit, omnia* (Amsterdam 1668). See WALLIS–OLDENBURG 5/[15].XII.1668.

128.
GIOVANNI ALFONSO BORELLI to WALLIS
second half of 1669 (i)

Transmission:

Manuscript missing.

Existence and date: Mentioned as one of two letters sent by Borelli in WALLIS–BORELLI 13/[23].I.1669/70.

129.
GIOVANNI ALFONSO BORELLI to WALLIS
second half of 1669 (ii)

Transmission:

Manuscript missing.

Existence and date: Mentioned as the second of two letters sent by Borelli in WALLIS–BORELLI 13/[23].I.1669/70.

This letter arrived with a package of books, some of which were destined for Boyle. By sending copies of his books to England in this way, Borelli evidently sought their distribution and sale by local booksellers.

130.
PIETER VAN SCHOOTEN to WALLIS
December 1669 – January 1669/70

Transmission:

Manuscript missing.

132. WALLIS to OLDENBURG, 7/[17] January 1669/70

Existence and date: Referred to in WALLIS–COLLINS 11/[21].I.1669/70.

Schooten apparently expressed support for the proposal that Wallis's anti-Hobbesian tracts be reprinted in the Netherlands. He also answered questions which Wallis had raised on behalf of Collins. See WALLIS–COLLINS 11/[21].I.1669/70.

131.
HENRY OLDENBURG to WALLIS
6/[16] January 1669/70

Transmission:

Manuscript missing.

Existence and date: mentioned in and answered by WALLIS-OLDENBURG 9/[19].I. 1669/70.

132.
WALLIS to HENRY OLDENBURG
Oxford, 7/[17] January 1669/70

Transmission:

W Letter sent: LONDON *Royal Society* Early Letters, W1, No. 99, 4pp. (our source). At top of p. 1 in Oldenburg's hand: 'A letter of Dr Wallis to M. Oldenburg Concerning some (*1*) Observations of (*2*) Thermometrical and Baroscopical Observations' and 'Entered LB. 3. 217.'—printed: OLDENBURG, *Correspondence* VI, 409–13.
w^1 Copy of letter sent: LONDON *Royal Society* Letter Book Original 3, pp. 217–22.
w^2 Copy of w^1: LONDON *Royal Society* Letter Book Copy 3, pp. 261–7.

E First edition of letter sent (with minor alterations): *Philosophical Transactions* No. 55 (17 January 1670), 1116–20 ('Some Observations Concerning the Baroscope and Thermoscope, made and communicated by Doctor I. Wallis at Oxford, and Dr. I. Beale at Yeovil in Somerset, deliver'd here according to the several dates, when they were imparted').

Oxford. Jan 7. 1669./70.
Sir,

To what you inquire[637] concerning my Thermoscope & Baroscope, (the former of which gives account of the Temper of the Ayr, as to Heat & Cold; the

---

[637]inquire: there is no record of this inquiry.

132. Wallis to Oldenburg, 7/[17] January 1669/70

other of its Weight;) I have some few particulars to adde to the Observations I sent you some years[638] agoe.

The first is, that whereas I did then observe, that in hot weather the Quick-silver in the Baroscope, did use to rise observably, especially in sun-shine & the Heat of the day; which might seem to argue the Ayr to be thereby made heavyer; (which I find was the case of some other Baroscopes as well as mine:) which put us to some thoughts concerning the reason of it: I do now find (having kept the same Baroscope for the space of five years unaltered) the case, for these two years last past, to be somewhat otherwise: & that in hot sun-shiny weather the Quick-silver doth rather subside a little; & in extreme cold & frosty weather it riseth.

Which makes mee judge the cause of these contrary Observations to be this: *viz.* That the Quicksilver at its first putting into the Tube or Baroscope, was not so perfectly clensed from Air, but that some small quantity of it did remain (undiscerned) in the Quick-silver: which latent particles of Air, though so small as not to be at all discernable to the Eye by bubbles, yet by the Externall heat (adding new strength, as it useth to do, to its elastick or springy power) were so much expanded as to give somewhat a greater bulk to the same quantity of Quicksilver with which it was mingled, & consequently to make it rise somewhat higher, as being specifically lighter (that is, having the same weight in a larger dimension; or, in the same dimensions, a lesser weight:) & upon the recesse of the externall heat, the spring again slackening, the air being more compressed suffered the quicksilver to be again contracted into its former lesser dimensions, & so to become heavyer & not

---

1 Weight;) *(1)* It is not much that I have now *(2)* I have some few particulars
2 agoe *(1)* save in *(a)* one or two *(b)* some few particulars, which in November last I gave you some notice of. *(2).* The first is,
5 sunshine & *add.*
8 for the space of five years *add.*
9 case, *(1)* to be *(2)* for these... somewhat
11 & |that *del.*| in
15 latent *add.*
16 by bubbles, *add.*

---

[638] some years agoe: i.e. Wallis's 'A Relation concerning the late Earthquake neer Oxford; together with some Observations of the sealed Weatherglass, and the Barometer both upon that Phaenomenon, and in General', printed in *Philosophical Transactions* No. 10 (12 March 1665/6), 166–71. Wallis reports on further observations with his baroscope in his letter to Oldenburg of 11/[21] August 1666; WALLIS, *Correspondence* II, 281–3.

132. WALLIS to OLDENBURG, 7/[17] January 1669/70

to rise so high as before when it was hotter. But now, the Quicksilver having continued in the Tube for five years & upwards, hath by its own weight clensed itself better from that little Air that was in it, & that Air freed from its intanglements with the quick-silver being got up into the voyd part of the Tube above the quick-silver, doth act contrarywise; that is, when it is by heat (upon the strengthening of its spring) expanded, it presseth downward upon the quicksilver & doth| a little depresse it; & contrary wise, when by frost or very cold weather this air (by the abatement of its spring) is contracted; the quick-silver (freed from that pressure) riseth a little. But this Rising and Sinking on this account; (as well that formerly, when this Air was in the quick-silver; as that now, when it is gotten above it;) is not very considerable; hardly exceeding the twelfth part of an Inch, or thereabouts.

This account I thought not amisse to give you; because it may possibly preserve yourself, or some others who make use of Baroscopes, from being imposed upon by such Observations, as if the Sun shining or Heat of the Weather did make the Air heavyer than before: whereas it seems to bee but an accidental operation upon that unobserved Air latent in the Quicksilver. And I thought it the more necessary, because it is not so easily discovered unlesse by keeping the Baroscope unaltered for a longer time than (perhaps) is done by some of those who make use of it; and, without which, I had not discerned it myself.

On this occasion, I shall adde another accident which I lately took notice of. Observing, in the late hard frost, that the quicksilver did not rise a little, at such time as, by reason of the fierce freezing, I expected it might; I did suspect (as it proved to bee) that a little drop of water, (which was at first made use of for the clensing of the quick-silver from the Air, and which hath ever since remained on the top of the quick-silver within the tube) was

3 from (*1*) the (*2*) that little
6 heat (*1*) expanded (*2*) (upon ... expanded
7 quicksilver (*1*) making it (*2*) & doth
8 very (*1*) hard (*2*) ⟨—⟩ (*3*) cold
11 is (*1*) gone (*2*) gotten
15 Sun shining or *add.*
16 it (*1*) is (*2*) seems to bee
17 latent *add.*
19 time (*1*) that (*2*) than (*a*) is pe *breaks off* (*b*) (perhaps)
20 by (*1*) many (*2*) some
20 of (*1*) it thems *breaks off* (*2*) it;
22 lately (*1* observed) (*2*) took notice of
24 little, (*1*) as (*2*) at such ... freezing,

132. WALLIS to OLDENBURG, 7/[17] January 1669/70

frozen fast to the glasse, so as to stop the quicksilver from ascending: &, to try whether it were so or no, [I] did a little shake the tube (by moving it up & down) so as to make the Quick-silver undulate. Whereupon I found the frozen drop of water to keep its place, while the undulating Quicksilver did several times beat against it. And (which is the thing for which I mention it) the noise upon these knocks, was not such a dull noise as Quicksilver or other Liquids use to make in the open Air, by dashing against Glasse, or Ice, or other such hard bodyes; but such a hard smart noise as hard metals use to make by knocking one against the other; or as if this Ice had been so knocked by a solid piece of Iron or other metal of such a bignesse. Which difference of noise from what would have been in the open air (where the intermediate air must first have been beat away before the quick-silver could strike the Ice, & thereby the stroke of the liquid body obtunded or broken) I attribute to that voydnesse of Air which was between the Ice & the distant Quick-silver. And I remembered presently that the Honorable Mr Boyle, had formerly shewed me an experiment very like this, upon| another occasion:[639] which made me [3] the readyer to take notice of this; & I did it severall days successively. But when, by applying the heat of a candle to the side of the Glasse, I had melted this Ice; I found (as I expected) that within a little time the quick-silver was risen, about a sixteenth part of an Inch above what it was before; which the freezing of that water had till then hindered it from doing.

My Thermoscope, or sealed Weather-glasse; (which, having no communication with the open air, & so not being affected with its weight, gives account but of its Heat or Cold;) hath, this last frost been much lower than I have ever known it, upon five years constant observation. Which proceedeth

---

1 fast to the glasse, *add.*
2 no, a did *corr. ed.*
3 undulate. (*1*) Upon wh *breaks off* (*2*) Whereupon
5 times (*1*) knock (*2*) beat
8 hard (*1*) mettles (*2*) metals
10 other (*1*) mettle (*2*) metal
11 the intermediate *add.*
15 presently *add.*
20 before; | when the frost began, *del.*| which
23 being *add.*
25 Which (*1*) arise (*2*) proceedeth

---

[639] occasion: there is evidence that Wallis and Boyle collaborated on investigations on air. See BOYLE–OLDENBURG *c.*10/[20]. March 1665/6 (OLDENBURG, *Correspondence* III, 55–6).

132. WALLIS to OLDENBURG, 7/[17] January 1669/70

partly from the extremity of the cold more then ordinary; & partly from the inclosed liquor (being spirit of wine, tinged with Cochineel,) growing less spirituous.

It was first made, in December 1664. In the monthes of January, & of February following, wee had very smart frosts, more cold then ordinary; when yet the lowest mark to which the liquor did subside (in extreme hard frosts, and very cold wind,) was at Inches $12\frac{1}{4}$: (at which time $14\frac{1}{2}$ was frost certain & sometimes at 15 and at $15\frac{1}{2}$:) The hight in summer following, 1665, was usually at 20, 21, 22, or thereabouts; but in some few very hot days at 25, 26, $26\frac{1}{2}$: (the whole hight of the smal cylindrick glasse, whose cavity was about $\frac{1}{8}$ of an Inch diameter, being about 28 inches; beside a small spherical bowle at the top, of about $\frac{3}{4}$ of an Inch diameter; & a bowl at the bottom, which contained the liquor, of about 2 Inches diameter: the space above the liquor being, at the first composure of it, voyd of air, save what it had out of the liquor, which, being warm at the first putting in, freed the whole cavity while the glasse was hermetically sealed.)

The Winter following; the liquor seemed to remain much about the same temper as before; for in December, January, & February, wee had at $14\frac{1}{2}$ Frost certain; sometimes at 15 or higher: & the lowest to which it did that winter descend, was $12\frac{3}{4}$: The hight in the following summer, 1666, was usually about 19, 20, 21; the highest of all at 25.

About the end of December 1666 & the beginning January following, it was, in hard frosty weather at 12, 11, & once $10\frac{1}{2}$, the weather being very cold, & the liquor (it should seem) becoming somewhat lesse spirituous, having evaporated some of its more subtile parts into the voyd cavity; & was Frost certain that winter about $13\frac{1}{2}$; (an inch lower than the years before;) sometimes at 14, or $14\frac{1}{2}$. The usual hight in summer following, 1667, was about 19, 20, 21, & the highest at $24\frac{1}{2}$.

4 made (*1*) about (*2*) in
6 subside (*1*) was (*a*) at (*b*) in (*2*) (in extreme ... wind,) was at
8 15 (*1*) or higher at (*2*) and
8 following, (*1*) being (*2*) 1665, was
12 spherical *add.*
15 what (*1*) was (*2*) it
20 1666, (*1*) being (*2*) was
22 1666 *add.*
22 following, *add.*
23 11, (*1*) or lower (*2*) & once at
24 somewhat *add.*
26 that winter *add.*
27 1667, (*1*) being (*2*) was

132. WALIIS to OLDENBURG, 7/[17] January 1669/70

The winter following; it was scarce certain frost at 13; but yet sometimes at 14 or a little higher.| The lowest to which it did descend that winter (being very mild after Christmas) was at 12: And the following summer, 1668, usually about 18, 19, 20; the highest of all (the heat of summer being but very moderate) at 22.

The next winter; it was frost certain, about $12\frac{1}{2}$; but sometimes at 13 or higher: the lowest of all, at $10\frac{1}{4}$. And, in the Summer following, 1669, the highest of all (being but a cold summer) not much above 20.

But now this Christmas, 1669, though I find it to be frost certain, about $12\frac{1}{4}$, & sometimes at higher then 13; yet hath it come sometimes lower then 8; & particularly Dec. 26. in the morning to $7\frac{3}{4}$, & did not all that day come so high as 8 inches. Which being so much lower, then ever it had been in any of the precedent years; though it may in part be attributed to the disspiriting of the liquor; yet principally to the extremity of the Cold.

It hath ever since been rising (but with some descent in the night time) & was on Jan. 1. when the frost seemed first to relent, some what higher then 9; & is Jan. 7. about $13\frac{1}{2}$. The Baroscope at 29 but for some days before, it was about $28\frac{3}{4}$ (the wether having been windy and rainy;) & so it was in the frost, about Dec. 25; but then continued to rise till about Jan. 2. to $29\frac{3}{8}$. But had been Dec. 13. at $30\frac{1}{8}$: which is the highest I have ever known it in my baroscope; $27\frac{7}{8}$ being the lowest that I have ever observed in it; (Octob. 26. 1665:) the most usual hight being about 29, or somewhat higher.

But, though mine have been very rarely, & but very little, above 30; or lower 28, (reconing from the surface of the stagnant quicksilver;) yet in

3 1668, *add.*
6 but *add.*
10 come (*1*) as low as $7\frac{3}{4}$ (*2*) sometimes lower than 8
13 of the (*1*) preceeding (*2*) precedent
15 (but ... time) *add.*
17 is (*1*) this day (*2*) Jan. 7.
17 Baroscope | now *add. and del.*| at
17 29 | (not withstanding the rain & very high wind this last night;) *del.*| but
18 (the wether ... rainy;) *add.*
19 was (*1*) about the beginning of the frost (*2*) in the frost, about
19 then (*1*) rose (*2*) continued to rise
21 baroscope; (*1*) the lowest (*2*) $27\frac{7}{8}$ being the lowest
22 in (*1*) my baroscope: (*2*) it; (Octob. 26. 1665.)
24 & but very little, *add.*

134. WALLIS to OLDENBURG, 9/[19] January 1669/[1670]

other places (according to the difference of Airs) it may by others have been found either higher or lower: & so likewise, according as the Quicksilver at the first filling of the inverted Tube, was more or lesse clensed of Air. For a very little air left in the Quicksilver, & undiscernable to the eye; will, when it gets free of it, & remains in the voyd space above the quicksilver, sensibly depresse the quicksilver: And in the mean time (before it so gets free) will, upon heat of wether, make it swell.

Yours

John Wallis.

For Mr. Henry Oldenburg, in the
Palmal London

133.
JOHN COLLINS to WALLIS
8/[18] January 1669/[1670]

Transmission:

Manuscript missing.

Existence and date: mentioned in and answered by WALLIS–COLLINS 11/[21].I.1669/70.

Collins apparently sought Wallis's advice on which books to send Borelli in return for those sent by him from Italy. To this end, Collins enclosed a list of possible titles.

134.
WALLIS to HENRY OLDENBURG
9/[19] January 1669/[1670]

Transmission:

W Letter sent: LONDON *Royal Society* Early Letters W1, No. 100, 2 pp. (our source). Underlining by Oldenburg of words and passages not to be included in the Letter Book copies. At top of p. 1 in Oldenburg's hand: 'Enter'd L.B.3.223.'. At foot of p. 1, again in Oldenburg's hand: 'Dr Wallis's Letter to M. Oldenburg concerning (*1*) his Answer (*2*)

5 it (*1*) comes to (*2*) gets free
6 quicksilver: (*1*) And ⟨—⟩ I have often observed ⟨—⟩ *del.* (*2*) And in the meantime

287

134. WALLIS to OLDENBURG, 9/[19] January 1669/[1670]

some copies of his Answer to M. Hobbes's Quadrature, to be presented to the R. Soc.'
On p. 2 beneath address in Oldenburg's hand: 'For more Copies, for Pell, Pope, Bishop
of |Chester del.| Salisbury. Given to President, Chester, Moray, Hook, Jeffreys.'—printed:
OLDENBURG, *Correspondence* VI, 419-20.
w^1 Copy of letter sent: LONDON *Royal Society* Letter Book Original 3, p. 223.
w^2 Copy of w^1: LONDON *Royal Society* Letter Book Copy 3, pp. 268-70.

Reply to: OLDENBURG–WALLIS 24.XII.1669/[3.I.1670] and OLDENBURG–WALLIS 6/[16].I.
1669/70.

Wallis sent this letter either accompanying or at the same time as a packet containing copies of his 'Brief account of Mr. Hobbes's fundamental mistake' and his *Thomae Hobbes quadratura circuli ... denuo refutata*. He also enclosed in the packet two sheets simply with the diagrams to 'A brief account of Mr Hobbes's fundamental mistake', which had been missing in the copies sent to Oldenburg and Collins in December (see WALLIS for OLDENBURG *c*.21/[31].XII.1669). Oldenburg's note on p. 2 was probably for a now missing reply to Wallis: that he had given copies of *Thomae Hobbes quadratura circuli ... denuo refutata* to Brouncker, Wilkins, Moray, Hooke, and Edward Jeffreys, besides that intended for the Royal Society itself, and that he was to request a further three copies (he neglected to correct 'for') for John Pell, Walter Pope, and Seth Ward.

Oxford Jan. 9. 1669

Sir,

I have yours[640] of Jan. 6. & that[641] of Dec. 24 (post free) & thank you for them. I had sooner sent the pacquet,[642] which comes with this of my Answer to Mr Hobbes's quadrature; but that it was some while before I could get the cuts wrought off at the Rolling Press (by reason of the frosty wether, & of the holy-days sticking in the workmens hands) though the graver had done his Work before I sent you the printed sheet.[643] And, since the cuts were finished, I deferred a little, that the two written papers might come with the bookes,[644] (to make one businesse of it,) considering the Society

4 sooner *add.*
4-5 of my ... quadrature *add.*
6 Press *add.*
9 I *(1)* stayd *(2)* deferred

---

[640]yours: i.e. OLDENBURG–WALLIS 6/[16].I.1669/70.
[641]that: i.e. OLDENBURG–WALLIS 24.XII.1669/[3.I.1670].
[642]pacquet: i.e. a packet containing copies of Wallis's *Thomae Hobbes quadratura circuli ... denuo refutata* together with two separate copies of the diagram contained in that work.
[643]printed sheet: i.e. the printed sheet of 'A brief account of Mr Hobbes's fundamental mistake', which Wallis sent Oldenburg (and Collins) in December 1669. See WALLIS for OLDENBURG *c*.21/[31].XII.1669 and WALLIS–COLLINS 11/[21].I.1669/70.
[644]bookes: i.e. copies of the first part of Wallis's *Mechanica: sive, de motu, tractatus geometricus*, delivered directly to the Royal Society from the printers in London.

135. WALLIS to COLLINS, 11/[21] January 1669/70

meet not till next Thursday.[645] You will then please to next deliver two of them, one to the Society, & the other to the President:[646] & the rest to any particular & mathematical friends[647] as far as they will go. And I have sent two cuts supernumerary to perfect the sheets I sent to you & Mr Collins at first: (In both which, let this erratum[648] be mended; pag. penult. lin. 4 a fine: for $\frac{25}{16}$, read $\frac{25}{32}$: which in all the rest I have mended myself;) but did not see it, being in hast, before the sheets were wrought off. No more ,but that I am,

Yours to serve you
John Wallis.

[2] For Mr Henry Oldenburg
in the Palmal near
St James's. London.

135.
WALLIS to JOHN COLLINS
Oxford, 11/[21] January 1670

Transmission:

W Letter sent: CAMBRIDGE *Cambridge University Library* MS Add. 9597/13/6, f. 211^r–211^v (our source). On f. 211^v in Collins's hand: 'Van Schootens Answer'. Postmark: 'IA/12'.—printed: RIGAUD, *Correspondence of Scientific Men* II, 518–20.

Reply to: COLLINS–WALLIS 17/[27].VI.1669 and COLLINS–WALLIS 8/[18].I.1669/70.

---

10 one *(1)* work *(2)* businesse

---

[645]Thursday: i.e. 13 January 1669/70 (old style). At the meeting on that day a copy of Wallis's *Thomae Hobbes quadratura circuli ... denuo refutata* was presented to the Society, together with the first part of his *Mechanica: sive, de motu, tractatus geometricus*. See BIRCH, *History of the Royal Society* II, 415.

[646]President; i.e. William Brouncker, q.v.

[647]friends: Oldenburg presented the remaining copies to Wilkins, Moray, Hooke, and Jeffreys.

[648]erratum: the Royal Society's copy of 'A brief account of Mr Hobbes's fundamental mistake' has this error corrected by hand.

135. WALLIS to COLLINS, 11/[21] January 1669/70

Oxford. Jan. 11. 1669./70.

Sir,

To yours[649] of the 8 instant. I think it would be convenient to acquaint Mr Boyle that wee are preparing to send,[650] & know whether hee think fit to send any or all of his own peeces, in return for those which Borelli hath sent[651] him: at lest those which he hath not. Of those in the catalogue[652] you send mee, I think it may be convenient to send, (beside those of Mr Boyle,) Hooks Micrography.[653] Lowers two bookes.[654] Wings[655] & Streets[656] Astronomy. Hobs[657] contra fastum Geometrarum. Merrets[658] Pinax. Willis's[659] Pathologia cerebri (his former bookes, Borrelli hath.) Needham's[660] disquisitio. Uleg Beig's[661] Catalogue. Wilkins[662] universal character. Evelins[663]

---

6 which (*1*) are (*2*) he hath
7 (beside ... Boyle,) *add.*
8 Wings & *add.*

---

[649] yours: i.e. COLLINS–WALLIS 8/[18].I.1669/70.

[650] send: i.e. the package of books Wallis and Collins were to send to Borelli, in return for those he had sent with a recent letter to Wallis: BORELLI–WALLIS, second half of 1669(ii).

[651] sent: cf. WALLIS–BORELLI 13/[23].I.1669/70.

[652] catalogue: i.e. a list of books to send to Borelli, probably enclosed in COLLINS–WALLIS 8/[18].I.1669/70. See appendix to WALLIS–BORELLI 13/[23].I.1669/70.

[653] Micrography: i.e. HOOKE, *Micrographia*, London 1665.

[654] bookes: i.e. LOWER, *Tractatus de Corde, item de Motu et Colore Sanguinis*, London 1669 and idem, *Diatribe Th. Willisii de Febribus vindicatio*, Amsterdam 1666.

[655] Wings: i.e. WING, *Astronomia Britannica*, London 1669.

[656] Streets: i.e. STREETE, *Astronomia Carolina*, London 1661.

[657] Hobs: i.e. HOBBES, *De principiis et ratiocinatione geometrarum: ubi ostenditur incertitudinem falsitatemque non minorem inesse scriptis eorum quam scriptis physicorum et ethicorum. Contra fastum professorum geometriae*, London 1666.

[658] Merrets: i.e. MERRET, *Pinax, rerum naturalium britannicarum continens*, London 1666.

[659] Willis's: i.e. WILLIS, *Pathologiae cerebri*, London 1667.

[660] Needham's: i.e. NEEDHAM, *Disquisitio anatomica de formato foetu*, London 1667.

[661] Beig's: i.e. [ULUG BEG], *Jadâwil-i mawâdi'-i thawâbit dar tûl u'ard kih bi-rasad yâftah ast Ulugh Baik Sive Tabulae long. ac lat. Stellarum fixarum*, ed. Thomas Hyde, Oxford 1665.

[662] Wilkins: i.e. WILKINS, *An Essay Towards a Real Character, And a Philosophical Language*, London 1668.

[663] Evelins: i.e. EVELYN, *Sylva, or A discourse of forest-trees, and the Propagation of timber ... to which is annexed Pomona; or An appendix concerning fruit-trees in relation to cider*, London 1664; idem, *Sylva*, second edition London 1670.

135. WALLIS to COLLINS, 11/[21] January 1669/70

Sylva & Pomona. Beverige's[664] Chronology. Glanvils[665] Progress, Mercators[666] Logarithmotechnia. Gregories[667] exercitationes. The Hystory of the R. Society.[668] (& what more of them you please.)

And moreover; A collection of the Transactions;[669] if Mr. Oldenburg have not sent them allready. And Barrow's pieces[670] al of them. (Which I find not in your list.)

Amongst mine, let my Grammatica[671] linguae Anglicanae be one.

I shall speedyly prepare a letter to send him. But consider what answere wee are to give to what hee moves, of sending over some copies of his[672] to bee sold; whether any booksellers here will take off any number of copies. Pray let Mr Oldenburg have the two bookes[673] hee desires. I like well inough of Flamstede's design;[674] but I would not have him too severe[675] with Street; who, I think, hath deserved well in Astronomy. Mr. Horro:[676] his name is

---

1 Evelins ... Progresse &c. *add.*
3 of them *add.*
3-5 And Barrow's ... list.) *add.*

---

[664]Beverige's: i.e. BEVERIDGE, *Institutionum chronologicarum libri II, una cum totidem arithmetices chronologicae libellis*, London 1669.

[665]Glanvils: i.e. GLANVILL, *Plus ultra: or, The progress and advancement of knowledge since the days of Aristotle*, London 1668.

[666]Mercators: i.e. MERCATOR, *Logarithmotechnia*, London 1668.

[667]Gregories: i.e. GREGORY, *Exercitationes geometricae*, London 1668.

[668]Hystory: i.e. SPRAT, *The history of the Royal-Society of London*, London 1667.

[669]Transactions: i.e. *Philosophical Transactions*.

[670]pieces: i.e. BARROW, *Lectiones XVIII, Cantabrigiae in scholis habitae; in quibus opticorum phaenomenωv genuinae rationes investigantur, ac exponuntur. Annexae sunt lectiones aliquot geometricae*, London 1669; idem, *Euclidis Elementorum libri XV. breviter demonstrati*, Cambridge 1655.

[671]Grammatica: i.e. WALLIS, *Grammatica linguae anglicanae*, Oxford 1653; idem, *Grammatica linguae anglicanae, Editio secunda, priore auctior*, Oxford 1664.

[672]his: i.e. BORELLI, *Historia et meteorologia incedii Aetnaei anni 1669. Accessit Responsio ad censuras rev. p. Honorati Fabri contra librum auctoris De vi percussionis*, Reggio di Calabria 1670.

[673]bookes: presumably Wallis's *Thomae Hobbes Quadratura Circuli ... denuo refutata* and *Mechanica, sive de motu tractatus geometricus; pars prima* At the meeting of the Royal Society on 13 January 1669/70, copies of both books were presented. See BIRCH, *History of the Royal Society* II, 415.

[674]design: see the report on a manuscript sent by Flamsteed to Brouncker at the meeting of the Royal Society on 13/[23] January 1669/70 in BIRCH, *History of the Royal Society* II, 415.

[675]severe: Flamsteed compared his results to those of Thomas Streete, as published in the latter's *Astronomia Carolina*, London 1661. See FLAMSTEED–OLDENBURG 26.II/[8.III].1669/70; OLDENBURG, *Correspondence* VI, 513–7, 514.

[676]Horro: i.e. Jeremiah Horrox (1618–41), *ODNB*.

135. WALLIS to COLLINS, 11/[21] January 1669/70

truly spelled Horrockes, not Horrox; which I could wish to be preserved, at lest in some places, in the printed bookes; though (since hee hath been pleased so to put it) it may in Latine elsewhere bee written with x.

I sent[677] yesterday to Mr Oldenburgh (by Bartlet's coach) some more of my last papers against Mr Hobs; (& a Cut supernumerary to perfect what[678] I sent you before.)

I had lately a letter[679] from Peter Van Schooten; who longs to hear of some here willing to part in the impression[680] of my things at Leyden. Hee tells me that young Golius[681] being very sick, & himself scarce so well as to go abroad, hee hath not yet inquired of the book[682] sent by Sir Alexander Humes to Golius his father; but wil do on the first occasion. That he knows of nothing of Huddens in the presse; but that he had long since spoken of a book[683] near finished, treating of Algebra a capite ad calcem; but beleeves that Hudden being ful of business, & but slow in writing bookes, (&, I may adde, who doth not so much as understand Latine,) he doth not think it to be finished or in the presse; &, if ever, it will be in dutch. That Martin Wilkens his Officina[684] Algebrae, hath nothing in it but vulgaria. Sclot vand Scloten[685] hee knows not. That, of vander Huyps, he knows nothing extant but the book[686] you mention. That he will very readyly serve you in buying

---

11 Golius *add.*

---

[677] sent: see WALLIS–OLDENBURG 9/[19].I.1669/70.

[678] what: Wallis had sent Collins and Oldenburg each a copy of his one-page refutation of Hobbes's method of quadrature, but without the diagram, in December 1669. To rectify this omission, supernumerary copies of the wood-cut were printed. See WALLIS for OLDENBURG it c.21/[31].XII.1669 and WALLIS–OLDENBURG 9/[19].I.1669/70.

[679] letter: i.e. SCHOOTEN–WALLIS ?.XII.1669.

[680] impression: Wallis clearly desired at this time, after the publication of Hobbes's *Opera philosophica* in Amsterdam in 1668, that his writings be made more widely available on the European continent. See the note appended to the review of *Thomae Hobbes quadratura circuli ... confutata* in *Philosophical Transactions* No. 48 (21 June 1669), 972, and COLLINS–VERNON 4/[14].IV.1671; RIGAUD, *Correspondence of Scientific Men* I, 160–5, 161: 'Dr. Wallis his former works are to be printed at Leyden, some of our booksellers joining with the Dutch in the impression'.

[681] Golius: presumably Theodorus van Gool (d. 1679), sometime burgomeister of Leiden and the eldest son of Jakob van Gool.

[682] book: i.e. the manuscript book of Alexander Anderson, which Alexander Hume had sent to Jakob van Gool. See COLLINS–WALLIS 17/[27].VI.1669.

[683] book: Collins had questioned Wallis about a mathematical book of Hudde's which was supposedly being printed in Amsterdam. See COLLINS–WALLIS 17/[27].VI.1669.

[684] Officina: i.e. WILKENS, *Officina algebrae*, Groningen 1636.

[685] Sclot vand Scloten: probably MARTINI, *Slot en sleutel van de navigation, ofte groote zeevaert*, Amsterdam 1659.

[686] book: i.e. HUIPS, *Algebra*, Amsterdam 1654.

135. WALLIS to COLLINS, 11/[21] January 1669/70

what bookes there you shal send for: (mathematick books being there, in their Auctions to be had very cheap:) & will, if you give him leave, desire of you the like favour at London, when he shal know how to direct a letter to you. That Letters will come to him safe, & post free, if directed thus:

> Mijn Heer Willem vander Cruck Procureur,
> in's Graven Hage om te behandigen aende
> Professor Van Schooten te Leyden.

And this in a cover thus directed:

> Mijn Heer mijn Heer Bisdommer Commis-
> saris in Schraven Hage.

And the letter given to the Secretary of Holland Ambassador at London to send.

I have here received 12 copies of myne de Motu;[687] but in one of them the sheet I is torn; & the last leaf of all is wanting. I wish (with the perfecting of this) hee would send mee all the printeds after Cc in the second alphabet, together with the first sheet. Hee hath made the year in the title page 1670, whereas I directed it to be 1669.

I wish they would make more hast at the presse. I have had (I think) but two sheets since I was in London. Of Mr Houghtons[688] son, I can say no more then I did: Dr Crosse[689] hath had mu[ch] occasion to know both the father & the son; being his next neighbours & of long acquaintance.

I am yours &c.
John Wallis.

[0ᵛ] For Mr John Collins
at the three Crowns in
Bloomsbury market
London.

---

2 leave, (1) put you to (2) desire
6 (1) in Schraven (2) in's Graven
20 had mu occasion *corr. ed.*

---

[687] De Motu: i.e. the first part of Wallis's *Mechanica: sive, de motu, tractatus geometricus*, London 1670.
[688] Houghtons: not identified. Cf. COLLINS–WALLIS 12/[22].VIII.1672.
[689] Crosse: probably Joshua Crosse (1614–76), sometime fellow of Magdalen College, who after marriage to Rachel Knight continued to live near the college. See WOOD, *Life and Times* II, 345.

136.
Wallis to Giovanni Alfonso Borelli
Oxford, 13/[23] January 1669/70

Transmission:

w^1 Copy (in Collins' hand) of missing letter sent: LONDON *Royal Society* Early Letters W1, No. 101, 1 p. (our source).

w^2 Copy (in Collins' hand) of missing letter sent: CAMBRIDGE *Cambridge University Library* MS Add. 9597/13/6, f. 212ʳ.—printed: RIGAUD, *Correspondence of Scientific Men* II, 520–2.

Enclosure: Catalogue of Books for Borelli.

As emerges from COLLINS–OLDENBURG ?.II/III.1669/70, the original letter was sent with a parcel of the books specified in the accompanying catalogue by boat to Palermo in February or March 1670. Collins clearly carried out the task of procuring or collecting the books and of having them packaged. He subsequently sent Oldenburg a copy of both the letter and the catalogue as an enclosure to COLLINS–OLDENBURG ?II/III.1669/70.

Clarissimo Doctissimoque Viro D Johanni Alphonso Borellio,
in Academia Pisana Matheseos Professori, Messanae Siculorum
agenti. Johannes Wallis S P D

Agnosco (vir Clarissime) me tibi pluribus nominibus obstrictum esse, qui
me et literis et libris tuis locupletasti: quem primum miseras Librum unum
et periisse putaveram, tandem postliminio accepi: Literas etiam binas,[690]
et cum secundis, Librorum molem majorem. Ex quibus, quos Honoratissimo Boylio designaveras, Londinum propediem ad illum remittendos curabam, qui plurimas tibi gratias rependit, quod eum hoc honore affeceris.
Unisque mihi reservatis Singulorum Exemplaribus: reliqua (quippe praeter
Apollonii[691] unicum erant singulorum si memini terna Exemplaria) Londinum mittebam ad D. Johannem Collins virum Mathematicum, et diligentem
Scriptorum Mathematicorum Exquisitorem. Eundem ego rogavi meo nomine
ad te remittere meorum omnium quae non habes Exemplaria (gratitudinis

---

3 Wallis *missing in* w^2
4 *In margin*: tradantur Messanae in Sicilia. w^2
6 Literas item binas w^2

---

[690]Literas ... binas: i.e. BORELLI–WALLIS, second half 1669(i), and BORELLI–WALLIS, second half 1669(ii).

[691]Apollonii: i.e. [APOLLONIUS], *Apollonii Pergaei Conicorum lib. V. VI. VII.*, paraphraste Abalphato Asphahanensi nunc primun editi ... ex codicibus arabicis m.ss. Abrahamus Ecchellensis ... Latinos reddidit. Io: Alfonsus Borellus ... curam in geometricis versioni contulit, & notas ... adiecit, Florence 1661.

136. WALLIS to BORELLI, 13/[23] January 1669/70

meae ob acceptos tuos qualecunque indicium,) una cum his Literis, et Catalogo[692] Librorum praecipuorum apud nos nuperis annis editorum, quae rem Mathematicam praesertim spectant atque etiam Physiologiam, (Nescio an non a D Boyle ad te missuros sit libros aliquot) simulque petit ut in posterum, librorum Mathematicorum, non tuorum tantum, sed et ab aliis editorum, Exemplar unum aut alterum, ad illum mittere dignari velis, quae vel numeratis pecuniis rependet, vel (si tu id malis) libris aliis hinc mittendis (quippe libros in Italia, et Oris adjacentibus editos, nos vix accipimus, aut etiam ne omnino) atque in eum finem hos nominatim recensendos a me petiit, quos ut mittas rogat

    Alexandri Marchetti[693] Exercitationes Mechanicas jam perfectas
    Eschinardi[694] Centuria Problematum Opticorum
    Reinaldini[695] Geometram promotam, seu tractatum de Curvis Maediceis determinationi et Solutioni aequationum inservientibus
    Riccii[696] Tractatum ejusdem Argumenti si extat
    Antanalisi di Grisio[697] against Maghetto
    Il compendio[698] delle regole Trigonometriche, e Centuria[699] di Problemi dal Cavaleirio
    Griembergeri de Luce[700] et Refractionibus Opus Posthumum

---

4 a *add.* w^1
13 promotam *missing in* w^2
17 Grisio, contra Maghetto w^2
19 di Cavaleirio w^2

---

[692]Catalogo: see COLLINS–OLDENBURG ?.II/III.1669/70.

[693]Marchetti: i.e. MARCHETTI, *Exercitationes mechanicae*, Pisa 1669.

[694]Eschinardi: i.e. ESCHINARDI, *Centuria problematum opticorum ... seu dialogi optici pars altera*, Rome 1666.

[695]Reinaldini: i.e. RENALDINI, *Geometra promotus*, Padua 1670.

[696]Ricci: i.e. Michelangelo Ricci. In the dedicatory epistle to his *Exercitatio geometrica de maximis et minimis* (1666), Ricci indicates that in the event of positive reviews of this work he would publish his earlier studies on analysis. This project was never realized. See HOFMANN, 'Über die Exercitatio geometrica des M.A. Ricci', *Centaurus* 9 (1963), 139–93, 145, 172. Cf. COLLINS–OLDENBURG *c*.12/[22].IX.1669; OLDENBURG, *Correspondence* VI, 226–9.

[697]Grisio: i.e. GRISIO, *Antanalisi a quesiti stampati nell'analisi di Benedetto Maghetti*, Rome 1641.

[698]compendio: i.e. CAVALIERI, *Directorium generale Uranometricum. In quo Trigonometriae Logarithmicae fundamenta, ac regulae demonstrantur ...*, Bologna 1632.

[699]Centuria: i.e. CAVALIERI, *Centuria di varii problemi*, Bologna 1639.

[700]de Luce: no work of this title by Grienberger is recorded. Cf. COLLINS–OLDENBURG *c*.15/[25].III.1669/70; OLDENBURG, *Correspondence* VI, 565–73, 570.

137. Wallis to Giovanni Alfonso Borelli, Books for Borelli

Ejusdem[701] Speculum Ustorium Ellipticum, cum appendice ad Praxin Sectionum Conicarum, et Consectariis de Circulorum Contactibus et Sectionibus angularibus
Ejusdem[702] Novam Coeli Perspectivam
Mengoli[703] tractatum de Additione fractionum, seu quadraturas Arithmeticas
Hodiernae Opera Mathematica omnia

Et si quid tuorum extat praeter illa quae jam scripsisti. Idemque ni fallor etiam hac vice tibi mittet libros aliquos alios post missurus, quos petieris. Vale

Scribebam Oxonii 13 Jan 1669/70 Stilo Angliae

Citius misissem nisi quod commoda Navigii occasio non occurrebat.[704]

137.
Wallis to Giovanni Alfonso Borelli
Oxford, 13/23 January 1669/70, enclosure:
Catalogue of Books for Borelli

Transmission:

C^1 Copy (in Collins's hand) of missing catalogue sent: London *Royal Society* Early Letters W1, No. 101, 1 p. (our source).

C^2 Copy (in Collins's hand) of part of catalogue: Cambridge *Cambridge University Library* MS Add. 9597/13/6, f. 212^v.—printed: Rigaud, *Correspondence of Scientific Men* II, 522–3.

Enclosure to: Wallis–Borelli 13/[23].I.1669/70.

The earliest version of this catalogue was apparently sent to Wallis as an enclosure to Collins–Wallis 8/[18].I.1669/70. Most of Wallis's suggestions for its augmentation, made in his reply (Wallis–Collins 11/[21].I.1669/70) were clearly taken up, as emerges from the comparison of C^1 and C^2. The final catalogue was sent, as stated in Collins–Oldenburg ?II/III.1669/70, with a parcel of the books specified and the original of

20 opus posthumum nuper editum w^2

---

[701] Eiusdem: i.e. Grienberger, *Speculum ustorium verae ac primigeniae suae formae restitutum*, ed. F. de Ghevara, Rome 1613.

[702] Eiusdem: i.e. Grienberger, *Catalogus veteres affixarum longitudines, ac latitudines conferens cum novis: imaginum caelestium prospectiva duplex*, Rome 1612.

[703] Mengoli: i.e. Mengoli, *Novae quadraturae arithmeticae: seu, De additione fractionum*, Bologna 1650.

[704] occurebat: cf. Collins–Oldenburg II–III.1669/70.

137. WALLIS to GIOVANNI ALFONSO BORELLI, Books for Borelli

WALLIS–BORRELLI 13/[23].I.1669/70 by ship to Palermo in February or March 1670. Collins subsequently sent copies of both the catalogue and Wallis's letter to Oldenburg as enclosures to COLLINS–OLDENBURG ?II/III.1669/70.

Catalogus Librorum missorum ad D Borellium
a Domino Boyle

A Continuation[705] of New Experiments Physico-Mechanicall, touching the Spring and Weight of the Ayre, and their effects the 1 part Oxford 1669 4°
Some Considerations[706] touching the usefulnesse of Experimentall Phylosophy Oxford 1664 4°
Of Absolute rest[707] in Bodies London 1669 4°
Hydrostaticall Paradoxes[708] Oxford 1666 8°
The Origine[709] of formes and qualities Oxford 1666

Sequentes libri a Clarissimo Wallisio et Joanne Collinsio
mittuntur

Wallisii Opera[710] Mathematica 2 Vol 4°
Eiusdem Tractatus[711] de motu et Libra
Eiusdem Elenchus[712] Geometriae Hobbianae
His Due Correction[713] for Mr Hobbs

1 Bookes sent by Mr Boyle C^2
10 Sent by John Collins Slusii Mesolabum De Respirationis usu Primario Auctore Malachia Thruston Londini 1670 8° Tractatus de Urim et Thummim Barrovii Optica C^2

---

[705]Continuation: i.e. BOYLE, *A continuation of New Experiments Physico-Mechanical, touching the spring and weight of the air, and their effects. The I. Part ... Whereto is annext a short discourse of the atmospheres of consistent bodies*, Oxford 1669.

[706]Considerations: i.e. BOYLE, *Some Considerations touching the Usefulnesse of Experimentall Naturall Philosophy*, Oxford 1663.

[707]rest: i.e. BOYLE, *Certain Physiological Essays, and other tracts; written at distant times, and on several occasions. The second edition ... increased by the addition of a discourse about the absolute rest in bodies*, London 1669.

[708]Paradoxes: i.e. BOYLE, *Hydrostatical Paradoxes, Made out by New Experiments*, Oxford 1666.

[709]Origine: i.e. BOYLE, *The Origine of Formes and Qualities (According to the Corpuscular Philosophy,) Illustrated by Considerations and Experiments ...*, Oxford 1666.

[710]Opera: i.e. WALLIS, *Operum mathematicorum pars prima*, Oxford 1657; idem, *Operum mathematicorum pars altera*, Oxford 1656.

[711]Tractatus: i.e. WALLIS, *Mechanica: sive, de motu, tractatus geometricus*, pars prima, London 1670.

[712]Elenchus: i.e. WALLIS, *Elenchus geometriae Hobbianae*, Oxford 1655.

[713]Correction: i.e. WALLIS, *Due Correction for Mr Hobbes*, Oxford 1656.

137. WALLIS to GIOVANNI ALFONSO BORELLI, Books for Borelli

Eiusdem Thomae Hobbs[714] quad Circuli Cubatio Sphaerae et Duplicatio Cubi confutata 1669 4°
Thomas Hobbs de Principiis[715] et ratiocinatione Geometrarum Londini 1666
Merrets Pinax[716] rerum Anglicarum Londini 1667 8°
Jacobi Gregorii Exercitationes[717] Geometricae
Pells Introduction[718] to Algebra
Wingi Astronomia[719] Britannica fo
Streets Astronomia[720] Carolina
Willis de Anatome[721] Cerebri
Loweri tractatus[722] de Corde
Beveregii Institutiones[723] Chronologicae
N Mercatoris Logarithmotechnia[724]
Barrovii Optica[725]

Spencerus de Urim[726] et Thummim

Mori Ethica[727]
Slusii Mesolabum[728]

Thruston de respiratione[729]

---

[714]Hobbs: i.e. WALLIS, *Thomae Hobbes Quadratura circuli, cubatio sphaerae, duplicatio cubi, confutata*, Oxford 1669.

[715]Principiis: i.e. HOBBES, *De principiis et ratiocinatione geometrarum*, London 1666.

[716]Pinax: i.e. MERRETT, *Pinax rerum naturalium Britannicarum*, London 1667.

[717]Exercitationes: i.e. GREGORY, *Exercitationes geometricae*, London 1668.

[718]Introduction: i.e. [RAHN], *An Introduction to Algebra, translated out of the High-Dutch into English, by Thomas Brancker. M. A. Much altered and augmented by D[r]. P[ell].*, London 1668.

[719]Astronomia: i.e. WING, *Astronomia Britannica*, London 1669.

[720]Astronomia: i.e. STREETE, *Astronomia Carolina. A new theory of the coelestial motions*, London 1661.

[721]Anatome: i.e. WILLIS, *Cerebri anatome: cui accessit, Nervorum descriptio et usus*, London 1664.

[722]tractatus: i.e. LOWER, *Tractatus de corde, item de motu & colore sanguinis et chyli in eum transitu*, London 1669.

[723]Institutiones: i.e. BEVERIDGE, *Institutionum chronologicarum libri II. Una cum totidem arithmetices chronologicae libellis*, London 1669.

[724]Logarithmotechnia: i.e. MERCATOR, *Logarithmotechnia*, London 1668.

[725]Optica: i.e. BARROW, *Lectiones XVIII, Cantabrigiae in scholis publicis habitae; in quibus opticorum phaenomenων genuinae rationes investigantur, ac exponuntur. Annexae sunt lectiones aliquot geometricae*, London 1669.

[726]Urim: i.e. SPENCER, *Dissertatio de Urim & Thummim*, Cambridge 1669.

[727]Ethica: i.e. MORE, *Enchiridion ethicum, praecipua moralis philosophiae rudimenta complectens*, London 1668.

[728]Mesolabum: i.e. SLUSE, *Mesolabum*, Liège 1659, 2nd ed., Liège 1668.

[729]respiratione: i.e. THRUSTON, *De respirationis usu primario, diatriba*, London 1670.

138.
ROBERT WOOD to WALLIS
8/[18] February 1669/[1670]

Transmission:

Manuscript missing.

Existence and date: mentioned in and answered by WALLIS–WOOD 10/[20].III. 1669/70.

As emerges from Wallis's reply, this letter enclosed two mathematical examples of approximation.

139.
HENRY HYRNE to WALLIS
Parson's Green, 28 February/[10 March] 1669/70

Transmission:

C Letter sent (including appendix): LONDON *Royal Society* Early Letters, H1, No. 107, 1 p. (our source). At top of page in Oldenburg's hand: 'Mr Henry Hyrnes, |Letter Containing A Copy of his *add.*| Objections to Dr Wallis's Hypothesis of Tydes' and 'Entered LB. 3. 326.'

c^1 Copy of letter sent: LONDON *Royal Society* Letter Book Original 3, pp. 326–8.

c^2 Copy of c^1: LONDON *Royal Society* Letter Book Copy 3, pp. 402–5.

Answered by: WALLIS–HYRNE 9/[19].III.1669/70.

Enclosure: Hyrne's additional objections to Wallis's hypothesis of tides (missing).

Wallis apparently received this letter on 8 March 1669/70 (old style), prompting him to send a reply the following day (see WALLIS–OLDENBURG 29.III/[8.IV].1670). Wallis sent Hyrne's letter together with a copy of his reply as an enclosure to WALLIS–OLDENBURG 24.III/[3.IV].1669/70. Both letters were produced at the meeting of the Royal Society on 14 April 1670 and ordered to be read at the following meeting on 21 April 1670. See BIRCH, *History of the Royal Society* II, 432–3.

Honour'd Sir, The last Octob. one[730] shew'd me the Conjectural Hypothesis,[731] as you are pleas'd to call it, about the Flux and Reflux of the Sea; and after I had perused it I ask'd of me what I thought of it. I having been for

2 Hypothesis, *(1)* about the *(2)* as you

---

[730] one: not identified.

[731] Hypothesis: i.e. WALLIS, 'An Essay of Dr John Wallis, exhibiting his Hypothesis about the Flux and Reflux of the Sea', *Philosophical Transactions* No. 16 (6 August 1666), 263–81; *Correspondence of John Wallis* II, 200–22.

139. Hyrne to Wallis, 28 February/[10 March] 1669/70

many years before, as fully satisfy'd in my Judgement concerning the Cause of this Phaenomenon as of any in Nature, and consequently have a little insight into the History of Tydes; returned this Answer: That this Hypoth. was very ingenious; and that it was great pitty, both that it came so far short of what it was intended for, and that the Judicious Author did not look for the Cause of the Seas Motion, where it was to be found. And that I might not seem to say this without reason; I then made several objections against it; and afterwards, at greater leisure, more.[732]

And because I find in the beginning of the Appendix[733] to the Publisher, That you are well contented, that Objections be made against the Hypothesis; I doe here send you in short, what I have to say against it; which to me seems to prove, that, although your Conjecture be witty, yet it is not solid truth.

First of all, I believe, that no Circular Motion of the Earth, as long as the Ambient Air is carried about with it (whatsoever the Hypothesis be,) can at all conduce to the causing of the Flux and Reflux of the Sea.

Secondly, The Circular motion, that you suppose, cannot; because the Acceleration & Retardation of a Point in the superficies of the earth is Inconsiderable; it being not above one inch in 2 miles about O in the 4th Figure, and not above half so much about M.

Thirdly, your Hypothesis will not solve the *Diurnal* Tydes: *i*. Because it would make but one Tyde and one Ebb in 24. hours. *ii*. It would never alter the Course of the Tydes; but they would always happen in the same time of the day.

Fourthly, It will not solve the *Menstrual* high waters: *i*. Because the highest Spring-tydes would be just at the New & Full Moon, whereas they are about 3. days after. *ii*. The Ebbe after a Spring-tyde would not be so great as after a Neap-tide; whereas it is greater. *iii*. We should have the Spring-tydes at the New & Full Moon at contrary hours; whereas they are at the same.

Fifthly, It will not solve the *Annual* high waters, because it doth not suppose, that the Earth moves faster at one time of the year, than it doth [at] another.

---

32 doth an another *corr. ed.*

[732] more: The additional objections to Wallis's hypothesis which the letter contained are now missing.

[733] Appendix: i.e. WALLIS, 'An Appendix, written by way of Letter to the Publisher; Being an Answer to some Objections, made by several Persons, to the precedent Discourse', *Philosophical Transactions* No. 16 (6 August 1666), 281–9; WALLIS, *Correspondence* II, 232–5, 240–5, 246–50.

139. HYRNE to WALLIS, 28 February/[10 March] 1669/70

Sixtly, By your Hypothesis should seem to follow: *i*. That under and near the Æquator the Rise & Fall of the Water would be greater than elsewhere; whereas it is lesse. *ii*. That the bay of Mexico would have as great a rise & fall of the water, as almost any place in the world; whereas it hath none. *iii*. That when it is high water at the East-side of America, it would be low water at the West; whereas it is high water at both the ends of the Streights of Magellan at the same time. *iv*. The Tydes and Ebbs would be at the same time, under the same Meridian, as farr as it reacheth from North to South in an Open Sea, such as the Atlantick is; whereas it is high water towards the North-pole, at the same time that it is low-water towards the South-pole, and at the Streights of Magellan. *v*. In places farr distant under the same Parallel, such as are Europe and the opposite parts of America, the Tydes would not happen at the same time: which they do.

I could instance in several other Phaenomena, which I cannot see, how your Hypothesis will salve; but these shall suffice at present.

Lastly, your Hypothesis adscribeth the Shooting forward and Casting back of the water to the Slowest and Swiftest motion of the several parts of the Earth's superficies at M & O. Fig. 4; Where, as the Instance of the loose incumbent weight upon a board or table seems to give it to the greatest Retardation, which is at $n.$, and Acceleration, which is at i.

I have prov'd all these Objections, and answer'd to your Instance of a Pendulum, why the highest waters may be after the New and Full Moon; but I forbear to send you the proof of them, because I would not seem to doubt of your acuteness to apprehend how they follow, as soon as you shall but look upon the Hypothesis again with an impartial Eye. But if upon consideration you shall beleive, that I do but think, I see what is not to be seen; and that what I assert does not follow from your Hypothesis, you may at your pleasure command the proof of these objections from

Sir

Parsons Green in the parish of Fulham,   your admirer,
Febr. 28. 1669/70.   and most humble servant
                                                                                       Harry Hyrne

---

19 incumbent *add.*

140.
RENÉ FRANÇOIS DE SLUSE to WALLIS
Liège, [28 February 1669/70]/10 March 1670

Transmission:

c^1 Copy of letter sent: LONDON *Royal Society* Early Letters S1, No. 62, 2 pp. (our source). At top of p. 1 in Oldenburg's hand: 'Apographum epistolae a domino Slusio scriptae ad D. Wallis: dab. d. 10. Martii 1670.'—printed PAIGE, *Correspondance de René François de Sluse*, 641–2.
c^2 Copy of letter sent: LONDON *Royal Society* Letter Book Original 3, pp. 295–6.
c^3 Copy of c^2: LONDON *Royal Society* Letter Book Copy 3, pp. 366–7.

Reply to: WALLIS–SLUSE 10/[20].IX.1669.
Answered by: WALLIS–SLUSE 29.III/[8.IV].1670.

After Wallis's original letter to Sluse of 10 September 1669 (old style), which was sent as an enclosure to OLDENBURG–SLUSE 14/[24].IX.1669, was lost in the post, Oldenburg sent a (slightly modified) copy, together with one of his own letters, with OLDENBURG–SLUSE 26.I/[5.II].1669/70. Sluse sent the present letter as an enclosure to one of the same date directed to Oldenburg. See SLUSE–OLDENBURG 28.II/[10.III].1669/70; OLDENBURG, *Correspondence* VI, 520–6.

Clarissimo et Celeberrimo Viro
D. Jo. Wallis, Geom. Prof. Savil. Oxoniens.
Ren. Franc. Slusius S.

Nescio, quid factum sit, Vir Celeberrime, ut literae,[734] quae Septembri praeterito ad me dare volueris, perierint in itinere; nisi quod plerumque accidat, ut quae maxime cupimus, ea vel sero vel nunquam consequamur. Sed bene est, quod earum Apographum,[735] Clarissimi Oldenburgi beneficio, tandem acceperim, ex quo videre licuit, non eruditionem tantum tuam orbi literato notissimam, verum humanitatem etiam singularem, qua me de invento[736] Clarissimi Wrenni, ejusque occasione certiorem esse voluisti. Equidem perlecta ejus demonstratione, quae ad parallelogrammum restricta erat, impulsus fueram ut quaererem etiam triangulum: sed ex erudita tua demonstratione, quae casum utrumque complectitur, video, rem vobis jam ante notam fuisse.

---

[734] literae: i.e. WALLIS–SLUSE 10/[20].IX.1669.

[735] Apographum: i.e. the copy of WALLIS–SLUSE 10/[20].IX.1669 sent as enclosure to OLDENBURG–SLUSE 26.I/[5.II].1669/70. See OLDENBURG, *Correspondence* VI, 447–8.

[736] invento: i.e. Wren's machine for grinding hyperbolic lernses. See OLDENBURG–SLUSE 26.I/[5.II].1669/70; OLDENBURG, *Correspondence* VI, 447–8.

Non destiti tamen, solidi illius sectiones ulterius explorare; et comperi (quod etiam tibi fortasse occurrit) non parallelogrammum tantum et triangulum in eo secari posse; sed parabolam quoque et Ellipsin, et Hyperbolam pluribus modis: imo et duo triangula juncta in communi vertice. [2] Ea nimirum sectione, cujus axis, axi solidi sit parallelus| et per verticem hyperbolae genetricis transeat. Unde sequitur, solidum illud torno elaborari posse; acie dolabrae (quam rectam supponimus) existente in plano, quod plano normali per axem torni parallelum sit, dummodo eadem acies ad horizontem inclinetur. Sed haec leviora sunt, quam ut te morari possint. Ut itaque finem faciam, gratias ago quam possum maximas humanitati tuae, ac vicissim obsequiorum meorum tenuitatem, si qua in re utilis esse possit, tibi ex animo addico. Vale, Vir praestantissime, et novis inventis, ut soles, rempubl. literariam ornare perge.

Dabam Leodii X Martii MDCLXX st. n.

141.
JOHN COLLINS to HENRY OLDENBURG
February/March ? 1669/70

Transmission:

C Letter sent: LONDON *Royal Society* Early Letters W1, No. 101, 1 p. At foot of page in Oldenburg's hand: 'I have written accordingly to Messina to [*blank*] April 4. 1670.'

Enclosures: WALLIS–BORELLI 13/[23].I.1669/70 (copy) and Catalogue of Books for Borelli (copy).

Collins notifies Oldenburg that a box of books for Borelli has been loaded onto a ship bound for Messina, and encloses the bill of lading together with copies of Wallis's letter to Borelli and the catalogue of books contained. Oldenburg's endorsement indicates that in accordance with Collins's suggestions he sent the bill of lading with instructions for forwarding of the books to a merchant in Messina.

Mr Oldenburgh

The abovesaid[737] Bookes are put up or nailed and Chorded up, in a Box, put on board the Alice and Francis, Stephen Dring Commander as appeares by the bill of Lading inclosed, which it may be convenient to send to a

---

16 put |up *add.*| or

[737] abovesaid Bookes: i.e. those listed in the *Catalogue of Books for Borelli*, drawn up by Boyle, Wallis, and Collins (APPENDIX to WALLIS–BORELLI 13/[23].I.1669/70).

141. COLLINS to OLDENBURG, February/March ? 1669/70

Merchant at Messina to take up the Box for the use of Borellius if he should be at Palermo, when the Ship arrives, the originall of Dr Wallis his Letter is on board the ship with a bill of lading, and Catalogue of the Bookes now sent in it; the other Catalogue mentioned[738] by the Doctor I have not yet in readinesse but intimate that I may send it hereafter within lesse then 3 Months, with Dr Wallis his booke[739] de Calculo Centri gravitatis, and Mr Barrows Elementa[740] Curvilineorum which will be finished by or before that time, you may likewise intreate Borellius to send Gotignies new Disposition[741] of Euclids Elements, and the 3. Decas of Camillus Gloriosus his Exercitationes[742] Geometricae printed at Naples 1635, and desire to know how the Tesoro[743] matematico di Gieronimo Pico fonticolano Romae 1645 fo, is approved, and what it handles what he disburses we shall order to be paid him by some Merchant in Messina, as soone as we know what it is, he is desirous to put off some of his owne workes here, you may signify that our Booksellers being much impoverished by the late fire, will not part with money, but will exchange for other Bookes, that may goe off there, either Dr Wallis or Mr Barrows workes or others, and if he please to send, not above 20 of a kind I shall use my indeavour to serve him without any reward for my paines, and what he sends to me let it be directed to be left for me with Mr George Cowart[744] Merchant well knowne in the Salters walke on the Exchange, and liveth in the Alley behind Token house yard.[745]

1 should *add.*
16 off *add.*

---

[738] mentioned: see WALLIS–BORELLI 13/[23].I.1669/70.

[739] booke: i.e. the second part of Wallis's *Mechanica: sive, de motu, tractatus geometricus*, which was published in July 1670.

[740] Elementa: i.e. BARROW, *Lectiones geometricae: in quibus, praesertim, generalia curvarum linearum symptomata declarantur*, London 1670.

[741] Disposition: i.e. GOTTIGNIES, *Elementa geometricae planae*, Rome 1669.

[742] Exercitationes: i.e. GLORIOSO, *Exercitationum mathematicarum Decas tertia*, Naples 1639. The *Decas prima* was published in Naples in 1627, the *Decas secunda* in Naples in 1635.

[743] Tesoro: PICO FONTICULANO, *Tesoro di matematiche considerationi dove si contiene la teorica e la prattica di tutta la geometria, il trattato della transformatione, circonscrittione, & riscrittione delle figure piane e solide*, Rome 1645.

[744] Cowart: not identified.

[745] Token house yard: yard built during the reign of Charles I by William Petty, taking its name from the house where farthing tokens were produced.

142.
JOSHUA CHILDREY to SETH WARD
Upwey, 4/[14] March 1669/70

Transmission:

C Letter sent: LONDON *Royal Society* Early Letters C1, No. 4, 8 pp. (our source). Words and passages to be omitted in printing have been underlined. Endorsed by Oldenburg at foot of p. 1: 'A Letter to the Bp of Salisbury containing some Animadversions on Dr Wallis's Hypothesis of Tides.' and 'Ent. LB. 3. 316.' On p. 8 Childrey's endorsement 'Post p[ai]d 6^d' and postmark: 'MR/7'.
c^1 Copy of letter sent: LONDON *Royal Society* Letter Book Original 3, pp. 316–25.
c^2 Copy of *c*^{1}: LONDON *Royal Society* Letter Book Copy 3, pp. 391–402.
E^1 First edition: *Philosophical Transactions* No. 64 (10. October 1670), 2061–8 ('A Letter of Mr. Joseph Childrey to the Right Reverend Seth Lord Bishop of Sarum, containing some Animadversions upon the Reverend Dr. John Wallis's Hypothesis about the Flux and Reflux of the Sea, publish't No. 16. of these Tracts').
E^2 Second edition: *The Philosophical Transactions and Collections ... Abridged and Disposed under General Heads* I, 516–20.

Enclosure to: OLDENBURG–WALLIS *c*.16/[26].III.1669/70.
Answered by: WALLIS–OLDENBURG 19/[29].III.1669/70.

Childrey sent his animadversions on Wallis's hypothesis of tides in a letter to Seth Ward, clearly intending that they be sent on by him to Wallis or communicated to the Royal Society. Ward, who was apparently staying at Westminster at the time, transmitted the letter to Oldenburg, who in turn sent it to Wallis as an enclosure to OLDENBURG–WALLIS 16?/[26?].III.1669/70. Wallis replied by means of his letter to Oldenburg of 19/[29].III.1669/70, with which he probably returned Childrey's letter. The letter was originally dated 3 March; Childrey subsequently altered the date to 4 March.

My Lord,

The last Summer I acquainted your Lordship that I had in my mind animadverted somethings upon Dr Wallis his Hypothesis[746] concerning the Flux & Reflux of the Sea, which I have at length gotten time to putt in writing, & (according to my duty & promise) here present inclosed to your Lordship. If upon perusall of them (so soon as the Weighty affaires of the Kingdome shall permitt) your Lordship shall thinke them worth transmitting to Dr Wallis, or communicating[747] to the R. Society, I onely desire your Lordships

---
[746]Hypothesis: i.e. Wallis's 'An Essay of Dr. John Wallis, exhibiting his Hypothesis about the Flux and Reflux of the Sea', *Philosophical Transactions* No. 16 (6 August 1666), 263–81.
[747]communicating: cf. CHILDREY–OLDENBURG 29.III/[8.IV].1670; OLDENBURG, *Correspondence* VI, 603–4.

142. CHILDREY to WARD, 4/[14] March 1669/70

& their favourable censure of my judgment & reasons, & in the meane time humbly beg your Lordships blessing for

Upway March 4th
1669.

My very good Lord
Your Lordships most obedient Son
& devoted Servant
Josh. Childrey.| [2]

Some briefe Animadversions upon the Reverend
Dr John Wallis's Hypothesis about the Flux & Reflux of the Sea.

My intention is not to argue against that part of the Hypothesis, that relates to the Common Center of gravity of the Earth & Moone; & the Diurnall & Menstruall vicissitudes of the Tides, the Authors discourse being (in my poor judgment) so rationall & satisfactory, as to those, that I cannot see, what cleare objection can be made against it. But that which I would beg his leave to except against, till better reason convince me; is his Opinion concerning the Annuall vicissitudes, & the true cause thereof, which he supposeth to be quite another thing from the Common Center of gravity; namely the Inequality of the Naturall Dayes. For I feare he may be mistaken in the time of the Annuall vicissitudes, which he contendeth to be about Allhollandtide, & Candlemas, & the reasons of my feare are these.

1. Because, if he dare stand to the generall judgment of Seamen, which I conceive is more to be trusted, than that of the inhabitants of Rumney Marsh, he will (I dare assure him) find very few of our English Seamen of that mind, who use to say either that the Time of the yeare signifieth nothing at all; or if it doe, that the highest tides of the yeare seeme to happen rather about the Equinoxes, then those 2 other assigned Times, when the Naturall dayes are longest, & shortest.

2. Whereas he gives an instance or two (pag. 276.) of very high Tides in the Thames in November 1660, & 1665; the trueth of which we need not question; & of which there are sundry other the like instances in our English Chronicles, I have reason to beleeve, that those high Tides may proceed from another cause; then he supposeth. For first if that, which he supposeth should be the cause, the like high Tides might be expected every November, & (which is more) they should happen as frequently about February, as about November; of which yet he gives not one instance. And (which is yet more, & very considerable) though I have perused throughly that perfect collection I have of all high Tides in the Thames, that our Chronicles take notice of since the Conquest, I can hardly find one such high Tide in the Thames

142. CHILDREY to WARD, 4/[14] March 1669/70

in February, or thereabout. Secondly those high Tides in the Thames in November, if we dare credit the London Watermen, are caused by the coming downe of the land waters after very great raines; which being encountred by the Tide of the floud from the mouth of the Thames, cannot but swell to [3] an unusuall height. To| induce us to beleeve which, we need onely consider, that the latter end of October, & the beginning of November (or rather both those whole moneths) are generally the rainiest part of the whole yeare. Now if the great raines fall so, that the land waters come downe to the flowing part of the Thames, just upon the Full or Change, when the Spring Tides happen; as they did (for Example) Septemb. 30. 1555, & Octob. 22. being Thursday, 1629. (Stow & Howes are my authors,[748]) those Spring Tides must be the higher, as proceeding from a double cause. But

3. (To say no more of the Thames, but to consider that, & other great Rivers joyntly with the Sea) there is another thing notoriously knowen by all Seamen to be a cause of high or low Tides, which I cannot but say, that I wonder the Author hath taken so little, or no notice of in his Essay; namely, the sitting of the wind at such or such a point of Compasse, & blowing hard. It is the constant saying of all Sea men in Kent, that ever I met with, that the Northwest wind makes the highest tides in the Thames, Medway, & all the coasts about the South & North forelands, & likewise on the Coasts of Holland & Flanders. And the reason they alledge for it, is because (say they) that wind doth with equall force blow in the tide of floud at both ends of this Island of Britaine; that is, from the Northward between the coasts of Scotland, Norway & Jutland; & also from the Westward by the coasts of Cornewall, Devonshire, Dorsetshire &c: up along the Sleeve.[749] And for the same reason they say (& I think truely) that a SE. wind deads & hinders the tides in the places before mentioned neare the Forelands. And agreeably to this (if the testimony of youth may be admitted) I very well remember, when I was a boy, & lived at home with my Father at Rochester, which is neare enough to Chatham, to observe, how the Tides run there; that when the Tides were unusually high, the wind was always at NW, & the ☽ neare the Full or Change. And so confident I am of my memory in

---

10 October 22. 1629. E^1

---

[748]authors: i.e. STOW, *Annales, or, A generall Chronicle of England. Begun by John Stow: Continued and augmented with matters Forraigne and Domestique, Ancient and Moderne, unto the end of this present yeere, 1631*, London 1631, 1045. Howes had first begun work on Stow's *Annales* three years before the latter's death. The first edition appeared in 1615.
[749]Sleeve: i.e. the English Channel.

142. CHILDREY to WARD, 4/[14] March 1669/70

this point, that if enquiry be made about Chatham, the Hundred of Hoo,[750] & the Ile of Graine;[751] I beleeve the inhabitants will with one voice say, that they never feare their low Marshes being overflowed by the Tide, but when the wind is at Northwest, or thereabout upon the Spring Tides. Here at Weymouth those able & ancient Seamen I have talked with, tell me that a SSE. wind makes the greatest Tides, & that according to the degree of the wind, caeteris paribus, the Tides rise more or less notably; But that they never observe any extraordinary swelling of the Tides about Allhollantide or Candlemas, unlesse the wind be about SSE. And the reason| they give [4] for that winds raising the Tides there, is in my opinion very convincing, if we consider the lying of that haven in the Map. And for the same reason I suppose the wind from the same point may make the highest tides at Southampton; a Westerly wind at Bristoll & in Severne; an Easterly wind at Hull, a NE. wind at Wisbich[752] & Lyn;[753] a Southerly wind upon the opposite coasts of England & Ireland &c: And as confident I am, that if more particular enquiry be made in Rumney-marsh, it will be found, that Dym-church-wall[754] is never in danger of being overflowed or broken by the Tides, but upon very stormy & tempestuous weather; especially when the wind either blowes right on upon the shore, or when it sitts in that point, that raiseth the Tides highest there. Whether the Northwest wind (because of the little distance of Rumney Marsh from the Forelands) be the raising wind or no; I cannot certainly affirme; But so much I beleeve, that were it not for the running out of the Nesse point[755] on this side, (which makes that coast a Bay) & the running out of Blacknesse point[756] in France on the other side; Dimchurch wall would be more secure, & need lesse constant reparation then it doth. And if we do but consider, that Allhollandtide & Candlemas are no more famous for the longest and shortest Naturall dayes, then they are generally infamous for stormy weather; especially the former season, (Wet & Windy weather being mostly concomitant) we have good ground to attribute high Tides, at those times of the yeare to another cause, then the Author supposeth, & make a more then probable conjecture at the occasion

---

2 inhabitants (*1*) would (*2*) will *C*

[750]Hundred of Hoo: much of the Hoo Peninsula lies in the original Saxon Hundred of Hoo.

[751]Ile of Graine: i.e. Isle of Grain, the outermost end of the Hoo Peninsula in north Kent.

[752]Wisbich: i.e. Wisbech.

[753]Lyn: i.e. King's Lynn.

[754]Dym-church-wall: i.e. Dymchurch wall, a sea defence originating in the first millenium and designed to prevent the sea from flooding Romney Marsh.

[755]Nesse point: i.e. Ness Point, St Margaret's Bay, Kent.

[756]Blacknesse point: i.e. Cap Gris-Nez, a cape on the Côte d'Opale in northern France.

142. CHILDREY to WARD, 4/[14] March 1669/70

of the mistake. Tis true, March is very often more stormy then February (though seldome so stormy as October & November) which possibly might occasion that opinion, that some hold, (of which number, Pliny[757] is one) that the highest Tides are about the Equinoxes. And if the thing were found to hitt pretty frequently in March, men might not be carefull to observe the other Equinox; Though yet it cannot be denied, that we have blustring weather may times before Michaelmas. In confirmation of all this that I have said concerning the influence of the Wind its being considerable on the Tides, I shall add these following Collections of my owne out of Histories, Chronicles, &c:

1250. Octob. 1. (saith Holinshead[758]) upon the change of the ☽ was a most dreadfull inundation of the Sea, that did exceeding much hurt to Holland beyond Sea, Holland in Lincolnshire; & the Marsh ground in Flanders, & drowned *Winchelsea*. But he tells us withall, that an unheard of tempest of wind accompanied it.|

1555. Septemb. 30. (saith Stow[759]) was a notable inundation of the Thames; But he saith withall, that it was by occasion of a great wind, & rain, that had fallen.

15$\frac{69}{70}$. March 10th, I find this manuscript note in Latine in an Ephemerides[760] for that yeare against the day; Septentrionis maxima saevitia. Nivis flocci magni, ingens frigus. Maxime tumescebat aestus maris die & nocte; nam excurrebat in agros late.

1592. Septemb. 6. Wednesday (saith Stow[761]) the wind being West & by South, as it had been for 2 dayes before very boistrous, the Thames was made so void of water, by forcing out the fresh & keeping backe the salt, that men in diverse places might goe 200 paces over, & then fling a stone to the land, &c:

1600. Decemb. 8. st. vet. I find this note written in another Ephemerides[762] for that yeare against the day, by an unknowen person, who (as it seemes) was then at Venice (where a SE. wind makes the highest tides) Inundatio Venetiis 6. pedes, temp. Sirocco.

---

[757]Pliny: i.e. PLINY THE ELDER, *Naturalis historia* II, 99.
[758]Holinshead: i.e. HOLINSHEAD, *The third volume of Chronicles, beginning at duke William the Norman*, London 1586, 243.
[759]Stow: i.e. STOW, *Annales*, 627.
[760]Ephemerides: not identified.
[761]Stow: i.e. STOW, *Annales*, 765.
[762]Ephemerides: not identified.

142. CHILDREY to WARD, 4/[14] March 1669/70

1601. In Aprill (saith Grimston[763] in his Netherland History) the Sea being forced in by a strong NW. wind, did some mischiefe to Ostend.

1601. Octob. 26. new style, Great tempest (saith the same Author[764]) & the wind W & NW. & the tide much higher then usuall at Ostend.

1602. Februar. 23. & 24. new Style blew a terrible North-west wind, which made the water rise higher than usual at Ostend. Idem.[765]

1604. *March* 1. (new style) the wind was very great at West & Northwest with a furious tempest, the Tyde at Ostend rising so high, as it had not done in 40 yeares before. Idem.[766]

4. There is yet another thing, which seemes to have (at least) some influence on the tides, & to make them swell higher then else they would doe; to witt, the Perigaeosis of the ☽. And this hath been my opinion (taken up first upon the consideration of the ☽'s coming nearer the Earth) ever since 1652, when living at Feversham in Kent neare the Sea, I found by observing of the Tides (as often as I had leisure) that there might be some trueth in my conjecture; & therefore in a little Pamphlet,[767] published in 1653. by the name of Syzygiasticon instauratum, I desired that others would observe that yeare, whether the Spring Tides after those Fulls & Changes, when the ☽ was in Perigaeo (the wind together considered) were not higher then usuall. And since that time I have found severall high Tides, & inundations| [6] (though I must not say all) to happen upon the ☽'s being in, or very neare her Perigaeum. For Example;

1. That famous inundation, mentioned before out of Holinshead, 1250. Octob. 1. was, when the ☽ was in Perigaeo; as appeares by calculation.

2. 1530. Novemb. 5. that inundation, on which was made the Distick, Anno ter deno post sesquimille, Novembris Quinta stat salsis Zelandia tota sub undis, was, when ☽ was in Perigaeo.

---

1 1601. (saith *Grimston*) E^1

---

[763]Grimston: i.e. [GRIMESTON], *A generall historie of the Netherlands. Newly renewed, corrected, and supplied with sundrie, necessarie observations omitted in the first impression*, London 1627, 1139. This work is a translation and augmentation of Jean-François le Petit's *La grande chronique ancienne et moderne, de Hollande, Zelande, West-Friese, Utrecht, Frise, Overyssel & Groeningen*, 2 vols, Dordrecht 1601.
[764]Author: i.e. [GRIMESTON], *A generall historie*, 1146–7.
[765]Idem: i.e. [GRIMESTON], *A generall historie*, 1154.
[766]Idem: i.e. [GRIMESTON], *A generall historie*, 1182.
[767]Pamphlet: i.e. CHILDREY, *Syzygiasticon instauratum, or, An ephemeris of the places and aspects of the planets*, London 1653.

142. CHILDREY to WARD, 4/[14] March 1669/70

3. Jan. 13. $155\frac{1}{2}$. the sea (saith Michell[768] in his Chronicle) brake in at Sandwich, & overflowed all the marshes thereabout, & drowned much Cattell: ☽ in Perigaeo.

4. 1570. Novemb. 1. was a dreadfull floud[769] at Antwerpe, & on all the coasts of Holland, that made infinite spoil. ☽ in Perigaeo.

5. 1600. Dec. 8. above mentioned, ☽ was in Perigaeo.

6. $160\frac{6}{7}$. Jan. 20th. was a great inundation in Severne, mentioned in Howes's Chronicle[770], that did much hurt in Somersetshire & Gloucestershire &c. ☽ in Perigaeo.

7. 1555. Septem. 30. (forgotten in its due place) ☽ was in Perigaeo.

8. 1643. Jan: 23. st. no. (saith a little Low-Dutch Chronicle[771] that I have) was a terrible high water floud in Frizland &c: whereby much hurt was done to the dykes, & at Gaes by Haerlingen the dead bodies streamed out of the Earth. ☽ in Perigaeo.

9. 1651. Feb. 23. st. no. (saith the same Chronicle[772]) was S. Peters high-floud, whereby much hurt was done to the Dykes in Frizland, Embderland, & elsewhere: And not far from Dockum[773] by Oudt-woudumer-zijl[774] is a breach of 42 roods long broken in the dyke: ☽ in Perigaeo.

10. August 2. 1657. old style, at Feversham (where I then lived) was a very high Spring tide, & yet the wind was at SE; which deads the tides there: ☽ in Perigaeo.

11. August. 22. 1658. old style, at Feversham, was a very high tide in the afternoone, though the wind was Southerly, & blew very stiff, which the Seamen there wondred at: ☽ in Perigaeo.

12. 1661. Upon Michaelmas day was a great overflowing of the Severne, that it drowned the low grounds lying by it. I lived then in Gloucestershire, & immediately, so soone as I heard of it, I noted it downe in my Memorandums. [7] ☽ in Perigaeo.|

13. The Scheme of the weather printed in the History[775] of the R. Society tells us, that May 24th. 1663. was a very great Tide at London. But it tells us withall, that the same day the Moone was in Perigaeo.

---

[768] Michell: i.e. MITCHELL, *A breviat cronicle, containing all the kynges from Brut to this daye*, [Canterbury] 1554, sig. Oi^{r}.
[769] floud: see [GRIMESTON], *A generall historie*, 341.
[770] Chronicle: i.e. [STOW], *Annales*, 889.
[771] Chronicle: not identified.
[772] Chronicle: not identified.
[773] Dockum: i.e. Dokkum, Friesland.
[774] Oudt-woudumer-zijl: i.e. Oudwoude, Friesland.
[775] History: i.e. SPRAT, *The History of the Royal-Society of London*, London 1667, 179.

142. CHILDREY to WARD, 4/[14] March 1669/70

14. 1669. Septemb. 1. here at Weymouth I observed my selfe a very high Tide, & so did severall Seamen in that Towne, who wondered at it; the weather being very calme, & that little wind that was being at NE, which uses to contribute nothing at all to the Tides in that haven. ☽ in Perigaeo.

Further, that which inclines me to thinke, that the Perigaeosis of ☽ is of some concernment in this matter, is because it is a maxime amongst our *Kentish* Seamen, that they never have 2 running Springs (as they call them) together, but that the next Spring tide after a high running Spring is proportionably weake & slacke. Which, if true, is very correspondent to my opinion, because if the ☽ be in Perigaeo at this Spring Tide, she will be in Apogaeo at the next.

But I conceive the best Touch stone to prove the soundness of my opinion, (which I confesse I never had the happy opportunity to doe yet) is to have it observed, whether those Neaptides be not apparently higher, (consideratis considerandis) that happen upon the ☽'s being in Perigaeo either at the First or Last quarter. Because it is a received & demonstrable trueth in Astronomy, that the ☽ being in Perigaeo at either Quarter comes then nearer the Earth, then when she is in Perigaeo at the Change or Full. And I could wish for the further clearing of this matter, that observation were made at Bristoll (because there is the most considerable flux & reflux of any Port of England) whether this yeare 1670 the Tides be not higher (consideratis &c:) when the ☽ passeth through ♓. ♈. & ♉, then when she passeth through the opposite signes ♍. ♎. & ♏. And particularly whether the the Spring Tides be not sensibly higher after the Change, then after the Full, in February, March, & Aprill; & higher after the Full then after the Change in August, September, & October; As also whether the Neaptides after the Last quarter in May & June rise not apparently higher then expected. I am promised, that observation shall be made here at at Weymouth for this whole yeare round; from whence I have already received this account, that this present February $16\frac{69}{70}$ the Spring Tides ran very high after the Change, though the weather were pretty calme, & that wind that was, not very favourable to the Tides, & that the Spring Tides after the Full were very low, & weake, which is exactly according to my conjecture.

6 mee to believe E^1
14 had the opportunity E^1
28 the Neap-tides in May E^1

[8] For the Right Reverend Father
in God, & my very good Lord
the Lord Bishop of Sarum at
Dr Perrenchefes lodgings neare the
little cloysters in Westminster Abbey
these humbly present
Westminster.

143.
WALLIS:
Reply to Collins's Question on Algebraic Roots, 8/[18] March 1669/70

Transmission:

W^1 Draft of paper sent: OXFORD *Bodleian Library* MS Don. d. 45, f. 149^r–149^v (our source).
W^2 Paper sent: CAMBRIDGE *Cambridge University Library* MS Add. 9597/13/6, f. 252^r–252^v (our source). On f. 252^v in Wallis's hand: 'For Mr John Collins', and the remark 'I should have sent this sooner (for I wrote it presently after I came home) but forgot it.' On f. 252^v also Collins' endorsement: 'Dr Wallis about Cubicks'.—printed RIGAUD, *Correspondence of Scientific Men* II, 601–4.

In reply to a question, possibly posed by Collins in a now missing letter, Wallis analyzes a sequence of cubic equations characterized by having as one root the arithmetic progression of integers 1, 2, 3, ..., and in the remaining quadratic equations the linear terms also running through an arithmetical progression $-14a$, $-13a$, $-12a$, ... The three highest terms of the cubic are identical ($a^3 - 15a^2 + 54a$); the variation of the constant term in both the cubic and the quadratic equation is explained by the construction of repeated differences: for the cubic, the third difference (-6) is constant, for the quadratic, the second difference ($+2$). The date and circumstances emerge from Wallis's remarks on the draft (W^1). The form of the paper sent (W^2) and the wording on the cover suggests that it was conveyed as an enclosure to a letter to Oldenburg, possibly WALLIS–OLDENBURG 10/[20].III.1669/70.

[W^1]

March. 8. 1669./70.

In answere to a question of Mr Joh. Collins.

143. WALLIS: Reply to Collins's Question on Algebraic Roots

Root. first.	Cubick Equation. #	Quadr: Æquat.*
$a-1)$	$a^3 - 15a^2 - 40$	$(a^2 - 14a + 40$
$a-2)$	$a^3 - 15a^2 - 56$	$(a^2 - 13a + 28$
$a-3)$	$a^3 - 15a^2 - 54$	$(a^2 - 12a + 18$
$a-4)$	$a^3 - 15a^2 - 40$	$(a^2 - 11a + 10$
$a-5)$	$a^3 - 15a^2 - 20$	$(a^2 - 10a + 4$
$a-6)$	$a^3 - 15a^2 \mp 0$	$(a^2 - 9a \pm 0$
$a-7)$	$a^3 - 15a^2 + 14$	$(a^2 - 8a - 2$
$a-8)$	$a^3 - 15a^2 + 16$	$(a^2 - 7a - 2$
$a-9)$	$a^3 - 15a^2 \pm 0$	$(a^2 - 6a \mp 0$
$a-10)$	$a^3 - 15a^2 - 40$	$(a^2 - 5a + 4$
$a-11)$	$a^3 - 15a^2 - 110$	$(a^2 - 4a + 10$
$a-12)$	$a^3 - 15a^2 - 216$	$(a^2 - 3a + 18$
$a-13)$	$a^3 - 15a^2 - 364$	$(a^2 - 2a + 28$
$a-14)$	$a^3 - 15a^2 - 560$	$(a^2 - 1a + 40$
$a-15)$	$a^3 - 15a^2 - 810$	$(a^2 \mp 0a + 54$
$a-16)$	$a^3 - 15a^2 - 1120$	$(a^2 + 1a + 70$
$a-17)$	$a^3 - 15a^2 - 1496$	$(a^2 + 2a + 88$

			That is,	
2^{d}Root.	3^{d}Root.	Root, 1st.	Root, 2^{d}.	Root, 3^{d}.
$7-3.$	$7+3.$	1.	4.	10.
$6\frac{1}{2} - \frac{1}{2}\sqrt{57}$	$6\frac{1}{2} + \frac{1}{2}\sqrt{57}$	2.	$2,7251-$	$10,2749+$
$6 - \sqrt{18}$	$6 + \sqrt{18}$	3.	$1,7575-$	$10,2425+$
$5\frac{1}{2} - 4\frac{1}{2}$	$5\frac{1}{2} + 4\frac{1}{2}$	4.	1.	10.
$5 - \sqrt{21}$	$5 + \sqrt{21}$	5.	$0,4194+$	$9,5826-$
0.	9.	6.	0.	9.
$4 - \sqrt{18}$	$4 + \sqrt{18}$	7.	$-0,2425+$	$8,2425+$
$3\frac{1}{2} - \frac{1}{2}\sqrt{57}$	$3\frac{1}{2} + \frac{1}{2}\sqrt{57}$	8.	$-0,2749+$	$7,2749+$
0.	6.	9.	0.	6.
$2\frac{1}{2} - 1\frac{1}{2}$	$2\frac{1}{2} + 1\frac{1}{2}$	10.	1.	4.
$2 - \sqrt{-6}$	$2 + \sqrt{-6}$	11.		
$1\frac{1}{2} - 1\frac{1}{2}\sqrt{-63}$	$1\frac{1}{2} + \frac{1}{2}\sqrt{-63}$	12.		
$1 - \sqrt{-27}$	$1 + \sqrt{-27}$	13.	Impossib.	Impossib.
$\frac{1}{2} - \frac{1}{2}\sqrt{-159}$	$\frac{1}{2} + \frac{1}{2}\sqrt{-159}$	14.		
$-\sqrt{-54}$	$+\sqrt{-54}$	15.		
$-\frac{1}{2} - \frac{1}{2}\sqrt{-279}$	$-\frac{1}{2} + \frac{1}{2}\sqrt{-279}$	16.		
$-1 - \sqrt{-87}$	$-1 + \sqrt{-87}$	17.		

143. WALLIS: Reply to Collins's Question on Algebraic Roots

#	Differences.			*	Differences.	
−40	−16			+40	−12	
−56	+2	+18	−6	+28	−10	+2
−54	+4	+12	−6	+18	−8	+2
−40	+20	+6	−6	+10	−6	+2
−20	+20	+0	−6	+4	−4	+2
∓0	+14	−6	−6	±0	−2	+2
+14	+2	−12	−6	−2	−0	+2
+16	−16	−18	−6	−2	+2	+2
±0	−40	−24	−6	∓0	+4	+2
−40	−70	−30	−6	+4	+6	+2
−110	−106	−36	−6	+10	+8	+2
−216	−148	−42	−6	+18	+10	+2
−364	−196	−48	−6	+28	+12	+2
−560	−250	−54	−6	+40	+14	+2
−810	−310	−60	−6	+54	+16	+2
−1120	−376	−66		+70	+18	
−1496				+88		

It appears by this: That, supposing an affected Cubick Equation, fitted to one series of Roots, arithmetically proportional; (as 1, 2, 3, &c.) the other two ranks of roots answering thereunto, will not be arithmetically proportional.

But the rank of Coefficients (in the middle terms of the Quadratick Equation, containing the two latter ranks of roots,) wil so be.

And allso, of the Absolute numbers in those quadratick Equations, the second differences will be equal; like as they would have been, if in all of them there had been the same coefficient of the middle terme, & one rank of rootes as 1, 2, 3, &c.

Note also, that if in any such series of Equations (of what degree so ever) one rank of roots be as 1, 2, 3, &c. arithmetically proportional; then will the series of the aggregate of all of rest (which is the coefficient of the second term in the subordinate Equation, as here in the Quadratick) be allso arithmetically proportional: but decreasing if those first did increase: & contrarywise, increasing if they did decrease.

And consequently, if in a series of Quadratic Equations, one rank of rootes be arithmetically proportional, the other will be so to: For the series of aggregates is but the other series of roots.|

---

46 series of the *add.*
50 in a (*1*) Quadratic Equation (*2*) series of Quadratic Equations

143. WALLIS: Reply to Collins's Question on Algebraic Roots

Note allso, that in such a rank of Equations (having one rank of Roots Arithmetically proportional) the rank of absolute numbers is so constituted, as that, if it be a Lateral Equation, their differences be equal; if a Quadratik Equation, their second differences (or differences of differences) are equall; if a Cubick, their 3^d differences; & so onward according as the degrees rise.

This hath been observed long since, as to simple Equations, or not-affected. As for instance

```
           In laterals,    1 · 2 · 3 · 4 · 5 · 6 · 7.
     The differences are ···  1 · 1 · 1 · 1 · 1 · 1 ·
             In Squares ···   1 · 4 · 9 · 16 · 25 · 36 · 49.
         The differences ···    3 · 5 · 7 · 9 · 11 · 13.
    The second differences ···    2 · 2 · 2 · 2 · 2.
              In Cubes ···    1 · 8 · 27 · 64 · 125 · 216 · 343.
         The differences ···    7 · 19 · 37 · 61 · 91 · 127.
    The second differences ···   12 · 18 · 24 · 30 · 36.
     The third differences ···    6 · 6 · 6 · 6.
```

And, in like manner, the fourth differences in Biquadratick, the fifth in those of the fifth degree, & the sixth in those of the sixth, (& so onwards,) will be found equal.

But I do not know, that the like hath been observed, till of late, in affected Equations.

Yet doth this depend upon the other; & is easily demonstrated from it. For, in a rank of Quadratick Equations, $x^2 \pm ax = b$. If x be successively interpreted of 1, 2, 3, &c (or any other rank of Arithmetical proportionals) the differences of ax must be equal; &, consequently, their second differences vanish: And therefore, the second differences of xx being equal, the second differences of both jointly ($x^2 \pm ax$,) that is, of b, must be equal also; and the very same with those of xx.

And, in like manner, in the Cubick Equation; $x^3 \pm ax^2 \pm bx = c$. For, the first differences of bx being equal, their second differences are nothing: and the second differences of ax^2 being equal, their third differences are nothing: and, consequently, the third differences of $x^3 \pm ax^2 \pm bx$, (that is, of c,) are the same with those of x^3 (the rest being vanished;) & must therefore be equal.

And, in like manner, in all superior Equations, (the difference of the inferior degrees vanishing) the last differences of the highest term, (be it

34 those |of *add.*| x^3

143. WALLIS: Reply to Collins's Question on Algebraic Roots

the fourth, fifth, sixth, or so onwards, according to its dimensions,) wil be same with the like differences of the absolute numbers, or (as *Vieta* calls[776] them) *Homogenea comparationis*.

But I observe further, in our rank of Quadratick Equations, though neither of the rank of Roots therein contained be Arithmetically proportional; yet are the second differences of the Absolute Numbers, in such proportion; in like manner as they would have been if one of their ranks of Roots had been so.

Which comes to pass by reason of the Coefficients of the second term being in Arithmetical proportion:

Which Coefficients are the aggregate of the two roots in each Equation. Which I do not know, that any have before observed.

[W^2]

Root. 1st.	Cubick Æquation. #	*
$a-1$)	$a^3 - 15a^2 + 54a - 40$	$(a^2 - 14a + 40$
$a-2$)	$a^3 - 15a^2 + 54a - 56$	$(a^2 - 13a + 28$
$a-3$)	$a^3 - 15a^2 + 54a - 54$	$(a^2 - 12a + 18$
$a-4$)	$a^3 - 15a^2 + 54a - 40$	$(a^2 - 11a + 10$
$a-5$)	$a^3 - 15a^2 + 54a - 20$	$(a^2 - 10a + 4$
$a-6$)	$a^3 - 15a^2 + 54a \mp 0$	$(a^2 - 9a \pm 0$
$a-7$)	$a^3 - 15a^2 + 54a + 14$	$(a^2 - 8a - 2$
$a-8$)	$a^3 - 15a^2 + 54a + 16$	$(a^2 - 7a - 2$
$a-9$)	$a^3 - 15a^2 + 54a \pm 0$	$(a^2 - 6a \mp 0$
$a-10$)	$a^3 - 15a^2 + 54a - 40$	$(a^2 - 5a + 4$
$a-11$)	$a^3 - 15a^2 + 54a - 110$	$(a^2 - 4a + 10$
$a-12$)	$a^3 - 15a^2 + 54a - 216$	$(a^2 - 3a + 18$
$a-13$)	$a^3 - 15a^2 + 54a - 364$	$(a^2 - 2a + 28$
$a-14$)	$a^3 - 15a^2 + 54a - 560$	$(a^2 - 1a + 40$
$a-15$)	$a^3 - 15a^2 + 54a - 810$	$(a^2 \mp 0a + 54$
$a-16$)	$a^3 - 15a^2 + 54a - 1120$	$(a^2 + 1a + 70$
$a-17$)	$a^3 - 15a^2 + 54a - 1496$	$(a^2 + 2a + 88$

4 rank of *add*.

---

[776]calls: i.e. VIÈTE, *In artem analyticem Isagoge*, chap. 3; *Opera mathematica*, ed. F. v. Schooten, 2–4.

143. WALLIS: Reply to Collins's Question on Algebraic Roots

			That is,	
Root.2^d.	Root.3^d.	Root.1.	Root.2.	Root.3.
$7 - 3$.	$7 + 3$.	1.	4.	10.
$6\frac{1}{2} - \frac{1}{2}\sqrt{57}$	$6\frac{1}{2} + \frac{1}{2}\sqrt{57}$	2.	$2,7251-$	$10,2749+$
$6 - \sqrt{18}$	$6 + \sqrt{18}$	3.	$1,7575-$	$10,2425+$
$5\frac{1}{2} - 4\frac{1}{2}$	$5\frac{1}{2} + 4\frac{1}{2}$	4.	1.	10.
$5 - \sqrt{21}$	$5 + \sqrt{21}$	5.	$0,4194+$	$9,5826-$
0.	9.	6.	0.	9.
$4 - \sqrt{18}$	$4 + \sqrt{18}$	7.	$-0,2425+$	$8,2425+$
$3\frac{1}{2} - \frac{1}{2}\sqrt{57}$	$3\frac{1}{2} + \frac{1}{2}\sqrt{57}$	8.	$-0,2749+$	$7,2749+$
0.	6.	9.	0.	6.
$2\frac{1}{2} - 1\frac{1}{2}$	$2\frac{1}{2} + 1\frac{1}{2}$	10.	1.	4.
$2 - \sqrt{-6}$	$2 + \sqrt{-6}$	11.		
$1\frac{1}{2} - 1\frac{1}{2}\sqrt{-63}$	$1\frac{1}{2} + \frac{1}{2}\sqrt{-63}$	12.		
$1 - \sqrt{-27}$	$1 + \sqrt{-27}$	13.		
$\frac{1}{2} - \frac{1}{2}\sqrt{-159}$	$\frac{1}{2} + \frac{1}{2}\sqrt{-159}$	14.	Impossib. Impossib. &c.	
$-\sqrt{-54}$	$+\sqrt{-54}$	15.		
$-\frac{1}{2} - \frac{1}{2}\sqrt{-279}$	$-\frac{1}{2} + \frac{1}{2}\sqrt{-279}$	16.		
$-1 - \sqrt{-87}$	$-1 + \sqrt{-87}$	17.		

# differences.				* differences.		
$+40$	16			40	-12	
$+56$	-2	-18	$+6$	28	-10	$+2$
$+54$	-14	-12	$+6$	18	-8	$+2$
$+40$	-20	-6	$+6$	10	-6	$+2$
$+20$	-20	∓ 0	$+6$	4	-4	$+2$
± 0	-14	$+6$	$+6$	0	-2	$+2$
-14	-2	$+12$	$+6$	-2	-0	$+2$
-16	$+16$	$+18$	$+6$	-2	$+2$	$+2$
∓ 0	$+40$	$+24$	$+6$	0	$+4$	$+2$
$+40$	$+70$	$+30$	$+6$	4	$+6$	$+2$
$+110$	$+106$	$+36$	$+6$	10	$+8$	$+2$
$+216$	$+148$	$+42$	$+6$	18	$+10$	$+2$
$+364$	$+196$	$+48$	$+6$	28	$+12$	$+2$
$+560$	$+250$	$+54$	$+6$	40	$+14$	$+2$
$+810$	$+310$	$+60$	$+6$	54	$+16$	$+2$
$+1120$	$+376$	$+66$		70	$+18$	
$+1496$				88		

It appears by this: That, supposing an affected Cubick Equation, fitted to one series of Roots, Arithmetically proportional, (as 1, 2, 3, &c:) the other two ranks of Roots answering thereunto, will not be Arithmeticall Proportional.

But the rank of Coefficients (in the middle termes of the Quadratick Equations, containing those two later ranks of Roots,) will so bee.

And allso, (of the Absolute numbers in those quadratick Equations,) the second differences will be equal; like as they would have been, if in all of them there had been the same coefficient of the middle term, & one ranke of rootes, as 1, 2, 3, &c.

Note allso, that if in any such series of Equations (as here in Cubicks) one rank of Rootes be arithmetically proportional; then will the series of Aggregates of all the rest, (which is the coefficient of the second term of the subordinate Equation, as here of the Quadratic,) bee arithmetically proportional also: but decreasing, if those first did increase; & contrarywise, increasing if they did decrease.

And consequently, if the proposed series be of Quadratick Equations, whereof one rank of rootes be Arithmetically proportional, the other will be so too: for the series of Aggregates is no other then the other series of roots.

144.
HENRY OLDENBURG to WALLIS
8/[18] March 1669/[1670]

Transmission:

Manuscript missing.

Existence and date: mentioned in and answered by WALLIS–OLDENBURG 10/[20].III. 1669/70.

As emerges from Wallis's reply, this letter enclosed an account of Roberval's new balance, based on a report contained in the *Journal des Sçavans* for 10 February 1670 (new style), 9–12 ('Nouvelle maniere de balance inventée par M. de Roberval').

145.
WALLIS to HENRY HYRNE
Oxford, 9/[19] March 1669/70

Transmission:

w^1 Copy (in Oldenburg's hand) of Wallis's copy of missing letter sent : LONDON *Royal Society* Early Letters W1, No. 102, 2 pp. (our source). At top of p. 1 in Oldenburg's hand: 'Enter'd LB.3.328.' and 'A Copy of Dr Wallis's Answer of March 9: 1669/70. Oxford. To Mr Harry Hyrne's Letter of Febr. 28. 1669 |, Parson's Green in Fulham *del.*| .' At foot of p. 1 deleted by Oldenburg: 'For Mr Henry Hyrne, to be left with Mr Fish, at the 3. Fishes at queen Hythe stairs, London.' At top of p. 2 in Oldenburg's hand: 'Copy of M. Hyrne's Objections against Dr Wallis's Hypothesis of the Tydes, together with the Doctors Answer.'
w^2 Copy of w^1: LONDON *Royal Society* Letter Book Original 3, pp. 328–30.
w^3 Copy of w^2: LONDON *Royal Society* Letter Book Copy 3, pp. 406–8.
Reply to: HYRNE–WALLIS 28.II/[10.III].1669/70.
Answered by: WALLIS–HYRNE 2/[12].IV.1670.

A copy of this letter was sent to Oldenburg together with the letter from Hyrne to which it replied as an enclosure to WALLIS–OLDENBURG 24.III/[3.IV].1669/70. Both letters were produced at the meeting of the Royal Society on 14/[24] April and ordered to be read at the following meeting on 21 April/[1 May]. See BIRCH, *History of the Royal Society* II, 432–3.

Sir,

Yours[777] of Febr. 28. I received this afternoon by the hands of Mr Savile,[778] containing some Objections against my Hypothesis[779] concerning Tydes. There be many[780] of you, and it would require a large discourse to make
5 particular Answers severally to you all; and which, when all is done, may perhaps give not much more satisfaction, than the Hypothesis itself well weighed (For the strength of an Argument is commonly couched in few words.) And if the foundations thereof be not sound, long altercations in

---

5 may *add.*

---

[777] Yours: i.e. HYRNE–WALLIS 28.II/[10.III].1669/70.
[778] Savile: not identified.
[779] Hypothesis: i.e. the letters WALLIS–BOYLE 25.IV/[5.V].1666, *Philosophical Transactions* No. 16 (6 August 1666), 263–81; WALLIS–OLDENBURG 18/[28].VII.1666, *Philosophical Transactions* No. 16 (6 August 1666), 281–9; WALLIS–OLDENBURG 7/[17].III.1667/8, *Philosophical Transactions* No. 34 (13 April 1668), 652–3.
[780] many: i.e. most recently Joshua Childrey. See CHILDREY–WARD 4/[14].III. 1669/70.

145. WALLIS to HYRNE, 9/[19] March 1669/70

writing will not make them so: if they be, I shall so far presume on your own Judgement (tho a person hitherto wholly unknown[781] to me) as to think you will be able not only to understand the consequence of your own arguments, but likewise what from this hypothesis is proper to be reply'd to each of you. I shall only suggest in generall; That I doe not take the Air to be more carried about with the Earth's motion, than the water is; and therefore not lesse liable to as great or greater inequalities than is the water; That, as the Accelerations and Retardations are but small, so are also the Rising and Falling of some few Fathoms but small motions compar'd with the Bigness of the whole Earth; That the Phaenomena, in this Hypothesis, are not deriv'd from one single Circular motion (as most of your Objections seem to suppose) but from a complication of divers; That Accelerations and Retardations do (both of them) occasion Accumulations of waters, and therefore two Tydes in one revolution; That the various positions of Shores, and other such inequalities, cause infinit variations from what would be, if there were one equal Channel round the world; which makes it not only difficult but impossible for me to undertake the solving of all particular phaenomena on each coast, without a better History of Tydes and Cosmographick Tables, than I am ever like to be furnish't with; That, upon the whole matter, though I have not yet met with any objections so cogent as to alter my opinion of this Hypothesis, as the most satisfactory to my own thoughts of any I have yet seen; yet am I not so fond of it, but that I can freely allow others to be of another mind, without thinking myself oblig'd to dispute them out on't, if they be not satisfy'd with my reasons; But lastly, That the greatest objection (with all) against this Hypothesis would be the proposing of a better and more satisfactory. And therefore, since you intimate, that you have spent thoughts about it (more perhaps than I have done) and with so good success as to satisfy yourself as fully therein as in any Phaenomenon of Nature; if you think fit to impart to me what Hypothesis that is, which hath given you such satisfaction; and if on me it have the like effect, I shall not be so fond of my own as not to be willing to exchange it for a better: Resting

Your friend and servant
J. Wallis.

---

[781]unknown: cf. OLDENBURG–WALLIS 26.III/[5.IV].1670.

146.
WALLIS to HENRY OLDENBURG
Oxford, 10/[20] March 1669/[1670]

Transmission:

W Letter sent (original figures, valedictory and signature inserted in w^1): LONDON *Royal Society* Early Letters W1, No. 103, 2 pp. (our source). At top of p. 1 in Oldenburg's hand: 'Dr Wallis's letter to M. Old. concerning M. Robervals New Ballance.' and 'Ent⟨ere⟩d LB. 3. 354.'—printed: OLDENBURG, *Correspondence* VI, 547–9.

w^1 Copy of letter sent with original figures inserted (with corrections in Oldenburg's hand): LONDON *Royal Society* Letter Book Original 3, pp. 354–5 (figures p. 355) (our source for the valedictory, signature, and original figures).

w^2 Copy of w^1: LONDON *Royal Society* Letter Book Copy 3, pp. 438–40.

Reply to: OLDENBURG–WALLIS 8/[18].III.1669/70.

Oxford, March 10. 1669./70.
Sir,

Yours[782] of I received this day; with the Paper[783] inclosed concerning M. Robervals new balance. Which, I suppose, hee intends rather for a curiosity, than for use. For (beside other inconveniences) it is not to be imagined, that such a Machine moving on six several Pins or Cylinders as so many centers, can be so nice or tender as the common Balance moving but upon one single Center: (& that more nice than any one of those six can bee.) And there seemes to mee nothing strange or surprising in it, save onely giving the name of Balance to an Engine so compounded. The name Balance hath been hitherto used to signify, A single Ruler moved on on[e] single Center, (or what is equivalent hereunto.) And of such a Balance it is that the known Laws of the Balance are to be understood. But if hee will extend the name Balance to Engines quite of another nature; it is not at al strange or surprising that these should not answere those Laws.

3 of | March 8, *del.*| I received
8 (& ... bee.) *add.*
11 hitherto (*1*) used (*2*) applyed (*3*) used
11 single *add.*
11 on on single *corr. ed.*
14 at al *add.*

---

[782]Yours: i.e. OLDENBURG–WALLIS 8/[18].III.1669/70.
[783]Paper: i.e. an account of the 'Nouvelle maniere de balance inventée par M. de Roberval', printed in the *Journal des Sçavans* for 10 February 1670 (new style), 9–12.

146. WALLIS to OLDENBURG, 10/[20] March 1669/[1670]

And this of his, is but one particular instance, of what is generally delivered in my Prop. 7. Cap. 2. Mechanicorum:[784] According to which, if any two weights be by any Engine so connected as that the one must just so much ascend as the other doth descend; these, if in one position they be equiponderant, will be so in any position (whether nearer or farther off, whether on the same or contrary sides of what you please to call the Center,) so long as that condition is preserved. Of which it will be easy to give many instances more simple than is that of M. Robervals Engines, which is he is pleased to call a Balance. As for example, If about a wheel or Pully, whose Center is C, the weights P, Q, be hanged by a string; (which Engine, if you please, you may as well cal a Balance, & C the Center of it; for a Circle may as well be so called, as a Parallelogram;) and P, Q, be supposed Equiponderant, if sited as in the first figure; they will be so allso if sited as in second, third, or fourth figures, (or in any other Position, wherein the Descents & Ascents of PQ are equal.)

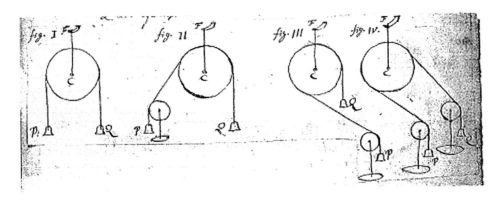

Whether they be equally or unequally distant from the Center C, or Perpendicular FC; & whether on the same or contrary sides: And, if in the first they bee not aequiponderant; they wil not be equiponderant in any of the rest. This in hast (lest I loose the Post) from

---

1 particular *add.*
3 one *(1)* doth *(2)* must
5 whether *(1)* farther or nearer *(2)* nearer or farther off
11 & C ... of it; *add.*
14 other *(1)* where *(2)* Position
17 Center C, or *add.*

---

[784] Mechanicorum: i.e. WALLIS, *Mechanica: sive, de motu, tractatus geometricus* I, 39; *Opera mathematica* I, 598–9.

147. WALLIS to WOOD, 10/[20] March 1669/[1670]

Sir

Your humble servant
J. Wallis.

147.
WALLIS to ROBERT WOOD
Oxford, 10/[20] March 1669/[1670]

Transmission:

W Letter sent: LONDON *Royal Society* Boyle Letters 7, No. 25, 4 pp. (p. 3 blank). On p. 4 endorsements in unknown hands: 'March 10. 1669/70 Dr Wallis to Mr Robert Wood.' and 'probably he that was afterward LLD & mat. of mathemat. School at Christ's Hospital & FRS. vid. Woods Athen: oxonienses Tom: II, p. 780'.

Reply to: WOOD–WALLIS 8/[18].II.1669/70.

Oxford. March. 10. 1669./70.

Sir,

I thank you for your very kind letter[785] of Febr. 8. and those two very ingenious Examples of approximation, which came with it. The former of them (I am told by one to whom I shewed your letter) is much the same as one in Albert Girard[786] his Treatise of Algebra[787] (if I mistake not the name of the book;) but, how near to it, I do not know; (not having the book by mee to consult.) The latter is no lesse ingenious. And your notes on both, are so too. I do not doubt but that in many other cases there may be such regular approximations: but it will not in all cases succeed so well.

I did some years since, draw up a little Tract[788] about Reducing Fractions or Proportions, to others in less numbers, the nearest that may bee, not *exceeding* numbers given; (suppose, whose Denominator or Consequent shal not exceed 3 places, or 4 places, or the like.) As for instance: The proportion of the Diameter to the Perimeter of a Circle, being as 1. to 3.1415926536 fore. The continual approches (the nearest that may be in terms no greater) are $\frac{1}{3} \cdot \frac{8}{25} \cdot \frac{15}{47} \cdot \frac{22}{69} \cdot \frac{29}{91} \cdot \frac{36}{113} \cdot \frac{43}{135} \cdot \frac{50}{157} \cdot \frac{57}{179} \cdot \frac{64}{201} \cdot \frac{71}{223} \cdot \frac{78}{245} \cdot \frac{85}{267} \cdot \frac{92}{289} \cdot \frac{99}{311} \cdot$

18-20 And, if ... any of the rest. *add.*

---

[785] letter: i.e. WOOD–WALLIS 8/[18].II.1669/70.
[786] Girard: i.e. Albert Girard (1595–1632), French mathematician and musician.
[787] Treatise of Algebra: i.e. GIRARD, *Invention nouvelle en l'algebre*, Amsterdam 1629.
[788] Tract: i.e. WALLIS, *Adversus M. Meibomii De proportionibus dialogum*, Oxford 1657.

147. WALLIS to WOOD, 10/[20] March 1669/[1670]

$\frac{106}{333} \cdot \frac{219}{688} \cdot \frac{332}{1043}$. &c. (the Diameter in these being still too big.) Or $\frac{7}{22} \cdot \frac{113}{355}$. &c. (the Diameter in these, being too little.) of which so onward to as great an exactness as you please. (All other designations of it, in numbers not greater than these, respectively, being farther from the true proportion.) And in this particular case, I have prosecuted all the approches so far as till they give the proportion within the most accurate limits yet assigned (that I know of) by any; that is, as 1. to 3.14159, 26535, 89793, 23846, 26433, 83279, 5028$\frac{8\pm}{9-}$.

The Methode is universal; (solving your two Problemes, by continual approches, & all others of the like kind:) but is too long to send in a letter.

By this Methode of mine, I find that, in the former of your two Examples, (about Extreme & Mean proportion,) you take-in all the approches: But in your latter (about the Side & Diagonal of a Square) you leave-out half; (but supply them by one of your observations.) For the continual next approches are these.

The Proportion of the Side to the Diagonal of a Square.

Too Great, but continually Decreasing.		Too Little, but continually Increasing.	
Side.	Diagonal.	Side.	Diagonal.
1	1	1	2
3	4	2	3
5	7	7	10
17	24	12	17
29	41	41	58
99	140	70	99
169	239	239	338
577	816	408	577
985	1393	1393	1970
3363	4756	2378	3363
5741	8119	8119	11482
19601	27720	13860	19601
33461	47321	47321	66922
	&c.		&c.

As to what you have in your Post-script; either I do not aright understand your meaning, or there must be some mistake in it. The series $\frac{1}{4} + \frac{1}{9} + \frac{1}{16} + \frac{1}{25}$ &c. is (as I call it) Reciproca Secundanorum: And $\frac{1}{8} + \frac{1}{27} + \frac{1}{64} + \frac{1}{125}$ [2] &c. is Reciproca Tertianorum: And $\frac{1}{16} + \frac{1}{81} + \frac{1}{256} + \frac{1}{625}$ &c.| Reciproca Quartanorum. (And so onward.) And every of these, infinitely continued, becomes infinitely great; or rather, more than so: viz. (to a quantity assigned) the

147. WALLIS to WOOD, 10/[20] March 1669/[1670]

first, as 1 to $(-2+1=)-1$; the second, as 1 to $(-3+1=)-2$; the third, as 1 to $(-4+1=)-3$, (& so onward:) that is, (not as 1 to 0; which is infinite; but) as 1 to a negative quantity, or to less than 0. (As I demonstrate[789] in my Arithm. Infin. prop. 104.) So that I do not understand, in what sense it is that you say, All of these

$$\left\{\begin{array}{cccc} \frac{1}{4}, & \frac{1}{9}, & \frac{1}{16}, & \&c. \\ \frac{1}{8}, & \frac{1}{27}, & \frac{1}{64}, & \&c. \end{array}\right\} \&c.$$

can be equal to an Unite.

I have had now, for about a year & an half, a treatise[790] in the press at London, about Mechanicks, or De Motu: one part of it, is lately gott out; & another part I suppose may be out about ten weekes hence.

I suppose you have heard of Mr Jeremy Horrockes[791], of Lancashire; a young man but a diligent Astronomer, who died about 1642. His excellent Treatise, De Venere in Sole Visa, 1639; is, some years since, (after it had long lyen by the walls) published[792] by Hevelius, amongst his own works. Wee have many other broken scattered Papers of his, worth preserving: But it is thought, the best of his Remaines, coming to his Brothers[793] hands (who was allso Mathematical) who in the time of the Wars went over into Ireland (I think as a Souldier; but I know not, in what capacity,) and died there, were by that meanes lost. If you could by any meanes find out, where in Ireland he died, & what became of what papers hee had: It might be a means to recover some of them, which have come to the hands of those who understand them not, or do not value them.

I have little more to adde, but that I am (with my family) in good health (through Gods mercy) &

Sir,
Your affectionate Friend & Servant
John Wallis.

---

[789]demonstrate: i.e. in WALLIS, *Arithmetica infinitorum*, 78–9; *Opera mathematica* I, 409.

[790]treatise: i.e. WALLIS, *Mechanica: sive, de motu, tractatus geometricus*, pars prima, London 1670. The second part of the work appeared later the same year. Wallis had declared his intention to publish a book on mechanics at the meeting of the Royal Society on 30 April 1668. See BIRCH, *History of the Royal Society* II, 275. In OLDENBURG–AUZOUT 2/[12].I.1668/9 (OLDENBURG, *Correspondence* V, 296–9, 298) Oldenburg informed Auzout that Wallis's work on mechanics was being printed.

[791]Horrockes: i.e. Jeremiah Horrox (1618–41), astronomer, former contemporary of Wallis at Emmanuel College, Cambridge.

[792]published: Hevelius published Horrox's *Venus in sole visa* as an appendix to his *Mercurius in sole visus*, Danzig 1662.

[793]Brothers: i.e. Jonas Horrox.

[4] For my worthy Friend
Mr Robert Wood, at his
house at Radans-town,
near Dublin
in Ireland.

148.
HENRY OLDENBURG to WALLIS
c.16/[26] March 1669/70

Transmission:

Manuscript missing.

Existence and date: mentioned in and answered by WALLIS–OLDENBURG 19/[29].III. 1669/70.
Enclosure: CHILDREY–WARD 4/[14].III.1669/70.

According to Wallis, Oldenburg's letter reached Oxford late on 17 March. It is therefore probable that Oldenburg had written the letter on the preceding day.

149.
WALLIS to HENRY OLDENBURG
Oxford, 19/[29] March 1669/[1670]

Transmission:

W Letter sent: LONDON *Royal Society* Early Letters W1, No. 104, 4 pp. (our source). At top of p. 1 in Oldenburg's hand: 'Dr Wallis's Answer to Mr Childreys Animadversions on his Theory of Tydes.' and 'Read Mar. 24. 69. Enter'd LB. 3. 355.'—printed: OLDENBURG, *Correspondence* VI, 578–84.
w^1 Copy of letter sent: LONDON *Royal Society* Letter Book Original 3, pp. 355–63.
w^2 Copy of w^1: LONDON *Royal Society* Letter Book Copy 3, pp. 440–50.
E^1 First edition: *Philosophical Transactions* No. 64 (10 October 1670), 2068–74 ('Dr. Wallis's Answer to the foregoing Animadversions, directed in a Letter to the Publisher, March 19. 1669/70.').
E^2 Second edition (partly): *The Philosophical Transactions and Collections ... Abridged and Disposed under General Heads* I, 520–3.

Reply to: CHILDREY–WARD 4/[14].III.1669/70.

This letter represents Wallis's reply to Childrey's animadversions on his hypothesis of tides, which had been forwarded to him by Oldenburg on 16 March or just before. Oldenburg

149. WALLIS to OLDENBURG, 19/[29] March 1669/[1670]

read the present letter at the meeting of the Royal Society on 24 March, when it was also ordered that a copy be sent to Childrey. See BIRCH, *History of the Royal Society* II, 431.

Oxford. March 19. 1669./70. Saturday.

Sir,

Yours[794] with the inclosed Animadversions[795] of Mr Childrey on my Hypothesis[796] of Tydes, came so late to hand on Thursday last,[797] (when the Post was allmost ready to be gone,) that I had not time, by that Post, to give you an account of them. And it is not much that I need to say now. For I do not find, that hee & I are like much to disagree.

That the Winds have a great influence on the Tides of particular Coasts & Havens; according as they are more or lesse stiff or slack, & do blow from this or that part: I do not at all question. But did always take for granted, as generally received, & upon good grounds.

And the like I say of Land-waters: which (though as to the Sea they doe not signify much in this Point) are, as to In-land Rivers, very considerable; especially as to Inundations upon rising of the Water: Which is rather by Checking than Promoting the Tides. For, certainly, these Land-waters, meeting the Tide of Floud, do hinder it from coming so far up the River as otherwise it would. And, consequently, if notwithstanding such Land-flouds, the Tide flow higher up the River than at other times, this must be derived from some other cause. But, that the Tyde & Land-floud should jointly make a greater Inundation than either singly would have done, is not to be doubted.

But hee need not wonder that, in my Essay,[798] though I grant both these, I sayd so little of either; Because it was wholly beside my business; which was, to give a statical account of Stated Periods (Diurnal, Menstrual, Annual,) arising from Regular Motions: not, of Accidental Extravagances, such as these are. And therefore I did, in the beginning of that discourse preclude the consideration of the Advantage or Disadvantage which should arise from such uncertain contingences, as extrinsecal to that busyness.

---

6 I (*1*) intend (*2*) need

---

[794]Yours: i.e. OLDENBURG–WALLIS *c*.16/[26].III.1669/70.

[795]Animadversions: i.e. CHILDREY–WARD 4/[14].III.1669/70.

[796]Hypothesis: i.e. WALLIS, 'An Essay of Dr John Wallis, exhibiting his Hypothesis about the Flux and Reflux of the Sea', *Philosophical Transactions* No. 16 (6 August 1666), 263–81; WALLIS, *Correspondence* I, 200–22.

[797]Thursday last: i.e. 17 March 1669/70.

[798]Essay: i.e. WALLIS–BOYLE 25.IV/[5.V].1666 (WALLIS, *Correspondence* II, 200–22), published in *Philosophical Transactions* No. 16 (6 August 1666), 263–81 ('An Essay of Dr John Wallis, exhibiting his Hypothesis about the Flux and Reflux of the Sea').

149. WALLIS to OLDENBURG, 19/[29] March 1669/[1670]

His third thing Suggested, The Moones Perigaeosis; is so far from being contrary to my Hypothesis, that it is a part of it. And (if I do not much mis-remember) it is, in one of my letters to you, expressely mentioned as such. But for as much as it doth not still fall out at the same time of the Day, Moneth, or Year; I could not make it a Component of any of those noted periods, Diurnal, Menstrual, or Annual; (& of more Periods than these, I did not know that there hath been any generall notice taken, of which I might think myself obliged to give an account:) But it may very well influence any or all of these, according as it falls out advantageous or disadvantageous for them.

And as I do so readyly concur with him in all the particulars by him suggested; so I think he will not be difficult in assenting to all the Materials of my Hypothesis.

The Account which I give of the Diurnal & Menstrual Periods (from the Common Center of Gravity, of the Earth & Moon,) hee doth allow as very Rational. And consequently (which is the foundation of it) that any Acceleration or Retardation of the Compound Motion of the particular parts in the Earths Surface, is to give such an Accumulation of waters as causeth a Tyde. And the complication of such Accelerations & Retardations, concurring or enterfeering one with another, doth occasion the perplex Varieties in them. Of which therefore there is no clear account to be given, without considering severally the proper Effects of each; from whence doth result the Compound Effect of all together.

Now as to the two most signal motions of the Earth, the Diurnal & Annual; if we suppose them each in themselves equal, & both perfectly Circular & upon Parallel Axes; though neither of them singly considered would give any Inequality of Motion; yet the Compound of both together, being Swiftest at Mid-night, & Slowest at Noon, (because the Compound of both is, in that, the Aggregate; in this, the Difference of them;) would give us two Tides in each Diurnal Revolution: But those allways at Noon, & Midnight.

If, to these, wee adde the Menstrual; whereby the Earth describes a smal Epicycle about the Common Center of gravity of the Earth & Moon; & suppose this allso Equal in itselfe, and Circular, about an Axe parallel to the rest: Neither would this, of itself, give any Inequality: But, compounded with the rest, it will. For this compounded with the Annual; doth, at the New-Moon, Increase; at the Full-moon, Abate, of that Motion; as to all

7 any *add.*
34 Axe (*1*) equal (*2*) parallel

parts of the Earths Surface: But, compounded with the Diurnall, (which, in this case, is| much the more considerable, as recurring every day,) it doth [2] most Adde to, or Abate of, that Motion, as to each particular place of the Earths Surface, when the Moon is in the Meridian of that place, Below or Above the Horizon; & would therefore, at those times, give us two Tides. (For which, & other particulars of like nature, that they may be the better apprehended, I refer myself to the inspection of the Schemes pertaining to my Hypothesis.)

Now because this coming of the Moon to the Meridian, above or below the Horizon, or (as the Seamen call it) the Moons *Southing & Northing*, doth, in a Moneths time, passe round the whole circle of 24 hours: hence it comes to pass, that the Time of the Tydes doth so allso. Which I take to be the true account of the Menstrual period. And because this composition of the Menstrual with the Diurnal (which seemes by the effect to be most predominant, though not to extinguish the other,) casts the time at the Moons being in the Meridian: and that of the Annual & Diurnal; when the Sun is in the Meridian: When both these happen at the same time, as at the Full & Change of the Moon; the Tides must needs be the Greater. Which I take to be the true account of the *Spring-tides, & Neap-tides*.

And thus far, (which is the Main of my Hypothesis,) hee concurs with mee, as having given at lest a very Rational & Probable account.

If therefore there be no other Periods of Tydes but these; or, no other remarkable: My work is done, & I need not be further solicitous. For then there will seem to be either no other inequalitie of motions, or none considerable. But if there be allso observable an Annual period, (as very many think there is;) or any other such Periode, (as perhaps there may bee:) then are wee to seeke for the cause thereof in somewhat of inequality which doth (for the Annual Period) Annually recur; or (for any other periode) which doth recur in such a time as that other periode doth require.

Now forasmuch as the three Motions above mentioned, are neither (as was above supposed) each equal in itself, or perfectly circular; nor, all on parallel Axes: there is, both as to the Sun, & as to the Moon, at lest a double inequality; the one by reason of the *Excentricity*, & (which depends thereon) the *Apogaeum* & Perigaeum; the other by reason of the *Obliquity* of the Zodiack & the Moons Orbit, with the Equinoctial, & with each other. From every of which doth proceed some little inequality of motion in the Earths Surface. But, whether so much as to make any remarkable alteration in the Tides; is hardly otherwise determinable than by Observation.

6 may *add*.

149. WALLIS to OLDENBURG, 19/[29] March 1669/[1670]

Now for that of the Moon, both as to its *Apogaeum* & *Perigaeum*, (with the inequality of motion depending on it:) and as to the Obliquity of its Orb, both with the Zodiack & the Equator, (which causeth another inequality both in the Motion of Longitude & Right-Ascension;) I have hitherto contented myself onely to insinuate it, in one of my letters on this subject, without further insisting on it: Because I did not know of any Periodical Vicissitude of Tydes consonant thereunto. When any such shal be discovered; we have here a foundation ready for the salving of it. But as to any Annual vicissitude, it is not of use; because it doth not Annually recurre.

But, because it hath been, allmost, generally received, That there is an observable Annual Period: I did, for the salving of that, apply, (not *the Inequality of the Natural days*, but those causes from whence that proceedes) the *Excentricity* of the Sun or Earths Orb, & the *Obliquity* of the Zodiack.

The Former of these, if singly considered, would cast those Annual Tides in *June* & *December*, (the times of the Suns *Apogaeum* & *Perigaeum*, or rather the Earths *Aphelium* & *Perihelium*, when are the slowest & the swiftest Annual Motions in the Zodiack:) The Latter, if considered alone, would cast them upon the two *Equinoxes*, & the two *Solstices*, (the times of the Lest & of the Greatest Right-Ascensions:) But if both be jointly considered; they must cast these (as they do the greatest Inequalitie of the Natural days) at some intermediate times, between the *Autumnal Equinox* in September, & the *Perigaeum* in December; and again, between this *Perigaeum*, & the following *Vernal Equinox* in March. As is more than probable (without the trouble of any new Computation) from the greatest inequalitie of the Natural Days, arising from the same causes. But whether precisely at the same time with that Inequality, or whether in all parts of the World at any one Time; I do not undertake there to determine: But do rather beleeve the contrary; because the different position of places, may very much alter the influence of both or either causes. I did onely mention, as a thing very notorious, that it doth so constantly fall out on the coasts of Kent, & particularly of Rumney-Marsh, about Allhollantide & Candlemas.|

This Account of the Annual vicissitude, is that onely to which hee doth except. Opposing, first the judgement of Sea-men, (more considerable than that of the Inhabitants of Rumney-Marsh,) who use to say, either that the time of the year signifies nothing; or, if at all, it is about the Equinoxes. Then, that if this be the cause, it will be constant; & that, in February as well as in November. And thirdly, that the Seamen about Weymouth have not observed anything signal about those times.

3 both *add.*

149. WALLIS to OLDENBURG, 19/[29] March 1669/[1670]

To the first, I answere; If not then, but at the Equinoxes: then so much of the Hypothesis as concerns the Excentricity; may be spared; (or allowed to be so little as not to be remarkable;) and that of the Obliquity alone, will give a sufficient account of it. Or if (to which hee seemes rather to incline) there be no such Annual Vicissitude at all; then may that of the Obliquity be spared allso; & the Hypothesis perfect without it. And, till some such be observed & acknowledged; it will be sufficient to say, that though both the excentricity & the Obliquity do cause some inequality in the motion; yet, so little, as that in the Tides it is not remarkable; they falling, just as if the three Motions (Annual, Menstrual, & Diurnal,) were all exactly circular, & on Parallell Axes.

To the second; which concerns matter of fact in Rumney-Marsh.[799] I say that (according to the best account I can there get, & the unanimous consent as well of Fisher-men, & other Water-men, as of other Inhabitants) it is constant; hardly missing (or very seldome) any one year; (be the weather fair or foul:) and as well, about Candlemas, as about Alhollantide, every year; though not then so high. Of which (though they do not pretend to give any reason of it) I think a cause may be very rationally assigned. For if you consult the tables of the Inequality of Natural days (which parallell I make use of for the explication of this,) you will find that about one of the extreemes (in January) the increase & decrease of the Natural days fluctuates very much; sometime increasing, sometime decreasing, according as this or that of the two causes, thwarting one another, doth prevail: But about the other extreme, (in October,) it is much otherwise; the increasings & decreasings going on in a continual course for a long time together. And the same causes, applyed to the business of Tides, may very rationally be supposed to produce as unequal effects.

To the third, That the Sea-men at Weymouth have not observed any such signal effects about Allhollandtide & Candlemas: It is very possible that they have not, & that nothing signal on those coasts doth use to happen at those times: For I fix that matter of fact, principally, on Rumney-Marsh, (& that it doth there constantly happen, I am pretty well out of doubt,) & do but by conjecture extend it to the River of Thames, (as having its mouth not far from those coasts,) where yet, I think, you can be my witness, that you have observed it several years to succeed accordingly. What variety is

---

7 to say *add.*
9-11 they falling ... Parallell Axes. *add.*

---

[799]Rumney-Marsh: i.e. Romney Marsh.

149. WALLIS to OLDENBURG, 19/[29] March 1669/[1670]

on other coasts, I am not certain: But (from an account[800] read in the R. Society in my hearing, about the end of the year 1667) I understand, that, about Chepstow-bridge (& consonantly, I suppose, on the Severn at other places,) they observe the like to happen about the beginning of March & end of September, (the one about as much before the Vernal, as the other is after the Autumnal Equinox, like as in our case it happens,) which they call by the name of *St Davids-stream*, and Michaelmas-stream; as wee do those in Kent, *Candelmas-stream* and *Allholland-stream*: (And, when Sea-men take so much notice of particular Tides as to give names to them; 'tis a great presumption, that it is for some remarkable accident usually happening at those seasons.) Of these different seasons at Chepstow-bridge from those of Rumney-marsh, I gave you my remarks in a letter of mine[801] to you in March following. And the like differences, I suppose, will be observable on other coasts, according as their positions be advantageous or disadvantageous to the one or the other of the two causes on which this Phaenomenon doth depend.

But since it is not yet (it seemes) agreed, whether such Annual Phaenomena do happen; or, if so, not at what time; (so that, for ought that appears, it may be at the seasons I design; that is, between the Winter Solstice & the two Equinoxes on either side of it; though, on severall coasts, severally remote:) I think it best to let this part of the Hypothesis stand as it is, [4] unrevoked.|

As that which, when it shal bee discovered & agreed on, stands ready inough to give a rational account of it; &, in the mean time, doth no hurt. And, in such a complication of causes so abstruse, scarce any thing, but observation, wil determine, which of the causes, & in what degree, is to be judged predominant.

And if to this of the Suns or Earths, be added that of the Obliquity & Excentricity of the Moons Orbit, (of which, for the reason above mentioned, I had taken so little notice,) it will, if it do no good, at lest do no hurt. And I do the rather think it may be considerable; because the Earth & Moons Appropinquation & Elongation, doth really alter the distance of the common Center of gravity (of the Earth & Moon) from the Earth; (rendering the

---

8 And *(1)* 'tis a great presumption, *(2)* when Sea-men
33 Earth; *(1)* & render *(2)* (rendering

---

[800]Account: Wallis refers to this account in WALLIS–OLDENBURG 7/[17].III.1667/8 (WALLIS, *Correspondence* II, 437). The meeting concerned had probably taken place more than a year earlier in December 1666.

[801]letter of mine: i.e. WALLIS–OLDENBURG 7/[17].III.1667/8 (WALLIS, *Correspondence* II, 435–9).

151. WALLIS to OLDENBURG, 24 March/[3 April] 1669/70

Earths Epicycle Elliptical:) and much to favour what Mr Childrey observes of the Moon in Perigaeo.

And this is the summe of what I thought proper to return you (upon those Animadversions) being

Sir,

Your friend to serve you,
John Wallis.

These
For Mr Henry Oldenburg, at his
house in the Palmal, near St
James's
London.

150.
JOSHUA CHILDREY to WALLIS
c.20/[30] March 1669/70

Transmission:

Manuscript missing.

Existence and date: mentioned in WALLIS–OLDENBURG 29.III/[8.IV].1670. Childrey clearly wrote to Wallis after 5/[15] March and probably not long before receiving Wallis's reply.

In this letter Childrey apparently apologised for not having sent his animadversions on Wallis's hypothesis of tides directly to Wallis rather than through the hands of Oldenburg. He also expressed his interest in seeing Wallis's response.

151.
WALLIS to HENRY OLDENBURG
Oxford, 24 March/[3 April] 1669/70

Transmission:

W Letter sent: LONDON *Royal Society* Early Letters W1, No. 105, 2 pp. (our source). At foot of p. 1 note appended by Oldenburg. On p. 2 beneath address in Oldenburg's hand:

4 Animadversions) (*1*) from (*2*) being

151. WALLIS to OLDENBURG, 24 March/[3 April] 1669/70

'Rec. March 25. 1670. with two letters, mentioned at the beginning, of which I had *(1)* copies) *(2)* a copy taken. Answ. March 26. 1670.' Beneath this the draft of Oldenburg's reply (OLDENBURG–WALLIS 26.III/[5.IV].1670).—printed OLDENBURG, *Correspondence* VI, 592–4.

Answered by: OLDENBURG–WALLIS 26.III./[5.IV].1670.
Enclosures: HYRNE–WALLIS 28.II/[10.III].1669/70 and WALLIS–HYRNE 9/[19].III. 1669/70.

Oxford March. 24. 1669./70.

Sir,

I send you, with this, Mr Hyrne's letter[802] & the copy of my Answere[803] to it. Both of which I desire you to return[804] to mee, because I have no other copies of either.

I have been looking over my letters to you about Tydes (those that I have any copies of) to see in which of them I had mentioned the Moones apogaeum & perigaeum (for that I had in some mentioned them I was very well assured.) And find in my Latine Translation[805] of the Printed Appendix[806] (which I sent[807] you in March $166\frac{6}{7}$.) At the first Break of the English printed in pag. 286, (ending with these words[808], *If the Hypothesis, for the main of it, be found rational; the Niceties of it are to bee adjusted, in time, from particular Observations.*) there is added,[809] as followeth.

8 was *(1)* certain *(2)* very well assured.)
10 which *(1)* I borrowed back from you |⟨earlier⟩ *add.*| when I was at London *(2)* I sent you in March 1666./7.)

---

[802] letter: i.e. HYRNE–WALLIS 28.II/[10.III].1669/70.
[803] Answere: i.e. WALLIS–HYRNE 9/[19].III.1669/70.
[804] return: Oldenburg returned the original of HYRNE–WALLIS 28.II/[10.III].1669/70 and Wallis's copy of WALLIS–HYRNE 9/[19].III.1669/70 with OLDENBURG–WALLIS 26.III/[5.IV].1670. Copies of both letters were made beforehand for the archival records of the Royal Society.
[805] Translation: i.e. LONDON *Royal Society* MS 368, No. 1, 1–44. Wallis's remarks indicate that his *De aestu maris* had largely already been written by March 1667, although it was not published until 1693. See WALLIS, *Opera mathematica* II, 737–56.
[806] Appendix: i.e. Wallis's 'An Appendix, written by way of a Letter to the Publisher; Being an Answer to some Objections, made by several Persons, to the precedent Discourse', published in *Philosophical Transactions* No. 16 (6 August 1666), 281–9. This is the printed version of WALLIS–OLDENBURG 18/[28].VII.1666 (WALLIS, *Correspondence* II, 251–63).
[807] sent: possibly as an enclosure to a now missing letter. Cf. WALLIS–OLDENBURG 21/[31].III.1666/7.
[808] words: i.e. Wallis's 'An Appendix', 286.
[809] added: see LONDON *Royal Society* MS 368, No. 1, 37–8, and (slightly diverging from this) WALLIS, *Opera mathematica* II, 752.

151. WALLIS to OLDENBURG, 24 March/[3 April] 1669/70

Et quidem si ad minutias descendere libuisset; habenda fuisset ratio tum Apogaei et Perigaei Lunae, tum Obliquitatis Orbitae Lunaris ad Ecclipticam comparatae, et Intersectionem quas Caput et Caudam Draconis dicunt. Sed et monendum fuisset, Terrae motum Annuum atque Diurnum, cum non eundem seu parallelos Axes habeunt, se mutuo nunc magis nunc minus intersecare. Item Inundationes Nili aliorumque Fluviorum, statis Anni temporibus factas, ad hujusmodi forte causam referri posse. Sed haec intacta plane praetereo, aliaque hujusmodi multa: quae tamen siquando ad particularia descendendum erit, consideranda venient.

Which I thought fit to transcribe out of the Latine Copy, (which I borrowed from you when I was lately[810] at London) because in the English they are not.

That the watermen, not knowing any other cause, should suppose the Tydes you mention,[811] to proceed from the Winds: is not at all strange. For so they do allways: And if they see no reason for it on Land here; they presume there be winds far off at Sea which occasion it. But it is the Phaenomenon which we are to have from them: the Cause may be, what they are not aware of.

I had, this week, letters[812] from Hythe in Kent (on the same coast with Rumney-Marsh) which say, that *This New Moon the Tydes have been high, & that on March. 13. was the highest Tyde they have had a good while.*

14 [*Note appended by Oldenburg:*] (*1*) In his (*2*) Dr Wall. in his letter to Mr Childrey (which he sent to me open) adds, that he is not at all averse from taking into consideration the Moons Perigaeum. But, because it falls not out at set times of the day, month or year, he saith, he would not take it into consideration as a component of the diurnal, menstrual or annual period. (And much lesse that of Winds and Landflouds occasionally happening.) And of more stated Periods than the diurnal, menstr. and Annual, he did not know there had been any publick notice taken, of which he might seem obliged to give an account; thinking it time inough, to give account of other stated inequalities, when they should be discover'd & agreed upon.

12-13 (which ... London) *add.*
21 from (1) Kent (2) Hythe in Kent

---

[810] lately: presumably at the beginning of March 1669/70. Cf. Wallis's 'Reply to Collins's Question on Algebraic Roots', 8/[18].III.1669/70.

[811] mention: possibly in OLDENBURG–WALLIS 16/[26]?.III.1669/70 or in another missing letter.

[812] letters: i.e. ?-WALLIS middle of.III.1669/70.

152. Oldenburg to Wallis, 26 March/[5 April] 1669

Which suites well inough with Mr Childrey's expectation.[813] It seemes, the moons Perigaeum, the Change, & the Vernal Equinox, falling all together, did jointly operate. And that those two did more promote the influence of the Equinox, than the disadvantageous position of the Earths being so far from both its Aphelium & Perihelium did disfavour it: which doth use to abate the Equinoxes Influence, & (on those coasts) draw it back to February.

The disquesi[814] of Slusius, I mean to take care of,[815] as soon as I get a little leisure. I am

Yours &c.
John Wallis.

[2] For Mr Henry Oldenburg
in the Palmal, near St James's
London.

152.
Henry Oldenburg to Wallis
London, 26 March/[5 April] 1670]

Transmission:

C Draft of (missing) letter sent: LONDON *Royal Society* Early Letters W1, No. 105 (beneath address of WALLIS–OLDENBURG 24.III/[3.IV].1669/70) (our source).—printed OLDENBURG, *Correspondence* VI, 600.

Reply to: WALLIS–OLDENBURG 24.III./[3.IV].1669/70.
Answered by: WALLIS–OLDENBURG 29.III/[8.IV].1670.
Enclosures: HYRNE–WALLIS 28.II/[10.III].1669/70 and WALLIS–HYRNE 9/[19].III. 1669/70.

Oldenburg's description of this letter as an endorsement indicates that he wrote it on Wallis's copy of his reply to Hyrne's objections (WALLIS–HYRNE 9/[19].III.1669/70) which Oldenburg returned to Wallis together with the original letter to which it replied (HYRNE–WALLIS 28.II/[10.III].1669/70).

---

[813] expectation: see CHILDREY–WARD 4/[14].III.1669/70.
[814] disquesi: see SLUSE–OLDENBURG 28.II/[10.III].1669/70; OLDENBURG, *Correspondence* VI, 520–3. Sluse sought Wallis's opinion on Riccioli's argument against the Copernican system based on the accelerated motion of heavy bodies.
[815] care of: Wallis gave his opinion on Riccioli's argument in WALLIS–SLUSE 29.III/[8.IV].1670. See also OLDENBURG–SLUSE 26.III/[5.IV].1670; OLDENBURG, *Correspondence* VI, 596–8.

152. OLDENBURG to WALLIS, 26 March/[5 April] 1669

Sir,

Not to increase the bulk of these letters[816] (return'd to you with many thanks) I shall only intimate to you, by way of endorsement on your Answer,[817] that, since the Objections of M. Hyrne (whom I yet can hear no news of) seem to strike at the foundations of your Hypothesis, it appears highly necessary, that the proofs of them as well as his owne Hypothesis, should be seen at large,[818] to enable you to discusse them as they shall merit.

As to your answer[819] to Mr Childrey, that was read[820] on Thursday last at the Society, where litle was said to it, because it required a more leisurable consideration of the whole hypothesis; only M.H.[821] was of opinion, that the Perigaeosis of the ☽ was contrary to your Theory. About March 13. there was, by the Observation of the Water-men, and others, a very high Tyde here also. I shall now, within a day or two, dispatch to M. Childrey your answer. What returne he shall make, shall be sent you, God willing, by

London March 26. 70.

Your serv.
Oldenb.

When you are ready[822] for Slusius, I am.

4 can *add*.
6 that *add*.

---

[816] letters: Oldenburg returned, together with the present letter, the original of HYRNE–WALLIS 28.II/[10.III].1669/70 and the copy, in Wallis's hand, of his reply to this, WALLIS–HYRNE 9/[19].III.1669/70.

[817] Answer: i.e. WALLIS–HYRNE 9/[19].III.1669/70.

[818] at large: Hyrne's objections and Wallis's reply, together with Hyrne's own hypothesis (HYRNE–WALLIS 2/[12].IV.1670) were produced at the meeting of the Royal Society on 14/[24] April 1670; at the following meeting on 21 April 1670 (new style), which Wallis attended, Hyrne's hypothesis was read. See BIRCH, *History of the Royal Society* II, 432–3.

[819] answer: i.e. WALLIS–OLDENBURG 19/[29].III.1669/70.

[820] read: i.e. at the meeting of the Royal Society on 24.III/[3.IV].1669/70. See BIRCH, *History of the Royal Society* II, 431.

[821] M.H.: i.e. Robert Hooke, who was himself working on an hypothesis of the tides. See BIRCH, *History of the Royal Society* II, 433.

[822] ready: cf. OLDENBURG–SLUSE 26.III/[5.IV].1670; OLDENBURG, *Correspondence* VI, 596–8, in which Oldenburg mentions having hinted to Wallis that Sluse would be interested in his opinion on Giovanni Battista Riccioli's demonstration against the movement of the earth. Oldenburg promises to convey Wallis's opinion with any written reply by Wallis to Sluse's latest letter to him (SLUSE–WALLIS [28.II]/10.III.[1669]/1670).

153.
WALLIS to RENÉ FRANÇOIS DE SLUSE
Oxford, 29 March/[8 April] 1670

Transmission:

w^1 Copy of missing letter sent: LONDON *Royal Society* Early Letters W1, No. 106, 4 pp. (our source). At top of p. 1 in Oldenburg's hand: 'Dr Wallis's Letter to Monsr Slusius. Concerning Ricciolo's Argument against the motion of the Earth, taken from the Accelerate motion of Descending Bodies.' and 'Enter'd LB. 3. 364.' On p. 4 in Oldenburg's hand: 'A copy of Dr Wallis's Answer to M. Slusius concerning Ricciolo's late argument against the motion of the Earth, mentioned in Slusius's letter to me of March 10. 1670.' Also in Oldenburg's hand: 'A Copy of Dr Wallis's Answer to M. Slusius concerning Ricciolo's argument against the motion of the Earth taken from the Accelerated Motion of Descending Bodies, mentioned in the sayd Slusius's letter of March 10. 1670.'
w^2 Copy of w^1: LONDON *Royal Society* Letter Book Original 3, pp. 364–6.
w^3 Copy of w^2: LONDON *Royal Society* Letter Book Copy 3, pp. 451–4.

Reply to: SLUSE–WALLIS [28.II]/10.III.[1669]/1670 and to SLUSE–OLDENBURG [28.II]/10. III.[1669]/1670.
Answered by: SLUSE–WALLIS [15]/25.VII.1670.
Enclosure to: WALLIS–OLDENBURG 29.III/[8.IV].1670.

It appears that this letter was forwarded by Oldenburg to Sluse without a covering letter of his own. Oldenburg had already replied himself to Sluse three days earlier: OLDENBURG–SLUSE 26.III/[5.IV].1670; OLDENBURG, *Correspondence* VI, 596–8.

Clarissimo Celeberrimoque Viro
D. Renato Francisco Slusio Canonico Leod.
Johannes Wallis, S.

Non multis Te morabor (Vir celeberrime) cum praeter gratias (quas rependo maximas) vix aliud habeam quod tuis reponam literis.[823] Ad quaesitum[824] autem tuum, quod per D. Oldenburg accepi,[825] ut ut tu eo sis judicio, ut quid ego cadere sententiam non opus sit ut admodum sis solicitus, cum Tu tamen id expetas, reponendum certe nonnihil erit. Quodquo rectius

---

[823] literis: i.e. SLUSE–WALLIS [28.II]/10.III.1669/70.

[824] quaesitum: i.e. the question concerning Riccioli's argument against the motion of the Earth, contained in SLUSE–OLDENBURG [28.II]/10.III.[1669]/1670; OLDENBURG, *Correspondence* VI, 520–3.

[825] accepi: possibly in OLDENBURG–WALLIS c.16/[26].III.1669/70, or in another, now missing letter.

153. Wallis to Sluse, 29 March/[8 April] 1670

possim Argumentum Riccioli, tum ut id in Almagesto[826] suo, tum ut in Astronomia Reformata[827] conspiciendum exhibet, perlegi: quo ex gravium descendentium motu accelerato, adeoque aucta vi percussionis (quod experimentis constat) probatum it, Tellurem non moveri: Eo quod, si ad Copernici mentem moveretur Tellus diurno motu (vel, et annuo simul et diurno) motus ille, qui ex Telluris motu communi et gravium descendentium peculiari componitur, erit vel non omnino acceleratus, vel certe tantillum, ut non possit sensu percipi; Adeoque, quae ex lapsu gravium, sive ex altiori loco sive ex humiliori decidentium oritur percussio, vel plane aequalis esset, vel ejus saltem inaequalitas inobservabilis. Quod quidem, siquid ego judico, est argumentum nihili, atque ex ignoratis motuum et percussionum legibus videtur proficisci. Quippe vis percussionis non aestimanda est simpliciter secundum velocitatem percutientis absolute consideratam, sed secundum ipsius excessum supra velocitatem percussi corporis ad eandem plagam moti, (aut etiam summam velocitatis utriusque, si contrariis motibus sibi occurrant invicem.) Et quidem, si in corpus quiescens impingat percutiens (seposita ponderis absoluta consideratione, de qua hic non agitur) percussio aestimanda erit secundum totam percutientis velocitatem, si vero corpus fugiens velocius moveatur, ad eandem plagam;| corpus insequens (quantacunque [2] velocitate) tantum abest ut feriat, ut ne assecuturum sit: Si eadem ferantur velocitate, sintque ab initio non contigua, necdum ad praecedens pertinget insequens, nedum id percutiet: sin ab initio contigua sint, fugiatque praecedens eadem velocitate qua sequatur insequens, manebit quidem contactus, sed nulla fiet pressio, nedum percussio: si autem fugientis velocitas minor sit quam insequentis, assequetur quidem illud, atque percutiet, sed non ea vi qua percuteret quiescens, hoc est, secundum totam suam velocitatem, sed tantummodo secundum illum velocitatis gradum, quo superat fugientis velocitatem (puta, fugienti ut 3, persequens ut 5, impinget ut 2:) quippe qui utrique communis est motus, atque ad easdem partes, instar nullius erit: Sin occurenti impingat percutiens contrario motu, percussio jam aestimanda erit secundum utriusque simul velocitatem. Quodque in simplici motu obtinet, obtinebit etiam (proportione servata) in compositis. Puta, si percutiens ferri intelligatur motu, ex binis (vel etiam pluribus) composito, quorum saltem unus est percussio communis; motus hic communis instar nullius erit, (cum alterum tantundem declinet ictum, quantum reliquum

---

[826] Almagesto: i.e. RICCIOLI, *Almagestum novum Astronomiam veterem novamque complectens*, Bologna 1651. The argument is contained in lib. II, cap. 3, 51–2: De terrae immobilitate.

[827] Reformata: i.e. RICCIOLI, *Astronomiae reformatae tomi duo*, Bologna 1665. The argument is contained in lib. I, cap. 17, appendix, 81–91.

intentat;) fietque percussio secundum motum illum, qui est percutienti proprius, atque eadem praecise vi qua foret si communis ille motus omnino non fuisset. (Ecquis enim dubitat, quin, si duo in eadem navi placido omne ferantur, possit alter alteri eadem vi colaphum impingere ac si uterque foret in littore.) Adeoque, sive sit, sive non sit, Telluris motus diurnus, Annuus, aliusve (vel plures alii,) qui et pavimento communis sit et corpori decidenti, omnino perinde est, percussionem quod spectat; ut quae in eadem ratione major minorve sit futura (caeteris paribus) qua major est minorve velocitas istius motus (seclusis reliquis) qui est percutienti proprius, hoc est, in lapsu [3] gravium, ipsius descensionis. Aliique| quotiunque componantur motus, corpori percusso et percutienti communes, ne hilum promovent impediuntve percussionem, quae tanta plane erit, atque ad eandem praecise plagam, ac, si communes illi omnes abfuissent. Ut mirum sit, Clarissimum virum, existimasse, ex hoc capite desumi posse Argumentum pro Telluris motu vel ponendo, vel tollendo. Atque haec sunt, Vir Clarissime, quae, sentententiam rogatus, hac in re licenda judico. Quae an tuis item sensis sint consona, lubens audivero. Vale

Dabam Oxonii 29 Martii 1640 st. v.

154.
WALLIS to HENRY OLDENBURG
Oxford, 29 March/[8 April] 1670

Transmission:

W Letter sent: LONDON *Royal Society* Early Letters W1, No. 107, 2 pp. (our source). On p. 2 beneath address in Oldenburg's hand: 'Rec. March. 30. 70. Answ. Apr. 9. 70.' Postmark on p. 2: 'MR/30'.—printed OLDENBURG, *Correspondence* VI, 601–2.

Answered by: OLDENBURG–WALLIS 9/[19].IV.1670.
Enclosures: WALLIS–CHILDREY *c*.29.III/[8.IV].1670 and WALLIS–SLUSE 29.III/[8.IV].1670.

Oxford Mar. 29. 1670.

Sir,

You have here my answere[828] to a letter[829] of Mr Childrey's, (which was onely an apology for not having sent his Animadversions[830] directly to my

---
[828]answere: i.e. WALLIS–CHILDREY *c*.29.III/[8.IV].1670.
[829]letter: i.e. CHILDREY–WALLIS *c*.20/[30].III.1669/70.
[830]Animadversions: i.e. CHILDREY–WARD 4/[14].III.1669/70.

154. WALLIS to OLDENBURG, 29 March/[8 April] 1670

self, but through other hands: &, a desire to see what Answere I give to them.) I know not how better to address it to him than through your hands; to whom (I understand by your last[831]) you are sending.

As to Mr Hyrne, I know not what he is, nor of what humour. But his objections[832] are meer mistakes of the Hypothesis: which hee seemes either not to have well considered, or not to understand it. Which made me give him that suddain & short answere[833] the next morning; (wherein yet all that is material in his exceptions is fully answered:) &, withal, to set him to work, to let us see what those his notions are which render this matter as clear to him as any Phaenomenon in nature. By which, if he attempt it, we shal be able to make some estimate of the man.

To Slusius, I have written as you see inclosed.[834] I am

Yours &c.
J. Wallis.

I have received the papers[835] returned concerning Mr Hyrn; with your indorsement: to which this is in answere.

I onely adde, that Mr Childrey's notion, of the Moon's Apogaeosis (who ever it were that made the Objection,) is not at all contrary to my Theory. It onely addes another Period, beside those three of which I gave an account (the diurnal, menstrual & annual,) it takes away none of those. And of this allso, you see, I was before aware; though, because it was not observed, I did not incumber my hypothesis with it.

These [2]
For Mr Henry Oldenburg,
in the Palmal near St James's,
London.

3 you are *(1)* writing *(2)* sending.
20 allso, *add.*

---

[831]last: i.e. OLDENBURG–WALLIS 26.III/[5.IV].1670.
[832]objections: i.e. HYRNE–WALLIS 28.II/[10.III].1669/70.
[833]answere: i.e. WALLIS–HYRNE 9/[19].III.1669/70.
[834]inclosed: i.e. WALLIS–SLUSE 29.III/[8.IV].1670.
[835]papers: i.e. HYRNE–WALLIS 28.II/[10.III].1669/70 and WALLIS–HYRNE 9/[19].III. 1669/70. Oldenburg wrote OLDENBURG–WALLIS 26.III/[5.IV].1670 as an endorsement on the latter.

155.
WALLIS to JOSHUA CHILDREY
c.29 March/[8 April] 1670

Transmission:

Manuscript missing.

Existence and date: mentioned in WALLIS–OLDENBURG 29.III/[8.IV].1670, with which it was sent as an enclosure.

After Childrey had sent his animadversions on Wallis's hypothesis of tides through Seth Ward, and Wallis his reply through Oldenburg, Childrey wrote to Wallis apologizing for not having written to him directly. Wallis sent this latest reply again through the hands of Oldenburg, but this time it was addressed to Childrey himself.

156.
? to WALLIS
mid March 1669/70

Transmission:

Manuscript missing.

Existence and date: mentioned in WALLIS–OLDENBURG 24.III/[3.IV].1669/70.

Wallis wrote to Oldenburg on Thursday, 24 March 1669/70 that he had received letters that week from Hythe in Kent. In these it was pointed out that the tides had been high during the then full moon and that on 13 March (old style) there had been the highest tide for some considerable time. There is no indication who might have been the author of this letter or letters.

157.
HENRY HYRNE to WALLIS
Parson's Green, 2/[12] April 1670

Transmission:

C Letter sent (original figures inserted in c^1): LONDON *Royal Society* Early Letters H1, No. 108, 17 pp. (our source). At top of p. 1 in Oldenburg's hand: 'Mr Hyrns Hypothesis of the Tydes written to Dr Wallis, and by him communicated to M. Oldenburg for the R. Society.' and 'Read April 21.70. Entered LB. 3. 330.'

157. HYRNE to WALLIS, 2/[12] April 1670

c^1 Copy of letter sent with original figures inserted: LONDON *Royal Society* Letter Book Original 3, pp. 330–49 (figures inserted between p. 348 and p. 349) (our source for the original figures).
c^2 Copy of c^1: LONDON *Royal Society* Letter Book Copy 3, pp. 408–32.
Reply to: WALLIS–HYRNE 9/[19].III.1669/70.
Answered by: WALLIS–HYRNE 4/[14].IV.1670.

With the present letter Wallis followed the same procedure he had adopted with Hyrne's objections to his theory of tides (HYRNE–WALLIS 28.II/[10.III].1669/70): he sent the original letter, in which Hyrne sets out his own hypothesis on tides, together with a copy of his reply (WALLIS–HYRNE 4/[14].IV.1670), to Oldenburg. The package, without a covering letter, was probably sent on 11 April 1670 (old style). See WALLIS–OLDENBURG 9/[19].III.1669/70. Both Hyrne's letter and Wallis's reply were produced at the meeting of the Royal Society on 14 April 1670, whereupon it was ordered that they should be read at the following meeting. Hyrne's 'Scheme' or diagram was evidently considered to be too small, for it was also ordered that it should be redrawn in larger scale by an amanuensis. The diagrams contained in the *Letter Book* are presumably the result of this re-drawing. Hyrne's hypothesis of the tides was duly read by Oldenburg at the meeting of the Royal Society on 21 April 1670. It was then recommended for further consideration by Wallis, who was present at that meeting. See BIRCH, *History of the Royal Society* II, 432–3.

Honoured Sir,

Yours[836] of March 9th came safe to my hands; to which although it be easy enough for mee to send you an answer to what you say in defence of your Hypothesis, (as having before made it in the proof of my objections,[837] yet) I shall at present only say that I had respect to the complication of all the motions in that which I objected; and research that further answer, which I suppose will be the most welcome to you, to wit the proposall of that Hypothesis, which hath given mee such satisfaction, though it be more troublesome to mee, as not having pen'd any thing about this subject before now.

My Hypothesis is this, that the Earth, besides the diurnall and annuall motion, hath another, directly from North to South, for the space of 6. hours and some odd minutes, and then again from South to North for the same time, and that in this motion, the Earth doth not allways move to the same points; but farther, when we have spring-tydes than at other times: And that the motion of the earth in each vibration (if I may so call it) from the spring-tide to the neap-tide, doth decrease, as that of a Pendulum will do;

6 in that ... answer, *add*.
7 to you, *add*.

---

[836] yours: i.e. WALLIS–HYRNE 9/[19].III.1669/70.
[837] objections: i.e. enclosure to HYRNE–WALLIS 28.II/[10.III].1669/70.

and from thence again encrease in the same proportion that it did decrease, till the tides bee at the highest: And that whilst the Earth thus moves to and fro, the center of gravity is not in the midst of the Earth, but in the midst of the Earths Vortex, equally distant from those points, to which the center of the Earth doth come in its motion towards the North and south. And that these points, to which the center of the earth doth come, are not farther distant from one another, than the difference between the high water and low-water under one of the Poles, if they be so far.

From this Hypothesis, the diurnall and menstruall tide with all those Phaenomena, which I said in my former letter,[838] I could not see how your Hypothesis would salve, will evidently follow. For let $SE\ NE$; be the terraqueous globe in the midst of its| Vortex, whose center is G; when it moves towards the North to the circle of pricks $AEBE$, whose center is F, the fluidity of water, which is a heavy body, and seeks to approach as near as it can to the center of gravity G, causeth a motion in all the waters in the world, that are not hindered, from B towards A: and consequently an ebbe in all the North part of the world, and a tide in all the South part. And then again when it moves towards the south, to the circle of the pricks $DECE$, whose center is H, the same fluidity of water causeth a return of all the waters in the world that are not hindered, from D towards C; and consequently an ebb in all the south part of the world, and a tide in all the north part.

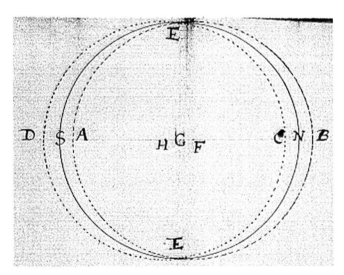

2 Earth (*1*) doth (*2*) thus
4 the center of *add.*

---

[838]former letter: i.e. HYRNE–WALLIS 28.II/[10.III].1669/70.

157. Hyrne to Wallis, 2/[12] April 1670

Now this motion from North to South and back again being finished in the same time that there is between two succeeding high waters, salves the diurnall tides, and the Earth moving farther towards the North and South, when wee have spring-tides, than at other times, it salves the menstruall tides. And because the center of gravity G, and the center of the Earth F or H are allmost of the same distance from E, the waters in their passage to and fro, will make a very inconsiderable, if any, rise and fall, under and near the Æquator. And because in the Bay of Mexico, which lyes all on one side of the Æquator, both the North and South-sides of it come towards the center of gravity at the same time, to wit, when the earth moves towards the South, and go from it again together, when the earth moves towards the North; the North-side doth not come so much nearer to, nor go so much farther from the center of gravity than the south side, as to cause an evident tide, and therefore navigators say, that there is none, as Sfondratus[839] reports[840] of one Philippus Folius,[841] de causa æstus Maris cap. 15. Perhaps some curious observers may find a little at the North and South shores, as Scaliger[842] says,[843] that by his care there was found at Marseils, subt. exer. 52 though none I beleive, was ever taken notice of there before. And because both ends of the Straits of Magellan| are on the same side of the Æquator, they have high water and low water at the same time. As Europe and the opposite parts of America allso have, for the same reason. And because, as the North-pole goes towards the center of gravity, the south pole goes from it, and vice versa, there must be of necessity high water towards the North-pole, at the same time that it is low water towards the South-pole; and high water there, when low here; And lastly, because when the earth is at the greatest distance from one side of the center of gravity, the vibration beyond it on the other side is also the greatest, the lowest ebbs will be after the spring-tides.

This Hypothesis will allso afford a reason, 1. why the spring-tydes are all over the world at the same time. 2. Why a place hath the greater tydes the farther it is distant from the Æquator, if nothing hinders them. 3. Why these Mediterranean seas, the Caspian, the Euxine, and the Baltick, and all others that are lesse, have no rise nor fall of the water. 4. Why the Mediterranean κατ' ἐξοχήν so called, in most places hath no ebbe, and not much in any.

---

[839]Sfondratus: i.e. Pandolfo Sfondrati *fl.* 1540.
[840]reports: i.e. SFONDRATI, *Causa aestus maris*, Ferrara 1590, 23^r–23^v.
[841]Folius: in fact, Sfondrati cites a certain Philippus Folcus.
[842]Scaliger: i.e. Julius Caesar Scaliger (1484–1558) Italian classical scholar, natural philosopher, and physician, mainly active in France.
[843]says: i.e. SCALIGER, *Exotericarum exercitationum liber quintus decimus de subtilitate, ad Hieronymum Cardanum*, Paris 1557, 84^r.

157. HYRNE to WALLIS, 2/[12] April 1670

5. Why the rise and fall of the water in the Adriatick, is greater at Venice than else-where. 6. Why at Corcyra, now Corfu, the water doth run to and fro from north to south, and back again, without any rise and fall. 7. Why the same happeneth between Rhegium[844] and Sicily, which motion being very swift there, through a chanell not a league wide, and causing the water to turn about on either side of it, gave occasion to the fiction of Scilla and Charibdes. 8. Why the red Sea ebbs and flows at both ends, and not in the middle.|

From this Hypothesis and the various positions and nature of shores, I can also frame to my selfe a satisfactory reason 1. For the notable tides at Cambaia[845] and Pegu,[846] and in the Pacifick Sea at Panama. 2. Why the water riseth so high at Bristol, and St Michaels mount[847] in Normandy, and the Magellan straits. 3. Why the tides are greater, at the very Isle of St Thomas, though it lyes under the line, than they are all along the coasts of Africk from Barbary[848] to Guiny;[849] and why in America at the mouth of the river of the Amazons, which is also under the Æquator, the tides are big enough to make a ship float, which did ly dry at low water. 4. Why the water runs with such violence into the Persian gulf, and so gently, if at all, into the red sea. 5. Why the Mediterranean flows in the north-part from East to West, and in the south part from West to East: why the Adriatick flows to the north by Dalmatia, & to the south by Italy: and why the German Ocean floweth unto the North by Holland, Danmark, Norway; and to the South by Scotland and England. 6. Why there is such a visible tide in Euripies, whereas, for ought that I could ever learn, there is scarce any in the Ægean Sea? I might instance in other Phaenomena, which maybe salved by this Hypothesis; but I chose rather to mention these, because there may possibly be objections drawn from every one of them against it.

As for that of Vossius[850] de Motu marium et ventorum,[851] cap. 16. that the spring tides, at Cambaia and Pegu are not at the new and full

---

[844]Rhegium: i.e. modern-day Reggio di Calabria, founded in 730 BC.

[845]Cambaia: i.e. Khambhat. The Gulf of Khambhat in northwest India experiences regularly extreme high tides.

[846]Pegu: i.e. Bago. The city in southern Burma (Myanmar) was founded on the silted-up Gulf of Martaban around 573 AD.

[847]St Michaels mount: i.e. Mont Saint-Michel, Normandy.

[848]Barbary: i.e. Barbary Coast; a term of reference to the middle and western coastal regions of North Africa.

[849]Guiny: i.e. Guinea, the forested coastal region of West Africa between the Tropic of Cancer and the Equator.

[850]Vossius: i.e. Isaak Vossius (1618–89), Dutch scholar and sometime librarian to Queen Christina of Sweden; from 1670 resident in England.

[851]de Motu...ventorum: i.e. VOSSIUS, *De motu marium et ventorum liber*, The Hague 1658, 71–2.

157. HYRNE to WALLIS, 2/[12] April 1670

moon, but at the quadratures, which was also formerly objected against your Hypothesis; I must crave leave not to beleive, till some eye witnesse confirmes it to me. Scaliger saying[852] de Subtil. exer. 52. In Calicuto mari crementa fiunt pleniluniis: contra ad Indi fluvii litus, noviluniis.| Out of [5] which words, though he seems to say, that the spring-tides happen but once in a month (which he alone I think asserts) yet I may conclude, that they do not happen at the quadratures at Cambaia, or the mouth of Indus.

As for that, which Scaliger saith[853] in the same exercitation, and many after him, that the sea flows 7. hours at the mouth of the Garonne,[854] and ebbs but 5; if it were true, there might possibly be an argument drawn from it against this Hypothesis. But Monsieur d'Arcons[855] (to whom I am beholden for the determination of the place, where the earth approacheth nearest to, and goes farthest from the center of gravity,) an Advocate of the Parliament of Bourdeaux, and living by the river, in his second part du Flux et Reflux de la mer cap. 15. paragr. 2. saith,[856] that he cannot imagine why Scaliger fancied such a thing, unlesse knowing the ebbe to continue 7. hours at Bordeaux, and the floud 5. he took the one for the other, and in the 3^d Paragr. he saith,[857] that the Flux and Reflux do allways happen at the Mouth of the river 6. hours and 12. Min. one after the other.

Besides the Solution of these Phaenomena by this Hypothesis, there is this to be said for it, that such a motion of the Earth was thought by a learned Mathematician, whom Fromondus[858] mentions,[859] Meteorol: lib. 5. cap. 1. Artic. 8. to be so probable a cause of the seas motion, that he did assert it to be so, though he could not tell under what Zenith the globe of the earth was lifted up and depressed. The same also, when it was propounded to Grandamicos[860] (so far as to the motion from North to South, and back again, for the Propounder Monsieur d'Arcons acknowledged[861] no

9 7. (*1*) times (*2*) hours
25 depressed. (*1*) It also, (*2*) The same also, when it

---

[852]saying: i.e. SCALIGER, *De subtilitate*, 82^v.

[853]saith: i.e. SCALIGER, *De subtilitate*, 82^v.

[854]Garonne: river in southwest France and northern Spain.

[855]d'Arcons: i.e. César d'Arçons (d. 1681), French lawyer and physician.

[856]saith: i.e. D'ARÇONS, *Le secret decouvert du flux et reflux de la mer et des longitudes*, Paris 1656, 121.

[857]saith: i.e. D'ARÇONS, *Le secret decouvert*, 124, 126.

[858]Fromondus: i.e. Libert Froidmont (1587–1653), Belgian theologian.

[859]mentions: i.e. FROIDMONT, *Meteorologicorum libri sex*, Antwerp 1627, 254.

[860]Grandamicos: i.e. Jacques Grandami (or Grandamy) (1588–1672), French physician and astronomer.

[861]acknowledged: i.e. D'ARÇONS, *Le secret decouvert*, 14–9.

157. HYRNE to WALLIS, 2/[12] April 1670

other motions of the earth) pleased him so much, that allthough he was praengaged against all motion, by his Nova Demonstratio immobilitatis terrae petita ex virtute magnetica,| yet he gave[862] the propounder the Answer, that if the Flux and Reflux of the sea were to be caused, and God should give him order concerning it, he would not do it otherwise than as he said it was done.

But that which is of greater force with mee, is the conformity of this motion of the globe of the earth from North to South, and from South to North, with that of the whole Vortex of the earth, and the rest of the planets. For following the Vortex of the earth such a motion from North to South, and from South to North, and to be carryed about by the Sun, not in the plain of the Zodiack, but in the plain of the Equator, and parallel to it (the sun with all the fixed stars not stirring out of their places,) all the Phaenomena, demonstrable by Copernicus his Hypothesis, may be demonstrated, and also the necessity of the variation of the obliquity of the Ecliptick; of which Bullialdus[863] Astronom. Philola: lib. 5. cap. 5. saith[864] it is impossible for any man to find out the cause: with the necessity of variation of the latitude of the fixed starrs; which Tycho Brahe saith he hath observed, and Bullialdus doubts[865] of, at prius cap. 4.

And allowing Saturn, Jupiter, Mars, Venus, and Mercury the same motion; and to be carryed about by the Sun in the plain of the Equator and parallel to it, the various declination of every one of them from the Ecliptick may be demonstrated, without the introducing of above one Axis for the Sun and all these planets to move about; which seems to mee a good argument for such an Hypothesis, because Frustra fit per plura quod potest fieri per pauciora. I shall not say, what may be demonstrated of the moon, because I am not yet so fully satisfyed about it, as I could wish to be. And that you may not think that this conformity of the motion of the globe of the earth to its Vortex and the planets, is the proof of an unknown thing by that which is more unknowne, or of one thing that is fals, by another that is more fals; I shall now prove that the Vortex of the| earth and the other planets have such a motion, as I ascribe to them; by a plain demonstration of some Phaenomena by this Hypothesis, than any other will afford.

---

[862]gave: i.e. GRANDAMI, *Nova demonstratio immobilitatis terrae petita ex virtute magnetica*, La Flèche 1645, 151–2.
[863]Bullialdus: i.e. Ismaël Boulliau (1605–94), French theologian and astronomer.
[864]saith: i.e. BOULLIAU, *Astronomia philolaica*, Paris 1645, 229.
[865]doubts: BOULLIAU, *Astronomia philolaica*, 226–7.

157. HYRNE to WALLIS, 2/[12] April 1670

Bullialdus saith[866] in his Astron. Philol. lib. 1. cap. 8. that at the Æquinoxes the spots in the Sun seem to describe right lines in the body of the Sun by their motion, and out of the Equinoxes, portions of circles. And Scheinerus[867] part. 2. lib. 4. cap. 10. Rosae Ursinae (as Andreas Cellarius[868] quotes[869] him in his Atlas Universalis page 127.) saith,[870] that the Poles of the Sun do appear above, and are hid under the Horizon of the Sun, first one and then another, for 6. months a peice. These 2. observations are in effect the same; for first, the spots of the Sun can not seem to describe portions of circles, unlesse the earth be nearer to one pole than the other; and the earth cannot be nearer to one pole than the other, but one pole will appear above the Horizon of the Sun, and the other be hid. Secondly the spots in the sun cannot seem to describe right lines, unlesse the earth be equally distant from both poles; nor first one pole of the Sun seem to appear, and then the other, but the earth must be equally distant from both the poles between the disappearing of the one and the appearing of the other. And it may be worth the while to consider which Hypothesis will best salve this Phaenomenon, with the rest that are more taken notice of.

Bullialdus saith,[871] that the Axis of the Sun hath such an inclination to the plain of the Zodiack, that the spots of the Sun do at the Æquinoxes describe right lines, and at other times, portions of circles. I know this possible; but what kind of motion must the annual motion of the earth then be? It is commonly ascribed to the Sun, who is thought to carry not only the earth, but the rest of the planets about with him;| which if he doth, it is not [8] consonant to what wee see every day, that he carry them about in a plain, which will make right angles with his Axis; and therefore not in the Zodiack, if this inclination of his Axis to the plain of the Zodiack be true. For to say, that he carryes them about in a plain to which his Axis hath any inclination, seems to mee but a pitifull shift. But Bullialdus will have[872] the planets not to be carryed about by the sun, lib. 1. cap. 12. but to be moved of themselves per propriam formam. But since there is such a continued proportion

---

[866]saith: BOULLIAU, *Astronomia philolaica*, 15.

[867]Scheinerus: i.e. Christoph Scheiner S.J. (1575–1650), German mathematician and astronomer.

[868]Cellarius: i.e. Andreas Cellarius (1596–1665), German–Dutch mathematician and cartographer.

[869]quotes: i.e. CELLARIUS, *Harmonia macrocosmica seu Atlas universalis et novus, totius universi creati cosmographiam generalem, et novam exhibens*, Amsterdam 1661, 127.

[870]saith: i.e. SCHEINER, *Rosa ursina sive sol ex admirando facularum & macularum suarum phaenomeno varius*, Braga 1630, 601 (I).

[871]saith: i.e. BOULLIAU, *Astronomia philolaica*, 15.

[872]will have: i.e. BOULLIAU, *Astronomia philolaica*, 21.

157. HYRNE to WALLIS, 2/[12] April 1670

in the motion of all the primary planets, from Mercury, which is nearest to the Sun, to Saturn, which is farthest off, the nearest allways finishing its revolution in lesse time than that which is immediatly next to it; I dare not exclude one common cause from their severall motions, neither can I think what Bullialdus saith to be any better evasion, than if I, discoursing of the ebbing and flowing of the sea, and not being able to give a reason for the different Tydes of each particular sea from one common cause, should say, that it is the nature of every sea to move as it doth.

Scheinerus will have[873] the Sun to bend first one pole & then the other, towards the earth. But the mutuall appearing and disappearing of the Suns poles being demonstrable, though its Axis allways continues in the same position to the Axis of the earth, I take this to be the worst shift of all.

My owne Hypothesis is, that the Sun being the center of all the Planets, moves them all about according to the succession of the signs in the plain of the Æquator, according as his owne body is moved: and that every planet hath a proper motion of its owne directly from north to south, and from South to North again; which proper motion, besides the solution of the [9] fore-mentioned Phaenomena, makes them at severall times| to appear in all their severall declinations from the Æquator, or in all the Signs of the Zodiack. Let AB be the Tropick of cancer, CD the Tropick of Capricorne, EF the Æquator, AEC the side of the Colurus Solstitiorum,[874] which passeth through the beginning of Capricorn. Whilst the Vortex of the earth moves from A to B by the motion of the Sun, which is common to all the planets, if it moves æqually from A to C by its owne proper motion, it will describe the line AD; and so be first in the Tropick of cancer, then in the Æquator, and lastly in the Tropick of capricorn; and in its return to cancer, it will pass by the Æquator; And this all the world must grant me.

But it will be said, that if the Vortex of the earth hath such a motion, then the declination of it from the Æquator will encrease and decrease *equally*, whereas wee see by the tables calculated for the Suns Declination, that it encreaseth and decreaseth allmost as much, whilst it passeth through one signe that is next to the Æquator, as it doth in the other two that are farthest from it. To which I answer, that the motion of a Pendulum in the descending arch doth continually encrease till it comes to the lowest point;

---

34 descending arch (*1*) and this motion of the earths vortex from North to South &c. being to and fro like that of a Pendulum in the descending arch (*2*) doth

---

[873] will have: i.e. SCHEINER, *Rosa ursina*, 601 (II).

[874] Colurus Solstitiorum: i.e. the solstitial colure which together with Colurus Aequinoctiorum forms the Coluri: two circles in the heavenly sphere, passing through the poles of the earth and cutting each other at right angles.

and from thence continually decrease till it comes to the top of the ascending Arch. And this motion of the earths Vortex from north to south &c. being to and fro like that of a Pendulum, it may with reason be supposed that it is swifter in the middle than at the beginning and ending of the vibration. And it is possible for it to encrease and decrease in such proportion, that in its compound motion it shall not describe the right line AD, but $AGHD$, in all points like to the ecliptick; And allowing this to be, I make no question, but that my Hypothesis will salve all Phaenomena.| [10

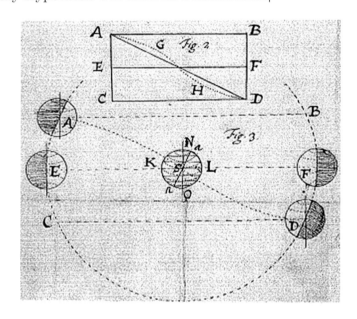

And if the apparent place of the sun as well as of the planets be different from what it should be; it may possibly be salved by this: that the motion compounded of that from West to East, with that from North to South &c: doth not cause the earths Vortex to describe a compleat circle.

I shall now proceed to shew, how my Hypothesis will salve the Phaenomena, 1. of the spots of the sun describing at one time right lines, and at another portions of circles. 2. of the mutuall appearing and disappearing of the poles of the Sun above the Horizon. To which I shall adde in the 3^d place the demonstration of the inequality of the nights and days all over the world, except under the Æquator, and of their æquality all over the world at the Æquinoxes. 4. of the necessity of the variation of the obliquity of

17 which (*1*) you (*2*) I

157. HYRNE to WALLIS, 2/[12] April 1670

the Ecliptick; and of the latitude of the fixed starrs. 5. of the other Planets motions.

Fig. 3. Let. S be the Sun, whose poles are NO; about which it turns from K, by S, to L and amongst the rest of the Planets carryes the earth about from A to B with the Axis parallell to his owne. And let the Earth in the time that it moves from A to B by the common motion of the Planets, move by its proper motion from A to C or from 23 deg. 30′ north declination to 23. gr. 30′ South declnation. The compound motion, as I said before, may cause the earth to bee in all the severall declinations, as if it were moved in a circle in the midst of the Zodiack. And from thence I infer, first, that whilst the earth is at E or F in the Æquator, then the spots in the Sun will all seem to describe right lines, like those drawne in S; that, whilst the earth is at A, at the Tropick of Cancer, they will seem to describe portions of circles, like those marked with pricks in S; and that, when the earth is between A and E, they will describe other portions of circles, lesse and lesse crooked than those here marked, answerable to the earths distance from A. And whilst the earth is at D, or between D and F, the spots will seem to describe the like portions of circles that they do, when the Earth is at A, and between A and E; only they will be turn'd the contrary way. Secondly, supposing that halfe the Sun appeares at a time when the earth is at E or F, it will appear from pole to pole, both the poles then touching the Horizon; but when the earth moveth towards A, the Pole N will begin to appeare, and be at the highest elevation, or seem to be farthest within the circle of the Sun, when the earth is at A, the Sun| appearing then from $a.$ to n. The same may be said of the pole O when the earth moves towards, and is at D.

[11]

Thirdly, the Sun enlightening that halfe of the earth that is next to it; when the earth is not at the Æquinoxiall points, the line that divideth the enlightened part from the dark, will cut cut all the parallells on either side of the Æquator into unaequall parts; and the Æquator into æquall parts, and so cause unaequall dayes and nights all over the world, except under the Æquator. And the parallele being divided most unequally, when the earth is at A or D, the greatest inequallity of the days and nights happens when the earth is at the Tropicks. Again, when the earth is at the æquinoxiall points, the line that divideth the enlightened part from the dark, will divide all the parallels into æquall parts, and so cause the days and nights to be equall all over the world.

Fourthly, if the motion of the terraqueous globe holds its conformity with that of the earths Vortex, not only in that it moves from North to

5 to (*1*) its (*2*) his

157. HYRNE to WALLIS, 2/[12] April 1670

South &c: but also, in that it moves sometimes farther to the North and South, than it does at other times: then the earths Vortex must not only move towards the North and South, but at one time farther to the North and South than it doth at another. For the Ecliptick is nothing else but a name given to that line wherein the earth moves because the Sun & moon are Eclipsed in it: And if the earth should decline 60 degrees on either side of the Æquator, or only on one side of it, wherever it went it would describe the Ecliptick. And lastly, the obliquity of the Ecliptick varying, and neither Sun nor fixed stars stirring out of their places; the latitude of those starrs especially that are neare to the Tropicks, must of necessity be varyed; their latitude being nothing else, but their distance from the Ecliptick where ever it be.

I shall now in the last place come to the Demonstration of the other Planets Motions; in which I shall demonstrate 1. Their latitude. 2. The variation of it. and 3. the Motion of their Nodi.

Suppose one of the other planets to move farther or lesse way towards the North, and then towards the South again, than the centers of A and D do move: it will by the motion compounded of that from West to East, and the other from North to South &c. describe a different line from that which the earth doth: which line| being on one side of the Ecliptick in the North, [12] and on the other in the South, must necessarily crosse it in 2 places. Now because this line is different from that in the ecliptick, it will therefore in the first place, bend from it and consequently have latitude.

Secondly, if the motion be at one time farther to the North and South than at the other, the latitude will alter.

And thirdly if the motion from north to south, and back againe, be not finished in the same time, that the planet moves the compleat circle by the common motion of the planets, the intersection of the line that the planet describes, and the Ecliptick, will alter, and the Nodi seem to move. For Example, if the planet in its motion towards the North, crosseth the Ecliptick in the first degree of Leo, and in the next returne towards the North, in the $26.^{gr}$ of Cancer, because it hath not finished one circle from West to East; or in the $6.^{gr}$ of Leo, because it hath gone more than a circle; In the first case the Nodi will seem to have moved 5 degrees in Antecedentia; in the second, as much in consequentia. I shall conclude this with the repetition of what I sayd before: that if the apparent place of any of the planets, be different from what it should be; it may possibly be salved by this, that its compound motion is not a compleat circle.

I do not know any thing that may be objected against the motion that I ascribe to the planets, with the Vortex of the earth; nor against that which

157. HYRNE to WALLIS, 2/[12] April 1670

I ascribe to the terraqueous globe, but that the moon, whose motion doth agree so exactly with the tides, is not made to be concerned in the causing of them: to which I answer, that although I confesse that I cannot, yet possibly some others that are very able, if they knew my Hypothesis, may give such a reason for the motion of the globe of the earth, as to concern the moon in it, it being most certain, that in whatsoever Meridian the moon was to day, for example, when the globe of the earth began to returne from the North towards the South; whensoever the Moon returns to the Meridian again, the earth will again begin to move from the North towards the South; and whensoever the Moon comes to the opposite side of the same Meridian, the earth will begin its motion from the South towards the North. In the meane time there being so much to be sayd against any other Hypothesis, that I know off, where the moon is concerned in the causing of the tides, I can satisfye my selfe; that this exact agreement between the Moons motion and the tides, is not in vain, though the moon be only an Index, and not an efficient cause of the Seas motion, as Vossius hath[875] it, de Motu marium et Ventorum. Cap. 18.

Thus much of the diurnall and menstruall tides; the annuall follow: concerning which that question must be premised, whether there be such things| or no? If by annuall tides be meant such tides exceeding the ordinary spring tides, as do constantly happen at certain times of the yeare all over the world, as the diurnall and menstruall tides do at a certain time of the day and month, it is a vulgar errour to assert them; but if such tides be meant as do usually, though not allways happen at such times of the yeare, I shall deliver my opinion of them in these assertions. 1. That there are such tides. 2. That the winds are the cheife causes of these tides. 3. That the raine and dissolved snow encreaseth them. 4. That where the winds are inconstant, these tides are not constantly at the same time of the yeare; though where the winds are constant, they may be constant too. 5. That these Tides do usually, but not allways, happen to those that live by the German Ocean, and the outlet of it into the British Sea, about Allhollantide and Candlemas. 6. That these tides may happen anywhere else, where there hath a strong wind blown for some considerable time, about the time of the Spring tides, from some large Sea towards the land. 7. That if there hath great store of raine happened some few days before, the tides will be much greater than they would have been without the raine.

7 North (*1*) to (*2*) towards
34 hath |some *del.*| great

---

[875]hath it: i.e. VOSSIUS, *De motu marium*, 79.

157. HYRNE to WALLIS, 2/[12] April 1670

As for my first assertion, it being no more than what every one grants, I shall not stand to prove it but proceed to my second, which I have knowne for a certaine truthe ever since the yeare of 1653: for having had occasion at that time, to be at the sea side at Warham,[876] about 8 miles distant from Yarmouth in Norfolk to the Westward, at least twice in a week all summer long, and many times the following winter, I did constantly observe that when the wind had been Northwest, at whatsoever time of the moon it was, the tides were mightily encreased, and I found that the countrey people, who suffered much by the breaking in of the Sea, never thought themselves in danger either at Allhollantide or Candlemas, or any time of the winter unlesse the wind had also been North-west. And I beleive, that your acquaintance at Rumney| marsh will tell you the same of themselves, [14 if they have observed the wind. The reason that these people had to be afraid of the sea at a spring tide after the Northwest wind, is very evident, though I beleive they did not at all know it; For if one doth but cast his eye on the map of Europe, he will find that from the German Ocean to the Northwest, there lyeth a great deale of Sea, for above 20 degrees without the interposition of any land: upon all which sea, when the wind bloweth from the Northwest, it hath power to drive it before it, and so to cause the waters in the German Ocean to swell hugely, especially towards the Southeast end of it, or between England and Holland, and at the outlet of it into the British Sea, where Rumney marsh lyeth; which getting out into the narrow passage between England and France, must of necessity cause the water to rise there, more than ordinary: And this rising of the water is also encreased by the Ness points reaching farther than the rest of the land, and bending somewhat towards the east, which must needs cast the water, that would else passe by, upon Rumney and the rest of the Marsh. And that the Northwest winds bring in the sea to the Inhabitants about the German Ocean, was not unknowne to Vossius, who in his 15 chap. saith,[877] that if the waves that beat upon Holland were not broaken by their banks and beds of Sand, there is no doubt but that they would be higher than the very hills; when there doth but a gentle Northwest wind blow. But what I observed by the Sea side in Norfolk, I have found to be true by above a dozen yeares experience in

---

11 I *add.*
11 that *add.*
13 that *add.*

---

[876]Warham: the village of Warham in north Norfolk is situated about three miles inland from the North Sea coastline.

[877]saith: i.e. VOSSIUS, *De motu marium*, 68.

157. Hyrne to Wallis, 2/[12] April 1670

the Thames, where a great tide doth not happen, without a Northwest wind going before. I have been further confirmed in my opinion of the winds power over the water to drive it before it, and to raise it, by what I have seene in Lough Neagh[878] in the North of Ireland. lat. 56.^{gr} long. 12.^{gr}30′. where I have seen the water 4 or 5 foot perpendicular higher at one time than another, only by the winds driving it to the shoar, which the inhabitants told mee allways happpened. And you may be satisfyed concerning the winds power to drive the water before it, by what Livy saith[879] dec. 3. lib. 6. cap. 45. about the taking[880] of Carthago nova, now Cartegena, by Scipio. For a stiffe north wind blowing at the tides going out, it drove so much water out of the lake on the west side of the towne that Scipio led his soldiers through it, some up to the Navell, others not above| the knees. Though I for my part beleive there is no other tide at Cartagena, but what the wind causeth, yet supposing there to be, every body must grant, that this North wind blew the waters out more than ordinary. For who can think that the subtill Carthagians would leave a place of such importance, as where they kept the Pledge of all Spain, their Treasury, Magazin, granary, and store house of all things, a brave haven, and the only harbour between the Piranaean mountains[881] and Cadiz, and which they had fortifyed on one side with walls too high for most of the Roman scaling ladders, and whose height made soldiers giddy, and fall downe from off their ladders; who, I say, can think, that the subtill Carthaginians would leave such a place no better fortifyed than with a lake, which twice in a day, or twice in a month was no deeper than to take one up to the knees or navell.

That the Annuall Tides are encreased by great quantityes of raine, no body can doubt, who doth consider, besides what falls in to the Sea, what a great deale runs into it, out of the rivers; which makes the Sea, I am sure, at our shoars, higher in winter than summer. And I could never yet observe, that there was an extraordinary tide without great rains preceding. And this is one reason, why the Easterly winds in March, though they drive the sea full upon the east side of England, never cause any great tide, the other, and that the principall one being, because these winds do not blow from so large a sea as the North west winds do.

2 confirmed (*1*) of the winds power in my opinion (*2*) in ... over

---

[878]Lough Neagh: the largest lake in the British Isles, about twenty miles west of Belfast.
[879]saith: i.e. LIVY, *Roman History* III, 6, §45.
[880]taking: Scipio Africanus (235–183 BC), Roman consul, destroyed the combined armies of the Carthaginians and Numidians in 203 BC.
[881]Piranaean mountains: i.e. the Pyrenees.

157. HYRNE to WALLIS, 2/[12] April 1670

That these tides are not constant, where the winds are inconstant, appeareth first, because we have not such huge tides every yeare about the same time, as experience witnesseth, and secondly because they sometimes happen about other times of the yeare, though never in summer. For Bodinus, as Fromondus quotes[882]| him, saith that in the yeare 1624 about [1⟨ twelfetide the Sea breaking the banks, took away from the Hollander above a hundred thousand acres of ground. For it drowned about 70 villages. And upon inquiry I have been informed by ancient watermen upon the Thames, that the bigest tide that they ever saw there in their lives, was upon new years day, and that the next to it, was much upon the same time of the yeare. But where the winds are constant, I beleive that that these yearely tides are constant too; for posita causa, ponitur effectus.

The reason why these tides do commonly happen to them that live by the German Ocean about Allhollandtide and Candlemas, is because the winds are commonly North west at those times of the yeare, after a great deale of raine. For all summer long the wind is commonly west, or Southwest with us; in the winter North and north east, in March east, and at the forementioned times northwest. And that we have commonly great store of raine after harvest about the end of October, every body knows, as also that usually wee have raine towards the end of January, which melting the snow, and opening the brooks and rivers causeth great flouds about that time: that such great tides do not happen allways at those times is before proved.

As for my sixth and seventh assertions, allthough I cannot prove the certainty of them by any particular instances, yet knowing to be true, that the wind coming from the sea, with a spring tide, makes a very great tide with the inhabitants of England and Holland, & that the tide is much the greater, if there hath great store of raine fallen before; I have not the least reason to doubte, but that it is so elsewhere.

Thus have I shewn you what hath given mee so much satisfaction about the motion of the sea or tides, and by the by,| what hath satisfyed me [17 more than anything else about the motion of the Planets. If my Hypothesis will salve all the Phaenomena, that any other will do, and some that no other pretends to do, without falling foule upon any absurdity: I have some reason for it. But whether it salves the Phaenomena, or escapes absurdityes; I submit to your judgment: and desiring that you would be pleased to let mee know your opinion both concerning the motion, that I ascribe to the planets, and my Hypothesis concerning the diurnall and menstruall tides.

---

13 commonly *add.*

[882]quotes: i.e. FROIDMONT, *Meteorologicorum libri sex*, 253.

For I am as sure, that what I say concerning the Annuall tides is true, as I am, that it is the light of the sun that causeth the day.

From this Hypothesis of the motion of the terraqueous globe, if it proves true, Seamen (as Monsieur d'Arcons observed[883]) may infallibly conclude in what longitude they are, being in the open sea, if they do but know the exact time of the day, and in what meridian the moon is, from one single observation, when the earth begins its motion to the North or South, because there cannot be any mistake, but of 180 degrees, which can never happen.

It may also be a great furtherance to the discovery of the North-East or North-West passages. For unlesse the North shore, whether to the East or West, be exactly parallel to the Æquator, men may proceed upon allmost demonstrable grounds in the searching out of those passages.

I pray pardon this long letter, as being writ not with an intent to trouble, but to satisfye you, by him, who is

Parsons green
April 2. 1670.

Honoured Sir
Your most humble servant
Harry Hyrne

158.
WALLIS to HENRY HYRNE
Oxford, 4/[14] April 1670

Transmission:

w^1 Copy of missing letter sent (with corrections entered by Oldenburg): LONDON *Royal Society* Early Letters W1, No. 108, 4 pp. (our source). At top of p. 1 in Oldenburg's hand: 'Read April: 14: 70. Ent⟨ere⟩d LB. 3. 350.' and 'Dr Wallis reply to Mr Hyrne's hypothesis of Tydes.'

w^2 Copy of w^1: LONDON *Royal Society* Letter Book Original 3, pp. 350–3.

w^3 Copy of w^2: LONDON *Royal Society* Letter Book Copy 3, pp. 433–7.

Reply to: HYRNE–WALLIS 2/[12].IV.1670.

This letter, the original of which was sent to Hyrne, was produced and read at the meeting of the Royal Society on 14 April 1670. See BIRCH, *History of the Royal Society* II, 433.

---

[883] observed: i.e. D'ARÇONS, *Le secret decouvert*, 213–4.

158. WALLIS to HYRNE, 4/[14] April 1670

Sir,

Yours[884] of the second of April I have received. Wherin you give your selfe the trouble of a large account of your Hypothesis, both as to the tides and as to the planets motions. Which though they be not wholly new to mee (for I have met with some of the like notions before) yet I must confesse I have no fansy to either of them. And though I do not think my selfe concerned to dispute against them, I cannot think either of them to be the true Hypothesis of Nature.

As what you say of rains and winds influencing the greatnesse of the tides: I do not question it. And particularly that of north west winds making high tides at the Kentish shore. Not only upon the account you give (of driving the Northerne Sea into the German Ocean) but (which is here considerable): because it doth at the same time, drive in the westerne sea by the Sleeve (as they call it,) or British Sea (between us and France;) for it is not the Northerne tide (which comes no farther than the Dogger sands, or there abouts, where the 2 tides meet), but the Western tide (by the British or narrow seas) which coming round by the Nesse point,[885] makes the tides on the coasts of Kent and Essex, and perhaps higher: (That of the Northerne Tide concurring no otherwise to it, than by hindring the other freely to passe on Northward.) And therefore the Nesse point standing far into the sea, (though it may upon another account) doth not upon the account you mention (of hindering the North-Tide, which comes not near it, from passing by) cast the waters on Rumny marsh.

But this of the raines and winds I took to be a thing so well knowne and agreed upon, as that I did not think it necessary to mention it in my Hypothesis, (because casuall not Periodicall,) as that neither of the Moons Apogaeum and Perigaeum, and the obliquity of its orb (which yet I take to bee very considerable,) because, though Periodical, it answers to none of those 3 (Diurnall, Menstruall, Annuall) of which I was giving an account (but doth rather disturb especially| that of the Annuall, what ever it bee, and [2] make it lesse observeable.) All which I did, in the beginning of my Hypothesis, exclude (among the casualls or extraordinaryes) from the present consideration then in hand.

But if there be not (as you would intimate) any Annuall period, but such only as depends on the uncertaintyes of wind and weather; then I am

---

22 comes not under, from *corr. Oldenburg.*

---

[884]Yours: i.e. HYRNE–WALLIS 2/[12].IV.1670.

[885]Nesse point: i.e. Ness Point, the most easterly point in the British Isles, in Lowestoft, Suffolk.

158. WALLIS to HYRNE, 4/[14] April 1670

easyed (as to my Hypothesis) of so much of the trouble (being not obliged to give any account of such;) for it is so fitted, as that, if there be any (as many have thought) it can give an account of it from the inequalityes arising from the obliquity of the Zodiack, and the excentricity of the Earths orb) but if there be none, these considerations may be spared, and the worke done without them. For then, these inequalityes will be found to be so little, as not to influence the tides, or so perplexed (by other accidents) as not to bee easily observed.

On the other hand, that (on the same side of the Æquator) all tides should be at the same time all the world over, seems to mee very strange, and contrary at least to the received tradition, that they are at or neare the Moons coming to the Meridian, which is not to all parts of the world at the same time: And I am not so much a seaman, as upon my owne knowledge to contradict that received Tradition.

Neither am I concerned that all the Annuall high tides should all the world over happen about Allhollandtide and Candlemas, though (I think) on the coast of Kent they do. For this happening not upon one single cause, but upon a complication of two (of which neither singly would cast it on those times) the position of severall coasts may so much favour, some the one, and some the other cause, as to make them fall in some nearer, in some farther off from the Æquinoxes, or the suns Perigaeum. And particularly about the Severne, where St David's streame and Michaelmas streame, are as famous, as are, in Rumney marsh, Candlemas streame, and Allholland streame. And no doubt, but that on other coasts, other periods may be observed. I only made choice of that of Rumny Marsh (as best known to mee) [3] to| give account, on that occasion, of the complication of those two causes, which may, by thwarting one another cast it upon some time intermediate between what either of them singly would have determined. But this (as I said before) is the least considerable part of my Hypothesis, and (if no such period be) may be spared.

And as to your observation, that the high spring-tides are never, but upon the account of the winds from such a coast, and never but after great rains, and never in summer: it is far otherwise.

For I have been very lately minded (upon another occasion) amongst other things, that August 2. 1657. was a very high spring tide, observed (by this relater who then lived there) at Feversham[886] in Kent, the wind at SE. And again Aug. 22. 1658. (at the same place) the wind blowing stiffely at

---

[886] Feversham: i.e. Faversham, market town in Kent.

158. WALLIS to HYRNE, 4/[14] April 1670

South; the wind not at all favouring the tide at other times. For (by your owne rule) if it's a Northwest-wind, that, on those coasts, makes high-tide, the South and SE both hindering it;) and so the seamen there do all agree. And at London, May 24. 1663, a very high tide. Yet all these were in summer; and I do not know, that either wind or raine did favour any of them. And march 13. now last past, was on the coast of Kent (as I am informed from thence) the highest tide they have had a great while (and so it was at London at the same time) when yet there had been neither wind nor raine for a great while before: And it is well knowne, how very little raine had fallen for 6 or 7 months before. All which are attributed, neither to wind or raine, but to the moons being in perigaeo at all those times. And in June 1667 it may be remembred by an accident too remarkable, that when the Dutch fired[887] our ships at Chattam,[888] they did upon the advantage of a high spring-tide (which there happened presently after the solstice) carry off the Charles[889] (at one tide) to sea, which was then thought to have been impossible, and at ordinary spring tides would have required 3 spring-tides to do it. Yet was that in summer & in a dry season, & I do not remember that the wind was either NW, or near that point. But I have never made it my businesse (nor have I had opportunity) to be a diligent| observer of those accidents (else I doubt not but many more might be produced:) only these were by chance at hand. I add no more but that I am

Sir

Oxford
April 4. 1670

your freind and servant
John Wallis.

For Mr Henry Hyrne
at Parsons green

---

1 at all *add. Oldenburg*
2 it's *add. Oldenburg*
2 that, | is *del. Oldenburg* | on those
3 there *add. Oldenburg*
5 any of *add. Oldenburg*
8 had neither been *corr. Oldenburg*

---

[887] fired: i.e. De Ruyter's surprise raid on English battleships laid up in the Medway during the Second Anglo-Dutch war.
[888] Chattam: i.e. Chatham Dockyard.
[889] Charles: i.e. HMS Royal Charles, the pride of the English fleet, which was taken to Hellevoetsluis in the United Provinces and later auctioned for scrap.

159.
HENRY OLDENBURG to WALLIS
9/[19] April 1670

Transmission:

Manuscript missing.

Existence and date: mentioned in Oldenburg's endorsement on WALLIS–OLDENBURG 29.III/[8.IV].1670.

Reply to: WALLIS–OLDENBURG 29.III/[8.IV].1670.

160.
WALLIS to HENRY OLDENBURG
Oxford, 9/[19] April] 1670

Transmission:

W Letter sent LONDON *Royal Society* Early Letters W1, No. 109, 2 pp. (our source). On p. 2 below address in Oldenburg's hand: 'Rec. Apr. 11. 70. Answ. 16. 70. that I will return his and Hyrnes papers next week.' Postmark on p. 2: 'AP/11'.—printed: OLDENBURG, *Correspondence* VI, 615–16.

Answered by: OLDENBURG–WALLIS 16/[26].IV.1670.

Oxford Apr. 9.1670.

Sir,

Since my last[890] to you, I have received another pacquet[891] from Mr Hyrn; which, with my Answere[892] to him, I mean to send you on Munday[893] morning by a friend[894] in the flying Coach from hence: & shall put it up with some things that I send to Mr. Faithorn;[895] where, I suppose, on Tuesday or Wednesday you may receive it. His Hypothesis supposeth, a motion of

---

5 a friend in *add.*

[890]last: i.e. WALLIS–OLDENBURG 29.III/[8.IV].1670.
[891]pacquet: i.e. containing HYRNE–WALLIS 2/[12].IV.1670.
[892]Answere: i.e. WALLIS–HYRNE 4/[14].IV.1670.
[893]Munday: i.e. on 11/[21].IV.1670.
[894]friend: not identified.
[895]Faithorn: probably William Fairthorne, the elder (1616–91), engraver and portrait painter in London, *ODNB*.

160. WALLIS to OLDENBURG, 9/[19] April 1670

the Earth from north to south every six hours (with some minutes) & back again in as long a time. And that (on the same side of the Æquator) the tydes are at the same time all the world over: without any reference to the Moones being at or near the Meridian. (Which is crosse to the generally received opinion of the Phaenomenon itself.) And that the Spring-tides at Change & Ful,[896] are no otherwise depending on the moons motion, then barely by a Synchronism. How ever; as it is I send it you; (which, when you have done with it, you will return mee again.) I return you allso, with it, the Latine version[897] of my hypothesis; which I borrowed from you.

In my letter[898] about Mr Childries objections: I have (I find) mis-dated one particular: That of the Narration[899] of the Tydes at Chepstow Bridge in the Society: which was about the end of the summer 1667, not 1666. Which you may mend.

You may remember, that, some years since; you sent (for me) a letter[900] to Leotaudus. Have you since heard any of it; or that it was delivered?

I am yours &c
J. Wallis.

For Mr Henry Oldenburg,
in the Palmal, near
St James's
London.

[2]

1 hours (1) & he (2) (with some

---

[896]Change & Ful: i.e. the passage from one 'moon' to another.

[897]version: cf. WALLIS–OLDENBURG 24.III/[3.IV].1669/70.

[898]letter: i.e. WALLIS–OLDENBURG 19/[29].III.1669/70.

[899]Narration: cf. WALLIS–OLDENBURG 7/[17].III.1667/8 and WALLIS–OLDENBURG 19/[29].III.1669/70. Wallis was most likely mistaken in wishing to have the date amended. The account of tides at Chepstow Bridge probably came from Henry Powle, FRS, in late Summer or Autumn 1666. He had announced his intention to visit Chepstow for this purpose in a letter to Oldenburg of late September 1666. See OLDENBURG, *Correspondence* III, 235–6.

[900]letter: i.e. WALLIS–LÉOTAUD 17/[27].II.1667/8 (WALLIS, *Correspondence* II, 412–18). Justel via whom Oldenburg had apparently forwarded the letter for Wallis had at the time informed Oldenburg that he intended to convey it to Léotaud. See JUSTEL–OLDENBURG [25.IV]/5.V. 1668; OLDENBURG, *Correspondence* IV, 333–4.

161.
JOSHUA CHILDREY to HENRY OLDENBURG
Upwey, 12 and 15/[22 and 25] April 1670

Transmission:

C Letter sent: LONDON *Royal Society* Early Letters C1, No. 10, 4 pp. (p. 3 blank) (our source). At top of p. 1 in Oldenburg's hand: 'Enter'd LB. 4. 04.' At top of p. 4, again in Oldenburg's hand: '(*1*) Mr Childreys letter (*2*) An Extract of Mr Childrey's letter to Mr Oldenburg, about the intercept of the Moons Perigaeum and Apogaeum at the Tydes.' On p. 4 also endorsement in Oldenburg's hand: 'Rec. April 18. 70.' and postmark: 'AP/18'.— printed: OLDENBURG, *Correspondence* VI, 625–7.
*c*¹ Copy of letter sent: LONDON *Royal Society* Letter Book Original 4, pp. 4–5.
*c*² Copy of *c*¹: LONDON *Royal Society* Letter Book Copy 4, pp. 4–6.
Reply to: OLDENBURG–CHILDREY 9/[19].IV.1670 and WALLIS–OLDENBURG 19/[29].III. 1669/70.

Sir,

Though I have much businesse lying upon me at present, yet I was willing (because you desired it) to send you word by the next Post, that I have received yours[901] of Aprill 9th with all the inclosed, for which & all your paines & kindnesse I returne you abundance of thankes, & shall endeavour to deserve what you have obliged me with, by hastening (as fast as I can) what you desire. And am sorry I have not leisure to say more in reply to Dr Wallis his answer,[902] then this, That the Annuall periods of the Tides will not yet downe with me, for a reason which weighes much with me upon second thoughts, & which I shall hereafter give you.[903] All I shall adde touching that affaire now, is, that the observations at Weymouth this March do still wonderfully correspond with the Perigaeosis & Apogaeosis of the ☽, the Spring tides being exceeding high after the Change, & low & very slacke after the Full. I have taken order with my friend there, who is a very able ancient Seaman, & as fit a person for observation as any Seaman there, to

---

12 wonderfully |(absit dicto &c.) *del.*| correspond

---
[901]yours: i.e. OLDENBURG–CHILDREY 9/[19].IV.1670. This letter enclosed Wallis's reply to Childrey's animadversions on his hypothesis of tides: WALLIS–OLDENBURG 19/[29].III.1669/70. Oldenburg had at first waited to send these through the hands of Seth Ward. See the memorandum of OLDENBURG–CHILDREY 9/[19].IV.1670; OLDENBURG, *Correspondence* VI, 621–2.
[902]answer: i.e. WALLIS–OLDENBURG 19/[29].III.1669/70.
[903]you: there appears to have been no further communication from Childrey before his death later in the year.

161. CHILDREY to OLDENBURG, 12 and 15/[22 and 25] April 1670

observe this whole yeare, whether my predictions doe not hitt every moneth, & to set downe the point of the wind, & the degree of the Tides rising. And, particularly yesterday I desired him to observe, whether this Aprill & in May, (because the ☽ comes to her Perigaeum some dayes before the Change:) the height of the Spring tides happens not sooner then ordinary, & possibly in May upon the very day of the Change. When he hath observed the yeare round, I shall then either send you a copy of the observations, or the very Originall. He seemes to be highly pleased with my fancy (if he doe not flatter me) & hath communicated my paper of predictions (as he tells me) to some of the Pilots of the Towne, because he saith (if it hit but as right for the future, as it hath done yet) it wil be of good use to them for bringing in, & carrying out of Vessells from that Harbour. And indeed if it doe but succeed for this yeare (as I have very great hope it will) it will cleare the onely difference[904] between the Dr & me.

Sir I give you free leave to insert my prediction of the shifting of the Tides in the Ph. transactions for the reason you give; but I have not time yet to give an account of my Hypothesis, whereon I grounded it. And for my animadversions of Dr W.'s Hypothesis, you have my leave to publish[905] them, upon condition you have this leave, whose the Hypothesis is; For, I have too great a value for his worth to doe anything that relates to him, without his consent. Sir I beseech you pardon the hast of

Upway Aprill 12th 1670.

Your very much obliged Servant
J. Childrey.| [2]

Sir,

I gave you an account in my last,[906] what ill successe I have had with the Copies of the Enquiries I dispersed about the County. I feare the multitude of the Questions to be answered is the reason why they are so cold in making returnes.

---

22 of |Sir *del.*| Your very

---

[904] difference: cf. WALLIS–OLDENBURG 19/[29].III.1669/70.

[905] publish: Childrey's animadverions were printed in *Philosophical Transactions* No. 64 (10 October 1670), 2061–8. Wallis's reply appeared in the same issue (on pages 2068–74).

[906] last: i.e. CHILDREY–OLDENBURG 29.III/[8.IV].1670; OLDENBURG, *Correspondence* VI, 603–4.

161. CHILDREY to OLDENBURG, 12 and 15/[22 and 25] April 1670

Sir,

I had not time to send this time enough to reach the Posts going by our Parish on Wednesday[907] last. But to recompense that failer, I have sent you the copy of the Observations of the tides taken at Weymouth by my friend there, in his owne words as followeth.

>   1669. Feb. 12 the tide very high, the wind Southeast.
>   1669. Feb. 27. 3 dayes after the full, low tide, the wind Easterly.
>   1669. March 11. Extraordinary high tide, the wind calme.
>   March 12. 13. 14. Very high tides, little wind, Southerly.
>   1670. March 26. to the 29th of March all the spring very low, the wind southerly.
>   1670. Aprill 9th & 10th. the tide very high, the wind Southwest.
>   Aprill 11. & 12. the tides lower, & the wind Southerly.
>
>   Memorandum, that March 11. 1669, being Friday morning, the tide was so high at Melcombe & Weymouth, that it came over the Key into the Streets.

This account I had from him, on Wednesday last in the afternoone.

Upway Aprill 15. 70.

[3] For my honoured friend Mr Henry
Oldenburg living in the Pal-mall
of St James's fields these present
Westminster.
Post p⟨ai⟩d 3^d.

---

[907] Wednesday: i.e. on 13 April 1670 (old style).

162.
ROBERT WOOD to WALLIS
15/[25] April 1670

Transmission:

Manuscript missing.

Existence and date: mentioned in and answered by WALLIS–WOOD 16/[26].V.1670.

As emerges from Wallis's reply, this letter was forwarded from Oxford to Kent, where Wallis evidently spent much of the time in April and May 1670. This circumstance also explains the delay in Wallis's reply.

163.
HENRY OLDENBURG to WALLIS
16/[26] April 1670

Transmission:

Manuscript missing.

Existence and date: mentioned in Oldenburg's endorsement to WALLIS–OLDENBURG 9/[19].IV.1670.
Reply to: WALLIS–OLDENBURG 9/[19].IV.1670.

From Oldenburg's endorsement to WALLIS–OLDENBURG 9/[19].IV.1670, it appears that he promised to return to Wallis with this letter the latest exchange in Wallis's epistolary debate with Hyrne: HYRNE–WALLIS 2/[12].IV.1670, and WALLIS–HYRNE 4/[14].IV.1670. These two letters had probably been sent to London on Monday, 11 April 1670 (old style). See WALLIS–OLDENBURG 9/[19].IV.1670.

164.
WALLIS to ROBERT WOOD
Hythe, 16/[26] May 1670

Transmission:

W Copy of (missing) letter sent: LONDON *Royal Society* Boyle Letters 7, No. 26, 4 pp. (on pp. 1–3). The copy is contained in WALLIS–WOOD 4/[14].VIII.1670).

Reply to: WOOD–WALLIS 15/[25].IV.1670.

164. WALLIS to WOOD, 16/[26] May 1670

As Wallis explains in WALLIS–WOOD 4/[14].VIII.1670, his son, John Wallis jr, was with him in Kent when he wrote the present letter; it was he who produced a copy of it. Recognizing that the original letter had failed to reach its destination, Wallis copied the text from that copy into WALLIS–WOOD 4/[14].VIII.1670.

Hythe in Kent, May. 16. 1670.

Sir,

(To begin where you end), I am very sorry for your losse,[908] of which I cannot but be very sensibly concerned, & pray God (who onely can) so to supply it to you, with comfort, as to him shal seem best. My last[909] to you was returned through the same hands by which yours came to mee, But, it seems, in passing from hand to hand, it mett with more rubs, than (probably) it would have done had I at first committed it to the Post. Yours[910] of April. 15 in answere to it, came to Oxford since I came thence (where I have not since been) & was from thence sent back after mee. Which is the reason why it came so late to my hands, & why you have no sooner an account of it. What I sayd of Albert Girard[911] (who hath severall smal pieces amongst those[912] of Stevinus,[913] at lest in the French Edition) was onely for your information, not supposing you had seen him or taken any thing out of him. And on the same account I thought it not amiss to acquaint you, that there is one Mengolus[914] (if I mistake not the name) who hath written a treatise[915] about divers such inquiries as that of your present letter. I think the title of it is Quadraturae Arithmeticae, but the book being not at hand I can (upon memory) say but little of it, having but slightly seen it, not read it over.

Your demonstration (which is short & clear) did a little surprise mee. But, upon consideration, I find the case different from those of my Arithm: Infin. For you here suppose the termes 1, 2, 3, &c, from the first term given to proceed onward to a term infinitely great: mine suppose them from a term infinitely small to proceed (infinitely, but) ending in a term given;

---

[908] losse: not identified. Wood had evidently reported a death in his immediate family.
[909] last: probably WALLIS–WOOD 10/[20].III.1670.
[910] Yours: i.e. WOOD–WALLIS 15/[25].IV.1670.
[911] Girard: Albert Girard (1595–1632), French mathematician and musician.
[912] those: i.e. STEVIN, *L'arithmetique de Simon Stevin de Bruges, reveuë, corrigee & augmentee de plusieurs traictez et annotations par Albert Girard Samielois mathematicien*, Leiden 1625; STEVIN, *Les œuvres mathématiques de Simon Stevin de Bruges [...]. Le tout reveü, corrigé, & augmenté par Albert Girard Samielois, mathematicien*, Leiden 1634.
[913] Stevinus: Simon Stevin (1548–1620), Flemish mathematician.
[914] Mengolus: Pietro Mengoli (1625–86), Italian priest and mathematician.
[915] treatise: i.e. MENGOLI, *Novae quadraturae arithmeticae, seu de additione fractionum*, Bologna 1650.

164. WALLIS to WOOD, 16/[26] May 1670

or (which is equivalent) from the last term given, infinitely decreasing (in such a proportion) to one infinitely smal. And though all your Aggregates if compleated ($R + P$.) would be a quantity infinitely great: yet, because the defalcations ($= R$) would be so too; it comes to passe, that the remainder ($= P$) comes to bee finite. But I have not time at present to inlarge so much on that subject as to make it cleare to you; possibly hereafter I may. And when I come home to my papers shal be able to furnish you with more instances of the approaches of the Diameter & Perimeter. As allso for the first & third of two mean proportionalls between 1 & 2 but at present can onely subscribe myselfe, &c.| [2]

By what means this letter came to miscarry, I know not; being sent by the Dover Post to London, as was that of two days after; & both with the same direction. And had this come before or with the other, that would not have seemed so abrupt.

In Mengolus, to whom I there rifer, there is (as I find by a note by mee, for the book I have not) such a proposition.

If an infinite rank of Fractions have an Unite for their common Numerator, & the squares of all numbers (Unite being included) + their roots for the Denominators; as $\frac{1}{2} \frac{1}{6} \frac{1}{12} \frac{1}{20} \frac{1}{30}$ &c, the summe of such a rank is equal to an Unite.

And if to each square, in stead of its root, be added the double of its root, as $\frac{1}{3} \frac{1}{8} \frac{1}{15} \frac{1}{24} \frac{1}{35}$ &c. the summe of such a rank is $\frac{3}{4}$ of an Unite.

And if to each square bee added the triple of its root, the summe of the rank is $\frac{11}{18}$ of an Unite.

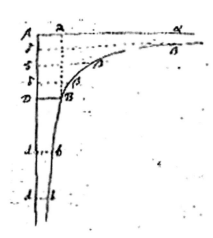

13 come (*1*) first (*2*) before

164. WALLIS to WOOD, 16/[26] May 1670

Next that I may a little explain the difference between your ranks & mine: In my reciprocal series, as of Numbers, $\frac{1}{0}, \frac{1}{1}, \frac{1}{2}, \frac{1}{3}$ &c: (infinitely,) of which the last is DB.

Of Squares, $\frac{1}{0}, \frac{1}{1}, \frac{1}{4}, \frac{1}{9}$, &c.
Of Roots, $\frac{1}{\sqrt{0}}, \frac{1}{\sqrt{1}}, \frac{1}{\sqrt{2}}, \frac{1}{\sqrt{36}}$, &c. $\Big\}$ of which still the last is DB.

& the like in other Powers & roots.

The particulars are expressed by $\delta B, \delta B$, &c.

And the proportion of the figure $ADB\beta\beta\alpha$, to the rectangle $ADBa$; is in the first case Infinite; in the second, more then infinite; in the third, lesse then Infinite; that is, finite. But if $ADBBd$ be infinite; then the somme (bounded that way but) infinitely continued the other way, $AaBbbd$; will be infinite allso. If that be more then Infinite; this lesse then infinite. If that lesse; this more. By Arithm. Infin[916]. prop. 102, 103, 104, 105.

This made me wonder at first, that your Aggregate of series $1+\frac{1}{4}+\frac{1}{9}+\frac{1}{16}$ &c: plus $1+\frac{1}{8}+\frac{1}{27}+\frac{1}{64}$ &c. plus &c being Reciprocalls of Squares, Cubes, &c; of which (DB the last, being supposed to be given,) each series must be infinitely great: should yet altogether be equall but to 1.

But, with you, the case is other wise: for, beginning at $\frac{1}{1}$ (suppose DB); you proceed to $\frac{1}{4}, \frac{1}{9}, \frac{1}{16}$, &c, (still decreasing;) as are the lines db db &c (not $\delta\beta, \delta\beta$, as in my case;) You taking the last of the squares (& so of cubes &c) to be determined; & d the greatest. So that your series are the continuation of mine the other way. So that in my way, the Reciprocalls of Powers be more then infinite; but of Rootes,

lesse then infinite: In your way, that of Powers (which (which is aequivalent to mine of Roots,) is to be less than infinite; but, of Roots are more than Infinites. Which are very well consistent.

---

3 the (*1*) greatest (*2*) last
4 of which still the (*1*) greatest (*2*) last
10 if (*1*) the infinite (*2*) $ADBBd$
11 infinitely *add.*
14 Aggregate of *add.*
17 should (*1*) be (*2*) yet
22 the | series of *del.*| Reciprocalls
24 (which ... Roots,) *add.*

---

[916] Arith. Infin.: i.e. WALLIS, *Arithmeticia infinitorum*, 76–9; *Opera mathematica* I, 408–10.

164. WALLIS to WOOD, 16/[26] May 1670

Because your figure for Powers continued one way; is the same with my figure for Rootes continued the other way. For with mee, if $\alpha ADB\beta\beta$ be a Reciprocal of Squares; then is $dAaBbb$ the Reciprocal of Square Roots: instead of which you (beginning, not at $\frac{1}{0}, \frac{1}{1}$, but at $\frac{1}{4}, \frac{1}{9}$ &c) take $dDBbb$.

I have now (after I have thanked you for your last, & the particulars in it:) But one thing remaining upon my hands to satisfye the defects of your letters. Which is; to give you some more particulars of the approaches for the Diameter & Perimeter of a Circle. My generall method of doing it; you had in my last: You have here the numbers themselves, in the next page. Which done; I have but to kisse your hands, & subscribe myself,

Your affectionate Friend & servant
John Wallis.| [3]

If the Diameter be 1.   the Perimeter is
   more   8.
   then   3.14159, 26535, 89793, 23846, 26433, 83279, 5028
   lesse   9.
                            Suppose,   50288, 5
Therefore
If the Perimeter be 1.   the Diameter is
   proxime, 0.31830, 98861, 83790, 67153, 77675, 26745, 02872, 4

The Approximation[917] is here continued in both Inquisitions (that is, where the Perimeter first, then the Diameter, is taken to little, but continually approaching to the truth) till the last numbers fall within the first proposed limits; (that is, more than 8, but lesse then 9, & nearer to $8\frac{1}{2}$ in

---

1 for Powers *add.*
7 it;) (*1*) which (*2*) But
9 generall *add.*
10 here (*1*) some examples (*2*) the numbers
14 Therefore *add.*
15 The (*1*) Inquisition (*2*) Approximation

---

[917] Approximation: a large part of this approximation is contained in chapter eleven of Wallis's later work *Treatise of Algebra* (Oxford 1685).

164. WALLIS to WOOD, 16/[26] May 1670

the last figure.) Which are Snellius[918] & Van-Culens[919] limits; & none have yet determined this proportion more exact.

I have only left some chasms, which are to be supplyed (if you please) by continuall addition of the continual Increment next preceding.

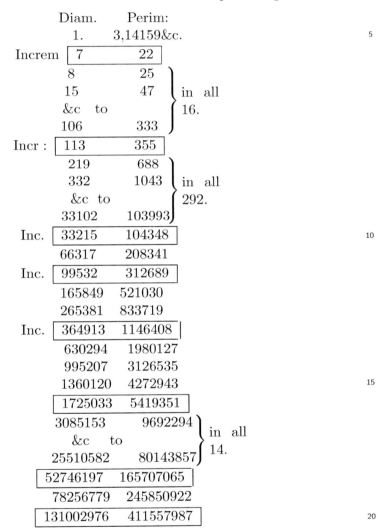

4 the (*1*) next (*2*) continual

---

[918]Snellius: i.e. Willebrord Snell (1580–1626), Dutch mathematician, former student of van Ceulen. His expansion of pi was published in *Cyclometricus de circuli dimensione secundum logistarum abacos* (Leiden 1621).

[919]Van-Ceulens: i.e. Ludolf van Ceulen (1540–1610), German mathematician, active in the Netherlands. He published his expansion of pi in *Van den Circkel* (Delft 1596).

164. WALLIS to WOOD, 16/[26] May 1670

209259755	657408909
340262731	1068966896
811528438	**2549491779**
1151791169	3618458675
1963319607	6167950454
4738167652	**14885392687**
6701487259	21053343141
567663097408	**1783366216531**
574364584667	1804419599672
1142027682075	3587785776203
1709690779483	**5371151992734**
2851718461558	8958937768937
44485467702853	**139755218526789**
47337186164411	148714156295726
91822653867264	288469374822515
136308121570117	428224593349304
1816491048114374	**5706674932067741**
1952799169684491	6134899525417045
9627687726852338	**30246273033735921**
115804868965336829	36381172559152966
21208174623389167	66627445592888887
136876735467187340	**430010946591069243**
158084910090576507	496638392183958130
&c	to
842468587426513207	2646693125139304345
979345322893700547	**3076704071730373588**
1821813910320213754	5723397196869677933.

	Perim.	Diam:
	1.	0.3483&c.
Increm :	**3**	**1**
	7	2
	&c. to	} in all 7.
	22	7
Increm :	**333**	**106**
	355	113
	103993	**33102**

374

164. WALLIS to WOOD, 16/[26] May 1670

104348	33215	
208341	66317	
312689	99532	
833719	265381	
1146408	364913	
4272943	1360120	
5419351	1725033	
80143857	25510582	
85563208	27235615	
165707065	52746197	
245850922	78256779	
411557987	131002976	
1068966896	340262731	
1480524883	471205707	
2549491779	811528438	
6167950454	1963319607	
8717442233	2774848045	
14885392687	4738167652	
21053343141	6701487259	
35938735828	11439654911	} in all 84.
&c to		
1783366216531	567663097408	
3587785776203	1142027682075	
5371151992734	1709690779483	
8958937768937	2851718461558	
14330089761671	4561409241041	} in all 15.
&c to		
13975548526789	44485467702853	
428224593349304	136308121570117	
567979811876093	180793589272970	} in all 13.
&c to		
5706674932067741	1816491048114374	
6134899525417045	1952799169684491	
11841574457484786	3769290217798865	} in all 4.
&c to		
30246273033735921	9627887726852338	
66627445592888887	21208174623389167	

164. Wallis to Wood, 16/[26] May 1670

$$\left.\begin{array}{ll} 96873718626624808 & 30835862350241505 \\ \&\text{c to} & \\ 430010946591069243 & 136876735467187340 \end{array}\right\} \text{in all 6.}$$

$$\boxed{\begin{array}{ll} 2646693125139304345 & 842468587426513207 \end{array}}$$

$$\overline{3076704071730373588 \quad 979345322893700547.}$$

If in either inquisition (for 'twil be same in both); if you take the continual Increments together with the numbers next before them; you will have the proportions alternately too big, & too little. As

$$\begin{array}{rl} 1. & 3\ + \\ 7. & 22\ - \\ 106. & 333\ + \\ 113. & 355\ - \\ 33102. & 103993\ + \\ 33215. & 104348\ - \quad \&\text{c.} \end{array}$$

which is a Series such as you ask: & if you can find in it, in your way, a regular approach; it will be a very noble discovery. For you may then continue, with ease, the approche for the same Geometrical Proportion, to what accurateness you please; without the troublesome processe of Von Culen or Snellius. Two methods of approch for this case you have in my Arith. Infin[920] prop. 191. viz.

1. Quadratum diametri ad Circulum;

$$\begin{cases} \text{ut } 9 \times 25 \times 49 \times 81 \times 121 \times 169\ \&\text{c} \\ \text{ad } 8 \times 24 \times 48 \times 80 \times 120 \times 168\ \&\text{c} \end{cases}$$

2. Item, Circ. ad quadratum diametri, ut
1 ad 1
$$\frac{1}{2}$$
$$\frac{9}{2}$$
$$\frac{25}{2}$$
$$+\frac{49}{2}$$
$$+\frac{81}{2}$$
$$+\frac{121}{2}$$
$$+\frac{169}{2}$$
$$+\&\text{c.}$$

But I would be glad to have a better than either.

---

5 (for 'twil be same in both) *add.*
15 120×1168 *corr. ed*
13-19 Two methods ... than either. *add.*

[920]Arith. Infin.: i.e. Wallis, *Arithmetica infinitorum*, 178–82; *Opera mathematica* I,

165.
WALLIS to ROBERT WOOD
18/[28] May 1670

Transmission:

Manuscript missing.

Existence and date: mentioned in WALLIS–WOOD 4/[14].VIII.1670.
Answered by: WOOD–WALLIS 23.VII/[2.VIII].1670.

166.
WALLIS to FRANCIS VERNON
early June? 1670

Transmission:

Manuscript missing.

Existence and date: mentioned in VERNON–OLDENBURG [9]/19.VII.1670; OLDENBURG, *Correspondence* VII, 60–2.
Enclosure to: OLDENBURG–VERNON 10/[20].VI.1670.

Oldenburg forwarded this letter to Vernon as an enclosure to his own letter of 10 June 1670 (old style). Since there is unlikely to have been any considerable time lapse between receipt and forwarding, Wallis's letter was probably written at the beginning of June, possibly at the end of May.

167.
WALLIS for HENRY OLDENBURG
11/[21] July 1670

Transmission:

W Draft note: LONDON *Royal Society* Early Letters W1, No. 110, 1 p. On reverse in Wallis's hand: 'To Mr Oldenburg. July 11. 1670. Oxford. Sir' (deleted).

The complete transmission and edition of the text is to be found in WALLIS, *Correspondence* II, 60–3.

467–76.

169. SLUSE to WALLIS, [15]/25 July 1670

Wallis sent Oldenburg this note as an appendix to his letter to Boyle of 14/[24] March 1661/2, which was published in *Philosophical Transactions* No. 61 (18 July 1670), pp. 1087–97.

168.
JOHN COLLINS to WALLIS
14/[24] July 1670

Transmission:

Manuscript missing.

Existence and date: mentioned in and answered by WALLIS–COLLINS 23.VII/[2.VIII].1670.

169.
RENÉ FRANÇOIS DE SLUSE to WALLIS
Liège, [15]/25 July 1670

Transmission:

c^1 Copy of (missing) letter sent: LONDON *Royal Society* Early Letters S1, No. 64, 4 pp. (p. 4 originally blank) (our source). At top of p. 1 in Oldenburg's hand: 'Ent⟨ere⟩d' LB. 4. 40.' Endorsements by Oldenburg on p. 4: 'A letter of Monsr. Slusius to Dr Wallis concerning Ricciolo's Argument against the Motion of the Earth, taken from the Acceleration of Falling Bodies.' and 'A Copy of Slusius's letter, of July 25. 70. to Dr Wallis, left open in another letter to myself.'—printed: PAIGE, *Correspondance de René François de Sluse*, 645–6.
c^2 Copy of c^1: LONDON *Royal Society* Letter Book Original 4, pp. 40–2.
c^3 Copy of c^2: LONDON *Royal Society* Letter Book Copy 4, pp. 51–4.

Reply to: WALLIS–SLUSE 29.III/[8.IV].1670.
Enclosure to: OLDENBURG–WALLIS 28.VII/[7.VIII].1670.

This letter was sent initially to Oldenburg as an enclosure to SLUSE–OLDENBURG [15/16]/25/26.VII.1670; OLDENBURG, *Correspondence* VII, 73–7. Oldenburg subsequently forwarded it to Wallis as an enclosure to OLDENBURG–WALLIS 28.VII/[7.VIII].1670. Since Sluse left the letter unsealed, Oldenburg felt free to have it copied for the archives of the Royal Society.

Clarissimo ac Celeberrimo Viro
Domino Johanni Wallisio Geometriae Professori
Ren. Franciscus Slusius Salutem.

169. SLUSE to WALLIS, [15]/25 July 1670

Quanquam scriptione inutili tempora tua morari verear, pudet tamen diutius differre gratiarum actionem quas Tibi debeo, ne silentio meo testari videar, me non agnoscere beneficium, quod a Te nuper accepi.[921] Plurimum enim contra debere me profiteor humanitati tuae, quod tam candide sententiam tuam mihi aperueris, quae me omnino confirmavit. In eadem enim eram, Vir Clarissime, et saepe tecum mirari subiit: qui fieri potuerit, ut Vir eruditus tam levi argumento, rem adeo difficilem conficere se posse in animum induxerit. Quid quod ejus solutionem vidit ipse, dum Almagesti[922] pag. 411. exemplum navis ab adversariis contra se adduci scripsit; dum pag. 310. Gassendi experimentum in navi factum retulit,[923] ac probare visus est.

Poterat autem, ut existimo, absque taedioso calculo, rem absolvere, supponendo lineam mobilis esse parabolam, cum ad eam non minus quam ad circulum accedat. Sint enim parabolae quotlibet circa eundem axem AB descriptae, et sumpta AC, ipsius AB subquadrupla (in ratione nimirum accelerationis,) applicentur CD, BE junganturque AD, DE. Evidens est, quo amplior erit| parabola, eo magis rectas AD, DE ad aequalitatem accedere. Quod, si tanti est, etiam theoremate locali ostendi potest.

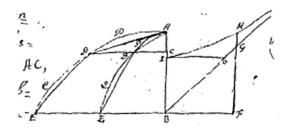

Sumpta enim BJ, quae possit differentiam quadratorum BA, CA; erigatur IO normalis ad AB, et aequalis BI, junctaque indefinita BO, vertice I, circa asymptoton BO, fiat semihyperbola CH. Patet, si in ea sumatur punctum quodlibet H, ex quo cadat in EB productam, normalis HF, secans, BO in G, ac FG supponatur aequalis AD, ipsam FH fore pariter aequalem DE. Cum autem rectae FH, FG, semper eo magis ad aequalitatem accedant, quo FG major est, patet in casu proposito (qui amplissimam parabolam requirit) rectas AD, DE fere aequales esse, et curvas quoque APD, DQE tanto magis, quanto ipsius APD, quam DQE curvatura major videtur.

---

[921]accepi: i.e. WALLIS–SLUSE 29.III/[8.IV].1670.
[922]Almagesti: i.e. RICCIOLI, *Almagestum novum Astronomiam veterem novamque complectens*, Bologna 1651, 411.
[923]retulit: i.e. RICCIOLI, *Almagestum novum*, 310.

Poterat igitur eodem quoque argumento concludere, motu ferme aequabili percurri lineam parabolicam, ideoque percussionem fere eandem in quolibet ejus puncto esse debere; cujus tamen contrarium, vel ipso fatente, experientia docet, et ratio. Sed instat[924] in Astronomia nova, quanquam ex duobus motibus, accelerato et aequabili componatur linea parabolica, id non impedit, quo minus motus per ipsam sit fere aequabilis; neque enim, ut ait pag. 84, motus localis corporum realiter fit per quodcumque spatium, per quod mensurari potest; ut ostendit exemplo lineae rectae| quae per duos [3] motus infinitis modis componi potest. Lubens equidem id admitto, et fateor quoque (si diversas accelerationis leges fingere quis velit) pluribus modis ex duobus motibus acceleratis componi posse parabolam: sed aio tecum, si qui Horizonti parallelus supponitur, sit communis, easdem in perpendiculari accelerationis et percussionis leges observatum iri, ac si prior non esset. Accedit, quod nec tangentium in curvis rationem habuerit, quarum respectu tamen percussio consideranda est. Sed haec nimium praesertim apud Te, qui motuum leges ex veris fundamentis jam ostendisti. Perge igitur, ut soles, Vir Celeberrime, meque semper virtutis tuae observantissimum esse Tibi persuade, ac vale.

Dabam Leodii 25. Julii Greg. MDCLXX.

170.
HENRY OLDENBURG to WALLIS
18/[28] July 1670

Transmission:

Manuscript missing.

Existence and date: partly quoted in and answered by WALLIS–OLDENBURG 22.VII/ .1670.

Wallis received this letter on the morning of Wednesday, 20 July, through the hands of a certain Dr. Upsall, travelling in the entourage of Karl II (1651–85), electoral prince of the Palatinate, who had arrived under a veil of secrecy in Oxford the night before. In the letter Oldenburg requested that Wallis look after the distinguished visitor, but without revealing his true identity.

---

[924]instat: i.e. RICCIOLI, *Astronomiae reformatae tomi duo*, Bologna 1665, lib. primus, appendix ad cap. XVII, I, 84.

171.
WALLIS to HENRY OLDENBURG
Oxford, 22 July/[1 August] 1670

Transmission:

W Letter sent: LONDON *Royal Society* Early Letters W1, No. 111, 4 pp. (our source). On p. 4 beneath address in Oldenburg's hand: 'Dr Wallis to be thanked for his Book. Answ. july 28. 1670.' Postmark:'JY/23'.—printed: OLDENBURG, *Correspondence* VII, 90–3.

Reply to: OLDENBURG–WALLIS 18/[28].VII.1670.
Answered by: OLDENBURG–WALLIS 28.VII/[7.VIII].1670.

The background to this letter is an unofficial and largely secret visit made to England by Karl II (1651–85), eldest son of the the Elector Palatine, Karl I Ludwig (1617–80), during his Grand Tour of Switzerland and France in 1670. For reasons of secrecy, the electoral prince used one of his lesser titles on his English excursion, namely that of the Graf von Schaumburg. Under this title and with a retinue of twenty-three persons he is reported as having departed from Dover for Calais on 2 August 1670 (old style) (*Calendar of State Papers Dom.* Car. II, 1670, 362). The electoral prince did not make an official visit to Oxford until September 1680, at which time he was created Doctor of physic. See WOOD, *Fasti Oxonienses* II, 378; WOOD, *Life and Times* II, 495.

Oxford July 22. 1670.

Sir,

I was just going to blame you for your last letter,[925] when upon a further view thereof I find it will (at lest in part) return upon myself. You intimate therein, but so obscurely that I did not understand it, that the Landgrave himself was coming hither. But your words being onely, that the bearer[926] was of *the suite of this prince now intending to see Oxford*; & himself onely acknowledging his relation to that Prince,[927] but that his business here, was to wait on a young Count whom he called the *Count of Schawmburg*; I apprehended no other sense of your words, but that *the suite* (not *the Prince*) was now intending to see Oxford; of whom I took this *Count of Schawmburg* to have been one. Nor was I delivered from that mistake till they were just going away. Otherwise, he should not have been so meanly attended at his own lodgings as by myself onely. 'Tis true, that they brought

---

[925]letter: i.e. OLDENBURG–WALLIS 18/[28].VII.1670.
[926]bearer: i.e. a certain Dr. Upsall.
[927]Prince: i.e. Karl II (1651–85), electoral prince of the Palatinate, Elector Palatine (1680–85).

171. WALLIS to OLDENBURG, 22 July/[1 August] 1670

a letter from Sir Charles Coterel[928] to the Vice-chancellor,[929] signifying his condition. But it was so late ere they delivered that Letter, as that hee was thereby hindered from giving such Academical Reception as would otherwise have been due to a Prince of his quality. They came hither late on Tuesday
5 night;[930] & the next morning your letter (with the inclosed papers for which I thank you) were delivered mee by Dr Upsall (if I mistake not the name:) whom I received with the Ordinary civility due to strangers; & (the rather because of your recommendation) understanding that hee did here attend a Count who came to see the place (though I knew not of what quality)
10 I offered with him to attend that Count at his lodgings, and wait on him to such places as hee had a desire to see. He signified, that the Count was then somewhat indisposed, & did not intend to goe forth that morning, nor would till about two hours after be in a condition to be attended. But (his indisposition, it seems, being pretty well over) when I came at the time
15 appointed, I found them not within, (being gone, as it afterward appeared, to see the Schooles, the Library, the Theater &c.) I left word that I had been there, & would again attend the Count after Dinner. At which time, I waited on his Highness from thence to Christchurch; where (missing of Dr Locky,[931] who speaks French, but was then not at home) I went with them
20 to the Deans[932] lodgings, who having treated them with a Glass of wine &c, did him self wait on them to shew his Highness their Church, their Library, their Hall, &c: and then at the College Gate left them to my further conduct. Thence, having taken a short view of the Colleges of Oriel, Corpus Christi, & Merton, wee went by Coach to the Physick Garden: where Bobart[933] (the

3 giving *(1)* that *(2)* such
21-22 Highness *(1)* the Church, the Library, the Hall *(2)* their Church, their Library, their Hall

---

[928]Coterel: i.e. Charles Cotterel (1615–1701), courtier and translator, *ODNB*. Cotterel had traveled to Heidelberg in 1652 in his capacity as steward to Elizabeth of Bohemia (1596–1662). Since then, close ties had existed between him and Elizabeth's son, the Elector Palatine, Karl I Ludwig (1617–80).

[929]Vice-chancellor: i.e. Peter Mews (1619–1706), President of St John's College, Oxford, *ODNB*.

[930]Tuesday night: Wallis wrote this letter on Friday, 22 July. The electoral prince therefore arrived on 19 July.

[931]Locky: i.e. Thomas Lockey (ca. 1602–79), Bodley's librarian (1660–65) and canon of Christ Church, *ODNB*.

[932]Deans: i.e. John Fell (1625–86), dean of Christ Church from 1660 until his death, vice-chancellor of the University of Oxford 1666–69, *ODNB*.

[933]Bobart: i.e. Jacob Bobart (or Bobert) the elder (*c.*1599–1680), botanist and superintendent of the Oxford physic garden from its foundation in 1632, *ODNB*.

171. WALLIS to OLDENBURG, 22 July/[1 August] 1670

[2] Gardiner) having in Dutch[934] given him an account of the Garden and some|
of his rarities there, we went cross the way to Magdalene College: where
it being then Prayer time, his Highness went into the Chappel (attended
with his Company) & stayd there all the time, to see the manner of the
Service, (Singing, Organs, &c;) which ended; the President, Dr Piers,[935] (to
whom I had intimated what I knew concerning the quality of the Person,)
treated him (in French) at his lodgings, with a Glass of Wine, Cidar, &c,
& then shewd him their Library, Walks, Hall, &c, & conducted him to his
Coach. His Highness being then weary, & not willing to see more Colledges,
I onely shewed him the outsides of University College, All-souls College,
& St. Mary's church, (as wee passed by,) returning to his lodgings at the
Bear. Understanding there, that the Vice-chancellor (with all his Bedles)
had been there whilst his Highness was absent to wait on him: I then waited
on him to the Vice-chancellors lodgings: who there received him (with the
attendance of his Bedles,) excused himself (by reason of the late reception
of Sir Charles Coterels letter that afternoon) that he had not waited on
him sooner: discoursed with him some time in French (which language his
Highness chose rather to speak than Latine, though he speaks this allso,)
treated him with a Glass of Wine &c, shewed him their Library, & walks,
& then attended him to his Coach; being desired by his Highness (as since
I understand) not to take notice of his quality: of which I was all this while
ignorant. Having then waited on him to his Inne, I there took my leave of
him: Yet afterwards waited on him again, to shew him a little curiosity[936]
of my own (of which there had before been mention made, by some of his
attendants who had seen it at my house,) and some of Mr. Birds[937] Stained
Marble. That night, & on Thursday morning (when I waited on him again
before he went out of Town) I began to discern my Error. And therefore,
before he went, I thought myself obliged to make an Apology; That, though

13 whilst his Highness was absent *add.*
20 desired *add.*

---

[934]in Dutch: i.e. in German.
[935]Piers: i.e. Thomas Pierce (or Peirse) (1621/2–91), President of Magdalen College (1661–72), *ODNB*.
[936]curiosity: probably Wallis's model of the roof design, involving an arrangement of interlocking wooden beams, which he had contrived, in 1644, while he was Fellow of Queens' College, Cambridge. Cf. LAMPLUGH–WILLIAMSON 5/[15].V.1669.
[937]Birds: i.e. William Byrd (d. 1690?), stonecutter and mason. In 1657/8 Byrd evidently discovered a technique of staining marble. He employed this technique in stonework which he carried out in the construction of the Sheldonian Theatre, and he presented an example of it to Cosimo III de' Medici during his ceremonial visit to Oxford in May 1669. See WOOD, *Life and Times* I, 241; II, 160, 213.

171. WALLIS to OLDENBURG, 22 July/[1 August] 1670

it were not civil for mee to be too inquisitive into what hee pleased to conceal: yet (by the Title of *Serenissimus* which I once discerned to fall unawares from one of his Attendance) I could not but beleeve him to be of a much higher quality than at first I apprehended, or did so much as suspect: beseeching his pardon as well in my own name as of the University; that hee had not been received with that Honor & Respect which his Quality, had it been known, would have required: & that what ever neglects or mistakes had been he would impute onely to our perfect ignorance of his condition; of which had wee been aware, his reception would have been much| otherwise, [3] & more sutable to his quality: But it being so late before I did so much as suspect it, it was now impossible (his Highness being just ready to take Coach for Salisbury) to make any other amends, than by professing our Ignorance & craving his Pardon. Which hee did not seem to take amisse. And it was then owned, that hee was indeed such a Person as I apprehended, (the Count of Schawmburg being allso one of his Titles,) but being willing to passe Incognito, it would rather be a favour to conceal his quality. After which he presently took Coach & went away towards Salisbury; when I had taken my leave of him at the Coach side. Had wee had timely notice of it, the University would doubtless have received him after another fashion, & somewhat answerable to his quality, as they did the Prince of Tuscany[938] the last year. (At lest it should have been my fault if they had not.) Or had the Vice-chancellors letter been delivered him the same night they came to Town, or betimes the next morning: there had been time to have called a Convocation, & given him therein some Academical Reception, & conferred some Degrees on those of his attendance (if himself sho⟨uld⟩ not do us the Honor to accept of any:) For though wee do not, in ord⟨i⟩nary course, grant Degrees without respect had to Time & Exercises: Yet wee are not so tyed up but that on such occasions our Honorary Degrees are conferred without such considerations.

Sir, having given you this Narrative, I have nothing further to adde, but to desire that you will, as there shal be occasion, make the best apology for any neglects of ours, & that the fairest constructions may be put upon

14 indeed *add.*
14 apprehended, *(1)* but *(2)* (the Count
26 though *add.*
26 course, *(1)* conf⟨er⟩ *(2)* grant
30 *(1)* Having *(2)* Sir, Having *corr. ed.*

[938]Tuscany: i.e. Cosimo III de' Medici. See WALLIS–OLDENBURG 24.IV/[4.V].1669, and WOOD, *Fasti Oxonienses* II, 310.

172. WALLIS to COLLINS, 23 July/[2 August] 1670

them; & to thank you for the honor you have given mee, in affording the opportunity of waiting on so Eminent a Prince. Resting,

Yours, to serve you,
John Wallis.|

[4] These
For Mr Henry Oldenburg,
in the Palmal near
St James's
London.

172.
WALLIS to JOHN COLLINS
Oxford, 23 July/[2 August] 1670

Transmission:

W Letter sent: CAMBRIDGE *Cambridge University Library* MS Add. 9597/13/6, f. 213^r – 213^v (our source). On f. 213^v postmark: 'JY/25'.—printed: RIGAUD, *Correspondence of Scientific Men* II, 523–4.

Reply to: Collins–Wallis 14/[24].VII.1670.

Oxford July. 23. 1670.

Sir,

I have now yours[939] of July 14. with the book[940] inclosed; & six bookes[941] from Mr Pits.[942] If Mr Bee[943] will Barter for the Book[944] you mention, you may please to make that bargain for mee. If there be need of taking any of the first part from Mr Pits to that purpose; I shal satisfy him either in mony or Bookes. I thank you for the book of that new Dutch Engine;[945] which

---

15 to that purpose; *add.*

[939] yours: i.e. COLLINS–WALLIS 14/[24].VII.1670.
[940] book: not ascertained.
[941] bookes: i.e. copies of part two of Wallis's *Mechanica: sive, de motu, tractatus geometricus*, London 1670.
[942] Pitts: i.e. Moses Pitt (1639–97), London printer and bookseller, *ODNB*.
[943] Bee: i.e. Cornelius Bee (d. 1671/2), bookseller in Little Britain, London.
[944] Book: not ascertained.
[945] engine: possibly Georg Christoph Werner's *Inventum novum, artis et naturae con-*

172. WALLIS to COLLINS, 23 July/[2 August] 1670

promiseth fair, but how it will perform wee cannot judge unlesse hee would open the covered box & shew what is within it.

I could wish, if it were not too much trouble; that, before the copies[946] be dispersed, these faults were mended with a pen;

$$\text{Pag. 560. lin. 28. after latus rectum } \frac{h^2}{h^2-\eta^2} \cdot L$$
$$\text{adde } +\frac{\eta^2}{h^2-\eta^2} \cdot T.$$
$$\text{Pag. 561. lin. 5. after Latus rectum } \frac{h^2}{\eta^2-h^2} \cdot L$$
$$\text{adde } +\frac{\eta^2}{\eta^2-h^2} \cdot T.$$

It should have been, in the first place, $\frac{h^2}{h^2-\eta^2} \cdot L + \frac{\eta^2}{h^2-\eta^2} \cdot T$: in the latter, $\frac{h^2}{\eta^2-h^2} \cdot L + \frac{\eta^2}{\eta^2-h^2} \cdot T$. And pag. 565, lin. 23. for $\frac{\eta^2-h^2}{L}$, make it $\frac{\eta^2-h^2}{L} \cdot T$. (the letter T, being omitted.) But the two former are the more considerable, & should have been put into the Table of Errata if I had continued it so far.[947]

Excuse the frequent trouble you have from

Your friend to serve you,
John Wallis.

It's very possible there may be some other like errors, either by the Printers mistake or mine; but I not yet had time to peruse the whole: If you find any, you will do mee a courtesy to mark them & give mee notice.

These,
For Mr John Collins at
the three Crownes in
Blomesbury-market
London

[2]

7-8 It should ... $\frac{\eta^2}{\eta^2-h^2} \cdot T$. add.

---

nubium, in copulatione levitatis cum gravitate & gravitatis cum levitate. Per artificium siphonis machinae aquaticae antliae, Augsburg 1670. Cf. OLDENBURG–LEIBNIZ 10/[20].VIII.1670; OLDENBURG, Correspondence VII, 110–2.

[946] copies: i.e. of the second part of Wallis's Mechanica: sive, de motu, tractatus geometricus, London 1670.

[947] so far: the corrections are contained in the table of errata at the end of the third part of Wallis's Mechanica: sive, de motu, tractatus geometricus, published in 1671.

173.
ROBERT WOOD to WALLIS
23 July/[2 August] 1670

Transmission:

Manuscript missing.

Existence and date: mentioned in and answered by WALLIS–WOOD 4/[14].VIII.1670.

174.
HENRY OLDENBURG to WALLIS
28 July/[7 August] 1670

Transmission:

Manuscript missing.

Existence and date: mentioned in and answered by WALLIS–OLDENBURG 4/[14].VIII. 1670. Also mentioned in Oldenburg's endorsement on WALLIS–OLDENBURG 22.VII/ [1.VIII]. 1670.

Reply to: WALLIS–OLDENBURG 22.VII/[1.VIII].1670.
Enclosure: SLUSE–WALLIS [15]/25.VII.1670 (cf. WALLIS–OLDENBURG 4/[14].VIII.1670).

In this letter, which enclosed Sluse's latest letter to Wallis, Oldenburg apparently thanked Wallis for the copy he had received of the second part of his *Mechanica: sive, de motu, tractatus geometricus.*

175.
JOHN COLLINS to WALLIS
end of July 1670

Transmission:

Manuscript missing.

Existence and date: As emerges from Wallis's reply (WALLIS–COLLINS 4/[14].VIII.1670), Collins reported on the meeting of the Council of the Royal Society on 26 July. His letter was therefore probably written sometime during the last days of the month.
Answered by: WALLIS–COLLINS 4/[14].VIII.1670.

176. WALLIS to COLLINS, 4/[14] August 1670

Collins evidently informed Wallis of the dispute between the Council of the Royal Society and Moses Pitt over the price for the first two volumes of Wallis's *Mechanica: sive, de motu, tractatus geometricus*.

176.
WALLIS to JOHN COLLINS
Oxford, 4/[14] August 1670

Transmission:

W Letter sent: CAMBRIDGE *Cambridge University Library* MS Add. 9597/13/6, f. 214^r–214^v (our source). On f. 214^v part of postmark:'AU/–'.—printed: RIGAUD, *Correspondence of Scientific Men* II, 524–5.

Reply to: COLLINS–WALLIS ?VII.1670.

Oxford. Aug. 4. 1670.

Sir,

I am sorry those of the Society have no better satisfyed[948] Mr Pits in the price of the book, which I am very sensible was both troublesome & chargeable to print. I have written[949] to Mr Oldenburg concerning it. You may forbear presenting any more[950] as from mee, save onely that to my Lord Brouncker & that to Mr Boyle, because I would not forestall his market.

The Theatrum Machinarum[951] you may let alone if they put it at such unreasonable terms.

My third part, I mean shall be very short: (And, when that is done;[952] I think I shall rest a while:) But something must bee done of a third part,

---

[948] satisfyed: there was disagreement between Moses Pitts, the printer of the *Mechanica: sive de motu, tractatus geometricus*, and the Royal Society over the price to be paid for the two volumes (i.e. pars prima and pars secunda) which had been published. Subscribers had agreed to pay the price set by Council, which was the sum of fourteen shillings for both volumes, whereas Pitts insisted on a price of fifteen shillings and six pence. See BIRCH, *History of the Royal Society* II, 446–7.

[949] written: i.e. WALLIS–OLDENBURG 4/[14].VIII.1670.

[950] more: i.e. copies of part two of the *Mechanica: sive, de motu, tractatus geometricus*.

[951] Theatrum Machinarum: probably BÖCKLER, *Theatrum Machinarum novum, exhibens aquarias, alatas, jumentarias, manuarias; pedibus, ac ponderibus versatiles, plures, et diversas molas* ..., Nuremburg 1662.

[952] done: part three of the *Mechanica: sive, de motu, tractatus geometricus* appeared in 1671.

because it is promised;⁹⁵³ &, without it, the other two are but an imperfect work.

I thank you for the book you send mee by the carrier; which I shal inquire after when the wagon comes in.

You may adde,⁹⁵⁴ if you please, in my book, pag. 565. l. 11. after conjugatus: adde, sin curvam tangat, perinde est ad utrumvis casum referas: quippe tum Hyperbolae degenerant in opposita Triangula, quorum communis vertex est O, punctum contactus; evanescente Axe transverso. (Which case, I wonder how I missed.) And pag. 556. lin. 24. after Genitricis; adde, aut etiam hyperbolam hanc ubivis tangat.

The post is going. I adde onely that I am

Your friend to serve you;
John Wallis.

177.
WALLIS to ROBERT WOOD
Oxford, 4/[14] August 1670

Transmission:

W Letter sent: LONDON *Royal Society* Boyle Letters 7, No. 26. 4 pp. On pp. 1–3 the text of WALLIS–WOOD 16/[26].V.1670 (see under that heading). On p. 4 in Oldenburg's hand: 'Aug. 4. 1670. Dr Wallis to Mr Robert Wood.' In an unknown hand the postal endorsement: 'forwarded'; Postmarks: 'AU/5' and 'AU/18'.

Reply to: WOOD–WALLIS 23.VII/[2.VIII].1670.

This letter was evidently sent first to London and then forwarded from there to Dublin. It contains a copy of WALLIS–WOOD 16/[26].V.1670, which Wallis suspected had not reached its addressee.

Oxford Aug. 4. 1670.

Sir,

By yours⁹⁵⁵ of July 23. which mentions mine⁹⁵⁶ of May 18, but not that⁹⁵⁷ of May 16, I perceive the former of them came not to your hand, to

---

⁹⁵³promised: in the scholion to Chapter 5, prop. 32.
⁹⁵⁴adde: these corrections (slightly modified) appear in the table of errata at the end of part three of the *Mechanica: sive, de motu, tractatus geometricus*.
⁹⁵⁵yours: i.e. WOOD–WALLIS 23.VII/[2.VIII].1670.
⁹⁵⁶mine: i.e. WALLIS–WOOD 18/[28].V.1670.
⁹⁵⁷that: i.e. WALLIS–WOOD 16/[26].V.1670.

which the latter was but an appendix; I therefore must needs seem to you very abrupt: the particulars of your letter being answered in the former of them.

But it happened, that my onely son[958] being then with mee in Kent, I caused him to copy that letter before I sent it away, by reason whereof I am now in abled to give you a duplicate of it. And it was allmost the last thing hee did before hee was taken sick of a very desperate malignant fever; (which made mee fear my next to you must have been of the like import with that ill news in yours to mee;) but which did after degenerate into a Chronick desease (of which part was a Kentish ague) the remainder of which hee is not yet quitt of. I say allmost the last: for hee had begun to copy that of May 18, when hee was taken so very ill that hee could not possibly goe on with it. After a considerable stay where I was; till I saw him begin (as wee hoped) to amend; I was fain (being called home) to leave him there behind mee; nor is hee yet in a condition to come after mee: but I hope will now in a short time.

That letter was in these words.

The text of WALLIS–WOOD 16/[26].V.1670, *which Wallis inserted here, is to be found under that heading.*

These [4]
For Mr Robert Wood at
Rathdans-town[959], near Dublin,
in
Ireland.

Post payd to London.

178.
WALLIS to HENRY OLDENBURG
Oxford, 4/[14] August 1670

Transmission:

W Letter sent: LONDON *Royal Society* Early Letters W1, No. 112, 2 pp. (our source). On p. 2 beneath address in Oldenburg's hand: 'Rec. Aug. 5. 70. Answ. Aug. 6. 70.' Postmark:

1 which (*1*) that of May 18 (*2*) the latter
10 the remainder *add.*

---

[958]son: i.e. John Wallis jr, q.v.
[959]Rathdans-town: i.e. Raddonstown, near Dublin.

178. WALLIS to OLDENBURG, 4/[14] August 1670

'AU/–'.—printed: OLDENBURG, *Correspondence* VII, 101–2.
Reply to: OLDENBURG–WALLIS 28.VII/[7.VIII].1670.
Answered by: OLDENBURG–WALLIS 6/[16].VIII.1670.

Oxford Aug. 4. 1670.

Sir,

I have yours[960] of July 28. with that[961] of M. Slusius inclosed. To whom I desire, when you next write, to return my thanks for it, & for the Theoreme therein contained: which is very ingenious. There being nothing in it which requires an answere I forbear to give him a letter merely of complements. The two books[962] for beyond sea, I therefore ordered you, because you desired them the last time: but you did not tell mee for whom: I suppose they may bee for Hugens[963] & Slusius.[964] Before you send them; be pleased with your pen make these additions[965]

pag. 560. lin. 28. after $\frac{h^2}{h^2-\eta^2}L$ adde $+\frac{\eta^2}{h^2-\eta^2}T$. pag. 561. lin. 5 after $\frac{h^2}{\eta^2-h^2}L$ adde $+\frac{\eta^2}{\eta^2-h^2}T$. pag. 565. lin. 11 after conjugatus, adde sin curvam tangat; perinde est ad utrumvis casum referas: quippe tum Hyperbolae degenerant in Opposita Triangula, quorum communis Vertex est O, punctum contactus; evanescente Axe transverso. pag. 556. lin. 24. after Genitricis, adde, aut etiam hyperbolam hanc ubivis tangat.

4-5 thanks *(1)* : There being *(2)* for it, & for the Theoreme ... ingenious. There being
15 evanescente Axe transverso. *add.*

[960]yours: i.e. OLDENBURG–WALLIS 28.VII/[7.VIII].1670.
[961]that: i.e. SLUSE–WALLIS 15/[25].VII.1670.
[962]books: i.e. copies of the second part of Wallis's *Mechanica: sive, de motu, tractatus geometricus*.
[963]Hugens: Oldenburg sent the copy of part two of the *Mechanica: sive, de motu, tractatus geometricus* to Huygens with OLDENBURG–HUYGENS 20/[30].IX.1670; OLDENBURG, *Correspondence* VII, 171–2. Huygens received the copy in Paris before his departure, but was too ill at the time to study it. See HUYGENS–OLDENBURG [5]/15.X.1670; OLDENBURG, *Correspondence* VII, 199–200, and HUYGENS–OLDENBURG [21]/31.X.1670; OLDENBURG, *Correspondence* VII, 216–8.
[964]Slusius: the copy of part two of the *Mechanica: sive, de motu, tractatus geometricus* for Sluse was sent by Collins through the hands of the Dutch seaman Johnson. See OLDENBURG–SLUSE 24.IX/[4.X].1670; OLDENBURG, *Correspondence* VII, 177–85. Oldenburg also sent a copy to Pierre Daniel Huet. See HUET–OLDENBURG [20]/30.X.1670; OLDENBURG, *Correspondence* VII, 206–7.
[965]additions: Wallis requested that Collins make the same corrections in WALLIS–COLLINS 23.VII/[2.VIII].1670 and WALLIS–COLLINS 4/[14][VIII].1670.

178. WALLIS to OLDENBURG, 4/[14] August 1670

I should have ordered more copies to my friends in the Society, but that the Book-seller[966] complained the last time that I thereby prevented those from buying who were most likely so to do; which in a book of so slow a sale as Mathematick bookes usually are, is a considerable prejudice to him: & therefore I hope my friends will not take it as any disrespect.

I find, (by a letter[967] from Mr Collins,) that Mr Pits, the Book-seller, is not satisfied in the price the Counsel of the Society hath put on the books for which their members have subscribed. I could wish (though I am not concerned in it) that they had satisfied him. For I know the printing of it hath been very troublesome & very chargeable, beyond what other books are of the same bulk: & the number of the Impression but smal, & the sale (as of other mathematick books, especially those more intricate, & not for every ones understanding,) but slow. (And, if I may say it, this is so closely penned as that if they have not bulk inough, they will at lest have matter inough for their mony; the same matter being inough to have filled large volumes.) And, considering how few are willing to undertake the printing of books that are a little out of the common road; I would not have those discouraged that are. I adde no more, but that I am

Yours to serve you
Joh: Wallis.

These [2]
For Mr Henry Oldenburg,
in the Palmal near
St James's
London.

2 the *(1)* Printer *(2)* Book-seller
11 & |the *add.*| number
12 of *add.*
17 road; *(1)* those *(2)* I would not

---

[966]Book-seller: i.e. Moses Pitt. Cf. WALLIS–COLLINS 4/[14].VIII.1670.
[967]letter: i.e. COLLINS–WALLIS ?VII.1670.

179.
HENRY OLDENBURG to WALLIS
6/[16] August 1670

Transmission:

Manuscript missing.

Existence and date: noted in Oldenburg's endorsement on WALLIS–OLDENBURG 4/[14].VIII.1670.
Reply to: WALLIS–OLDENBURG 4/[14].VIII.1670.

180.
FRANCIS VERNON to WALLIS
?[25 August]/4 September 1670

Transmission:

Manuscript missing.

Existence and date: mentioned in VERNON–OLDENBURG [25.VIII]/4.IX.1670; OLDENBURG, *Correspondence* VII, 139–141.
Enclosure to: VERNON–OLDENBURG [25.VIII]/4.IX.1670.

As Vernon reports in the letter to Oldenburg, in which this letter was sent as an enclosure, he thanked Wallis for the book, i.e. the second part of the *Mechanica: sive, de motu, tractatus geometricus*, which Oldenburg had sent him through the hands of a certain Dr Williams.

181.
JOHN AUBREY to WALLIS
Broad Chalke, 27 August/[6 September] 1670

Transmission:

W Copy of missing letter sent: LONDON Royal Society Early Letters W1, No. 113, 2 pp. (on p. 1; the copy is contained in WALLIS–OLDENBURG 25.X/[4.XI].1670) (our source). Above Chart⟨r⟩es in Wallis's hand: 'Carnutes'.—printed: OLDENBURG, *Correspondence* VII, 224–6.

182. Wallis to Neile, August/September 1670

In the course of research on antiquities in France, Aubrey had received a mathematical problem forwarded to him by one of his Jesuit contacts in Paris. He enclosed the problem in this letter, seeking Wallis's help in finding its solution.

Rev. Sir,

I received a letter a little while since from a Jesuite[968] of Paris, in answere to some Quaere's of mine concerning some Antiquitie at Chartes (the metropolis of the Druides) & withall he sent mee the inclosed probleme,[969] which was sent to them from one of their order in the Southern part of France.

If you please to honour mee with the Solution; I shal send it to Paris as done by you: which will be kindly taken by them, & you will much oblige,
Sir,

Your most humble servant
Jo: Awbrey.

Broad-Chalke, Aug. 27. 1670.

182.
Wallis to William Neile
August/September 1670

Transmission:

Manuscript missing.

Existence and date: referred to in Wallis–Oldenburg 4/[14].X.1673.
Answered by: ?–Wallis VIII/IX.1670.

Wallis tells Oldenburg in Wallis–Oldenburg 4/[14].X.1673 that he had requested by letter that Neile send him his proof of the rectification of the semi-cubic parabola. Unknown to him at the time, Neile had died on 24 August 1670 (old style). See Birch, *History of the Royal Society* II, 460–1.

---

[968] Jesuite: i.e. Richard Thimbleby (*alias* Ashby), (1614–80), admitted S.J. 1632, minister of the English College at St Omer from 1642. Between 1666 and 1672 he gave spiritual exercises at the convent of the blue nuns in Paris.

[969] probleme: the enclosure with the mathematical problem referred to has not survived, but cf. Wallis–Oldenburg 3/[13].XI.1670 and Morehouse–Aubrey 15/[25].X.1670; *Bodleian Library* MS Aubrey 12, f. 340^r–341^v.

183.
? to WALLIS
August/September 1670

Transmission:

Manuscript missing.

Existence and date: referred to in WALLIS–OLDENBURG 4/[14].X.1673.

Reply to: WALLIS–NEILE VIII/IX.1670.

Wallis tells Oldenburg in WALLIS–OLDENBURG 4/[14].X.1673 that in reply to his request for Neile's original proof of his rectification of the semi-cubic parabola he received the information that Neile had died. This letter notifying him of this would have been sent in August or September 1670.

184.
COSIMO III DE' MEDICI to WALLIS
3/[13] October 1670

Transmission:

Manuscript missing.

Existence and date: mentioned in and answered by WALLIS–COSIMO III MEDICI 9/[19].XI.1670.

185.
WALLIS to JOHN AUBREY
24 October/[3 November] 1670

Transmission:

Manuscript missing.

Existence and date: mentioned in WALLIS–OLDENBURG 25.X/[4.XI].1670.
Reply to: AUBREY–WALLIS 27.VIII/[6.IX].1670.
Enclosure to: WALLIS–OLDENBURG 25.X/[4.XI].1670.

As emerges from WALLIS–OLDENBURG 25.X/[4.XI].1670, this letter to Aubrey, which was sent with it as an enclosure, had been written the previous day. In it Wallis re-stated the mathematical problem which Aubrey had passed on to him and provided a solution to it. Cf. WALLIS–OLDENBURG 3/[13].XI.1670.

186.
[WALLIS] to HENRY OLDENBURG
Oxford, 25 October/[4 November] 1670

Transmission:

W Letter sent: LONDON *Royal Society* Early Letters W1, No. 113, 2 pp. (our source). On p. 2 endorsement in Oldenburg's hand: 'Rec. Oct. 26. 70.' and postmark: 'OC/25'.—printed: OLDENBURG, *Correspondence* VII, 224–6.

Enclosure: WALLIS–AUBREY 24.X/[3.XI].1670.

Wallis's decision to conceal his identity as the writer of this letter evidently reflects the sensitive nature of its subject matter: the support which was being given to Henry Stubbe in his attacks on the Royal Society by the president of Magdalen College. Oldenburg would, however, have recognized Wallis's handwriting immediately. Wallis enclosed his reply to Aubrey which also contained his solution to the mathematical problem which Aubrey had forwarded to him earlier.

Oxford. Octob. 25. 1670.

Sir,

The inclosed[970] is an answere to one[971] from Mr Awbrey to mee, in these Words.

The text of AUBREY–WALLIS 27.VIII/[6.IX].1670, *which Wallis inserted here, is to be found under that heading.*

    The Probleme you have recited in my Answere. His letter (by reason of my absence from home) came not to my hand till Sunday[972] last; which being not a day for business, I wrote the inclosed on the morrow, which this Morning I sent to you to transmit, because you are willing to see such transactions.

    I was told, about 3 days since, from two several persons, that Dr Piers[973] (our President of Magdalene College here) had sent to Mr. Stubbe[974] for his

---

9 inclosed *(1)* yesterday *(2)* on the morrow

[970] inclosed: i.e. WALLIS–AUBREY 24.X/[3.XI].1670. This letter to Aubrey is now missing.
[971] one: i.e. AUBREY–WALLIS 27.VIII/[6.IX].1670.
[972] Sunday: i.e. 23 October 1670.
[973] Piers: i.e. Thomas Pierce (or Pierse) (1622–91), religious controversialist, president of Magdalen College from 1661, *ODNB*.
[974] Stubbe: i.e. Henry Stubbe (1632–76), physician and author, *ODNB*.

186. [WALLIS] to OLDENBURG, 25 October/[4 November] 1670

good service,⁹⁷⁵ a piece of Plate: &, that it might be the more acceptable, by a Gentleman of quality. I was loth to take up such a report too hastyly, & therefore desired one of them that told it mee, to inform himself certainly & particularly about it; who assuring mee of the truth of it, did withall promise to give me a particular account from the person who told him. And the next day brought mee this. That a Fellow of a College here (whom I could name if it were convenient) tells him; that about 7 or 8 weekes since (he knows not whether) on a Friday in August [which must therefore by computation be Aug. 26] hee was at the Angell in Oxford with a Gentleman of Warwickshire, Mr Thomas Wagstaffe⁹⁷⁶ (sometime of Magdalene hall) heir to Sir Combe Wagstaf ⁹⁷⁷ (a gentleman of a good estate;) & whilst he was there, Dr Piers's man came to Mr Wagstaf (then ready to goe out of town) from his master desiring him to go back to the College (for hee had been there a little before, but had not found the Dr then at lesure to be spoken with) for that his Mr had a great desire to speak with him about some business. Hee did so. And when he returned back to this person at the Inne, hee told him the businesse was, that Dr Pierce had desired him to carry that piece of Plate (which he shewed this person, wrapped up in a paper, who took it in his hand & judged it by the weight to bee about 5 or 6$^£$ price) to Mr Stubs for a present.

My relater tells mee further, that another Fellow of the same house, (whom I could name allso,) tells him, that hee hath since mett Mr Stubs in the country, who told him, that hee had received a piece of Plate from Oxford; & that he expected another piece of Plate from London. These are the particulars of matter fact which I meet with of certainty. I am told some others, but the relaters have not the like evidence as of these. I am

---

13 from his *(1)* Mr *(2)* master
18 of Plate *add.*
24 Plate from *(1)* Dr Piers *(2)* |from *del. ed.*| Oxford

---

⁹⁷⁵service: evidently a reference to Stubbe's series of pamphlets, issued in 1670, attacking the virtuosi of the Royal Society. See for example STUBBE, *Legends no Histories: or, A specimen of some animadversions upon the History of the Royal Society*, London 1670; idem, *A censure upon certaine passages contained in the History of the Royal Society, as being destructive to the established religion and church of England*, Oxford 1670; idem, *Campanella revived, or an Enquiry into the history of the Royal Society, whether the virtuosi there do not pursue the projects of Campanella for the reducing England unto Popery*, London 1670.
⁹⁷⁶Wagstaffe: i.e. Thomas Wagstaffe (1645–1712), deacon, later canon of Lichfield; from 1694 non-juring bishop of Ipswich cathedral, *ODNB*. After studying at Pembroke College, Cambridge, Wagstaffe in fact migrated to New Inn Hall, Oxford.
⁹⁷⁷Combe Wagstaf: not identified.

Yours &c
Anonymus

These [2]
For Mr Henry Oldenburg,
in the Palmal, near
St James's
London.

187.
HENRY OLDENBURG to WALLIS
1/[11] November 1670

Transmission:

Manuscript missing.

Existence and date: mentioned in and answered by WALLIS–OLDENBURG 3/[13].XI.1670.

From Wallis's reply it appears that this letter was either sent with or at the same time as a package enclosing two books for Wallis: Erasmus Bartholin's *Experimenta crystalli Islandici dis-diaclastici*, Copenhagen 1669, and Honoré Fabri's *Dialogi physici*, Lyon 1669.

188.
WALLIS to HENRY OLDENBURG
Oxford, 3/[13] November 1670

Transmission:

W Letter sent: LONDON *Royal Society* Early Letters W1, No. 114, 2 pp. (our source). On p. 2 beneath address in Oldenburg's hand: 'Rec. Nov. 4. 70. Answ. Nov. 8. 70.' and postmark: 'NO/4'.—printed: OLDENBURG, *Correspondence* VII, 235–6.

Reply to: OLDENBURG–WALLIS 1/[11].XI.1670.
Answered by: OLDENBURG–WALLIS 8/[18].XI.1670.

Oxford. Novemb. 3. 1670.

Sir,

Yours[978] of Nov. 1. I receive this morning. What concerns Mr Awbrey is answered allready.[979] The Probleme is either a pitiful one, or pitifully

---
[978] Yours: i.e. OLDENBURG–WALLIS 1/[11].XI.1670.
[979] allready: i.e. in WALLIS–AUBREY 24.X/[3.XI].1670.

188. Wallis to Oldenburg, 3/[13] November 1670

worded. 'Tis possible hee may mean that the 10 Squares should be all equal: & the 4 Cubes likewise: & perhaps that those & these should have the same root. But there is no such thing in the words of the Probleme. And had it been so, I should scarce have meddled with it, as being a busyness of more time than the thing is worth. And those of France have sufficiently examined mee allready: 'tis time now that they give mee leave to mind my own busyness.

What concerns Mr Hugens,[980] I do not much question but hee will satisfy himself. But the thing is intricate, & will require attention: & will need to be read more then once.

But the truth is, I have been detained abroad the greatest part of the summer by the sickness of friends;[981] & now at home by an ill companion, a severe quartan Ague, which seised me presently after I came home; (& I know not whether it wil give mee leave to make an end of this letter, for I expect a 7th visite by & bye.) So that I have not had yet so much time as to read it[982] over since it was printed, to collect the typographicall errors.

The two books I have. That[983] of Bartholine I have not so much as looked into. That[984] of Fabry, I have part of (somewhat of what concerns Borelli:) but not all; onely some, part of the 2^d dialogue: wherein I am not at all satisfyed with his objections against Borelli, so far as I have read.[985]

But my Ague takes up so much time, & doth render me so indisposed for busyness, that I can scarce either write or read an hour together with attention: so that it will be the longer ere I can give you a full account

11 detained *(1)* the great⟨est⟩ *(2)* abroad the greatest
15 So that I have *(1)* ha⟨d⟩ *(2)* not had

---

[980]Hugens: cf. HUYGENS–OLDENBURG 21/[31].X.1670; OLDENBURG, *Correspondence* VII, 216–18. Huygens had remarked that he had difficulty in understanding some of the demonstrations in Wallis's *Mechanica: sive, de motu, tractatus geometricus*, such as that of the centre of gravity of a spiral. Oldenburg passed on Wallis's response in OLDENBURG–HUYGENS 8/[18].XI.1670; OLDENBURG, *Correspondence* VII, 239–40.

[981]friends: i.e. close relatives.

[982]it: i.e. the second part of his *Mechanica: sive, de motu, tractatus geometricus*. Since its publication Wallis had discovered innumerable typographical errors.

[983]That: i.e. BARTHOLIN, *Experimenta crystalli Islandici dis-diaclastici*, Copenhagen 1669.

[984]That: i.e. FABRI, *Dialogi physici*, Lyon 1669. This volume was apparently sent by Vernon from Paris together with other books by Fabri. See VERNON–OLDENBURG [9]/19.VII.1670 (OLDENBURG, *Correspondence* VII, 60–4).

[985]read: see WALLIS–OLDENBURG 24.XI/[4.XII].1670.

188. Wallis to Oldenburg, 3/[13] November 1670

of it. If you print that[986] of Mr Childrey, with my answere;[987] it will be expedient (where I refer to what I had written to you in a former letter,[988] about the Moones Perigaeum &c) to insert in the margin, somewhat to this purpose: [viz. in the Latine copy[989] of the letter of ... (the English of which is printed[990] in the Transact. ...) in which are these words, ...] for in the English they are not: & you will find them in a marginal insertion in my Original Latine. You may if you have occasion present my humble service to Mijn Heer Constantine Hugenius van Zulichem: who when hee was heretofore[991] in Oxford, gave mee a very civil visite with testimony of great respect upon the account of my acquaintance with his son.

Sir, I must break off when I have told you that I am

Your most affectionate friend & servant
Joh. Wallis.

When you se Mr Collins, desire that he would get the Printers to make hast:[992] For I would not willingly have it a Posthumous work; but see the end of it. But possibly, if this ague continue, I must fain to crave his help to draw my schemes for mee; for it renders mee allready very indisposed for busyness; & will do dayly more & more. I would write to him myself; but writing begins to grow troublesome.

These [2]
For Mr Henry Oldenburg, in the
Palmal near St James's
London.

---

3 about the *(1)* Suns *(2)* Earths Apheliu⟨m⟩ *(3)* Moones Perigaeum &c
6 insertion *add.*

---

[986] that: i.e. CHILDREY–WARD 4/[14].III.1669/70. This letter was printed in *Philosophical Transactions* No. 64 (10 October 1670), 2061–8.

[987] answere: i.e. WALLIS–OLDENBURG 19/[29].III.1669/70. This letter was likewise printed in *Philosophical Transactions* No. 64 (10 October 1670), 2068–74.

[988] letter: i.e. in WALLIS–OLDENBURG 18/[28].VII.1666; WALLIS, *Correspondence* II, 251–63.

[989] copy: i.e. LONDON *Royal Society* MS 368, No. 1, 1–44.

[990] printed: i.e. in *Philosophical Transactions* No. 16 (6 August 1666), 281–9.

[991] heretofore: i.e. Constantijn Huygens's most recent visit to Oxford in 1664. Cf. CONSTANTIJN HUYGENS–WALLIS [21.VI]/1.VII.1668; WALLIS, *Correspondence* II, 467–9.

[992] hast: i.e. with the third part of his *Mechanica: sive, de motu, tractatus geometricus*.

189.
HENRY OLDENBURG to WALLIS
8/[18] November 1670

Transmission:

Manuscript missing.

Existence and date: mentioned in Oldenburg's endorsement on WALLIS–OLDENBURG 3/[13].XI.1670 and in WALLIS–OLDENBURG 15/[25].XI.1670.
Reply to: WALLIS–OLDENBURG 3/[13].XI.1670.
Answered by: WALLIS–OLDENBURG 15/[25].XI.1670.

In this letter Oldenburg apparently announced the imminent arrival in Oxford of a visitor from Switzerland. He also informed Wallis that he was sending him a copy of the latest issue of *Philosophical Transactions*.

190.
WALLIS to COSIMO III DE' MEDICI
Oxford, 9/[19] November 1670

Transmission:

W Letter sent: FLORENCE *Biblioteca Nazionale* Gal. 286, f. 56^r–57^v (f. 56^v and f. 57^v blank) (our source).—partly printed in Italian translation: TENCA, *Giovanni Wallis e gli Italiani*, 413–15.

Reply to: COSIMO III MEDICI–WALLIS 3/[13].X.1670.

Oxonii Nov. 9. 1670.

Serenissime Magne Dux.

Non par est ut ego Serenissimi Principis negotia nimis morer. Neque permittendum tamen quin devotissima veneratione agnoscam quam antehac expertus sum Clementiam Vestram; et speciatim Literas[993] Clementissimas 5° Nonas Octobris Serenissima manu signatas. Et simul gratuler Serenissimae Vestrae Celsitudini, qui Vestras scribendi Methodos tanta felicitate administras.

Nec omnino erit incongruum, si grata recolam memoria, quantum debet Italiae Vestrae, et Vestrae praesertim Serenissimae Familiae, (quae mea est Professio) Mathesis.

---

[993]Literas: i.e. COSIMO III MEDICI–WALLIS 3/[13].X.1670.

190. WALLIS to COSIMO III DE' MEDICI, 9/[19] November 1670

Utut enim Architecturam taceam, qua Vos prae caeteris Europae partes celebres estis; Taceamque, inter alios, Commandinum,[994] qui tot ex situ et pulvere Scriptores Veteres in lucem produxit, Commentariis illustravit, atque ex suis multa addidit; Guid–Ubaldum[995] etiam, Virum Principem, et celebrem scriptorem; Cardanum[996] item, Tartagleam,[997] aliosque sua aetate celebres:

Certe Galilaeum[998] Vestrum (Magnum Magni Ducis Mathematicum) tacere non debeo; quem ut Novae Philosophiae Parentem veneror. Qui non modo Medicea Sydera, Optici Tubi sui beneficio, Orbi ostendit primus; aliaque in Coelis Phaenomena visu digna; ipsumque Mundi Systema, feliciter (si per ingratos licuisset) illustravit: Sed et Motus Physicos ad leges Mathematicas reduxit primus; eoque Verae Philosophiae viam aperuit; quam, ex eo tempore, Philosophi plures et Mathematici, eadem Schola oriundi, et Serenissimis Magnorum–Ducum auspiciis animati, dici non potest quantum promoverunt:

Addo Cavalerium;[999] qui, magno demonstrandi compendio, Methodum quam vocant Indivisibilium in Geometriam introduxit.[1000] Quae quanquam reapse non alia sit quam celebris illa Veterum, per Inscriptiones et Circumscriptiones Figurarum, (ad quam haec facile revocatur,) ad novam formam reducta; (non enim ille novum in Geometria Monstrum induxit, quod Recentiores aliqui reformident;) mira tamen ille brevitate et perspicuitate praestat, quod non nisi longis ambagibus illi; nec minori tamen certitudine, si caute administretur. Sicut nec Vieta,[1001] dum Arithmeticam Speciosam introduxit,[1002] induxit novas Demonstrandi leges, sed veteres expeditius administravit.[1003] Quem feliciter secuti sunt Oughtredus[1004] noster; et

---

1 taceam, *add.*

[994]Commandinum: i.e. Federigo Commandino (1509–75), Italian humanist and mathematician.

[995]Guid–Ubaldum: i.e. Guidobaldo, Marchese del Monte (1545–1607), Italian mathematician and astronomer.

[996]Cardanum: i.e. Girolamo (Geronimo) Cardano (1501–76), Italian mathematician and physician.

[997]Tartagleam: i.e. Niccolò Tartaglia (1499?–1557), Italian mathematician and engineer.

[998]Galilaeum: i.e. Galileo Galilei (1564–1642), Italian mathematician and astronomer.

[999]Cavalerium: i.e. Bonaventura Cavalieri (1598?–1667), Italian mathematician.

[1000]introduxit: e.g. in CAVALIERI, *Geometria indivisibilibus continuorum nova quadam ratione promota*, Bologna 1635.

[1001]Vieta: i.e. François Viète (1540–1603), French mathematician.

[1002]introduxit: e.g. in VIÈTE, *In artem analyticem isagoge*, Paris 1631.

[1003]administravit: e.g. in VIÈTE, *Apollonius Gallus, seu, Exsuscitata Apollonii Pergaei Περὶ ἐπαφῶν geometria*, Paris 1600.

[1004]Oughtredus: i.e. William Oughtred (1575–1660), mathematician and divine, *ODNB*.

190. WALLIS to COSIMO III DE' MEDICI, 9/[19] November 1670

Harriotus[1005] item noster, ex quo Cartesius (celato nomine) praecipua suae Geometriae Fundamenta mutuatus est.[1006] Nam ex Harrioti Algebra, (opere posthumo, Anno 1632 edito,) desumpsit ille, non modo Specierum mutationem a literis Majusculis in Minusculas; et potestatum designationem, per $q.\ c.\ qq.\ qc.$ &c. in Speciem toties positam quot sunt dimensiones, ut $a.\ aa.\ aaa.$ &c. (quae ubi numero nimis turgent, possunt appensa figura numerali designari, ut $a^4.\ a^5.$ &c.) omissa nomenclatura per Quadrata, Cubos, Surdesolida, &c: Sed, quod majoris est momenti, Aequationum Reductionem illam, qua tota ad unas partes posita| Nihilo aequetur: et, quod inde ortum ducit, Aequationum simpliciorum (sic reductarum) invicem ductu, Compositionem Altiorum: atque, quod hinc dependet, Multitudinis Radicum in qualibet Aequatione (vel possibilium vel saltem imaginarium) pro numero dimensionum in potestate suprema determinationem. Quae omnia, qui utrumque legerit, nulli dubium esse poterit, quin inde desumpta sint, sintque praecipua Geometriae Cartesianae Fundamenta; unde reliqua calculo eliciuntur. Sed ad Vestros redeo.

Cavallerii Methodum Indivisibilium, Torricellius[1007] Vester (Magni Ducis item Mathematicus) promovit[1008] feliciter, et illustravit. Quidque eidem superaddidit, mea Infinitorum Arithmetica,[1009] aliorum esto judicium, qui illam vel probe perpenderint et in usum redegerint, vel ea quae illius ope praestiterim ego consideraverint.

Eidem Torricellio debemus (praeter multa ab ipso ingenio se scripta) celebre illud quod ab eo nomen ducit, Experimentum Torricellianum;[1010] de Hydrargyro inverso Tubo suspenso. Quod tamen ipsum, si non originem, saltem ansam, debere poterit magno Galilaeo. Quam enim ille, ex eo quod Aquam ultra certam altitudinem nec suctione nec siphonum ope protrahi

---

12 imaginariarum *corr. ed.*
14 poterit, *add.*

[1005] Harriotus: i.e. Thomas Harriot (1560–1621), mathematician, astronomer, and ethnographer, *ODNB*.
[1006] mutuatus est: Wallis was convinced that Descartes had borrowed heavily from Harriot and argued for this in his *Treatise of Algebra*, London 1685.
[1007] Torricellius: i.e. Evangelista Torricelli (1608–47), Italian mathematician.
[1008] promovit: i.e. in TORRICELLI, *De sphaera et solidis sphaeralibus libri duo*, Florence 1644.
[1009] mea ... Arithmetica: i.e. WALLIS, *Arithmetica infinitorum*, Oxford 1656.
[1010] Experimentum Torricellianum: Torricelli first reported his experiment on the vacuum in his letter to Michelangelo Ricci of 11/[21] June 1644. This and other items from the correspondence with Ricci on the barometric experiment was first published by Carlo Dati (under the pseudonym Timauro Antiate) in *Lettera a Filaleti di Timauro Antiate della vera Storia della Cicliode e della famosissima esperienza dell' Argento vivo*, Florence 1663.

190. WALLIS to COSIMO III DE' MEDICI, 9/[19] November 1670

posse observaverit, suspicatus est Aeris Gravitatem Pressumque: Torricellius, liquore feliciter mutato, (quo Experimenta commodius administrari possent,) extra dubium posuit: eaque omnia, quae ad Fugam Vacui (celebre κρησφύγελον) relata prius fuerant, ad Pressum Aeris fuisse referenda.

Atque ex hoc uno; hem, quanta seges novorum indies succrescit Experimentorum: quae totam fere Naturalem Philosophiam ita penitus immutavit, ut Harvaei[1011] nostratis Circulatio Sanguinis,[1012] Anatomen Animalium: Atque, prae aliis, celeberrima illa Florentina, ante aliquot annos, sumptibus Mediceis instituta et edita.[1013] Sed et eidem felix accessio facta est ab Honoratissimi nostri Boylii[1014] (illustri familia et meritis nobilissimi) Organo Pneumatico[1015] ad exsugendum Aerem: quo innumeris Experimentis ab eo praestitis, confirmatur tum Aeris Gravitas, tum et Vis Elastica.

Addo insuper, hac occasione, rem plane non indignam, sed optandam potius, ut qui ante plures annos, sub ficto nomine Timauri Antiatis, prodiit in Torricellii Vindicias Tractatus[1016] Italicus (cujus praeter unicum exemplar, quod ad me transmittendum curavit Vir Nobilissimus Carolus Dati,[1017] quod aliis impertiendo non parcus fui, nescio an aliud ullum in Angliam advectum fuerit,) Latina etiam lingua (si non et Gallica) ederetur, quo in totam Europam spargeretur.

Taceo alios, adhuc in vivis, Viros magnos; Datum, Vivianum,[1018] Borellum,[1019] caeterosque, qui vel nominum celebritate, vel scriptis editis, literato Orbi innotescunt; Magalottum[1020] item et Falconerum,[1021] qui Oxonii

---

6 totam *(1)* ita *(2)* fere Naturalem Philosophiam ita

[1011] Harvaei: i.e. William Harvey (1578–1657), English physician, *ODNB*.

[1012] Circulatio Sanguinis: i.e. in HARVEY, *De motu cordis & sanguinis in animalibus, anatomica exercitatio*, Leiden 1639.

[1013] edita: i.e. [MAGALOTTI], *Saggi di naturali esperienze fatte nell' Accademia del Cimento*, Florence 1667.

[1014] Boylii: i.e. Robert Boyle (1627–1691), q.v.

[1015] Organo Pneumatico: e.g. in BOYLE, *New Experiments physico–mechanicall, Touching the Spring of the Air and its Effects*, Oxford 1660. Also published in Latin as *Nova experimenta physico–mechanica de vi aeris elastica, ex Angl. in Lat. conversa* [by R. Sharrock], Oxford 1661.

[1016] Tractatus: i.e. [DATI], *Lettera a Filaleti di Timauro Antiate*, Florence 1663.

[1017] Dati: i.e. Carlo Roberto de Cammillo Dati (1619–76),Italian scholar and member of the Accademia del Cimento.

[1018] Vivianum: i.e. Vincenzo Viviani (1622–1703), Italian mathematician.

[1019] Borellum: i.e. Giovanni Alfonso Borelli (1608–79), Italian physiologist and mathematician.

[1020] Magalottum: i.e. Lorenzo Magalotti (1637–1712), Italian scholar and diplomat, secretary to the Accademia del Cimento in Florence from 1660.

[1021] Falconerum: i.e. Paolo Falconieri (1638–1704), Italian architect and mathematician.

191. WALLIS to OLDENBURG, 15/[25] November 1670

aliquando dignati sunt me salutare:[1022] Ut quos omnes Serenissima Vestra Celsitudo rectius aestimare novit, quam mea tenuitas describere.

Unicum superest, ne nimius sim, ut exorare liceat Serenissimum Magnum–Ducem, quod facis, porro facere; hoc est, ut literis et literatis favere pergas; solidae praesertim Philosophiae instauratoribus: Ut quam coepit Philosophiam Lynceorum Academia, eandem ipsa perficiat; et non modo Medicea Sydera, sed et Medicea Philosophia, literato Orbe celebretur.

Serenissimae Vestrae Celsitudini
Devotissimus,
Johannes Wallis. S. T. D.
Geom. Prof. Oxon.

191.
WALLIS to HENRY OLDENBURG
Oxford, 15/[25] November 1670

Transmission:

W Letter sent: LONDON *Royal Society* Early Letters W1, No. 115, 2 pp. (our source). On p. 1 endorsements in Oldenburg's hand. At top of page: 'Read November 24:70 Entered L.B. 4. 117.', and at foot of page: 'Copy of a letter, recommending an Experiment to decide (*1*) a (*2*) the Controversy which is between Fabri and Borelli, touching the synchronisme in the fall of a (*a*) Body, moving (*b*) stone shot horizontally, and |of *add. and del.*| another (*aa*) moving (*bb*) descending perpendicularly.' On p. 2 beneath address, also in Oldenburg's hand: 'Rec. Nov. 23. 70.' Postmark: 'NO/23'.—printed: OLDENBURG, *Correspondence* VII, 283–5.

w^1 Copy of part of letter sent: LONDON *Royal Society* Letter Book Original 4, p. 117.

w^2 Copy of w^1: LONDON *Royal Society* Letter Book Copy 4, pp. 161–2.

Reply to: OLDENBURG–WALLIS 8/[18].XI.1670.

In its original form this letter was returned to Wallis undelivered. He subsequently sent it again with additional remarks added beneath his signature. Since the expanded (but not re-dated) letter was received by Oldenburg on 23 November, a Wednesday, it is probable that it was sent on the previous day, i.e. on 22 November. The letter was read at the meeting of the Royal Society on 24 November. The experiment which Wallis had proposed earlier and which he saw as a means of resolving the dispute between Borelli and Fabri

---

5-6 quam coepit... perficiat; et *add.*

[1022]salutare: Magalotti and Falconieri visited London in February and March 1667/8, during which time they attended meetings of the Royal Society and apparently also went to Oxford. See OLDENBURG–BOYLE 10/[20].III.1667/8 (OLDENBURG, *Correspondence* IV, 234–6). The following year they were members of the the entourage of the Duke of Tuscany when he visited Oxford.

191. WALLIS to OLDENBURG, 15/[25] November 1670

was finally carried out, after numerous postponements, on 26 January 1670/1 (old style). See Birch, *History of the Royal Society* II, 454, 461, 464–5.

Oxford Nov. 15. 1670.
Sir

I am sorry by yours[1023] of Nov. 8. to find my advertisement[1024] came too late. But your Swisse Balleville[1025] is not yet come at mee, nor the Transaction you mention. (As neither those of April, nor August.) It may be inserted in your next, with a particular advertisement,[1026] of the great mischief at Dover (on the Kentish coast) at the spring tides which happened a few days before All-hollantide this year. Of which the News letter[1027] (&, I think, the Gazette[1028]) took notice. Which I forgot to mind you of in my last. How it was then at London, I know not.

I have read over again that Proposition with the Demonstration, which Monsieur Hugens mentions;[1029] but find nothing, which upon a second reading will (probably) stick; if I know what it is, I shal be ready to satisfy him.

The Peruvian bark[1030] (or Jesuits powder as it is called) hath putt by one fitt of my Ague;[1031] but it is supposed (as it is usual with it) that it may in a weeks time return.

I am
Your friend to serve you
John Wallis.

6 mischief (*1*) do⟨ne⟩ (*2*) at Dover
15 it (*1*) will (*2*) may

[1023] yours: i.e. OLDENBURG–WALLIS 8/[18].XI.1670.

[1024] advertisement: i.e. Wallis's request for a margin note from the Latin copy of WALLIS–OLDENBURG 19/[29].III.1669/70 to be inserted in the English version of this letter, printed in *Philosophical Transactions*. He had made this request in WALLIS–OLDENBURG 3/[13].XI.1670.

[1025] Balleville: not identified.

[1026] advertisement: no notice to this effect was added to the printed version.

[1027] News letter: presumably the newsletter entitled *Newes*, edited by the London journalist Henry Muddiman (1628/9–92), *ODNB*.

[1028] Gazette: i.e. the *London Gazette* (formerly the *Oxford Gazette*), produced by the office of the government official and intelligence gatherer Joseph Williamson (1633–1701), *ODNB*. The *Gazette* was originally published in collaboration with Henry Muddiman and Roger L'Estrange (1616–1704), but Williamson transformed it into a government newspaper under his sole direction.

[1029] mentions: i.e. in HUYGENS–OLDENBURG [21]/31.X.1670; OLDENBURG, *Correspondence* VII, 216–18.

[1030] bark: i.e. cinchona.

[1031] Ague: Wallis suffered repeated attacks of quartan fever at this time.

191. WALLIS to OLDENBURG, 15/[25] November 1670

Since I wrote this, I have yours with the inclosed Transactions of October. But have not seen the man; who I hear is not well.

Sir

This letter hath been once at London allready but by reason of a mistake came back again. It now comes a second time with this Addition.

I find between Fabri & Borelli a controversy[1032] about matter of Fact or Experiment. Borelli supposeth (with many other) that a Stone thrown or Bullet shot Horizontally, as in HO, doth by reason of its gravity sink downwards according to a curve line HQ, (& so far they agree;) but, sayth Borelli, it will in the same time come at the Horizontal plain PQ, at Q, as if (without the motion of projection) it had fell directly down in the perpendicular HP. This Fabri denies; (& allegeth an experiment of Mersennus to that purpose;) & will have the motion of descent retarded by the additional Horizontall motion; supposing the descent in the curve to be, by the obliquity of the motion, hindered, as in Sloping Plains. I remember I have once formerly suggested[1033] the making of some Experiment by the Society, for the clearing this matter: And I could be content now to renew the same motion. For though I suppose most of us be rather of Borelli's then Fabri's opinion in it: Yet (especially since it is denyed) I think it might well deserve to be experimented.

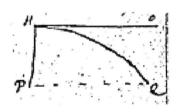

[1032]controversy: see WALLIS–OLDENBURG 3/[13].XI.1670.

[1033]suggested: there is no record of this earlier suggestion. Members of the Royal Society present at the meeting on 24 November 1670 would no doubt have have been able to recollect it. See BIRCH, *History of the Royal Society* II, 454.

One thing more. I find in Transact. numb. 46. (for Apr. 1669.) Mr Hugen's laws of motion printed;[1034] but not his Demonstration[1035] of them, which (as I remember) came with them. I desire you would favour mee with a Copy[1036] thereof to peruse; & I shall (if you so desire) return it to you, from Sir

Your friend to serve you,
J.W.

I have (upon taking the Peruvian Cortex) missed 3 or 4 fits of my Ague: And could hope myself rid of it, were it not frequent (after that Medicine) to have it return again.

These for Mr Henry Oldenburgh [2]
at his house in the Palmal
near St James's
London.

192.
WALLIS to HENRY OLDENBURG
Oxford, 24 November/[4 December] 1670

Transmission:

W Letter sent: LONDON *Royal Society* Early Letters W1, No. 116, 2 pp. (our source). On p. 2 beneath address endorsement in Oldenburg's hand: 'Rec. Nov. 25: 70. Answ. Dec. 1: 70. promised Diophantus and communication of Slusius letter, when he comes to London.' Postmark: 'NO/–'.—printed: OLDENBURG, *Correspondence* VII, 285–6.

Answered by: OLDENBURG–WALLIS 1/[11].XII.1670.

[1034] printed: i.e. 'A Summary Account of the Laws of Motion, communicated by Mr. Christian Hugens in a Letter to the R. Society', printed in *Philosophical Transactions* No. 46 (12 April 1669), 925–8; HUYGENS, *Œuvres complètes* V, 431–3. This Latin translation of Huygens's article, which had originally been published in French in the *Journal des Sçavans*, had been carried out by Oldenburg himself.

[1035] Demonstration: i.e. 'De motu corporum ex mutuo impulsu hypothesis Christiani Hugenii de Zulichem', which Huygens sent to Oldenburg with HUYGENS–OLDENBURG [26.XII.1668]/5.I.1669; OLDENBURG, *Correspondence* V, 282–3. It is now *Royal Society* Classified Papers III (i), No. 45; printed in HUYGENS, *Œuvres complètes* VI, 336–43. The *Tractatus de motu corporum ex percussione*, of which this enclosure was the beginning, was not published until after his death in *Christiani Hugenii . . . Opuscula posthuma*, ed. B. de Volder and B. Fullenius, Leiden 1703.

[1036] Copy: cf. WALLIS–OLDENBURG 24.XI/[4.XII].1670.

192. WALLIS to OLDENBURG, 24 November/[4 December] 1670

Oxford Nov. 24. 1670.

Sir,

Since mine[1037] by the last post, I find (amongst my papers) a Copy of M. Hugens's demonstration (as it first came) of which I sent[1038] to you for a copy: that, in case hee added no more afterward, you may save yourself the labour of sending the Copy I desired. I find his propositions printed,[1039] to be somewhat different (in words, not in sense,) from those, with the demonstrations, in my written paper: And that, in the printed Copy (without demonstrations) there bee divers added. Whether you have any further demonstrations of those added, I know not. My paper begins with: De Motu corporum ex mutuo impulsu Hypothesis. 1. Corpus quodlibet semel motum; &c. And ends with: Quod autem in navi contingit, idem in terra consistenti, uti diximus, evenire certum est. Igitur constat propositum.

But I had, a little before, Quae celeritates cum sint in proportione reciproca ipsarum magnitudinum, necesse est ut corpora AB ejusdem spectatoris respectu resiliant a contactu iisdem celeritatibus CA, CB; *Hoc enim postea demonstrabitur*.[1040] But of this promise, in my paper, there is no performance. Which makes mee suppose, there was somewhat to come after.

I have not yet read over all of Fabri[1041] against Borelli. That I have read gives mee no satisfaction. And hee doth manifestly cavil, very often, without any just cause. If you think fit to give any character[1042] of the book: I think it best to bee some such purpose as this. That he doth therein (in

---

19 over *(1)* ⟨—⟩ *(2)* all of
22 therein *(1)* by way of *(2)* (in six Dialogues)

[1037] mine: i.e. WALLIS–OLDENBURG 15/[25].XI.1670, which was probably re-sent on 22 November after having been returned to Oxford through postal error.

[1038] sent: i.e. in WALLIS–OLDENBURG 15/[25].XI.1670. The copy of Huygens's demonstrations which Wallis discovered among his papers after sending that request is now missing.

[1039] printed: i.e. 'A Summary Account of the Laws of Motion, communicated by Mr. Christian Hugens in a Letter to the R. Society', printed in *Philosophical Transactions* No. 46 (12 April 1669), 925–8.

[1040] Quae ... demonstrabitur: this passage is also found in Oldenburg's copy: *Royal Society Classified Papers* III (i), No. 45; HUYGENS, *Œuvres complètes* VI, 343. The demonstration was first published after Huygens's death in the complete tract *De motu corporum ex percussione*, prop. 8, in *Christiani Hugenii ... Opuscula posthuma*, ed. B. de Volder and B. Fullenius, Leiden 1703, II, 381; HUYGENS, *Œuvres complètes* XVI, 53.

[1041] Fabri: i.e. FABRI, *Dialogi physici*, Lyons 1669.

[1042] character: Oldenburg published a review of Fabri's *Dialogi physici* in *Philosophical Transactions* No. 67 (16 January 1670/1), 2057–9.

six Dialogues) write against, Grimaldi,[1043] Alfonsus Borelli,[1044] & Montanarius:[1045] who in divers things differ from what Fabri hath written. Against the first:[1046] *concerning Light*: [Against] the second,[1047] About *Motion & Percussion*: Against the third,[1048] About the *Ascent of Liquors in Tubes*, (as in the Torricellian Experiment, &c:) But whether hee have the better of those against whom hee writes; I shal not take upon mee to judge: but leave it to the Reader to think as hee shall see cause.

This is the present thought of

Yours to serve you
John Wallis.

These [2]
For Mr Henry Oldenburg,
in the Palmal near
St James's
London.

193.
GIOVANNI ALFONSO BORELLI to WALLIS
Messina, [26 November]/6 December 1670

Transmission:

C Letter sent: UPPSALA *Universitetsbibliotek* Ms it–00197, f. 001ᵃ–001ᵇ. Some textual damage through breaking of seal. Endorsement in Collins's hand on f. 001ᵇ: 'The Italian

2 Montanarius: *(1)* Against the *(2)* who in divers
3 first: *(1)* De Lumine *(2) concerning Light*:
3 Again the second, *corr. ed.*
6 writes: *(1)* I *(a)* do *(b)* shall not think fit to deliver my opi⟨nion⟩ *(2)* I shal not

---

[1043] Grimaldi: i.e. Francesco Maria Grimaldi (1618–63), S.J., sometime teacher of rhetoric and humanities at the College of Santa Lucia Bologna and close associate of Riccioli.
[1044] Borelli: i.e. Giovanni Alfonso Borelli, q.v.
[1045] Montanarius: i.e. Geminiano Montanari (1633–87), professor of mathematics at Bologna since 1664, later (from 1679) professor at Padua.
[1046] first: i.e. GRIMALDI, *Physico-mathesis de lumine, coloribus, et iride ... libri duo*, ed. H. Bernia, Bologna 1665.
[1047] second: i.e. BORELLI, *De vi percussionis liber*, Bologna 1667.
[1048] third: i.e. MONTANARI, *Pensieri fisico-matematici sopra alcune esperienze fatte in Bologna ... intorno diversi effetti de liquido in cannuccie di vetro, & altri vasi*, Bologna 1667.

193. Borelli to Wallis, [26 November]/6 December 1670

merch[ant] Capt. David Lambert at Redriffe, at Cherry Garden Chaires Prudence.'—
printed: Beretta, *A History of Non-Printed Science*, 114–15.
c Copy of letter sent: Cambridge *Cambridge University Library* MS Add. 4007 (B), f. 4ʳ–6ʳ (f. 4ᵛ and 5ᵛ originally blank). At top of f. 4ʳ in unknown hand: 'Borellius to Wallis Dec 6. 1670'.
Reply to: Wallis–Borelli 13/[23].I.1669/70.

Clarissimo Doctissimoque viro D. Johanni Wallisio in
Academia Oxoniensi
Matheseos professore Saviliano.
Jp. Alpons. Borellus S.

Accepi tandem (vir clarissime) diu desideratam tuam epistolam[1049] quae diuturnam moram longe majoribus beneficiis compensavit; quippequae dona amplissima[1050] et amicitiam pretiosam clarissimorum virorum D. Boylei, et D. Collinsii una cum tuis libris attulit, pro qua beneficentia summopere me tibi obstrictum profiteor. Doleo tamen libris Clar. Boyle frui non posse cum Anglici idiomatis prorsus ignarus sim; verum non despero auxilio anglorum hic degentium posse licet tardius eorum aliquam interpretationem consequi: interim ex titulis, ex figuris et ex genio authoris ex operibus ejus latinis praecognito conjecisse me puto quid in hisce contineatur. Quae non videntur recedere ab ea philosophia, quae valde mihi arridet.

Tua vero opera exprimere non possum quantopere me delectarunt, ob ingenii acumen et perspicaciam, inventionum copiam, et judicium adaequatum quo veriora a dubiis seligere conaris. Dolet tamen quod etiam apud vos non desint scioli qui livore aut ignorantia veritate[m] [i]nsectantes doctos distrahant, et in apologiis conficiendis bonas horas terere inutiliter cogant: hoc malum apud nos minus grave videtur, quia assidue nos vexat. Prodiit nuper ex Gallia dialogus[1051] religiosi cujusdam e societate Jesu Honorati Fabri qui integrum meum opus[1052] de vi percussionis superbissimo contemp[t]u rejicit. Huic profecto nil respondissem, nisi furti turpissimo nota me affecisset. Ergo coactus brevissime ei respondeo in prohemio opusculi[1053] de nupero Ætnae

---

22 contempu *corr. ed.*

[1049] epistolam: i.e. Wallis–Borelli 13/[23].I.1669/70.
[1050] dona amplissima: i.e. the parcel of books sent with Wallis–Borelli 13/[23].I.1669/70.
[1051] dialogus: i.e. Fabri, *Dialogi physici*, Lyon 1669.
[1052] meum opus: i.e. Borelli, *De vi percussionis liber*, Bologna 1667.
[1053] opusculi: i.e. Borelli, *Historia, et meteorologia incendii Ætnaei anni 1669, ... accessit Responsio ad censuras Rev. P. Honorati Fabri contra librum auctoris De vi percussionis*, Reggio di Calabria 1670.

193. BORELLI to WALLIS, [26 November]/6 December 1670

incendio quod sub praelo sudat, et quam primum ad vos mittam. Circa libros quos D. Boyle a me petit, video vos a titulis moveri dissitorum scriptorum, sicuti nos sepe decipimur; hujus generis sunt libri in scheda vestra adnotati, aliqui eorum sunt adeo pueriles ut contemptus ipsae raritatem eorum effecerit. Exercitationes mechanicae Marchetti[1054] non respondebunt tuae expectationi. Tractatus Rinaldini[1055] de curvis mediceis non est editus; sicut nec tractatus ejusdem argumenti Doctiss. Riccii;[1056] qui negociis Ecclesiasticis Romae implicitus mathematicos otiari sinit. Grimbergeri[1057] de luce et refractionibus opus postumum certe non extat. Sed ejus vice prodiit Grimaldi[1058] opus eodem titulo. Suspicor a nominis similitudine vos decipi. Hodiernae opera[1059] non sunt tanti facienda, is fuit parum matheseos peritus, edidit[1060] aliqua astronomica. Amicis tamen Romae, Pisis, Florentiae et Bononiae scripsi ut predictos libros undique perquirerent. Dumque eos expecto se obtulit repentina hujus navigii occasio. Non eam censui omittendam, mitto hinc qui apud me extant libros, reliquos liburni forsan transmittent amici. Interea, vir clarissime me amare perge et clariss. D. Boyleum et D. Collinsium salutes rogo. Vale.

Messanae 6. Decembris 1670.

Clarissimo Doctissimoque Viro D. Johanni Wallisio
in Oxoniensi Academia
Matheseos Professore Saviliano
Oxoniae.

Commendate Clarissimo D. Johanni Collins. Londini
Cum fasciculo librorum hoc nota signato. I. C. L.

[1054]Marchetti: i.e. MARCHETTI, *Exercitationes mechanicae*, Pisa 1669.

[1055]Rinaldini: i.e. RENALDINI, *Geometra promotus*, Padua 1670. This book had evidently not appeared by the time this letter was written.

[1056]Riccii: on account of his office in the Papal court, Ricci had increasingly little time for his scientific activity. Collins, who apparently read the present letter, reported the same to James Gregory in a letter probably written in November 1671. See COLLINS–GREGORY ?.XI.1671; TURNBULL, *James Gregory*, 193–205, 194; HOFMANN, 'Über die Exercitatio geometrica', 143.

[1057]Grimbergeri: no posthumous work on light and refraction by Grienberger is recorded as having existed, as Borelli correctly asserts.

[1058]Grimaldi: i.e. GRIMALDI, *Physico-mathesis de lumine, coloribus, et iride, aliisque adnexis libri duo ... opus posthumum*, ed. H. Bernia, Bologna 1665.

[1059]Opera: presumably HODIERNA, *Opuscoli del dottor Don Gio. Battista Hodierna*, Palermo 1644.

[1060]edidit: e.g. HODIERNA, *Protei caelestis vertigines seu Saturni systema*, Palermo 1657; idem, *De admirandis phasibus in sole et luna visis*, Palermo 1656; idem, *De systemate orbis cometici; deque admirandis coeli characteribus, opuscula duo*, Palermo 1654.

194.
HENRY OLDENBURG to WALLIS
1/[11] December 1670

Transmission:

Manuscript missing.

Existence and date: mentioned in Oldenburg's endorsement on WALLIS–OLDENBURG 24.XI/[4.XII].1670 and in OLDENBURG–BERNARD 1/[11].XII.1670; OLDENBURG, *Correspondence* VII, 292–3.
Reply to: WALLIS–OLDENBURG 24.XI/[4.XII].1670.
Answered by: WALLIS–OLDENBURG 12/[22].XII.1670.

As Oldenburg remarks in his letter to Bernard, written on the same day, he intended to congratulate Wallis on his recovery from illness and 'to acquaint him with some things, I have lately received from Slusius, and from Fermat'. His endorsement on WALLIS–OLDENBURG 24.XI/[4.XII].1670 indicates that he promised to show Wallis the copy of *Diophanti Alexandrini arithmeticorum libri sex ... cum commentariis C. G. Bacheti, & observationibus D. P. de Fermat*, which had recently arrived, and Sluse's recent letter, i.e. SLUSE–OLDENBURG 12/[22].XI.1670 (OLDENBURG, *Correspondence* VII, 246–52), when he next came to London.

195.
WALLIS to HENRY OLDENBURG
Oxford, 12/[22] December 1670

Transmission:

W Letter sent: LONDON *Royal Society* Early Letters W2, No. 34, 2 pp. (our source). On p. 2 in left margin in Oldenburg's hand: 'Dr Wallis' sheets to be procured by Lord Br. is that at large which his Hypothesis of motion printed in the Transactions, is an Epitome. There is somewhat peculiar in it, which makes me desire your Lordships opinion in it, both as to the truth of the proposition and strength of the demonstration. As to the agreement or disagreement with those of Dr Wren & M. Hugens your Lordship need not be sollicitous; For if they will allow their *dura* to be also Elastica, we shall agree well enough.' Beneath address, again in Oldenburg's hand: 'Received Dec. 14. 70. Answ. dec. 17. 70.' Postmark: 'DE/1–'.—printed: OLDENBURG, *Correspondence* XIII, 424–5.

Reply to: OLDENBURG–WALLIS 1/[11].XII.1670.
Answered by: OLDENBURG–WALLIS 17/[27].XII.1670.
Enclosure: WALLIS–BROUNCKER 12/[22].XII.1670.

Oxford. Dec. 12. 1670.

195. WALLIS to OLDENBURG, 12/[22] December 1670

Sir,

I have this morning sent away your Book[1061] of Fabry's Dialogues, with directions for it to be left for you at Mrs Lichfields[1062] lodging at a Poulterers house next door to Exeter house, towards the West: as being just in your common rode, & no trouble to call for it as you go by. I have read his 2^d & 3^d Dialogue, against Borelli; & his 4th against Montanarius: the two first, with some diligence; comparing them both with Borelli all along; & with Fabri's Tractatus Physicus de Motu Locali (being one of his 3 Volumes in $4°$,) which hath made the task the harder, & the time longer. But I am not satisfyed that hee hath at all the better of Borelli: And though I do in somethings differ from him, yet not at all upon the account of what Fabri says; to whom I do much lesse assent. And, so far as I can guesse by what he says of Montanarius (who's book[1063] I have not) I should take [Montanarius] to bee the more considerable by much. For, his whiffling with a few schole termes & notions, that Accidens non transit de subjecto in subjectum; that omne ens habet determinatam magnitudinem metaphysicam, & therefore Celeritas cannot a Quiete continue procedere, but must begin a certo gradu celeritatis, & go on per saltus; that Natura abhorret a vacuo, that natura nihil agit frustra, & therefore Impetus cessat quando finis ejus cessat, & upon no other account, (with such others,) are to mee very insignificant. I know not therefore what other character to give than what I formerly advised. For, to declare for him will not bee fit, unless there were more reason: And, to declare against him, would sett him a-wrangling; being (if I mistake him not by writings) a conceited quarrelsome man. As to what I wrote formerly,[1064] that hee dissented from Borelli in matter of fact; who affirms that a body Horizontally projected will in the same time come to the ground as if it had

3 for you at *add.*
13 Montarius *corr. ed.*
16-18 & therefore ... per saltus; *add.*
20 & upon ... account, *add.*
23 sett |him *add.*| a-wrangling
23 mistake |him *add.*| not
24 a *(1)* conceiting *(2)* conceited

[1061] Book: i.e. FABRI, *Dialogi physici*, Lyon 1669. Cf. WALLIS–OLDENBURG 24.XI/[4.XII].1670.

[1062] Lichfields: presumably Margaret Lichfield, daughter of Ann Lichfield (d. 1671), the Oxford printer.

[1063] book: i.e. MONTANARI, *Pensieri fisico-matematici sopra alcune esperienze fatte in Bologna*, Bologna 1667.

[1064] formerly: i.e. in WALLIS–OLDENBURG 24.XI/[4.XII].1670.

195. WALLIS to OLDENBURG, 12/[22] December 1670

of itself fallen right down; which Fabri denies: I sayd it from what I find in his Tractatus Physicus[1065] lib. 4. Theorem. 17. 18. 21. 26. 27. 28. 30. 31. 39. 46. Whether hee have since changed his mind or not, I know not: But when I expected to find the same asserted against Borelli, when he should speak to Borelli's 23^d chapter (where the contrary is positively affirmed against him) hee silently waves it, onely reciting that amongst some other particulars, without any opposition, but seeming in the grosse to admitt them all; onely slighting them as not new but formerly proved by Galileo, & not worth the taking notice of. (I cannot site the page, because the book is gone: but you could easily find it by the order.) Yet this makes me not the lesse of opinion; that the Society may do well to settle the truth by a just experiment.[1066]

I have not had time to read the rest of his book; & did not think fit to keep it longer, having now an oportunity of sending it.

But I have sent with it some sheets[1067] of what I am printing, to have my Lord Brouncker's opinion of it; which I desire you will, with your first opportunity deliver to him with the inclosed letter:[1068] & when he hath considered them, to return them back.

Monsieur de Monceaux[1069] was here this morning with your letter (since I had sent away your book.) My Ague's returning on mee, made mee not in a condition to wait on them abroad; & therefore I committed him & his friends to Mr Bernard's[1070] conduct.

---

1 had *add.*
3 I know not *add.*
4 But |when *add.*| I
6 that *(1)* which *(2)* amongst
9-10 (I cannot ... order.) *add.*

[1065]Tractatus Physicus: i.e. FABRI, *Tractatus physicus de motu locali*, Lyon 1646.
[1066]experiment: see BIRCH, *History of the Royal Society* II, 454.
[1067]sheets: i.e. proof sheets of third part of his *Mechanica: sive, de motu, tractatus geometricus.*
[1068]letter: i.e. WALLIS–BROUNCKER 12/[22].XII.1670.
[1069]Monceaux: i.e. André de Monceaux (17th century), F.R.S., son of one of Louis XIV's councilors and an acquaintance of Huygens. A former traveler in the Levant, he was elected F.R.S. on 15 December 1670. See BIRCH, *History of the Royal Society* II, 461, and HUYGENS–OLDENBURG [21]/31.X.1670; OLDENBURG, *Correspondence* VII, 216–8.
[1070]Bernard's: i.e. Edward Bernard q.v.

196. WALLIS to BROUNCKER, 12/[22] December 1670

At St Andrews day[1071] last, I presume they continued their President, Secretary, & Treasurer as before:[1072] the change of other members of the counsel is less material. No more at present but that I am,

Yours to serve you,
John Wallis.

For Mr Henry Oldenburg [2]
in the Palmal near St James's
London.

196.
WALLIS to WILLIAM BROUNCKER
12/[22] December 1670

Transmission:

Manuscript missing.

Existence and date: Mentioned in WALLIS–OLDENBURG 12/[22].XII.1670.
Enclosure to: WALLIS–OLDENBURG 12/[22].XII.1670.

In the morning of 12 December 1670 (old style), Wallis returned the copy of Fabri's *Dialogi physici* which Oldenburg had lent him. In the package he also enclosed some sheets of the third part of *Mechanica: sive, de motu, tractatus geometricus*, i.e. Chapter XI (*De percussione*), containing his general laws of motion, in order that he might obtain Brouncker's opinion on them. Later that day he sent WALLIS–OLDENBURG 12/[22].XII.1670, in which this letter for Brouncker, undoubtedly of the same date, was enclosed. In WALLIS–OLDENBURG 12/[22].XII.1670 he instructed Oldenburg to forward WALLIS–BROUNCKER 12/[22].XII.1670 to its addressee together with the printed sheets which he was to collect from the house of Margaret Lichfield. Oldenburg evidently attached a note of his own to the letter to Brouncker, corresponding to his notes on the cover of WALLIS–OLDENBURG 12/[22].XII.1670, in which he sought Brouncker's views on Wallis's laws of motion in relation both to their earlier presentation ('A Summary Account given by Dr. John Wallis, of the General Laws of Motion', *Philosophical Transactions* No. 43 (11 January 1668/9), 864–6) and to those of Huygens and Wren.

[1071] day: i.e. 30 November, the Anniversary Day of the Royal Society.
[1072] as before: as Wallis presumed, Brouncker was re-elected President, Daniel Colwall (?–1690) was re-elected Treasurer, and, alongside Oldenburg, Thomas Henshaw (1618–1700) was re-elected Secretary. See BIRCH, *History of the Royal Society* II, 456.

197.
HENRY OLDENBURG to WALLIS
17/[27] December 1670

Transmission:

Manuscript missing.

Existence and date: mentioned in Oldenburg's endorsement on WALLIS–OLDENBURG 12/[22].XII.1670.
Reply to: WALLIS–OLDENBURG 12/[22].XII.1670.

198.
WALLIS:
Lecture for the Prince of Orange
20/[30] December 1670

Transmission:

W Manuscript paper: OXFORD *Bodleian Library* MS Don. e. 12, f. 1^r–1^v. At top of f. 1^r endorsement in Wallis's hand: 'A lecture intended for the Prince of Aurange Dec. 20. 1670, if the Professors had read'.

William Henry, Prince of Orange and Nassau, the future King William III, visited England in late 1670. He was received by Charles II in Whitehall on 30 October (old style) and on 20 December he was created doctor of civil law at Oxford, in a convocation held in his honour in the Sheldonian Theatre. Wallis's endorsement suggests that the ceremonial visit was originally intended to include lectures in the Schools. See WOOD, *Life and Times* II, 206–11; *Fasti Oxonienses* II, 323–4.

Celsissime Princeps,
Proceres Illustros,
Celeberrima Corona,

Non mirabitur, Celsitudo Vestra, nobis Mortalibus tanta distantia infra positis oculos nictare, tanti splendoris luce perstrictos. At mirari forte posset, nisi ea esset Celsitudo Vestra ut cui nihil esse possset mirum; si ego ex Geometriae principiis demonstravero, (quod quidem si quod aliud inter Geometriae mirandum haberi solet,) Duas lineas, certa lege ductas, continuo sibi invicem appropinquare, seu propius accedere, nec tamen unquam, utcunque producantur, occursuras.

6 cui (*1*) mirum (*2*) nihil esse posset mirum
8-9 ductas, (*1*) ita continuo sibi invicem appropinquare, ut quantum (*2*) continuo ... seu propius

198. WALLIS: Lecture for the Prince of Orange, 20/[30] Dec. 1670

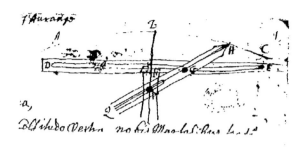

Mirum inquam hoc videre possit; nec immerito. Quippe illud fore videtur, minus attente rem intuentibus, instar communis notionis, sua luce clarum; Quod duo, continue appropinquantia, tandem coibunt.

At vero, Illustrum Nassoviorum Familiam, ut virtutibus aliis, et rei speciatim militaris peritia, ita et severioris matheseos cognitione jam olim instructos esse, testantur saltem Stevini opera.[1073] Quae ut multa continent in Mathesi cognitu dignissima; ita speciatim indicant quos in ea progressus fecerat, qui tum erat, Princeps Auriacus; in cujus gratiam conscripta erant. Quod facit ne existimem Auriaco Principi ea mira esse posse, quae aliis certe mortalibus videntur mira.| [1ᵛ]

Sunt autem in Geometria hujusmodi lineae, (quas *Asymptotas* vocant,) non unius generis; imo vero jam a veteribus plures observatae fuerunt, atque plures indies excogitantur. Asymptotas autem eas dicimus, quae utut continuo propius accedant, atque ita quidem ut minus tandem distent dato quovis intervallo, nunquam tamen sint coiturae.

Prae reliquis autem maxime celebres fuere, Hyperbola cum sua Asymptota recta, et Conchoides veterum cum regula sua.

Hyperbolae Asymptotam rectam, jam non attingo; propter ipsius naturam paulo intricationem: Conchoidis cum sua Regula $Ἀσυμπτωσίαν$ tradere contentus.

---

3 intuentibus, (*1*) prae (*2*) tanquam (*3*) instar
6 speciatim *add.*
7-8 continent in Mathesi *add.*
9 fecerat, (*1*) Auriacus Princeps (*2*) qui tum erat, Princeps Auriacus;
13 imo vero (*1*) indies plures (*2*) jam
15 quidem ut minus tandem distent *add.*

[1073]opera: i.e. STEVIN, *Les œuvres mathematiques de Simon Stevin, où sont insérées les Memoires mathematiques esquelles s'est exercé Maurice de Nassau, prince d'Aurenge. Le tout revue par A. Girard*, 6 vols. (in 2 parts), Leiden 1634. Maurice of Nassau, Prince of Orange (1567–1625), was Stadholder of the United Provinces of the Netherlands from 1585.

198. WALLIS: Lecture for the Prince of Orange, 20/[30] Dec. 1670

Hanc ut explicem; Intelligenda primum erit Recta Regula DE; atque in ea Canalis rectus quantumlibet productus. Intelligenda deinde erit Regula secunda FP, ad primam Normalis, seu ad Angulos rectos atque in ea Clavus aliquis ubivis fixus, ut in P, quem Polum vocant. Intelligenda demum Regula Tertia, quam Lineantem dicas; atque in ea clavum fixum ut G; qui per canalem DE ita hac illac moveatur ut inde non exeat; dum interim canalis alius in hac tertia regula, quantum opus est producta, polum P continuo amplectatur, eique circumequitet: Atque dum haec fiunt, haec regula lineans extremo suo puncto H circumducto, quod punctum lineans dicimus, describat ABC curvam. Atque tandem quos jam supponimus (phantasiae juvandae causa) canales DE, PQ, intelligantur latitudinis quantumvis exiguae, adeoque in lineas tandem degenerare; et clavos G, P, quantumvis exiguos; in duo itidem puncta degenerare. Ut saltem intelligatur rectae PGH punctum G, a DE recta non recedere; nec rectae BP punctum P a recta HGQ.

His positis; manifestum est 1°, Punctum lineans H, ad medium canalem DE (quam pro linea recta hic habemus) pervenire non posse, quantumvis protrahatur ABC curva. Cum enim rectae HGP Punctum G a recta DE non recedat; nec possit ea recta recedere a polo P: (per constructionem,) necesse est ut PGH recta rectam DE secet in G. Cum enim rectae PGH punctum P sit infra rectam DE, et punctum G, in ea recta, necesse est punctum ⟨reced⟩ere, supra illam.

Manifestum est 2°; propter G punctum in regula lineante fixum, adeoque eandem semper longitudinem GH; ipsius extremum H, quo longius a medio in utramvis partem recedit, eo propius ad DE accedere. Obliquius enim secat rectam DE, recta GH, quo longius abest a sectione recta seu perpendiculari. Idque eo (quod facile [demonstratum] est) ut puncti H altitudo supra DE rectam data quavis minor sit.

Est igitur puncti H, adeoque et Curvae ABC quam hoc describit, accessus ad DE regulam talis, ut data quavis distantia minus distet, nec tamen

---

4-5 Intelligenda (1) tertio Regula (2) demum Regula Tertia
6 hac illac add.
9-10 circumducto, (1) describat (2) quod punctum lineans (a) dici⟨—⟩ (b) dicimus, describat
10-15 Atque tandem quos jam supponimus |(phantasiae juvandae causa add.)| canales ... a recta HGQ. add.
20-22 Cum enim ... supra illam .add.
26 abest a (1) perpendiculari (2) sectione recta seu perpendiculari
27 demonstratu corr. ed.

unquam (quantumvis producantur) occurrat ABC curva Conchoidis cum recta DE regula. Q.E.D.

199.
WALLIS to JEAN BERTET
late 1670

Transmission:

Manuscript missing.

Existence and date: mentioned in COLLINS–VERNON 7/[17].II.1670/1; RIGAUD, *Correspondence of Scientific Men* I, 139–42, 140.

Collins reports in COLLINS–VERNON 7/[17].II.1670/1 that Wallis had written to the Jesuit mathematician Jean Bertet (1622–97) concerning the procurement through purchase or exchange of scientific books from France. Wallis had evidently been asked to assure Bertet that he could expect 'candid and upright dealing' from Collins, no doubt in view of the quantity of books he was being asked to obtain. After Bertet had initiated the correspondence with Collins in September 1670, Collins had sent Bertet a packet of books and written to ask him 'to procure many more'. Wallis's letter would have been sent sometime between the beginning of October and the end of the year. See COLLINS–GREGORY 29.IX/[9.X].1670; TURNBULL, *James Gregory*, 105–8, 106, and FLAMSTEED–COLLINS 1/[10].XII.1670; RIGAUD, *Correspondence of Scientific Men* I, 103–5, 104.

200.
CHRISTIAAN HUYGENS for WALLIS
late 1670–late 1671

Transmission:

C Draft of (missing) demonstration sent: LEIDEN, *Bibliotheek der Rijksuniversiteit*, Hug. 45, No. 483, 2 pp. (our source).—printed: HUYGENS, *Œuvres complètes* II, 170–3.
E^1 First edition: WALLIS, *Mechanica: sive, de motu, tractatus geometricus* III, 754–6.
E^2 Second edition: WALLIS, *Opera mathematica* I, 906–8 (our source).

Huygens first announced his successful measurement of the area between the cissoid and the asymptote in HUYGENS–WALLIS [27.VIII]/6.IX.1658 (WALLIS, *Correspondence* I, 522-31, 527). Wallis subsequently worked out a proof of his own employing the method for quadratures he had perfected in *Arithmetica infinitorum*. This proof was published as part of the epistolary tract addressed to Huygens contained in *Tractatus duo*, published in 1659: WALLIS, *Tractatus duo*, 81–90; *Opera mathematica* I, 545–50. Huygens refers to his own proof again in HUYGENS–WALLIS [21]/31.III.[1659]/1660 (WALLIS, *Correspondence* II, 11–12), but he did not send it to Wallis until after the publication of the second part of his *Mechanica: sive, de motu, tractatus geometricus* in the summer of 1670. There is no indication as to whether the demonstration was sent as an enclosure to a letter to Wallis or by some other means. After receiving Huygens' proof, Wallis published it as an appendix to the third part of his *Mechanica: sive, de motu, tractatus geometricus* in late 1671, this work providing the terminus ad quem for our dating. See WALLIS, *Mechanica: sive, de motu, tractatus geometricus* II, 532; *Opera mathematica* I, 905.

[C]

1658. Aprili.[1074]

APE est Cissoides Dioclis.[1075] ACB circulus a quo genita est. Semper nimirum $AC \infty EF$. Dico spatium $AVPEFB$ aequari triplo segmento $CBT + \triangle° ACB$. $\angle MBH \infty BAC$. $BM \infty AB$. $\mathcal{E}$[1076] triplum segm. $CBT + \triangle° ACB \infty$ sectori BKM + spatio; $HXMK$. nam segm. CBT seu $HMX \infty$ spatio; $HXMK$. $\angle HZM \infty 2\angle HBM$. $\mathcal{E}$ sector $ZHM \infty \frac{1}{2}$, sectori, BKM. $\mathcal{E}$ sector $ZHM \infty \triangle BHZ$ + spatio; $HXMK$. auferantur aequalia hinc $\triangle BHZ$, inde $\triangle ZHM$. fit spatium $HXMK \infty$ segmento HMX.

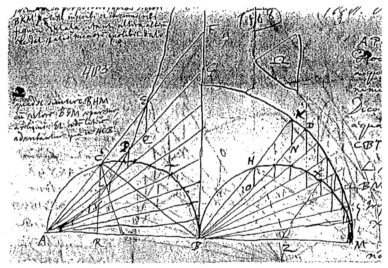

3 *[At top left of page in Huygens's hand:]* Premittenda propositio, quod sectori BKM potest inscribi et circumscribi figura dentata ita ut altera alteram excedat spatio minore quolibet dato.

7 $HMX \infty$ (*1*) segm (*2*) spatio
7-9 $\angle HZM$ ... segmento HMX. *add.*

[1074] 1658. Aprili: it is probable that Huygens first completed this proof in April 1658 (new style). See HUYGENS– SLUSE [26.III]/5.IV.1658; HUYGENS, *Œuvres complètes* II, 163–4.

[1075] Cissoides Dioclis: on the cissoid of Diocles see HEATH, *History of Greek Mathematics* I, 264–6.

[1076] $\mathcal{E}$: i.e. ergo.

200. Huygens for Wallis, late 1670–late 1671

Ergo ostendendum quod spatium $AVPEFB \infty$ sectori $BKM+$ spatio $HXMK$. si dicatur spatium cissoidis hisce minus esse; sit excessus horum Ω. Et inscribatur sectori BKM figura ordinate, ut duplum omnium trilineorum KND sit minus quam Ω. et spatio Cissoidis figura inscribatur ex totidem trapeziis. Ostenditur trapez. $EFGQ \infty \triangle BKN +$ trapezio HN. est enim trapez. EG ad $\triangle ACL$ ut $FG + EQ$ ad CL, (quia eandem habent altitudinem) hoc est ut $FA + AE$ ad AC. hoc est ut $AF + FC$ ad AC. h.e. ut $AB + BR \, . \, RA$.[1077] Ergo componendo, trapez. $EG + \triangle ACL$ ad $\triangle ACL$ ut $2AB$ ad AR. hoc est ut 2 quadratum AB ad qu. AC. sive ut 2 qu. BK ad qu. BH. hoc est ut $2 \triangle BKN$ ad $\triangle BHO$. Sed $\triangle BHO \infty \triangle ACL$. Ergo trapez. $EG + \triangle ACL$ aequale erit $2 \triangle BKN$. Et ablato hinc $\triangle° BHO$, inde $\triangle° ACL$, manet $\triangle BKN +$ trapez. HN, aequale trapez°. EG. et similiter de caeteris. Ergo figura in sectore inscripta $+$ trapeziis omnibus $HN \infty$ figurae in spatio cissoidis inscriptae. Sed figura in sectore assumens omnia KND, itemque trapezia HN assumentia omnia KND; ista omnia simul addita superant sectorem $+$ spatio $HXMK$. Ergo figura in sectore $+$ trapeziis HN hoc est figura in cissoide assumens spatium Ω, longe superabit sectorem $+$ spatio $HXMK$. sed ipsum Cissoidis spatium $+ \Omega$ aequatur ex hypoth. sectori $BKM+$ spatio $HXMK$. Ergo figura in Cissoide ipso Cissoidis spatio major erit. quod abs.| [2]

1 *[Alongside diagram in Huygens's hand:]* quidsi semicirc. BHM cum sectore BGM separetur, a reliquis. Et eaedem literae adscribantur quae in ACB? [For BGM read BKM.]

4 horum Ω. |Vel propius sic. Si dicatur primum, spatium Cissoidis minus esse sectori $BKM+$ spatio HMK. Ergo habemur jam duas magnitudines inaequales, quarum *add. and del.*| Et
5-6 et spatio ... trapeziis. *add.*
6-14 HN. |(*1*) quia (*2*) est enim trapez. ... ad $\triangle BHO$. (*a*) Ergo cum trapez. $EG+ \triangle ACL$ sit ad $\triangle ACL$ seu BHO, sicut $2 \triangle BKN$ ad $\triangle BHO$, aequale (*b*) sed $\triangle BHO \infty \triangle ACL$; Ergo trapez. $EG+ \triangle ACL$ ... trapez°. EG. *add.*| et similiter
16 ista omnia *add.*
18 hoc est figura in cissoide *add.*
19-20 sed ipsum ... $+$ spatio $HMXK$. *add.*

[1077] $AB + BR \, . \, RA$: i.e. $AB + BR$ ad RA.

Dicatur jam spatium $AEYGB$ majus sectore $BDM+$ spatio $IHMD$ sitque excessus Ω. Circumscribam jam sectori figuram. ut omnia DKN bis sumpta sint minora excessu Ω. Et ex totidem trapeziis circumscribam spatio Cissoidis. Sicut antea ostendetur trapez. $EG \infty \triangle BKN +$ trapez. HN. Ergo tota figura circumscripta Cissoidi aequalis circumscriptae sectori $+$ omnibus trapeziis HN. Sed ab his si demantur bis omnia trilinea DNK, residuum minus erit quam sector $BDM+$ spatio $IHMD$. (nam primum auferendo omnia DNK a circumscripta figura. sectori, relinquitur sector BDM: at eadem DNK auferendo a trapeziis HN, residua omnia simul, minora sunt spatio $IHMD$: quia additis rursus OHI, omnia simul aequantur demum

    2 *[To the left of diagram in Huygens's hand:]* $B\alpha\beta AB \infty 3 B\alpha\gamma$. Sed $\triangle A\alpha B - B\alpha\beta AB - B\alpha\gamma \infty AV\alpha\beta A$. $\mathcal{E} \triangle A\alpha\beta - 4B\alpha\gamma \infty AV\alpha\beta A$. Sed $\triangle A\alpha B - 4B\alpha\gamma \infty \square$, $\zeta B \div \triangle A\alpha B$. Ergo &c.

    2 *[To the right of diagram in Huygens's hand and encircled:]* Literae F et E, ubi G et Y, ponendae, quia rursus (*1*) vocandam figuram (*2*) vocandum spatium ut prius. item K et H in locum N et O.

    6 *[In left margin in Huygens's hand:]* NB. quod ultimum in figura circumscripta. Cissoidi est $\triangle APB$ (et non trapezium) aequale $\triangle°$ BSM in sectore.

6 spatio Cissoidis *add.*
11 trapeziis HN, (*1*) relinquiratur $DKHO$, (*2*) residua omnia simul
12 quia (*1*) $DKHO + OHI$ (*2*) additis rursus OHI

spatio $IHMD$.) Ergo si ab his ipsis, a figura nimirum circa sectorem + trapeziis NH hoc est a figura circumscripta spatio Cissoidis, auferatur Ω, reliquum multo minus erit sectore BDM + spatio $IHMD$. Sed spatium ipsum Cissoidis dempto Ω aequale dicebatur his ipsis. Ergo Cissoidis spatium majus erit figura sibi circumscripta, quod absurdum.

Hoc demonstrato quod spatium $AEYGB \backsim 3$ segmento $BVC+\triangle AVB$, facile ostendetur spatium infinitum $AEYBG$ aequale triplo semicirculo AVB.

Item quod spatium $AEYB \backsim$ triplo segmento BVC.

J'ay envoie[1078] a M. Sluze la demonstration de cecy; mais cellecy est plus belle. Je l'ay envoiée a M. Wallis qui l'a fait imprimer dans son traitè[1079] de Mechanique.

[E^2]

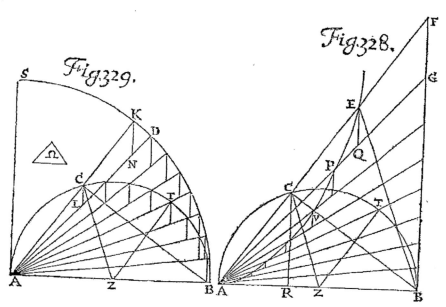

11 Hoc demonstratio ... belle *add. later*
11-12 Je l'ay envoiée ... Mechanique *add. later in crayon*.

[1078] envoie: probably HUYGENS–SLUSE [18]/28.V.1658; HUYGENS, *Œuvres complètes* II, 178–80.
[1079] traitè: i.e. WALLIS, *Mechanica: sive, de motu, tractatus geometricus* III, 754–6; idem, *Opera mathematica* I, 906–8.

200. HUYGENS for WALLIS, late 1670–late 1671

Sit ACB, semicirculus, (cui Centrum Z, tangens BF) & $AVPE$ Cissoides Dioclis inde genita: Cujus haec proprietas, ut sit ubique $AC = EF$. Atque eicdem Semicirculo (seorsum transcripto, quo vitetur linearum confusio,) circumponatur ABS circuli Quadrans (Centro A radio AB descriptus;) cui AC producta occurrat in K.|

Dico, *Spatium $AVPEFB$ aequari Triplo segmento CBT una cum Triangulo ACB*: Hoc est, Sectori AKB una cum Spatio $CTBK$. Est enim Segm. CBT = Spat. $CTBK$. Nam (juncta CZ) erit Ang. $CZB = 2$ Ang. CAB. Adeoque Sect. $ZCB = \frac{1}{2}$ Sect. AKB. Ergo Sect. ZCB = Triang. ACZ + Spat. $CTBK$. Auferantur aequalia, hinc Triang. ACZ, inde Triang. ZCB: Fit, Spat. $CTBK$ = Segm. CBT.

Ostendendum ergo, quod Spat. $AVPEFB$ = Sect. AKB + Spat. $CTBK$.

(Praesumitur autem, tanquam facile demonstratu, per notas exhaustionum methodos, *Sectori AKB Inscribi posse & Circumscribi figuram Dentatam, ita ut altera alteram excedat spatio minore quolibet dato: Et similiter, Spatio Cissoidali $AVPEFB$*.)

Si dicatur Cissoidis Spatium $AVPEFB$, minus esse quam Sect. AKB + Spat. $CTBK$: Sit horum excessus Ω. Et inscribatur sectori AKB figura ordinate, ut duplum omnium Trilineorum KND sit minus quam Ω. Et Cissoidis Spatio $AVPEFB$, figura inscribatur ex totidem trapeziis. Ostendetur Trapezium $EFGQ$ = Triang. AKN + Trapez. CN. Est enim Trapez. EG ad Triang. ACL, ut $FG + EQ$ ad CL, (quia eandem habent altitudinem;) Hoc est, ut $FA + AE$ ad AC; Hoc est, ut $AF + FC$ ad AC; Hoc est, (demissa perpendiculari CR) ut $AB + BR$ ad RA. Ergo, componendo, Trapez. EG + Triang. ACL ad Triang. ACL, ut $2AB$ ad AR; Hoc est, ut 2 Quadrat. AB ad Quadrat. AC; Hoc est, ut 2 Qu. AK ad Qu. AC; Hoc est, 2 Triang. AKN ad Triang. ACL. Ergo, Trapez. EG + Triang. ACL = 2 Triang. AKN. Et, ablato utrinque Triang. ACL, manet Triang. AKN + Trapez. CN = Trapez. EG. Et similiter de caeteris. Ergo, figura in Sectore Inscripta + Omn. Trapez. CN, = Figurae spatio Cissoidis Inscriptae. Sed figura in Sectore assumens omnia KND, item Trapezia CN assumentia omnia KND, ista inquam omnia simul sumpta superant Sectorem AKB + Spat. $CTBK$. Ergo, figura in Sectore + trapeziis CN (hoc est, figura in Cissoide,) assumens spatium Ω, longe superabit Sectorem AKB + spat. $CTBK$. Sed ipsum Cissoidis spatium $AVPEFB + \Omega$ aequatur ex hypothesi Sectori AKB + spat. $CTBK$. Ergo figura in Cissoide ipso Cissoidis spatio major erit. Quod est absurdum.

201. BERTET to WALLIS, beginning of 1671

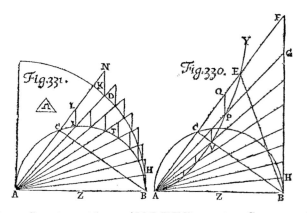

Dicatur jam Spatium idem $AVPEFB$, majus Sectore AKB + Spat. $CTBK$. Sitque excessus Ω. Et circumscribatur Sectori figura, ut omnia KND bis sumpta sint minora excessu Ω. Et Cissoidis spatio, figura ex totidem Trapeziis; (nisi quod, pro ultimo Trapezio, habeatur in Cissoide Triangulum AHB = Triang. AHB in Sectore.) Ostendetur, ut supra, Trapez. $PQFG$ = Triang. ADN + Trapez. LD. Ergo, tota figura circumscripta Cissoidi, aequalis circumscriptae| Sectori + omnibus Trapeziis LD. Sed ab his si demantur bis omnia Trilinea KDN, residuum minus erit quam Sector AKB + spat.$CTBK$. (Nam primum auferendo omnia KDN, a figura circumscripta Sectori, relinquitur Sector AKB: At eadem KDN auferendo a Trapeziis LD, residua omnia simul minora sunt spatio $CTBK$: Quin additis rursus spatiis LIC, omnia simul aequantur demum spatio $CTBK$.) Ergo, Si ab his ipsis, a figura nimirum circa Sectorem + Trapeziis LD; Hoc est, a figura Spatio Cissoidis circumscripta; Auferatur Ω: Reliquum multo minus erit Sectore AKB + spat.$CTBK$. Sed spatium ipsum Cissoidis dempto Ω aequale dicebatur his ipsis. Ergo Cissoidis spatium majus erit figura sibi circumscripta. Quod est absurdum.

Hoc itaque demonstrato, Quod Spat. $AVPEFB$ = 3 Segm. CBT + Triang. ACB: facile ostendetur, Quod *Spatium infinitum* $AVPEYFB$ = 3 Semicirc. ACB.

Item, *Quod Spatium* $AVPEB$ = 3 Segment. CBT.

201.
JEAN BERTET to WALLIS
beginning of 1671

Transmission:

Manuscript missing.

202. BROUNKER to WALLIS, 15/[25] February 1670/1

Existence and date: As reported by Collins in COLLINS–BERNARD 16/[26].III.1670/71, Bertet had recently written letters to him and to Wallis, both of which had miscarried. Since Collins and Wallis had written to Bertet in late 1670, it is probable that Bertet's replies were sent around the beginning of the new year.

202.
WILLIAM BROUNCKER to WALLIS
London, 15/[25] February 1670/1

Transmission:

C Letter sent: LONDON *Royal Society* Early Letters B2, No. 9, 4 pp. (p. 4 originally blank) (our source). Endorsement on p. 1 in Oldenburg's hand: 'Enter'd L.B. 4. 207.' On p. 4 further endorsement in Oldenburg's hand: 'The Lord Brouncker's letter to Dr Wallis concerning (*1*) some (*2*) a Difference about the time of the publication of some writings of the Doctor.'
c^1 Copy of letter sent: LONDON *Royal Society* Letter Book Original 4, pp. 207–10.
c^2 Copy of c^1: LONDON *Royal Society* Letter Book Copy 4, pp. 280–4.

Evidently at Wallis's behest, Brouncker confirms historical details concerning three of his writings, namely that his *Tractatus duo* had been written before the arrival in England of Pascal's *Lettres de A. Dettonville*, that his letter to Brouncker of 17 December 1657 had been written before the arrival of Frenicle's *Solutio duorum problematum circa numeros cubos et quadratos*, and that he had sent the results of his investigations on Fermat's negative theorem, carried out in the summer of 1668, in response to an anonymous letter to Brouncker which he had seen shortly beforehand.

Ad Clarissum Virum Dn. Johannem Wallisium SS Th. D.
et Geometriae Professorem Savilianum Oxoniae,
Honoratissimi Domini Vice-Comitis Brounckeri Epistola.

Duo sunt aut tria, Vir Clarissime, in quibus fidem appellas meam, sed ea sunt, in quibus, et fide optima possum et lubentissime Tibi testimonium perhibere.

    Primum illud est, quod in praefatione ad tuum de Cycloide Tractatum[1080] (et alibi) occurrit; Nempe jam diu scriptum fuisse tractatum illum antequam prodierit Dettonvillii liber.[1081] Quod mihi optime constare oportet, quoniam jam per aliquot menses, antequam hic comparuit Dettonvillii

---

[1080] Tractatum: i.e. WALLIS, *Tractatus duo*, Oxford 1659.
[1081] liber: i.e. PASCAL, *Lettres de A. Dettonville contenant quelques-unes de ses inventions de Geometrie*, Paris 1659.

202. Brounker to Wallis, 15/[25] February 1670/1

Tractatus, Tractatum illum tuum perlustrandum ad me miseris, et examinandum, (ne forte calculi lapsus alicubi obrepserit) quem et summa cura examinavi, nec ullum vel in Demonstrationibus, vel in calculo lapsum deprehendi emendandum; nec, credo, multo ante Parisiis prodiit, Domino Carcavio[1082] tum mittente tanquam recens editum. Et quidem cum paulo post, eodem anno, prodierit liber tuus (dum recens adhuc erat memoria) nihil observare potui (saltem quod ullius momenti sit) vel additum vel immutatum ab eo quod in scriptis ante legeram. Te autem ex Tractatu illo Dettonvillii in Tuis adjutum esse, aut tua inde desumpta, ne suspicandum quidem esse (quicquid insimulare velint, qui opus illud extenuatum eunt) res ipsa clamat, ne testimonio opus sit. Est enim Tua methodus ab ejus plane diversa. Nam quod Tu totius negotii fundamentum ponis, unde caetera deducis, nempe distributionem Cycloidis in portiones, quae correspondentium circuli portionum triplae sint, ille ne attinget quidem; et quidem videtur non considerasse; neque enim existimandum est, eum tam bellam speculationem, et praesenti negotio tam utilem, celare voluisse, si novisset.

Quod eo adhuc evidentius patet, quoniam ea quae solidum circa basin spectant, tanquam difficiliora proponit, quam sunt ea quae spectant solidum circa Axem; cum interim, qui distributionem illam noverint, inveniant, ea his longe faciliora; quod ex tua methodo patet.|

Sed et Figuram sinuum versorum (ipsi quidem inutilem, propter non consideratam illam Cycloidis distributionem,) ille non attinget, quam Tu in toto processu nusquam non adhibes.

Quapropter etiam ipsius, quam vocat[1083] *Cycloidis Sociam*, non nisi Sectione Sphaerae, et Cylindri, constructam exhibet, quae tamen alia non est quam sinuosa curva Figuram sinuum versorum terminans, Sectione Plani et Cylindri simplicius exhibenda; Quod Tu alicubi ostendis.

Verum rem extra omne dubium ponit, si (quod ais[1084]) Tractatus Tui Sceleton, artic. 55. primoribus contentum, Parisios ad eos jam ante annum miseras, ne suspicioni ullus locus sit, cum ex Dettonvillii Tractatu, post annum edito, mutuatum esse.

[1082] Carcavio: i.e. Pierre de Carcavi (1600?–84), French mathematician, custodian of the royal library in Paris from 1663, and founding member of the Académie royale des sciences.
[1083] vocat: i.e. Gilles Personne de Roberval (1602–75). See PASCAL, *Historia trochoïdis sive cycloïdis; gallicè la roulette*, Paris 1658, 5–6.
[1084] ais: WALLIS, *Tractatus duo*, 16; *Opera mathematica* I, 507; *Correspondence* I, 522. According to Wallis, the solution to Pascal's first challenge which he sent, as instructed to Carcarvi, in his letter of 19/[29] August 1658, consisted in the first fifty-five paragraphs of the solution he eventually published.

202. BROUNKER to WALLIS, 15/[25] February 1670/1

Alterum est, Literas[1085] tuas Decemb. 17. 1657. datas (commercii Epistolici, Epist. 17.) jam ante scriptas fuisse, quam ad nos pervenit Freniclii Tractatus.[1086] Quod ambigere nonnunquam videri volunt Fermatius, Frenicliusque, quo solutiones nostras extenuent.

Id autem ego fide optima possum affirmare. Nam literas eas jam aliquandiu hic habueram in manibus meis; atque ipsum autographum Parisios misissem, si Dominum White[1087] (cujus opera in transmittendis utenda erat) convenire potuissem, antequam Freniclii liber ille ad me pervenit (jam nuperrime Parisiis allatus) qui et serius adhuc aliquanto ad Te pervenit, hinc mittendus.

Sed nihil est utcunque, quod inde mutuatum suspicentur, cum nihil ibidem sit quod mutuari possit, quod non jamdudum ante, nos exhibueramus. Nam quod ad Fermatii quaestionem de Numero quadrato inveniendo, qui in non quadratum ductus unitate addita faciat quadratum, (de qua illic agitur;) exhibet quidem Freniclius exempla aliquam-multa, praeter illa duo quae exhibuerat Fermatius; de methodo vero, qua ad illa pervenit, nihil habet quod nos nondudum ostenderamus. Nam una illa Regula, nempe quoties $na^2 \sim c^2$ dividat $2ae$ (seu, quod eodem recidit, $nr^2 \sim s^2$ dividat $2rs$,) Quotiens est Radix Quadrati quaesiti, (Epist. 14. Oct. 22. 1657.exhibita,[1088] et epist. 16. Nov. 21. 1657. ad eos missa,[1089]) aequipollet| praeceptis ejus omnibus, quae per continuas decem paginas habet de usu numerorum suae quartae columnae; ut utraque comparanti liquido constabit. Saltem si illud alterum addas (eadem epist. 14. traditum) inventis quadratis pro quovis exhibito non-quadrato, haberi etiam quadratos pro ejusdem non-quadrati multiplo per quadratum quemvis; nempe, quadratos illos (quotquot sunt hujus divisionis capaces) per hunc dividendo. Ut omnino frustra causentur, Freniclii librum nobis visum fuisse ante scriptam epistolam 17. (quod tamen secus est;) aut causentur inde in hanc desumptum quicquam; cum totum

---

[1085]Literas: i.e. WALLIS–BROUNCKER 17/[27].XII.1657 (WALLIS, *Correspondence* I, 342–57).

[1086]Tractatus: i.e. FRENICLE DE BESSY, *Solutio duorum problematum circa numeros cubos et quadratos*, Paris 1657.

[1087]White: i.e. Thomas White (1593–1676). White conveyed many of the letters between England and France in the course of exchanges on Fermat's challenges on number theory. His name ('Mr le Blanc') is also appended to the copies of Pascal's challenges on the cycloid contained in the Savile Collection of the Bodleian Library (Savile G 8 i,2 and i,3), suggesting that White conveyed these, too.

[1088]exhibita: i.e. BROUNCKER–WALLIS 22.X/[1.XI].1657 (WALLIS, *Correspondence* I, 317–8).

[1089]missa: i.e. WALLIS–DIGBY 21.XI/[1.XII].1657 (WALLIS, *Correspondence* I, 320–42).

illud quod inde desumi posset, nos ante habueramus, atque indicaveramus ipsis.

Tertium est, de epistola[1090] ad me scripta, qua respondetur Anonymi epistolae,[1091] aliquot annos post editum commercium Epistolicum editae. De qua forsan imposterum non minus erunt suspicaces quam de reliquis, saltem siquid posthac prodierit, quod eo spectare possit. Verum ego demonstrationes inibi contentas, singulatim ad me transmissas, jam ante plures annos apud me habui, ipsamque illam responsoriam (in Epistolae formam redactam) aliquandiu. At nondum prodiit quicquam, quod scimus, vel a Fermatio, vel a Freniclio, vel quoquam alio, quod eo spectet, vel typis editum vel scripto transmissum, unde quicquam lucis ad Demonstrationes illas formandas habere posses. Verum quidem est rumorem aliquem esse de Fermatii operibus a filio suo edendis[1092] (in quibus quid contineatur ignoramus:) At certe (sive illa brevi edenda sint, sive, jam apud Gallos edita,) ipsorum exemplar nullum, in Angliam, hactenus allatum est; ut nullus sit suspicioni locus, Te inde adjutum fuisse; saltem cum Demonstrationes ipsae, dudum scriptae, jam ante plures annos fuerint ad me transmissae, et depositae mecum.

Atque haec sunt, quae de hoc negotio scripto consignare visum est, ut siquando occasio fuerit testimonio utendi meo, praesto habeas. Vale.

Londini Febr. 15. 1670/1.
Tuus amicus fidelissimus
& observantissimus
Brouncker

203.
John Collins to Wallis
February/early March? 1671

Transmission:

Manuscript missing.

Existence and date: Referred to in Collins–Bernard 16/[26].III.1670/1.

[1090]epistola: i.e. Wallis–Brouncker VIII.? 1668 (Wallis, *Correspondence* II, 573–92).
[1091]epistolae: not ascertained.
[1092]edendis: Fermat's eldest son and executor, Clément-Samuel de Fermat (c.1630–90), republished Bachet's edition of Diophantus's *Arithmetica* with his father's *Observationes*, some letters, and Jacques de Billy's *Doctrinae analyticae inventum novum* in 1670. In 1679 he published what he had been able to gather of his father's remaining papers under the title *Varia opera mathematica*.

204. COLLINS to BERNARD, 16/[26] March 1670/71

Collins reports to Bernard in COLLINS–BERNARD 16/[26].III.1670/1 that exactly three weeks earlier, i.e. on 23 February 1670/1, he had sent Wallis proofs of part of Horrox's *Opera posthuma* by carrier to Oxford, and that he had since written to him but received no answer. Correctly, Collins supposes in his letter to Bernard that Wallis had been incapacitated by another bout of quartan ague.

204.
JOHN COLLINS to EDWARD BERNARD
London, 16/[26] March 1670/71

Transmission:

C Letter sent: OXFORD *Bodleian Library* MS Smith 45, pp. 61–4 (p. 63 blank). On p. 64 note in Collins's hand: 'I have heard from Dr Wallis'. Postmark: 'MR/16'.

Answered by: BERNARD–COLLINS 3/[13].IV.1671 (RIGAUD, *Correspondence of Scientific Men* I, 158–60).

Mr Barnard
Worthy Sir

I will not goe about to detaine you with a Discourse to intimate how happy it is for a Man inferioris Subsellii, and a Non-Academick to have the honour of the Acquaintance with the learned, such as you are, but to come a little nearer in another respect I make knowne, that by vertue of your friendship with Mr Vernon[1093] at Paris, I wrote[1094] to him, and at my request he went to visit Pere Bertet (with whome I correspond) from whome he received a Box of Bookes, to be transmitted to me, with the first Conveniency that shall happen, and when arrived, you that are so great a Master of Literature, and willing to oblige the Republik of Learning by your Labours, may expect to heare a further Account of them, and perchaunce of another Box at Sea sent by Borellius to Dr Wallis with whome I ioyned[1095] in sending Borellius a present of Bookes, I have remitted to the said Mr Vernon 50 Corones to pay for the said Bookes, and buy others, out of which he is also intreated to pay Monsieur Duhamel[1096] now absent at his Abbey in Normandy, And

---

[1093] Mr Vernon: i.e. Francis Vernon, q.v.
[1094] wrote: i.e. COLLINS–VERNON 7/[17].II.1670/1; RIGAUD, *Correspondence of Scientific Men* I, 139–42.
[1095] ioyned: i.e. WALLIS–BORELLI 13/[23].I.1669/70.
[1096] Duhamel: i.e. Jean Baptiste Duhamel (1624–1706), French astronomer.

204. COLLINS to BERNARD, 16/[26] March 1670/71

I have Conveniency to enorder more at any time to be paid Mr Vernon in Paris, and I hope he will be willing to take that paines, and I may the rather beleive he will especially being thereto inclined by your next Letters and ere long he promised to write to you, In your Answer you may be pleased if he be unwilling to be at the trouble himselfe, or in case of his Returne from Paris, to intreat him to provide us some diligent correspondent there, that may informe us of, and furnish us with, (by Pere Bertets Assistance) such new Bookes of Italy, and France, as we shall desire, and money shall not be wanting for that purpose, Inclosed in Mr Vernons Letter I received (from Pere Bertet) an elegant Manuscript[1097] intituled Cogitata de Acceleratione motus gravium, by the Noble and Learned Geometer Reignault[1098] of Lyons, with leave and direction to get the Lord Brouncker and Dr Wallis, to peruse the same, and give their Censure thereof, the said MS &c shall be at your perusall, when you returne to London, a Letter[1099] in which P. Bertet wrote to Dr Wallis as likewise to my selfe is miscarried, Dr Wallis his Comment[1100] on the Astronomicall remaines of Horrox, is to goe into the Presse here, and there is a new type provided for the same, the Doctor desired to peruse it first, that he might adde a running title to the Topp, I sent it on this day three Weekes by Dobbins Moores Coachman, giving notice to the Doctor thereof by the Post and since wrote[1101] to the Doctor, but receiving no answer am afraid the Doctor is by his Disease incapacitated, or under some great affliction, If you were lately with him he could accquaint you that these Mathematicall Bookes were lately come out in Italy more than my former Letters mentioned

Borellius de Liquidis[1102] (his Comments on Archimedes are to be printed at Lyons)

6 to (1) provide (2) intreat him to provide
19 notice (1) thereof (2) to the Doctor thereof

[1097] Manuscript: not identified. Cf. VERNON–OLDENBURG 8/[18].III.1670/1; OLDENBURG, *Correspondence* VII, 496–8, 498. There is no record of this manuscript ever being sent to Brouncker or Wallis.

[1098] Reignault: i.e. François de Raynaud (Regnauld) (17th century) S.J., former pupil of Honoré Fabri and mathematician in Lyon.

[1099] Letter: i.e. the now missing letter BERTET–WALLIS beginning of 1671.

[1100] Comment: i.e. Wallis's Epistola nuncupatoria, addressed to William Brouncker, which prefaced his edition of Jeremiah Horrox's *Opera posthuma*, published in 1673.

[1101] wrote: i.e. COLLINS–WALLIS II/III.1670/1.

[1102] de Liquidis: i.e. BORELLI, *De motionibus naturalibus a gravitate pendentibus*, Reggio di Calabria 1670.

204. COLLINS to BERNARD, 16/[26] March 1670/71

Mengolus his Body of Musick[1103]
Gottignies Dioptricks[1104]
Honorato Fabri's Comment on Archimedes[1105]
In France a Capucine hath lately publisht Dioptricks Speculative[1106] and Practicall in fo.
Fermats Diophantus[1107] is not yet to be bought in Paris

[2] One Mr Walter Long[1108] a late Student in Oxford, and one of his| Majesties Pensioners being enioyned to reside in Holland to learne Fortification, is come over from thence, and suddainly returnes, I formerly remited him some Monies to buy Bookes, and after his arrivall he promiseth to send over Kinckhuysens[1109] Workes, and other good Dutch Mathematicall Bookes, Mr Streete[1110] presumes you may much oblige him, in relation to a treatise of Astronomicall tables he intends to publish, by communicating some ancient Observations, if such are to be found in your Libraries, a Transcript of what he desires followeth.

I have left with Mr Pitts[1111] to be sent you, according as you desire Seneschel[1112] about the time[1113] of our Saviours Nativity and Passion, any

---

[1103] Body of Musick: i.e. MENGOLI, *Speculazioni di musica*, Bologna 1670.

[1104] Dioptricks: there was considerable talk of a book on dioptrics by Gottignies in 1671. See DODINGTON–OLDENBURG [20]/30.I.[1670]/1671; OLDENBURG, *Correspondence* VII, 405–6; COLLINS–GREGORY 15/[25].XII.1670; TURNBULL, *James Gregory*, 137–41. Almost a year later there was still no news of the work. See OLDENBURG–SLUSE 4/[14].III.1671/2; OLDENBURG, *Correspondence* VIII, 571–4.

[1105] Comment on Archimedes: mention is made of Fabri's work on Archimedes in Bertet's letter to Collins of late 1670, copied in COLLINS–GREGORY 15/[25].XII.1670; TURNBULL, *James Gregory* 137–41. Dodington confirmed the existence of this work in DODINGTON–OLDENBURG [20]/30.I.[1670]/1671; OLDENBURG, *Correspondence* VII, 405–6.

[1106] Dioptricks: i.e. CHERUBIN, *La dioptrique oculaire, ou la theorique, la positive, et la mecanique, de l'oculaire dioptrique en toutes ses especes*, Paris 1671.

[1107] Diophantus: i.e. DIOPHANTUS, *Diophanti Alexandrini arithmeticorum libri sex*, ed. and transl. C.G. Bachet de Méziriac, with notes by P. Fermat, ed. S. Fermat, Toulouse 1670.

[1108] Long: possibly Walter Long (*c*.1648–1731) of Wraxall, Wiltshire, who matriculated at Trinity College, Oxford in April 1664. Cf. FLAMSTEED–COLLINS 1/[10].VIII.1671; RIGAUD, *Correspondence of Scientific Men* II, 118–22.

[1109] Kinckhuysens: i.e. Gerard Kinckhuysen (1625–66), Dutch mathematician.

[1110] Streete: i.e. Thomas Streete (1621–89), Irish-born astronomer and astrologer, active in London, *ODNB*.

[1111] Pitts: i.e. Moses Pitt (1639–97), London printer and bookseller.

[1112] Seneschel: i.e. Michael Seneschal (1606–73), S.J., Dutch theologian, professor at the University of Douai.

[1113] about the time: i.e. SENESCHAL, *Trias evangelica, sive, Quaestio triplex de anno, mense, et die Christi nati, baptizati, et mortui*, Liège 1670.

204. COLLINS to BERNARD, 16/[26] March 1670/71

Bookes I have, you may commaund the use of, and on the other side If I now and then trouble you for a Booke from Oxford, that we cannot have at London, I doubt not of your Courtesy and friendship, at present I want (to send to the Jesuite[1114]) Mr Boyles[1115] Treatise de Origine formarum lately turned into Latin and printed in 12º, the which is not here to be had, and if you can meete with Walker[1116] of Iustification[1117] against John Goodwin[1118] a quarto booke printed about 1643 the 12º booke intituted Socinianisme[1119] discovered is not intended) I should be glad of the same, but much more of your returne to London, that you may viva voce receive the thankes of

Sir
Your most humble
affectionate Servitor.
John Collins

from my house next the
three Crownes in Bloomsbury
Market 16 March 1670/71

Auncient Observations are desired besides those related by Ptolomy,[1120] the older the better to witt

Of the Appules of the Moone and Planets to fixed starres

In those of the Moone the exact time by the rising setting or Altitude of the Moone, or some starr will be requisite, with the Place viz in or neare what Citty the observation was made

And if at the time of the beginning or end &c of a Lunar Eclipse it will be so much the better, ♀ her neare Coniunctions with fixt Starres &c, at or neare the time of her greatest Elongation from ☉ and at other times

Observations of ♄ his close Coniunctions &c with fixt starres and of ☿ he being within his greatest distance from ☉ the nearer the Sun the better

---

[1114] Jesuite: presumably Jean Bertet, q.v.

[1115] Boyles: i.e. BOYLE, *Origo formarum et qualitatum*, Oxford 1669.

[1116] Walker: i.e. George Walker (1582?–1651), Church of England clergyman, *ODNB*.

[1117] Iustification: Walker attacked John Goodwin's views on justification in *A Defence of the true sence and meaning of the words of the Holy Apostle, Rom. chap. 4, ver. 3, 5, 9*, [London] 1641. Goodwin set out his position fully in *Imputatio fidei, or, A treatise of justification*, London 1642.

[1118] Goodwin: i.e. John Goodwin (*c*.1594–1665), Independent minister, *ODNB*.

[1119] Socinianisme: i.e. WALKER, *Socinianisme in the Fundamentall point of Justification discovered and confuted*, London 1641.

[1120] Ptolemy: i.e. in Ptolemy's *Almagest*.

If any such are in the Arabick Manuscripts The Arabick texts (with their translation into Latin) are desired without such the middle Motions cannot be certainly determined

To the Worthy and Learned
Mr Edward Barnard fellow
of St Johns Colledge
In Oxford

205.
WALLIS to JOHN COLLINS
?March 1670/1

Transmission:

Manuscript missing.

Existence: mentioned in and answered by COLLINS–WALLIS 21/[31].III.1670/1.

206.
JOHN COLLINS to WALLIS
[London], 21/[31] March 1670/1

Transmission:

C Draft of letter sent: CAMBRIDGE *Cambridge University Library* MS Add. 9597/13/6, f. 215^r–216^v (our source). On f. 216^v in Collins's hand: 'To Dr Wallis the 21 of March 1671 About drawing of tangents to Curves for Æquations'.—printed: RIGAUD, *Correspondence of Scientific Men* II, 525–7.

Reply to: Wallis–Collins ?III.1670/71.

Collins evidently sent as enclosures to this letter a copy of a letter from James Gregory on his method of equations and a copy of SLUSE–OLDENBURG [27.II]/9.III.1670/1 (OLDENBURG, *Correspondence* VII, 477–81).

Reverend Sir

I have yours[1121] of the ... instant wherein you mention the Printing of Mr

---
9 the ... instant *Space left for inserting date*

[1121] yours: i.e. WALLIS–COLLINS ?III.1670/71.

206. COLLINS to WALLIS, 21/[31] March 1670/1

Merries[1122] Exposition[1123] of Huddens rule[1124] (the MS which he is willing to communicate) about reducing Compound Æquations into their Components, concerning which I have this to say that Mr Merry did explaine only some of those rules to witt ... omitting the rest, that I doe beleive that if a treatise of that nature, and about finding the rootes and limits of Æquations collecting what is scattered in Hudden[1125] Bartholinus[1126] Dulaurens[1127] and the Dutch writers Kinckhuysen[1128] Furguson[1129] &c were well digested in Latin it would be very acceptable, I am sure to many here (especially in English into which it might afterwards be translated) and therefore might be another Treatise apart to be sold with his whereof is now in Agitation, for of the two Mathematicall Clubbs here, one is a large one, consisting of Diverse ingenious Mechanicks Guagers Carpenters Shipwrights some Seamen Lightermen &c whose whole Discourse is about Æquations, nor doth Mr Kersy[1130] speake of the rootes or Limits of æquations of any high degree

Neither may we doubt of considerable Improovements concerning Limits. I send you two Papers the one[1131] of Slusius (which he sent to Mr Oldenburgh[1132] who need not know that I have imparted the same) the other[1133]

1-2 (the MS ... communicate) *add.*
4 witt ... omitting *Space left for inserting rules*
4-15 that (*1*) a treatise of that nature, and about finding the rootes |and limits *add.*| of

---

[1122] Merries: i.e. Thomas Merry (d. 1682), mathematical practitioner. See the biographical details in *Bodleian Library* MS Aubrey 8, f. 82.

[1123] Exposition: i.e. Merry's 'Invention and Demonstration of Hudden's Rules for Reducing Equations', written sometime after 1659, and now *Bodleian Library* MS Savile 33. Merry gave Collins the book manuscript; Collins later passed it on to Wallis, who deposited it in the Savile Library. See COLLINS–GREGORY 24.XII.1670/[3.I.1671]; TURNBULL, *James Gregory*, 153–9; WALLIS, *Algebra*, 142, 'Additions and Emendations', 157–62; *Opera mathematica* II, 150.

[1124] rule: i.e. HUDDE, *De reductione aequationum*, in: DESCARTES, *Geometria*, ed. Frans van SCHOOTEN, I, Amsterdam 1659, 401–506.

[1125] Hudden: i.e. Jan Hudde (1628–1704), Dutch mathematician.

[1126] Bartholinus: i.e. Rasmus Bartholin (Berthelsen) (1625–98), Danish physician and mathematician.

[1127] Dulaurens: i.e. François Dulaurens, q.v.

[1128] Kinckhuysen: i.e. Gerhard Kinckhuysen (1625–66), Dutch mathematician.

[1129] Furguson: i.e. Johan Jacob Ferguson (*c.*1630–before 1706), Dutch mathematician.

[1130] Kersy: i.e. John Kersey (1616–77), mathematical practitioner, *ODNB*.

[1131] the one: i.e. SLUSE–OLDENBURG [27.II]/9.III.[1670]/1671; OLDENBURG, *Correspondence* VII, 477–81.

[1132] Oldenburgh: i.e. Henry Oldenburg, q.v.

[1133] the other: not identified. Cf. GREGORY–COLLINS 23.XI/[3.XII].1670; TURNBULL, *James Gregory*, 118–22.

206. COLLINS to WALLIS, 21/[31] March 1670/1

of Mr Gregory,[1134] neither of them come up to what I have heard discoursed by Dr Pell,[1135] to witt that he finds the limits of high Æquations made by the Multiplication of knowne rootes ascendendo first precisely limiting each degree in the order of the Scale as first the Quadratick æquation, then the Cubick and so on

After he hath the limits of an Æquation then giving any Homogeneum he affirmes he can fall upon the Logarithm of the roote quam accuratissime, 5ᵛ] now I shall speak my sense of it|

This figure may Represent the Curve of a Cubick Æquation you formerly Calculated, in which if AQ be a roote when the Homogeneum is $= 0$ and DA the Homogeneum when DC represents a paire of equall rootes If we shall Suppose another Homogeneum given as AE, and a roote found thereto by tryall somewhat neare the truth as FG, it seemes Probable that Parabolasters may be made to passe through the Points C and G so as the one to fall within AG and the other without, so that the ordinate HI shall be lesse and the Ordinate HK greater than the Complement or difference betweene the roote sought, and the root or Limit DC The performance whereof depends cheifly on the drawing of a line to touch the curve AOC at the point O, if the touch lines of those Parabolasters doe the one fall

Æquations would be very acceptable, I am sure to many here, for of two Mathematicall Clubbs here, one is a large one, consisting of Diverse ingenious Mechanicks Guagers Carpenters Shipwrights some Seamen Lightermen &c whose whole Discourse is about Æquations. Nor doth Mr Kersy speake of the rootes or Limits of Æquations of any high degree. If what is (*a*) said (*b*) scatterd in Hudden Bartholinus Dulaurens and (*aa*) Fergus *breaks off* (*bb*) the Dutch ... were well digested I beleive it would be (*2*) if a treatise ... of Æquations |collecting what is scatterd ... were digested in Latin it *add.*| would be ... to many here |(*a*) and peo *breaks off* (*b*) (especially ... translated) *add.*| |and therefore ... Treatise apart to (*aa*) goe (*bb*) be sold with his (*aaa*) one (*bbb*) the Printing (*ccc*) whereof ... Agitation *add.*| for of two ... high degree. Neither
3 made ... rootes *add.*
3 limiting (*1*) the Quad *breaks off* (*2*) each
4 æquation *add.*
7 can (*1*) find (*2*) fall
9 figure *add.*
13 that |other *del.*| Parabolasters
16 or difference *add.*
17 or Limit *add.*

[1134] Gregory: i.e. James Gregory, q.v.
[1135] Pell: i.e. John Pell (1611–85), q.v.

within, the other without the said touch line, the Quaesitum is thence easily obtained

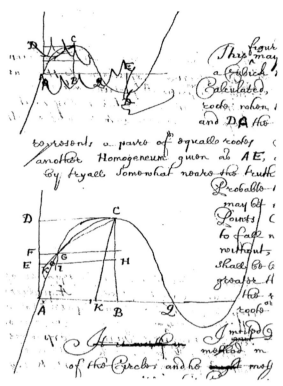

I incited Dr. Barrow[1136] to this method in relation to the quadrature of the Circle, and he mett with good successe in it, as you may see in his Geom Lectures[1137] page 103

If AQ be bisected by the line CK it is scarce worth Inquiry whether the parallells to AQ be bisected by the curve ACQ but I thinke they are not

---

17-2 The performance ... of a |touch del.| line ... touch line, the (1) matter is (2) Quaesitum ... obtained add.
4 (1) It is worth in breaks off (2) I (a) put (b) incited
5 he (1) might (2) mett
7 scarce add.
8 the curve add.

[1136] Barrow: i.e. Isaac Barrow (1630–77), mathematician and theologian, Lucasian professor of mathematics in the University of Cambridge from 1663, ODNB.
[1137] Geom Lectures: i.e. BARROW, Lectiones geometricae, London 1670, 103.

Another Inquiry may be whether these Curves that are described by making the rootes of affected Æquations, Perpendiculars or Ordinates to their Homogenea may not be described by ayd of a ranke of continuall Proportionalls as in Dr Barrows Curves[1138] Page 135 or by the Pure Powers of an Arith Progression seeing the pure Biquad Parabolaster at the end of your former workes[1139] was described from an affected Cub Æq Lastly that which is most desirable in Algebra is an easy method for obtaining the rootes of high affected Æquations, and that will be performed by one Series for all Cubick æquations, another for all Biquadraticks, &c ad inf. only varying the Signes and using due caution And those series are found or made by extracting the rootes of adfected æquations in Species and not in Numbers

207.
WALLIS to HENRY OLDENBURG
Oxford, 23 March/[2 April] 1670/1

Transmission:

W Letter sent: LONDON *Royal Society* Early Letters, W1, No. 118, 2 pp. (our source). At foot of p. 1 in Oldenburg's hand: 'Observed these directions in my letter to Vernon, to whom I sent Dr. Wallis's latin letter to Lord Brouncker, assuring him, we had seen any thing yet of Fermat's works; omitted those line of Dr. Wallis, which are included in [ ]. By my neveu March 25. 1671.' On p. 2 in Oldenburg's hand: 'Rec. March. 25. 71. Answ. March. 30. Sent his owne Copy of letter to Lord Brouncker by Oxf. Coach recomm. to Geffreys. And Leibniz's book to examine.' Postmark on p. 2: 'MR/24'.—printed: OLDENBURG, *Correspondence* VII, 530–1, 536.

Answered by: OLDENBURG–WALLIS 30.III/[9.IV].1671.
Enclosure?: Pelzhofer's Problem.

Wallis gives instructions to Oldenburg on the manner in which his Latin letter to Brouncker, containing his investigations on Fermat's negative theorem (WALLIS–BROUNCKER VIII?.1668; WALLIS, *Correspondence* II, 573–92) is to be forwarded to the mathematical community in Paris. Evidently, Wallis was concerned to provide evidence of the success of his own work on this topic in anticipation of the arrival of Jacques de

2 affected *add.*
5 pure *add.*
8 affected *add.*
10 or made *add.*

[1138]Curves: i.e. BARROW, *Lectiones geometricae*, 135.
[1139]workes: see WALLIS, *Arithmetica infinitorum*, 100–1; *Opera mathematica* I, 222.

207. WALLIS to OLDENBURG, 23 March/[2 April] 1670/1

Billy's new edition of Bachet's *Diophantus* with Fermat's commentaries. Oldenburg sent a copy of WALLIS–BROUNCKER VIII?.1668 (with bracketed lines omitted according to Wallis's instruction) as an enclosure to his next letter to Vernon (OLDENBURG–VERNON 25.III/[4.IV].1671; OLDENBURG, *Correspondence* VII, 536), which was conveyed to the French capital by his nephew, Heinrich von Coccejus. Evidently, Oldenburg also returned to Wallis the original copy of his letter to Brouncker, as Wallis had requested. The present letter possibly contained as enclosure Wallis's account of the mental arithmetic he had performed for Johann Georg Pelzhofer.

Sir,
                 Oxford March. 23. 1670./1.

If I had so well considered the trouble of transcribing, & that it had not been into a Book, I should have saved your scribe that trouble. As it is, if
5 your Nephew[1140] stay but so long as that there be time so to do; if you send mee the Copy[1141] that your man hath done, I will by it cause another to be transcribed here & sent you for France, & you may keep my original; I have no copy of it here at all; that which I sent you being the onely one that I had written. If your Nephew stay not so long, you may send the transcript
10 you have (provided it bee well compared;) & let me have my original back so long as to copy it. But I would have those words in the preamble (beginning, as I remember, with *praesertim,*) which concern a person there not named (but a blank left for it) to bee blotted out: For Monsieur *Carcavi*[1142] is the person there meant (who, amongst a great deal more of uncivil language
15 had in expresse terms given me the Ly; *il menti:*) but so far as is general; (*quod in materia de Cycloide expertus sum,* or to that purpose,) may stand. But if you send it, let it be as from yourself, & as the copy of a thing you have had some time in your hand, and the demonstrations much longer in my Lo. Brounckers[1143] hand, but which I did not til lately consent to have
20 sent over; till now hearing of Fermat's works[1144] ready to come abroad (of which wee have yet seen none in England,) I was content it be sent, that I

8 that (*1*) was (*2*) I
17 the copy of *add.*

[1140]Nephew: i.e. Heinrich von Coccejus (1644–1719), sometime professor of law at Heidelberg, and thereafter professor of law at Frankfurt an der Oder.
[1141]Copy: i.e. of WALLIS–BROUNCKER VIII?.1668; WALLIS, *Correspondence* II, 573–92.
[1142]Carcavi: i.e. Pierre de Carcavi (1600?–84), French mathematician, custodian of the royal library in Paris from 1663.
[1143]Brounckers: i.e. William Brouncker, q.v.
[1144]works: i.e. *Diophanti Alexandrini Arithmeticorum libri sex, et de numeris multangulis liber unus. Cum commentariis C.G. Bacheti v.c. & observationibus D.P. de Fermat,* ed. J. de Billy, Toulouse 1670.

208. WALLIS to OLDENBURG, 23 March/[2 April] 1670/1, enclosure

might not bee supposed to have taken any light from thence. No more (the post being going) but that I am

Yours to serve you
John Wallis.

If you publish the other thing[1145] of the numbers, let it be as in a Letter from that Monsieur Pelshofer[1146] to yourself, & in latine; & in terms modest & not too extravagant.

[2] For Mr Henry Oldenburg,
in the Palmal near St
James's
London.

208.
WALLIS to HENRY OLDENBURG
23 March/[2 April] 1670/1, enclosure:
Pelzhofer's problem

Transmission:

W Draft/copy of paper sent: OXFORD *Bodleian Library* MS Don. d. 45, f. 316^v (our source). Above text in Wallis's hand the report of a similar operation performed over a year earlier: 'Dec. 22. 1669. I did by memory (without pen ink or paper &c) extract the square root of $3,00000,00000,00000,00000,00000,00000,00000,00000$. Which I found to be, $1,73205,08075,68877,29353$, fere.'
E First edition: *Philosophical Transactions* No. 178 (December 1685), 1269–71.

Wallis reports on a task in mental arithmetic which Johann Georg Pelzhofer (*fl*.1666–71), who had been recommended to him by Oldenburg, set him during a visit to Oxford in February 1670/1. Wallis dictated the result of his calculation to Pelzhofer during a subsequent visit in March. As an enclosure either to WALLIS–OLDENBURG 23.III/[2.IV].1670/1, or to a letter preceding this, Wallis sent Oldenburg an account of his carrying out this task for possible inclusion in the *Philosophical Transactions*. Cf. BIRCH, *History of the Royal Society* IV, 389, and the minutes of the Philosophical Society of Oxford for 31 March 1685, GUNTHER, *Early Science in Oxford* IV, 134–6.

[1145] thing: evidently Wallis sent either with this letter or with one preceding it an account of the mental arithmetic he had performed for Johann Georg Pelzhofer (enclosure).
[1146] Pelshofer: i.e. Johann Georg Pelzhofer (*fl*. 1666–71), former student of the University of Königsberg.

210. WALLIS to OLDENBURG, 7/[17] April 1671

Feb. 18. 1670./1. Stilo Angliae; Johannes Georgius Pelshover (Regiomontanus Borussus) desiring it of me; I did the same night (by dark, in bed,) extract the root of this number of 53 places;

2, 4681, 3579, 1012, 1411, 1315, 1618, 2017, 1921, 2224, 2628, 3023, 2527, 293
1. Finding its square root (of 27 places) 157, 1030, 1687, 1482, 8058, 1715, 2171, fere. And upon another visit March. 11. following (not having before committed them to writing) I dictated to him, from my memory, both numbers; which he wrote & took with him to examine.

209.
HENRY OLDENBURG to WALLIS
30 March/[9 April] 1671

Transmission:

Manuscript missing.

Existence and date: mentioned in Oldenburg's endorsement to WALLIS–OLDENBURG 23.III/[2.IV].1670/71.
Reply to: WALLIS–OLDENBURG 23.III/[2.IV].1670/71.
Answered by: WALLIS–OLDENBURG 7/[17].IV.1671.

As emerges from the endorsement on WALLIS–OLDENBURG 23.III/[2.IV].1670/71, Oldenburg returned to Wallis at or about this time by carrier the Savilian professor's own copy of WALLIS–BROUNCKER VIII?.1668 (WALLIS, *Correspondence* II, 573–92). Oldenburg also sent him a copy of Leibniz's *Hypothesis physica nova* (Mainz 1671) for review, as decided at the meeting of the Royal Society on 23 March 1671 (old style). See BIRCH, *History of the Royal Society* II, 475.

210.
WALLIS to HENRY OLDENBURG
Oxford, 7/[17] April 1671

Transmission:

W Letter sent: LONDON *Royal Society* Early Letters W1, No. 119, 4 pp. (p. 3 blank) (our source). At top of p. 1 in Oldenburg's hand: 'Dr Wallis's Opinion concerning the printed Hypothesis Physica nova Leibnitii, which had been desired of him in the name of

6 2171, (*1*) proxime (*2*) fere.
6 fere. (*1*) Which (*2*) And

210. WALLIS to OLDENBURG, 7/[17] April 1671

the R. Society.' and 'Read April: 20: 71. Entered LB. 4. 264'. On p. 4 beneath address in Oldenburg's hand: 'Rec. d. 8. April 1671.' Postmark illegible.—printed: OLDENBURG, *Correspondence* VII, 559–62 (Latin original); 562–4 (English translation).
w^1 Copy of letter sent: LONDON *Royal Society* Letter Book Original 4, pp. 264–8.
w^2 Copy of w^1: LONDON *Royal Society* Letter Book Copy 4, pp. 352–7.
E First edition of letter sent: *Philosophical Transactions* No. 74 (14 August 1671), 2227–30 ('Dr. Wallis's opinion concerning the Hypothesis Physica Nova of Dr. Leibnitius, promised in Numb. 73, and here inserted in the same tongue, wherein it was written to the Publisher, April. 7. 1671.').

Reply to: OLDENBURG–WALLIS 30.III/[9.IV].1671.

At the meeting of the Royal Society on 23 March 1670/1 it was decided that Wallis, together with Boyle, Wren, and Hooke, should be asked to report on Leibniz's *Hypothesis physica nova* (Mainz 1671). Shortly afterwards, Oldenburg sent Wallis a copy of the book. See BIRCH, *History of the Royal Society* II, 475, and Oldenburg's endorsement on WALLIS–OLDENBURG 23.III/[2.IV].1671. The present letter, containing Wallis's views on the *Hypothesis physica nova*, was read at the meeting of the Royal Society on 20 April 1671. See BIRCH, *History of the Royal Society* II, 477. Oldenburg subsequently incorporated Wallis's report on the *Hypothesis physica nova* (with his shorter report on Leibniz's *Theoria motus abstracti*, Mainz 1671) in his letter to Leibniz of 12 June 1671 (old style). See OLDENBURG, *Correspondence* VIII, 99–103; LEIBNIZ, *Sämtliche Schriften und Briefe* II, 1 (2006), 216–21. Oldenburg also published Wallis's report in *Philosophical Transactions* No. 74 (14 August 1671), 2227–30 ('Dr. Wallis's opinion concerning the Hypothesis Physica Nova of Dr. Leibnitius ... here inserted in the same tongue, wherein it was written to the Publisher, April. 7. 1671.').

Oxoniae Aprilis 7. 1671.

Clarissime Vir,

Legi ego semel atque iterum, quam impertiisti D. Leibnitzii[1147] Hypothesim[1148] novam; de qua opinionem meam petis.

Authorem quod spectat, utut de nomine (quod memini) mihi ignotum prius, aestimare tamen debeo, ut qui, in loco magno inter magna negotia positus, vacare tamen potest liberae Philosophiae, et rerum causis investigandis, quique ad multa respexisse videtur.

Opus quod attinet, multa inibi reperio summa cum ratione dicta, et quibus ego plane assentior, ut quae sint sensis meis consona. Talia sunt, *Debere Physicum ad Mechanicas rationes, quam fieri potest, omnia accommodare,* §15, *Nihil seipsum, ex abstractis motus rationibus, in lineam priorem*

[1147] Leibnitzii: i.e. Gottfried Wilhelm Leibniz (1646–1716), German polyhistor who made important contributions to many branches of learning, especially philosophy, mathematics, natural science, history, and politics. At this time he was councillor and legal advisor to the Elector of Mainz, Johann Philipp von Schönborn.
[1148] Hypothesim: i.e. LEIBNIZ, *Hypothesis physica nova*, Mainz 1671.

210. WALLIS to OLDENBURG, 7/[17] April 1671

restituere, etiam sublato impedimento nisi accedat nova vis. §22. *Omnia corpora sensibilia, saltem dura, esse Elastica; Atque, Ab Elatere oriri Reflexionem,* §21. (Quae meis de Motu hypothesibus,[1149] Transactionibus Philosophicis jam antehac a te insertis omnino congruunt; quaeque in Mechanicis seu de Motu tractatu fusius prosequor[1150] Cap. 11. et 13.) Item, *Attolli gravia non metu Vacui, sed propter Atmosphaerae aequilibrium,* §25. *Levitatem vero per accidens tantum sequi ex Gravitate* (gravioribus minus gravia sursum pellentibus) §24. *Irruptionem aeris (sed et aquae &c) in vas exhaustum, ob aeris Gravitatem et Elaterem fieri;* §26. Item, *Exhausti atque Distenti* (ut loquitur) *effectus, (unde fermentationes, deflagrationes, et displosionum omne genus,) nempe, displodente altero quod alterum absorbet,* (seu admittit potius,) §27, 39, 40. Nam et haec etiam ab Elatere fiunt; vel in Contento vel in Continente, vel utroque; illic, se explicante quod nimis fuerat, compressum; hic, se contrahente quod nimis distentum fuerat; quippe utrovis modo, nedum utroque fiet irruptio vel explosio, dummodo locus sit quo, sine impedimento, recipi possit quod projiciendum erit. Suntque haec plane consona traditis nostris,[1151] Mechan. cap. 14.

Sed et illud, *Gravitatem in inferioribus oriri ex motu* (vel pressu) *superioris aetheris,* §13, 16. magna saltem verisimilitudine dicitur: quanquam enim Gravitatis causa (ut et Elateris) tam sit in abscondito ut mihi nondum usque quoque satisfactum sit quid ea in re statuam, naturae tamen phaenomena pulsione quam tractione felicius ut plurimum explicantur. Aliaque multa sunt, quae repetitu non est opus, quae magna verisimilitudine, si non et certitudine, dicta judico; quaeque per se satis consistunt independenter ab aliis: neque enim ita inter se sunt connexa omnia, ut uno vacillante caetera simul ruent.

1 *nisi ... vis* add.
2 *Omnia (1) , saltem dura corpora (2) corpora ... esse*
7-8 *Levitatem ... §24.* add.
10 *deflagrationes,* add.
15 *irruptio* add.
19 *aetheris,* §(*1*) 13. (*2*) 13, 16.

[1149] Quae ... hypothesibus: i.e. Wallis's 'A Summary Account given by Dr. John Wallis, of the General Laws of Motion, by way of Letter written by him to the Publisher, and communicated to the R. Society, Novemb. 26. 1668', published in *Philosophical Transactions* No. 43 (11 January 1668/9), 864–6.

[1150] prosequor: i.e. WALLIS, *Mechanica: sive, de motu, tractatus geometricus* III, 660–82; *Opera mathematica* I, 1002–15, 1018–31.

[1151] traditis nostris: i.e. WALLIS, *Mechanica: sive, de motu, tractatus geometricus* III, 708–46; *Opera mathematica* I, 1032–55.

210. WALLIS to OLDENBURG, 7/[17] April 1671

De tota vero hypothesi nequid statim pronuntiem, id saltem facit, quod non sim pronus ego (in rebus saltem pure Physicis, non Mathematicis,) assensum novis traditis adhibere, donec vel eruditorum sententiis in utramque partem ventilatis quid statuendum sit rectius constet, vel ipsa sui evidentia (quod in veris hypothesibus non raro fit) veritas eluceat. Fundamentum Hypotheseos novae repetit ex *Abstracta sua motus theoria*, (quam non vidi, ut nec hujus tractatus posteriora, quae passim citantur,) nempe, *Quod nulla sit cohaesio quiescentis, sed omnis consistentia seu cohaesio oriatur a motu*, §7, 12, 34. (Quod cum Guil. Nelii nostri placitis coincidit.[1152]) Contra vero, Honoratissimus Boylius *Consistentiam in particularum quiete, et Fluiditate in earundem continuo motu*, collocat.[1153] Alii ad varias *Atomorum figuras, hamatas et varie implicitas*, rem referunt. Neque ego is sum qui in tanta sententiarum variete me velim arbitrum interponere. Sed tempori res permittenda est, et doctorum in utramque partem rationibus. Quippe idem fere obtinet in novis Hypothesibus atque in Pendulorum oscillationibus; ubi, post crebras hinc inde reciprocationes factas, tandem in perpendiculo fit quies. Id vidimus in Hypothesi Copernicana, quae utut fuerit Veteribus cognita,[1154] tamdiu tamen jacuit sepulta ut pro nova haberetur: Et quamvis erat optima ratione suffulta, non tamen statim obtinuit, sed a variis fuit variis modis impetita, et acriter disputata, donec tandem rationibus authoritati praevalentibus ita jam universim admittitur, ut vix quispiam harum rerum gnarus de ea dubitet nisi quibus Cardinalium decretum praejudicio est: Et quanquam Tycho[1155] novam illius loco sub|stituerit quae illi aequipolleret, ea tamen tot incommodis onerata est ut existimandus videatur potius ad frangendam invidiam id fecisse (quoniam Telluris motus ita vulgi opinionibus horribilis videbatur) quam quod Copernici Hypothesin ex animo repudiaverit.

1 saltem *add.*
12 *varias add.*
25 id fecisse *add.*
27 ex animo *add.*

[1152] coincidit: see for example NEILE–OLDENBURG 18/[28].XII.1668; OLDENBURG, *Correspondence* V, 263–4.
[1153] collocat: i.e. Boyle's 'The History of Fluidity and Firmnesse', published in BOYLE, *Certain Physiological Essays*, London 1661, 137–249.
[1154] cognita: according to Archimedes, Aristarchus of Samus (310–230) developed heliocentric hypotheses in a work which has not survived. Nicholas Copernicus's (1473–1543) *De revolutionibus orbium coelestium* was not published until the year of his death.
[1155] Tycho: i.e. Tycho Brahe (1546–1601), Danish astronomer.

210. WALLIS to OLDENBURG, 7/[17] April 1671

Idem dicendum est de Circulatione Sanguinis Harvaeana; quae utut optime stabilita fuerit et oculorum αὐτοψίᾳ comprobata, disceptata tamen fuit inter Londinenses Medicos viginti plus minus annis antequam in publicum prodiret;[1156] et ab aliis postea: Quae tamen post maturam rei pensitationem (quod tempori dandum erat) ab omnibus ut indubitata recipitur. Sic Galilaei[1157] hypothesis (ob antlias aquam non ultra certam altitudinem attrahentes primum excogitata) quam Torricellius[1158] in graviori liquido adeoque magis tractabili promovit, Æquilibrium Atmosphaerae pro Veterum fuga vacui substituens, non nisi post diutinas hinc inde disputationes eum apud viros doctos locum obtinuit quem jam habet.

Idem dicendum de Jolivii[1159] nostri vasis Lymphaticis, ante multos annos Medicis Londinensibus ab illo indicatis atque ab eis admissis et approbatis, dicendum erit; Quae tamen ita rationi consona reperta sunt et oculari inspectioni manifesta ut tandem longo post tempore inter alios aliquot acriter disputatum est quis eorum primus inventor fuerit.

Similiter Whartoni[1160] nostri ductus salivales, quos libro edito[1161] indicaverat; quo libro quasi dissimulato, tandem inter Stenonem et Bartholinum disputatum est, uter horum prior invenerit.[1162] Item Infusio liquorum in venas animalium jam ante viginti fere annos a Wrennio[1163] nostro excogitata

11 dicendum *add.*
17 quasi *add.*

[1156]prodiret: i.e. in HARVEY, *Exercitatio anatomica de motu cordis et sanguinis in animalibus*, Frankfurt am Main 1628. Already in 1616 Harvey had presented his theory of the circulation of blood to the Royal College of Physicians in London.

[1157]Galilaei: i.e. Galileo Galilei (1564–1642), Italian natural philosopher, astronomer, and mathematician.

[1158]Torricellius: i.e. Evangelista Torricelli (1608–47), Italian natural philosopher and mathematician.

[1159]Jolivii: i.e. George Joyliffe (1621–58), anatomist and physician, *ODNB*. A former student of Pembroke College, Oxford, Joyliffe studied later under Francis Glisson in Cambridge. His observations of lymphatic vessels are described in Glisson's *Anatomia hepatis*, London 1654.

[1160]Whartoni: i.e. Thomas Wharton (1614–73), physician, from 1657 at St Thomas's Hospital, London, *ODNB*.

[1161]libro edito: i.e. WHARTON, *Adenographia, sive, glandularum totius corporis descriptio*, London 1656.

[1162]invenerit: the parotid and lachrymal ducts were discovered by Niels Stensen (1638–86) in 1660. Gerard Blaes (*c.*1600–82), and not, as Wallis incorrectly asserts, Thomas Bartholin (1616–80) disputed Stensen's right to priority in discovering of the parotid duct.

[1163]Wrennio: i.e. Christopher Wren (1632–1723), architect, mathematician, and astronomer, *ODNB*. Wren entered Wadham College, Oxford, in June 1650, and became a member of the experimental philosophy circle which for a time was centred there and to which Wallis also belonged.

210. WALLIS to OLDENBURG, 7/[17] April 1671

Oxoniae, ibique postmodum crebro administrata; et sanguinis Transfusio ab animali uno in alterum, a Lowero[1164] nostro feliciter aliquoties administrata Oxoniae, atque Londini postea coram Societate Regia, ab exteris interim, et quidem Parisiensibus, non credita sed pro impossibili habita et tantum non irrisa; tandem tamen ita illis placuit (post totum processus tenorem illis indicatum) ut illius authores haberi cupiant,[1165] atque apud illos nescio quem jam ante aliquid obiter dixisse quod eo jam trahant. Idemque in hoc negotio, aliisque novis hypothesibus, expectandum erit, quae nec oculi inspectione nec certa demonstratione probari possunt, ut, si veris rationibus fundatae sint, tandem, sed non nisi post velitationes utrinque factas, in libere philosophantium animis locum obtinebunt; interea pendulae mansurae.

Clarissimo interim Viro habendae gratiae, qui eam de Societate nostra opinionem concepit, ut dignatus fuerit sensa sua cum illis communicare, novamque suam Hypothesin exhibere, quibus certe res erit non ingrata. Tu vero Vale.

Tuus,
Johannes Wallis.

[4] These For Mr Henry Oldenburg,
at his house in the Palmal
near St James's
London.

7 obiter *add.*
10 sint *add.*
13 cum illis *add.*
14 res erit *add.*

[1164]Lowero: i.e. Richard Lower (1631–91), physician and physiologist, *ODNB*. After studying at Christ Church, Oxford, Lower established a medical practice in Oxford at the end of the 1650s. At the meeting of the Royal Society on 20 June 1666 Wallis reported on the successful experiment, conducted by Lower in Oxford, of 'transfusing the blood of one animal into the body of another'. See BIRCH, *History of the Royal Society* II, 98.

[1165]cupiant: on the controversy between the English and the French over priority in conducting the first experimental blood transfusion see OLDENBURG–BOYLE 24.IX/[4.X].1667; OLDENBURG, *Correspondence* III, 480–2.

211.
HENRY OLDENBURG to WALLIS
9/[19] May 1671

Transmission:

Manuscript missing.

Existence and date: mentioned in and answered by WALLIS–OLDENBURG 13/[23].V. 1671.

212.
WALLIS to HENRY OLDENBURG
9/[19] May 1671

Transmission:

Manuscript missing.

Existence and date: mentioned at the beginning of Wallis's succeeding letter to Oldenburg (WALLIS–OLDENBURG 13/[23].V.1671).

213.
WALLIS to HENRY OLDENBURG
Oxford, 13/[23] May 1671

Transmission:

W Letter sent: LONDON *Royal Society* Early Letters W1, No. 120, 2 pp. (our source). At top of p. 1 in Oldenburg's hand: 'Read May 18: 71. Entered LB. 4. 303.' and 'An Extract of Dr Wallis's Letter to M. Old. concerning the Experiment, wherein the (*1*) Tube (*2*) Mercury (*a*) in the Tube (*aa*) at most (*bb*) being inverted remaines (*b*) in the inverted Tube falls not at all.'—printed: OLDENBURG, *Correspondence* VIII, 50–1.
w^1 Copy of letter sent: LONDON *Royal Society* Letter Book Original 4, pp. 303–5.
w^2 Copy of w^1: LONDON *Royal Society* Letter Book Copy 4, pp. 403–6.

Reply to: OLDENBURG–WALLIS 9/[19].V.1671.

This letter contains Wallis's thoughts on barometric experiments and was evidently sent with a now missing enclosure. It was read at the meeting of the Royal Society on 18 May 1671 (old style). See BIRCH, *History of the Royal Society* II, 482.

213. WALLIS to OLDENBURG, 13/[23] May 1671

Oxford. May 13. 1671.

Sir

Yours[1166] of Tuesday last was in part answered before it came by one of mine[1167] of the same date; desiring you to return me the papers[1168] My Lo. Br.[1169] last had as soon as you could; which are not yet come to hand, perhaps they may come by this nights coach. To the experiment His Lordship mentions,[1170] I confesse I have sayd nothing in them: & the truth is, it will not be solved by the same onely principle with the rest, but hath somewhat singular in it, which may in time give some light to the nature of Gravitation:[1171] but 'tis rather Physicall than Mathematical, I was not satisfyed what clearly to say in it. Yet since his Lordship thinks it proper that I should say something; I shal do so; & it will be to the same purpose that I have some time spoken (as I remember) in the Royal Society at Arundel house, (whether any great notice were taken of it or no, I know not:) that whereas heretofore, in the old Philosophy, Earth &c were thought heavy, & the Ayr light; it would here seem contrary, & that all actual Gravitation amongst us proceeds from the Air or Æther's pressure & Spring, without which these Dull bodies, which wee call Heavy, would rest in quiet without any actuall gravitation or descension; & be no more apt to move downward, than sidewise; & accordingly, the quicksilver (perfectly freed from Air, & then resting suspended in an inverted Tube to the height of 40, 50, 60, or more inches,

---

3 last |(if I do not misremember the date) del.| was
8 onely add.
12 same (1) time (2) purpose
13 (as I remember) add.
17 Spring, |(which seemes allso to be from somewhat incumbent on it,) del.| without
19 to (1) presse (2) move
20-1 quicksilver |(perfectly freed ...ordinary hight,) add.| being (1) freed (2) freed (3) unmolested

---

[1166]Yours: i.e. OLDENBURG–WALLIS 9/[19].V.1671.
[1167]mine: i.e. WALLIS–OLDENBURG 9/[19].V.1671.
[1168]papers: Wallis had evidently given Brouncker part of Chapter XIV (De hydrostaticis) of his *Mechanica: sive, de motu, tractatus geometricus* for perusal before submitting it to the printer.
[1169]Lo. Br.: i.e. Lord Brouncker, q.v.
[1170]mentions: i.e. the anomalous suspension of mercury at heights above some 30 inches. It is probable that Oldenburg conveyed Brouncker's remarks on this phenomenon in the now missing letter OLDENBURG–WALLIS 9/[19].V.1671.
[1171]Gravitation: Wallis published *A Discourse of Gravity and Gravitation, grounded on experimental observations* under the imprimatur of the Royal Society in November 1674.

213. WALLIS to OLDENBURG, 13/[23] May 1671

much above the ordinary hight,) being unmolested by the Air's pressure or spring, (that without, pressing it upward if at all, & there being none in it or above it to presse downward,) rests at quiet in that position: But if either by a concussion of the Tube, or by an inward disturbance from the spring of
5 the Ayr left within the Tube or now suffered to enter, it be put into motion, then (in proportion to its quantity of matter, denseness of parts, or what ever that bee which is vulgarly or rudely expressed by the word Weight,) it pursues that motion & will compresse downwards, as it would do sideways in case of a laterall impulse given it: For, not onely downwards, but sidewise
10 allso, the heavyer a body is, the harder it strikes when put in motion.

So that whatever it bee that wee call Weight in Quicksilver (or the like) though without such pressure or spring of air it would not begin a motion, yet being put into motion obeys the Statick laws. And it was this consideration that made mee, (in the account[1172] I gave you of Monsieur
15 Lebnitzs Hypothesis, April.7.1671.) so inclinable to attribute Gravitation to the Motion or Pressure of the incumbent Æther. And I think My Lo. Brouncker will not much differ in opinion from mee in this point. But this account of it, though I know not how to give a better, I can but hesitantly deliver, not with a full confidence; being not forward to be positive in new
20 Hypotheses. But if, such as it is, it bee thought proper to be there published or his Lo. wil vouchsafe to help correct & perfect it, I shall find a place somewhere in those papers to insert it. Yet not omitting that other consideration, of the little inequalities or asperities even in the most polite surface, which cause somewhat of friction or cohaesion or parts whereby motion is more or
25 lesse hindered, (of which I have oft occasion to speak, as the cause of Rolling rather then Sliding of a Body moved on the contiguous surface of another Body;) which may be inough to counterpoise the action of some very little

5 or now ... to enter, *add.*
6 its *(1)* weight *(2)* quantity
7 or rudely *add.*
9 laterall *(1)* impulsive *(2)* impulse
17 point. *(1)* Yet I can say no *(2)* But this
20-21 or his ... perfect it, *add.*
25 lesse *(1)* empeded *(2)* hindered
25-26 of *(1)* rotation *(2)* Rolling rather then Sliding of a *(a)* Weight *(b)* Body
26 contiguous *add.*
26 another *(1)* cont *breaks off* *(2)* Body
27 to *(1)* answere *(2)* counterpoise

[1172] account: i.e. in WALLIS–OLDENBURG 7/[17].IV.1671.

Ayr which may (after the greatest diligence) be yet remaining amongst the Quicksilver. But in order hereunto, I must desire you, if you can, (out of your Notes or Memory,) to inform me what day or time this experiment was first exhibited[1173] in the Society then at Gresham College; which I think was about the year 1661 or 1662, by Monsieur Huygens, but with the assistance of the Ayr-pump, which hath since by my Lo. Brouncker & Mr Boyl been done without it. For it will be very convenient (for asserting the Experiment to the true Authors) to mention the date of it, & (if it be so) its being Registered in the Societies Journal.[1174] I shal adde no more, but that I am

Sir
Your affectionate friend & servant.
John Wallis.

I shal be glad to hear of the reception of this, because of the inclosed[1175] which is an Original.

These
For Mr Henry Oldenburg
in the Palmal near St.
James's
London.

214.
HENRY OLDENBURG to WALLIS
mid/end May 1671

Transmission:

Manuscript missing.

Existence and date: Mentioned in and answered by WALLIS–OLDENBURG 10/[20].VI.1671.

[1173] exhibited: the phenomenon of anomalous suspension was discovered by Huygens in an experiment using an air pump which he carried out in the Netherlands in July 1662. The experiment was successfully repeated, in Huygens's presence, at a meeting of the Royal Society in August 1663. On 7 August 1663, Boyle and Brouncker reported achieving the same effect without the aid of an air pump. See BIRCH, *History of the Royal Society* I, 287; BOYLE–OLDENBURG 29.X/[8.XI].1663; OLDENBURG, *Correspondence* II, 123–6.
[1174] Journal: i.e. the Royal Society's Journal Book or Register.
[1175] inclosed: this enclosure is now missing.

216. WALLIS to OLDENBURG, 2/[12] June 1671

In this letter, Oldenburg probably offered to arrange that John Wallis jr witness the installation by proxy of the King of Sweden and the Duke of Saxony as Knights of the Garter at Windsor on 29 May 1671 (old style). See 'News-letter of the Earl of Exeter', *Calendar of State Papers Dom.* Car. II, 1671, 287 (No. 67).

215.
JEAN BERTET to WALLIS
May/June 1671

Transmission:

Manuscript missing.

Existence: Referred to in BERTET–WALLIS [21.XI]/1.XII.1671.

Evidently this letter miscarried just as the previous one had done, which Bertet wrote to Wallis in February/March 1671.)

216.
WALLIS to HENRY OLDENBURG
Oxford, 2/[12] June 1671

Transmission:

W Letter sent: LONDON *Royal Society* Early Letters W1, No. 121, 2 pp. (our source). At top of p. 1 in Oldenburg's hand: 'Enter'd L.B. 4. 312.' and beneath signature: 'Dr Wallis's opinion of Monsieur Leibnitius his Theoria Motus abstracti.' On p. 2 beneath address in Oldenburg's hand: 'Acc. d. 5 jun. 1671.' Postmark illegible.—printed: OLDENBURG, *Correspondence* VIII, 72–3 (Latin original); 73–4 (English translation).
w^1 Copy of letter sent: LONDON *Royal Society* Letter Book Original 4, p. 312.
w^2 Copy of w^1: LONDON *Royal Society* Letter Book Copy 4, pp. 414–15.
E First edition of letter sent: *Philosophical Transactions* No. 74 (14 August 1671), 2231 ('After that the other part of this Tract was (a great while after) come to hand, namely De Abstracta Motus Theoria, and sent also to the same Dr. Wallis, he made this return to it in a Letter of June 2. 1671.').

At the meeting of the Royal Society on 25 May 1671, Oldenburg was asked to send Leibniz's *Theoria motus abstracti* (Mainz 1671) to Wallis in order to obtain his opinion on the work. Shortly afterwards Oldenburg sent Wallis a copy of the book. See BIRCH, *History of the Royal Society* II, 482. Oldenburg incorporated Wallis's report on the *Theoria motus abstracti* (together with his longer report on Leibniz's *Hypothesis physica nova* in his letter

216. WALLIS to OLDENBURG, 2/[12] June 1671

to Leibniz of 12 June 1671 (old style). See OLDENBURG, *Correspondence* VIII, 99–103; LEIBNIZ, *Sämtliche Schriften und Briefe* II, 1 (2006), 216–21. He also published Wallis's report in *Philosophical Transactions* No. 74 (14 August 1671), 2227–30 ('Dr. Wallis's opinion concerning the Hypothesis Physica Nova of Dr. Leibnitius ... here inserted in the same tongue, wherein it was written to the Publisher, April. 7. 1671.').

Oxoniae. Junii. 2. 1671.

Clarissime Vir,

Accepi nuperrime, a te transmissam,[1176] D. Leibnitzii *Theoriam Motus Abstracti*: de qua judicium meum expetis. Duo autem sunt quae suadeant ne illud praestem. Alterum; quod res invidiosa videatur de aliorum scriptis censuram agere: Alterum; quod occupatissimo tempore huc advenerit, quo aegre tempus obtinuerim semel atque iterum attentius legendi, nedum omnia pensiculatius expendendi. Quoniam vero tu id expetis, haec pauca dicam. Multa scilicet mihi contenta, ego plane approbo, ut subtiliter et solide dicta; quaeque Virum curiosum et cogitabundum indicant. Si pauca sint quibus non statim assentiar, ignoscet spero Vir humanissimus. Et speciatim, fateor mihi nondum satisfactum esse, ut, primis saltem cogitationibus, statim assentiar, Cohaesionem omnem ex continuo celerique sed inobservabili particularum motu fieri, (quod ille Theoriae motus concreti fundamentum ponit;) uti nec pridem, cum, ante aliquot annos, similem quietis et cohaesionis causam assignaverit[1177] Nelius noster. Quid olim aliquando fiet, post rem accuratius perpensam, nec dicere possum nec praevidere. Interim ego $\overset{\text{᾿}}{\alpha}\pi\acute{\epsilon}\chi\omega$, nec quicquam in aliorum praejudicium pronuntio; quin liberum quique sit eam quam rationi magis consentaneam judicaverit sententiam amplecti. Vale.

Tuus
Johannes Wallis.

I desire, when you send mee my Lords sense of my letter[1178] you shewed him, that you will not forget to send mee word when first that experiment was made amongst us, & how, & whether in the Society at Gresham College.

[2] These
For Mr Henry Oldenburg

9 ut (*1*) acute (*2*) subtiliter

[1176]transmissam: i.e. LEIBNIZ, *Theoria motus abstracti*, Mainz 1671.
[1177]assignaverit: see for example NEILE–OLDENBURG 18/[28].XII.1668; OLDENBURG, *Correspondence* V, 263–5.
[1178]letter: i.e. WALLIS–OLDENBURG 13/[23].V.1671.

in the Palmal near St
James's
London.

217.
WALLIS to HENRY OLDENBURG
Oxford, 10/[20] June 1671

Transmission:

W Letter sent: LONDON *Royal Society* Early Letters W1, No. 122, 2 pp. (our source). On p. 2 beneath address in Oldenburg's hand: 'not to be entered'. Postmark: 'IU/11'.—printed: OLDENBURG, *Correspondence* VIII, 88–9.

Reply to: OLDENBURG–WALLIS ?.V.1671.
Enclosure: Scholium to proposition 13 of chapter XIV (De hydrostaticis) of *Mechanica: sive, de motu, tractatus geometricus*.

Oxford June 10. 1671.

Sir,

I thank you for your intimation[1179] of readynesse to do my son the kindness I desired. If hee come too late, or that otherwise there were no opportunity; wee must be content. I have since the time of that solemnity heard nothing from him of what passed. The inclosed paper[1180] I had written presently after I sent my letter[1181] to the same effect & before I received your Answere. I had sooner sent it, but that I have been in continual expectation[1182] of receiving my Lords sense upon it; & an account of matter of fact how & when that experiment began & had its progresse with the Royal Society or the members of it: that so I might accordingly have altered the narrative in

---

10 & before ... Answere *add.*

[1179] intimation: Oldenburg had probably offered to arrange that Wallis's son, John Wallis jr, witness the installation by proxy of the King of Sweden and the Duke of Saxony as Knights of the Garter at Windsor on 29 May 1671 (old style). See 'News-letter to the Earl of Exeter', 30 May 1671; *Calendar of State Papers Dom.* Car. II, 1671, 287 (No. 67).

[1180] paper: evidently the scholium to proposition 13 of chapter XIV (De hydrostaticis) of his *Mechanica: sive, de motu, tractatus geometricus* for perusal by Oldenburg.

[1181] letter: presumably WALLIS–OLDENBURG 13/[23].V.1671.

[1182] expectation: cf. the postscript to WALLIS–OLDENBURG 2/[12].VI.1671. The experiment concerned is described in the long scholium to proposition 13 of chapter XIV of his *Mechanica: sive, de motu, tractatus geometricus* (WALLIS, *Mechanica* III, 732–42; *Opera mathematica* I, 1046–52).

the beginning of it, before I had sent it. I desire you will furnish mee with the history of fact; in which none can better inform you (I think) than my Lord himself; & return mee the paper as soon as you can. For the presse stays for Copy, having not inough to finish the sheet now in hand. And what I have now sent them (two or three leaves) will but just help out one sheet; & this paper is next to follow. I have no more at present, but that I am

Sir
Your affectionate friend to [serve you]
John Wallis.

[2] These
For Mr Henry Oldenburg,
in the Palmal near St James's
London.

218.
WALLIS to HENRY OLDENBURG
Oxford, 27 June/[7 July] 1671

Transmission:

W Letter sent: LONDON *Royal Society* Early Letters W1, No. 123, 4 pp. (our source). At top of p. 1 in Oldenburg's hand: 'Enter'd LB. 4. 329.' and on left margin of same page at 90°, again in Oldenburg's hand: 'Dr Wallis's Letter, containing an Answer to Mr Hobbes's Rosetum'. Postmark: 'JU/28'.—printed: OLDENBURG, *Correspondence* VIII, 128–31.
w^1 Copy of letter sent: LONDON *Royal Society* Letter Book Original 4, pp. 329–31.
w^2 Copy of w^1: LONDON *Royal Society* Letter Book Copy 4, pp. 438–41.

This letter constitutes the first draft in English of Wallis's reply to Hobbes's *Rosetum geometricum*, London 1671. Oldenburg was unable to include it in the June issue of *Philosophical Transactions*, but referred to it after the expanded Latin version of the reply, which he published in the July issue. See WALLIS–OLDENBURG 16/[26].VII.1671; *Philosophical Transactions* No. 73 (17 July 1671), 2002–9, 2009.

Oxford, June 27. 1671.

Sir,

On Saturday June 24, I had a sight of Mr. *Hobs's Rosetum*,[1183] (he should

---

5 now *add*.

[1183] *Rosetum*: i.e. HOBBES, *Rosetum geometricum, sive propositiones aliquot frustra ante-*

218. Wallis to Oldenburg, 27 June/[7 July] 1671

have called it *Fimetum*.) It comes forth with its Keeper (that it might do no hurt.) For the Animadversions[1184] printed with it (from I know not what hand) are a sufficient Confutation of it. On Munday, that is, yesterday, I had leisure to peruse so much as may serve to confute the whole. For his Construction of the first Probleme or Proposition, (on which the rest depend,) is False; &, with it, the rest therefore come to nought.

To cut a line in extreme and mean proportion, was taught by *Euclide* & demonstrated;[1185] *prop.* 30. *lib.* 6. with whom all Geometers hitherto have agreed.

The result of which amounts to thus much, That if the line to be so cut be $1R$, the greater segment thereof will be $\frac{\sqrt{5}-1}{2}R$; And, consequently, the lesser $\frac{3-\sqrt{5}}{2}R$ which multiplied or drawn into the whole $1R$, produceth $\frac{3-\sqrt{5}}{2}R^2$, equal to the square of Greater segment.

But Mr Hobs gives us here another construction; according to which, the greater segment should be $\frac{\sqrt{:5-2\sqrt{3}:}}{2}R$.

But whether you will rather trust *Euclide*'s geometry or Mr *Hobs*'s, I leave to your judgement. If you will trust neither; then try Mr *Hobs*'s, as we have done *Euclide*'s. To cut a line *in extreme & meane proportion*, is, so to cut it as that, the *Rectangle of the whole & the lesser segment* be equal to the *Square of the greater*. Now Mr *Hobs*'s greater segment being $\frac{\sqrt{:5-2\sqrt{3}:}}{2}R$, the lesse must be $\frac{2-\sqrt{:5-2\sqrt{3}:}}{2}R$ that so both together may be equal to the whole $1R$. Which lesser segment drawn into the whole, produceth $\frac{2-\sqrt{:5-2\sqrt{3}:}}{2}R^2$. Which should be equal to the square of the greater segment, but is not. For the square of $\frac{\sqrt{:5-2\sqrt{3}:}}{2}R$, is $\frac{5-2\sqrt{3}}{4}R^2$. But, in *Euclide*'s case, the Rectangle was equal to the Square, as it ought to be. This perhaps with Mr *Hobs* (who's

---

1-4 *Fimetum.* (*1*) of which on Munday, that is yesterday, I had leisure (*2*) It comes ... of it. (*a*) I had (*b*) On Munday ... leisure
4 as (*1*) serve *breaks off* 2 affords matter for the confutation of the whole (*3*) may serve to confute the whole
4 For (*1*) the (*2*) his
7 & demonstrated; *add.*
11 $\frac{\sqrt{5}-1}{2}R$; (*1*) But Mr Hobs gives us here another construction (*2*) And, consequently
12 or drawn *add.*
16 will (*1*) trust (*2*) rather trust
17 try (*1*) To cut a line in extrem *breaks off* (*2*) Mr *Hobs*'s

---

hac tentatae, cum censura brevi doctrinae Wallisianae de Motu, London 1671. It is noticed in *Philosophical Transactions* No. 72 (19 June 1671), 2185–6.
[1184] Animadversions: not identified.
[1185] demonstrated: i.e. Euclid, *Elements* VI, prop. 30.

218. WALLIS to OLDENBURG, 27 June/[7 July] 1671

Arithmetick & Geometry cannot agree,) will not pass for a demonstrative confutation of his construction, (nor do I care whether it will or not;) with others, it will.

The falts in Mr *Hobs*'s construction & (pretended) demonstration of it, (beside a great many little ones) are at lest these three great ones.

First, in his Construction, in stead of *Describatur centro D quadrans DAC secans FE et GH in K et X*. He might as well have sayd, *sumatur ubivis in IG puncto X*; (For so his whole demonstration following will just as well agree, as now it doth, without altering one word or letter;) or, *ubivis in IG utcunque in utramque partem producta*; (For then he need onely in like manner produce EA; or, in stead of *secans AE*, say, *secans AE saltem productam pag*. 2. *lin*. 17.) So that, X being taken arbitraryly, his EX (which he will have to be the greater segment) may be of what length hee please; & his demonstration for it, stand just as now it doth. As any who will take the pains to apply *verbatim* his Demonstration to this Construction will presently discern.

Next, in his Demonstration, having shewed, that *the two right angles mXl, IXl, are equal to the five angles mXF, FXy, yXz, zXE, EXl, taken alto-gether*: *pag*. 2. *lin*. 18. 19; (without proving, or so much as saying, whether they be or be not all equal to each other;) He then (silently supposing them to be all equal) goes to prove from hence, that *the angles zEX, zXE, are equal*, because *otherwise the sayd two right angles were not so divided* (quinquifariam) *into five equal parts*, *pag*. 3., *lin*. 4, 5. The truth is; the first, third, & fifth of those angles are equal each to other; & the second & fourth equal (not to any of the other three, but) each to other; But, whether one of these two, be equal, lesse, or greater than one of these

---

3 will. (*1*) You'll ask, where lyes the falt (*2*) The falts
4 of it (*1*) are (*2*) (beside
7 et X (*1*) ducatur denique FX (*2*). He might
10 onely (*1*) in stead of secans AE in l *breaks off* (*2*) in like manner
11 or (*1*) pag. 2. lin. 17. (*2*) in stead of
12 pag. 2. lin. 17. add.
17 having (*1*) proved (*2*) shewed
19 pag. 2. lin. 18. 19; add.
20 other;) (*1*) He |then add.| a silently taking |it as add.| for granted withou *breaks off* (*2*) He then
23 pag. 3., lin. 4, 5. add.
26 whether (*1*) |one of add.| these be equal, lesse, or gre *breaks off* (*2*) one of these

218. Wallis to Oldenburg, 27 June/[7 July] 1671

three, his Demonstration shows not, nor doth so much as attempt it; though on this depend the whole strength of his argument.| [2]

The third is, *pag.* 3. *lin.* 17, 18. where; Because *the angles yXz, Xyr are equal; & the angles yXr, yrX, equal; and the angles yXz, zEX, equal*: he infers, *that the angles zXE, zEX are allso equal.* Which doth not follow from thence. Yet the want of this, destroys his demonstration.

If all this be not inough; His second Proposition confutes his first. For, by his second, *As EF to FC*, that is, as 2 to $\sqrt{5}-1$; *so is, the greater segment AB, to the lesser BG*; & consequently, *the whole AG, to the greater segment AB*: As Euclide's construction would have it; not, as Mr *Hobs prop.* 1. makes it, as 2 to $\sqrt{:5-2\sqrt{3}}$.

His first Proposition thus failing, it will not be necessary for the confuting of his book, to trouble you with what faults are in the rest. For that failing, all that depends upon it must needs fall with it. But if you desire more; I refer you to the Latin Paper.[1186]

What in the end of his book is intended against mee,[1187] (beside some childish trifling with words, & indeavouring to pervert the sense of them,) amounts mostly to this, That *he doth not like Algebra; he doth not understand Symbols; hee likes not the doctrine of Indivisibles; nor the doctrine of Infinites*; (nor do I care whether hee do or do not:) and the whole is so weak& frothy, that it neither needs nor is worth answering: But I may safely trust the Reader with it; who if he do but confer what Mr *Hobs* says, with the places to which it doth refer, will easyly see its emptynesse without my help. I am Sir

Yours to serve you
John Wallis.| [3]

June 27. 1671

Sir,

---

10 *prop.* 1. *add.*
12 Proposition (*1*) being (*2*) thus failing
13 you *add.*
16 mee, (*1*) is so empty & childish, that it neither needs nor is worth answering (*2*) (beside some
21 weak & (*1*) emp *breaks off* (*2*) frothy
23 refer, |hee *del.*| will
23 easyly (*1*) answer it (*2*) see its

[1186] Latin Paper: i.e. WALLIS–OLDENBURG 16/[26].VII.1671.

[1187] against mee: i.e. Hobbes' criticisms of the first two parts of Wallis's *Mechanica: sive, de motu, tractatus geometricus.*

219. WALLIS to OLDENBURG, 16/[26] July 1671

I send you this, as that which you may, if it be not quite to late, publish in this months transactions : & refer the Latine paper to which this refers (which I shall send you very suddenly) to the next. If it be quite too late to insert it, you may at lest intimate that you have it, but want room for it.

The Post is going, so that I can say no more but that I am

Yours &c John Wallis.

[4] For Mr Henry Oldenburg,
in the Palmal near St James's
London.

219.
WALLIS to HENRY OLDENBURG
Oxford, 16/[26] July 1671

Transmission:

W Draft of (missing) paper sent: OXFORD *Bodleian Library* MS Savile 104, f. 3^r–5^v (f. 5^v blank) (our source).
E First edition of paper sent: *Philosophical Transactions* No. 73 (17 July 1671), 2202–9 ('An Answer of Dr. Wallis to Mr. Hobbes's Rosetum Geometricum in a Letter to a friend in London, dated July 16. 1671.').

Answered by: HOBBES–ROYAL SOCIETY end of VII.1671, HOBBES–ROYAL SOCIETY first half of VIII. 1671, HOBBES–ROYAL SOCIETY early IX.1671, and HOBBES–ROYAL SOCIETY IX.1671.

Cl. Viro H. Oldenburg.
Clarissime vir,

Perlegi *Hobbii* sive *Rosetum*, sive *Fimetum*, (nam utrumque olet;) in quo, antiquum obtinet. Mirumque est, ut nec sibi in animum inducere possit, nec ab amicis suaderi ne sic delirando persistat se contemptui exponere.

2 to ... refers *add.*
16 delirando *add. W*

219. WALLIS to OLDENBURG, 16/[26] July 1671

Primae Propositionis, sive Problematis, constructio, (utut in re facili,) falsa est.

Rectam extrema et media ratione secare; docuerat *Euclides* et demonstraverat,[1188] prop 30. El. 6. (cui et alii hactenus consenserunt.) Secundum quem, posita recta secanda $1R$, erit majus segmentum $\frac{\sqrt{5}-1}{2}R$; adeoque segmentum reliquum $\frac{3-\sqrt{5}}{2}R$; quod in totam $1R$ ductum, efficit $\frac{3-\sqrt{5}}{2}R^2$, quod est ipsum majoris quadratum.

Hobbius autem hic novam proponit constructionem; secundum quam, segmentum majus erit $\frac{\sqrt{:5-2\sqrt{3}:}}{2}R$. Nam, posita secanda recta, seu quadrantis radio, $DA = 1R$, adeoque $DH = \frac{1}{2}R$, erit $HX = \sqrt{\frac{3}{4}}R^2$, et $IX = \sqrt{\frac{3}{4}}R^2 - \frac{1}{2}R$, cujus quadratum $R^2 - R^2\sqrt{\frac{3}{4}}$, et (propter quadratum $EI = \frac{1}{4}R^2$) Quadratum $EX \frac{5}{4}R^2 - R^2\sqrt{\frac{3}{4}}$, hoc est $\frac{5-2\sqrt{3}}{4}R^2$; ergo ipsa $EX = \frac{\sqrt{:5-2\sqrt{3}:}}{2}R$, Segmentum majus (si *Hobbio* credas) secandae $DA = 1R$; Adeoque segmentum minus, $1R - \frac{\sqrt{:5-2\sqrt{3}:}}{2}R$.

Num vero *Euclidi* (atque post illum aliis hactenus) an *Hobbio* credendum sit, tuum esto judicium. Sin neutrius authoritati credendum putes, sed rationi; examinemus (ut jam *Euclidis*) *Hobbii* constructionem.

Rectam extrema et media ratione secare, est, ita secare, ut quadratum segmenti majoris aequetur rectangulo ex minore segmento et tota secanda. (Quod norunt omnes, nec *Hobbius* diffitetur.) Sed factum ex tota $1R$, et minori segmento $1R - \frac{\sqrt{:5-2\sqrt{3}:}}{2}R$, est $R^2 - \frac{\sqrt{:5-2\sqrt{3}:}}{2}R^2$: quod aequale esset majoris EX quadrato. Sed non est. Quippe hoc jam inventum est $\frac{5-2\sqrt{3}}{4}R^2$.

16-1 exponere. Notata quaedam hic tibi mitto: non quasi metuerim, te talibus ratiociniis seduci posse, sed ut tu, aliique, quibuscum haec forte communicaveris, sine anxia consideratione denuo instituenda, statim videatis ubi potissimum peccatur. —— *Primae Propositionis* E

1 (*1*) Propositionis (*a*) primae (*b*) Primae (*2*) Primae Propositionis W
8 quam (ad calculum redactam) segmentum E
9-10 recta, seu quadrantis radio, *add*. W
10 erit $HX = \sqrt{\frac{3}{4}}R^2 = \frac{1}{2}R\sqrt{3}$, & IX E
14 $1R$; (*1*) Ergo segmentum (*2*) Adeoque segmentum W
16 authoritati *add*. W
17 Euclidis,) sic *Hobbii* E
18 media (*1*) recta (*2*) ratione (*a*) tum sect *breaks off* (*b*) secare W
22 aequale (*1*) esse deberet quadrato (*2*) esset majoris EX quadrato. (*a*) $\frac{5-2\sqrt{3}}{4}R^2$ (*b*) Sed W
22 hoc jam (*1*) deprensum (*2*) inventum W

[1188] demonstraverat: i.e. EUCLID, *Elements* VI, prop. 30.

219. WALLIS to OLDENBURG, 16/[26] July 1671

Falsa igitur est *Hobbii* constructio. Et (propter hanc falsam) ruunt etiam quae annectit *Corollarium* et *Consectarium*.

Menda in ipsius qua constructione *qua* (praetensa) demonstratione, (praeter minora multa) sunt haec saltem tria grandia.

1. In Constructione, pro *Describatur centro D quadrans DAC secans FE et GH in K et X*: dici non minus potuisset, *sumatur X ubivis in IG recta*. Nam et sic non minus procederet, *Ducatur denique EX*, (et quae sequuntur omnia,) ne una quidem vel voce vel syllaba mutata. Vel etiam *ubivis in IG recta utcunque in utramvis partem producta*: Nam etiam hoc posito, si *pag. 2. lin.* 17. pro *secans AE*, ponatur *secans AE saltem productam*, omnia similiter procedent. (Quod legenti statim patebit.) Ut possit esse, per ipsius demonstrationem, segmentum majus quantumvis longum.

2. In Demonstratione Cum ostenderat *pag. 2. lin.* 18, 19, *duos rectos mXI, IXl, aequales quinque angulis $[m]XF$, FXy, yXz, zXE, EXl*, (ne insinuato quidem, nedum probato, hos omnes esse inter se aequales;) Hinc probatum it (quasi jam probasset, omnes illos quinque invicem aequales esse) *angulos zEX, zXE, esse invicem aequales*: nempe quia, *si secus, duo illi anguli recti non sic dividerentur quinquifariam*, seu in quinque partes *invicem aequales*. pag. 3. lin. 4, 5. Sunt quidem tres, mXF, yXz, EXl, invicem aequales; item duo, FXy, zXE, invicem aequales: sed utrumvis horum utrivis illorum aequalem esse, neque demonstratum est, neque verum.

3. Ubi, ex eo quod *anguli yXz, Xyr, sint aequales; item yXr, yrX, aequales; et yXz, zEX aequales;* infert, (pag. 2. lin. 17, 18) *Quare anguli X et E trianguli zEX sunt aequales*: Nulla est consequentiae vis. Quod attendenti patebit.

1-2 Et ... *Consectarium. add. W*
5 1. (*1*) Ubi ait Co *breaks off* (*2*) In Constructione *W*
9 etiam *add. W*
10 *pag. 2. lin. 17. add. W*
11 esse, (*1*) ipsius segmentum (*2*) per ipsius demonstrationem, segmentum *W*
14 *angulis nXF, FXy, yXz W corr. ed.*
16-17 (quasi ... aequales esse) *add. W*
17 *aequales*, quia *E*
18 *illi add. W*
18 *sic add. W*
19 (*pag. 3. lin.* 4, 5.) de quo in praecedentibus nihil dictum est. Sunt *E*
20 duo, *add. W*
21 neque (*1*) ostensum est (*2*) demonstratum est *W*
23 (pag. 2. lin. 17, 18) *add. W*

219. WALLIS to OLDENBURG, 16/[26] July 1671

Praeter haec omnia; *Propositionis secundae* constructio, hanc *primae* refutat. Nam, si ponamus illic, totam AG vel $CE, = 1R$, segmentum majus AB vel AC vel EF, erit $= \frac{\sqrt{5}-1}{2}R$. Nam ut $AG = CE$ ad $CA = AB$, sic CA ad CF. Est autem CA ad CF, ut 2 ad $\sqrt{5} - 1$, (nam propter $AC = AB = 2AD$, Quadr. $CB = 5$ Quadr. AD, ipsaque $CB = AD\sqrt{5}$, unde dempta AD vel DF, erit $CF = AD\sqrt{5} - AD$: ergo $CA = 2AD$ ad CF, ut 2 ad $\sqrt{5} - 1$.) Ergo et AG ad AB, ut 2 ad $\sqrt{5} - 1$, (quod vult *Euclides*,) non, ut 2 ad $\sqrt{} : 5 - 2\sqrt{3}$: quod vult *Hobbii prop.* 1.| [3ᵛ]

Tam turpiter autem titubasse *Hobbium* in ipso limine, eo magis mirandum est, et minus condonandum, quod problematis constructio vera (et facilis) in ipsis Elementis extet (pr. 30. El. 6.) estque pueris nota.

Propositio Tertia (multimembris,) *de Polygonis Regularibus*, (cum *Consectariis* suis,) dependet tota ex hac consequentia, *Quoniam chordae Cb, bc, ad chordas Ci, ic* (in eodem circulo,) *sunt ut 8 ad 7*, propterea etiam arcus $Cc = Cb + bc$, ad arcum $CE = Ci + iE$ ut 8 ad 7. (*pag.* 11. *lin.* 2.) atque in aliis proportionibus similiter, *p.* 13. *l*: 23, 26. *p.* 14. *l.* 20, 24. *p.* 15. *l.* 4 &c. Quasi quidem, in eodem circulo, *Arcus essent Chordis proportionales*. Quod quam ridiculum sit non dictu opus est.

Hinc infert, EH (subtensam octantis) ad EF (subtensam sextantis) esse ut 3 ad 4, (quia arcus sunt in ea ratione,) *p.* 13. *l.* 23. Item, EF (subtensam partis duodecimae) ad EH (partis decimae subtensam,) ut 5 ad 6 (in ratione arcuum,) *p.* 14. *l.* 19, &c. Satis crude.

Prop. Quarta; postquam *Circuli peripheriam curvam esse* ostenditur, *curvedinemque a flexione oriri*, diciturque *Curvedinum aliam alia majorem esse*: Ostensum it, primo, quod, *Quam rationem habet in eodem circulo angulus in circumferentia* (major) *ad angulum in circumferentia* (minorem,)

---

2 refutat. (*1*) Est enim illic (*2*) Nam, *W*
2 vel CE, add. *W*
3 Nam, (*1*) ut AC hoc est CE, ad AC hoc est EF, sic A (*2*) ut $AG = CE$ *W*
4-7 CA ad CF ut 2 ad $\sqrt{5} - 1$, (quod vult *Euclides*:) non *E*
5 $2AD$, (*1*) rectos (*2*) CB quadratum (*a*) erit (*b*) volebit 5 quadrata AD (*3*) Quadr. CB *W*
10 quod (*1*) vera hujus problematis constructio (*2*) problematis constructio vera *W*
15-16 (*pag.* 11. *lin.* 2.) ... *p.* 15. *l.* 4 &c. *add. W*
19-22 Hinc infert ... crude. *add. W*
23 curvam *add. W*
24 curvedinemque ... oriri *add. W*
25 quod, *add. W*
26 (major) *add. W*
26 (minorem,) *add. W*

219. WALLIS to OLDENBURG, 16/[26] July 1671

eandem habet: curvedo majoris arcus ad curvedinem minoris. (Puta, curvedo arcus quadrantalis ad curvedinem semiquadrantalis, est ut 2 ad 1; propter duplo plures in illo quam in hoc flexiones.) Deinde, quod, *In diversis circulis curvedo majoris perimetri minor est curvedine minoris.* Quasi quidem non tot essent in majori perimetro quot in minori Flectiones. Quod absurdum est. Utut enim in arcubus longitudine aequalibus pauciores essent in circulo majori quam in minori flexiones (eo quod ille minorem angulum subtendat:) Certe in tota perimetro majore (aut etiam partibus proportionalibus, ut quae aequales angulos subtendunt,) non pauciores erunt flexiones quam in minori. Quique totam circuli circa Terram maximi curvedinem simul conspiciat, (vel hujus partem aliquotam;) non minus curvedinis cernet quam in Annulo, vel hujus parte proportionali. Haec itaque cum praecedentibus non satis cohaerent. Si dicat, se alio sensu illic alio hic majoritatem curvedinis intelligere; Æquivoce loquitur.

Propositio Quinta, (quae exhibet rationem Radii ad Perimetrum circuli, ut R ad $10R\sqrt{\frac{2}{5}}$; hoc est, ut 10000 ad plusquam 63245; quam alii faciunt ut 10000 ad minus quam 62832;) dependet ex hac consequentia, (*pag. 18. lin. 5.*) *Quoniam ut DC ad DR* (radius ad [radium]) *sic arcus CA ad arcum RS* (similem;) *ita Quadrantalis arcus descriptus radio DC, id est Arcus CA, ad arcum descriptum* (radio DR, hoc est arcum RS, sic dicendum erat, sed ille) *radio RS extenso in rectitudinem.* Quod absurde dictum esse, per se liquet.

---

1 curvedo (*1*) arcus majoris ad (*2*) majoris arcus ad *W*
2 curvedo arcus (*1*) Quadrantale semicirculi ad quadr *breaks off* (*2*) quadrantalis *W*
2-3 propter ... flexiones.) *add. W*
4-5 Quasi quidem (*1*) non essent in majori perimetro tot flectiones (*2*) non tot essent ... perimetro quot in minori Flectiones. *W*
6 absurdum est |nec cum Accidentibus rite convenit. *add. and del.*| . Utut *W*
6-7 aequalibus (*1*) minores (*2*) pauciores essent (*a*) flexiones major *breaks off* (*b*) in circulo ... in minori flexiones (*aa*) (quoniam ille ad totam per (*bb*) (eo quod *W*
12 Annulo, hujusve parte *E*
13 cohaerent. Sin dicat, se alio sensu hic, alio illic *E*
14 intelligere; (*1*) Hoc est aequivoce loqui *breaks off* (*2*) Æquivoce loquitur. *W*
16 plusquam *add. W*
18 (*pag. 18. lin. 5.*) *add. W*
18 ad radius *W corr. ed.*

219. WALLIS to OLDENBURG, 16/[26] July 1671

Propositio Sexta, cum ejusdem *Scholio* et *Consectario;* item *Propositio Septima,* cum ejus *Corolario,* et *Consectariis* quatuor; item *Propositio Octava,* quaeque hac nituntur; dependent a prop. 5. (ut patet, *pag.* 20. *lin.* 4, 6, 8, 10. *pag.* 21. *lin.* 6. *p.* 22. *l.* 5. *p.* 23. *l.* 8, 14, 28. *p.* 24. *l.* 2. *p.* 25. *l.* 1, 14, 16. *p.* 26. *l.* 18. *p.* 27. *l.* 5, 17. *p.* 28. *l.* 4. *p.* 30. *l.* 21, 22. nec diffitebitur *Hobbius*) ergo, cum illa ruunt.

Propositio Nona, (de sectione Anguli in ratione data) eodem misero tibicine fulcitur cum prop. 3. nempe, in eodem circulo *Chordas arcubus proportionales esse.* Adeoque juxta cum illa cadit.

Propositionis Decimae Corollarium, verum est si sumatur P in producta Db, non autem in producta AK. Cum vero haec duo P habeat Hobbius pro eodem; hallucinatur. non enim coeunt AK, Db, in eodem rectae BC puncto P: ut post dicetur.

Proposito Undecima, falsa est: Nempe *Tangentes grad. 30, et grad.* $22\frac{1}{2}$. *simul aequari Radio.* Hoc est (per canonem Tangentium) in numeris absolutis quam proxime, $5773503 + 4142136 = 10000000$: vel (accurate) in surdis, $\frac{1}{3}\sqrt{3} + \sqrt{2} - 1 = 1$. (Satis absurde.) Nec probat ille (quod in demonstratione assumitur) *rectas AK, Db, productas, incidere in* (rectae BC) *punctum P.* Potest utique punctum concursus P (non obstante probatione sua) vel supra vel infra rectam BC contingere. Quod enim in probationem adducit, *pag.* 37. *lin.* 21. *Cum enim,* &c non sequitur. Utut enim angulus, quem faciunt (productae) AK, Db, sit $\frac{7}{12}$ (*septem duodecimae*) unius recti, et quem cum BC facit (producta) AK, $\frac{8}{12}$ (*octo duodecimae,*) et quem cum eadem BC, facit (producta) Db, $\frac{9}{12}$ (*novem duodecimae*) unius recti; non tamen hinc

2 *Corolario,* et *(1)* quatuor *(2) Consectariis* quatuor *W*
2-3 item *Propositio Octava* ... nituntur; *add. W*
4 *lin.* 4, 6, 8,|10. *add.*| *pag.* 21. *W*
5 *p.* 28. *l.* 4. *(1)*) adeoque cum illa ruunt *pag.* *(2) p.* 30. *l.* 21, 22. *W*
7 (de sectione ... data) *add. W*
10-13 *Propositionis Decimae* ... post dicetur *add. W*
11 non autem, si in producta *E*
14-17 falsa est: *(1)* Quod si probationem adducitur, *p.* 38. *l.* 1. *Cum ergo,* &c non sequitur *(2)* Nempe *Tangentes* ... Tangentium) |in numeris ... proxime, *add.*| 5773503 ... (Satis absurde.) *W*
18-20 *punctum P. (1)* in Quod *(2)* Potest utique ... Quod enim in probationem *W*
21 *Cum (1)* ergo *(2)* enim, &c *W*
21 angulus *(1) CPD* sit septem duodecimae *(2)* , quem faciunt ... *Db*, sit *W*
22 et quem *(1)* cum *BC AK* ⟨—⟩ *(2)* cum *BC* facit *W*
23 et quem *(1)* facit *(2)* cum eadem *BC*, facit *W*

219. Wallis to Oldenburg, 16/[26] July 1671

magis sequitur punctum concursus P in recta BC contingere, quam in recta GH. Nam hic eadem verbatim dicenda essent; *Cum enim angulus CPD vel HRD sit novem, et angulus GSA quem facit tangens 30 graduum cum sua secante, sit octo, (duodecimae unius recti) et reliquus angulus* (quem faciunt productae AK, Db, nempe septem duodecimae) *complementum ad duos rectos;* Ergo quid? Num| *Ergo rectarum AK, Db, punctum concursus est in recta BC?* Imo non minus sequitur, *Ergo est in recta GH.* Sed hoc non sequi, fatebitur *Hobbius.* Ergo nec illud.

Consectarium una ruit cum propositione.

Propositionis Duodecimae falsum est Corollarium. Non enim sunt aequales CO et $\frac{1}{2}AT$. Sed (posito Radio DA, vel $DC, = R,$) erit graduum 30 Tangens $AT = \frac{1}{3}R\sqrt{3}$, (utpote dimidia secantis $\frac{2}{3}R\sqrt{3}$, cujus quadratum complent quadrata AT et AD,) adeoque $\frac{1}{2}AT = \frac{1}{6}R\sqrt{3}$: Sed $CO = R - \frac{1}{2}R\sqrt{2}$ (excessus Radii DC supra DO sinum graduum 45:) Non sunt ergo CO, $\frac{1}{2}AT$, aequales. Sed nec ille aequales esse demonstrat. Et ubi hoc aggreditur, (coroll. prop. 10.) hallucinatur. Supponit enim (quod non probat, ut ad prop. 11. ostensum est) rectas AK et Db in eodem rectae BC puncto P concurrere.

Propositio Decimatertia subvertit primam. Nam hic agnoscitur et adhibetur sectionis rectae in extrema et media ratione constructio *Euclidaea*; ubi recta secanda ad segmentum majus est, ut 2 ad $\sqrt{5} - 1$, seu 1 ad $\sqrt{\frac{5}{4}} - \frac{1}{2}$; non (ut in *Hobbiana* prop. 1.) ut 2 ad $\sqrt{u : 5 - 2\sqrt{3}}$.

2-6 *Cum* (1) *ergo* (2) *enim angulus* |*CPD vel add.*| *HRD sit novem,* (a) *et angulus DAK octo duodecimae, erit reliquus angulus APD septem, et omne simul vigintiquatuor duodecimae et GSA* (b) *et angulus GSA ... reliquuus angulus* (aa) *APD* (sumpto (bb)) (quem faciunt ... AK, Db, (aaa) ⟨—⟩ septem (bbb) nempe septem ... duos rectos; (aaaa) id est, ad tres angulos (bbbb) Ergo quid? W

6 Num (1) Ergo Ak, Db, concurrent in rectae BC puncto. Imo non (2) Ergo rectarum W

9-10 propositione. (1) *Propositionis Duodecimae* Corollarium falsum est. Non sunt |aequales *add.*| enim CO et $\frac{1}{2}AT$. Sed (posito radio $= R,$) est $CO = R - R\sqrt{\frac{1}{2}}$ |(differentia radii et sinus graduum 45;) *add.*| et $\frac{1}{2}AT = \frac{1}{4}R$ (seu tangens graduum 30) (a) sive quadratus subtensae grad. 60 ⟨—⟩ (b) adeoque non invicem aequales. Sed nec ille aequales esse demonstrat. Sic ubi illud aggreditur, est ex falsis praemissis. (2) *Propositionis Duodecimae* falsum W

11-13 erit (1) AT Tangens grad. 90 (secantis secans) (2) graduum 90. Tangens $AT = \frac{1}{3}R\sqrt{3}$, (a) secantis (b) (utpote dimidia secantis (aa) secans autem (bb) $\frac{2}{3}R\sqrt{3}$, cujus quadratum (aaa) $\frac{4}{3}R^2$ aequit quadratum AT (bbb) complent W

15-16 Et (1) sicubi (2) ubi hoc aggreditur, (a) est ex falsis praemissis (b) (coroll. prop. 10.) hallucinatur. W

21-22 ut 2 ad $\sqrt{5} - 1$, non (ut in *Hobbiana*, prop. 1) ut 2 ad $\sqrt{u : 5 - 2\sqrt{3}}$. E

219. WALLIS to OLDENBURG, 16/[26] July 1671

Propositio Decimaquarta falsa est. Est enim graduum 30 secans $\frac{2}{3}R\sqrt{3}$; adeoque extrema et media ratione sectae segmentum majus $\frac{\sqrt{5}-1}{2} \times \frac{2}{3}R\sqrt{3} = \frac{\sqrt{5}-1}{3}R\sqrt{3}$. Sed quadrati ex Radio semidiagonalis, $\frac{1}{2}R\sqrt{2}$. Non sunt igitur aequales. Ubi autem (in probatione) dicit, *Ostensum enim est prop. 10. rectam BP duplam esse rectae CO:* ludit in ambiguo. Ostenderat enim *BP residuam tangentis CP* (grad. $22\frac{1}{2}$) ad Radium, duplam esse *CO*; sed non *BP tangentem* grad. 30. Quas cum ille pro eadem habet; hallucinatur: ut ad prop. 11. ostendimus.

Prop. Decimaquinta, suasum it (non probatum) *In omni quaestione Geometrica, multo prudentius esse, certe Mechanice mensurando magnitudinem quaesitam quam potest fieri veritati proximam assequi, et deinde causam inquirere propinquitatis; quam credens incertae Logicae vel Logisticae pronunciare. Et multo credibilius a Mensura pronunciare Mensorem diligentem, quam Algebristam seu Arithmeticum.* (Non itaque mirandum est *Hobbium*, his methodis utentem, talem nobis producere Geometriam.) Sed et *studiosum veritatis non putandum esse, qui sententiam videns suae contrariam, verisimilibus argumentis fultam, contentus sit pugnare contra solam demonstrationem.* (Quasi quidem, in re explorata et saepius demonstrata, non sufficeret praetensae in contrarium demonstrationis imbecillitatem ostendere. Sed neque hic desumus; nam sua non modo Indemonstrata, sed Falsa esse demonstramus.) Hoc est, audiendum potius esse *Hobbium*, ex mechanica mensura probabiliter (sine demonstratione) pronunciantem; quam demonstrative arguentes alios.

Ego, contra, *Hobbio* suaderem potius ut Verisimilia sua, et Mensurationem Mechanicam, Demonstrationi (sive ex Arithmetica, sive ex Geome-

---

2 sectae *add.* W

9 *Decima quinta,* novam commendat Geometrizandi methodum, quam (credo) ipse sequitur. Suasum utique it (non probatum) *E*

9 it (non (*1*) demonstrat *breaks off* (*2*) probatum) W

16 Geometriam; utpote cui Circinus est Calculo accuratior.) Sed & *E*

17 *argumentis* (puta, circini indicio) *contentus E*

19-20 sufficeret impugnantis paralogismos detegere. Sed neque *E*

20-21 Sed neque ... demonstramus. *add.* W

21-24 *Hobbium,* verisimiliter (ex circini indicio in angusto Schemate) sine demonstratione pronunciantem; quam demonstrative ratiocinantes alios: Nec satis, eum redargutum esse, ostensis ratiocinii sui *paralogismis* & demonstrationum defectibus; aut argumentis in contrarium sive ex *Logica* sive *Logistica* petitis: Quippe haec omnia cedere vult *Circino* & suis *Verisimilibus*. Bellum equidem Geometram! Ego, contra *E*

24 Ego, (*1*) potius, Hobbio suaderem et (*2*) contra, ... sua, et W

24-25 sua & *Mechanicam mensuram* (ubi ἀκρίβεια Geometrica spectatur) Demonstra-

219. WALLIS to OLDENBURG, 16/[26] July 1671

tria petitae) posthabeat; et Demonstrationi magis quam Circino credat; quo (quae sua vox est) minus *irrideatur*.

Propositione Decimasexta, affert (si credes) *demonstrationem, quae nisi confutetur, audiendi ulterius non erunt arguentes a potestate linearum*: (ergo, nec *Hobbius*, nam ille ubi potest sic arguit.) Sed confutatio facilis est. Quod enim assumit (*pag. 46. lin. 29.*) *Manifestum est tum Kd tum ba esse ad ae ut 3 ad 1*; falsum est. Verum quidem est Kd ad ae sic esse; sed non ba. Neque affert ille quicquam quo probet, vel Kd, ba, invicem aequales esse; vel, ba ad ae esse ut 3 ad 1. Sed neque verum est. Est enim Kd ad AD, ut $\frac{3}{4}\sqrt{5} - \frac{3}{2}$ ad 1: sed ba ad AD, ut $\frac{1}{12}\sqrt{5}$ ad 1. Non sunt igitur (quod ille gratis et falso assumpsit) invicem aequales.

Et propterea falsum est, ba ad Hc $esse$ ut 3 ad 2; item, fa ad ae ut 3 ad 2; item, *junctam cd esse parallelam ID*; et dH, Hb, *invicem aequales*; item, bK, da, *aequales esse*: (Nam haec| omnia supponunt, Kd aequalem rectae ba, rectaeque KH trientem:) item (quae hinc dependent) Md esse quater duo quorum MK est ter tria; adeoque MK ad MD sive MF ut 9 ad 8.

Sed et mox, dum ait, MK *quintuplum potest* FK *sive* Mb; supponit (gratis quidem et falso) FK, Mb, *invicem aequales esse*. *Gratis*, inquam; nam ne hilum affert quo probet, (nisi forte, circino rem explorans, hoc inde collegerit:) et *Falso*; est enim FK ad AD ut $\frac{3}{4}$ ad 1; sed Mb ad AD ut $\frac{1}{3}\sqrt{5}$ ad 1. Adeoque falsum porro est, *si detrahatur FK a recta MK, reliquam esse bK*. Sed et (propter dK, ba, non invicem aequales) falsum etiam est, bK aequalem esse da.

---

tioni E
1 posthabeat; quo (quae sua vox est) E
1 et Demonstrationi . . . credat; *add.* W
5 nam ille *(1)* quoties *(2)* ubi W
5 *Hobbius*, qui, ubi potest, sic arguit; ut & aliis methodis quas alibi damnat.) Sed confutatio E
10 sunt igitur *(1)* aequales *(2)* (quod ille . . . aequales. W
11 invicem aequales KD & ba. Et propterea E
12 falsum est (quod porro habet) ba E
12 ae ut 3 ad 2; *(1)* (quod supponit Kd aequalem rectae ba, adeoque trientem rectae KH, (quae falsa sunt,) *(2)* item W
14-15 supponunt, Kd *(1)*, ba, invicem aequales *(2)* aequalem . . . trientem W
14 omnia praesumunt, E
17 mox, *(1)* subsumit FK et Mb aequales (gratis quidem et falso,) *(2)* dum ait W
19 forte *add.* W
20 sed Mb ad |ad *del. ed.*| AD W
22 (propter . . . aequales) *add.* W

219. WALLIS to OLDENBURG, 16/[26] July 1671

Falsum igitur est quod hinc infert, *MK esse* 9, *quorum Md est* 8, *Ma* 3, et *da* 5. Neque hinc *Euclidi, (vel arguentibus a potestate linearum,* contradicitur. Quippe *illi* falsum esse pronunciarent, nec probat *Hobbius* esse verum.

Propositio Decima septima, item falsa est. Dependet utique tum a *prop.* 14. (ut *pag.* 49. *lin.* 14.) tum a *Consect.* 3. *prop.* 7. (ut *pag.* 50. *l.* 15.) quorum utrumque falsum esse supra ostendimus; et utriusvis falsitas sufficeret huic propositioni subvertenda.

Sed et alia subsumit falsa, ut (*pag.* 49. *l.* 22) *rectam Bf quintae partis lateris AB potentia quintuplam esse;* hoc est (posita $AC = 1$,) $\frac{1}{5}\sqrt{5}$; cum tamen sit $\sqrt{\frac{6-\sqrt{5}}{20}}$. Item *pag.* 50. *lin.* 4. ponit *latus Icosaedri* $\alpha\zeta$; sed idem facit $\alpha\gamma$, *lin.* 10, 11. Item, *lin.* 9. vult ut sumamus *arcus Pθ*, $\gamma\zeta$, (in circulis inaequalibus,) invicem aequales, (non similes,) quod quomodo faciamus nec docuit ille, nec docebit. Item (*lin.* 10.) *Arcus Pθ, radio BP descripti, radium alterum Bθ* (utpote ipsi *BP* aequalem) probat inde *aequalem rectae $\alpha\gamma$* (quae ex constructione est ipsi *BP* aequalis, *pag.* 49. *lin.* 6, 12;) sed mox (*p.* 50. *l.* 11, 12,) vult *eandem Bθ, minorem esse quam BP* (arcumque radio *Bθ* descriptum *secare* rectam *BP* in *e*,) ipsique (non *BP* sed) *Be aequalem*, hoc est *tertiae parti rectae Bb* (quam rectam ille perperam supponit aequalem arcui *AC*:) Et *(lin. 19) eidem Be aequalem esse* vult rectam $\alpha\gamma$ (quae tamen ex constructione fuerat aequalis *BP*, cujus pars est *Be* ex constructione, pag. 48. l. 22.) Adeoque omnia ponit in confuso.

---

1 hinc *add.* W
2-3 Neque hinc *Euclides, vel arguentes a potestate linearum*, evincuntur. Quippe *E*
4-5 verum. (1) *Propositio Decima septima*, dependet ex prop. 14. (ut liquet *pag.* 49. *lin.* 14) quem falsam deprehendimus. Ergo et haec simul ruit. (2) *Propositio Decima septima,* item *W*
7 utrumque falsum *(1)* est *(2)* esse |supra ostendimus; *add.*| et *W*
5-8 *Propositio Decima septima*, dependet ex prop. 14. (ut liquet *pag.* 49. *lin.* 14.) quam falsam esse deprehendimus: Ergo & haec simul ruit. Sed & a *consect.* 3. *prop.* 7. dependet, (*pag.* 50. *l.* 15) quod falsum esse ostendimus: Adeoque duplici ruina labitur. *E*
11-12 *pag.* 50. *lin.* 4. *(1)* posuerat *(2) latus Icosaedri* posuerat $\alpha\zeta$; ⟨—⟩ sed $\alpha\gamma$ *(3)* ponit ...sed idem facit $\alpha\gamma$ *W*
14 docebit. *(1)* Tum |lin. *add.*| arcus *Pθ*, centro *B* descripti, radium *Bθ* probat |inde *add.*| aequalem (*(a)* ipsi $\alpha\gamma$ *(b) BP* ideoque ut) ipsi $\alpha\gamma$ (quae ex constructione ponitur |ipsi *add.*| *BP* aequalis,) lin *(2)* Tum Arcus *(3)* Item (*lin.* 10.) Arcus *W*
15 alterum *add.* W
22 *pag. 48. l. 22.)* Adeoque, nunc $\alpha\zeta$ nunc $\alpha\gamma$ ponens pro Icosaedri latere, rectamque $\alpha\gamma$ seu *Bθ* nunc aequalem nunc minorem recta *BP*, omnia ponit in confuso. *E*

219. WALLIS to OLDENBURG, 16/[26] July 1671

Postque haec ita confusa omnia, et falsis suffulta, sibique invicem opposita; *rationem suam* assignatum it, *ex motu et mensura, verisimilem*; quam *demonstrationem* vocat, sed quam (recte conjicit) *Algebristas damnaturos*, sed et *alios*, (quippe qui ne *verisimilitudinem* ullibi deprehendunt ullam;) Qua quidem, si demonstrandi vim haberet ullam, probaretur, (non quod erat ab initio positum, *rectam* $\alpha\zeta$ *latus Icosaedri*, ut pag. 50. l. 3, 5; sed) *rectam* $\alpha\gamma$ *latus Icosaedri* (ut lin. 9, 10,) *aequalem esse tertiae parti* (arcus) *semicirculi*, (adeo illi indifferens est, sive $\alpha\zeta$, sive $\alpha\gamma$, sit latus Icosaedri; et de utrisque perinde procedit demonstratio; quas tamen invicem aequales esse, ne ipse dixerit:) adeoque *subvertit* id quod erat *probandum*. Qua tamen oscitantia stupidissima non obstante; male habet quod pro legitima demonstratione non simus habituri.

Propositio Decima octava, (de circuli quadratura,) est Crambe (non bis tantum, sed) saepius recocta; atque haec eadem constructio jam tertio saltem in cassum introducta. In Demonstratione, prout nunc prodit, illud (pag 54. lin. 17. 18.) *Quare sector reliquus DVP duplus est, tum trilinei CYP,* (non, sed sectoris *Dbi*) *tum quadrilinei VPbi,* nullam habet vel speciem consequentiae. Dum enim pro *Sectore Dbi* (quod dicendum erat) substituit *Trilineum CYP*, quasi haec essent aequalia; praesumit id quod erat probandum.

Quae hinc dependet *Duplicatio Cubi*, nihilo itaque firmior est. Quam ut defendat, admittit (tanquam non inconmodum) 10. *decimas sextas, et* 16. *vicesimas quintas aequales esse; (pag.* 55. *lin.* 10, 13.) ponitque *pag* 56.

---

2 assignatum it *verisimilem* E
3 vocat (ineptam satis,) sed E
4 ullam *add.* W
4-5 alios, (quippe qui *verisimilitudinem* ullam inibi deprehendunt;) Qua E
5 demonstrandi *add.* W
6 sed) (*1*) latus (*2*) rectam W
5-7 probaretur, non, (quod erat ab initio propositum,) *rectam* ... sed, (quo jam per oscitantiam delapsus est,) *rectam* $\alpha\gamma$ E
8-10 Icosaedri (*1*) nec majus de hoc quam de illo (*2*) et de utrisque perinde procedit demonstratio; (*a*) planaque $\alpha\gamma$, nunc ipsi rectae BB, nunc rectae Be (ejusdem parti) ponit aequalem: (*b*) quas tamen ... probandum W
10 tamen (*1*) stupiditate (*2*) stupidissima negligentia (*3*) oscitantia stupidissima W
13 (de ... quadratura,) *add.* W
15 introducta. (*1*) Demonstratio (*2*) In Demonstratione W
16 *duplus est add.* W
20 hinc (*1*) dependent non attingo (*2*) dependet W
20 firmior (*1*) Quodque hinc insent (*2*) est. Quam W

219. WALLIS to OLDENBURG, 16/[26] July 1671

lin. 2. (ut suis effatis consequens) *non modo* 50, 40, 32, *sed etiam* 50, 40, $31\frac{1}{4}$, *continue proportionales.* Quae pueris ridenda relinquo.

Propositio Decima Nona, dependet a prop. 5. (ut liquet *pag.* 58. *lin.* 8.) quam falsam deprehendimus. Item, falsum illud (*pag.* 58. *l.* 12.) *circulum centro l, radio lF descriptum, transiturum per G simul et C.* (Transibit quidem per C, propter bisectam CF in l, sed non per G.) Probatio ejus (*pag.* 58. *lin.* 20.) dependet a *prop.* 6, 7, quas falsas deprehendimus. Deinde, *pag.* 59. *l.* 9. praesumit rectas AH et DE productas,| ad F (punctum in CB [5ʳ producta assignatum) pertingere: Quorum neutrum probatum est; imo ne affirmatum quidem, sed tacite praesumptum; idque falso: Adeoque et falsa quae sequuntur.

Propositio Vicesima falsa est; utpote quae (ut *pag.* 61. *lin.* 22.) dependet a prop. 18. quam falsam esse ostendimus. Sed, utut hoc non esset in fundamento vitium, quod sequitur (*pag.* 62. *lin.* 1, 2, 3, 4) est lepida designatio centri gravitatis: Unde, qui res has intelligit, facile percipiet quam *Hobbius* eas non intelligit.

Consectarium (utpote inde dependens) est ejusdem commatis. Sed et alio nomine vitiosum, eo quod dependeat etiam a prop. 5, quae est itidem falsa.

Propositio Vicesima prima, (quae et *ultima*,) item falsa est. In probatione; falsum illud (*pag.* 63. *lin.* 14) *Gnomon YBM est Quinta pars quadrati DYQM*: (Nam differentia duorum quadratorum quae sunt inter se ut 5 ad 4, est quadrati Minoris pars *Quarta*, non quinta; Quinta vero Majoris.) Sed demus hoc: Falsum quod sequitur, *id est, pars quinta quadrantis DAC*: Dependet enim a prop. 18. (quam falsam ostendimus) ubi putat se probasse, Quadrati $DYQM$ et Quadrantis DAC aequalitatem. Item falsum illud, *Quare etiam Trilineum ABCLA est quinta pars quadrati ABCD*: pluribus nominibus. Nam, primo, dicendum erat, ad mentem suam, *quadrati DYQM*, non quadrati $ABCD$; (quippe $\frac{1}{5}$ Quadrantis DAC ab eo ponitur aequalis $\frac{1}{5}DYQM$, non $\frac{1}{5}ABCD$.) Sed neque de $DYQM$ verum est;

1 *pag* 56. *lin.* 2. *add.* W
9 pertingere: (*1*) Quod probatum non est (*2*) Quorum neutrum probatum est W
10 imo ne affirmatum, sed E
12 *Vicesima* (*1*) dependet (*2*) falsa est ...dependet W
21-22 probatione; illud ...$DYQM$; falsum est. (Nam E
27 pars |quadrati add.| ABCD: (*1*) Supponit ⟨—⟩ (*2*) pluribus W
28 erat, (*1*) quadrati D *breaks off* (*2*) ad mentem W
29 (quippe (*1*) illud, nec hoc, ab eo ponitur aequale quadranti DAC) (*2*) $\frac{1}{5}$ Quadrantis DAC ...$ABCD$.) W

219. WALLIS to OLDENBURG, 16/[26] July 1671

Praesumit enim ex prop. 18 (quod falsum est) *Trilinea CYP, PQL, esse inter se aequalia.* Sed et qualis est ea consequentia; Quoniam Gnomon est pars quinta quadrati Minoris, Ergo Trilineum (quod gnomoni praesumitur aequale) est pars quinta Majoris? Sed tales ejus esse solent consequentiae. Quod autem inter duo quadrata $DYQM$, $ABCD$, subsultim ludat, (nunc de hoc, nunc de illo, idem affirmans;) pro solita sua oscitantia factum est. Caeteraque quae sequuntur, tanquam ab his pendentia, falsa sunt.

Adeoque percurrimus olens *Rosetum*, brevibus stricturis menda ex innumeris praecipua notantes; alia quamplurima consulto praetereuntes, ut vel minoris momenti, vel quae opus non erant ad subvertendas propositiones. Sed talis expectanda est Geometria, ab eo qui, magnitudines circino dimensus in angusto schemate, quas ita non deprehendit inaequales eas pro aequalibus tuendas existimat, etiam contra demonstrationum authoritatem, quam ille circino postponit.

Quae autem de me habet, sive ad operis calcem sive ad frontem, contemnenda sunt. Quippe, praeter pueriles quasdam circa voces ineptias, quas conatur in sensum perversum detorquere; caetera fere huc tendunt, Se *Symbola* non intelligere, *Arithmeticam Speciosam*, sibi non placere, sed nec *Geometriam Indivisibilium*, aut *Arithmeticam Infinitorum*. (Et quidem mihi perinde est, sive sic, sive secus. Nam jamdiu est, quod Hobbii Authoritas in Mathematicis ne hilum significaverit; ejusque Rationes, tantundem.) Sunt autem ea omnia tam pueriliter ab eo dicta, ut quicunque rerum harum intelligens ad loca notata respicit, pro me facile etiam non monitus responderit.

Tuus Johannes Wallis

Oxoniae, Julii 16. 1671.

3 Ergo (*1*) (quod praesumit (*2*) Trilineum *W*
10 propositiones. (*1*) Est |haec *add.*| Geometria principio illo suo admodum consona, quo (*2*) Sed talis *W*
12 dimensus, quas ita *E*
13 etiam indemonstratas, & contra demonstrationum *E*
14 postponit. Gloriatum tamen audio (sic sua deperire solet,) *ex omnibus ab eo editis, hunc librum esse optimum. E*
15 habet, sive ad libri Calcem sive ad Frontem *E*
18 *Arithmeticam speciosam, & Logisticam* sibi non placere *E*
21 ne hilum valuerit, ejusque ratiocinium, tantundem.) *E*
21-22 tantundem.) (*1*) Et quidem (*2*) Sunt autem ... pueriliter |et ridicule *del.*| ab eo dicta *W*
22 ea omnia tam crude, insulse, pueriliter *E*
23 ad loca notita respiciat, pro me facile, etiam non monitus, sit responsurus. *E*

220.
THOMAS HOBBES to the ROYAL SOCIETY
end of July 1671

Transmission:

E^1 Printed paper: HOBBES, *To the Right Honorable and others, the Learned Members of the Royal Society for the Advancement of the Sciences*. Exemplar: OXFORD *Bodleian Library* Savile Ee 1 (14), 1 p. (our source).
E^2 Partly reprinted: HOBBES, *Three Papers Presented to the Royal Society Against Dr. Wallis. Together with Considerations on Dr. Wallis his Answer to them*, London 1671.—printed: HOBBES, *English Works* VII, 431–3.

Reply to: WALLIS–OLDENBURG 16/[26].VII.1671.
Answered by: WALLIS–HOBBES/ROYAL SOCIETY IX(i).1671

Hobbes published this first rejoinder to Wallis's refutation of his *Rosetum geometricum* at the end of July 1671. Referred to as Hobbes' 'first paper', it was, like those succeeding it, addressed to the members of the Royal Society.

To the Right Honorable and others, the Learned Members of the Royal Society, for the Advancement of the Sciences.

Presenteth to your Consideration, your most humble servant *Thomas Hobbes*, (who hath spent much time upon the same Subject) two Propositions, whereof the one is lately published by Dr. Wallis a Member of your Society, and Professor of Geometry; which, if it should be false, and pass for truth, would be a great obstruction in the way to the designe you have undertaken. The other is a Probleme, which, if well demonstrated, will be a considerable Advancement of Geometry, and though it should prove false, will in no wise be an impediment to the growth of any other part of Philosophy.

Dr. Wallis, *de Motu*, Cap. 5. Prop. 1.

If there be understood an infinite row of Quantities beginning with 0, or $\frac{1}{0}$, and increasing continually according to the natural order of numbers, 0, 1, 2, 3, &c. or according to the order of their squares, as, 0, 1, 4, 9, &c. or according to the order of their cubes, as, 0, 1, 8, 27, &c. whereof the last is given; the proportion of the whole, shall be to a row of as many, that are equal to the last (in the first case) as 1 to 2; (in the second case) as 1 to 3; (in the third case) as 1 to 4, &c.

This Proposition is the ground of all his doctrine concerning the Centers of Gravity of all Figures. Wherein it may please you to consider,

220. HOBBES to the ROYAL SOCIETY, end of July 1671

First, whether there can be understood an infinite row of Quantities, whereof the last can be given. Secondly, whether a Finite Quantity can be divided into an Infinite number of lesser quantities, or a Finite quantity can consist of an Infinite number of parts, (which he buildeth on as received from Cavallieri[1189].)

Thirdly, whether (which in consequence he maintaineth) there be any Quantity greater then Infinite. Fourthly, whether there be (as he saith) any Finite Magnitude of which there is no Center of Gravity. Fifthly, whether there be any number Infinite. For it is one thing to say, that a Quantity may be divided perpetually without end, and another thing to say that a Quantity may be divided into an infinite number of parts. Sixthly, if all this be false, whether that whole book of *Arithmetica Infinitorum*, and that Definition which he buildeth on, and supposeth to be the doctrine of Cavallieri, be of any use for the confirming or confuting of any propounded doctrine.

Humbly praying you would be pleased to declare herein your Judgement.

The Examination thereof being so easie, that there needs no skill either in Geometry, or in the Latine Tongue, or in the Art of Logick; but onely of the common understanding of Mankinde to guide your Judgement by.

Thomas Hobbes, *Roset*. Prop. 5.

To finde a straight line equal to two fifths of the Arc of a Quadrant.

I describe a Square $ABCD$, and in it a Quadrant DAC. Supppose DT be $\frac{2}{5}$ of DC; then will the Quadrantal Arc TV be $\frac{2}{5}$ of the Arc CA. Again, let DR be a mean proportional between DC and DT; then will the Quadrantal Arc RS be a mean proportional between the Arc CA, and the Arc TV.

Suppose further a right line were given equal to the Arc CA, and a quadrantal Arc therewith described; then will DC, CA, the Arc on CA be continually proportional, Set these proportionals in order by themselves.

DC. CA. Arc on CA $\quad\div$ ⎫
DR. RS. Arc on RS $\quad\div$ ⎬ which are in continual proportion to the Semidiameter of
DT. TV. Arc on TV $\quad\div$ ⎭ the Arc.

[1189] Cavallieri: i.e. Bonaventura Cavalieri (1598?–1647), Italian mathemaician. Wallis claims in numerous places that the method he employed in *Arithmetica infinitorum* represents an arithmetical development of the method of Cavalieri. See WALLIS–OUGHTRED 19/[29].VII.1655; WALLIS, *Correspondence* I, 151–9, 152. That letter also prefaced his *Arithmetica infinitorum*.

220. HOBBES to the ROYAL SOCIETY, end of July 1671

And *DC*, *DR*, *DT* are in continual proportion by construction, and therefore also *CA*, *RS*, *TV*, and Arc on *CA*, Arc on *RS*, Arc on *TV* in continual proportion.

Therefore as *DC* to *RS*, so is *RS* to the Arc to the Arc on *TV*. And *DC*, *RS*, the Arc on *TV* will be continually proportional. And because *DC*, *CA*, the Arc on *CA* are also continually proportional, and have the first Antecedent *DC* common; the proportion of the Arc on *CA* to the Arc on *TV* is (by Eucl.[1190] 14. 28.) duplicate of the proportion of *CA* to *RS*, and the Arc on *RS* a mean proportional between the Arc on *CA*, and the Arc on *TV*.

Now if *DC* be greater then *RS*, also *RS* must be greater then the Arc on *TV*; and the Arc *CA* greater then the Arc on *RS*. Therefore seeing *DC*, *CA*, Arc on *CA*, are continually proportional; the Arc on *TV*, the Arc on *RS*, the Arc on *CA* cannot be continually proportional, which is contrary to what has been demonstrated. Therefore *DC* is not greater then *RS*.

Suppose then *RS* to be greater then *DC*, then will the Arc on *RS* be a mean proportional between the Arc on *TV*, and a greater Arc then that on *CA*; and so the inconvenience returneth. Therefore the Semidiameter *DC* is equal to the Ard *RS*, and *DR* equal to *TV*, that is to say, to $\frac{2}{5}$ of the Arc *CA*. Which was to be demonstrated. Nor needeth there much Geometry for the examining of this Demonstration. Therefore I submit them both to your Censure, as also the whole *Rosetum*, a copie whereof I have caused to be delivered to the Secretary of your Society.

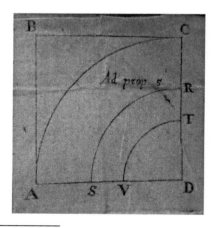

[1190]Eucl.: i.e. [EUCLID], *Elementorum libri XV*, ed. C. Clavius, third edition, Cologne 1591, 302–3; *Elements* XIV, theor. 28, prop. 28.

221.
HENRY OLDENBURG to WALLIS
1/[11] August 1671

Transmission:

Manuscript missing.

Existence and date: mentioned in and answered by OLDENBURG–WALLIS 5/[15].VIII.1671.
Enclosure: HOBBES–ROYAL SOCIETY end of VII.1671.

As Oldenburg notes in his subsequent letter (OLDENBURG–WALLIS 5/[15].VIII.1671), he enclosed in this letter a copy of the printed paper which Hobbes had addressed to the Royal Society (HOBBES–ROYAL SOCIETY end of VII.1671).

222.
WALLIS to HENRY OLDENBURG
Astrop, 4/[14] August 1671

Transmission:

W Letter sent: LONDON *Royal Society* Early Letters W1, No. 124, 2 pp. (our source). On p. 2 beneath address in Oldenburg's hand: 'Rec. Aug. 7. 71.' Postmark illegible.—printed: OLDENBURG, *Correspondence* VIII, 181.

Reply to: OLDENBURG–WALLIS 1/[11].VIII.1671.

At the time of writing this short reply to Hobbes' first printed paper addressed to the Royal Society (i.e. Hobbes' 'first paper'), Wallis was taking waters at Astrop Spa in Northamptonshire, following his recent severe bout of quartan ague.

Sir,

To your letter[1191] which I received last night, with Mr *Hobs*'s English paper[1192] inclosed, had I not been sick in bed I had presently writt you an Answere. But it is time inough this morning. As to the first part of his paper,

---

1 |Astrop. Aug. 4. 1671 *del.*| Sir,
2 with (*1*) an English paper of (*2*) Mr *Hobs*'s English paper
3 had (*1*) returned you (*2*) presently writt
4 it (*1*) will be (*2*) is ... morning (*a*) to tell you, That (*b*). As

---

[1191]letter: i.e. OLDENBURG–WALLIS 1/[11].VIII.1671
[1192]paper: i.e. HOBBES–ROYAL SOCIETY end of VII.1671.

222. WALLIS to OLDENBURG, 4/[14] August 1671

which concerns my *Prop I. Cap. 5. De Motu*:[1193] when he understands it better, he will be of another mind: In the meantime, I am not bound to give him understanding nor is it worth the answering. As for the Royal Society, whom hee would set on work: I presume, they may find wherein to imploy their time better, than in answering his idle questions. As to the second part of his paper, which concerns his own *Prop. 5. Roset.*[1194] it was answered in the Latine before it was written in English, in my Latine Confutation of his *Rosetum*. As he doth now repeat it in English, his demonstration is peccant in those words (*col. 2. lin.* 3, 33, 34,), *Therefore — the Arc on TV, the Arc on RS, the Arc on CA, cannot be continually proportional*, (with all that follows:) there being no strength in that inference, whether DC or RS be supposed the greater; for it may be either notwithstanding his Demonstration. And as to his whole paper; it was very well known by what he had written before, that *Mr Hobs is no Mathematician*; and therefore he might have spared the pains of publishing this paper further to demonstrate it.

This you may (if you think fit) insert[1195] in the transactions for July, onely leaving out the date.

Yours
Joh: Wallis.

Astrop. Aug. 4. 1671.

---

2 mind: (*1*) And (*2*) In the
3 nor ... answering. *add.*
3-4 answering. (*1*) And (*2*) As for ... Society, (*a*) I presum *breaks off* (*b*) whom hee
4 find (*1*) where about (*2*) wherein
6-7 answered |in the Latine *add.*| before it was written |in English, *add.*| in
8 English, |out of his Rosetum *del.*| his
9 words (*1*) (line, 32, 33, 34, of his second colu *breaks off* (*2*) (*col. 2.*
10 (with ... follows:) *add.*
11 DC (*1*) be supp *breaks off* (*2*) or RS
13 by ... written *add.*
14 *Mathematician*; (*1*) without the (*2*) and therefore ... pains of
15 paper (*1*) to demonstrate it afresh (*2*) further ... it.

[1193] *Prop I. Cap. 5. De Motu*: i.e. WALLIS, *Mechanica: sive, de motu, tractatus geometricus* II, 148–9; *Opera mathematica* I, 667–8.

[1194] *Prop. 5. Roset.*: i.e. HOBBES, *Rosetum geometricum*, London 1671, 17–19.

[1195] insert: this reply by Wallis to Hobbes's 'first paper' appeared heavily re-worked as part of his article 'An Answer to Four Papers of Mr. Hobs, lately published in the Months of August, and this present September, 1671' in *Philosophical Transactions* No. 75 (18 September 1671), 2241–50, 2244–5.

[2] These
For Mr Henry Oldenburg in
the Palmal near St James's
London.

223.
HENRY OLDENBURG to WALLIS
London, 5/[15] August 1671

Transmission:

C Letter sent: OXFORD *Bodleian Library* MS Savile 104, f. 8^r–9^v (9^r blank) (our source). Text loss through opening of seal. Postmark illegible.—printed: OLDENBURG, *Correspondence* VIII, 184–5.

Answered by: WALLIS–OLDENBURG 10/[20].VIII.1671.
Enclosure: Philosophical Transactions No. 73 (17 July 1671).

This letter was sent before Wallis's reply (WALLIS–OLDENBURG 4/[14].VIII.1671) to Oldenburg's previous letter (OLDENBURG–WALLIS 1/[11].VIII.1671) had arrived. Oldenburg informs Wallis that Hobbes has sent a copy of his printed paper (HOBBES–ROYAL SOCIETY end of VII.1671) to the king and encloses a copy of the July 1671 issue of *Philosophical Transactions*, containing Wallis's reply to Hobbes's *Rosetum geometricum*.

Sir,

On Tuesday last, being the 1st of August, I sent to Oxford for you a letter,[1196] with an inclosed printed Paper,[1197] by M. Hobbes addressed to the R. Soc. about two propositions, whereof one is deliver'd by you, the other by him, desiring the judgement of the Society concerning them. I hope, you have received[1198] it ere this, and are ready to give us your thoughts thereupon. Now I thought good to convey to you a copy of the Transactions, wherein your answer[1199] to his Rosetum is printed; wherein I can find but two or three inconsiderable faults of the presse (marked by me in

---

[1196]letter: i.e. OLDENBURG–WALLIS 1/[11].VIII.1671.
[1197]Paper: i.e. HOBBES–ROYAL SOCIETY end of VII.1671.
[1198]received: Wallis received OLDENBURG–WALLIS 1/[11].VIII.1671 at Astrop Spa on 3 August 1671 (old style); he replied the following day (WALLIS–OLDENBURG 4/[14].VIII.1671.)
[1199]answer: i.e. Wallis's article 'An Answer of Dr. Wallis to Mr. Hobbes's Rosetum Geometricum', published in *Philosophical Transactions* No. 73 (17 July 1671), 2202–9.

223. OLDENBURG to WALLIS, 5/[15] August 1671

this printed copy), whereof one is found in the original written copy, viz. that of p. 2207. l. 11. where it should be, *p. 49.* instead of *50.* (which I found by chancing to compare that quotation with the *Rosetum.*) In the same original there was omitted the number of the line quoted in *p. 49*; which number, I hope, you will find rightly supplied in the print, where *p. 2207. l. 24.* it is *49. lin. 12.* Whereas in the written copy there was no more than, *p. 49. lin.* (without the number of the line.)

I had nobody to read with me, when I review'd [for the] presse, which I was fain to doe twice, and yet could not [get] it done altogether to my mind.

[M. H]obbes hath sent me another printed Copy of his Rosetum, wherein I find some words blotted out; and some alter'd; which I found not done in the first book, I received. I know not, whether you found any such changes, made with the pen, in the Copy, you had of that Rosetum. As, p. 35. l. 8. for MC he hath made it NC; p. 46. l. ult. he hath made it fa ad ae; p. 47 l. 10. he hath made it MF sive Md. p. 50. l. 10 he hath made it, Ducta ergo $B\theta$ aequalis erit $\alpha\zeta$; ibid. l. 19. it must be, $\alpha\zeta$; p. 51 l. ult. he hath quite blotted out the words between CK, and, *aequales.* And again p. 52. l. 1. 2. he hath struck out the words between, *aequales erunt*, and, *Quare tres.* Again p. 59.| l. 5. it must be, ita BG ad BC. There is one alteration more [8ᵛ p. 62. to that Consectarium, but I know not what to make of it.

I hear, that Hobbes sent one of the printed English sheets of his to the king on munday last,[1200] and that the next day[1201] he came himself to his Majesty, and there vapour'd of his abilities, and that Dr. Wallis had no Geometry, but what he got from him. Which, if true, is so intolerable an impudence, that he ought to be soundly lash't for it; and I was very sorry, that none of our Mathematicall Society men, that know both you and him, were present, to tell him, that he durst not say such words to your face.

I think, he ought to be perstringed ad vivum and it cannot be too smart, so it be but with observing that decorum, which becoms members of the R. Society.

Sir, you see my freedom; whereupon you may act as shall seem good to your discretion. I remain.

Yours

London August 5. 71.

---

1 printed *add.*
4 I (*1*) thin *breaks off* (*2*) hope
21 English *add.*

[1200] Munday last: i.e. Monday, 31 July 1671 (old style).
[1201] next day: i.e. 1 August 1671 (old style).

[9ᵛ] To his honor'd friend
Dr John Wallis, Savilian
Professor of Geometry, and one
of his Majesties Chaplains.
Oxford.

224.
WALLIS to HENRY OLDENBURG
5–9/[15–19] August 1671

Transmission:

Manuscript missing.

Existence: mentioned in WALLIS–OLDENBURG 10/[20].VIII.1671.

As emerges from WALLIS–OLDENBURG 10/[20].VIII.1671, Wallis wrote to Oldenburg shortly after sending his brief reply to Hobbes' 'first paper' (WALLIS–OLDENBURG 4/[14].VIII.1671), indicating his intention to expand that reply through a number of insertions. He sent these insertions in his letter of 10 August 1671 (old style).

225.
WALLIS to HENRY OLDENBURG
Oxford, 10/[20] August 1671

Transmission:

W^1 First draft: OXFORD *Bodleian Library* MS Savile 104, f. 7ʳ.
W^2 Second draft: OXFORD *Bodleian Library* MS Savile 104, f. 6ʳ.
W^3 Letter sent: LONDON *Royal Society* Early letters W1, No. 126, 2 pp. (our source). On p. 2 beneath address in Oldenburg's hand: 'Rec. Aug. 11. 71.' Postmark illegible.—printed: OLDENBURG, *Correspondence* VIII, 195–7.

Reply to: OLDENBURG–WALLIS 5/[15].VIII.1671.

This letter contains the insertions to WALLIS–OLDENBURG 4/[14].VIII.1671 to which Wallis apparently referred in the now missing letter WALLIS–OLDENBURG 5–9/[15–19].VIII.1671. The text of the former was heavily revised before it finally appeared as part of 'An Answer to Four Papers of Mr. Hobbes, lately Published in the Months of August, and this present September, 1671' in *Philosophical Transactions* No. 75 (18 September 1671), 2241–50. His reply also appeared as a separate publication entitled *An Answer to three papers of Mr. Hobbes, lately published in the months of August and . . . September*. Wallis sent more insertions and corrections three days later in WALLIS–OLDENBURG 13/[23].VIII.1671.

225. WALLIS to OLDENBURG, 10/[20] August 1671

Oxford. Aug. 10. 1671.

Sir,

The insertions mentioned in my last,[1202] (which I had not then time, without loosing the post, to write) I now send you. Had I a copy of the brief answere[1203] I sent from Astrop, I would have transcribed it & inserted these in their proper places; but you may do it by these directions. I had once intended in the first paper[1204] to have put in somewhat to the same purpose; but left it out that it might not bee too long to insert in the last Transactions had it come time inough.[1205] But, for the next, I think with these it will not be too long. If any thing in it seem too severe, you may either mollify it, or leave it out. As, for *idle questions* you may put *impertinences*. And, of the last addition, you may (if it bee thought fit) leave out either one or both clauses: The first of which respects his sending abroad; the second, the words mentioned in your last letter.[1206] No more at present but thanks for the Transactions you sent me, & that I am

Your most humble servant
John Wallis.

After the fourth of fift period, ending with these words,—*idle questions.* Adde

He may do as well, the next time, to ask their Judgement concerning *Euclide's second Postulate*: wherein he requires, *A Streight line given, to be continued* (εἰς ἄπειρον) *Infinitely, either way*: Whether, the length of that line so continued one way, be not Infinitely great: Whether, if so continued both ways, it be not yet greater: Whether hereupon there bee a quantity Greater than Infinite: Whether the same given line may not as well (by his 10th Proposition) be continually Bisected, Infinitely: Whether, upon such Bisection Infinitely continued, will not arise a Series of Equal parts Infinitely many, (that is, more than any finite number assignable:) Whether, if this line be made the side of a Triangle, & therein from every point of division

---

9 had (*1*) they (*2*) it
22 (εἰς ἄπειρον) add.
28 many, (*1*) (or more than any |finite *add.*| number assigned: (*2*) (that is, ... assignable:

[1202]last: i.e. WALLIS–OLDENBURG 5–9/[15–19].VIII.1671.
[1203]brief answere: i.e. WALLIS–OLDENBURG 4/[14].VIII.1671.
[1204]paper: i.e. WALLIS, 'An Answer of Dr. Wallis to Mr. Hobbes's Rosetum Geometricum', in *Philosophical Transactions* No. 73 (17 July 1671), 2202–9.
[1205]time inough: cf. last line of WALLIS–OLDENBURG 4/[14].VIII.1671.
[1206]letter: i.e. OLDENBURG–WALLIS 5/[15].VIII.1671.

225. WALLIS to OLDENBURG, 10/[20] August 1671

be inscribed streight lines parallel to one of the other sides, there will not be a series of lines, infinitely many, in arithmetical proportion, as 1, 2, 3, 4, &c, of which the Last or Greatest (viz. that other side) is given; (and the Squares of those, as 1, 4, 9, 16, &c; the Cubes, as 1, 8, 27, 64, &c; and so of other powers:) Whether the Whole Doctrine of Euclide depending on this Postulate, must not therefore be rejected, as not of any use for the confirming or confuting of any propounded doctrine: rather than Mr Hobs's paralogisms not take place.

As to the later part, &c.

Next before the last periode, insert

And the thing is Manifest. For DC, DR, DT, being (by construction) in continual proportion, as $1, \sqrt{\frac{2}{5}}, \frac{2}{5}$: And

DC. CA. Arc on CA extended $\div$ ⎫
DR. RS. Arc on RS extended $\div$ ⎬ in the continual proportion of r to q (the Radius to the Quadrantal Arc:)
DT. TV. Arc on TV extended $\div$ ⎭

It is evident that (putting $r = DC$, and $q = CA$,) the quantities will be these,

$DC = r \times 1$. $CA = q \times 1$. Arc on $CA = \frac{q^2}{r} \times 1$ ⎫
$DR = r \times \sqrt{\frac{2}{5}}$. $RS = q \times \sqrt{\frac{2}{5}}$. Arc on $RS = \frac{q^2}{r} \times \sqrt{\frac{2}{5}}$ ⎬ And therefore ($\frac{q^2}{r}$ being the same in all) the Arcs on TV, on RS, on CA,
$DT = r \times \frac{2}{5}$. $TV = q \times \frac{2}{5}$. Arc on $TV = \frac{q^2}{r} \times \frac{2}{5}$ ⎭

in continuall proportion, (viz: as $\frac{2}{5}, \sqrt{\frac{2}{5}}, 1$,) whatever bee the proportion of r to q or to $q\sqrt{\frac{2}{5}}$, that is, of DC to CA or to RS (greater, lesse, or equal;) notwithstanding Mr *Hobs's* pretended Demonstration. And Mr *Hobs* must needs be (as he is) a very weak Demonstrator who doth not discern it.

Lastly, as to his whole paper, &c.
At the end, adde
: And the more hee Divulgeth it, the more he *proclaimeth or layeth open his folly*. And, if (as, I hear, he sayth) I have no Mathematicks but what I

1-2 sides, (*1*) these will not be a series of so *breaks off* (*2*) there will not hence arise a series (*3*) there will ... series
17 notwithstanding ... Demonstration *add*.
21 hee (*1*) proclaime (*a*) the (*b*) it (*2*) Divulgeth it

had from him: It seems I have had from him so much that he hath none left for himself. Farewell.

Yours
John Wallis.

5 Query *As to* &c: make a Break, for the better distinction.

For Mr Henry Oldenburg in   [2]
the Palmal near St James's
London.

226.
HENRY OLDENBURG to WALLIS
10/[20] August 1671

Transmission:

Manuscript missing.

Existence and date: mentioned in and answered by WALLIS–OLDENBURG 13/[23].VIII.1671.

227.
WALLIS to HENRY OLDENBURG
Oxford, 13/[23] August 1671

Transmission:

W Letter sent: LONDON *Royal Society* Early Letters W1, No. 128, 2 pp. (our source). On p. 2 beneath address in Oldenburg's hand: 'Rec. Aug. 14. 71.' Postmark: 'AU/14'.—printed: OLDENBURG, *Correspondence* VIII, 202–3.

Reply to: OLDENBURG–WALLIS 10/[20].VIII.1671.

This letter contains further insertions to WALLIS–OLDENBURG 4/[14].VIII.1671, the text of which was heavily revised before it finally appeared as part of 'An Answer to Four Papers of Mr. Hobbes, lately Published in the Months of August, and this present September, 1671' in *Philosophical Transactions* No. 75 (18 September 1671), 2241–50. Wallis's reply also appeared as a separate publication entitled *An Answer to three papers of Mr. Hobbes, lately published in the months of August and ...September.*

227. WALLIS to OLDENBURG, 13/[23] August 1671

Oxford. Aug. 13. 1671

Sir,

To yours[1207] of Aug. 10 (beside thanks for it & for the receit it,) I have little to reply save what mine[1208] by the last post brought you, which I suppose by this time is come to hand: If that be not thought clear inough, you may to those insertions adde these that follow.

In the first Insertion, after these words, *as 1, 8, 27, 64, &. and so of other powers*; Adde

(Which is the proposition at which he cavils:)

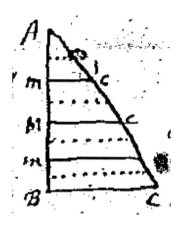

In the end of that Insertion, after, *of any doctrine propounded.* Adde[1209]

But if he can (without Latine, Logick, or Mathematicks,) solve These: he need not ask help of the Royal Society to solve His *Quaere's*.

Or (which wil be harder to solve than all those) Mr *Hobs* may ask himself, who will not allow, that there is any Argument (beside the Magistrates authority, who commands us to beleeve the Scripture,) to prove, that *the World had a beginning*: Whether, in case it had not, there must not have passed an Infinite number of years before Mr *Hobs* was born, (for, if but a finite number, how great so ever, it must have had a beginning so many years before:) Whether, now, there have not passed more years, (that is,

---

1 years *add*.

[1207] yours: i.e. OLDENBURG–WALLIS 10/[20].VIII.1671.

[1208] mine: i.e. WALLIS–OLDENBURG 10/[20].VIII.1671; this letter was received on 11/[21] August.

[1209] Adde: a considerably shorter version of this addendum was published.

227. WALLIS to OLDENBURG, 13/[23] August 1671

more than that Infinite number:) Whether, in that Infinite (or more than Infinite) number of Years, there have not been yet a greater number of Days & Hours, (&, of which the last is given:) Whether, if this be an Absurdity, wee have not (contrary to what Mr *Hobs* would perswade us) an Argument (beside the authority of Scripture or of the Magistrate) to prove, that the World had a beginning. And these will be more hard to solve, than all those before, or those of Mr. *Hobs*; Because these *more than Infinite* numbers of Years, Days, & Hours must have *Actually Existed*, & been passed allready; whereas it serves *Euclide*, & the Mathematicians, if their *Infinite* be but *Imaginable*; with whom it is frequent, upon impossible Suppositions, to infer usefull Truthes; (and, by *Infinite*, to mean onely, *more than any Finite assignable*; As, If we *suppose* the side of a triangle AB to bee bisected in M, & each of those Halves again in m, & so onwards *Infinitely*; we must withall *suppose* the number of the parts resulting to be *Infinite*, or more than any assignable *Finite* number: And if we *suppose* further, from every of those Infinite points of division to be drawn as many lines mc parallel to the Base BC, these lines must be *supposed*, *Infinitely many*, & those in an Arithmetical progression, as 1, 2, 3, &c., of which BC the *last and greatest is given*. Nor is it necessary that this *Infinite Bisection* be actually performed (or possible so to be;) it is inough, that if wee *suppose* that, we must *suppose* the other allso. But, in Mr *Hobs's* notion, of the *Worlds Eternity*, there must have been, not onely *supposed*, but *actually existing*, those *Infinites*, and *More than Infinites*.

I am

yours &c.
John Wallis.

For Mr Henry Oldenburg, [2]
Secretary to the Royall Society,
in the Palmal near St James's
London.

1 in (*1*) those (*a*) Infinitely many, or more than (*b*) years, (*2*) that Infinite
10 *Imaginable*; (*1*) who do |frequently *add*.| upon (*2*) with whom it is frequent upon
11 mean (*1*) no more but (*2*) onely, *more*
12 As, (*1*) supposing (*2*) If we *suppose*
13 *Infinitely*; (*1*) As must needs (*2*) we must withall
16 Infinite *add*.
18 BC *add*.

228.
Thomas Hobbes to the Royal Society
first half of August 1671

Transmission:

E^1 Printed paper: HOBBES, *To the Right Honourable and others, the Learned Members of the Royal Society, for the Advancement of Science.* Exemplar: OXFORD *Bodleian Library* Savile Ee 1 (10), 1 p. On reverse in unknown hand: 'For Dr Wallis Math. proffessor in Oxford'. Postmark: 'AU/–'.
E^2 Reprinted: HOBBES, *Three Papers Presented to the Royal Society Against Dr. Wallis. Together with Considerations on Dr. Wallis his Answer to them,* London 1671.—printed: HOBBES, *English Works* VII, 435–6.

Reply to: WALLIS–OLDENBURG 16/[26].VII.1671.
Answered by: WALLIS–HOBBES/ROYAL SOCIETY IX.1671(I) and WALLIS–HOBBES/ROYAL SOCIETY IX.1671(II).

Hobbes published this second rejoinder to Wallis's refutation of his *Rosetum geometricum* in early August 1671. Oldenburg received a copy of Hobbes's 'second paper' addressed to the Royal Society on 14 August and sent it with his letter to Wallis the following day.

To the Right Honourable and others, the Learned Members of the Royal Society, for the Advancement of the Sciences.

Presenteth to your Consideration, your most humble servant *Thomas Hobbes*, a Confutation of a Theoreme which hath a long time passed for Truth; to the great hinderance of Geometry, and also of Natural Philosophy, which thereon dependeth.

The Theoreme.

The four sides of a Square being divided into any number of equal parts, for example into 10; *and straight lines drawn through the opposite points, which will divide the Square into* 100 *lesser Squares; The received Opinion, and which* Dr. Wallis *commonly useth, is, that the root of those* 100, *namely* 10, *is the side of the whole Square.*

The Confutation.

The Root 10 *is a number of those Squares, whereof the whole containeth* 100, *whereof one Square is an Unitie; therefore the Root* 10, *is* 10 *Squares:*

Therefore the Root of 100 *Squares is* 10 *Squares, and not the side of any Square; because the side of a Square is not a Superficies, but a Line. For as the root of* 100 *Unities is* 10 *Unities, or of* 100 *Souldiers* 10 *Souldiers: so the root of* 100 *Squares is* 10 *of those Squares. Therefore the Theoreme is false; and more false, when the root is augmented by multiplying it by other greater numbers.*

Hence it followeth, that no Proposition can either be demonstrated or confuted from this false Theoreme. Upon which, and upon the Numeration of Infinites, is grounded all the Geometry which Dr. *Wallis* hath hitherto published.

And your said servant humbly prayeth to have your Judgement hereupon: And that if you finde it to be false, you would be pleased to correct the same; and not to suffer so necessary a Science as Geometry to be stifled, to save the Credit of a Professor.

229.
HENRY OLDENBURG to WALLIS
[London], 15/[25] August 1671

Transmission:

C Letter sent: LONDON *Royal Society* Early Letters W1, No. 130, 1 p. (our source). Address on p. 2 now Early Letters W1, No. 129. On p. 2 Wallis has written his reply. Postmark: 'AU/16' and '2/Off'.—printed: OLDENBURG, *Correspondence* VIII, 204–5.

Reply to: WALLIS–OLDENBURG 13/[23].VIII.1671.
Answered by: WALLIS–OLDENBURG 16/[26].VIII.1671.
Enclosure: HOBBES–ROYAL SOCIETY VIII.1671.

With this letter Oldenburg sent Wallis a copy of Hobbes' latest paper addressed to the Royal Society (i.e. his 'second paper'), which he had received the day before.

Sir,

By your last,[1210] which I received yesterday, me thinks you have pressed your Adversary home; but I have not yet been able to shew it to My Lord Brouncker, who told me, when I saw him last, that the king had commanded

---

[1210]last: i.e. WALLIS–OLDENBURG 13/[23].VIII.1671.

229. Oldenburg to Wallis, 15/[25] August 1671

him to take care of getting the printed sheet[1211] answer'd in such another sheet, and that in such a manner, that every body of common understanding and good sense might be able to apprehend where the error lyeth. This made his Lordship wish very much, that you, that have already so well satisfaied Mathematicians, would take a litle more pains to satisfy those also, that are no Mathematicians, but otherwise intelligent men; and particularly all those Queries about your Proposition, which he hath proposed so plausibly, as some doe think. What My Lord shall say after the view of your last, you will soon know.

But, I believe, you will be surprised to see this other very bold paper[1212] of the Author, which came yesterday to my hands, and of which, I heare, a Copy hath been also by him presented to the king $\mu\epsilon\tau\grave{\alpha}$ $\pi o\lambda\lambda\tilde{\eta}\varsigma$ $\phi\alpha\nu\tau\alpha\sigma\acute{\iota}\alpha\varsigma$.[1213]

I am very sorry, he should give occasion to torment you at such an unseasonable time as this is; though you will soon despatch this also, in the persuasion of

Sir
Your faithf. servant
Aug. 15. 71.   Oldenb.

[2] To his much honord friend
Dr John Wallis Savillian
Professor of Geometry, and one
of his Majesties Chaplains
Oxford

---

1 of (1) having (2) getting
3 be able to add.
6 all add.
8-9 What My Lord ... know. add.
10 see (1) another (2) this other

[1211] printed sheet: i.e. HOBBES–ROYAL SOCIETY end of VII.1671—Hobbes's 'first paper'.
[1212] paper: i.e. HOBBES–ROYAL SOCIETY first half of VIII.1671–Hobbes's 'second paper'.
[1213] $\mu\epsilon\tau\grave{\alpha}$ $\pi o\lambda\lambda\tilde{\eta}\varsigma$ $\phi\alpha\nu\tau\alpha\sigma\acute{\iota}\alpha\varsigma$: Acts 25, 23.

230.
WALLIS to HENRY OLDENBURG
Oxford, 16/[26] August 1671

Transmission:

W Letter sent: LONDON *Royal Society* Early Letters W1, No. 129, 1 p. (our source). Written on cover of OLDENBURG–WALLIS 15/[25].VIII.1671.—printed: OLDENBURG, *Correspondence* VIII, 205–6.

Reply to: OLDENBURG–WALLIS 15/[25].VIII.1671.

Henry Oldenburg. Oxford. Aug. 16. 1671.

Sir

I received from you this night, yours[1214] of yesterday; in which was a paper[1215] of Mr *Hobs*, wherein hee pretends to confute a Theorem which, hee says, hath long time passed for a Truth.

'The Theoreme'.

'*The four sides of a Square being divided into any number of equal parts, for example into 10; and streight lines drawn through the opposite points; which will divide the Square into 100 lesser Squares; The received Opinion, and* which Dr Wallis *commonly useth, is, that the root of those 100, namely 10, is the side of the whole Square.*'

My Answere.

It is not the Opinion of *Dr Wallis*, (and Mr *Hobs* knows, it is not; having been oft told the contrary;) and it is so far from being a *received Opinion*, that he doth not know it to be the Opinion of one Person (unlesse Mr *Hobs*.) But hee is of opinion (& so are all Mathematicians, so far as hee knows: that if (for instance), the Square do contain 100 *Square Feet*, the side doth 10 *Long Feet*, (because the *Number* 10 being the root or side of the *number* 100; and a *Foot in Length* of an *Inch Square*; the product of the Rootes, that

---

14-15 and (*1*) he beleeves, (*2*) it is ... *received Opinion*, (*a*) thinks it is not the (*b*) that he
17 *Square* (*1*) Inches (*2*) Feet
18 *Long* (*1*) Inches (*2*) Feet
19 and (*1*) an *Inch* (*2*) a *Foot*

[1214] yours: i.e. OLDENBURG–WALLIS 15/[25].VIII.1671.
[1215] paper: i.e. HOBBES–ROYAL SOCIETY first half of VIII.1671—Hobbes's 'second paper'.

is a *Foot long, multiplied by 10*; is the Root of the product of the Squares, that is, of *a Foot Square, multiplied by 100*). And so in all other Squares. The Root of the Number of *Squares* in the Plain, is the number of *Lengths* in the side. And this Mr *Hobs* hath been often told; & particularly in my *Hobbius Heautontimenos*[1216] pag. 142, 143, 144. (&, long before that, in my *Opus Arithmeticum*[1217] pag. 196, 197, 198. and elsewhere,) by

Yours
John Wallis.

231.
WALLIS to the ROYAL SOCIETY
August 1671

Transmission:

W Draft paper: LONDON *Royal Society* Early Letters W1, No. 124a, 4 pp. (our source).
E Printed: *Philosophical Transactions* No. 75 (18 September 1671), pp. 2241–2250 ('An Answer to Four Papers of Mr. Hobs, lately Published in the Months of August, and this present September, 1671').

Reply to: HOBBES–ROYAL SOCIETY end of VII.1671 and HOBBES–ROYAL SOCIETY first half of VIII. 1671.

This is apparently an early draft reply to the first two papers presented by Hobbes to the Royal Society: HOBBES–ROYAL SOCIETY end of VII.1671 and HOBBES–ROYAL SOCIETY first half of VIII.1671. It was eventually published as part of 'An Answer to Four Papers of Mr. Hobs, lately published in the Months of August, and this present September' in the issue of *Philosophical Transactions* for September 1671. See WALLIS–HOBBES/ROYAL SOCIETY IX.1671(ii).

An Answere to two Papers of Mr Hobs
published this present Moneth of August, 1671.
The former of which is in these Words.

1 is (*1*) an Inch (*2*) a Foot
2 of (*1*) an Inch (*2*) a Foot
2 Squares. (*1*) For instance: Because the root of 25 is 5, therefore (*2*) The Root
5 144. (*1*) by (*2*) &, long
11 which (*1*) begins (*2*) is in

[1216] *Hobbius Heautontimenos*: i.e. WALLIS, *Heauton-Timorumenos*, 142–4.
[1217] *Opus Arithmeticum*: i.e. WALLIS, *Mathesis universalis*, 196–8; *Opera mathematica* I, 120–1.

231. Wallis to the Royal Society, August 1671

To the Right Honourable &c. ──
── but onely of the common understanding of Mankinde to guide your Judgement by.

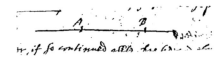

To this former part of that Paper (allowing the Proposition therein mentioned to be for sense the same with one of mine, though otherwise worded,) I say; First, Hee might as well have asked their Judgement concerning *Euclide's second Postulate*; which requires *A streight Line given, to be produced Infinitely, either way*: As 1. Whether it be possible for any man to *produce a line Infinitely*. 2. Whether, if *AB* be so produced forward from *B*; the Length will not become *Infinitely great*. 3. Whether, if so continued allso backward from *A*, it will not be yet *Greater*; that is, *Greater than that Infinite*. 4. Whether the same *AB* may not as well (by *Euclide's* 10th Proposition) be *Bisected* in *M*, & each of the halves again in *m*; and so onward *Infinitely*. 5. Whether, upon such Bisection Infinitely continued, will not arise a number of equal parts *Infinitely many*. 6. Whether, if this line be made the side of a triangle *ABC* & therein from every point of division be inscribed streight lines parallel to the Base *BC*, there will not be a series of lines *Infinitely many*, & those in *Arithmetical progression*, as 1, 2, 3, &c. of which *BC* the *Last or Greatest is given*: (and the *Squares* of these, as 1, 4, 9, &c; their *Cubes*, as 1, 8, 16, &c.; & so of other Powers:) Which is the proposition at which he cavills. 7. Whether the *whole Doctrine of Euclide*, depending on this *Postulate*, must not therefore be rejected, as not *of any use for the confirming or confuting of any propounded doctrine*. And this I say, to shew, that the *supposition of Infinites* is not so new, or so peculiar to *Cavallerius* or Dr *Walllis*, but that *Euclide* himself presumes it, & the same

6 this (*1*) first (*2*) former
14 by (*1*) his (*2*) *Euclide's* add.
18 *ABC* & (*1*) from each of those points of section lines therein in *breaks off* (*2*) therein from
21 of which |*BC* add.| the *Last*
27 same (*1*) supposed (*2*) pretended

pretended inconveniences of Mr *Hobs* as well follow from *Euclide*'s doctrine. But may both in His & Theirs, be easyly solved, as wee shall see by and by.

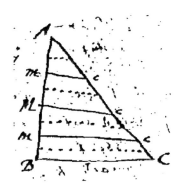

Or, secondly, Mr *Hobs* may ask himself, if still of opinion that there is not any Argument (beside the Magistrates authority; who commands us to beleeve the Scripture) to to prove *the World had a beginning*: 1. Whether, in case it had not, there must not have passed an *Infinite number of Years* before Mr *Hobs* was born; For, if but a *finite number*, how great soever, it must have had a beginning so many years before:) 2. Whether, since, there have not passed *more years*; that is *more than that Infinite number*. 3. Whether in that *Infinite* (or *more than Infinite*) *number of years*, there have not been yet a *Greater number of Days & Hours* (and, *of which, the last is given*). 4. Whether, if this be an Absurdity, we have not then (contrary| to what Mr *Hobs* would perswade us) a good Argument (beside the Authority of Scripture or the Magistrate) to prove, that the World had a beginning: Or, if hee think it no Absurdity, why doth he cavil at the like in the doctrine of *Euclide*, of *Cavallerius*, or of Dr *Wallis*? And this is sayd, to shew that Mr *Hobs* (however he please to prevaricate) is at lest as much concerned, as any of those, to solve these *Quaere*'s about *Infinites*.

But I say further, That 'twill be a harder task for Mr *Hobs* to solve these, than for them to solve those *quaere*'s: Because this *Infinite, & more*

2 easly (*1*) answered (*2*) solved
4 secondly, (*1*) (which will be *inadvertently not deleted* (*a*) hard for him to solve (*b*) more (*c*) harder to solve than will the (*2*) Mr *Hobs*
4 himself, (*1*) who will not allow (*2*) if still of opinion
12 yet *add*.
14 us) (*1*) an (*2*) a good
21 Because (*1*) those *Infinitely many* (*2*) this *Infinite,*

231. WALLIS to the ROYAL SOCIETY, August 1671

than Infinite number of Years, Days, & Hours, must have *actually existed* & been passes allready; whereas it serves *Euclide* & the *Mathematicians* well inough, if their *Infinite* be but *Imaginable*, (though never actually existent, nor possible so to bee,) with whom it is frequent, upon impossible Suppositions, to infer usefull Truthes; and, by *Infinite*, to mean onely, *more than any assignable Finite*. But with Mr. *Hobs* it will be necessary to affirm his *Infinites* not onely to be *Possible*, but *actually to have been*, in case the World had no beginning.

Leaving therefore Mr *Hobs* to answer for his *Infinites*, (*which must have actually existed and be already past*:) I answer to the *Quaere's* concerning *Euclide's* Infinites, (in which are included those Quaere's of Mr *Hobs*,) 1. That, by *Infinite*, *Euclide* & other Mathematicians after him hitherto, understand but *more than any assignable Finite*, though not *absolutley Infinite,,* or that *Greatest possible*: And this they do not require *actually to be*: but onely to be *imagined, supposed*, or (as Mr *Hobs* words it) *understood*; whether possible to exist or not: And when *Euclide* requires to *produce a line Infinitely*; it was not his meaning that this should be *actually performed* (for that were impossible for any man to do) but that it might be *supposed*; there being no repugnance *in the nature of a Line*, but that it may be produced *beyond any assigned Finite length*. 2. That, if *AB* be *supposed* so produced *one way*, as forward from *B*, its length must be *supposed* to become *Infinitely great*, or *greater than any Finite length assignable*. For if but *Finite*, a *Finite production* would have served. 3. If *supposed* so produced *both ways*; it must needs be (*supposed*) yet *Greater*, that is; *more than* that *that Infinite*, (for

---

9-10 Infinites, (*1*) actually existing (*2*) (*which* (*a*) have (*b*) must have actually existed |and be already past:) add.|
11 concerning (*1*) *Euclide* (*2*) *Euclide's* Infinites
11 Quaere's *add*.
12 Hobs, |concerning those of Dr *Wallis*.) del.| 1. That
15 imagined, (*1*) or supposed, whether (*2*) suposed, or (as Mr *Hobs* words it) *understood*;
16 not: (*1*) And nor can any believe, that (*2*) And when
20 *one way*, as *add*.
21 from *B*, (*1*) it must (*2*) its length must
21 to (*1*) be (*2*) become
24 (*supposed*) *add*.
24 Greater; (*1*) (for such Addition must needs increase what was before,) (*2*) that is, (*a*) Greater than that former Infinite (*b*) more than that Infinite (*c*) Greater than that Infinite, (for such addition must needs increase it;) or (*d*) Greater than was necessary to have made it (*e*) more than |that add.| Infinite

231. WALLIS to the ROYAL SOCIETY, August 1671

such addition must needs increase it,) or *more than was necessary* to make it *greater than any assignable finite length*: And whosoever *supposeth* Infinites, must withal *suppose* One Infinite Greater than Another. 4. That it may as well be *supposed* to be continually *Bisected Infinitely.* 5. That upon such *supposed*, infinite Bisection, the number of *parts* must be *supposed* to be *Infinite*, (for no *finite* number of *parts* could be sufficient for *infinite sections*;) and all those parts, *infinitely many*, are in the undivided *Whole*, (Else, where should they be hiden) 6. If, from each of those (infinitely many) points of section, be *supposed* as many lines inscribed parallel to *BC* the base of the Triangle *ABC*, those lines (as *mc*) must allso be *supposed* to be *infinitely many*; and those in *arithmeticall progression, or according to the natural order of* | *Numbers* 1, 2, 3, &c. (each surpassing the next foregoing, as much as that did the next before it,) and (taking *BC* into the number) *whereof the last is given: BC being the last & greatest* of that series of parallels, *infinitely many*, appertaining to *AB* or *ABC, the Line of Triangle given.* 7. There is therefore no reason upon the account of these *Supposed Infinites*, to reject the doctrine either of *Euclide* or of Dr *Wallis*. But as to Mr *Hobs*'s *Infinites* (and *more than Infinites*) which must have *really existed* & be *actually passed*, I shall leave that to him to solve. Lastly, as to his question (of another nature) *Whether there can be any Finite Magnitude of which there* is no Center of Gravity; I say, as before, That, whether it *be* or *be not*; yea, whether it *can* or *cannot be*; it may at lest bee *Imagined* or *Supposed*, which is inough to the present purpose: That is, A Plain or Solide may be *Supposed* so constituted, as to be *Infinitely Long* but *Finitely Great* (the Breadth continually Decreasing in greater proportion than the Length Increaseth;) and, so as *not to have any Center of Gravity*:

[3]

1 it (*1*) more (*2*) greater
2 And whosoever ... Another *add.*
4 continually *add.*
9 lines (*1*) drawn (*2*) inscribed
14 of (*1*) those (*2*) that series of parallels, (*a*) *respecting AB the line given* (*b*) |Yet *del.*| *infinitely many*, (*aa*) respecting the *A* breaks off (*bb*) appertaining to
17 either *add.*
17 But (*1*) whether (*2*) as to
18 *Infinites*) (*1*) actually (*2*) really existing & actually (*3*) which must
21 there (*1*) be (*2*) is no Center of Gravity;

231. WALLIS to the ROYAL SOCIETY, August 1671

(Of which see *Toricellius*[1218] in his *Solidum Hyperbolicum Acutum*, and Dr *Wallis* in his *Arithmetica Infinitorum, prop.*[1219] And the letters between Mons. Fermat and him, in his *Commercium Epistolicum, Epist.*[1220] And, in his Treatise *De Motu, Cap. 5. prop. 7, 8, 9.*[1221]) But to solve this, requires more *of Logick & Geometry*, (whatever it do, of the *Latine Tongue*,) than Mr *Hobs* is master of; and therefore, no wonder if he doubt of it.

In brief therefore, to his Six Questions wee may thus Answer 1. There may be *Supposed* (or understood) a Series (or Row) of Quantities *infinitely many* of which *the last may be given*. (As if, in a Triangle ABC, the number of Parallels be *Supposed* infinitely many, arising from a continuall bisection of AB infinitely continued; of these; (reconing from A downwards, *the Last is given, viz. BC.*) 2. There may be, at lest *Supposed*, in a Finite Quantity, (as in AC) a number of parts *Infinitely many*, or *more than any assignable Finite number*, into which it is divisible: (There being no stint or bound beyond which it may not be *Supposed* further divisible.) 3. That there may be *Supposed* one Infinite greater than another, or, more than what was necessary to make it infinite: (As, a *Supposed* Infinite number of *Men*, may be *supposed* to have a greater number of Eyes.) 4. There may be *Supposed* a Surface or Solide so constituted, as to be *Infinitely Long*, but *Finitely Great*, (the Breadth stil Decreasing faster than the Length Increaseth,) and so as to have *no Center of Gravity*: (but this, not to be proved, *without Logick or Geometry*.) 5. There may be *Supposed* a Number *Infinite* or *greater than any assignable Finite*: (because no Finite number can be assigned so Great, but that we may still *imagine* a Greater.) 6. These things being so, there is no reason from hence to reject Dr *Wallis's Arithmetick of Infinites*, or *the Doctrine of Cavallieri*.

Thomas Hobbes Roset. prop. 5. *To find a streight line* &c.

This latter part of his first paper was, before it was printed in English,

---

7-26 In brief therefore ... *the Doctrine of Cavallieri* add. on following page
8 understood) (*1*) an Inf breaks off (*2*) a Series
12 (as in AC) add.

[1218] *Toricellius*: i.e. TORRICELLI, *De solido hyperbolico acuto*, published in his *Opera geometrica* Florence 1644.

[1219] *Arithmetica infinitorum, prop.*: Wallis has left space for numbers of the propositions. Cf. WALLIS, *Arithmetica infinitorum*, 74–5, 83, 198; *Opera mathematica* I, 407–8, 412, 478.

[1220] *Commercium Epistolicum, Epist.*: Wallis has left space for numbers of the letters. Cf. WALLIS, *Commercium epistolicum*, 4–6, 6–9, 31–3, 33–56; *Opera mathematica* II, 760–1, 762–3, 776, 777–89; *Correpondence* I, 281–4, 287–90, 318–20, 320–42.

[1221] *De Motu, Cap. 5. prop. 7, 8, 9*: i.e. WALLIS, *Mechanica: sive, de motu, tractatus geometricus* II, 165–89; *Opera mathematica* I, 679–93.

231. WALLIS to the ROYAL SOCIETY, August 1671

allready answered in the Latine Confutation[1222] of his *Rosetum*, published in the *Philosophical Transactions* of the last Month. As it now stands in English (not altogether so absurd as it was in the Latine) His Demonstration is peccant in those words, *Therefore —— the Arc on TV, the Arc on RS, the Arc on CA, cannot be continually proportional,* (with all that follows,) there being no strength in that inference.

And the thing is manifest. For *DC, DR, DT*, being (by construction) in continual proportion, as $1, \sqrt{\frac{2}{5}}, \frac{2}{5}$: And

$$\left. \begin{array}{lll} DC. & CA. & \text{Arc on } CA \;\;\div \\ DR. & RS. & \text{Arc on } RS \;\;\div \\ DT. & TV. & \text{Arc on } TV \;\;\div \end{array} \right\} \text{In the continual proportion of the Radius to the Quadrantal Arc,}$$

(suppose, as r to q: It is evident that (putting $r = DC$, and $q = CA$) the quantities will be these,

$$\left. \begin{array}{lll} DC = r \times 1. & CA = q \times 1. & \text{Arc on } CA = \frac{q^2}{r} \times 1. \\ DR = r \times \sqrt{\frac{2}{5}}. & RC = q \times \sqrt{\frac{2}{5}}. & \text{Arc on } RS = \frac{q^2}{r} \times \sqrt{\frac{2}{5}}. \\ DT = r \times \frac{2}{5}. & TV = q \times \frac{2}{5}. & \text{Arc on } TV = \frac{q^2}{r^2} \times \frac{2}{5}. \end{array} \right\} \text{And therefore } (\frac{q^2}{r} \text{ being in all the same)}$$

the Arcs on *TV*, on *RS*, on *CA*, (wil be in continual proportion, (viz. as $\frac{2}{5}$, $\sqrt{\frac{2}{5}}$, 1,) whatever be the proportion of r to q or to $q\sqrt{\frac{2}{5}}$, that is of *DC* to *CA* or to *RS* (greater, less, or equal) notwithstanding Mr *Hobs* pretended demonstration: & Mr *Hobs* must needs be a very weak Demonstrator if he do not discern it.

And indeed, at this rate, he may put what proportion hee pleaseth, and his Demonstration will equally prove it. For, at the first proposal, in those words, *Suppose DT be $\frac{2}{5}$ of DC then will the Quadrantal Arc TV be $\frac{2}{5}$ of the Arc CA*; if for $\frac{2}{5}$, he say $\frac{1}{2}$, or $\frac{1}{10}$, or what he please; all the rest

2 in the (*1*) last months *Transactions* (*2*) *Philosophical Transactions*
11 (wil be ... 1,) *add*.
14 Mr *Hobs* ... discern it. *add*.
16 may |as well *del*.| put
17 proposal, (*1*) in stead of (*2*) in those words
19 please; (*1*) his Demonstration (without (*a*) alteration) (*b*) altering one word) will conclude,) (*2*) all the

---

[1222] Latine Confutation: i.e. Wallis's article 'An Answer of Dr. Wallis to Mr. Hobbes's Rosetum Geometricum in a Letter to a friend in London, dated July 16. 1671', published in *Philosophical Transactions* No. 73 (17 July 1671), 2202–9.

will follow as now it doth (without altering one word) to the conclusion, *Therefore* —— *DR equal to TV, that is to say to* $\frac{2}{5}$ *of the Arc CA*; where in stead of $\frac{2}{5}$, he is to put $\frac{1}{2}$, or $\frac{1}{10}$, or whatever proportion he pleased in the first proposal to assign. That is, he may assign the proportion of the Radius to the Quadrantal Arc *what hee please*, and his Demonstration will *equally prove it of any.*| [4]

His later Paper in these words,

> To the Right Honourable &c
>
> For Answere to this, I say, That it is neither the opinion of Dr *Wallis*, nor of any other that I know, (so far is it from being a *Received opinion*, which Mr *Hobs* would here insinuate as such,) that 10 is the Side of *100 Squares*. For no man (who doth but understand the terms) can think, that a *bare number* can be the Side of a *Square Figure*. 10 is indeed the Root of 100, but not of *100 Squares*.
>
> Neither is Dr *Wallis* (nor, I beleeve, any one else) of Mr *Hobs*'s opinion (in his Confutation) that the *Root* of *100 Squares* is 10 *Squares*. For if the Root be *10 Squares*, the Square of that Root will bee (not 100 *Squares*, but) 100 *Squared Squares*. Like as if the Root be 10 *Fower*'s, the Square will be (not 100 *Fower*'s, but) 10 *Sixteens*, or *Squares of Fower*. For to have the Square of *10 Fowers*, you must not onely multiply 10 *into* 10, which makes 100; but allso *Fower into Fower* which makes *Sixteen*.
>
> But the Dr is of opinion (and all Mathematicians with him) That *10 Lengths* is the Root of *100 Squares*: because 10 *into* 10, makes 100; and *Length into Length*, makes a *Square*; and consequently *10 Lengths* into *10 Lengths*, makes *100 Squares*. And so in all other proportions, the *number of Squares* in the *Plain*, is the Square of the *number of Lengths* in the *Root* or *Square side*. As if the Side of a Square be *2 Foot in Length*, the Plain of it will be *4 Foot in Square*, (because there will be 2 rows of 2 in a row:)

12 that (*1*) *the number 10, is the side of a Square figure* (*2*) 10 is the
12 man (*1*) can be so simple (who doth but understand the terms) as to think (*2*) (who ...terms) can think
18 Root (*1*) must (*2*) will
31 be (*1*) (twice two, that is) Four times Four (*2*) *4 Foot in Square*
31 (because ...row:) *add.*

If that be *3 in Length*, this will be 9 (or 3 times 3) *in Square*, (because there are 3 rows of 3 in a row, and therefore the number of square feet, 3 *times* 3, that is the *square* of 3:) If that be 4 *in Length*, this will be 16 (or 4 times 4) *in Square*. Of which if any man doubt, let him beleeve his own Eys.

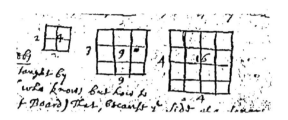

And this is what I thought fit to say of Mr *Hobs*'s two Papers; (rather to satisfy the desire of some friends, than because I thought it necessary;) And do submit the whole to the Judgement of the *Royal Society*; to whom Mr *Hobs* makes his appeal.

232.
THOMAS HOBBES to the ROYAL SOCIETY
early September 1671

Transmission:

E^1 Printed paper: HOBBES, *To the Right Honourable and others, the Learned Members of the Royal Society, for the Advancement of the Sciences.* Exemplar: OXFORD, *Bodleian Library* Savile Ee 1 (11).
E^2 Reprinted as part of: HOBBES, *Three Papers Presented to the Royal Society Against Dr. Wallis. Together with Considerations on Dr. Wallis his Answer to them*, London 1671.—printed: HOBBES, *English Works* VII, 437–42.

2 because (*1*) of (*2*) there are
3 number of (*1*) little Squares will bee (*2*) square feet,
5 Eys. |And this Mr *Hobs* might have been taught by the next Carpenter (who knows but how to measure a Foot of Board) That, because the *Side* of a *Square foot* is *12 Inches in Length*, therefore the *Plain* of it will be *12 times 12 Inches in Squares*; because there will be (*1*) twelve (*2*) 12 Rows of 12 in a Row. del.| And this
9 the (*1*) Reverend (*2*) Royal Society |*of London for the* (a) *Advancement* (b) *Improvement of Natural Knowledge del.*|; to whom

232. HOBBES to the ROYAL SOCIETY, early September 1671

Reply to: WALLIS–OLDENBURG 16/[26].VII.1671.
Answered by: WALLIS–HOBBES/ROYAL SOCIETY IX.1671(i) and WALLIS–HOBBES/ROYAL SOCIETY IX.1671(ii).

Hobbes's third rejoinder to Wallis's refutation of his *Rosetum geometricum*, the 'third paper', was printed at the beginning of September 1671.

To the Right Honourable and others, the Learned Members of the Royal Society, for the Advancement of the Sciences.

Your most humble servant *Thomas Hobbes* presenteth, That the quantity of a Line calculated by extraction of Roots, is not to be truely found. And further presenteth to you the Invention of a Straight Line equal to the Arc of a Circle.

Definition.

A Square Root is a number which multiplied into it self produceth number. And the number so produced is called a Square number. For example: Because 10 multiplied into 10 makes 100; the Root is 10, and the Square number 100.

Consequent.

In the natural row of Numbers, as 1, 2, 3, 4, 5, 6, 7, 8, 9, 10, 11, 12, 13, 14, 15, 16, &c. every one is the Square of some number in the same row. But Square numbers (beginning at 1) intermit first two numbers, then four, then fix, &c. So that none of the intermitted numbers is a Square number, nor has any Square root.

Prop. I.

A Square root (speaking of quantity) is not a Line, such as *Euclide* defines,[1223] without Latitude, but a Rectangle.

Suppose $ABCD$ be the Square, and AB, BC, CD, DA be the sides; and every side divided into 10 equal parts, and Lines drawn through the opposite points of division; there will then be made 100 lesser Squares, which taken all together are equal to the Square $ABCD$. Therefore the whole Square is 100, whereof one Square is an Unit; therefore 10 Units, which is the Root, is ten of the lesser Squares, and consequently has Latitude; and therefore it cannot be the side of a Square, which according to Euclide is a Line, without Latitude.

[1223] defines: i.e. EUCLID, *Elements* I, def. 2.

232. HOBBES to the ROYAL SOCIETY, early September 1671

Consequent.

It follows hence, that whosoever taketh for a Principle, That a Side of a Square is a meer Line without Latitude, and That the Root of a Square is such a Line, (as Dr. *Wallis* continually does) demonstrates nothing.

But if a Line be divided into what number of equal parts soever, so the Line have bredth allowed it, (as all Lines must, if they be drawn) and the length be to the bredth as Number to an Unite, the Side and the Root will be all of one length.

Prop. II.

Any Number given is produced by the greatest Root multiplied into it self, and into the remaining Fraction. Let the Number given be two hundred Squares, the greatest Root is $14\frac{4}{14}$ Squares. I say, that 200 is equal to the product of 14 into it self, together with 14 multiplied into $\frac{4}{14}$. For 14 multiplied into it self, makes 196. And 14 into $\frac{4}{14}$ makes $\frac{56}{14}$, which is equal to 4. And 4 added to 196 makes 200; as was to be proved.

Or take any other Number 8, the greatest Root is 2; which multiplied into it self is 4, and the Remainder $\frac{4}{2}$ multiplied into 2 is 4; and both together 8.

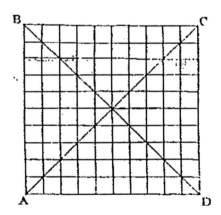

Prop. III.

But the same Square calculated Geometrically by the like parts, consisteth (by *Eucl.* 2. 4.)[1224] of the same numeral great Square 196, and of the two Rectangles under the greatest side 14, and the Remainder of the side, or (which is all one) of one Rectangle under the greatest side, and double

---

[1224] by *Eucl.* 2. 4.: i.e. EUCLID, *Elements* II, prop. 4.

232. Hobbes to the Royal Society, early September 1671

the Remainder of the side; and further of the Square of the less Segment; which all together make 200, and moreover $\frac{1}{49}$ of those 200 Squares, as by the operation it self appeareth thus.

The side of the greater Segment is $14\frac{4}{14}$.

Which multiplied into it self, makes $14\frac{4}{14}$. 200

The product of 14 the greatest segment, into the two Fractions $\frac{4}{14}$, that is, into $\frac{4}{14}$ (or into twice $\frac{2}{14}$) is $\frac{56}{24}$ (that is 4) and that 4 added to 196 makes 200.

Lastly, the product of $\frac{2}{14}$ into $\frac{2}{14}$, or $\frac{1}{7}$ into $\frac{1}{7}$, is $\frac{1}{49}$.

And so the same Square calculated by Roots, is less by $\frac{1}{49}$ of one of those two hundred Squares, then by the true and Geometrical Calculation; as was to be demonstrated.

Consequent.

It is hence manifest, That whosoever calculates the length of an Arc or other Line by the extraction of Roots, must necessarily make it shorter then the truth, unless the Square have a true Root.

The Radius of a Circle is a Mean Proportional between the Arc of a Quadrant and two fifths of the same.

Describe a Square $ABCD$, and in it a Quadrant DCA. In the side DC take DT two fifths of DC; and between DC and DT a Mean Proportional DR; and describe the Quadrantal Arcs RS, TV.

I say, the Arc RS is equal to the streight line DC.

For seeing the proportion of DC to DT is duplicate of the proportion of DC to DR, it will be also duplicate of the proportion of the Arc CA to the Arc RS; and likewise duplicate of the proportion of the Arc RS to the Arc TV.

Suppose some other Arc less or greater then the Arc RS to be equal to DC, as for example rs: Then the proportion of the Arc rs to the streight line DT will be duplicate of the proportion of RS to TV, or DR to DT. Which is absurd; because Dr is by construction greater or less then DR.

Therefore the Arc RS is equal to the side DC. Which was to be demonstrated.

Corol.

Hence it follows that DR is equal to two fifths of the Arc CA. For RS, TV, DT being continually proportional; and the Arc TV being described by *DT*, the Arc *RS* will be described by a streight line equal to *TV*. But *RS* is described by the streight line *DR*. Therefore *DR* is equal to *TV*, that is, to two fifths of *CA*.

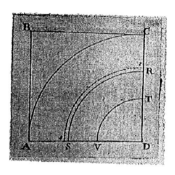

And your said servant most humbly prayeth you to consider (if the demonstration be true and evident) whether the way of objecting against it by Square Roots, used by Dr. *Wallis*; and whether all his Geometry, as being built upon it, and upon his supposition of an Infinite Number, be not false.

233.
EDWARD POCOCKE to WALLIS
21 September/[1 October] 1671

Transmission:

Manuscript missing.

Existence and date: mentioned in and answered by WALLIS–POCOCKE 23.IX/[3.X]. 1671.

234.
WALLIS to EDWARD POCOCKE
London, 23 September/[3 October] 1671

Transmission:

W Letter sent: LONDON *British Library* MS Sloane 4025, f. 310 and f. 306 (address) On f. 306 in unknown hand: 'Dr J: Wallis DD to Dr Pocock Hebrew Professor of Oxford Sept 23. 1671.' and Postmark: 'SE/23'.

Reply to: POCOCK–WALLIS 21.IX/[1.X].1671.

234. WALLIS to POCOCKE, 23 September/[3 October] 1671

Talbot in Fleet street. Sept. 23. 1671.

Sir,

I have neither matter nor time (my Ague coming on) to write much at present. But onely, in answere to yours of the 21th instant, I think it not amisse to send a quarter of a hundred of your books.[1225] Directed to Mr Oldenburg & a letter to him with it; & another Letter to Mr Haak;[1226] signifying that half thereof (or what proportion he shall have occasion to use,) he may have from Mr Oldenburg: The latter I presume I shal speak with in a few days, but whether I shal see Mr Haak again, or not, I cannot tell; for his lodgings & mine are a great way a sunder:) If they have opportunities of putting them all of, it will not be amisse: If not, however it will be no disadvantage to have them bye in their hands, from whence you may have them if there shal be occasion. After your Spider medecine, I had remission of some fits; but neither intermission, nor a riddance of them: I have since used some other medecines; from which yet (though greatly promising) I have little more confidence than the intermission of some fits (if so much) with a return afterwards. Most telling mee that there is little expectation of riddance before the Spring. Excuse the abrupt of

Sir

Your affectionate friend to
serve you
John Wallis.

These   [30
For the Reverend Dr Pocock
Canon of Christchurch
in
Oxon.

---

3 time (1) at present (2) (my Ague
5 of your books *add*.

[1225] books: i.e. the translation of Ibn Tufayl's 'philosophical fable' under the title *Philosophus autodidactus sive epistola Abi Jaafar, Ebn Tophail*, Oxford 1671, which was probably largely the work of the writer of this letter, the oriental scholar Edward Pocock, but which he nominally ascribed to his son, Edward Pococke junior.

[1226] Haak: i.e. Theodore Haak (1605–90), translator and natural philosopher, *ODNB*.

235.
WALLIS to THOMAS HOBBES/ROYAL SOCIETY
September 1671 (i)

Transmission:

E^1 Printed paper: WALLIS, *An Answer to Three Papers of Mr. Hobs, Lately Published in the Months of August, and this present September, 1671*. Exemplars OXFORD, *Bodleian Library* Savile Ee 1 (15) (with hand-written comment by Wallis) (our source); KEW *The National Archives* PRO SP29/293, No. 67.
E^2 Reprinted: *Philosophical Transactions* No. 75 (18 September 1671), pp. 2241–50 ('An Answer to Four Papers of Mr. Hobs, lately Published in the Months of August, and this present September, 1671').

Reply to: HOBBES–ROYAL SOCIETY end of VII.1671, HOBBES–ROYAL SOCIETY first half of VIII.1671 and HOBBES–ROYAL SOCIETY early IX.1671.
Answered by: HOBBES–ROYAL SOCIETY IX.1671.

Despite the title, this rejoinder by Wallis is essentially concerned with the first two papers addressed by Hobbes to the Royal Society, the third paper coming into Wallis's hands while the rejoinder was already at the press; presumably, the title was changed at the last minute. As in the earlier version of his reply to Hobbes (WALLIS–ROYAL SOCIETY) VIII.1671, Wallis incorrectly suggests that Hobbes' first paper was published in August rather than July.

An Answer
to
Three Papers of Mr. *Hobs*,
Lately Published in the Months of *August*, and
this present *September*, 1671.

In the former part of his first Paper;

By reason of a Proposition of Dr. *Wallis (Prop. 1. Cap. 5. De Motu)*[1227] to this purpose (for he doth not repeat it *Verbatim*): *If there be supposed a row of Quantities infinitely many, increasing according to the natural Order of Numbers, 1, 2, 3, &c. or their Squares, 1, 4, 9, &c. or their Cubes, 1, 8, 27, &c. whereof the last is given. It will be to a row of as many, equal to the last, in the first case, as 1 to 2; in the second case, as 1 to 3; in the third, as 1 to 4,* &c. (Where all that is affirmed, is but; *If we SUPPOSE That; This will Follow.* Which Consequence Mr. *Hobs* doth not deny. and therefore all that he saith to it, is but Cavelling.)

[1227] *De motu*: i.e. WALLIS, *Mechanica: sive, de motu, tractatus geometricus* II, 148–9; *Opera mathematica* I, 667–8.

235. WALLIS to HOBBES/ROYAL SOCIETY, September 1671 (i)

Mr. *Hobs* moves these Questions, (and proposeth them to the *Royal Society*, as not requiring any skil in *Geometry*, *Logick*, or *Latin*, to resolve them:) 1. *Whether there can be understood* (he should rather have said, *supposed*) *an infinite row of Quantities, whereof the last can be given.* 2. *Whether a Finite Quantity can be divided into an Infinite Number of lesser Quantities, or a Finite quantity consist of an Infinite number of Parts.* 3. *Whether there be any Quantity greater than Infinite.* 4. *Whether there be any Finite Magnitude of which there is no Center of Gravity.* 5. *Whether there be any Number Infinite.* 6. *Whether the Arithmetick of Infinites be of any use, for the confirming or confuting any Doctrine.* For answer. In general, I say, 1. Whether those things *Be* or *Be not*; yea, whether they *Can* or *Cannot be*; the Proposition is not at all concerned, (which affirms nothing either way;) but, whether they can be *supposed*, or made the *supposition, in a conditional Proposition*. As when I say, *If* Mr. Hobs *were a Mathematician, he would argue otherwise*: I do not affirm that either *he is*, or ever *was*, or *will be* such. I only say (upon supposition) *If he were*, what he is not; he would not do as he doth. 2. Many of these *Quaere's* have nothing to do with the Proposition: For it hath not one word concerning *Gravity*, or *Center of Gravity*, or *Greater than Infinite*. 3. That usually in *Euclide*, and all after him, by *Infinite* is meant but, *More than any assignable Finite*, though not Absolutely Infinite, or the greatest possible. 4. Nor do they mean, when Infinites are proposed, that they should *actually Be*, or be *possible to be performed*; but only, that they be *supposed*. (It being usual with them, upon *supposition* of things *Impossible*, to infer useful Truths.) And *Euclide* (in his second *Postulate*)[1228] requiring, *the producing a streight line Infinitely, either way*; did not mean, that it should be *actually performed*, (for it is not possible for any man to produce a streight line Infinitely;) but, that it be *supposed*. And if AB be *supposed* so produced, though but one way; its length must be *supposed* to become *Infinite* (or *more than any Finite length assignable*;) For, if but *Finite*, a Finite production would serve. But, if so produced both ways; it will be yet *Greater*, that is, *Greater than that Infinite*, or Greater than was necessary to make it more than any Finite length assignable. (And whoever doth thus *suppose Infinites;* must consequently *suppose, One Infinite greater than another*.) Again, when (by *Euclide's* tenth Proposition)[1229] the same AB, may be *Bisected* in M, and each of the halves in m, and so onwards, *Infinitely:* it is not his meaning (when such continual section is proposed)

1-3 to the *Royal Society*, to pass a judgment on them.) *1. Whether E*²

[1228] second *Postulate*: i.e. EUCLID, *Elements* I, post. 2.
[1229] tenth Proposition: i.e. EUCLID, *Elements* I, prop. 10.

that it should be *actually done*, (for, who can do it?) but that it be *supposed*. And upon such *(supposed)* section *infinitely continued*, the parts must be *(supposed) infinitely many*; for no *Finite* number of parts would suffice for *Infinite* sections. And if further, the same AB so divided, be *supposed* the side of a Triangle ABC; and, from each point of division, *supposed* lines (as *mc*, *Mc*, &c.) parallel to BC: these parallels (reckoning downward from A to BC) must consequently be *(supposed) infinitely many*; and those, in *Arithmetical progression*, as 1, 2, 3, &c. (each exceeding its Antecedent as much as that exceeds the next before it;) and, *whereof the last* (BC) *is given*. Nor is the *supposition of Infinites* (with these attendants) so new, or so *Peculiar* to *Cavallerius* or Dr. *Wallis*, but that *Euclide* admits it, and all Mathematicians with him; as at least *supposable*, whether *Possible* or not.

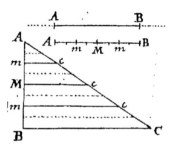

In particular, therefore, to his *Quaere's*, I answer, 1. There may be *supposed* a row of Quantities *Infinitely many*, and *continually increasing*, (as the supposed parallels in the Triangle ABC, reckoning downwards from A to BC,) *whereof the last* (BC) *is given*. 2. A Finite Quantity (as AB) may be *supposed* (by such continual Bisections) divisible into a number of parts *Infinitely many* (or, more than any Finite number assignable:) For there is no stint beyond which such division may not be *supposed* to be continued; (for still the last, how small soever, will have two halves;) And, all those Parts *were in* the Undivided whole; (else, where should they be had?) 3. Of *supposed* Infinites, one may be *supposed* greater than another. As a, *supposed*, infinite number of *Men*, may be *supposed* to have a *Greater* number of eyes. 4. A surface, or solid, may be *supposed* so constituted, as to be *Infinitely Long*, but *Finitely Great*, (the Breadth continually Decreasing in greater

---

10-11 *is given*: (and their Squares,a s 1, 4, 9, &c. their Cubes, as 1, 8, 27, &c.) And this I say, to shew that the supposition of Infinites (with these attendants) is not so new, or so *Peculiar* E^2

235. WALLIS to HOBBES/ROYAL SOCIETY, September 1671 (i)

proportion than the Length Increaseth,) and so as to have *no Center of Gravity*. Such is *Toricellio's Solidum Hyperbolicum acutum*[1230] and others innumerable, discovered[1231] by Dr. *Wallis*, Monsieur *Fermat*, and others. But to determine this, requires more of *Geometry, and Logick* (whatever it do of the *Latin Tongue*) than Mr. *Hobs* is Master of. 5. There may be *supposed* a number *Infinite*; that is, greater than any assignable Finite:

As the *supposed* number of parts, arising from a *supposed* Section *Infinitely continued*. 6. There is therefore no reason, on this account, why the Doctrin of *Euclide, Cavallerius*, or Dr. *Wallis*, should be rejected as of no use.

But having solved these *Quaere's*, I have some for Mr. *Hobs* to answer, which will not so easily be dispatched by him. For though *Supposed Infinites* will serve the Mathematicians well enough: yet, howsoever he please to prevaricate (which, he saith, is *for his Exercise*,) Mr. *Hobs* himself is more concerned than they, to solve such *Quaere's*. Let him ask himself therefore, if he be still of opinion, that *there is no Argument in nature to prove, the World had a Beginning*:

1. Whether, in case it had not, there must not have passed an *Infinite number of years* before Mr. *Hobs* was born. (For, if but *Finite*, how many soever, it must have begun so many years before.) 2. Whether, now, there have not passed *more*; that is, *more than that infinite* number. 3. Whether, in that *Infinite* (or *more than infinite*) number of *Years*, there have not been a *Greater* number of *Days* and *Hours*: and, of which hitherto, *the last is given*. 4. Whether, if this be an Absurdity, we have not then (contrary to what Mr. *Hobs* would perswade us) an *Argument in nature* to prove, *the world had a beginning*. Nor are we beholden to Mr. *Hobs* for this Argument; for it was an Argument in use before Mr. *Hobs* was born. Nor can he serve himself (as the Mathematicians do) with *supposed Infinites*; For his Infinites, and more than Infinites of Years, Days, and Hours, *already past*, must be *Real Infinites*, and which have *actually existed*, and whereof the *last is given*; (and yet there are more to follow.) Mr. *Hobs* shall do well (for his Exercise) to solve *these*, before he propose more *Quaere's* of *Infinites*.

25 *[Comment in Wallis's hand:]* If no Absurdity, what is then he quarrels at.

4-5 more of *Geometry, and Logick* than E^2
31 *Infinites*. And this I say, to shew that Mr. Hobs is, as much as any, concerned to solve the Quaere's by himself proposed. E^2

[1230] *Solidum Hyperbolicum acutum*: i.e. TORRICELLI, *De solido hyperbolico acuto*, pubished in his *Opera geometrica*, Florence 1644.
[1231] discovered: cf. WALLIS, *Arithmetica infinitorum*, 198; *Opera mathematica* I, 478; FERMAT–DIGBY [10]/20.IV.1657; WALLIS, *Correspondence* I, 281–4.

235. WALLIS to HOBBES/ROYAL SOCIETY, September 1671 (i)

In the latter part of his first Paper,

He gives us (out of his *Roset. Prop.* 5.[1232]) this Attempt of *Squaring the Circle. Suppose* DT *be* $\frac{2}{5}$ DC, *and* DR *a mean proportional between* DC *and* DT: *the Semidiameter* DC *will be equal to the Quadrantal Arc* RS, *and* DR *to* TV.

That the thing is false, is already shewed in the Latin Confutation of his *Rosetum*, published[1233] in the *Philosophical Transactions* for *July* last past.

As it is now in the English; his Demonstration is peccant in these words, (*Col.* 2. *lin.* 31, 32, 33.) *Therefore — the Arc on* TV, *the Arc on* RS, *the Arc on* CA, *cannot be in continual proportion*; (with all that follows:) There being no ground for such Consequence.

And the thing is manifest; for since that, by his construction,

$$\left.\begin{array}{l}\text{DC. CA. Arc on CA extended} \div \\ \text{DR. RS. Arc on RS extended} \div \\ \text{DT. TV. Arc on TV extended} \div\end{array}\right\} \begin{array}{l}\text{are in the same continual} \\ \text{proportion, of the Semi-} \\ \text{diameter to the Quadran-} \\ \text{tal Arc;}\end{array}$$

Let that proportion be *what you will*; suppose, as 1 to 2; and consequently, DC to CA being as 1 to 2, it will be to the Arc on CA, as 1 to 4: And by the same reason, DR to the Arc on RS, and DT to the Arc on TV, must also be as 1 to 4: And therefore the Arcs on TV, on RS, on CA; that is, 4 DT, 4 DR, 4 DC; will be in the same proportion to one another, as (their singles) DT, DR, DC: But these (by construction) are in continual proportion; therefore those Arcs also, as they ought to be. Indeed, if (by changing some one of the terms) you destroy (contrary to the Hypothesis) the continual proportion of DT, DR, DC, you will destroy that of the Arcs also (which are still proportional to these:) but so long as DT, DR, DC, be in *any* continual proportion (whether that by him assigned or any other) those will be in the same continual proportion with them. As if for DT, DR, DC, be taken Dt, Dr, DC, in any continual proportion (greater, less, or equal to his) the Arcs on *tu*, on *rs*, on CA, (extended) will be in the same continual proportion.

[1232] *Roset. Prop.* 5: i.e. HOBBES, *Rosetum geometricum*, 17–19.

[1233] published: i.e. Wallis's article 'An Answer of Dr. Wallis to Mr. *Hobbes's Rosetum Geometricum* in a Letter to a friend in London, dated July 16. 1671', published in *Philosophical Transactions* No. 73 (17 July 1671), 2202–9.

235. Wallis to Hobbes/Royal Society, September 1671 (i)

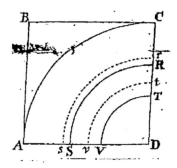

But (which is the common fault of Mr. *Hobs's* Demonstration) if this Demonstration were good, it would serve as well for any proportion as that for which he brings it. For if, instead of $\frac{2}{5}$, he had said $\frac{4}{9}$, $\frac{1}{2}$, $\frac{1}{100}$, or what else he pleased; the Demonstration had been just as good as now it is, without changing one syllable: That is, it will equally prove the proportion of the Semidiameter to the Quadrantal Arc, to be, *what you please.*

In his second Paper.

He pretends to *confute a Theorem, which hath a long time passed for truth*; (and therefore doth no more concern Dr. *Wallis*, than other men.) And 'tis this, *The four sides of a square being divided into any number of equal parts, for example, into* 10; *and streight lines drawn through the opposite points, which will divide the Square into* 100 *lesser Squares: The received opinion* (saith he) *and which* Dr. Wallis *commonly useth, is, that the Root of those* 100, *namely* 10, *is the side of the whole Square.* Which to confute, he tells us, *The Root* 10 *is a number of Squares, whereof the whole contains* 100; *and therefore the Root of* 100 *Squares is* 10 *of those Squares, and not the side of any Square; because the side of a Square is not a Superficies, but a Line.*

For Answer; I say, that 'tis neither the opinion of Doctor *Wallis*, nor (that I know) of any other (so far is it from being a *Received Opinion*, which Master *Hobs* insinuates as such) that 10 is the Root of 100 *Squares* (For surely a *Bare Number* cannot be the side of a *Square Figure*:) Nor yet (as Master *Hobs* would have it) that 10 *Squares* is the Root of 100 *Squares*: But that 10 *Lengths* is the Root of 100 *Squares*. 'Tis true that the *Number* 10 is the Root of the *Number* 100, but not, of a 100 *Squares*: and, that 10 *Squares* is the Root (not of 100 *Squares*, but) of 100 *Squared Squares*: Like as 10 *Dousen* is the Root, not of 100 *Dousen*, but of 100 *Dousen dousen*,

---

7 *please.* As any may presently see, who doth but read over his Paper. E^2
12 *into* 100; *corr. ed.*

235. WALLIS to HOBBBES/ROYAL SOCIETY, September 1671 (i)

or *Squares of a Dousen*. And, as, there, you must multiply not only 10 *into* 10, but *Dousen into Dousen*, to have the Square of 10 *Dousen*; so here 10 *into* 10 (which makes 100) and *Length into Length* (which makes a *Square*) to obtain the Square of 10 *Lengths*, which is therefore 100 *Squares*, and 10 *Lengths* the Root or side of it. But, says he, the Root of 100 *Soldiers*, is 10 *Soldiers*. *Answer*. No such matter: For 100 *Soldiers* is not the product of 10 *Soldiers into* 10 *Soldiers*, but of 10 *Soldiers into* the *Number* 10; And therefore neither 10, nor 10 *Soldiers*, the Root of it. So 10 *Lengths* into the *Number* 10, makes no Square, but 100 *Lengths*; but 10 *Lengths* into 10 *Lenghts* makes (not 100 *Lengths*, but) 100 *Squares*.

So in all other proportions: As, if the number of *Lenghts* in the *Square side* be 2; the number of *Squares* in the *Plain* will be *twice two*, (because there will be *two* rows of *two* in a row:) If the number of *Lengths* in the *side*, be 3; the number of *Squares* in the *Plain*, will be 3 times 3, or the Square of 3: If that be 4, this will be 4 times 4: And so in all other proportions. Of which, if any one doubt he may believe his own eyes.

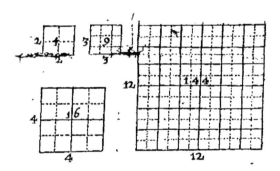

And this Mr. *Hobs* might have been taught by the next Carpenter (that knows but how to measure a Foot of Board) who could have told him, that because the *side* of a Square Foot, is 12 *inches in Length*, the Plain of it will be 12 *times* 12 *Inches in Squares*: Because there will be 12 Rows of 12 in a Row.

His third Paper,

Which came out just as the Answer to the two former was going to the Press, contains, for substance, the same with his Second, and the Latter part of the First: And so needs no farther Answer.

Only I cannot but take notice of his usual trade of contradicting himself. His second Paper says, *The side of a Square is not a Superficies, but a*

Line: His Third says the quite contrary, (Prop. 1.) *A Square root (speaking of Quantity) is not a Line, but a Rectangle*. Other faults, falsities, and contradictions, there are a great many; which I omit, as too gross to need an Answer.[1234]

And this is what I thought fit to say to Mr. *Hobs's Three Papers* (rather to satisfie the importunity of others, than because I thought them worth Answering:) And submit the whole, with all Respects, to the *Royal Society*, to whom Mr. *Hobs* makes his Appeal.

236.
THOMAS HOBBES to the ROYAL SOCIETY
September 1671

Transmission:

E^1 Printed paper: HOBBES, *Considerations upon the Answer of Dr. Wallis to the Three Papers of Mr. Hobbes*. Exemplar: OXFORD *Bodleian Library* Savile Ee 1 (12).
E^2 Reprinted as part of: HOBBES, *Three Papers Presented to the Royal Society Against Dr. Wallis. Together with Considerations on Dr. Wallis his Answer to Them*, London 1671 (our source).—printed: HOBBES, *English Works* VII, 443–8.

Reply to: WALLIS–HOBBES/ROYAL SOCIETY IX.1671 (i).
Answered by: WALLIS–HOBBES/ROYAL SOCIETY IX.1671 (ii).

Soon after the appearance of Wallis's printed paper entitled *An Answer to Three Papers of Mr. Hobs* (WALLIS–HOBBES/ROYAL SOCIETY IX.1671 (i)), Hobbes wrote and published the present paper in response.

Considerations upon the Answer of Dr. Wallis to the Three Papers of Mr. Hobbes.

Dr. Wallis sayes, All that is affirmed, is but, *If we SUPPOSE That, This will follow*.

But it seemeth to me, that is the Supposition be impossible, then that which follows will either be false, or at least undemonstrated.

[1234] Answer: Wallis's extensive reply to Hobbes's 'third paper' was published in 'An Answer to Four Papers of Mr Hobs, lately published in the Months of August, and this present September, 1671', *Philosophical Transactions* No. 75 (18 September 1671), 2241–50; 2246–50. See WALLIS–HOBBES/ROYAL SOCIETY IX.1671 (ii).

236. HOBBES to the ROYAL SOCIETY, September 1671

First, this Proposition being founded upon his *Arithmetica Infinitorum*, If there he affirm an absolute Infiniteness, he must here also be understood to affirm the same. But in his 39th Proposition he saith[1235] thus: *Seeing that the number of terms increasing, the excess above sub-quadruple is perpetually diminished, so as at last it becomes less than any proportion that can be assigned; If it proceed in* infinitum *it must utterly vanish. And therefore if there be propounded an Infinite row of quantities in triplicate proportion of quantities Arithmetically proportional (that is, according to the row of Cubical numbers) beginning from a point or 0; that row shall be to a row of as many, equal to the greatest, as 1 to 4.*

It is therefore manifest that he affirms, That in an Infinite row of quantities the last is given; and he knows well enough that this is but a shift.

Secondly, he says, That usually in *Euclide* and all after him, by Infinite is meant but, more than any assignable *Finite*, or the greatest possible.

I am content it be so interpreted. But then from thence he must demonstrate those his conclusions, which he hath not yet done. And when he shall have done it, not only the Conclusions, but also the Demonstration will be the same with mine[1236] in *Cap.* 14. *Art.* 2, 3, &c. of my Book *De Corpore*. And so he steals what he once condemn'd. A fine quality.

Thirdly, he says[1237] (by *Euclides* 10th Proposition, but he tells not of what Book) That a Line may be bisected, and the halves of it may again be bisected, and so onwards infinitely; and that upon such supposed Section Infinitely continued, the parts must be supposed Infinitely many.

I deny that; for *Euclide*, if he says a Line may be divisible into parts perpetually divisible, he means, That all the divisions, and all the parts arising from those divisions, are perpetually Finite in number.

Fourthly, he sayes, That there may be supposed a row of quantities infinitely many, and continually increasing, whereof the last is given.

'Tis true, a man may say (if that be supposing) that white is black; but if *Supposing* be *Thinking*, he cannot suppose an Infinite row of quantities whereof the last is given. And if he say it, he can demonstrate nothing from it.

Fifthly, He sayes (for one absurdity begets another) *That a Superficies or Solid may be supposed so constituted, as to be* Infinitely long, *but* Finitely great (*the breadth continually decreasing in greater proportion than the length increaseth*) *and so as to have no center of gravity.* Such is Toricel-

---

[1235] saith: i.e. WALLIS, *Arithmetica infinitorum*, 31–2; *Opera mathematica* I, 382.
[1236] mine: i.e. HOBBES, *De corpore*, 107–8.
[1237] sayes: Wallis refers to EUCLID, *Elements* I, prop. 10.

236. Hobbes to the Royal Society, September 1671

lio's Solidum Hyperbolicum acutum,[1238] *and others innumerable discovered by* Dr. Wallis, *Monsieur* Fermat, *and others. But to determine this, requires more of Geometry and Logick (whatsoever it do of the Latine Tongue) than* Mr. Hobbes *is master of.*

I do not remember this of *Toricellio*, and I doubt Dr *Wallis* does him wrong, and Monsieur *Fermat* too. For to understand this for sense, 'tis not required that a man should be a Geometrician or a Logician, but that he should be mad.

In the next place he puts to me a Question as absurd as his Answers are to mine. Let him ask himself (saith he) if he be still of opinion, *That there is no Argument in Natural Philosophy to prove that the World had a beginning*: First, whether in case it had no beginning, there must not have passed an Infinite number of years before Mr. *Hobbes* was born. Secondly, whether at this time there have not passed more, that is, more than that Infinite number. Thirdly, whether, in that Inifnite (or more than Infinite) number of years, there have not been a greater number of dayes and hours, and of which hitherto the last is given. Fourthly, whether, if this be an Absurdity, we have not then (contrary to what Mr. *Hobbes* would perswade us) an Argument in Nature to prove the World had a beginning.

To this I answer, not willingly, but in service to the Truth, that by the same Argument he might as well prove that God had a beginning. Thus: in case he had not, there must have passed an Inifnite length of time before Mr. *Hobbes* was born; but there have passed at this day more than that Infinite length (by eighty four years). And this day, which is the last, is given. If this be an Absurdity, have we not then an Argument in Nature to prove that God had a beginning? Thus 'tis when men intangle themselves in a Dispute of that which they cannot comprehend. But perhaps he looks for a Solution of his Argument to prove that there is somewhat greater than Infinite which I shall do so far, as to shew; it is not concluding, If from this day backwards to Eternity be more than Inifnite, and from Mr. *Hobbes* his birth backwards to the same Eternity be Infinite, then take away from this day backwards to the time of *Adam*, which is more than from this day to Mr. *Hobbes* his birth, then that which remains backwards must be less than Infinite. All this arguing of Infinites is but the ambition of School-boyes.

[1238]Solidum Hyperbolicum acutum: i.e. TORRICELLI, *De solido hyperbolico acuto*, published in his *Opera geometrica*, Florence 1644.

236. HOBBES to the ROYAL SOCIETY, September 1671

To the Latter part of the first paper.

There is no doubt, if we give what Proportion we will of the Radius to the Arc, but that the Arc upon that Arc will have the same Proportion. But that is nothing to my Demonstration. He knows it, and wrongs the Royal Society in presuming they cannor find the Impertinence of it.

My proof is this; That if the Arc on *TV*, and the Arc *RS*, and the streight Line *CD*, be not equal, then the Arc on *TV*, the Arc on *RS*, and the Arc on CA, cannot be proportional. Which is manifest by supposing in *DC* a less than the said *DC*, but equal to RS, and another streight Line, less than *RS*, equal to the Arc on *TV*; and any body may examine it by himself.

I have been asked by some that think themselves Logicians, Why I proceeded upon $\frac{2}{5}$ rather than any other part of the Radius. The reason I had for it was, That long ago some *Arabians* had determined, That a streight Line whose square is equal to 10 squares of half the Radius, is equal to a quarter of the Perimiter; but their demonstrations are lost. From that Equality it follows, that the third proportional to the Quadrant and Radius, must be a mean proportional between the Radius and $\frac{2}{5}$ of the same. But my answer to the Logicians was, That though I took any part of the Radius to proceed on, and lighted on the Truth by chance, the Truth it self would appear by the Absurdity arising from the denial of it. And this is it that *Aristotle* meant, where he distinguisheth between a Direct demonstration, and a demonstration leading to an absurdity. Hence it appears, that Dr. *Wallis* his objections[1239] to my *Rosetum*[1240] are invalid, as built upon Roots.

To the second Paper.

Firstly, he sayes, That it concerns him no more than other men. Which is true. I meant it against the whole Herd of them who apply their *Algebra* to *Geomtry*.

Secondly, He sayes, That a bare Number cannot be the Side of a Square figure.

I would know what he means by a Bare Number. Ten Lines may be the side of a Square figure. Is there any Number so bare, as by it we are not to conceive or consider any thing numbred? Or by ten Nothings understands he Bare 10? He struggles in vain, his Conscience puzzles him.

[1239] objections: i.e. WALLIS, 'An Answer of Dr. *Wallis* to Mr. *Hobbes's Rosetum Geometricum* in a Letter to a friend in London, dated July 16. 1671', *Philosophical Transactions* No. 73 (17 July 1671), 2202–9. See WALLIS–OLDENBURG 16/[26].VII.1671.

[1240] *Rosetum*: i.e. HOBBES, *Rosetum geometricum*, London 1671.

Thirdly, He sayes, Ten Squares is the Root of 100 Square-squares. To which I answer, first, That there is no such Figure as a Square-square. Secondly, That it follows hence that a Root is a Superficies, for such is 10 Squares.

Lastly, He sayes, That neither the Number 10 nor 10 Souldiers is the Root of 100 Souldiers; because 100 Souldiers is not the Product of 10 Souldiers into 10 Souldiers.

That last I grant, because nothing but Numbers can be multiplied into one another. A Souldier cannot be multiplied by a Souldier. But no more can a Square-figure by a Square-figure, though a Square-number may. Again, If a Captain will place his hundred Men in a square Form, must not he take the root of 100 to make a Rank or File? And are not those 10 Men?

To the third Paper

He objects nothing here, but that, *The Side of a Square is not a Superficies but a Line*, and that *a Square Root (speaking of quantity) is not a Line but a Rectangle*, is a contradiction. The *Reader* is to judge of that.

To his Scoffings I say no more, but that they may be retorted in the same words, and are therefore childish.

And now I must submit the whole to the Royal Society, with confidence that they will never ingage themselves in the maintenance of these Unintelligible Doctrines of Dr. *Wallis*, that tend to the suppression of the Science which they endeavour to advance.

237.
WALLIS to THOMAS HOBBES/ROYAL SOCIETY
September 1671 (ii)

Transmission:

E First edition: *Philosophical Transactions* No. 75 (18 September 1671), pp. 2241–50 ('An Answer to Four Papers of Mr. Hobs, lately Published in the Months of August, and this present September, 1671').

Reply to: HOBBES–ROYAL SOCIETY end of VII.1671, HOBBES–ROYAL SOCIETY first half of VIII.1671, HOBBES–ROYAL SOCIETY early IX.1671 and HOBBES–ROYAL SOCIETY IX.1671.

While Wallis's paper entitled *An Answer to Three Papers of Mr. Hobs, Lately Published in the Months of August, and the present September, 1671* was in the final stages of being prepared for printing, Hobbes published his third rejoinder to Wallis's refutation of his

237. WALLIS to HOBBBES/ROYAL SOCIETY, September 1671 (ii)

Rosetum geometricum. Wallis's paper prompted Hobbes to write a 'fourth paper', his *Considerations upon the Answer of Dr. Wallis to the Three Papers of Mr. Hobbes.* The present extensive reply by Wallis to Hobbes' 'third' and 'fourth paper', which concluded the latest debate between the two men, was added to the content of Wallis's paper in reply to Hobbes' two earlier papers and published as 'An Answer to Four Papers of Mr. Hobs, lately Published in the Months of August and this present September, 1671' in the September issues of *Philosophical Transactions.* See WALLIS–HOBBES/ROYAL SOCIETY IX.1671 (i) for the text of Wallis's reply to Hobbes' 'first' and 'second paper'.

An Answer to Four Papers of Mr. *Hobs*, lately Published in the Months of *August*, and this present *September, 1671.*

[...]

His third Paper,[1241]

Which came out just as the Answer to the two former[1242] was going to the Press, contains, for substance, the same with his second, and the Latter part of the first: And so needs no farther Answer.

Only I cannot but take notice of his usual trade of contradicting himself. His second Paper says, *The side of a Square is not a Superficies, but a Line*: His third says the quite contrary, (Prop. 1.) *A Square root, (speaking of Quantity) is not a Line, but a Rectangle.* Other faults, falsities, and contradictions, there are a great many.

As for Instance: He tells us first, *In the natural Row of Numbers, as 1, 2, 3, 4, 5, 6, &c. every one is the Square of some number in the same Row;* (that is, of some Integer number; which is notoriously false.) This he contradicts in the very next words, *But Square numbers (beginning at 1) intermit first two numbers, then four, then six, &c; so that none of the intermitted numbers is a Square number, nor hath any Square root.* (If these *intermitted numbers*, between 1, 4, 9, 16, &c. be not *Squares* how is it that *every one* in the whole row is a *Square*, and that of some Integer number?) But this again is contradicted *prop. 2.* where 200 (one of such intermitted numbers) is made a *Square*, and $14\frac{4}{14}$ the *Root* of it.

Again; in his *Definition* he tells us, that *a Square Root multiplied into it self produceth a Square:* But (*prop. 2.*) he multiplieth *the Root* $14\frac{4}{14}$ (not into it self, but) into 14 (a part thereof,) to make 200, which he will have to be *the Square* of that Root. Nor is it a meer slip of negligence in the computation, but his Rule directs to it; *Any number given is produced by*

[1241] *third Paper*: i.e. HOBBES–ROYAL SOCIETY early IX.1671.
[1242] two former: i.e. HOBBES–ROYAL SOCIETY end of VII.1671 and HOBBES–ROYAL SOCIETY first half of VIII.1671.

237. Wallis to Hobbbes/Royal Society, September 1671 (ii)

the greatest Root multiplied into it self, and into the remaining Fraction. Whereof he gives this instance: *Let the number given be* 200 *Squares, the greatest Root is* $14\frac{4}{14}$ *Squares* (he should rather have said *Lengths*; but that is a small fault with him;) *I say, that* 200 *is equal to the product of* 14 *into it self (which is* 196,*) together with* 14 *multiplied into* $\frac{4}{14}$ *(which is equal to* 4:*)* that is $14\frac{4}{14}$ multiplied into 14. But this calculation is again contradicted in his third proposition, where he calculates *the same Square* otherwise, as we shall see by and by. In the mean time let's consider this alone, and see the contradictions within it self. His Rule bids us multiply *the greatest Root into it self,* &c. This *greatest Root* he says is $14\frac{4}{14}$; yet doth he not muliply this, but 14 (a part thereof) *into it self and into the Fraction* $\frac{4}{14}$. Again; if $14\frac{4}{14}$ be *the greatest Root,* what shall be *the remaining Fraction?* Doth he take the Root of 200 to be more than $14\frac{4}{14}$ by some further *remaining Fraction?* If so, he should have told us what that Fraction is; for $\frac{4}{14}$ it is not, this being part of his *greatest Root* $14\frac{4}{14}$. But if we should allow (as I think we must,) that by *the greatest Root* he means sometimes $14\frac{4}{14}$, sometimes 14, (that is, if we allow him to contradict himself,) yet how comes he by the Fraction $\frac{4}{14}$? For, $\frac{2}{14}$ is too much (the square of $14\frac{2}{14}$ being more then 200, as by multiplying $14\frac{2}{14}$ into it self will appear;) which destroys his whole design; for 14, multiplied into $14\frac{2}{14}$, will not make 200, but 198; contrary to his rule. But further, it is so gross a mistake, to make 200 the Square of $14\frac{2}{14}$, that every Apprentice boy, (that can but multiply whole numbers, and fractions,) could have informed him better, who would first have reduced the fraction to smaller terms, putting $14\frac{2}{7}$ for $14\frac{4}{14}$, and then multiplying $14\frac{2}{7}$ into it self, would have shew'd him, that the Square of $14\frac{4}{14}$, that is, $14\frac{4}{14}$ multiplied into it self, is (not 200, but) $204\frac{4}{49}$.

But the Root of 200, is the [surd] number $10\sqrt{2}$, which is less than $14\frac{2}{14}$, and bigger than $14\frac{2}{15}$: the Square of that being somewhat more than 200; and, of this, somewhat less; but either of them within an unite of it.

$$14\frac{2}{7}$$
$$14\frac{2}{7}$$
$$\overline{56}$$
$$14$$
$$4$$
$$4$$
$$\frac{4}{49}$$
$$\overline{204\frac{4}{49}}$$

27 said number *corr. ed.*

237. WALLIS to HOBBBES/ROYAL SOCIETY, September 1671 (ii)

But this second Proposition, is (as I said) contradicted by his third, which makes the Square of $14\frac{4}{14}$ to be $200\frac{1}{49}$, (by what computation, we shall see by and by;) and then finds fault, that this and the former do not agree. (But 'tis no wonder they should disagree, when both are false.) *The same Square* (saith he) *calculated Geometrically, consisteth (by Euclid.*[1243] *2. 4.) of the same numeral great Square* 196, *and of two Rectangles under the greatest side* 14 *and the Remainder of the side, and further of the Square of the less segment; which altogether make* $200\frac{1}{49}$. (He might have learned to reckon better; but let us see how he makes it out.) *As by the operation it self* (saith he) *appeareth thus: The side of the greater segment is* $14\frac{4}{14}$ (this was, but now, the side of the whole square, how comes it now to be but the side of the greater Segment?) *which multiplied unto it self* (saith he) makes 200: (no; but $204\frac{4}{49}$:) *The product of* 14 *the greatest Segment into the two Fractions* $\frac{4}{14}$ *is* 4, *and that added to* 196 *makes* 200: (if by two fractions $\frac{4}{14}$, he mean, as he ought by his Rule, the Fraction 4 twice taken, or the double of it, it will be not 4, but 8, and this added to 196 make 204; But all this he puts in his pocket, for it comes not into account at all.) *Lastly, the product of* $\frac{2}{14}$ *into* $\frac{2}{14}$, *or* $\frac{1}{7}$ *into* $\frac{1}{7}$ *is* $\frac{1}{49}$; which with the first 200 makes $200\frac{1}{49}$: (But he forgets himself, for his lesser segment was not $\frac{2}{14}$, but $\frac{4}{14}$; he should therefore have said $\frac{4}{14}$ into $\frac{4}{14}$, or $\frac{2}{7}$ *into* $\frac{2}{7}$, is $\frac{4}{49}$.) His calculation therefore should have been this: The greater segment is (not [$14\frac{4}{14}$], but) 14; which multiplied into it self makes (not 200 but) 196: The Rectangle of the greater segment 14, into the lesser $\frac{4}{14}$, is 4: And this taken a second time, is another 4: The lesser segment (not $\frac{2}{14}$, but) $\frac{4}{14}$, or $\frac{2}{7}$, multiplied into it self, is (not $\frac{1}{49}$, but) $\frac{4}{49}$: All which added together make not $200\frac{1}{49}$, but $196+4+4+\frac{4}{49}=) 204\frac{4}{49}$, which is just the same with $14\frac{4}{14}$ multiplied into it self. So that, had he known how to multiply a number into a number, especially when incumbred with fractions (which it is manifest he doth not,) he would have found no disagreement between the *Arithmetical calculation*, and what he calls the *Geometrical*. But I am ashamed (for him) that so great a pretender to such high things in Geometry, should be so miserably ignorant of the common operations of practical Arithmetick.

His repeated Quadrature he now expresseth thus, *The Radius of a Circle is a mean Proportional between the Arc of a Quadrant and two fifths of the same*. But instead of *two fifths*, he might as well have said the *half*, or *tenth*, or *hundredth part*, &c; or (taking T in DC produced beyond C,) the *double, decuple, centuple, &c.* or *what you please*: For his Demonstration

---

21 is not $14\frac{2}{14}$ *corr. ed.*

[1243] *Euclid*: i.e. EUCLID, *Elements* II, prop. 4.

would have proved it, which is this. *Describe a Square ABCD, and in it a Quadrant DCA. In the side DC* (continued if need be,) *take DT two fifths of DC,* (or its Half, Double, Hundredth part, or what you please;) *and between DC and DT a mean proportional DR; and describe the Quadrantal Arcs RS, TV. I say, the Arc RS is equal to the streight line DC. For seeing the proportion of DC to DT is duplicate of the proportion of DC to DR, it will be also duplicate of the Proportion of the Arc CA to the Arc RS, and likewise duplicate of the Proportion of the Arc RS to the Arc TV. Suppose some other Arc, less or greater than the Arc RS, to be equal to DC, as for example* rs; *Then the proportion of the Arc* rs *to the streight line DT will be duplicate of the proportion of RS to TV, or DR to DT, which is absurd; because* Dr *is by construction greater or less than DR. Therefore the Arc RS is equal to the side DC; which was to be demonstrated.* Which demonstration therefore proving indifferently *every* proportion, doth not indeed prove *any*. In brief: The force of his Demonstration is but this; *DT being to DC as* 2 *.to* 5 (or in an other proportion) *and DR a mean proportional between them; RS will be so between TV and CA; and therefore* rs *(greater or less than RS,) will not be a mean proportional between TV and CA*: which is true; *but why it may not be equal to DC*, we have nothing but his word for it; there being nothing to shew that *DC is equal to such a mean proportional.* Again; though rs be not a mean proportional between TV and CA, yet it may be between tv and CA, which serves his Demonstration as well; which is indifferent to any three continual proportionals, as was shewed before. So that now we have had three Demonstrations of this Quadrature, (in his *Rosetum,* in his *first* paper, and in his *third,*) and this common fault in all of them, that they equally prove the proportion by him proposed, or any other what you please. But such his Demonstrations use to be.

And this is what I thought fit to say to Mr. *Hobs's [Three] Papers* rather to satisfie the importunity of others, than because I thought them worth Answering:) And submit the whole, with all Respects, to the *Royal Society,* to whom Mr. *Hobs* makes his Appeal.

His Fourth Paper;[1244]

Which came out since the *Three former* were answer'd, (containing some faint endeavors to re-assert some

---

28 Hob's Four Papers corr. ed.

[1244] *Fourth Paper*: i.e. HOBBES, *Considerations upon the Answer of Dr. Wallis to the Three Papers of Mr. Hobbes.* See HOBBES–ROYAL SOCIETY IX.1671.

What he would therein insinuate concerning *God* (that we may as well prove *Him* to have had a Beginning, as that the World had) smells too rank of Mr. *Hobs*. We are not to measure Gods *Permanent* Duration of Eternity, by our *successive* Duration of Time: Nor, his Intire *Ubiquity*, by Corporeal *Extension*.

What in it concerns *Mathematicks*, (whether his own or others,) is so weak and trivial, (and said only, that he may seem to say something, though nothing to the purpose,) that I shall trust it with those to whom he makes his appeal, without thinking it to need any Reply; The view of what he writeth against, being a sufficient Answer to all he saith.

238.
HENRY OLDENBURG to WALLIS
4/[14] November 1671

Transmission:

Manuscript missing.

Existence and date: mentioned in and answered by WALLIS–OLDENBURG 6/[16].XI.1671.

239.
WALLIS to THOMAS LAMPLUGH
6/[16] November 1671

Transmission:

Manuscript missing.

Existence and date: mentioned in WALLIS–OLDENBURG 6/[16].XI.1671.

As emerges from WALLIS–OLDENBURG 6/[16].XI.1671, Wallis at this time owed £2-3s-6d to Oldenburg. In order to offset his debt, he asked Lamplugh to purchase from Oldenburg a copy of the recently-printed third part of his *Mechanica: sive, de motu, tractatus geometricus* for 16s. The rest of the debt was to be settled by Crew (see WALLIS–CREW 6/[16].XI.1671) and by John Wallis jr.

240.
WALLIS to NATHANIEL CREW
6/[16] November 1671

Transmission:

Manuscript missing.

Existence and date: mentioned in WALLIS–OLDENBURG 6/[16].XI.1671.

As emerges from WALLIS–OLDENBURG 6/[16].XI.1671, Wallis at this time owed £2-3s-6d to Oldenburg. In order to offset his debt, he asked Crew to purchase from Oldenburg a copy of the recently-printed third part of his *Mechanica: sive, de motu, tractatus geometricus* for 16s-6d. The rest of the debt was to be settled by Lamplugh (see WALLIS–LAMPLUGH 6/[16].XI.1671) and by John Wallis jr.

241.
WALLIS to HENRY OLDENBURG
Oxford, 6/[16] November 1671

Transmission:

W Letter sent: LONDON Royal Society Early Letters W1, No. 131, 2 pp. (our source). At top of p. 1 in Oldenburg's hand: 'Entered L.B. 5. 33.' On p. 2, again in Oldenburg's hand, 'Novemb. 8. 71. Ans. Nov. 11: 71.' and 'Dr Wallis's sentiment about Dr Leibnitius his assertion, that Motion is the Principle of Cohesion |and not *del.*|.' Postmark: 'NO/8'.—printed: OLDENBURG, *Correspondence* VIII, 341–3.
w^1 Copy of letter sent: LONDON Royal Society Letter Book Original 5, pp. 33–5.
w^2 Copy of w^1: LONDON Royal Society Letter Book Copy 5, pp. 36–8.

Reply to: OLDENBURG–WALLIS 4/[14].XI.1671.
Answered by: OLDENBURG–WALLIS 11/[21].XI.1671.

November 6. 1671. Oxford.

Sir,

I have yours[1245] of Nov. 4. this morning. *I have written*[1246] to the Bishop of
Oxford to send to you for one of the 2 Copies in your hand, paying 16^s. 6^d.
and to Dr Lamplugh to send to you for the other paying 16^s. (They have each them in their letters an order to you to deliver the books.) & I send

---

[1245] yours: i.e. OLDENBURG–WALLIS, 4/[14].XI.1671.
[1246] written: i.e. the two letters evidently written on the same day to Nathaniel Crew and Thomas Lamplugh, after receiving that from Oldenburg, WALLIS–CREW 6/[16].XI.1671 and WALLIS–LAMPLUGH 6/[16].XI.1671.

241. WALLIS to OLDENBURG, 6/[16] November 1671

you herewith an order on my son[1247] to pay you 11^s. Which with the two former summes makes up the $2^{\pounds}$. 3^s. 6^d. for which I was Debitor to you on the foot of your last account: And I am further obliged to you for your pains in the busyness.

D. Leibnitium quod attinet:[1248] num quies an motus sit principium cohaesionis, non libet ultra disputare; neque de conciliatione sententiarum DD. Boylii et Leibnitii ea de re, aut utriusvis cum Cartesiana. Quippe in quaestionibus Physicis non ea certitudo demonstrationis esse solet quae in pure Mathematicis: Adeoque in illis dissentiendi locus. Motum esse cohaesionis principium, jam ante Leibnitium (uti nosti) per aliquot annos, contendebat G. Nelius noster. De quo quid senserim jam olim dictum est in literis[1249] quae inter me et ilium te mediante intercesserunt. Malim adhuc hac de re ἀπέχειν, quam in utramvis partem determinare. Hactenus tamen assentio, vix aut ne vix quietem ullam absolutam in corporibus reperiri, saltem iis quibuscum nos versamur. Spatium item Vacuum quin dari possit, nihil in contrarium video; quicquid sentiant vel Peripatetici vel Cartesiani. Huic autem, non minus quam Corpori, si Extensio concedatur, concedenda videtur et Magnitudo, (quippe, Extensionem esse ubi nulla est Magnitudo, non satis assequor quo possit concipi;) sed et, si Vacuum illud Terminatum sit, etiam Figuram habet, non minus quam terminatum Corpus quod propterea erit et Figuratum; sin Vacuum intelligatur non Terminatum sed in infinitum extensum, Figuram quidem non habebit; sed nec Corpus, si concipiatur sic extendi, habebit Figuram. Non video igitur quomodo Figura et Magnitudo, magis quam Extensio, Corpus (seu Spatium plenum) a Spatio Vacuo distinguat. Dixerim ego potius, eo distingui, quod alterum Materialem sit, alterum non sit. Sin quaeretur, quid illud sit quod Materiam dicimus; dico, illud magis a communi apprehensione quam sibi concipit animus ex vocis illius in communi sermone usurpatione animo innotescere quam ulla definitione explicari; idemque de Spatio, Loco, Tempore, aliisque notionibus simplicibus

9 in illis *add.*
23 igitur *add.*
29 idemque de *(1)* Loco, Tempore, Spatio *(2)* Spatio, Loco, Tempore

---

[1247]son: i.e. John Wallis jr, q.v.
[1248]Leibnitium quod attinet: Oldenburg had evidently reported to Wallis on the letter he had recently received from Leibniz, LEIBNIZ–OLDENBURG 15/25.X.1671; OLDENBURG, *Correspondence* VIII, 292–5. In that letter Leibniz discusses at length the questions of motion and rest, fluidity and firmness which Wallis addresses in the present letter evidently at Oldenburg's request.
[1249]literis: see for example NEILE–OLDENBURG 13/[23].V.1669, NEILE–OLDENBURG 20/[30].V.1669 and NEILE–OLDENBURG 1/[11].VI.1669.

non paucis dicendum existimo. Alicubi enim sistendum erit in vocum significationibus, et rerum incomplexarum notionibus, non minus quam in notionibus complexis seu propositionibus: utrobique scilicet dicendum, esse κοινὰς ἐννοίας. Nam, ut nullas concedentibus praemissas (tamquam per se notas, aut a notis ante probatas,) nulla potest probari conclusio seu propositio; ita nulla verba quasi primitus nota concedentibus, nihil potest verbis explicari; nullasque rerum notiones simplices quasi primitus notas concedentibus, nulla res poterit definiri. Quidni itaque dicamus conceptus Simplices *Temporis; Spatii; Materiae, Motus*, &c, non minus esse menti Apprehendenti congenios; quam conceptus complexos, *Totum esse majus sua parte, Æqualia aequalibus addita conficere aequalia, &c*, menti *Judicanti*? *Motus* autem quomodo Corpus a Vacuo distingueret non video. Utut enim concesserim nullum forte corpus de facto esse quod non vel ipsum Totum, vel Partes saltem ipsius, moveantur: non tamen id concesserim ita ad Materiae Corporisve naturam spectare, ut, siquando quiesceret, desinerit esse corpus materiave. Habes itaque (quod petis) quid de his sentiam (ea libertate animi quam ab aliis expeto eisque vicissim concedo) raptim expositum.

Tuus
Joh. Wallis.

These [2]
For Mr Henry Oldenburg,
in the Palmal near St. James's,
London.

242.
HENRY OLDENBURG to WALLIS
11/[21] November 1671

Transmission:

Manuscript missing.

Existence and date: mentioned in and answered by WALLIS–OLDENBURG 23.XI/[3. XII]. 1671, and also mentioned in Oldenburg's endorsement to WALLIS–OLDENBURG 6/[16].XI. 1671.
Reply to: WALLIS–OLDENBURG 6/[16].XI.1671.
Answered by: WALLIS–OLDENBURG 23.XI/[3.XII].1671.

4 ut (*1*) nihil ⟨—⟩ concedentibus (*2*) nullas concedentibus
6 verbis (*1*) definiri (*2*) explicari
7 quasi (*1*) per se (*2*) primitus
13 de facto *add.*

243.
WALLIS to CHARLES SCARBOROUGH
16/[26] November 1671

Transmission:

W Letter sent: in private possession. Referred to, and partly quoted from, in Sotheby's Catalogue *The Library of the Earls of Macclesfield removed from Shirburn Castle*, Part Two: Sciences A–C, p. 56.

In this letter, Wallis discusses a missing passage of the dedication to Dositheus of Pelusium in Archimedes' *Dimensio circuli*. The letter is endorsed: 'These for my worthy Friend Sir Charles Scarbrough ... London'.

244.
WALLIS to CHARLES SCARBOROUGH
Oxford, 21 November/[1 December] 1671

Transmission:

W Letter sent: in private possession. Facsimile in Sotheby's Catalogue *The Library of the Earls of Macclesfield removed from Shirburn Castle*. Part Two: Science A–C, p. 56.

Enclosure: Demonstrations of three propositions of Archimedes.

Wallis evidently wrote the present letter and that preceding it (WALLIS–SCARBOROUGH 16/[26].XI.1671) after a meeting with Scarborough during a recent visit to London. At Scarborough's request, he enclosed a demonstration of three propositions of Archimedes.

Oxford. Novemb. 21. 1671.

Sir,

I send you here the Demonstrations of those two Propositions of Archimedes which you desired, sett down so distinctly as that I hope you will find no difficulty in them: viz. Prop. 10 et 11 $\pi\epsilon\rho\grave{\iota}$ $\accentset{\sim}{E}\lambda\iota\kappa\tilde{\omega}\nu$.[1250] The Demonstrations are those of Archimedes, and (as near as I could) in his own Methode. They need no Schemes, (being equally true in other quantities as well as lines:) or, if you please, you make use of those in Archimedes. For which end I use, in the former of them, the same Letters that he doth; but in the latter (where

---

4 which you desired *add.*

[1250] $\pi\epsilon\rho\grave{\iota}$ $\accentset{\sim}{E}\lambda\iota\kappa\tilde{\omega}\nu$: i.e. ARCHIMEDES, *De lineis spiralibus*, prop. 10, 11.

he changeth the letters & makes use of others in their stead) I chose to retain those of the precedent, that the dependence of this upon the former might be the more clear. It differs not in substance from that I gave you at London, but in the order of setting it down. The fault in the Greek copy (which the Latine Translations retain,) where, in the Ecthesis of the latter, wee have ἴσα τᾶ ὑπεροχᾶ in stead of ἴσα τᾶ ἐλαχίστα, I find to have been observed, before me, by Sir Henry Savile,[1251] and noted as an error in the Margin of that which was his Book with his own hand; which doubtless was not an error of the Author, but of the Transcribers. I sent[1252] you last week what I find written by the same hand (out of some Manuscript, I suppose,) in the beginning of his Archimedes De Sphaera et Cylindro, viz. the Dedication[1253] of that Piece by Archimedes to Dositheus[1254] which in the Printed Copies is all wanting except the two last Lines of it. Nor have I more to adde now, but, that I am

Sir,

Your very humble servant,
John Wallis.

Since I had written the rest, I thought fit to subjoin this other,[1255] (which you allso desired,) somewhat altered in form, though not in the substance of the Demonstration, from what I gave you at London.

245.
Wallis to Charles Scarborough
21 November/[1 December] 1671, enclosure:
Demonstrations of Three Propositions of Archimedes

Transmission:

W^1 Paper sent: in private possession. Partial facsimile in Sotheby's Catalogue *The Library of the Earls of Macclesfield removed from Shirburn Castle.* Part Two: Science A–C, p. 56. W^2 Copy of paper sent: Oxford *Bodleian Library* MS Don. d. 45, f. 151^r–152^v (f. 152^v blank) (our source).

[1251] Savile: i.e. Henry Savile (1549–1622), classical scholar and mathematician, *ODNB*.
[1252] sent: i.e. Wallis–Scarborough 16/[26].XI.1671.
[1253] Dedication: in the prefatory letter ot Dositheus, Archimedes summarizes the main results of Book I of *De sphaera et cylindro*.
[1254] Dositheus: i.e. Dositheus of Pelusium (*fl.* 230 BC), Greek mathematician.
[1255] other: i.e. the enclosed demonstrations of three propositions of Archimedes.

245. WALLIS to SCARBOROUGH, 21 Nov./[1 Dec.] 1671, enclosure

E^1 WALLIS, *Algebra*, [299]–304.
E^2 WALLIS, *Opera mathematica* II, 324–330.

Enclosure to: WALLIS–SCARBOROUGH 21.XI/[1.XII].1671.

The Three following Propositions of Archimedes were, at the desire of Sir Charles Scarbrough, thus by mee demonstrated, & sent to him in a Letter of November 21. 1671.

Propositio IX Lib. II. Ἰσορροπικῶν Archimedis.[1256]

Sint continue proportionales $\alpha^3.\alpha^2\epsilon.\alpha\epsilon^2.\epsilon^3 \div\div.$ et $\epsilon^3.\alpha^3 - \epsilon^3 ::$
$\beta\alpha^2.\alpha^3 - \alpha\epsilon^2$ in $\frac{2}{5}$.
Et $2\alpha^3 + 4\alpha^2\epsilon + 6\alpha\epsilon^2 + 3\epsilon^3.5\alpha^3 + 10\alpha^2\epsilon + 10\alpha\epsilon^2 + 5\epsilon^3 ::$
$\gamma\alpha^2.\alpha^3 - \alpha\epsilon^2$.
Erunt $\beta\alpha^2 + \gamma\alpha^2 = \frac{2}{5}\alpha^3$.

Demonstratio.

Hoc est, (propter Analogias,)
$\frac{\alpha^3-\alpha\epsilon^2}{\alpha^3-\epsilon^3}$ in $\frac{3}{5}\epsilon^3$ ($=\beta\alpha^2,$) +
$(\gamma\alpha^2 =) \frac{\alpha^3-\alpha\epsilon^2,\text{in},2\alpha^3+4\alpha^2\epsilon+6\alpha\epsilon^2+3\epsilon^3}{5\alpha^3+10\alpha^2\epsilon+10\alpha\epsilon^2+5\epsilon^3}, = \frac{2}{5}\alpha^3$.

Hoc est, (dividendo utrinque per $\frac{\alpha^3-\alpha\epsilon^2}{5}$,)
$\frac{3\epsilon^3}{\alpha^3-\epsilon^3} +$
$\frac{2\alpha^3+4\alpha^2\epsilon+6\alpha\epsilon^2+3\epsilon^3}{\alpha^3+2\alpha^2\epsilon+2\alpha\epsilon^2+\epsilon^3} = \frac{2\alpha^3}{\alpha^3-\alpha\epsilon^2}$.

Hoc est, (reductis omnibus ad communem Denominatorem, eoque sublato,)

$$\left.\begin{array}{r}3\epsilon^3 \text{ in } \alpha^3 - \alpha\epsilon^2 (= 3\alpha^3\epsilon^3 - 3\alpha\epsilon^5) \text{ in } \alpha^3 + 2\alpha^2\epsilon + 2\alpha\epsilon^2 + \epsilon^3 \dots \\ \text{pl}: 2\alpha^3 + 4\alpha^2\epsilon + 6\alpha\epsilon^2 + 3\epsilon^3, \text{ in } (\alpha^3 - \epsilon^3 \text{ in } \alpha^3 - \alpha\epsilon^2 =) \alpha^6 - \alpha^4\epsilon^2 - \alpha^3\epsilon^3 + \alpha\epsilon^5\end{array}\right\} = a$$

[1256] Ἰσορροπικῶν: i.e. ARCHIMEDES, *De aequiponderantibus* II, prop. 9.

245. Wallis to Scarborough, 21 Nov./[1 Dec.] 1671, enclosure

$$
\begin{array}{llllllllll}
& & & & +3\alpha^6\epsilon^3 & +6\alpha^5\epsilon^4 & +6\alpha^4\epsilon^5 & +3\alpha^3\epsilon^6 & & \\
2\alpha^9 & +4\alpha^8\epsilon & +6\alpha^7\epsilon^2 & +3\alpha^6\epsilon^3 & & -3\alpha^4\epsilon^5 & -6\alpha^3\epsilon^6 & -6\alpha^2\epsilon^7 & -3\alpha\epsilon^8 \\
& & -2\alpha^7\epsilon^2 & -4\alpha^6\epsilon^3 & -6\alpha^5\epsilon^4 & -3\alpha^4\epsilon^5 & & & \\
& & & -2\alpha^6\epsilon^3 & -4\alpha^5\epsilon^4 & -6\alpha^4\epsilon^5 & -3\alpha^3\epsilon^6 & & \\
& & & & & +2\alpha^4\epsilon^5 & +4\alpha^3\epsilon^6 & +6\alpha^2\epsilon^7 & +3\alpha\epsilon^8 \\
\hline
a = \; 2\alpha^9 & +4\alpha^8\epsilon & +4\alpha^7\epsilon^2 \ldots & & \ldots & -4\alpha^5\epsilon^4 & -4\alpha^4\epsilon^5 & -2\alpha^3\epsilon^6 & &
\end{array}
$$

$$
2\alpha^3 \text{ in } \alpha^3 - \epsilon^3 \; (= 2\alpha^6 - 2\alpha^3\epsilon^3) \text{ in } \alpha^3 + 2\alpha^2\epsilon + 2\alpha\epsilon^2 + \epsilon^3 = b
$$

$$
\begin{array}{llllllll}
2\alpha^9 & +4\alpha^8\epsilon & +4\alpha^7\epsilon^2 & +2\alpha^6\epsilon^3 & & & & \\
& & & -2\alpha^6\epsilon^3 & -4\alpha^5\epsilon^4 & -4\alpha^4\epsilon^5 & -2\alpha^3\epsilon^6 & \\
\hline
b = \; 2\alpha^9 & +4\alpha^8\epsilon & +4\alpha^7\epsilon^2 & \ldots & -4\alpha^5\epsilon^4 & -4\alpha^4\epsilon^5 & -2\alpha^3\epsilon^6 & = a.
\end{array}
$$

Quod erat ostendendum.

|Propositio X. Archimedis περί Ἑλικῶν.[1257]

Sint, quotlibet continue aequaliter decrescentium, $\alpha, \beta, \gamma, \delta, \epsilon, \zeta, \eta, \theta$, Maxima α, Minima θ, idemque Excessus communis, Numerum omnium π. Erunt, $\pi\alpha^2, +\alpha^2, +\theta$ in $\alpha + \beta + \gamma + \delta + \epsilon + \zeta + \eta + \theta, = 3$ in $\alpha^2 + \beta^2 + \gamma^2 + \delta^2 + \epsilon^2 + \zeta^2 + \eta^2 + \theta^2$.

Demonstratio.

Nam, $\pi\theta = \alpha$. Adeoque $\pi\alpha\theta = \alpha^2$.
Item, $\alpha = \beta + \theta = \gamma + 2\theta = \delta + 3\theta = \epsilon + 4\theta = \zeta + 5\theta = \eta + 6\theta = \theta + 7\theta$.
Hoc est, $\alpha = \beta + \theta = \gamma + \eta = \delta + \zeta = \epsilon + \epsilon = \zeta + \delta = \eta + \gamma = \theta + \beta$.
Ergo, $\pi\alpha = \alpha + 2\beta + 2\gamma + 2\delta + 2\epsilon + 2\zeta + 2\eta + 2\theta$.

Item,
$$
\left\{
\begin{array}{llllllll}
\alpha^2 = & \beta^2 = & \gamma^2 = & \delta^2 = & \epsilon^2 = & \zeta^2 = & \eta^2 = & \theta^2 \\
+\theta^2 & +\eta^2 & +\zeta^2 & +\epsilon^2 & +\delta^2 & +\gamma^2 & +\beta^2 & \\
+2\theta\beta & +2\eta\gamma & +2\zeta\delta & +2\epsilon\epsilon & +2\delta\zeta & +2\gamma\eta & +2\beta\theta &
\end{array}
\right\}
$$

[1257] περί Ἑλικῶν: i.e. Archimedes, *De lineis spiralibus*, prop. 10.

245. Wallis to Scarborough, 21 Nov./[1 Dec.] 1671, enclosure

Ergo,

$$\begin{cases} +\alpha^2 & = & \alpha^2 & \cdots & \cdots & \cdots & \cdots & \cdots & \cdots & \cdots & \cdots \\ & & \alpha^2 & +\beta^2 & +\gamma^2 & +\delta^2 & +\epsilon^2 & +\zeta^2 & +\eta^2 & +\theta^2 \\ +\pi\alpha^2 = & \begin{cases} & +\theta^2 & +\eta^2 & +\zeta^2 & +\epsilon^2 & +\delta^2 & +\gamma^2 & +\beta^2 \end{cases} \Big\} = f \\ & & & +2\theta\beta & +2\eta\gamma & +2\zeta\delta & +2\epsilon\epsilon & +2\delta\zeta & +2\gamma\eta & +2\beta\theta & = a \\ & a & = & & 2\theta\beta & +4\theta\gamma & +6\theta\delta & +8\theta\epsilon & +10\theta\zeta & +12\theta\eta & +14\theta\theta & = b \\ & b & = \theta \text{ in } 2\beta & +4\gamma & +6\delta & +8\epsilon & +10\zeta & +12\eta & +14\theta \\ & +\theta \text{ in } \alpha & +\beta & +\gamma & +\delta & +\epsilon & +\zeta & +\eta & +\theta \end{Big\} = c \\ & c = \theta \text{ in, } \alpha & +3\beta & +5\gamma & +7\delta & +9\epsilon & +11\zeta & +13\eta & +15\theta & = d \end{cases}$$

$$d = \begin{cases} \theta \text{ in,} & \alpha +2\beta +2\gamma +2\delta +2\epsilon +2\zeta +2\eta +2\theta(=\pi\alpha\theta) = \alpha^2 \\ +\theta \text{ in } \ldots \beta & +2\gamma +2\delta +2\epsilon +2\zeta +2\eta \quad +2\theta = \beta^2 \\ +\theta \text{ in } \ldots \ldots & \gamma +2\delta +2\epsilon +2\zeta +2\eta \quad +2\theta = \gamma^2 \\ +\theta \text{ in } \ldots \ldots & \ldots \delta +2\epsilon +2\zeta +2\eta \quad +2\theta = \delta^2 \\ +\theta \text{ in } \ldots \ldots & \ldots \ldots \epsilon +2\zeta +2\eta \quad +2\theta = \epsilon^2 \\ +\theta \text{ in } \ldots \ldots & \ldots \ldots \ldots \zeta +2\eta \quad +2\theta = \zeta^2 \\ +\theta \text{ in } \ldots \ldots & \ldots \ldots \ldots \ldots \eta \quad\quad +2\theta = \eta^2 \\ +\theta \text{ in } \ldots \ldots & \ldots \ldots \ldots \ldots \ldots \quad\quad \theta = \theta^2 \end{cases} = e$$

$$\begin{aligned} e & = & \alpha^2 & +\beta^2 & +\gamma^2 & +\delta^2 & +\epsilon^2 & +\zeta^2 & +\eta^2 & +\theta^2 \\ f & = & 2\alpha^2 & +2\beta^2 & +2\gamma^2 & +2\delta^2 & +2\epsilon^2 & +2\zeta^2 & +2\eta^2 & +2\theta^2 \end{aligned} \Bigg\} = g$$

$$g = \quad\quad 3\alpha^2 +3\beta^2 +3\gamma^2 +3\delta^2 +3\epsilon^2 +3\zeta^2 +3\eta^2 +3\theta^2 = h$$
$$h = \quad 3 \text{ in } \alpha^2 +\beta^2 +\gamma^2 +\delta^2 +\epsilon^2 +\zeta^2 +\eta^2 +\theta^2 = i$$
$$i = \pi\alpha^2, +\alpha^2, +\theta \text{ in } \alpha \quad +\beta \quad +\gamma \quad +\delta \quad +\epsilon \quad +\zeta \quad +\eta \quad +\theta.$$

Quod erat demonstrandum.

Corollarium

Ergo,

$$\pi\alpha^2 \begin{cases} < 3 \text{ in} & \alpha^2 +\beta^2 +\gamma^2 +\delta^2 +\epsilon^2 +\zeta^2 +\eta^2 +\theta^2 \\ > 3 \text{ in} & \ldots \quad \beta^2 +\gamma^2 +\delta^2 +\epsilon^2 +\zeta^2 +\eta^2 +\theta^2, \end{cases} \text{propter } k$$

$k\ [=]\ \theta$ in $\alpha+\beta+\gamma+\delta+\epsilon+\zeta+\eta+\theta < (\theta$ in $\alpha+2\beta+2\gamma+2\delta+2\epsilon+2\zeta+2\eta+2\theta = \alpha^2 <)\ 2\alpha^2$.

245. WALLIS to SCARBOROUGH, 21 Nov./[1 Dec.] 1671, enclosure

|Propositio XI. ejusdem.[1258]

Sint quotlibet continue aequaliter decrescentium $\alpha + v$, $\beta + v$, $\gamma + v$, $\delta + v$, $\epsilon + v$, $\zeta + v$, $\eta + v$, $\theta + v$, v, Maxima $\alpha + v$, Minima v, Communis excessus θ, numerus omnium $\pi + 1$.

Erit $\pi Q : \alpha + v$.

$$\left\{\begin{array}{l} Q:\alpha+v, +Q:\beta+v, +Q:\gamma+v, +Q:\delta+v, \ +Q:\epsilon+v \ +Q:\zeta+v, +Q:\eta+v, +Q:\theta+v,\ldots\ldots < \\ \ldots\ldots\ldots\ Q:\beta+v, +Q:\gamma+v, +Q:\delta+v, +Q:\epsilon+v, +Q:\zeta+v, +Q:\eta+v, +Q:\theta+v,\ +Qv,\ > \end{array}\right\} ::$$

$$Q : \alpha + v. \ v \text{ in } \alpha + v, +\frac{1}{3}\alpha^2, :: \alpha$$

Demonstratio

$a :: \pi Q : \alpha + v.\ \pi v \text{ in } \alpha + v, +\frac{1}{3}\pi\alpha^2 = \pi v^2 + \pi\alpha v, +\frac{1}{3}\pi\alpha^2 = b$

$b = \left\{\begin{array}{l} < Q:\alpha+v, +Q:\beta+v, +Q:\gamma+v, +Q:\delta+v, +Q:\epsilon+v, +Q:\zeta+v, +Q:\eta+v, +Q:\theta+v,\ldots\ldots = c \\ > \ldots\ldots\ldots\ Q:\beta+v, +Q:\gamma+v, +Q:\delta+v, +Q:\epsilon+v, +Q:\zeta+v, +Q:\eta+v, +Q:\theta+v, +Qv, = d \end{array}\right\}$

$c = \left\{\begin{array}{llllllll} +v^2 & +v^2 & +v^2 & +v^2 & +v^2 & +v^2 & +v^2 & +v^2 = \pi v^2. \\ +\alpha^2 & +\beta^2 & +\gamma^2 & +\delta^2 & +\epsilon^2 & +\zeta^2 & +\eta^2 & +\theta^2 > \frac{1}{3}\pi\alpha^2. \quad \text{per Corol: prop. praeced.} \\ +2\alpha v & +2\beta v & +2\gamma v & +2\delta v & +2\epsilon v & +2\zeta v & +2\eta v & +2\theta v > \pi\alpha v. \quad \text{propter e.} \end{array}\right.$

$d = \left\{\begin{array}{llllllll} \ldots & +v^2 & +v^2 & +v^2 & +v^2 & +v^2 & +v^2 & +v^2 = \pi v^2. \\ \ldots & +\beta^2 & +\gamma^2 & +\delta^2 & +\epsilon^2 & +\zeta^2 & +\eta^2 & +\theta^2 < \frac{1}{3}\pi\alpha^2. \quad \text{per Cor. prop. praec.} \\ \ldots & +2\beta v & +2\gamma v & +2\delta v & +2\epsilon v & +2\zeta v & +2\eta v & +2\theta v > \pi\alpha v. \quad \text{propter e.} \end{array}\right.$

e. $\alpha + 2\beta + 2\gamma + 2\delta + 2\epsilon + 2\zeta + 2\eta + 2\theta = \pi\alpha$, ut ad praeced. ostensum est.

Corollarium

Si intelligantur α, β, γ, &c, in propositione praecedente, et $\alpha+v$, $\beta+v$, $\gamma+v$, &c, in hac propositione, pro totidem Rectis: Quae de harum Quadratis demonstrantur, pariter obtinent in Figuris Similibus quibusvis super eas rectas constitutis; Propter Figuras Similes, Quadratis Homologorum Laterum, proportionales.

[1258] ejusdem: i.e. ARCHIMEDES, *De lineis spiralibus*, prop. 11.

246.
JEAN BERTET to WALLIS
Pontoise, near Paris, [21 November]/1 December 1671

Transmission:

c^1 Copy of missing letter sent: LONDON *Royal Society* Early letters B2, No. 14, 7pp. At top of p. 1 in Oldenburg's hand: 'A Copy of P. Bertets letter to Dr. Wallis' and 'Enter'd LB. 5. 54.' In margin, also in Oldenburg's hand: 'A Copy of a letter of |the *del.*| P. Bertet, a French Jesuit, to Dr Wallis, concerning divers nice Geometrical matters, and the present endeavors and labors abroad in mathematicks.'
c^2 Copy of missing letter sent (in Wallis's hand): OXFORD *Bodleian Library* MS Add. D. 105, f. 159^r–161^r (our source). On f. 159^r in Wallis's hand instruction for printer of the first edition of the *Algebra*: 'Pray that this and the other Latine Epistle which follows, be carefully corrected. The last Latine piece was left very full of faults.' On f. 161^v beginning of draft of Wallis's reply, i.e. WALLIS–BERTET 19/[29].XII.1671.
c^3 Copy of missing letter sent: LONDON *Royal Society* Letter Book Original 5, pp. 54–61.
c^4 Copy of c^3: LONDON *Royal Society* Letter Book Copy 5, pp. 60–7.
E^1 First edition of c^2: WALLIS, *Algebra*, 355–8.
E^2 Second edition of c^2: WALLIS, *Opera mathematica* II, 410–13.

Answered by: WALLIS–BERTET 19/[29].XII.1671.

Viro Clarissimo, Eruditissimo Matheseos
Professori, D. J. Wallis. Oxonium.
Pontisarae ad Parisios, 1. Dec. 1671.

Vir Clarissime,

Scripseram[1259] ante sex menses Epistolam ad D. V. in qua Meditationes quasdam Geometricas Tibi Geometrarum hujus aetatis facile principi, proponebam; ut aliquid lucis exquirerem circa ea quae per me ipse assequi non potueram. Factum est, nescio quo casu, ut haec Epistola intercideret.

    Nunc, quoniam ruri vitam hic dego quamquam brevi rediturus Parisios cum Eminentissimo Cardinale Bullionio,[1260] nactus tantisper otii, scripsi opusculum[1261] contra hypothesin Cartesii, quod brevi prodibit in lucem.

[1259] Scripseram: i.e. BERTET–WALLIS V./VI.1671. This letter evidently miscarried.
[1260] Bullionio: i.e. Emmanuel Théodose Bouillon (1643–1715), canon of Liège from 1658, created cardinal, 1669; made Grand Almoner of France, 1671.
[1261] opusculum: not identified. Dechales's *Cursus seu mundus mathematicus*, 3 vols, Leiden 1674, contains a section with this title. Cf. BERTET–COLLINS [16]/26.I.1671; RIGAUD, *Correspondence of Scientific Men* I, 157–8.

246. Bertet to Wallis, [21 November]/1 December 1671

Quamquam autem careo et libris et scriptis, occurrit cogitatum Geometricum, quod mihi inciderat priusquam divinum tuum opus Arithmet. Infinitorum percurrissem. Hoc vero est hujusmodi.

Si dentur Figurae quaelibet super eadem Basi et ejusdem Altitudinis, intelligaturque Axis divisus in quapiam ratione Geometrica continua, cujus terminatio sit Vertex figurae: Dico, Figuram unam toties aliam continere, quoties ratio partium unius figurae inter se comparatarum continet rationem partium alterius figurae inter se. Verbi gratia, Spatium Asymptoticum, Rectangulum, Parabola, Triangulum, et rectilineum quodlibet cuilibet comparatum, sunt unum totuplex alterius, quoties ratio A ad B unius est Multiplicata rationis Partis A ad partem B alterius; sed reciproce sumpta.| [15 Verbi gratia. Ratio partium Trianguli, est duplicata rationis partium Rectanguli: ergo, reciproce, Rectangulum duplum est Trianguli. Ita potest Parabolae cujuslibet et Trilinei quantitas exquiri. Quod a Te in illo mirabili opere facillime praestitum est.

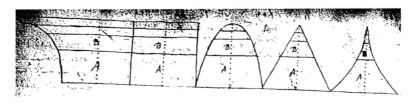

In hoc tamen haec mea methodus non omnino videtur contemnenda; Quod ipse figurarum quantitatem dimetiaris per rationem quam habent singulae ad Rectangulum; ego vero, independenter a Rectangulo, statim earum rationem inter se inveniam. Verbi gratia. Spatii Asymptotici, non quidem verae Hyperbolae, sed pseudo-hyperbolae, in qua Ordinatae in Asymptoto sunt reciproce in ratione subduplicata partium a Centro Hyperbolae sumptarum in Asymptoto seu in Axe AB. Et sic de caeteris figuris. (Nota, Rectangula inscripta aut circumscripta in figura, se habere ut seriem partium ipsius figurae.)

Alterum est; Quod quamcunque tandem surdam rationem habeant Ordinatae in qualibet Figura quae applicantur partibus Axis divisi ut supra in aliqua ratione continua, non minus reperire possum quantitatem figurae, quam si Ordinatae sint in ratione aliqua quae exprimi possit. Ipse vero duntaxat methodum innueris dimetiendi figuras quarum Ordinatae ab 0 incipientes sint in serie Arithmetica 1, 2, 3, 4. vel ut eorum radices aut potestates.

9 Triangulum, & Trilineum, quodlibet E^2

246. BERTET to WALLIS, [21 November]/1 December 1671

Haec cum nuper animo versarem, succurrit mihi subobscura cogitatio; Qua via reperiri posset, non solum figuras Directas metiendi, sed forte etiam eas quas Gregorius a S. Vincentio vocat[1262] Subcontrarias. (In quo, si bene memini, relicta est Circuli Quadratura, a Gregorio a S. Vinc. et a Te pariter.) Ut, quemadmodum Trilinei, Trianguli, et Parabolae; et Conoeidis Trilinearis, Triangularis (seu Coni,) et Parabolici ratio nullo negotio reperitur; ita etiam, si reperiri posset ratio Solidorum quae fiunt ex ductu Trilinei, Trianguli et Parabolae subcontrarie positorum, absoluta esset Circuli Quadratura.| Ratio autem ordinatarum in Trilineo, nempe EF, CD, GH, toties est multiplicata rationis ordinatarum in Triangulo, FI, KL, MN, quoties haec ultima est multiplicata rationis ordinatarum in Parabola IO, PQ, RS: Ratio vero partium vel Rectangulorum AB in Trilineo, ad rationem rectangulorum AB in Triangulo; et rursus ratio partium AB in Triangulo, ad rationem partium AB in Parabola; non sunt aeque multiplicatae una alterius, quamvis denominatores rationis inter lineas ordinatas servent eandem proportionem quam habent denominatores rationis inter rectangula vel partes AB istarum figurarum

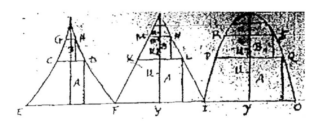

Observavi praeterea; quoties in figuris directis tribus, ratio ordinatarum unius toties est multiplicata rationis ordinatarum alterius, quoties ratio ordinatarum secundae est multiplicata rationis ordinatarum tertiae; tunc figuras esse arithmetice proportionales; ut Trilineum, Triangulus, et Parabola: At comparatae inter se rationes partium sive Rectangulorum AB, sunt Harmonice multiplicatae alicujus rationis simplicis.

Ut redeam ad propositum. Demonstravit[1263] Gregorius a S. Vincentio Rectangulum ACB ad rectangulam DEF habere rationem aeque multiplicatam rationis rectanguli GHI ad rectangulum KLM, ut haec est alterius rationis multiplicata inter rectangula NOP, QRS. Quod si lineae CX HY

[1262] vocat: cf. SAINT–VINCENT, *Opus geometricum*, 1217–19. It does not appear that Saint-Vincent expressly used this term, the underlying concept of which can be traced to APOLLONIUS, *Conics* I, 5.

[1263] demonstravit: see SAINT–VINCENT, *Opus geometricum*, 37–50.

246. BERTET to WALLIS, [21 November]/1 December 1671

OZ aequales, divisae sint in aliqua ratione continua cujus terminatio sit XYZ; intelliganturque fieri solida per ductum rectangulorum ACB in altitudinem CE, et Rectanguli DEF per altitudinem ET, et sic de caeteris; non dico haec Solida eandem rationem habitura quam habebant Rectangula; sed, ut supra observavi, denominatores harum rationum Solidorum erunt proportionales denominatoribus rationum quas habebant Rectangula: At rationes Rectangulorum erant aeque multiplicatae una alterius; quod non amplius convenit Solidis.

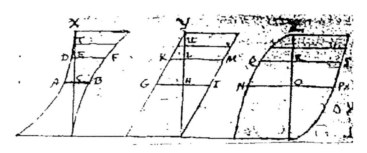

Quemadmodum autem, si subdivideretur supra, in figuris directis, Axis YZ proportionaliter in punctis UUU, ratio applicatarum| esset semper una aeque multiplicata rationis applicatarum alterius figurae directae; ita pariter si subdivideretur in aliqua ratione continua Axis CX vel HY, in subcontrariis esset ratio rectangulorum aeque multiplicata invicem in tribus figuris in infinitum.

Sed duo videntur obstare mensurae Solidi totius geniti ex ductu subcontrario. Primo, quod rationes linearum et Solidorum in figuris directis sit semper una eademque ratio: At, in figuris subcontrariis, semper sit mutata. Unde, quamvis in hoc conveniant, ut rationes baseon partium unius, rationibus baseon partium alterius comparatae, sint singulae aeque multiplicatae singularum; non tamen sunt omnes similes rationes continuae. Praeterea, in figuris directis, ita se habent portiones figurae interceptae inter ordinatas, ut rectangula inscripta aut circumscripta inter ordinatas intercepta: At idem non demonstratur de solidis interceptis inter ordinatas in figuris subcontrariis, quae non habent eandem rationem ac parallelepipeda quae inscribi aut circumscribi intelligerentur.

Tamen quia (ut alicubi D. V. observat[1264]) aliquando infinita series tollit irrationalitatem singularum partium; dubitavi an vere tres illae figurae subcontrariae essent inter se Arithmetice proportionales, ut sunt directae, modo reducantur ad eandem basin.

[1264] observat: i.e. WALLIS, *Arithmetica infinitorum*, 132; *Opera mathematica* I, 439.

246. BERTET to WALLIS, [21 November]/1 December 1671

Igitur, quandoquidem in postrema figura, nota est quantitas Solidi Y et solidi X, quorum assumo solam medietatem, sed bases seu rectangula ACB GHI, non sunt aequalia rectangulo NOP; intelligantur ergo poni aequalia, et continuetur eadem prorsus ratio infinita quae antea erat, in rectangulis ACB DEF etc. in trilineo; et pariter posita basi seu rectangulo GHI trianguli aequali rectangulo seu basi NOP, intelligatur fieri applicatio rectangulorum in punctis L et U solidi Triangularis subcontrarii, servata eadem ratione quae erat inter rectangula priora ejusdem figurae triangularis. Certe solida ista duo nova, habebunt semper in quibuscunque suis rectangulis ad axem applicatis, rationes aeque multiplicatas ut erant antea inter istas tres figuras.

Jam vero inveniatur quantitas novorum istorum solidorum; Si ut est rectangulum ACB ad rectangulum NOP, ita fiat solidum AXB trilineare ad novum solidum trilineare; rursusque fiat ut Rectangulum GHI ad rectangulum NOP, ita solidum GYI triangulare ad novum solidum triangulare fictum; denique inventa quantitate duorum novorum solidorum trilinearis et triangularis fictorum: Dico, tertiam quantitatem Arithmetice Proportionalem fore Solidum Parabolicum subcontrarium quaesitum.|

Dico, inquam, optando et divinando, non demonstrando. Tuum, erit, Vir illustrissime, detegere paralogismum sine dubio latitantem.

Expectamus avidissime opera tua in Hollandia recens typis edita.[1265] Nihil hic novi extudit Mathesis.

Lugduni, Pater Claudius Franciscus de Chales[1266] edit cursum integrum[1267] Mathematicum; non, ut est Cursus Scotti[1268] interpolatoris; sed cursum absolutum, Clarum, ac ubique novis demonstrationibus illustratum. Delineavit[1269] idem Author Nauticas Mediterranei Chartas reformatas, quae Anglis Vestris necessariae essent: Sed nondum invenit artificem qui laminis aeneis velit incidere.

[1265]edita: see COLLINS–VERNON 4/[14].IV.1671; RIGAUD, *Correspondence of Scientific Men* I, 160–5, 161: 'Dr. Wallis his former works are to be reprinted at Leyden, some of our booksellers joining with the Dutch in the impression'. This plan was evidently never realized.

[1266]de Chales: i.e. Claude François Milliet Dechales, S.J. (1621–78), French mathematician and missionary.

[1267]cursum integrum: i.e. DECHALES, *Cursus seu mundus mathematicus*, 3 vols., Leiden 1674.

[1268]Scotti: i.e. Caspar Schott S.J. (1608–66), teacher of mathematics in Palermo, Rome, and later Würzburg; author of numerous mathematical-technical works, including *Cursus mathematicus*, Würzburg 1661.

[1269]delineavit: possibly DESCHALES: *l'Art de naviger demontré par principes & confirmé par plusieurs observations tirées de l'experience*, Paris 1677.

246. BERTET to WALLIS, [21 November]/1 December 1671

Aedificatur[1270] hic Parisiis Observatoria Arx insignis Regiis Sumptibus; jam in eo domicilio habere incipit[1271] D. Cassini.

Missus est[1272] Uraniburgum ut loci situm observaret, et Eclipsin nuperam aspiceret in eo loco ubi Ticho Brahe[1273] suas omnes Observationes peregit. Nondum rescivi quid novi attulerit.

Non hic habeo ad manum observationes ultimae illius Eclipseos, quae variis in locis peractae sunt; quas mittam quamprimum ad D. Collins tibi communicandas. Nos hic Parisiis ob nebulosum Coelum nihil observare potuimus, praeter Eclipseos[1274] finem. Tabulae Wingii[1275] et Prutenicae[1276] satis quadrarunt cum Observationibus. Sed, quod mirum est, Rudolfinae,[1277] alias exactissimae, immane quantum aberrarunt.

Advocavi huc Lugduno Adolescentem Analyseos peritissimum D. Hozanam,[1278] discipulum P. Jacobi de Billy.[1279] Proponit ille nostris Geometris

---

2 in eo domicilio habitare incipit E^2

[1270] Aedificatur: members of the newly founded Académie royale des sciences decided on the need for an observatory at their first meeting in December 1666. Work on the Observatoire royale, designed by Claude Perrault (1613–88), was started in 1667 and completed in 1671.

[1271] incipit: the Italian astronomer Giovanni Domenico Cassini (1625–1712) moved to Paris in 1669 in order to help set up the Observatoire royale; he became its first director when it opened two years later.

[1272] missus est: the French astronomer Jean Picard (1620–82) travelled to the site of Tycho Brahe's former observatory Uraniborg in order to determine its exact position and to observe from there the September lunar eclipse. Cf. CASSINI–OLDENBURG [10]/20.VIII.1671; OLDENBURG, *Correspondence* VIII, 193–4, and HEVELIUS–OLDENBURG [27.IX]/7.X.1671; OLDENBURG, *Correspondence* VIII, 271–2.

[1273] Brahe: i.e. Tycho Brahe, born Tyge Ottesen Brahe, (1546–1601), Danish nobleman and astronomer. His observatory Uraniborg, on the island of Hven, was built around 1576–80 and destroyed in 1601.

[1274] Eclipseos: i.e. the lunar eclipse on 8 September 1671 (old style). Cf. OLDENBURG–HEVELIUS 9/[19].XI.1671; OLDENBURG, *Correspondence* VIII, 349–51.

[1275] Wingii: i.e. Vincent Wing (1619–68), astronomer and land surveyor, *ODNB*. Wing published various collections of astronomical tables, including those in *Harmonicon coeleste: or, the coelestiall harmony of the visible world*, London 1651.

[1276] Prutenicae: the German astronomer Erasmus Reinhold (1511–53) published various collections of astronomical tables, including *Prutenicae tabulae coelestium motuum*, Tübingen 1551.

[1277] Rudolfinae: with Tycho Brahe's backing, the German mathematician and astronomer Johannes Kepler (1571–1630) produced the Rudolphine Tables as a replacement for Reinhold's Prutenic Tables. See KEPLER, *Tabulae Rudolphinae, quibus astronomiae scientiae, temporum longinquitate collapsae restauratio continetur*, Ulm 1627.

[1278] Hozanam: i.e. Jacques Ozanam (1640–1718), French mathematician, teacher of mathematics in Lyon.

[1279] de Billy: i.e. Jacques de Billy, S.J. (1602–79), French mathematician and astronomer. Contrary to Bertet's assertion, there is no evidence that Billy had been Ozanam's teacher.

varias Quaestiones, sed hactenus insolutas, quod ad earum solutionem nolint, ut aiunt, animum adjicere.

P. Pardies[1280] Catalogum[1281] Stellarum novum aereis laminis incidit: Sed novo projectionis genere in plano descriptarum, et ad usum facillima et accomodatissima methodo.

Nihil audio novi ex Italia praeter mirabilia Telescopia Patris Gotignes:[1282] Cujus Opticam Manuscr. amicus quidam meus qui profectus est in Angliam ostendet Domino Collins; cui reddidit librum Geometricum Patris Pardies, quamquam eum domi nactus non fuerit.

Audiam libenter siquid novi prodierit apud Vestros; quorum inventa tanti facio, ut Anglicae Linguae addiscendae operam navarim ut possem opera vestra perlegere, quae nunc fere sine adjutore intelligo, et Gallice reddo.

Proponam interdum dubia mea D. V. Ac siquid hac in parte quae ad Mathesim spectat possim D. V. inservire, habebit nutibus suis

Addictissimum Servum
Johannem Bertet.

247.
WALLIS to HENRY OLDENBURG
Oxford, 23 November/[3 December] 1671

Transmission:

W Letter sent: LONDON *Royal Society* Early Letters W1, No. 132, 2 pp. (our source). On p. 2 beneath address in Oldenburg's hand: 'Rec. Nov. 24. 71. Answ. Nov. 26. 71.' Postmark illegible.—printed: OLDENBURG, *Correspondence* VIII, 372–3.

Reply to: OLDENBURG–WALLIS 11/[21].XI.1671.
Answered by: OLDENBURG–WALLIS 26.XI/[6.XII].1671.

[1280]Pardies: i.e. Igance-Gaston Pardies, S.J. (1636–73), French mathematician and astronomer.
[1281]Catalogum: probably PARDIES, *Globi coelestis in tabulas planas redacti descriptio*, Paris 1674.
[1282]Gotignes: i.e. Gilles-François Gottignies, S.J. (1630–89), Belgian mathematician and astronomer. A manuscript copy of his tract on the telescope was on sale in Rome in 1670. There was much talk of his treatise of dioptrics in contemporary correspondence. See NEWTON–? 23.II/[6.III].1668/9; NEWTON, *Correspondence* I, 3–5; COLLINS–GREGORY 25.III/[4.IV].1671; TURNBULL, *James Gregory*, 178–81, and COLLINS–VERNON 4/[14].IV.1671; RIGAUD, *Correspondence of Scientific Men* I, 160–5.

247. WALLIS to OLDENBURG, 23 November/[3 December] 1671

Oxford. Nov. 23. 1671.

Sir,

To yours[1283] of Nov. 11. I had sooner answered but that I was willing to have answered more fully than yet I can. By *Pocockii folium de Caave et She*, (that is, of Coffee & The) I suppose they mean a paper of half a sheet,[1284] printed here in Arabick and Latin, out of an Arab Physician concerning their *Kawha* or (as wee call it) *Coffee*, (but nothing of *The* or *Tea*,) I have indeavoured to get one of these papers, both from the Dr[1285] and from the Printer, but I cannot yet light upon one: when I do, I will send it to you.

I have attempted to meet with Dr Morrison,[1286] once or twice, but have missed of him; & so can give you as yet no account of those Trees.

Concerning Abulfeda:[1287] where Mr Grave's[1288] copy[1289] is, I cannot tell, nor how much paines hee took about it. But Dr Pocock hath two excellent Copies of it. And Mr Clark[1290] had (before his death) taken a great deal of paines in comparing them,[1291] & other Geographers with them; in order to an Edition of it. But his death, as in other regards, (he being a person of very good worth & industry,) so particularly in reference to this happened unluckyly, before it was ordered & digested, so that I doubt that labour will be lost.

---

17 very *add.*

[1283] yours: i.e. OLDENBURG–WALLIS 11/[21].XI.1671.
[1284] sheet: i.e. POCOCKE (ed.), *The Nature of the drink Kauhi, or Coffe, and the Berry of which it is made*, Oxford 1659. This was largely a translation from the Arabick, presented on facing pages; Pococke did not write on tea.
[1285] the Dr: i.e. Edward Pococke, q.v.
[1286] Morrison: i.e. Robert Morison (1620–83), physician to Charles II, and professor of botany in the University of Oxford from 1669, *ODNB*.
[1287] Abulfeda: i.e. Abu'l Fedā (Abu Al-fidā) (1273–1331), Kurdish geographer and historian. His 'Geography' held considerable appeal among contemporary scholars as a means to perfecting that subject.
[1288] Mr Grave's: i.e. John Greaves (1602–52), astronomer and orientalist, Fellow of Merton College, 1624, and Savilian professor of astronomy in the University of Oxford, 1643–48, *ODNB*. See WALLIS, *Correspondence* I, passim.
[1289] copy: Greaves worked on Abulfeda's 'Geography' and printed some of the text in Arabic with Latin translation – GREAVES (ed.), *Chorasmiae et Mawaralnahrae hoc est regionum extra fluvium Oxum descriptio ex tabulis Abulfedae Ismaelis*, London 1650.
[1290] Clark: i.e. Samuel Clarke (1624–69), orientalist and architypographus to the University of Oxford, *ODNB*.
[1291] comparing them: among Clarke's papers at his death was a complete transcript of Abulfeda's 'Geography' collated with several manuscripts. See WOOD, *Athenae Oxonienses* III, 885.

248. OLDENBURG to WALLIS, 26 November/[6 December] 1671

Dr Pocock's copies (I think, both of them,) are not pointed for the pronunciation; but so accurate are they that (in a distinct columne) the names of those points are in words at length set down. But as to Mr Grave's coppy, I doubt there may be some mistake, (for if any such were, Mr Clark surely would have indeavoured to have gott a sight of it, but I do not remember that hee ever spake of other than Dr Pococks to Copies:) But if any such bee I suppose it is in his Brothers hands Mr Edward Graves,[1292] a Minister (I think) in Lecestershire (or Lincolnshire.) But I mean to inquire further of it.

Dr Pocock desires that when you have occasion or opportunity of sending anything to Mr Vernon[1293] at Paris, that you would send to him one of those Books[1294] you have of his, to Mr Vernon to present from the Author to Monsieur Thevenot.[1295] No more at present but that I am

Your very humble servant
John Wallis.

[2] For Mr Henry Oldenburg
in the Palmal near St
James's
London.

248.
HENRY OLDENBURG to WALLIS
26 November/[6 December] 1671

Transmission:

Manuscript missing.

Existence and date: mentioned in WALLIS–OLDENBURG 27.XI/[7.XII].1671 and in Oldenburg's endorsement on WALLIS–OLDENBURG 23.XI/[3.XII].1671.

[1292]Graves: i.e. Edward Greaves (1613–80), Linacre superior reader of physic at Oxford from 1643, Fellow of the College of Physicians, physician in ordinary to Charles II, *ODNB*. See WALLIS–OLDENBURG 27.XI/[7.XII].1671.

[1293]Vernon: Cf. OLDENBURG–VERNON 4/[14].XII.1671; OLDENBURG, *Correspondence* VIII, 392.

[1294]Books: i.e. copies of POCOCKE, *Philosophus autodidactus sive epistola Abi Jaafar, Ebn Tophail*, Oxford 1671. Cf. WALLIS–POCOCKE 23.IX/[3.X].1671.

[1295]Thevenot: i.e. Melchisédech Thévenot (c.1620–92), French orientalist, inventor, and diplomat. Thévenot published a small fragment of Abulfeda's 'Geography' in his *Relations de divers voyages curieux*, Paris 1663; his plans to publish a fuller edition were never realized. See DEW, *Orientalism in Louis XIV's France*, Oxford 2009, 107–30.

249. WALLIS to OLDENBURG, 27 November/[7 December] 1671

Reply to: WALLIS–OLDENBURG 23.XI/[3.XII].1671.
Answered by: WALLIS–OLDENBURG 27.XI/[7.XII].1671.

249.
WALLIS to HENRY OLDENBURG
Oxford, 27 November/[7 December] 1671

Transmission:

W Letter sent: LONDON *Royal Society* Early Letters W1, No. 133, 2 pp. (our source). On p. 2 beneath address in Oldenburg's hand: 'Rec. Nov. 28. Answ. Dec. 5. 71.' Postmark illegible.—printed: OLDENBURG, *Correspondence* VIII, 387–9.

Answered by: OLDENBURG–WALLIS 5/[15].XII.1671.

Oxford November 27. 1671.

Sir,

Since I sent away my last,[1296] I have received from Dr Pocock the inclosed Paper[1297] in Arabick & (not Latine, but) English, which therefore (because you desire it for a forraigner[1298] I have, in two blank pages, put into Latine.[1299] He tells mee that Mr John Grave's Abulfeda,[1300] (much inferiour to those copies of his,) Mr Clark had the use of, as allso of that at Cambridge which had been Erpenius's,[1301] written (he thinks) by Erpenius own hand. That of Mr John Grave,[1302] he supposeth is still in the hands of his brother

---

[1296] last: i.e. WALLIS–OLDENBURG 23.XI/[3.XII].1671.
[1297] paper: i.e. POCOCKE (ed.), *The Nature of the drink Kauhi, or Coffe, and the Berry of which it is made*, Oxford 1659.
[1298] forraigner: i.e. Martin Vogel (Fogelius) (1634–75), medical practitioner, and professor of logic and metaphysics at the Akademisches Gymnasium in Hamburg. See VOGEL–OLDENBURG 1/[11].XI.1671; OLDENBURG, *Correspondence* VIII, 330–2; OLDENBURG–VOGEL 14/[24].XI.1671; OLDENBURG, *Correspondence* VIII, 356–7.
[1299] put into Latine: Wallis's translation has not survived.
[1300] Abulfeda: cf. WALLIS–OLDENBURG 23.XI/[3.XII].1671.
[1301] Erpenius's: i.e. Thomas van Erpe (Erpenius) (1584–1624), professor of oriental languages at Leiden from 1613.
[1302] John Grave: i.e. John Greaves. Cf. WALLIS–OLDENBURG 23.XI/[3.XII].1671.

249. WALLIS to OLDENBURG, 27 November/[7 December] 1671

Dr Thomas[1303] (not Edward) Grave, to whom, if you have occasion to send, it may be done by his Brother Sir Edward Grave,[1304] a Physition, about Henrietta Street in Covent-Garden. As to what I wrot[1305] on occasion of your Letter[1306] concerning Leibnitius,[1307] you may use your discretion. That (which you conjecture, & not unlikely, may be his meaning,) the Matter, which, when in Motion, is Body, should become, when at rest, but Voyd Space; seems hard: and though he should so distinguish the Words, I doubt the Thing will be the same: & if, contrary to the old opinion in appearance, he affirm a Vacuity, & mean nothing by it but Quiescent Matter, he will not be thought to overthrow their sentiments, who never denyed Quiescent Matter: but did allways allow this, even when they most denied Vacuity. So that if you send him what I wrote last, you may adde this to it; *Sin forte velit, (quod insinuas,) Corpus, a Vacuo, eo tantum distingui, quod eadem Materia, quae, dum in Motu, Corpus erat, fiat, Quiete, Vacuum: Hoc durum dictu videtur: nec Vacuum negantibus adversatur. Quippe, qui hoc negant vel maxime, minime tamen negaverint Materiam Quiescentem: Sed, Spatium ea refertum, non Vacuum, sed Plenum vocant: Ut, de solo Nomine, non Re, sit controversum.*

I desired, in my last, that you would send, for Dr Pocock, one of his Bookes[1308] to Mr Vernon[1309] for Monsieur Thevenot.[1310] You may intimate allso to Mr Vernon, that since they are so greedy of those books (as appears)

---

10 to *(1)* contradict *(2)* overthrow
15 nec *(1)* iis adversatur *(2)* Vacuum
19 would *add*.

[1303] Thomas: i.e. Thomas Greaves (1611–76), oriental scholar, younger brother of the Savilian professor of astronomy, John Greaves, *ODNB*. Fellow of Corpus Christi College, Oxford 1636. Ejected from the University 1648; from 1664 rector of Benefield, Northamptonshire.
[1304] Edward Grave: i.e. Edward Greaves (1613–80), Linacre superior reader of physic at Oxford from 1643, Fellow of the College of Physicians, physician in ordinary to Charles II, *ODNB*.
[1305] wrot: i.e. WALLIS–OLDENBURG 6/[16].XI.1671.
[1306] your Letter: i.e. OLDENBURG–WALLIS 4/[14].XI.1671.
[1307] Leibnitius: i.e. Gottfried Wilhelm Leibniz (1646–1716), German polyhistor.
[1308] Bookes: i.e. copies of POCOCKE, *Philosophus autodidactus sive epistola Abi Jaafar, Ebn Tophail*, Oxford 1671. Cf. WALLIS–POCOCKE 23.IX/[3.X].1671.
[1309] Vernon: i.e. Francis Vernon, q.v. Oldenburg wrote to Vernon a week later, indicating that he would send him a copy of *Philosophus autodictatus*. See OLDENBURG–VERNON 4/[14].XII.1671; OLDENBURG, *Correspondence* VIII, 392.
[1310] Thevenot: i.e. Melchisédech Thévenot. See WALLIS–OLDENBURG 23.XI/[3.XII]. 1671.

at Paris; if hee think fit, some number of them (a dousen or twenty) may be sent him to put off to some Book-seller there at a price; but move it as from your self, or mee, not as from Dr Pocock, who is loth to take the confidence of giving him that trouble, unlesse he should intimate his willingnesse to undertake it. If the Bishop of Oxford[1311] have not yet sent for his book, I suppose it is but his forgetfullness. For it was his own desire when he was here, to have that delivered him at London rather then at Oxford: &, if you think fit, you may send it to his lodgings in White-hall, & receive the mony there. I had written thus far, when I received from Dr Pocock the inclosed[1312] to Mr Vernon, which he desires you to transmit. You may to him allso present the service of

Sir

Your humble servant
John Wallis.

For Mr. Henry Oldenburg [2]
at his house in the
Palmal near
St. James's
London.

250.
HENRY OLDENBURG to WALLIS
5/[15] December 1671

Transmission:

Manuscript missing.

Existence and date: mentioned in Oldenburg's endorsement on WALLIS–OLDENBURG 27.XI/[7.XII].1671.
Reply to: WALLIS–OLDENBURG 27.XI/[7.XII].1671.

---

5 yet *add.*

[1311] Bishop of Oxford: i.e. Nathaniel Crew, q.v. Cf. WALLIS–OLDENBURG 6/[16].XI.1671.
[1312] inclosed: probably Pococke's letter to Vernon, which Oldenburg sent with OLDENBURG–VERNON 11/[21].XII.1671; OLDENBURG, *Correspondence* VIII, 397.

251.
WALLIS to ROBERT BOYLE
Oxford, 13/[23] December 1671

Transmission:

W Letter unsent: OXFORD *Bodleian Library* MS Add. D. 105, f. 38ʳ–38ᵛ (our source). On f. 38ᵛ, memorandum by William Wallis: 'This letter of the great and learned Dr. Wallis was found amongst writings unopened December 10. 1785. by me his greatgreatgrandson, and opened the same day in the presence of my Brother Alex. Wallis, being 114 years within 3 days after it was written. W. Wallis. Whitechurch Oxfordshire Dec. 10ᵗʰ 1785. Witness Alex[ander] Wallis.' Some text lost through breaking of seal.—printed: BOYLE, *Correspondence* IV, 235–7.
E First edition: *Gentleman's Magazine* 61 (1791), 404.

Evidently Wallis decided against sending this letter only after it had been signed and sealed.

Oxford. Dec. 13. 1671.

Sir,

The China Almanach,[1313] which you shewed mee at London, and have since sent to Oxford, I have further considered of. I find one leaf at the beginning, to give account of some generals of the several months following. Which are 12 Lunar months; the greater of 30 days which are thus marked. ★ ; the lesser of 29 days, thus marked ⸸ The numeral figures, from 1 to 30, are there observable: & I suppose the same methode is pursued in those that follow. Amongst the days in the several Lunar months; that in a white character within the black, (which wee looked on as some festivalls,) are the Solar months, or half months; that is, the suns enterance into the beginning & middle of the 12 signes; and are at 15 days distance, sometimes 16 days, one from another: And in most of the Lunar months, two of them happen; In one ⟨—⟩ there happens but one of them (the next before it, being ⟨at the⟩ end of the former month: & the next after it, in the ⟨beginning⟩ of the following month;) And in the epitome in the first leaf, there is notice given in the column of each Lunar month, in what days of it those happen; & for that month wherein but one happens, the column is but of half the

3 London, (*1*) I since considered (*2*) and have
9 Lunar *add.*
13 And in (*1*) those most (*2*) most of the Lunar
16 And (*1*) column (*2*) in the

[1313]China Almanach: not identified.

251. WALLIS to BOYLE, 13/[23] December 1671

length: which makes mee think that the other things in that breviate are concerning those Solar months fitting to the Lunar. The Almanach, I take to bee for the year 1666. And comparing this with Mr John Grave's[1314] *Epochae Celebriores*,[1315] (printed in Persian and Latin, out of Uleg-beg, I find it very well sute with the *Cathayan* year there described; (where we have the names both of the Lunar months, & Solar months or half-months.) And, consulting Golius's[1316] Appendix[1317] to the *Atlas Sinicus*, I am confirmed in it; who thence concludes the *Cathaians* to be the same with the *Chineses*. Their manner of writing (as wee bind their books) is column-wise from the top of the page to the bottom, beginning at (what wee should call) the End of Book. But, if we hold the book side-long (with the binding from us) it is just as the Lawyers Briefs are wont to bee read; And if their leaves (which, as wee bind them, are usually doubled back,) were out at length, a book of theirs would be just as one of those Briefs, & so to be read; & the leaves written but upon one page, as theirs are wont to bee. A more particular account of their marks would be useless unlesse you had the book withall in your view: which since you have not I adde no more but that I am,

Sir,

Your Honors very humble servant,
John Wallis.

These
For the Honorable Robert
Boyle Esquire, at the Lady
Ranaleigh's house in the
Palmal near St James's,
London.

[38ᵛ]

---

2 are |but *del.*| concerning
5-6 (where we ... months.) *add.*

[1314] Grave's: i.e. John Greaves (1602–52), astronomer and orientalist, *ODNB*.
[1315] *Epochae Celebriores*: i.e. GREAVES, *Epochae celebriores, astronomis, historicis, chronologis, Chataiorum, Syro-Graecorum, Arabum, Persarum, Chorasmiorum, usitatae: ex traditione Ulug Beigi*, London 1651.
[1316] Golius's: i.e. Jacobus Golius (Jakob van Gool) (1596–1667), Dutch orientalist and mathematician.
[1317] Appendix: i.e. GOLIUS, *De regno Catayo additamentum*. This text was appended to MARTINI, *Novus atlas Sinensis*, [1655].

252.
COLLINS to FRANCIS VERNON
[London], 14/[24] Decenber 1671

Transmission:

C Draft of letter sent: CAMBRIDGE *Cambridge University Library* MS Add. 9597/13/5, f. 64^r–65^v (our source). At top of f. 64^r in unknown hand: 'To Mr Vernon'. On f. 65^v in Collins's hand: 'To Mr Vernon the 14 Decemb 1671'.—printed: RIGAUD, *Correspondence of Scientific Men* I, 176–9.

Reply to: VERNON–COLLINS 4/[14].XII.1671.

Fearing that this letter and his previous letter to Vernon of 23 November had miscarried, Collins wrote again to Vernon on 26 December 1671, giving a fuller account of Newton's new telescope. See HALL, *John Collins on Newton's Telescope*, 73–5.

Your communications[1318] I imparted[1319] to Mr. Bernard and when I had wrote hitherto, I received his answer[1320] which I shall give you, As to Harriots[1321] Papers he sayth nothing as to those of Rawlinson,[1322] he saith that Mr Carre[1323] one of the Proctors hath those that did not goe along with his Bookes, but in short there is nothing in them desirable being but   5
Excerpta from Euclid Herigon[1324] and Oughtreds[1325] Clavis He saith that

---

2 to *(1)* Dr Rawlinsons Papers *(2)* |Mr *del.*| Harriots
5 desirable being *add.*

[1318]communications: Vernon probably wrote this letter in the second half of November. Cf. VERNON–OLDENBURG 23.XI/[3.XII].1671, OLDENBURG, *Correspondence* VIII, 383–6.
[1319]imparted: Collins probably wrote this letter to Bernard at the beginning of December 1671.
[1320]answer: Bernard's reply to Collins probably arrived shortly before the present letter was written.
[1321]Harriots: i.e. Thomas Harriot (*c.*1560–1621), mathematician and natural philosopher. Harriot bequeathed his voluminous mathematical papers to Henry Percy (1564–1632), the ninth earl of Northumberland.
[1322]Rawlinson: i.e. Richard Rawlinson (d. 1668), mathematician and Fellow of the Queen's College, Oxford.
[1323]Carre: i.e. Alan Carr (d. 1676), Fellow of All Souls College, Oxford, 1667; his election to proctor in May 1671 caused protests on account of his youth. See WOOD, *Life and Times* II, 222.
[1324]Herigon: i.e. HÉRIGONE, *Les six premiers livres des Elements d'Euclide*, Paris 1639.
[1325]Oughtreds: i.e. OUGHTRED, *Arithmeticae in numeris et speciebus institutio: quae tum logisticae, tum analyticae, atque adeo totius mathematicae, quasi clavis est*, London 1631. A further four Latin editions of Oughtred's *Clavis* were published up to 1693, an English edition appeared in 1647.

252. COLLINS to VERNON, 14/[24] December 1671

Mr Gale[1326] of Cambridge writes him word that Mr Newton[1327] (Barrows[1328] Successor) hath abbreviated a 16 foot tube to the length of a Spann, which is a most happy Invention[1329] I further adde, that the eye glasse is placed towards the Object the object glasse from it, the eye lookes in through the middle of the side, and sees all by reflection, as 'tis said in the same perfection as, and certainly takes in much more than, when the glasses are placed in their long tube

Sir Samuell Morlands[1330] Trumpett[1331] is now publish, oh that you had one of them wherewith to display its owne fame, and the due praise of this Tellescope, however I shall endeavour to send you the Booke by the first oppor As to the book[1332] of Pere Poterius[1333] de Ponderibus ac Mensuris compared with the Standards (Mr Bernard[1334] writes thus) it is a necessary treatise, and if it were accurately done, I doubt not of the Printing of it here at Oxford, for I designe[1335] to print Mr Greaves foot[1336] and Denarius

---

4 it, (1) you look in (2) the eye
6 perfection (1) as of the (2) as, and (a) for a much greater (b) doubtless (c) certainly
9 due add.
11 oppor (1) Mr Barnard writes (2) As
12 (Mr Bernard ... thus) add.

[1326] Gale: i.e. Thomas Gale (1635/6–1702), Fellow of Trinity College, and from 1672 regius professor of Greek in the University of Cambridge, secretary of the Royal Society 1679–81, *ODNB*.

[1327] Newton: i.e. Isaac Newton (1642–1727), natural philosopher and mathematician, Fellow of Trinity College, and Lucasian professor of mathematics in the University of Cambridge, 1669–1702, *ODNB*.

[1328] Barrows: i.e. Isaac Barrow (1630–77), mathematician and theologian, Lucasian professor of mathematics in the University of Cambridge, 1663–9, *ODNB*.

[1329] Invention: i.e. Newton's reflecting telescope, which was examined by the Royal Society in late December 1671. Oldenburg reported on this telescope in OLDENBURG–WALLIS 30.XII.1671/[9.I.1672].

[1330] Morlands: i.e. Samuel Morland (1625–95), natural philosopher and diplomat, *ODNB*.

[1331] Trumpett: i.e. MORLAND, *Tuba Stentoro-Phonica, an instrument of excellent use, as well at sea, as at land*, London 1671.

[1332] book: not identified. Cf. COLLINS–GREGORY 23.II/[4.III].1671/2, TURNBULL, *James Gregory*, 218-20; COLLINS–GREGORY 14/[24].III.1671/2, TURNBULL, *James Gregory*, 224-5; WALLIS–COLLINS 27.III./[6.IV].1672.

[1333] Poterius: not identified.

[1334] Bernard: i.e. Edward Bernard, q.v. Bernard himself wrote a tract on ancient weights and measures, which was appended to POCOCKE, *A Commentary on the Prophecy of Hosea*, Oxford 1685. An amended and substantially enlarged version of this was later published as BERNARD, *De mensuris et ponderibus antiquis libri tres*, Oxford 1688.

[1335] designe: Collins's plan for a Latin edition was evidently not realized.

[1336] foot: i.e. GREAVES, *A Discourse of the Romane Foot, and Denarius* London 1647.

252. COLLINS to VERNON, 14/[24] December 1671

rendred into Latin with some Additions, and ioyne with you in desiring to know of Mr Vernon his opinion (which is sufficient) of the worth of the Booke

I adde that I mooved the Printing of Poterius to Horne[1337] a Stationer at the Royall Exchange who is the Principall person that prints Mercantile affaires, and the cheife of them that is concerned in the late Impression of Roberts[1338] his Map of Commerce,[1339] which treats of the Trade Weights Measures Coines and Exchanges of all the most Principall Places in the World

He said if the Treatise be good, and were here, he would be at the charge of translating and printing it in| English, but would not give above 25 Printed Coppies for the Manuscript, Querie how much it may make printed

As to Laloveras[1340] workes Mr Bernard writes thus Father Lalovera was an excellent Person, and if father Pardiez[1341] would but send some small treatise of the said Laloveras to you that buisinesse would be soone at an End: I strangely came to know that Lalovera had a Booke[1342] in print intituled Geometria veterum promota in 7 de Cycloide Libris, una cum Appendicibus, after much sending and long Expectation one was procured from France for Dr Wallis, who finding the Author and himselfe Wonderfully to agree, wrote a Letter[1343] to Lalovera and sent it to me to transmitt, but it being rumoured he was dead, I never sent it, having it still by me, the Dr hath a great esteeme for Lalovera, and if you please I shall send you a Coppy of that Letter, I have wrote and sent often to P. Bertet[1344] and others to procure that Booke of Laloveras (which one Mr Hoot[1345] assures me is very

4 Royall *add.*
6 Trade (*1*) Coines (*2*) Weights
9 Treatise (*1*) were here and (*2*) be

[1337] Horne: i.e. Robert Horne (*fl.* 1660–85), bookseller in London at the south entrance to the Royal Exchange, Cornhill.
[1338] Roberts: i.e. Lewes Roberts (1596–1641), London merchant and author, *ODNB*.
[1339] Map of Commerce: i.e. ROBERTS, *The Merchants Map of Commerce*, London 1671. The second edition of this work (and those subsequent to it) was printed for Horne; the first, which appeared in 1638, was printed by R. Oulton for Ralph Mabb.
[1340] Laloveras: i.e. Antoine de Laloubère (Lalovera), S. J. (1600–64), French mathematician, professor in the Jesuit college in Toulouse.
[1341] Pardiez: i.e. Ignace-Gaston Pardies.
[1342] Booke: i.e. LALOUBÈRE, *Veterum geometria promota in septem de cycloide libris*, Toulouse 1660.
[1343] Letter: i.e. WALLIS–LALOUBÈRE *c.*10/[20].IX.1668, WALLIS, *Correspondence* II, 603. This letter is now missing.
[1344] Bertet: i.e. Jean Bertet, q.v.
[1345] Hoot: i.e. Mr Hoot (Hoote) (*fl.* 1670–71), son of an eminent London merchant. While residing in France for health reasons, Hoot brought about the correspondence between

252. COLLINS to VERNON, 14/[24] December 1671

common in Paris) and his Appendices[1346] Polemicae contra Maignanum but could never prevaile

Now as to the printing of Laloveras remaines,[1347] I thinke it were best they should come out in small treatises one after another, and when the Royall Societie begin to have a place for the late Arundelian Library bestowed on the said Society by the bounty of the Lord Howard,[1348] with what else they have, the Project I hope will be so laid that I may buy 50 or more of any Booke they would encourage to be printed for ready money at a Shop rate, and take 50 Bookes more to barter for others that may be desire hence, and when these 100 Bookes are here, to sell as many of them as I can to private persons and Barter away the rest for other Bookes for the Library, and if a Losse ensue not to fall on my selfe, and if the Royall Academy have the like Agency we might be mutually help of all, I hope to say more hereafter concerning this

A Bookseller here will print any Booke if he but sure to sell 80 or 100 Bookes for ready money

I wrote[1349] a Discourse to P Bertet to shew the ill consequences that ensue the not encouraging good Bookes to be printed, giving him instances of many good treatises that are by this meanes either quite lost, or never come to| be publique, I received Billys Diophantus[1350] Redivivus that you sent by [65r

---

5 place (1) for their owne and the (2) for ... Library |and what they *add. and del.*| bestowed
6 said *add.*
7 Project (1) is so laid (2) I hope
9 to barter *add.*
9 others (1) they may desire (2) that may be desire *corr. ed.*
13 all (1). (2), but (3), I hope ... this.
20 to (1) Leight (2) be publique
20 received (1) the (2) Billys

Bertet and Collins. See COLLINS–VERNON late 1670; RIGAUD, *Correspondence of Scientific Men* I, 139–41.

[1346] Appendices: the appendixes to *Veterum geometria promota* contain criticisms of the Minim friar and natural philosopher Emmanuel Maignan (1601–76).

[1347] remaines: cf. COLLINS–GREGORY 14/[24].III.1671/2; TURNBULL, *James Gregory*, 224–5; WALLIS–COLLINS 25.I/[4.II].1671/2.

[1348] Howard: i.e. Henry Howard (1628–84), the sixth duke of Norfolk. At the persuasion of John Evelyn (1620–1706), Howard presented a valuable collection of books and manuscripts from the library at Arundel House to the Royal Society and was subsequently elected fellow on 28 November 1666. See BIRCH, *History of the Royal Society* II, 128.

[1349] wrote: this letter from Collins to Bertet has apparently not survived.

[1350] Diophantus: i.e. BILLY, *Diophantus redivivus*, Toulouse 1670.

252. COLLINS to VERNON, 14/[24] December 1671

Mr Nott[1351] and immediatly left it at Mr Kersies[1352] house for him to peruse, When Dr Wallis was last here, a Physician alleadged he did not doubt to cure him of his Ague and that he should have but one fitt more, which should be but a faint one and come 12 houres after its wonted time, the event prooved true, and the Dr hath been freed from that distemper diverse Months and is in good health he hath much enlarged his Commercium Epistolicum which handles such unlimited Problems and if you be minded to present the Dr or Mr Bernard with a Booke I beleive one of Billys said bookes would be very acceptable

I have the Catalogue[1353] of the last tearme to send you, I am glad to heare of three Bookes mentioned in it, those are James Calvert[1354] a learned Minister of Yorke his Colluctationes[1355] Theologicae cum tribus ingentibus Dubiis, de reditu decem tribuum, de Conversione Judaeorum et sacris mensuris Ezekielis, he affirmes the 10 tribes returned with the 2, and that there shall be no future Nationall Conversion of the Jewes, the Booke I would willingly present you with might it find admittance in France, Another is Sir Robert Cottons[1356] Posthuma[1357] reprinted about State affaires, the last Dr John Newtons[1358] Arith[1359] wherein all the possible Answers

2 peruse (1) Dr Wallis ha *breaks off* (2) when
2 Physician (1) assured him he is (2) alleadged
5 been (1) ever since (2) freed
6 which ... Problems *add.*
8 said bookes *add.*
10 you, (1) three Bookes mentioned (2) I am
15 Nationall (1) Church (2) Conversion

[1351] Nott: not identified.
[1352] Kersies: i.e. John Kersey, the elder (1616–77), surveyor and teacher of mathematics in London, *ODNB*.
[1353] Catalogue: i.e. the *Catalogue of Books*, published (since Easter 1670) by the London bookseller Robert Clavell (in or before 1633–1711), *ODNB*. Collins is referring to the catalogue for Michaelmas Term, in which the books he cites are listed.
[1354] Calvert: i.e. James Calvert (1631–98), moderate Presbyterian minister. Probably working as a merchant in York from 1664, Calvert was licensed to preach at his house there in July 1672, *ODNB*.
[1355] Colluctationes: i.e. CALVERT, *Naphtali: seu Colluctationes theologicae cum tribus ingentibus dubiis*, London 1672.
[1356] Cottons: i.e. Robert Bruce Cotton (1571–1631), antiquary and politician, *ODNB*.
[1357] Posthuma: i.e. COTTON, *Cottoni posthuma: divers choice pieces of that renowned antiquary Sir Robert Cotton*, London 1651.
[1358] Newtons: i.e. John Newton.
[1359] Arith: i.e. NEWTON, *The art of natural arithmetick: in whole numbers and fractions vulgar and decimal*, London 1671. This work was advertised in the term catalogue for

which he had from Mr Dary[1360]) in Alligation, are found, Tartalea[1361] gloried that in a certaine question he had found two Answers in whole Numbers, afterwards Bachet[1362] in his Problems[1363] Plaisans found about 7000 Answers to the same question in whole Numbers

This conveighs you Flamsteads[1364] Calculations[1365] of the Lunar Apulses to the fixed starres (which I thought unfitt to delay) and an Account of the late excellent Dutch Booke[1366] of Navall Architecture

Mr Hodges[1367]

253.
WALLIS to JEAN BERTET
Oxford, 19/[29] December 1671

Transmission:

W Part copy of (missing) letter sent: OXFORD *Bodleian Library* MS Add. D. 105, f. 161^v (on reverse of BERTET–WALLIS [21.XI]/1.XII.1671) (our source).

w^1 Copy of (missing) letter sent: LONDON *Royal Society* Early Letters W1, No. 134, 8 pp. At top of p. 1 in Oldenburg's hand: 'Entered LB. 5. 76.' and in left margin, again in Oldenburg's hand: 'Doctoris Wallisii Responsio ad literas Domini Johannis Bertet, &c vide supra'.

w^2 Copy of (missing) letter sent: OXFORD *Bodleian Library* MS Smith 3, pp. 77–9.

w^3 Copy of (missing) letter sent: LONDON *Royal Society* Letter Book Original 5, pp. 76–82.

1 (which ... Dary) *add.*
2 certaine *add.*

Michaelmas 1671. Collins makes similar references in a contemporary letter: COLLINS–GREGORY 24.XII.1670/[3.I.1671], TURNBULL, *James Gregory*, 153–9.

[1360] Dary: i.e. Michael Dary (1613–79), mathematical teacher in London.

[1361] Tartalea: i.e. Niccol Fontana Tartaglia (c.1499–1557), Italian mathematician and engineer.

[1362] Bachet: i.e. Claude-Gaspard Bachet de Méziriac (1580–1639), French mathematician, pupil of Jacques de Billy.

[1363] Problems: i.e. BACHET, *Problemes plaisans et delectables, qui se font par les nombres*, Lyon 1612.

[1364] Flamsteads: i.e. John Flamsteed (1646–1719), astronomer; first Astronomer Royal *ODNB*.

[1365] Calculations: presumably FLAMSTEED, 'Lunae ad fixas appulsus visibiles, nec non arctiores juxta eas transitus, observabiles A. 1672', *Philosophical Transactions* No. 77 (20 November 1671), 2297–3001.

[1366] Booke: not identified.

[1367] Hodges: i.e. Hodges (*fl.*1671). Collins refers to a Mr Hodges coming over by packet boat from France in his next letter to Vernon. See COLLINS–VERNON 26.XII.1671/[5.I.1672]; HALL, *John Collins on Newton's Telescope*, 73.

253. WALLIS to BERTET, 19/[29] December 1671

w^4 Copy of (missing) letter sent: LONDON *Royal Society* Letter Book Copy 5, pp. 84–91.
E^1 First edition of letter sent: WALLIS, *Algebra*, 358–62.
E^2 Second edition of letter sent: WALLIS, *Opera mathematica* II, 414–17 (our source).
Reply to: BERTET–WALLIS [21.XI]/1.XII.1671.

Clarissimo Doctissimoque Viro,
D. Joanni Bertet, Parisiis

Oxoniae, Dec. 19. 1671.

Clarissime Vir,

Quod meam Infinitorum Arithmeticam[1368] tanti aestimaveris, gratias habeo.
Literas[1369] tuas quod attinet (Decemb. 1. Pontisarae datas;) Parabolas quidem ego, Paraboloeides, aliasque ad eandem familiam spectantes Figuras, concipio (in Arithm. Infin.) tanquam (parallelis) sectas in partes aeque altas. Non quod alias Sectiones respuerim (nam & alias passim adhibeo) sed hanc ut simplicissimam elegerim. Adeoque Figurae partes habeam (nempe Parallelogramma interjecta) ipsis (quibus adjacent) Ordinatis proportionales: Ut non sit opus, propter Altitudinum considerationem (cum sit in omnibus eadem) calculum perplexiorem reddere; quod omnino faciendum esset, si Altitudines sumerentur inaequales.

Et quidem sive serierum Indices sint $-\frac{1}{2}$. 0. $\frac{1}{2}$. 1. 2. &c. sive numerus utcunque surdus, ut $\sqrt{2}$. &c. perinde est: quod monuimus[1370] ad Arithm. Infin. prop. 64. De Motu Cap. 5. def. 1, 2, & prop. 1. 28, & alibi. Utut tu id forte non animadverteris.

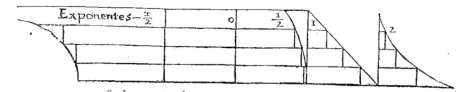

11 (ut simplicissimam) *W*
14 (quod ... faciendum esset,) *W*

[1368] Arithmeticam: i.e. WALLIS, *Arithmetica infinitorum*, Oxford 1656.
[1369] Literas: i.e. BERTET–WALLIS [21.XI]/1.XII.1671.
[1370] monuimus: see WALLIS, *Arithmetica infinitorum*, 52–3; *Opera mathematica* I, 395; WALLIS, *Mechanica: sive, de motu, tractatus geometricus* II, 144–9, 511–31; *Opera mathematica* I, 665–8, 892–904.

253. WALLIS to BERTET, 19/[29] December 1671

Nempe semper, posito Indice, verbi gratia, s vel t, (numero Integro, Fracto, Surdove, aut Negativo;) erit tota series $\frac{1}{s+1}$ vel $\frac{1}{t+1}$, correspondentis seriei aequalium; seu, in hoc casu, Parallelogrammi Circumscripti (si Index seu Exponens sit Affirmativus,) vel (si Negativus) Inscripti.

Earumque ad invicem rationes nullo negotio obtinentur: Nempe ut $\frac{1}{s+1}$ ad $\frac{1}{t+1}$, sive ut $t+1$ ad $s+1$; quicunque fuerint s, t, numeri, (Integri, Fracti, Surdive, aut Negativi.)

Sed & porro advertendum erit; Me non de Figuris tantum (nedum sic ad Axem positis,) sed de omne genus Quantitatibus indiscriminatim, (puta, Lineis, Superficiebus, Solidis, Rectis, Planis, Curvis, Ponderibus, Momentis, Temporibus, Celeritatibus, &c.) tractationem illam instituisse; quibus omnibus, pro re nata, series illae pariter accommodandae erunt. Adeoque nulla erat mihi, in Propositionibus generalibus, vel Axis, vel Sectionum Axis, facienda mentio.

Quod si secuissem Axem Figurarum earundem in alia proportione, puta, ut partes Axis essent, verbi gratia, ut 1, 3, 5, 7, &c. Arithmetice Proportionales; adeoque, distantiae a vertice ut 1, 4, 9, 16, &c. series Secundanorum, (ut Parabolam speciatim Sectam videas,[1371] Arith. Infin. prop. 24, 38, 55, 56; & de Curvarum Ἐυθύνσει, Fig. 24; & de Motu, Cap. 5. Prop. 28. Cap. 10. Pr. 7, 8, Cap. 15. Pr. 1. & alibi, idemque generaliter moneo[1372] ad Def. Cap. 4. de Motu,) Forent earundem Figurarum Series admodum ab illis modo dictis diversae, & magis compositae.

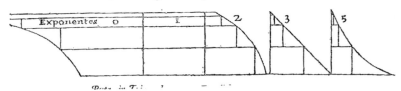

Puta, in Triangulo, propter Parallelogrammorum Latitudines (ut distantias a vertice) in Serie Secundanorum; & Altitudines (ut quadratorum differentias) in Serie Primanorum; erunt illa interjecta Parallelogramma (seu partes Figurae) Series (ex duabus illis composita) Tertianorum. Similiter ostendetur (in hujusmodi Sectione) Parallelogrammum, Series Primanorum, (quae, apud me, est Aequalium:) Parabola, Secundanorum, (quae est, apud

---

[1371] videas: i.e. WALLIS, *Arithmetica infinitorum* 18–19, 27–31, [44–5]; *Opera mathematica* I, 375, 380–2, 390–1; WALLIS, *Tractatus duo*, 93; *Opera mathematica* I, 552; WALLIS, *Mechanica: sive, de motu, tractatus geometricus* II, 511–31, III, 656–9, 747–53; *Opera mathematica* I, 892–904, 999–1001, 1056.

[1372] moneo: i.e. WALLIS, *Mechanica: sive, de motu, tractatus geometricus* II, 110–12; *Opera mathematica* I, 645–6.

253. WALLIS to BERTET, 19/[29] December 1671

me, Subsecundanorum:) Reciproca Parabolae, Aequalium, (quae est, apud me, Reciproca subsecundanorum:) Et universaliter, Index est mei duplus uno auctus. Nempe, si Index in Sectione mea sit s; erit in hac $2s + 1$. Adeoque Series, est $\frac{1}{2s+2}$ seriei Aequalium. Duaeque Series comparatae, quae habeant indices Ordinatarum s, t; erunt ad invicem ut $\frac{1}{2s+2}$ ad $\frac{1}{2t+2}$; hoc est, ut $2t+2$ ad $2s+2$; seu ut $t+1$ ad $s+1$, ut prius.

Neque est quod haereas, si Seriem (verbi gratia) Primanorum, nunc dicamus ut 0, 1, 2, 3, 4, &c; nunc ut 1, 2, 3, 4, 5, &c; nunc ut $\frac{1}{2}$, $1\frac{1}{2}$, $2\frac{1}{2}$, $3\frac{1}{2}$, $4\frac{1}{2}$, &c; seu 1, 3, 5, 7, 9, &c. Quippe haec omnia, in Serie infinita, coincidunt; pariter atque Expositae Figurae coincidunt Inscripta, Circumscripta, & Intermediae omnes; Sectionibus ad minutissima continuatis. Ut moneo[1373] ad Schol. Prop. 182. Arithm. Infin. & ad Prop. 1. Cap. 5. de Motu.

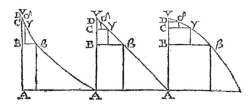

Ponamus jam (quod tu vis) Sectos esse Axes, earundem Figurarum secundum rationem aliquam Geometricam continuam; puta Triplam: Nempe, VA, VB, VC, VD, &c. ut a, $\frac{1}{3}a$, $\frac{1}{9}a$, $\frac{1}{27}a$, &c. seu a, $\frac{a}{3}$, $\frac{a}{9}$, $\frac{a}{27}$ &c. seu a, $\frac{a}{3}$, $\frac{a}{3\times 3}$, $\frac{a}{3\times 3\times 3}$ &c. Adeoque AB, BC, CD, &c. ut $\frac{2}{3}a$, $\frac{2}{3}\times\frac{a}{3}$, $\frac{2}{3}\times\frac{a}{3\times 3}$, &c. seu $\frac{2a}{3}$, $\frac{2a}{3\times 3}$, $\frac{2a}{3\times 3\times 3}$. Et Bases, in Triangulo quidem (Diametris Proportionales) ut b, $\frac{b}{3}$, $\frac{b}{3\times 3}$, $\frac{b}{3\times 3\times 3}$, &c. In Parabola (in subduplicata ratione Diametrorum) ut b, $b\sqrt{\frac{1}{3}}$, $b\sqrt{\frac{1}{3\times 3}}$, $b\sqrt{\frac{1}{3\times 3\times 3}}$, &c. In Parabolae complemento (in duplicata ratione diametrorum) ut b, $\frac{b}{3\times 3}$, $\frac{b}{3\times 3\times 3\times 3}$, $\frac{b}{3\times 3\times 3\times 3\times 3\times 3}$, &c. Adeoque interjecta Parallelogramma, (seu, ut tu loqueris, parets Figurae) $A\beta$, $B\gamma$, $C\delta$, &c. in ratione ex basium & altitudinum composita) in Triangulo, ut $\frac{2ab}{3\times 3}$, $\frac{2ab}{3\times 3\times 3\times 3}$, $\frac{2ab}{3\times 3\times 3\times 3\times 3\times 3}$, &c: In Parabola, ut $\frac{2ab}{3\sqrt{3}}$, $\frac{2ab}{3\times 3\sqrt{3}\times 3}$, $\frac{2ab}{3\times 3\sqrt{:3\times 3\times 3}}$, &c. In Parabolae Complemento, ut $\frac{2ab}{3\times 3\times 3}$, $\frac{2ab}{3\times 3\times 3\times 3\times 3}$, $\frac{2ab}{3\times 3\times 3\times 3\times 3\times 3\times 3}$, &c. Adeoque Figura inscripta, in Triangulo, erit (circumscripti Parallelogrammi ab,) $\frac{2}{9-1} = \frac{2}{8}$ seu $\frac{1}{4}$; in Parabola, $\frac{2}{3\sqrt{3}-1}$; in Parabolae Complemento, $\frac{2}{27-1} = \frac{2}{26} = \frac{1}{13}$.

[1373] moneo: i.e. WALLIS, *Arithmetica infinitorum*, 146–61; *Opera mathematica* I, 447–58; WALLIS, *Mechanica: sive, de motu, tractatus geometricus* II, 148–9; *Opera mathematica* I, 667–8.

253. WALLIS to BERTET, 19/[29] December 1671

Et universaliter, si sit ratio VB ad VA ut 1 ad z (Integrum, Fractum, Surdumve quemlibet,) Figura ex Parallelogrammis inscripta, erit, in Triangulo $\frac{z-1}{z^2-1}ab$; in Parabola $\frac{z-1}{z\sqrt{z,-1}}ab$; in Complemento Parabolae, $\frac{z-1}{z^3-1}ab$. Quod, alibi demonstratum, [Nempe, ad prop. 8, & 15, cap. praeced. hujus,] longius est quam ut hic commode inseratur.

Hoc est, Posito $z = 4$; erit in Parabolae complemento, $\frac{3}{63}$; in Triangulo, $\frac{3}{15}$: in Parabola, $\frac{3}{7}$. Posito $z = 9$; erit $\frac{8}{728}, \frac{8}{80}, \frac{8}{26}$. Posito $z = 100$; erit $\frac{99}{999999}, \frac{99}{9999}, \frac{99}{999}$. Et in aliis casibus similiter.

Sed, quo propius z superat 1, eo propius accedet Figura inscripta ad expositam. Adeoque, posito, verbi gratia, $z = 1.00020001$, (quadrato numero, quo sit $\sqrt{z}$ non sit Surdus; nempe $\sqrt{,}z = 1.0001;$) erit, in Parabolae Complemento, $\frac{0.0002,0001,}{0.0006,0015,0020,0015,0006,0001}$, In Triangulo $\frac{0.0002,0001,}{0.0004,0006,0004,0001}$: In Parabola $\frac{0.0002,0001,}{0.0003,0003,0001}$. Quae proxime accedunt, ad $\frac{2}{6}, \frac{2}{4}, \frac{2}{3}$; quae est vera ratio Figurarum expositarum. Quod magis adhuc patebit, si, pro 1.0001, sumeretur 1.00000001; aut si plures adhuc Ciphrae quotlibet interponerentur locis Fractionum decimalium.

Verum si hujusmodi Figurae concipiantur in semet inverse positae (seu, ut tu loqueris subcontrarie,) ductae: Ducentur, non quidem planum in planum, quo fiat plano-planum, sed Ordinatae in Ordinatas respective sumptas, (manente ubique quae prius erat communi altitudine.) Hoc est, $bB\beta$ Rectangulum, in Altitudinem BA, & sic in caeteris.

(Quae itaque Altitudo, si foret, ut apud me, ubique eadem; negligi jure posset; ut sola haberetur Rectangulorum consideratio: Hic vero, quoniam est alibi alia, ad calculum revocanda est.)

Vel etiam $A\beta$ Parallelogrammum, in Altitudinem Bb, & sic in reliquis. Hoc est, $AB \times B\beta \times Bb$, $BC \times C\gamma \times Cc$, $CD \times D\delta \times Dd$, &c.

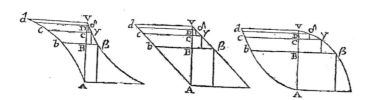

Adeoque partes Solidi adjacentis Triangulo; $\frac{z-1}{z} \times \frac{1}{z} \times \frac{z-1}{z}, +\frac{z-1}{z^2} \times \frac{1}{z^2} \times \frac{z^2-1}{z^2}, +\frac{z-1}{z^3} \times \frac{1}{z^3} \times \frac{z^3-1}{z^3}$, &c. seu $\frac{z-1}{z^3}$ in $z-1, +\frac{z-1}{z^6}$ in $z^2-1, +\frac{z-1}{z^9}$ in z^3-1, &c. Hoc est, $\frac{z-1}{z^3} \times z, +\frac{z-1}{z^6} \times z^2, +\frac{z-1}{z^9} \times z^3, (= \frac{z-1}{z^2}, +\frac{z-1}{z^4}, +\frac{z-1}{z^6}$, &c.$)$ minus, $\frac{z-1}{z^3} + \frac{z-1}{z^6} + \frac{z-1}{z^9}$, &c. Hoc est, omnino, $\frac{z-1}{z^2-1}$ minus $\frac{z-1}{z^3-1}$.

Adjacentis Complemento Parabolae; $\frac{z-1}{z} \times \frac{1}{z^2} \times \mathrm{Q}\frac{z-1}{z}, +\frac{z-1}{z^2} \times \frac{1}{z^4} \times \mathrm{Q}{:}\frac{z^2-1}{z^2}, +\frac{z-1}{z^3} \times \frac{1}{z^6} \times \mathrm{Q}{:}\frac{z^3-1}{z^3}$, &c. seu $\frac{z-1}{z^5}\mathrm{Q}{:}z-1. +\frac{z-1}{z^{10}}\mathrm{Q}{:}z^2-1. +\frac{z-1}{z^{15}}\mathrm{Q}{:}z^3-1.$

253. WALLIS to BERTET, 19/[29] December 1671

&c. Hoc est, $\frac{z-1}{z}$ in $z^2 - 2z + 1$, $+\frac{z-1}{z^{10}}$ in $z^4 - 2z^2 + 1$, $+,\frac{z-1}{z^{15}}$ in $z^6 - 2z^3 + 1$,
&c. Hoc est, $\frac{z^3-z^2-2z^2+2z+z-1}{z^5}$, $+\frac{z^5-z^4-2z^3+2z^2+z-1}{z^{10}}$, $+\frac{z^7-z^6-2z^4+2z^3+z-1}{z^{15}}$,
&c. Quorum summa $\frac{z-1}{z^3-1} - \frac{2z-2}{z^4-1} + \frac{z-1}{z^5-1}$.

Adjacentis Parabolae $\frac{z-1}{z} \times \frac{1}{\sqrt{z}} \times \sqrt{\frac{z-1}{z}}$, $+\frac{z-1}{z^2} \times \frac{1}{\sqrt{z^2}} \times \sqrt{\frac{z^2-1}{z^2}}$, $+\frac{z-1}{z^3} \times \frac{1}{\sqrt{z^3}} \times \sqrt{\frac{z^3-1}{z^3}}$. &c. seu $\frac{z-1}{z^2}\sqrt{\,} : z-1.$ $+\frac{z-1}{z^4}\sqrt{\,} : z^2-1.$ $\frac{z-1}{z^6}\sqrt{\,} : z^3-1.$ &c. Quae summa aequatur Surdo alicui novi generis adhuc Anonymo.

In casu Triangulorum, pro $\frac{z-1}{z^2-1} - \frac{z-1}{z^3-1}$, habemus nos $\frac{1}{6}$.

In casu Complementorum Parabolae; pro $\frac{z-1}{z^3-1} - \frac{2z-2}{z^4-1} + \frac{z-1}{z^5-1}$, habemus nos $\frac{1}{30}$. (Ad quas, rationes Sectioni tuae accommodatae, eo propius accedunt, quo z propius superat 1.)

In casu Parabolarum; pro Surdo illo Anonymo, habemus $\frac{1}{2\square}$; proque hoc utcunque designando, quos progressus fecimus, videas[1374] ad Arithm. Infin. prop. 166, & sequentes.

Sin tibi adhuc spes est, in seriebus adhuc perplexioribus (propter interjectorum Solidorum altitudines inaequales,) rem felicius assequendam, quam ego in simplicioribus (propter sumptas altitudines aequales) assecutus fuerim; per me liceat inquiras bonis avibus. Eique inquisitioni hactenus ego tibi viam praeparavi, interjectorum Solidorum in expositis a te Figuris quas quaerebas rationes (pro quacunque Axis Sectione in Geometrica ratione continua) exhibendo.

Metuo autem ne eo semper res redeat (quocunque te vertas) ut incidas in Seriem Radicum Universalium, Apotomarum (si Circulum aut Ellipsin secteris,) vel (si Hyperbolam) Binomiorum. Et quidem cum ego, simplicioribus insistens, in avia devenerim, haud sperandum videtur ut perplexiores sectanti felicius res succedat.

Quam autem istiusmodi Series Radicum Universalium, ad congruam Seriem Aequalium, rationem habeat; nec numeris veris, nec etiam hactenus receptis Surdis Radicibus, explicari posse; jam satis demonstrasse videamur[1375], ad Prop. 189, 190, Arithm. Infin. Quippe, quo id fiat, ostendimus, dividi oportere numerum imparem in duos aequales integros: Et Aequationum ordinem inquirendum, qui sit Lateralibus & Quadraticis intermedius, (habeatque radices plures quam unam, & pauciores quam duas:) Quorum utrumque est ἀδύνατον. Sed aliusmodi Surdum cogitandum, adhuc Anonymum.

[1374]videas: i.e. WALLIS, *Arithmetica infinitorum* 133–6; *Opera mathematica* I, 440–2.
[1375]videamur: i.e. WALLIS, *Arithmetica infinitorum*; 169–78, *Opera mathematica* I, 462–7.

Qui cujusmodi sit, illic explicavimus: Et (prop. seq.) ostendimus, quomodo possit, continua approximatione, veris numeris quam proxime explicari ejus valor; (non minus quam valor Radicis Surdae $\sqrt{2}$.)

Ut id solum supersit, ut inter se conveniat Mathematicis, quo velint charactere Surdum illum designare, puta, illo quem ibidem exhibemus, ♍ $1|\frac{3}{2}$; vel alio quovis, (prout jam convenit medium proportionalem inter 1 & 2, charactere $\sqrt{1 \times 2}$, vel $\sqrt{2}$ insinuare.)

Atque eo saltem rem illam ulterius provexi quam a Gregorio San Vincentiano factum est,[1376] (qui illic, si memini, ubi tu dicis, consistit.) Qui multa quidem habet nobiscum communia; quamquam eum non ante viderim, quam scripserim Tractatum illum. Sed neque dum evolvere contigit; utut eum satis aestimo, saepiusque statuerim evolvendum. Nec dubito quin quae de continue proportionalium terminationibus habet, nostris consentiant; utut ad manum non sit liber quem consulam, atque ex inventis meis propriis ea deduxerim.

Atque haec sunt, Vir Clarissime, quae ad quaesita tua reponenda censui.

Tuus ad Officia,
Johannes Wallis.

254.
HENRY OLDENBURG to WALLIS
30 December 1671/[9 January] 1672

Transmission:

Manuscript missing.

Existence and date: mentioned in and answered by WALLIS–OLDENBURG 14/[24].I. 1671/72 and WALLIS–OLDENBURG 18/[28].I.1671/72.

[1376] factum est: i.e. in SAINT-VINCENT, *Opus geometricum*, Antwerp 1647.

BIOGRAPHIES OF CORRESPONDENTS

Aubrey, John (1626–97). — Born in Easton Pierse near Kington St Michael, Wiltshire, son of the landowner Richard Aubrey and his wife Deborah, née Lyte. Entered Trinity College, Oxford, May 1642. Left Oxford during first Civil War. Entered Middle Temple 1646. Inherited estates and debts of father, following his death in 1652. From 1671 largely reliant on hospitality of friends and patrons. Elected fellow of the Royal Society 1663. Antiquarian interests led to important archaeological studies, particularly in Wiltshire; further studies on regional and natural history, educational reform, astrology, and mathematical science. Produced biographical sketches of many great 16th and 17th century figures in *Brief Lives*. Donated his extensive collection of books and papers to the Ashmolean Museum 1689. Died during journey from London to Wiltshire. Buried in church of St Mary Magdalene, Oxford on 28 May 1697.

Bernard, Edward (1638–97). — Born in Paulerspury, near Towcester, Northamptonshire, son of the clergyman Joseph Bernard and his wife Elizabeth, née Lenche (Linche). Admitted Merchant Taylors' School 1647. In 1655 elected Sir Thomas White scholar at St John's College, Oxford. Fellow of St John's 1658. Graduated B.A. 1659. Studied Arabic with Edward Pococke; took private tuition in mathematics with John Wallis. Proceeded M.A. 1662. College reader in mathematics 1663. In 1667 chosen University proctor. Graduated B.D. in 1668. In December 1668 travelled to Leiden with John Wallis jr to work on transcription of the *Conics* of Apollonius. Deputized for Wren as Savilian professor of astronomy from 1669. Presented to rectory of Cheam, Surrey, in 1672. On 9 April 1673 appointed successor to Wren as Savilian professor. On same day elected Fellow of the Royal Society. Subsequently resigned rectorship. Soon grew unhappy with Savilian professorship. In 1676–7 resided in Paris, teaching two illegitimate sons of Charles II. Revisited Leiden in 1683 to attend auction of Heinsius's library; unsuccessfully sought election to vacant professorship in Arabic at Leiden. Proceeded D.D. 1684. Resigned Savilian professorship 1691; presented to rectory of Brightwell, Berkshire, but continued to live in Oxford. Married

Eleanor Howell (b. 1667) in August 1693. In September 1696 again travelled to Leiden, to attend auction of Golius's manuscripts. Died from consumption in Oxford on 2 January 1697. Buried in the chapel of St John's College.

Bertet, Jean (1622–92). — Born in Tarascon. Admitted S.J. 1637. Studied under Honoré Fabri at Collège de la Trinité in Lyon. Taught philosophy and mathematics at University of Grenoble. By 1659 professor of mathematics at Aix. In 1689 conspired with Leibniz and Antonio Baldigiani S.J. in Rome to have ban on Copernicanism lifted. Died in Paris on 29 June 1692 (new style).

Borelli, Giovanni Alfonso (1608–79). — Born in Castelnuovo, near Naples, son of a Spanish infantryman, Miguel Alonso, and a local woman, Laura Borelli. Studied mathematics under Galileo's pupil and disciple Benedetto Castelli in Rome. Professor of mathematics at Messina 1639. Appointed to chair of mathematical science at University of Pisa 1656. Founder member of the Accademia del Cimento in 1657. Returned to Messina in 1667, but fled to Rome on suspicion of involvement in political conspiracy. Lived under protection of Queen Christina of Sweden in Rome. Spent last years of his life teaching mathematics in a convent. Produced important work in the mathematical sciences, biology, and astronomy, including *De vi percussionis liber* (1667) and *De motionibus naturalibus a gravitate pendentibus, liber* (1670). His two volume *De motu animalium* (1680–81) was published posthumously at the expense of Christina. Died in Rome on 31 December 1679 (new style).

Boyle, Robert (1627–91). — Born at Lismore Castle, Munster, Ireland, son of Richard Boyle, first earl of Cork, and his wife Catherine, née Fenton. Attended Eton College 1635–38; thereafter privately tutored. Spent years 1639 to 1644 on a continental tour, mainly in Geneva. In 1645 settled in the manor at Stalbridge, Dorsetshire. Established close contacts with the intelligencer Samuel Hartlib. From summer 1652 to summer 1654 spent most of the time in Ireland; contact to William Petty in Dublin. In late 1655/early 1656 moved to Oxford and joined the experimental philosophy club around John Wilkins in Wadham College. After Wilkins's departure, meetings took place in his lodgings. Conducted extensive correspondence, particularly with Henry Oldenburg. From 1661 onwards frequently in London. Became leading representative of experimental philosophy and one of the most active members of the early Royal Society. In September 1665 created doctor of physic at Oxford. In 1668 moved permanently to London; lived with his sister Katherine, Lady Ranelagh, in Pall Mall. Published numerous works,

Biographies of correspondents

including *Sceptical Chymist* (1661), *New Experiments Physico-Mechanicall, Touching the Spring of the Air, and its Effects* (1660), and *A Continuation of New Experiments Physico-Mechanical* (1669). Died in London a week after his sister on 21 December 1691. Buried in St Martin's-in-the-Fields.

Brouncker, William (1620–84). — Born at Newcastle Lyons, county Dublin, Ireland, son of Sir William Brouncker, Viscount Brouncker of Lyons, and his wife Winifred, née Leigh. Succeeded to his father's titles 1645. Studied at Oxford from 1636. Awarded M.D. 1646/7. Settled in London. First president of the Royal Society after its incorporation by royal charter in 1662. Deposed from office 1677. Published translation of Descartes's *Musicae compendium* into English (1653). His most important mathematical work carried out during the Commonwealth and the Protectorate; especially intensive correspondence with Wallis in connection with Fermat's challenges on number theory. In 1662 appointed chancellor to Queen Catherine and made keeper of the great seal. President of Gresham College 1664–67. Commissioner for the navy 1664–79, 1681–84. Controller of the navy accounts 1668–79. Master of the Hospital of St Katharine, London, 1681–4. Did not marry and had no children. Died in his house on St James's Street, Westminster, on 26 March 1684. Buried in chapel of St Katharine's.

Childrey, Joshua (1625–70). — Son of Robert Childrey of Rochester, Kent. Matriculated Magdalen College, Oxford 1640/1. Studies interrupted by Civil Wars. Graduated B.A. 1646. Expelled from Magdalen by parliamentary visitors 1648. Kept a school in Faversham, Kent, until the Restoration. Proceeded M.A. 1660/1. Rector of Shepton Beauchamp, Somerset 1661. Rector of Upwey, Dorset 1664. Presented to archdeaconry of Sarum and made prebendary of Yetminster Parum 1664. Published *Syzygiasticon instauratum*, a heliocentric ephemeris in 1653, and compiled *Britannia Baconica* in 1660. Married Elizabeth Todd in 1665. Died in Upwey on 16 August 1670.

Collins, John (1625–83). — Born in Wood Eaton, Oxfordshire, son of a non-conformist minister. Around 1638 apprenticed to the bookseller Thomas Allen (or Allam) in Oxford. Employed later as clerk by John Marr, clerk of the kitchen to the Prince of Wales. During 1642–49 served on board an English merchant ship engaged by the Venetian republic. Devoted leisure time to the study of mathematics and merchants' accounts. On leaving the service, worked as a mathematics teacher in London. After the Restoration, appointed successively as accountant to the excise office, accountant in chancery, and secretary to the council of plantations. Around 1670 married Bellona Austen, daughter of head cook to Charles II. Became manager of the farthing office in 1672. Not long afterwards, he became accountant to the Royal Fishery Company and remained in this post until his death. Elected fellow of the Royal Society in October 1667. Often advised Oldenburg on mathematical topics. Alongside numerous mathematical publications of his own, he assisted in seeing the works of others through the press, including Brancker's translation of Rahn's *Teutsche Algebra* and Wallis's *A Treatise of Algebra*. As a prolific scientific intelligencer, whose correspondents included Newton, Leibniz, Wallis, Flamsteed, and Sluse, he was styled by Isaac Barrow as the 'English Mersenne'. Died at his lodging on Garlick Hill, London on 31 October 1683. Buried in the parish church of St James Garlickhythe.

Cosimo III de' Medici (1642–1723). — Elder son of Ferdinando II de' Medici, grand duke of Tuscany, and the grand duchess, Vittoria delle Rovere of Urbino. Taught privately by Volunnio Bandinelli, a Siennese theologian. In 1661 married by proxy Marguerite Louise d'Orléans, a cousin of Louis XIV. Marriage unhappy from the outset; Marguerite Louise eventually returned to France in 1674. Travelled on first tour of Europe (Austria, United Provinces, Germany) 1667–68; second tour 1668–69 took him to Spain, Portugal, England, Ireland, and France. Assumed title of grand duke, following death of Ferdinando II in May 1670. Later years marked by economic decline and introduction of repressive legislation in Tuscany. Died in Florence on 31 October 1723 (new style). Interred in the Basilica of San Lorenzo in Florence.

Crew, Nathaniel (1633–1721). — Born in Steane, Northamptonshire, son of John Crew (later first Baron Crew) and his wife Jemimah, née Waldegrave. Admitted to Gray's Inn 1652. Matriculated at Lincoln College, Oxford 1653. Graduated B.A. 1656. Elected Fellow of Lincoln College 1656. Proceeded M.A. 1658. Fellow in canon law at Lincoln College 1659. Senior university proctor 1663. Made D.C.L. 1664. Ordained deacon 1665. King's chaplain 1666. Preferred to rectory of Gedney, Lincolnshire 1667. Elected rector of Lincoln College 1668. Developed close ties to future James II. In 1671 elected bishop of Oxford; subsequently resigned rectorship of Lincoln. Elected bishop of Durham 1674. Member of Inner Temple and lord lieutenant of Durham 1674. Made privy councilor 1676. In 1685 under James II became dean of the Chapel Royal (dismissed 1689). Married Penelope, widow of Hugh Tynte 1691. Following her death married Dorothy, daughter of William Forster. Enjoyed partial revival of political standing towards end of Queen Anne's reign. Died in Steane on 7 September 1721. Buried in Steane.

Dulaurens, François (d. c.1675). — Probably originated from Montpellier. Evidently met Oldenburg, with whom he later corresponded, in Paris.

Author of *Solutiones aliquot quaestionum* (1663) and *Specimina mathematica duobus libris comprehensa* (1667).

Ellis, John (*c*.1646–1738). — Eldest son of the clergyman John Ellis and his wife Susannah, née Welbore. Entered Westminster School 1660. Matriculated at Christ Church, Oxford 1664. Recommended by John Fell to Joseph Williamson at State Paper Office. In 1674 sent by Williamson to France to further his education. Following Williamson's promotion to secretary of state, employed by Llewelyn Jenkins as secretary during peace negotiations at Nijmegen, 1675–78. Attended the prince of Orange in military campaigns between 1676 and 1678. Upon return to England employed as secretary to Thomas Butler (1634–80), sixth earl of Ossory. With help of James Butler (1610–88), first duke of Ormond, Henry Bennet, and Jenkins appointed secretary to the Irish revenue commissioners 1682. Returned to England 1688; became secretary to James Butler (1665–1745), second duke of Ormond. Appointed commissioner for transports 1691. Made under-secretary to secretary of state William Trumbull in May 1695. Retained this post under successive secretaries until his dismissal in 1705. Appointed comptroller of the Royal Mint 1701; relieved of this post 1711. Member of parliament for Harwich 1702–08. In later years a man of independent means. Died in his house in Pall Mall on 27 June 1738.

Gott, Samuel (1613/4–71). — Son of Samuel Gott, a London merchant. Attended Merchant Taylors' School. Matriculated as pensioner from St Catharine's College, Cambridge 1629/30. Graduated B.A. 1632/3. Proceeded M.A. 1644. Admitted as student of Gray's Inn 1632/3. Called to the Bar 1640. Appointed reader 1657; ancient 1658. Author of religious texts, including *Novae Solymae libri sex sive institutio christiani* (1648) and *The Divine History of the Genesis of the World* (1670). Member of parliament for Winchelsea 1645–48 and 1660–61, for Sussex 1656–58, and for Hastings 1659. Died in 1671 and buried in Battle, Sussex.

Gregory, James (1638–75). — Born at Drumoak near Aberdeen, son of the local minister John Gregory and his wife Janet. Studied at Marischal College, Aberdeen. In 1663 published *Optica promota*, a description of the reflecting telescope he had invented in 1661. During 1664–67 pursued mathematical studies at Padua, where his *Vera circuli et hyperbolae quadratura* (1667) was printed. The same work was reprinted in Padua in 1668, appended by *Geometriae pars universalis*. Elected fellow of the Royal Society in June 1668, following his return to England. Published his *Exercitationes geometricae* in London in 1668. Later the same year appointed professor of mathematics in the University of St Andrews. Married Mary Jameson, daughter of the painter George Jameson, in 1669. Elected professor of mathematics in the University of Edinburgh in 1674. Suffered a stroke in October 1675, a few months after his arrival in Edinburgh. Died in Edinburgh shortly afterwards.

Hevelius (or Hewelcke), Johannes (1611–87). — Born in Danzig, son of the wealthy brewery owner Abraham Hewelcke and his wife Cordula, née Hecker. Educated at the Akademisches Gymnasium in Danzig by the mathematician Peter Krüger. Between 1630 and 1634 studied law at University of Leiden, and made extended visits to England and France. Developed close ties to Gassendi and Boulliau in Paris. In 1634 returned to Danzig at father's wish. In 1635 married Katharine Rebeschke (d. 1662). In 1649, following death of father, built observatory above his own and adjoining houses. Many of his instruments he designed and engraved himself. In 1663 married Elisabetha Koopmann (Kaufmann) (1647–93), who later assisted him in his observations. Published several important astronomical works, including *Selenographia* (1647), *Cometographia* (1668), *Machina coelestis*, 2 parts (1673–9). Elected fellow of the Royal Society in March 1664. Conducted extensive correspondence with the leading scientific figures of his day. In September 1679 his observatory was destroyed by fire. Thereupon published *Annus climactericus* (1685). Rebuilding of observatory incomplete at death. Died in Danzig on 28 January 1687 (new style).

Hobbes, Thomas (1588–1679). — Born in Westport near Malmesbury, Wiltshire, son of Thomas Hobbes, curate of Brokenborough and his wife Catherine, née Middleton. Entered Magdalen Hall, Oxford, 1602/3. Graduated B.A. 1607/8. Tutor to several members of the Cavendish family, including William (1591?–1628), the second earl of Devonshire, and William (1617–84), the third earl of Devonshire. During 1629–31 mainly in Paris. Received in the circles of Mersenne and Descartes. Meetings with Galileo around 1636. After returning briefly to England, fled to France 1640. Taught elements of mathematics to the Prince of Wales in exile in Paris 1646–48. Uproar following publication of *Leviathan* (1651) led him to return to England. In 1653 rejoined and remained in the Cavendish household, first in London, and then at Hardwick Hall and Chatsworth, Derbyshire. Other publications include *De corpore* (1655), *De homine* (1658), and *Problemata physica* (1662). His *Opera philosophica quae latine scripsit, omnia* was published in Amsterdam in 1668. Attempts to produce solutions to ancient mathematical problems in the wake of the Webster–Ward debate led to a long-drawn-out war with Wallis. Died at Hardwick Hall on 24 November 1679. Buried in the church of St John the Baptist in Ault Hucknall, Derbyshire.

Biographies of correspondents

Huygens, Christiaan (1629–95). — Born in The Hague, son of the diplomat, poet, and Latin scholar Constantijn Huygens (1596–1687) and Suzanna van Baerle. Educated largely by his father and private tutors, including Jan Stampioen. Entered University of Leiden in 1645 to study mathematics and law; taught (also privately) by Frans van Schooten. In 1647–49 studied law at Collegium Auriacum in Breda. In 1655 bought doctorate of law in Angers. From 1654 onwards worked on lenses, and constructed microscopes and telescopes. Invented pendulum clock 1655. Carried out numerous astronomical investigations, especially on the rings of Saturn. In 1659, published *Systema Saturnium*. Obtained important results concerning curves and surfaces (tangents, quadratures, cubatures, rectifications). Applied theoretical mathematics to problems in physics, astronomy and technology. Numerous stays in Paris in 1650s and 1660s, as well as visits to England. Elected fellow of the Royal Society 1663. From 1666 to 1681 almost exclusively in Paris. Salaried member of Académie royale des sciences. Published *Horologium oscillatorium* in 1673. Final years spent in The Hague, where he died on 8 July 1695 (new style).

Hyrne, Henry (or Harry) (*c.*1626–after 1672). — Probably identical with Henry Hearne, who matriculated at Magdalene College, Cambridge in 1642. Graduated B.A. 1648/9. Proceeded M.A. 1652. Private secretary to Edward Conway (*c.*1623–83), third viscount Conway and Killultagh, husband of the philosopher Anne Conway, née Finch. Mentioned often in Conway papers. Lived at Parson's Green, near Fulham. Visited Ireland and Norfolk. Corresponded with the philosopher and theologian Henry More on Cartesianism, Latinity, and the flux and reflux of the sea.

Lamplugh, Thomas (1615–91). — Born in Little Riston, Yorkshire, son of Christopher Lamplugh and his wife Anne, née Roper. Matriculated at the Queen's College, Oxford 1634. Graduated B.A. 1639. Proceeded M.A. 1642. Elected fellow of the Queen's College 1642. Proceeded B.D. 1657; thereafter minister of Holy Rood, Southampton. Created D.D. 1660. In 1663 married Katherine Davenant, daughter of the mathematician Edward Davenant. Presented to archdeaconry of London 1664. Made prebendary of Worcester 1669. Installed dean of Rochester 1673. Elected bishop of Exeter 1676. Entered House of Lords 1677. Installed archbishop of York 1688. Died at archiepiscopal residence, Bishopthorpe, on 25 April 1691. Buried in York Minster.

Langham, James (*c.*1621–99). — Son of John Langham, first Baron Cottesbrooke, of Cottesbrooke, Northamptonshire, and Crosby Place, Bishopsgate, London. Ascended to baronetcy on his father's death in 1671.

Sheriff of Northampton 1664. Member of parliament for Northamptonshire and Northampton. In 1647 married Mary Alston (c.1628–60), daughter of Edward Alston, president of Royal College of Surgeons. In 1662 married Elizabeth Hastings (1635–64), daughter of Ferdinando Hastings, later epitomized as godly gentlewoman by Samuel Clarke. In 1667 married Penelope Holles (c.1643–84), daughter of John Holles. Following her death, married Dorothy Pomeroy (1631/2–before 1713), daughter of John Pomeroy. Elected fellow of the Royal Society in January 1677/8, but remained inactive. Died in August 1699.

Neile, William (1637–70). — Born in Bishopthorpe, son of the courtier and patron of scientific learning Paul Neile and his wife Elizabeth. Entered Wadham College, Oxford 1652. Matriculated at Oxford 1655. Admitted as student of Middle Temple 1657. Discovered an exact rectification of semi-cubical parabola 1657. Member of privy council of Charles II. Elected fellow of the Royal Society in January 1663, and member of council in April 1666. Died at his father's residence, Hill House, White Waltham, Berkshire on 14 August 1670. Buried at White Waltham.

Oldenburg, Henry (1618?–77). — Born in Bremen, son of the schoolmaster Heinrich Oldenburg. Studied theology at the Gymnasium Illustre in Bremen. Matriculated at the University of Utrecht 1641. During the following years apparently worked as a tutor and travelled on the Continent. Returned to Bremen 1652. In 1653 sent to England on diplomatic mission. Around 1655 became tutor to Richard Jones, the son of Lady Ranelagh and nephew of Robert Boyle. Matriculated at the University of Oxford 1656. Devoted time to the study of experimental philosophy. In 1657–60 travelled in continental Europe with Richard Jones. In 1660 returned to London. Founder member of the Royal Society. In 1663 married Dorothy West (c.1623–65). In April 1663 appointed second secretary of the Royal Society. Prolific scientific intelligencer with correspondents at home and abroad. Instituted and published the journal *Philosophical Transactions* 1665–77. Collaborated with Joseph Williamson in procuring political and military information from France and the Low Countries for the State Paper Office from 1666. In June 1667 accused of espionage and imprisoned in the Tower of London for two months. 1668 married Dora Katherine Durie, daughter of John Durie, and his former ward. Died in Charlton, Kent. Buried in St Mary the Virgin, Bexley, on 28 August 1677.

Pell, John (1611–85). — Born in Southwick, Sussex, son of the local vicar John Pell and his wife Mary née Holland. Entered Trinity College, Cambridge 1624. B.A. 1628. Proceeded M.A. 1630. Incorporated at Oxford 1631. In 1632 married Ithumaria Reginolles of London. During 1638–43 apparently teacher of mathematics in London. In 1643 succeeded Hortensius as professor of mathematics at the Athenaeum Illustre in Amsterdam. Removed to the college at Breda 1646. Publications include *Controversiae de vera circuli mensura* (1647), based on his dispute with Longomontanus, and *An Idea of Mathematics*, which appeared as an appendix to Durie's *The Reformed Librarie-Keeper* (1650). Returned to England 1652. Appointed by Cromwell to lecture on mathematics. During 1654–8 served as Commonwealth agent in Zurich. Took holy orders on return to England. Rector of Fobbing, Essex 1661–85. Vicar of Laindon, Essex 1663–85. Elected fellow of the Royal Society in May 1663. Created D.D. at Lambeth in November 1663. Collaborated with Brancker in producing *An Introduction to Algebra* (1668), a translation of Rahn's *Teutsche Algebra* (1659). Resided for some years at Brereton Hall, Cheshire, as guest of his former pupil William Brereton. Financially impoverished towards the end of his life, he died in London on 2 December 1685. Buried in the vault of St Giles-in-the-Fields.

Pococke, Edward (1604–91). — Born in Oxford, eldest child of the clergyman Edward Pococke (d. 1636) and his wife, Hester, née Shepard. Matriculated Magdalen Hall, Oxford 1619. Admitted scholar Corpus Christi College, Oxford 1620. Graduated B.A. 1622. Proceeded M.A. 1626. Attended lectures on Arabic at Oxford and later taught privately by William Bedwell at Tottenham High Cross. Acquaintance with John Selden, who became his patron and protector. Admitted probationer fellow at Corpus Christi 1628. Ordained in Oxford by Richard Corbet. In 1630 appointed chaplain to the Levant Company. During 1630–36 resided in Aleppo. In 1636 returned to Oxford; took B.D. and was appointed professor of Arabic. In 1637 travelled to Constantinople (for part of time accompanied by John Greaves) to procure oriental manuscripts for Archbishop Laud. Returned to Oxford via Paris 1640–41. Presented to rectory in Childrey, Berkshire 1642. Marriage to Mary Burdet of West Worldham, Hampshire 1646. Submitted to authority of visitors charged with reforming University of Oxford in 1647. Appointed professor of Hebrew 1648, but deprived of associated canonry of Christ Church. At Restoration canonry restored; moved with family into canon's quarters. Advanced D.D. by royal command 1660. From 1662 delegate of University Press. Developed permanent lameness in later years. Died on 31 August 1691 in Oxford and buried in Christ Church cathedral.

Scarborough (or Scarburgh), Charles (1615–94). — Born in London, son of Edmund Scarborough and his wife Hannah, née Butler. Attended St

Paul's School. Entered Gonville and Caius College, Cambridge 1633. Graduated B.A. 1637. Proceeded M.A. and elected fellow of Caius 1640. Studied mathematics with Seth Ward. Both men tutored privately by William Oughtred at Albury, Surrey. Ejected from Caius around 1644. Entered Merton College, Oxford; assisted William Harvey. Created M.D. at Oxford 1646. Thereafter moved to London. Admitted as candidate to College of Physicians 1648. Helped arrange for Ward to be appointed John Greaves's successor as Savilian professor of astronomy 1649. Elected reader in anatomy at Surgeon's Hall 1649. Elected fellow of College of Physicians 1650. Succeeded Harvey as Lumleian lecturer 1656. Following Restoration appointed First Physician. Knighted by Charles II 1669. Served later as physician to James II and to William and Mary. Member of parliament for Camelford, Devon, 1685–87. Died in London on 16 February 1694. Buried at Cranford, Middlesex.

Schooten, Pieter van (1634–79). — Born in Leiden in February 1634, younger half-brother of Frans van Schooten (1615–60). Studied mathematics at University of Leiden. Appointed successor to Frans van Schooten on chair of mathematics in the engineering school of the University of Leiden 1660. Edited Frans van Schooten's *Tractatus de concinnandis demonstrationibus geometricis ex calculo algebraico*, Amsterdam 1661. Died in Leiden 1679.

Sluse, René François Walter de (1622–85). — Born in Vise, in the principality of Liège, son of René François de Sluse, a notary, and his wife Catherine. Studied law at the University of Louvain 1638–42. Awarded L.L.D. at La Sapienza, Rome 1643. Remained in Rome to study ancient languages, astronomy, and mathematics. In 1651 appointed prebendary in the chapter of St Lambert, Liège by Innocent X. Published first edition of *Mesolabum* in 1659. Became member of privy council of bishop of Liège 1659. Appointed abbot of Sainte-Ode at Amay 1666. Councilor ordinary 1673. Elected fellow of the Royal Society on nomination by Oldenburg in April 1674. Vice-provost of Liège cathedral 1676. Died in Liège in March 1685.

Vernon, Francis (1637–77). — Born near Charing Cross in London, son of Francis Vernon and his wife Anne, née Smithes. Entered Westminster School 1649. Matriculated Christ Church, Oxford 1654. Graduated B.A. 1658. Proceeded M.A. 1660. Thereafter extensive travel on the continent. In 1669 sent as diplomat to Paris with the embassy of Ralph Montagu. Apart from brief visits to England in 1670 and 1671, remained in Paris until 1672. Developed close ties to Parisian virtuosi, especially Giandomenico Cassini and Christiaan Huygens. Elected fellow of the Royal Society 1672. Began extensive journey to Persia 1673. Killed during a dispute with local men in Esfahan in 1677.

Wallis jr, John (1650–1717). — Born in Oxford, only son of John Wallis and his wife Susanna, née Glyde. Matriculated Trinity College, Oxford, 1666. Travelled to Leiden with Edward Bernard in December 1668. Graduated B.A. 1669. Admitted Inner Temple 1669. Called to the Bar 1676. Initial career in law. Trained by his father as potential successor as decipherer; gave occasional assistance to him. In February 1682 married Elizabeth Harris (d. 1693), daughter of John Harris of Soundess House, Nettlebed, Oxfordshire. Lived at first in Wallingford, Oxfordshire. In 1685, following death of Elizabeth's brother, Taverner Harris, sometime member of parliament for Wallingford, inherited Soundess. From then on devoted himself largely to politics and to management of estate. Member of parliament for Wallingford from 1689/90 to 1695. Recommended William Trumbull as his successor. Marriage produced three children: John, Mary, and Elizabeth. His daughter Mary (1699–1761) married the barrister Edward Filmer (1683–1755). Died on 14 March 1717.

Ward, Seth (1617–88). — Born in Buntingford, Hertfordshire, son of the attorney John Ward and his wife Martha Dalton. Entered Sidney Sussex College, Cambridge 1632. Graduated B.A. 1637. Proceeded M.A. 1640. Fellow of Sidney Sussex 1640. Appointed lecturer in mathematics to the University 1643. Together with Scarborough visited William Oughtred in Albury, Surrey, to discuss questions arising from his *Clavis mathematicae*. Deprived of fellowship for refusing to take the covenant 1644. With Oughtred's help became tutor to the sons of his friend Ralph Freeman in Aspenden until 1649. On Scarborough's recommendation and with assistance of John Greaves and John Trevor appointed Savilian professor of astronomy at Oxford 1649. From April 1650 fellow-commoner at Wadham College, Oxford. Became member of the scientific circle around the warden, John Wilkins. As astronomer devised an alternative to Kepler's second law of planetary motion, as formulated in Boulliau's *Astronomia philolaica* (1645). Presented his own position in *In Ismaelis Bullialdi astronomiae philolaicae fundamenta, inquisitio brevis* (1653); developed more fully in *Astronomia geometrica* (1656). In collaboration with Wilkins wrote *Vindiciae academiarum* (1654), a defence of the English universities. From 1650 also influential work on development of universal language schemes. Elected president of Trinity College, Oxford in September 1659; resigned at restoration in favour of ejected predecessor Hannibal Potter. Resigned Savilian professorship in April 1660; presented to vicarage of St Lawrence Jewry, London. Elected dean of Exeter, December

1661; consecrated bishop, July 1662. Translated to see of Salisbury, September 1667. Died at his home in Knightsbridge on 27 December 1688. Buried in Salisbury Cathedral.

Williamson, Joseph (1633–1701). — Born in Cumberland, son of Joseph Williamson (d. 1634), vicar of Bridekirk, and his wife Agnes, née Bowman. Admitted to Westminster School 1648. Entered the Queen's College, Oxford as servitor 1650. Graduated B.A. 1654. Proceeded M.A. by diploma 1657. In later years generous benefactor to his college. Private tutor in France 1655–58. Elected fellow of Queen's 1658. In 1660 appointed under-secretary to secretary of state for the south, Edward Nicholas. In December 1661 appointed keeper of the king's library and of the State Paper Office. Effectively became head of Restoration government's system of intelligence and information. Elected fellow of the Royal Society 1663. Member of parliament for Thetford 1669–85. Election defeat after revolution of 1688. Member of parliament for Rochester 1690–1701. Elected member of parliament for Thetford 1695 and 1698. Clerk of privy council 1672. Received knighthood 1672. Plenipotentiary (with Llewelyn Jenkins) at congress of Cologne 1673. Appointed secretary of state 1674. Elected president of the Royal Society 1677. Following the Popish Plot replaced as secretary of state 1679. Plenipotentiary at congress of Rijswijk (Ryswick) 1697. Thereafter ambassador to The Hague. Died at Cobham Hall on 22 September 1701. Buried in Westminster Abbey.

Wood, Robert (1621/2–85). — Born in Peper Harow, near Godalming, Surrey, son of Robert Wood, rector of Peper Harow. Educated at Eton College. Received instruction in mathematics from William Oughtred at Albury, Surrey. Matriculated at New Inn Hall, Oxford 1640. Obtained postmastership at Merton College, Oxford 1642. Graduated B.A. 1647. Proceeded M.A. 1649. Elected fellow of Lincoln College, Oxford 1650. Member of Oxford experimental philosophy club. Licensed to practice physic 1656. Wrote economic tract *Ten to One* 1655–56. Sent to Ireland by Henry Cromwell 1656; collaborated with William Petty. Appointed professor of mathematics at Durham College 1657. At restoration dismissed from fellowship of Lincoln College. Returned to Ireland as chancellor of diocese of Meath. On return to England appointed mathematical master at Christ's Hospital. Later returned to Ireland as commissioner for revenue; later appointed accountant-general. Elected fellow of the Royal Society 1681. Married a Miss Adams. Died in Dublin in April 1685.

LIST OF MANUSCRIPTS

CAMBRIDGE *Cambridge University Library*
MS Add. 9597/13/5, f. 64^r–65^v: Letter 252.
MS Add. 9597/13/6, f. 204^r–204^v, 205ra–205^{vd}, 206^r–207^v, 208^r–208^v, 209^r–209^v, 210^r–210^v, f. 210*, f. 211^r–211^v, 213^r–213^v, 214^r–214^v, 215^r–216^v: Letters 15, 58, 59, 70, 91, 94, 96, 135, 172, 176, 206.

FLORENCE *Biblioteca Nazionale*
Gal. 286, f. 56^r–57^v: Letter 190.

KEW *The National Archives*
SP 29/259, No. 183: Letter 75.
SP 29/262, No. 174: Letter 100.

LEIDEN *Bibliotheek der Rijksuniversiteit*
Hug. 45, Nos. 483, 1676: Letters 19, 200.

LONDON *British Library*
Add. MS 4278, 127^v–128^r, f. 344^r–344^v, 345^r–345^v: Letters 6, 10, 20.
Add. MS 32499, f. 32^r–32^v: Letter 114.
MS Sloane 4025, f. 306, f. 310: Letter 234.

LONDON *Royal Society*
Boyle Letters 5, f. 174^r–175^v: Letter 102.
Boyle Letters 7, Nos. 25, 26: Letters 147, 164, 177.
Classified Papers III (i), No. 53: Letter 77.
Early Letters W1, Nos. 15, 64, 65 , 66, 68, 69, 70, 71, 72, 73, 74, 75, 76, 77, 78, 79, 80, 81, 82, 83, 84, 85, 86, 87, 88, 89, 90, 91, 92, 93, 94, 95, 96, 97, 98, 99, 100, 101, 102, 103, 104, 105, 106, 107, 108, 109, 111, 112, 113, 114, 115, 116, 118, 119, 120, 121, 122, 123, 124, 124a, 126, 127, 128, 130, 129, 131, 132, 133: Letters 1, 16, 17, 21, 22, 25, 27, 28, 29, 30, 34, 35, 37, 38, 43, 47, 48, 52, 60, 66, 71, 73, 78, 81, 84, 87, 90, 92, 101, 104, 109, 110, 113, 116, 117, 121, 122, 123, 125, 132, 134, 136, 137, 141, 145, 146, 149, 151, 152, 153, 154, 158, 160, 171, 178, 181, 186, 188, 191, 192, 207, 210, 213, 216, 217, 218, 222, 225, 227, 229, 230, 231, 241, 247, 249.

List of manuscripts

Early Letters W2, No. 34: Letter 195.
Early Letters B2, No. 9: Letter 202.
Early Letters C1, Nos. 4, 10: Letters 142, 161.
Early Letters G1, No. 21: Letter 8.
Early Letters H1, Nos. 58, 107, 108: Letters 14, 139, 157.
Early Letters N1, Nos. 8, 21, 22, 23, 24: Letters 79, 82, 85, 89, 93.
Early Letters S1, Nos. 62, 64: Letters 140, 169.
Letter Book Original 3, between pp. 222 and 223, p. 355, between pp. 348 and 349: Letters 125, 146, 157.

OXFORD *Bodleian Library*
MS Add. D. 105, f. 22^v, f. 34^r–35^v, f. 37^r, f. 159^r–161^r, f. 38^r–38^v, f. 161^v: Letters 2, 3, 95, 111, 112, 246, 251, 253.
MS Don. d. 45, f. 149^r–149^v, f . 151^r–152^v, f. 316^v: Letters 143, 208, 245.
MS Don. e. 12, f. 1^r–1^v: Letter 198.
MS Smith 45, pp. 61-4: Letter 204.
MS Savile 104, f. 3^r–5^v, f. 8^r–9^v: Letters 219, 223.

OXFORD *Oxford University Archives*
SP/D/5/37: Letter 119.

PARIS *Bibliothèque Nationale*
Nouv. acq. latines 1641, f. 114^r–118^v: Letters 54, 55.

PARIS *Observatoire de Paris*
C 1 (9), No. 1357: Letter 68.

UPPSALA *Universitetsbibliotek*
Waller Ms gb-01783, f. 1^r–2^v: Letter 9.
Ms it-00197, f. 001^a–001^b: Letter 193.

BIBLIOGRAPHY

ANDERSON, ALEXANDER: *Exercitationum mathematicarum decas prima.* Paris, 1619.

ANDERSON, ROBERT: *Gaging Promoted. An Appendix to Stereometrical Propositions.* London, 1669.

ANDERSON, ROBERT: *Stereometrical Propositions variously appicable, but particularly intended for Gageing.* London, 1668.

ANDREWS, CHARLES MCLEAN: *British Committees, Commissions, and Councils of Trade and Plantations, 1622–1675*, Baltimore, 1908.

APOLLONIUS OF PERGA: *Apollonii Pergaei conicorum Libri quattuor*, ed. F. COMMANDINO. Bologna, 1566.

APOLLONIUS OF PERGA: *Apollonii Pergaei Conicorum lib. V. VI. VII. paraphraste Abalphato Asphahanensi ... ex codicibus Arab. MSS. Abrahamus Ecchellensis Lat. reddidit. I. A. Borellus curam in geometricis versioni contulit et notas adiecit.* Florence, 1661.

ARBER, EDWARD, ed.: *The Term Catalogues 1668–1709 A.D.; with a number for Easter Term, 1711 A.D.* 3 vols. London 1903–06.

ARCHIMEDES: *Archimedis opera omnia: cum commentariis Eutocii*, ed. J. L. HEIBERG. 2 vols. Leipzig, 1880–1; 2nd edn., 3 vols. Leipzig, 1910–15.

AUBREY, JOHN: *Brief Lives, chiefly of contemporaries, set down by John Aubrey, between the years 1669 and 1696*, ed. A. CLARK. 2 vols. Oxford, 1898.

BACHET DE MÉZIRIAC, CLAUDE GASPAR: *Problemes plaisans et delectables, qui se font par les nombres.* Lyon, 1612.

BACHET DE MÉZIRIAC, CLAUDE GASPAR: *Diophanti Alexandrini Arithmeticorum libri sex, et, De numeris multangulis liber unus.* Paris, 1621.

Bibliography

BACHET DE MÉZIRIAC, CLAUDE GASPAR: *Diophanti Alexandrini Arithmeticorum libri sex... Accessit... inventum novum, collectum ex variis D. de Fermat epistolis.* Toulouse, 1670.

BARROW, ISAAC: *Euclidis elementorum libri XV breviter demonstrata.* Cambridge, 1655.

BARROW, ISAAC: *Lectiones geometricae: in quibus (praesertim) generalia curvarum linearum symptomata declarantur.* London, 1670.

BARROW, ISAAC: *Lectiones XVIII, Cantabrigiae in scholis habitae; in quibus opticorum phaenomenωv genuinae rationes investigantur, ac exponuntur. Annexae sunt lectiones aliquot geometricae.* London, 1669.

BARTHOLIN, ERASMUS: *Experimenta crystalli Islandici dis-diaclastici.* Copenhagen, 1669.

BARTHOLIN, ERASMUS: *Specimen recognitionis nuper editarum observationum astronomicarum N. V. Tychonis Brahe.* Copenhagen, 1668.

BEELEY, PHILIP and CHRISTOPH J. SCRIBA: 'Controversy and Modernity. John Wallis and the Seventeenth-Century Debate on the Nature of the Angle of Contact.' *Acta Historica Leopoldina* 54 (2008), 431–50.

BEELEY, PHILIP and CHRISTOPH J. SCRIBA: 'Disputed Glory. John Wallis and some questions of precedence in seventeenth-century mathematics.' In: *Kosmos und Zahl. Beiträge zur Mathematik- und Astronomiegeschichte, zu Alexander von Humboldt und Leibniz*, ed. H. HECHT, R. MIKOSCH, I. SCHWARZ, et al. Stuttgart, 2008, 275–99.

BEELEY, PHILIP: 'Eine Geschichte zweier Städte. Wallis, Wilkins und der Streit um die wahren Ursprünge der Royal Society. *Acta Historica Leopoldina* 49 (2008), 135–62.

BEELEY, PHILIP: 'Infinity, Infinitesimals, and the Reform of Cavalieri: John Wallis and his Critics.' In: *Infinitesimal Differences: Controversies between Leibniz and his Contemporaries*, ed. U. GOLDENBAUM and D. M. JESSEPH. Berlin and New York, 2008, 31–52.

BEELEY, PHILIP and SIEGMUND PROBST: 'John Wallis (1616–1703): Mathematician and Divine.' In: *Mathematics and the Divine*, ed. T. KOETSIER and L. BERGMANS. Amsterdam, 2005, 441–57.

Bibliography

BEELEY, PHILIP: 'Un de mes amis. On Leibniz's Relation to the English Mathematician and Theologian John Wallis.' In: *Leibniz and the English-Speaking World*, ed. P. PHEMISTER and S. BROWN. Dordrecht, 2007, 63–81.

BEELEY, PHILIP and CHRISTOPH J. SCRIBA: 'Wallis, Leibniz und der Fall von Harriot und Descartes. Zur Geschichte eines vermeintlichen Plagiats im 17. Jahrhundert.' In: *Physica et historia—Festschrift für Andreas Kleinert zum 65. Geburtstag. Acta Historica Leopoldina* 45 (2005), 115–29.

BENNETT, JAMES ARTHUR: *The Mathematical Science of Christopher Wren.* Cambridge, 1982.

BERNARD, EDWARD: *De mensuris et ponderibus antiquis libri tres.* Oxford, 1668.

BERETTA, MARCO: *A History of Non-Printed Science. A select catalogue of the Waller Collection.* Uppsala, 1993.

BEVERIDGE, WILLIAM: *Institutionum chronologicarum libri II, una cum totidem arithmetices chronologicae libellis.* London, 1669.

BILLY, JACQUES DE: *Diophantus redivivus.* Toulouse, 1670.

BIRCH, THOMAS: *History of the Royal Society.* 4 vols. London, 1756–57; repr. Hildesheim, 1968.

BOECKLER, GEORG ANDREAS: *Theatrum Machinarum novum, exhibens aquarias, alatas, jumentarias, manuarias; pedibus, ac ponderibus versatiles, plures, et diversas molas.* Nuremburg, 1662.

BORELLI, GIOVANNI ALFONSO: *De motionibus naturalibus a gravitate pendentibus.* Reggio di Calabria, 1670.

BORELLI, GIOVANNI ALFONSO: *De vi percussionis liber.* Bologna, 1667.

BORELLI, GIOVANNI ALFONSO: *Euclides restitutus, sive, Prisca geometriae elementa.* Pisa, 1658.

BORELLI, GIOVANNI ALFONSO: *Historia et meteorologia incedii Aetnaei anni 1669. Accessit Responsio ad censuras Rev. P. Honorati Fabri contra librum auctoris De vi percussionis.* Reggio di Calabria, 1670.

Bibliography

BORELLI, GIOVANNI ALFONSO: *Theoricae mediceorum planetarum ex causis physicis deductae.* Florence, 1666.

BOULLIAU, ISMAËL: *Astronomia philolaica.* Paris, 1645.

BOYLE, ROBERT: *A continuation of New experiments physico-mechanical, touching the spring and weight of the air. Whereto is annext a short discourse of the atmospheres of consistent bodies.* Oxford, 1669.

BOYLE, ROBERT: *Certain physiological essays, written at distant times, and on several occasions. 2nd ed., increased by the addition of a discourse about the absolute rest in bodies.* London, 1669.

BOYLE, ROBERT: *Hydrostatical paradoxes, made out by new experiments.* Oxford, 1666.

BOYLE, ROBERT: *New Experiments and Observations touching Cold, or an Experimental History of Cold begun. To which are added, an Examen of Antiperistasis, and an Examen of Mr. Hobbes's Doctrine about Cold. Wereunto is annexed an Account of Freezing brought into the Royal Society by the learned Dr. C. Merret, a Fellow of it. Together with an Appendix containing some promiscuous Experiments and Observations relating to the precedent History of Cold.* London, 1665.

BOYLE, ROBERT: *New Experiments Physico-Mechanicall.* Oxford, 1660.

BOYLE, ROBERT: *Nova experimenta physico–mechanica de vi aeris elastica, ex Angl. in Lat. conversa.* Oxford, 1661.

BOYLE, ROBERT: *Origo formarum et qualitatum.* Oxford, 1669.

BOYLE, ROBERT: *Some considerations touching the usefulnesse of experimentall naturall philosophy.* Oxford, 1663.

BOYLE, ROBERT: *The Correspondence of Robert Boyle*, ed. M. HUNTER, A. CLERICUZIO, and L. M. PRINCIPE. 6 vols. London, 2001.

BOYLE, ROBERT: *The origine of formes and qualities, according to the corpuscular philosophy.* Oxford, 1666; 2nd edn., Oxford, 1667.

BOYLE, ROBERT: *The Works of Robert Boyle*, ed. M. HUNTER and E. B. DAVIS. 14 vols. London, 1999–2000.

BOYLE, ROBERT: *The Works of the honourable Robert Boyle*, ed. T. BIRCH. 5 vols. London, 1744.

Bibliography

BOYLE, ROBERT: *The Works of the honourable Robert Boyle*, ed. T. BIRCH. 6 vols. London, 1772.

BRAHE, TYCHO: *Historia coelestis*, ed. A. VON CURTZ. Augsburg, 1666.

Calendar of State Papers, Domestic Series, preserved in the Public Record Office, 1547–1704. 92 vols. London, 1856–1972.

CASSINI, GIOVANNI DOMENICO: *Ephemerides Bononienses mediceorum syderum ex hypothesibus et tabulis Jo. Dominici Cassini... ad observationum opportunitates praemonstrandas deductae*. Bologna, 1668.

CAVALIERI, BONAVENTURA: *Centuria di varii problemi*. Bologna, 1639.

CAVALIERI, BONAVENTURA: *Directorium generale Uranometricum. In quo Trigonometriae Logarithmicae fundamenta, ac regulae demonstrantur*. Bologna, 1632.

CAVALIERI, BONAVENTURA: *Geometria indivisibilibus continuorum nova quadam ratione promota*. Bologna, 1635; 2nd edn., Bologna, 1653.

CELLARIUS, ANDREAS: *Harmonia macrocosmica seu Atlas universalis et novus, totius universi creati cosmographiam generalem, et novam exhibens*. Amsterdam, 1661.

CHILDREY, JOSHUA: *Britannia Baconica: or, the natural rarities of England, Scotland, & Wales. According as they are to be found in every shire*. London, 1660.

CHILDREY, JOSHUA: *Syzygiasticon Instauratum. Or, an Ephemeris of the places and aspects of the planets, as they respect the ☉ as center of their orbes, calculated for the year of the incarnation of God, 1653*. London, 1653.

CLAVIUS, CHRISTOPH: *Euclidis elementorum libri XV. Accessit liber XVI*. 2 vols. Frankfurt, 1654.

CLAVIUS, CHRISTOPH: *Euclidis Elementorum libri XV*. Cologne 1591.

COTTON, ROBERT BRUCE: *Cottoni posthuma: divers choice pieces of that renowned antiquary Sir Robert Cotton*. London, 1651.

CRINÒ, ANNA MARIA, ed.: *Un principe di Toscana in Inghilterra e in Irlanda nel 1669. Relazione ufficiale del viaggio di Cosimo de' Medici*

Bibliography

tratta dal 'Giornale di L. Magalotti', con gli acquerelli palatini. Rome, 1968.

D'ARÇONS, CÉSAR: *Le secret decouvert du flux et reflux de la mer et des longitudes.* Paris, 1656.

DATI, CARLO ROBERTO DE CAMMILLO: *Lettera a Filaleti di Timauro Antiate.* Florence, 1663.

DECHALES, CLAUDE FRANÇOIS MILLIET: *Cursus seu mundus mathematicus.* Lyon, 1674.

DECHALES, CLAUDE FRANÇOIS MILLIET: *l'Art de naviger demontré par principes & confirmé par plusieurs observations tirées de l'experience.* Paris, 1677.

DESCARTES, RENÉ: *Geometria, a Renato DesCartes Anno 1637, Gallicè edita*, ed. F. VAN SCHOOTEN. 2nd edn., 2 vols. Amsterdam, 1659–61.

DEW, NICHOLAS: *Orientalism in Louis XIV's France.* Oxford, 2009.

DICKINSON, HENRY WINRAM: *Sir Samuel Morland. Diplomat and Inventor, 1625–1695.* Cambridge, 1970.

DIOPHANTUS OF ALEXANDRIA: *Diophanti Alexandrini arithmeticorum libri sex, et de numeris multangulis liber unus*, Gr. & Lat. ed., atque comm. illustr. auctore C. G. Bacheto. Cum observationibus P. de Fermat. Accessit doctrinae analyticae inventum novum collectum ex variis ejusdem D. de Fermat epistolis. Toulouse, 1670.

DSB: *Dictionary of Scientific Biography*, ed. C. C. GILLISPIE. 16 vols. New York, 1970–80.

DULAURENS, FRANÇOIS: *Specimina mathematica duobus libris comprehensa. Quorum primus syntheticus agit de genuinis matheseos principiis in genere, in specie autem de veris geometriae elementis hucusque nondum traditis. Secundus vero de methodo compositionis, atque resolutionis fuse disserit, & multa nova complectitur, quae subtilissimam analyseos artem mirum in modum promovent.* Paris, 1667.

EGMOND, W. VAN: 'A Catalog of François Viète's Printed and Manuscript Works.' In: *Mathemata. Festschrift für Helmuth Gericke*, ed. M. FOLKERTS and U. LINDGREN. Stuttgart, 1985, 359–96.

Bibliography

ESCHINARDI, FRANCESCO: *Centuria problematum opticorum ... seu dialogi optici pars altera.* Rome, 1666.

EUCLID: *Euclidis elementorum lib. XV: Accessit XVI De solidorum regularium comparatione. Omnes perspicuis demonstrationibus, accuratisque scholiis illustrati*, ed. C. CLAVIUS. Rome, 1574.

EUCLID: *Euclidis opera omnia*, ed. J. L. HEIBERG and H. MENGE. 9 vols. Leipzig, 1883–1916.

EVELYN, JOHN: *Sylva, or A discourse of forest-trees, and the Propagation of timber ... to which is annexed Pomona; or An appendix concerning fruit-trees in relation to cider.* London, 1664; 2nd edn., London, 1670.

EVELYN, JOHN: *The Diary of John Evelyn*, ed. E. S. DE BEER. 6 vols. Oxford, 1955.

FABRI, HONORÉ: *Dialogi physici.* Lyon, 1669.

FABRI, HONORÉ: *Tractatus physicus de motu locali.* Lyon, 1646.

FAUVEL, JOHN, RAYMOND FLOOD, and ROBIN WILSON, eds.: *Oxford Figures. 800 Years of the Mathematical Sciences.* Oxford, 2000.

FEINGOLD, MORDECHAI: 'Jesuits: Savants.' In: *Jesuit Science and the Republic of Letters*, ed. M. FEINGOLD. Cambridge, Mass. and London, 2003, 1–45.

FEINGOLD, MORDECHAI: 'Oriental Studies.' In: *The History of the University of Oxford. IV. Seventeenth-Century Oxford*, ed. N. TYACKE. Oxford, 1997, 449–503.

FEINGOLD, MORDECHAI: 'The Mathematical Sciences and New Philosophies.' In: *The History of the University of Oxford. IV. Seventeenth-Century Oxford*, ed. N. TYACKE. Oxford, 1997, 359–448.

FERGUSON, JOHANN JAKOB: *Labyrinthus algebrae.* The Hague, 1667.

FERMAT, PIERRE DE: *Diophanti Alexandrini arithmeticorum libri sex*, ed. and transl. C. G. BACHET DE MÉZIRIAC, with notes by P. FERMAT, ed. S. FERMAT. Toulouse, 1670.

FERMAT, PIERRE DE: *Varia opera mathematica D. Petri de Fermat: Accesserunt selectae quaedam ejusdem Epistolae.* Toulouse, 1679.

FLAMSTEED, JOHN: 'Lunae ad fixas appulsus visibiles, nec non arctiores juxta eas transitus, observabiles A. 1672'. *Philosophical Transactions* No. 77 (20 November 1671), 2297–3001.

FOSTER, JOSEPH, ed.: *Alumni Oxonienses: The Members of the University of Oxford 1500–1714; Their parentage, birthplace, and year of birth, with a record of their degrees, being the matriculation register of the university arranged, revised, and annotated.* Early series, 4 vols. Oxford and London, 1891–92.

FRANK, ROBERT G.: *Harvey and the Oxford physiologists. A study of scientific ideas.* Berkeley, Los Angeles, London, 1980.

FRENICLE DE BESSY, BERNARD: *Solutio duorum problematum circa numeros cubos et quadratos.* Paris, 1657.

FROIDMONT, LIBERT: *Meteorologicorum libri sex.* Antwerp, 1627.

GALILEI, GALILEO: *Opere.* 20 vols. Florence, 1929–39; repr. Florence, 1964/5, 1968; 2nd rev. edn., Turin, 1980ff.

GIRARD, ALBERT: *Invention nouvelle en l'algebre.* Amsterdam, 1629.

GLANVILL, JOSEPH: *Plus ultra: or, The progress and advancement of knowledge since the days of Aristotle.* London, 1668.

GLORIOSI, GIOVANNI CAMILLO: *Exercitationes mathematicae decas tertia.* 3 vols. Naples, 1627–39.

GOLIUS, JACOBUS: *De regno Catayo additamentum.* (Published with MARTINI: *Novus atlas Sinensis.*) 1655.

GOODWIN, JOHN: *Imputatio fidei, or, A treatise of justification.* London, 1642.

GOTT, SAMUEL: *The Divine History of the Genesis of the World explicated and illustrated.* London, 1670.

GOTTIGNIES, GILLIS FRANÇOIS: *Elementa geometricae planae.* Rome, 1669.

GRANDAMI (GRANDAMY), JACQUES: *Nova demonstratio immobilitatis terrae petita ex virtute magnetica.* La Flèche, 1645.

GREAVES, JOHN: *A Discourse of the Romane Foot, and Denarius.* London, 1647.

Bibliography

GREAVES, JOHN: *Chorasmiae et Mawaralnahrae hoc est regionum extra fluvium Oxum descriptio ex tabulis Abulfedae Ismaelis.* London, 1650.

GREAVES, JOHN: *Epochae celebriores, astronomis, historicis, chronologis, Chataiorum, Syro-Graecorum, Arabum, Persarum, Chorasmiorum, usitatae: ex traditione Ulug Beigi.* London, 1651.

GREGORY, JAMES: 'An Extract of a Letter of Mr. James Gregory to the Publisher, containing some Considerations of his, upon Mr. Hugens his Letter, printed in Vindication of his Examen of the Book, entitled Vera Circuli & Hyperbolae Quadratura.' *Philosophical Transactions* No. 44 (15 February 1668/9), 882–6.

GREGORY, JAMES: 'Appendicula ad veram Circuli & Hyperbolae Quadraturam' (Appendix to *Exercitationes geometricae*). London, 1668.

GREGORY, JAMES: 'Mr. Gregories Answer to the Animadversions of Mr. Hugenius upon his Book, De Vera Circuli & Hyperbolae Quadratura.' *Philosophical Transactions* No. 37 (13 July 1668), 732–5.

GREGORY, JAMES: *Exercitationes geometricae.* London, 1668.

GREGORY, JAMES: *Geometriae pars universalis.* Padua, 1668.

GREGORY, JAMES: *Vera circuli et hyperbolae quadratura.* Padua, 1667.

GRIENBERGER, CHRISTOPH: *Catalogus veteres affixarum longitudines, ac latitudines conferens cum novis: imaginum caelestium prospectiva duplex.* Rome, 1612.

GRIENBERGER, CHRISTOPH: *Speculum ustorium verae ac primigeniae suae formae restitutum,* ed. F. DE GHEVARA. Rome, 1613.

GRIENBERGER, CHRISTOPH: *Prospectiva nova coelestis.* Rome, 1612.

GRIFFITHS, JOHN, ed.: *Statutes of the University of Oxford, codified in the year 1636 under the authority of Archbishop Laud, Chancellor of the University.* Oxford, 1888.

GRIMALDI, FRANCESCO MARIA: *Physico-mathesis de lumine, coloribus, et iride, aliisque adnexis libri duo ... opus posthumum,* ed. H. BERNIA. Bologna, 1665.

GRISIO, SALVATOR: *Antanalasi a quesiti stampati nell'analisi di Benedetto Maghetti.* Rome, 1641.

Bibliography

GUNTHER, ROBERT THEODORE: *Early Science in Oxford. Volume IV. The Philosophical Society*. Oxford, 1925.

HALL, A. RUPERT: 'John Collins on Newton's Telescope.' *Notes and Records of the Royal Society of London* 49 (1995), 71–8.

HALL, MARIE BOAS: *Henry Oldenburg: shaping the Royal Society*. Oxford, 2002.

HARRIOT, THOMAS: *Artis analyticae praxis*. London, 1631.

HARTLEY, HAROLD, ed.: *The Royal Society: its Origins and Founders*. London, 1960.

HARVEY, WILLIAM: *De motu cordis et sanguinis in animalibus, anatomica exercitatio*. Leiden, 1639.

HENDERSON, FELICITY: 'Putting the dons in their place: A Restoration Oxford Terrae filius speech.' *History of Universities* 16 (2000), 32–64.

HÉRIGONE, PIERRE: *Les six premiers livres des Elements d'Euclide*. Paris, 1639.

HEVELIUS, JOHANNES: *Cometographia*. Danzig, 1668.

HEVELIUS, JOHANNES: *Descriptio Cometae anno aerae Christ. MDCLXV... cui addita est Mantissa prodromi cometici*. Danzig, 1666.

HEVELIUS, JOHANNES: *Dissertatio de nativa Saturni facie ejusque variis phasibus, certa periodo redeuntibus*. Danzig, 1656.

HEVELIUS, JOHANNES: *Eclipsis solis observata Gedani anno...1649, die 4 Novembris st. Greg*. Danzig, 1650.

HEVELIUS, JOHANNES: *Epistolae II. Prior: De motu lunae libratorio, in certas tabulas redacto. Ad...Iohannem Bapt. Ricciolum...Posterior: De utriusque luminaris defectu anni 1654. Ad...Petrum Nucerium*. Danzig, 1654.

HEVELIUS, JOHANNES: *Machina Coelestis*, part I. Danzig, 1673.

HEVELIUS, JOHANNES: *Machina Coelestis*, part II. Danzig, 1679.

Bibliography

HEVELIUS, JOHANNES: *Mercurius in Sole visus Gedani, Anno Christiano MDCLXI. d. III. Maj, St. n. Cum aliis quibusdam rerum Coelestium observationibus, rarisque phaenomenis. Cui annexa est Venus in Sole pariter visa, Anno 1639, d. 24 Nov. St. V. Liverpoliae, A Jeremia Horoxio, nunc primum edita, notisque illustrata. Quibus accedit succincta novae Historiola illius, ac mirae Stellae in collo Ceti, certis annis temporibus clare admodum affulgentis, rursus omnino evanescentis. Nec non Genuina Delineatio, Paraselenarum et Paraliorum quorundam rarissimorum.* Danzig, 1662.

HEVELIUS, JOHANNES: *Prodromus cometicus, quo historia, cometae anno 1664 exorti cursum, faciesque, diversas capitis ac caudae accurate delineatas complectens, nec non dissertatio, de cometarum omnium motu, generatione, variisque phaenomenis, exhibitur.* Danzig, 1665.

HEVELIUS, JOHANNES: *Selenographia: sive, lunae descriptio: atque accurata, tam macularum ejus, quam motuum diversorum, aliarumque omnium vicissitudinum, phasiumque, telescopii ope deprehensarum, delineatio.* Danzig, 1647.

HOBBES, THOMAS: *De principiis & ratiocinatione Geometrarum. Ubi ostenditur incertitudinem falsitatemque non minorem inesse scriptis eorum, quam scriptis physicorum & ethicorum. Contra fastum professorum Geometriae.* London, 1666.

HOBBES, THOMAS: *Dialogus physicus, sive de natura aeris.* London, 1661.

HOBBES, THOMAS: *Elementorum philosophiae sectio prima De corpore.* London, 1655.

HOBBES, THOMAS: *Examinatio et emendatio mathematicae hodiernae.* London, 1660.

HOBBES, THOMAS: *Lux mathematica.* London, 1672.

HOBBES, THOMAS: *Opera philosophica quae latine scripsit in unum corpus nunc primum collecta*, ed. W. MOLESWORTH. 5 vols. London, 1839–45.

HOBBES, THOMAS: *Quadratura circuli, cubatio sphaerae, duplicatio cubi, breviter demonstrata.* London, 1669.

HOBBES, THOMAS: *Rosetum geometricum, sive propositiones aliquot frustra antehac tentatae, cum censura brevi doctrinae Wallisianae de Motu.* London, 1671.

HOBBES, THOMAS: *Six Lessons to the Professors of the Mathematiques, one of geometry, the other of astronomy: in the chaires set up by the noble and learned Sir Henry Savile, in the University of Oxford*. London, 1656.

HOBBES, THOMAS: *The Correspondence of Thomas Hobbes*, ed. N. MALCOLM. 2 vols. Oxford, 1994.

HOBBES, THOMAS: *The English works of Thomas Hobbes of Malmesbury*, ed. W. MOLESWORTH. 11 vols. London, 1839–45.

HOBBES, THOMAS: *Three Papers Presented to the Royal Society Against Dr. Wallis. Together with Considerations on Dr. Wallis his Answer to Them*. London, 1671.

HODIERNA, GIOVANNI BATTISTA: *De admirandis phasibus in sole et luna visis*. Palermo, 1656.

HODIERNA, GIOVANNI BATTISTA: *De systemate orbis cometici; deque admirandis coeli characteribus, opuscula duo*. Palermo, 1654.

HODIERNA, GIOVANNI BATTISTA: *Opuscoli*. Palermo, 1644.

HODIERNA, GIOVANNI BATTISTA: *Protei caelestis vertigines seu Saturni systema*. Palermo, 1657.

HOFMANN, JOSEPH EHRENFRIED: 'Über die Exercitatio geometrica des M. A. Ricci.' *Centaurus* 9 (1963), 139–93.

HOFMANN, JOSEPH EHRENFRIED: *Ausgewählte Schriften*, ed. C. J. SCRIBA. 2 vols. Hildesheim, 1990.

HOOKE, ROBERT: *Micrographia*. London, 1665.

HORROX, JEREMIAH: *Opera Posthuma, viz. Astronomia Kepleriana defensa.... Excerpta ex epistolis ad Crabtraeum suum. Observationum coelestium catalogus. Lunae theoria nova. Accedunt G. Crabtraei... Observationes Coelestes... Adjiciuntur J. Flamstedii de Temporis Aequatione Diatriba. Numeri ad Lunae Theoriam Horrocianam*, ed. J. WALLIS. London, 1673.

HORROX, JEREMIAH: *Venus in Sole... visa Anno 1639, d. 24 Nov. St. V. Liverpoliae*. (Published in HEVELIUS: *Mercurius in Sole visus*.) 1662.

Bibliography

HUDDE, JAN: *Epistola prima de reductione aequationum.* In: DESCARTES, *Geometria,* Amsterdam, 1659, I, 401–506.

HUDDE, JAN: *Epistola secunda de maximis et minimis.* In: DESCARTES: *Geometria,* Amsterdam, 1659, I, 507–16.

HUIPS, FRANS VAN DER: *Algebra.* Amsterdam, 1654.

HUNTER, MICHAEL: *John Aubrey and the Realm of Learning.* New York, 1975.

HUNTER, MICHAEL: *The Boyle Papers. Understanding the Manuscripts of Robert Boyle.* Aldershot and Burlington, VT, 2007.

HUNTER, MICHAEL: *The Royal Society and its Fellows, 1660–1700. The morphology of an early scientific institution.* 2nd edn. Chalfont St Giles, 1994.

HUYGENS, CHRISTIAAN: 'A Summary Account of the Laws of Motion, communicated by Mr. Christian Hugens in a Letter to the R. Society.' *Philosophical Transactions* No. 46 (12 April 1669), 925–8.

HUYGENS, CHRISTIAAN: *De circuli magnitudine inventa.* Leiden, 1654.

HUYGENS, CHRISTIAAN: 'Examen de Vera Circuli & Hyperboles Quadratura, in propria sua proportionis specie inventa & demonstrata a Jacobo Gregorio Scoto, in 4°. Patavii.' *Journal des Sçavans* (2 July 1668), 52–6.

HUYGENS, CHRISTIAAN: *Œuvres complètes,* ed. D. BIERENS DE HAAN, J. BOSSCHA, D. J. KORTEWEG, and J. A. VOLLGRAFF. 22 vols. The Hague, 1888–1950; repr. Amsterdam, 1967–78.

HUYGENS, CHRISTIAAN: *Systema saturnium, sive de causis mirandorum Saturni phaenomenon.* The Hague, 1659.

HUYGENS, CHRISTIAAN: *Theoremata de quadratura hyperboles, ellipsis et circuli, ex dato portionum gravitatis centro.* Leiden, 1651.

HYDE, THOMAS: *Catalogus impressorum librorum bibliothecae Bodlijanae in academia Oxoniensi.* Oxford, 1674.

JESSEPH, DOUGLAS M.: *Squaring the Circle: the War between Hobbes and Wallis.* Chicago and London, 1999.

Bibliography

KEPLER, JOHANNES: *Gesammelte Werke*, ed. M. CASPAR, F. HAMMER, and W. VON DYCK. Munich, 1937ff.

KEPLER, JOHANNES: *Tabulae Rudolphinae, quibus astronomiae scientiae, temporum longinquitate collapsae restauratio continetur*. Ulm, 1627.

KERSEY, JOHN: *The Elements of that Mathematical Art commonly called Algebra*. 4 vols. London, 1673.

KIESSLING, NICOLAS K.: *The Library of Anthony Wood*. Oxford, 2002.

KINCKHUYSEN, GERARD: *Algebra ofte stelkonst*. Haarlem, 1661.

LALOUBÈRE, ANTOINE DE: *De cycloide Galilaei et Toricellii propositiones viginti*. Toulouse, 1658.

LALOUBÈRE, ANTOINE DE: *Quadratura circuli et hyperbolae segmentorum*. Toulouse, 1651.

LALOUBÈRE, ANTOINE DE: *Veterum geometria promota in septem de cycloide libris*. Toulouse, 1660.

LEIBNIZ, GOTTFRIED WILHELM: *Hypothesis physica nova*. Mainz 1671; repr. (with *Theoria motus abstracti*) London, 1671.

LEIBNIZ, GOTTFRIED WILHELM: *Sämtliche Schriften und Briefe*, ed. Prussian Academy of Sciences (and its successors). Series I–VIII. Darmstadt, Leipzig, Berlin, 1923ff.

LEIBNIZ, GOTTFRIED WILHELM: *Theoria motus abstracti*. Mainz 1671; repr. (with *Hypothesis physica nova*) London, 1671.

LE TENNEUR, JACQUES-ALEXANDRE: *Traité des quantités incommensurables, où sont décidées plusieurs belles questions des nombres rationaus et irrationaus. Les erreurs de Stevin réfutées. Et le Dizième livre d'Euclide, illustré de nouvelles démonstrations plus faciles et plus succinctes que les ordinaires, et réduit à 62 propositions. Avec un Discours de la manière d'expliquer les sciences en français*. Paris, 1640.

LOWER, RICHARD: *Diatribe Th. Willisii de Febribus vindicatio*. Amsterdam, 1666.

LOWER, RICHARD: *Tractatus de corde, item de motu et colore sanguinis*. London, 1669.

Bibliography

MADDISON, ROBERT EDWIN WITTON: *The life of the honourable Robert Boyle F.R.S.* London, 1969.

MAGALOTTI, LORENZO: *Saggi di naturali esperienze fatte nell' Accademia del cimento, descritte dal segretario.* Florence, 1667.

MAIERÙ, LUIGI: *John Wallis. Una vita per un progretto.* Soveria Mannelli, 2007.

MALCOLM, NOEL: 'An unpublished letter from Henry Oldenburg to Johann Heinrich Rahn.' *Notes and Records of the Royal Society of London* 58 (2004), 249–66.

MALCOLM, NOEL: *Aspects of Hobbes.* Oxford, 2002.

MALCOLM, NOEL and MIKKO TOLONEN: 'The correspondence of Thomas Hobbes: some new items.' *The Historical Journal* 51 (2008), 481–95.

MALLET, CHARLES EDWARD: *A History of the University of Oxford.* 3 vols. London, 1924–27; repr. New York and London, 1968.

MALYNES, GERARD: *Consuetudo, vel, Lex mercatoria, or The Ancient Law Merchant.* London, 1622.

MARCHETTI, ALESSANDRO: *Exercitationes mechanicae.* Pisa, 1669.

MARTINI, CHRISTIAAN: *Slot en sleutel van de navigation, ofte groote zee-vaert.* Amsterdam, 1659.

MARTINI, MARTINO: *Novus atlas Sinensis.* [Amsterdam] 1655.

MENGOLI, PIETRO: *Novae quadraturae arithmeticae: seu, De additione fractionum.* Bologna, 1650.

MENGOLI, PIETRO: *Speculazioni di musica.* Bologna, 1670.

MERCATOR, NICOLAUS: *Logarithmotechnia: sive methodus construendi Logarithmos nova, accurata, & facilis; scripto antehac communicata, Anno Sc. 1667. nonis Augusti: Cui nunc accedit. Vera Quadratura Hyperbolae, & Inventio Summae Logarithmorum.* London, 1668.

MERRET, CHRISTOPHER: *Pinax rerum naturalium Britannicarum.* London, 1667.

MERSENNE, MARIN: *Correspondance du P. Marin Mersenne, religieux minime*, ed. C. DE WAARD, R. PINTARD, B. ROCHOT, A. BEAULIEU, et al. 17 vols. Paris, 1932–88.

MITCHELL, JOHN: *A breviat cronicle, containing all the kynges from Brut to this daye.* [Canterbury], 1554.

MONTANARI, GEMINIANO: *Pensieri fisico-matematici sopra alcune esperienze fatte in Bologna ... intorno diversi effetti de liquido in cannuccie di vetro, & altri vasi.* Bologna, 1667.

MORE, HENRY: *Enchiridron ethicum, praecipua moralis philosophiae rudimenta complectens.* London, 1668.

MORLAND, SAMUEL: *Tuba Stentoro-Phonica, an instrument of excellent use, as well at sea, as at land.* London, 1671.

NEEDHAM, WALTER: *Disquisitio anatomica de formato foetu.* London, 1667.

NEWMAN, JOHN: 'The Architectural Setting.' In: *The History of the University of Oxford. IV. Seventeenth-Century Oxford*, ed. N. TYACKE. Oxford, 1997, 135–77.

NEWTON, ISAAC: *The Correspondence of Isaac Newton*, ed. H. W. TURNBULL, A. R. HALL, L. TILLING, J. F. SCOTT. 7 vols. Cambridge, New York, 1959–77.

NEWTON, JOHN: *The Art of natural arithmetick: in whole numbers and fractions vulgar and decimal.* London, 1671.

NEWTON, JOHN: *The Art of Practical Gauging.* London, 1669.

OLDENBURG, HENRY: *The Correspondence of Henry Oldenburg*, ed. A. R. HALL and M. B. HALL. 13 vols. Madison, London, 1965–86.

OUGHTRED, WILLIAM: *Arithmeticae in numeris et speciebus institutio, quae ... totius mathematicae quasi clavis est.* London, 1631. From the 2nd Latin edition (London 1648) onwards under the title *Clavis mathematicae*; 3rd edn., ed. J. WALLIS, Oxford, 1652; 4th edn., ed. J. WALLIS, Oxford, 1667.

ODNB: *Oxford Dictionary of National Biography*, ed. C. MATTHEW and B. HARRISON. 60 vols. Oxford, 2004.

Bibliography

PALMER, JOHN: *The Catholique planisphaer. Which Mr. Blagrave calleth the mathematical Jewel.* London, 1658.

PARDIES, IGNACE GASTON: *Élémens de géométrie.* Paris, 1671.

PARDIES, IGNACE GASTON: *Globi coelestis in tabulas planas redacti descriptio.* Paris, 1674.

PASCAL, BLAISE: *Historia trochoïdis sive cycloïdis; gallicè la roulette.* Paris, 1658.

PASCAL, BLAISE: *Lettres de A. Dettonville contenant quelques-unes de ses inventions de Geometrie.* Paris, 1659.

PASCAL, BLAISE: *Œuvres*, ed. L. BRUNSCHVICG, P. BOUTROUX, and F. GAZIER. 14 vols. Paris, 1904–14.

PEPYS, SAMUEL: *The Diary of Samuel Pepys*, ed. R. LATHAM and W. ANDREWS. 11 vols. London, 1970–83.

PETERS, CHRISTIAN HEINRICH F. and EDWARD B. KNOBEL: *Ptolemy's Catalogue of Stars. A revision of the Almagest.* Washington, 1915.

PICO, GERONIMO: *Tesoro di matematiche considerationi dove si contiene la teorica e la prattica di tutta la geometria, il trattato della transformatione, circonscrittione, & riscrittione delle figure piane e solide.* Rome, 1645.

PLOMER, HENRY R.: *A Dictionary of the Booksellers and Printers who were at Work in England, Scotland and Ireland from 1641 to 1667.* London, 1907.

PLOMER, HENRY R.: *A Dictionary of the Printers and Booksellers who were at Work in England, Scotland and Ireland from 1668 to 1725.* Oxford, 1922.

POCOCKE, EDWARD: *A Commentary on the Prophecy of Hosea.* Oxford, 1685.

POCOCKE, EDWARD, ed.: *Philosophus autodidactus sive epistola Abi Jaafar, Ebn Tophail.* Oxford, 1671.

POCOCKE, EDWARD, ed.: *The Nature of the drink Kauhi, or Coffe, and the Berry of which it is made.* Oxford, 1659.

Bibliography

POOLE, WILLIAM: *John Aubrey and the Advancement of Learning.* Oxford, 2010.

POPE, WALTER: *The Life of the Right Reverend Father in God Seth, Lord Bishop of Salisbury.* London, 1697.

POWELL, ANTHONY: *John Aubrey and his Friends.* London, 1948.

POWICKE, MAURICE and E. B. FRYDE, eds.: *Handbook of British Chronology.* 2nd edn. London, 1961.

PRAG, ADOLF: 'John Wallis, 1616–1703. Zur Ideengeschichte der Mathematik im 17. Jahrhundert.' *Quellen und Studien zur Geschichte der Mathematik, Astronomie und Physik, Abteilung B: Studien* 1 (1931), 381–412.

PROBST, SIEGMUND: *Die mathematische Kontroverse zwischen Thomas Hobbes und John Wallis.* Hanover, 1997 (Universität Regensburg doctoral thesis, 1994).

PTOLEMY, CLAUDIUS: *Ptolemy's Almagest,* ed. G. J. TOOMER. Princeton, 1998.

PTOLEMY, CLAUDIUS: *The history of Ptolemy's star catalogue,* ed. G. GRASSHOFF. New York, 1990.

PTOLEMY, CLAUDIUS: *Claudii Ptolemaei Opera quae exstant omnia.* Leipzig, 1898–1954. Reprinted 1961ff.

RAHN, JOHANN HEINRICH: *An introduction to algebra, translatedout of the High-Dutch into English by Thomas Brancker, ... Much altered and augmented by D[r] P[ELL]. Also a table of odd numbers less than one hundred thousand, shewing those that are incomposit.* London, 1668.

REINHOLD, ERASMUS: *Prutenicae tabulae coelestium motuum.* Tübingen, 1551.

RENALDINI, CARLO: *Ars analytica mathematum.* 3 vols. Florence, 1665–9.

RENALDINI, CARLO: *Geometra promotus.* Padua, 1670.

RICCI, MICHEL ANGELO: *Exercitatio geometrica de maximis et minimis.* [Rome, 1666.]

RICCIOLI, GIOVANNI BAPTISTA: *Almagestum novum Astronomiam veterem novamque complectens*. Bologna, 1651.

RICCIOLI, GIOVANNI BAPTISTA: *Astronomiae reformatae tomi duo*. Bologna, 1665.

RIDER, ROBIN E.: *A Bibliography of Early Modern Algebra 1500–1800*. Berkeley, 1982.

RIGAUD, STEPHEN J., ed.: *Correspondence of Scientific Men of the Seventeenth Century*. 2 vols. Oxford, 1841; repr. Hildesheim, 1965.

RUDERMAN, ARTHUR: *The Personal Life and Family of Doctor John Wallis F.R.S., born in Ashford, 1616*. Folkestone, 1997.

ROBERTS, LEWES: *The Merchants Map of Commerce*. London, 1671.

SAINT-VINCENT, GRÉGOIRE DE: *Opus geometricum quadraturae circuli et sectionum coni decem libris comprehensum*. Antwerp, 1647.

SAINTY, JOHN CHRISTOPHER: *Officials of the Secretaries of State, 1660–1782*. London, 1973.

SAINTY, JOHN CHRISTOPHER: *Treasury Officials, 1660–1782*. London, 1972.

SCALIGER, JULIUS CAESAR: *Exotericarum exercitationum liber quintus decimus de subtilitate, ad Hieronymum Cardanum*. Paris, 1557.

SCHEINER, CHRISTOPH: *Rosa ursina sive sol ex admirando facularum & macularum suarum phaenomeno varius*. Braga, 1630.

SCHOOTEN, FRANS VAN: *Appendix de cubicarum aequationum resolutione*. In: DESCARTES, *Geometria*. Amsterdam, 1659, I, 345–68.

SCHOOTEN, FRANS VAN: *Tractatus de concinnandis demonstrationibus geometricis ex calculo algebraico*, ed. P. VAN SCHOOTEN. Amsterdam, 1661.

SCHOTT, CASPAR: *Cursus mathematicus*. Würzburg, 1561.

SCOTT, JOSEPH FREDERICK: *The Mathematical Work of John Wallis, D.D., F.R.S. (1616–1703)*. London, 1938; repr. New York, 1981.

SCRIBA, CHRISTOPH J., 'A Conjecture on Mr Gott's Proposal of an Artificiall Spring (1668).' *Algorismus* 60 (2007), 389–96.

SCRIBA, CHRISTOPH J.: 'Gregory's Converging Double Sequence.' *Historia mathematica* 10 (1983), 274–85.

SCRIBA, CHRISTOPH J.: *James Gregorys frühe Schriften zur Infinitesimalrechnung.* Gießen, 1957.

SCRIBA, CHRISTOPH J.: *Studien zur Mathematik des John Wallis (1616–1703). Winkelteilungen, Kombinationslehre und Zahlentheoretische Probleme.* Stuttgart, 1966.

SCRIBA, CHRISTOPH J.: 'The Autobiography of John Wallis, F.R.S.' *Notes and Records of the Royal Society of London* 25 (1970), 17–46.

SENESCHAL, MICHAEL: *Trias evangelica, sive, Quaestio triplex de anno, mense, et die Christi nati, baptizati, et mortui.* Liège, 1670.

SFONDRATI, PANDOLFO: *Causa aestus maris.* Ferrara, 1590.

SHAKERLEY, JEREMY: *Tabulae Britannicae.* London, 1653.

SLUSE, RENÉ FRANÇOIS DE: *Mesolabum.* Liège, 1659. 2nd edn. Liège, 1668.

SNELLIUS, WILLIBRORD: *Cyclometricus de circuli dimensione secundum logistarum abacos.* Leiden, 1621.

SORBIÈRE, SAMUEL: *Relation d'un voyage en Angleterre, où sont touchées plusieurs choses, qui regardent l'estat des Sciences, de la Religion, & autres matieres curieuses.* Paris, 1664.

SOTHEBY: *Catalogue: The Library of the Earls of Macclesfield removed from Shirburn Castle.* Part Two: Science A–C. [London, 2004]

SPENCER, JOHN: *Dissertatio de Urim & Thummim.* Cambridge, 1669.

SPRAT, THOMAS: *The History of the Royal Society of London, For the Improving of Natural Knowledge.* London, 1667.

SPURR, JOHN: *England in the 1670s. 'This Masquerading Age'.* Oxford and Malden, Mass., 2000.

STEDALL, JACQUELINE A.: *A Discourse Concerning Algebra. English Algebra to 1685.* Oxford, 2003.

STEDALL, JACQUELINE A.: 'Ariadne's Thread: The Life and Times of Oughtred's *Clavis*.' *Annals of Science* 57 (2000), 27–60.

Bibliography

STEVIN, SIMON: *Les œuvres mathematiques de Simon Stevin, où sont inserées les Memoires mathematiques esquelles s'est exercé Maurice de Nassau, prince d'Aurenge. Le tout revue par A. Girard*, 6 vols in 2 parts. Leyden, 1634.

STOW, JOHN: *The annales of England untill 1592. Continued unto the ende of 1631*, by E. HOWES. With an appendix. London, 1631.

STREET, THOMAS: *Astronomia Carolina. A new theorie of the celestial motions*. London, 1661.

STUBBE, HENRY: *A censure upon certaine passages contained in the History of the Royal Society, as being destructive to the established religion and church of England*. Oxford, 1670.

STUBBE, HENRY: 'An Enlargement of the Observations, Formerly Publisht Numb. 27.' *Philosophical Transactions* No. 36 (15 June 1668), 699–709.

STUBBE, HENRY: *Campanella revived, or an Enquiry into the history of the Royal Society, whether the virtuosi there do not pursue the projects of Campanella for the reducing of England to popery*. London, 1670.

STUBBE, HENRY: *Legends on histories: or, A specimen of some animadversions upon the history of the Royal Society*. London, 1670.

STUBBE, HENRY: 'The Remainder of the Observations made in the formerly mention'd Voyage to Jamaica, publisht Numb. 36.' *Philosophical Transactions* No. 37 (13 July 1668), 717–22.

TAYLOR, EVA GERMAINE RIMINGTON: *The Mathematical Practitioners of Tudor & Stuart England 1485–1714*. Cambridge, 1967.

TENCA, LUIGI: 'Giovanni Wallis e gli Italiani.' *Bollettino della unione matematica Italiana* 19 (1955), 412–18.

THÉVENOT, MELCHISÉDECH: *Relations de divers voyages curieux*. Paris, 1663.

THRUSTON, MALACHI: *De respirationis usus primario, diatriba*. London, 1670.

TORRICELLI, EVANGELISTA: *De sphaera et solidis sphaeralibus libri duo*. Florence, 1644.

Bibliography

TORRICELLI, EVANGELISTA: *Opera geometrica.* Florence, 1644.

TURNBULL, HERBERT WESTREN, ed.: *James Gregory Tercentenary Memorial Volume.* London, 1939.

TYACKE, NICHOLAS, ed.: *The History of the University of Oxford*, vol. 4: *Seventeenth-Century Oxford.* Oxford, 1997.

ULUG BEG [BEIG]: *Jadâwil-i mawâdi'-i thawâbit dar tûl u'ard kih bi-rasad yâftah ast Ulugh Baik Sive Tabulae long. ac lat. Stellarum fixarum. Ex tribus invicem collatis MSS. Persicis jam primum luce ac Latino donavit, & Commentariis illustravit, Thomas Hyde.* Oxford, 1665.

VENN, JOHN AND J. A. VENN, eds.: *Alumni Cantabrigienses: A biographical list of all known students, graduates and holders of office at the University of Cambridge, from the earliest times to 1900.* Part 1 (from the earliest times to 1751), 4 vols. Cambridge, 1922–27.

VIÈTE, FRANÇOIS: *In artem analyticam isagoge.* Tours, 1591.

VIÈTE, FRANÇOIS: *Opera mathematica, in unum Volumen congesta, ac recognita*, ed. F. VAN SCHOOTEN. Leiden, 1646.

VOSSIUS, GERARDUS JOANNES: *De quatuor artibus popularibus, de philologia et scientiis mathematicis, cui opera subjungitur: chronologia mathematicorum libri III, editio nova.* Amsterdam, 1660.

VOSSIUS, ISAAC: *De motu marium et ventorum liber.* The Hague, 1663.

WALKER, GEORGE: *A Defence of the true sence and meaning of the words of the Holy Apostle, Rom. chap. 4, ver. 3, 5, 9.* [London], 1641.

WALKER, GEORGE: *Socinianisme in the Fundamentall point of Justification discovered and confuted.* London, 1641.

WALLIS, JOHN: 'A Continuation of Dr. Wallis his second Letter, to the printed Paper of Mr. Du Laurens.' *Philosophical Transactions* No. 41 (16 November 1668), 825–32.

WALLIS, JOHN: *A Defense of the Treatise of the Angle of Contact.* London: Printed by John Playford, for Richard Davis, Bookseller, in the University of Oxford, 1684.

Bibliography

WALLIS, JOHN: *A Discourse of Combinations, Alternations, and Aliquot Parts.* London: Printed by John Playford, for Richard Davis, Bookseller, in the University of Oxford, 1685.

WALLIS, JOHN: *A Discourse of Gravity and Gravitation, grounded on experimental observations.* London: Printed for John Martyn, 1675.

WALLIS, JOHN: *Adversus Marci Meibomii de proportionibus dialogum tractatus elencticus.* Oxford: Typis Leon. Lichfield, Academiae Typographi, Impensis Tho. Robinson, 1657 (in *Operum mathematicorum pars prima*).

WALLIS, JOHN: 'A Letter written by Dr. John Wallis to the Publisher, concerning the Variety of the Annual High-Tydes, as to several places; with respect to his own Hypothesis, deliverd No. 16, touching the Flux and Reflux of the Sea.' *Philosophical Transactions* No. 34 (13 April 1668), 652–3.

WALLIS, JOHN: *Algebra*, 'Additions and Emendations'. London, John Playford for Richard Davies, 1685.

WALLIS, JOHN: 'An Accompt of a small Tract, entituled, Thomae Hobbes Quadratura Circuli, Cubatio Sphaerae, Duplicatio Cubi, secundo edita,) Denuo Refutata, Auth. Joh. Wallis. S. T. D. Geom. Prof. Saviliano. Oxoniae, 1669.' *Philosophical Transactions* No. 55 (17 January 1670), 1121–2.

WALLIS: 'An Answer to Three Papers of Mr. Hobs, lately published in the Months of August, and this present September, 1671.' [1671]

WALLIS, JOHN: 'An Answer to Four Papers of Mr. Hobs, lately published in the Months of August, and this present September, 1671.' *Philosophical Transactions* No. 75 (18 September 1671), 2241–50.

WALLIS, JOHN: 'An Answer of Dr. Wallis to Mr. Hobbes's Rosetum Geometricum in a Letter to a friend in London, dated July 16. 1671.' *Philosophical Transactions* No. 73 (17 July 1671), 2202–9.

WALLIS, JOHN: 'An Appendix, written by way of a Letter to the Publisher; Being an Answer to some Objections, made by several Persons, to the precedent Discourse.' *Philosophical Transactions* No. 16 (6 August 1666), 281–9.

Bibliography

WALLIS, JOHN: 'An Essay of Dr. John Wallis, exhibiting his Hypothesis about the Flux and Reflux of the Sea.' *Philosophical Transactions* No. 16 (6 August 1666), 263–81.

WALLIS, JOHN: 'Animadversions of Dr. Wallis, upon Mr. Hobs's late Book, De Principiis et Ratiocinatione Geometrarum.' *Philosophical Transactions* No. 16 (6 August 1666), 289–94.

WALLIS: 'Another Letter Written by the same Hand, concerning some Mistakes. to be found in a Book lately publishd under the Title of Specimina mathematica Francisci Du Laurens, especially touching a certain Probleme, affirmd to have been proposed by Dr. Wallis, to the Mathematicians of all Europe, to solve.' *Philosophical Transactions* No. 34 (13 April 1668), 654–[655].

WALLIS, JOHN: 'A Relation concerning the late Earthquake neer Oxford; together with some Observations of the sealed Weatherglass, and the Barometer, both upon that Phaenomenon, and in General.' *Philosophical Transactions* No. 10 (12 March 1665/6), 166–71.

WALLIS, JOHN: *Arithmetica infinitorum, sive nova methodus inquirendi in curvilineorum quadraturam, aliaque difficiliora matheseos problemata.* Oxford: Typis Leon. Lichfield, Academiae Typographi, Impensis Tho. Robinson, 1656 (in *Operum mathematicorum pars altera*).

WALLIS, JOHN: 'A second Letter... on the same printed Paper of Franciscus Du Laurens.' *Philosophical Transactions* No. 39 (21 September 1668), 775–9.

WALLIS, JOHN: 'A Summary Account given by Dr. John Wallis, of the General Laws of Motion, by way of Letter written by him to the Publisher, and communicated to the R. Society, Novemb. 26. 1668.' *Philosophical Transactions* No. 43 (11 January 1668/9), 864–6.

WALLIS, JOHN: *Autobiography.* See: SCRIBA, 'Autobiography of John Wallis'.

WALLIS, JOHN: *Commercium epistolicum, de quaestionibus quibusdam mathematicis nuper habitum.* Oxford: Excudebat A. Lichfield, Acad. Typographi; Impensis Tho. Robinson, 1658.

WALLIS, JOHN: *Cono-Cuneus: or, the Shipwright's circular wedge.* London: Printed by John Playford, for Richard Davis, Bookseller, in the University of Oxford, 1684.

Bibliography

WALLIS, JOHN: *De aestu maris; hypothesis nova.* In *Opera mathematica I.* Oxford: e Theatro Sheldoniano, 1695; repr. Hildesheim and New York, 1972, 737–56.

WALLIS, JOHN: *De angulo contactus et semicirculi disquisitio geometrica.* Oxford: Typis Leon. Lichfield Academiae Typographi, Impensis Tho: Robinson, 1656 (in *Operum mathematicorum pars altera*).

WALLIS, JOHN: *Defence of the Royal Society, and the Philosophical Transactions, Particularly those of July, 1670. In Answer to the Cavils of Dr. William Holder.* London: Printed by T. S. for Thomas Moore, 1678.

WALLIS, JOHN: *Defense of the Treatise of the Angle of Contact.* London: John Playford for Richard Davis, Bookseller, in the University of Oxford, 1684.

WALLIS, JOHN: *Due Correction for Mr Hobbes. Or Schoole Discipline, for not saying his Lessons right. In Answer To His Six lessons, directed to the Professors of Mathematicks. By the Professor of Geometry.* Oxford: Leonard Lichfield for Thomas Robinson, 1656.

WALLIS, JOHN: *Elenchus geometriae Hobbianae. Sive, geometricorum, quae in ipsius Elementis philosophiae a Thoma Hobbes Malmesburiensi proferuntur, refutatio.* Oxford: Excudebat H. Hall; Impensis Johannis Crook, 1655.

WALLIS, JOHN: *Grammatica linguae Anglicanae. Cui praefigitur, De loquela, sive sonorum formatione, tractatus grammatico-physicus.* Oxford: Excudebat Leon. Lichfield Acad. Typographus. Veneunt apud Tho. Robinson, 1653; *Editio secunda, priore auctior*, Oxford: Lichfieldianis, Acad. Typog., 1664.

WALLIS, JOHN: *Hobbiani Puncti Dispunctio. Or The Undoing of Mr Hobss Points; In Answer to M. Hobss ΣΤΙΓΜΑΙ, Id est, Stigmata Hobbii.* Oxford: Leonard Lichfield for Thomas Robinson, 1657.

WALLIS, JOHN: *Hobbius Heauton-timorumenos. Or a consideration of Mr. Hobbes his dialogues. In an epistolary discourse, addressed to the Hon. R. Boyle.* Oxford: A. & L. Lichfield, for Samuel Thomson, 1662.

WALLIS, JOHN: *Mathesis universalis: sive, Arithmeticum opus Integrum, tum philologice, tum mathematice traditum.* Oxford: Typis Leon. Lichfield Academiae typographi, impensis Tho. Robinson, 1657 (In *Operum mathematicorum pars prima*).

Bibliography

WALLIS, JOHN: *Mechanica: sive, De motu, tractatus geometricus. Pars prima.* London: Typis Gulielmi Godbid; impensis Mosis Pitt, 1670.

WALLIS, JOHN: *Mechanica: sive, De motu, tractatus geometricus. Pars secunda.* London: Typis Gulielmi Godbid; impensis Mosis Pitt, 1670.

WALLIS, JOHN: *Mechanica: sive, De motu, tractatus geometricus. Pars tertia.* London: Typis Gulielmi Godbid; impensis Mosis Pitt, 1671.

WALLIS, JOHN: *Opera mathematica* I. Oxford: e Theatro Sheldoniano, 1695; repr. Hildesheim and New York, 1972.

WALLIS, JOHN: *Opera mathematica* II. Oxford: e Theatro Sheldoniano, 1693; repr. Hildesheim and New York, 1972.

WALLIS, JOHN: *Opera mathematica* III. Oxford: e Theatro Sheldoniano, 1699; repr. Hildesheim and New York, 1972.

WALLIS, JOHN: *Operum mathematicorum pars altera.* Oxford: Typis Leon. Lichfield Academiae typographi, Impensis Tho. Robinson, 1656.

WALLIS, JOHN: *Operum mathematicorum pars prima.* Oxford: Typis Leon. Lichfield Academiae typographi, Impensis Tho. Robinson, 1657.

WALLIS, JOHN: 'Some Animadversions...on...Responsio Francisci Du Laurens.' *Philosophical Transactions* No. 38 (17 August 1668), 744–50.

WALLIS, JOHN: 'Some Observations Concerning the Baroscope and Thermoscope, made and communicated by Doctor J. Wallis at Oxford, and Dr. J. Beale at Yeovil in Somerset, deliver'd here according to the several dates when they were inparted.' *Philosophical Transactions* No. 55 (17 January 1669/70), 1113–20, 1116–20.

WALLIS, JOHN: *The Correspondence of John Wallis (1616–1703)*, ed. P. BEELEY and C. J. SCRIBA. Oxford, 2003ff.

WALLIS, JOHN: *Thomae Hobbes Quadratura circuli, cubatio sphaerae, duplicatio cubi, confutata.* Oxford: Typis Lichfieldianis; Impensis Tho. Gilbert, 1669.

WALLIS, JOHN: *Thomae Hobbes Quadratura Circuli...denuo refutata.* Oxford: Typis Lichfieldianis; Impensis Tho. Gilbert, 1669.

Bibliography

WALLIS, JOHN: *Tractatus duo, prior de cycloide et corporibus inde genitis, posterior epistolaris in qua agitur de cissoide et corporibus inde genitis, et de curvarum, tum linearum Εὔθυνσις, tum superficierum Πλατυσμός.* Oxford: Typis Academicis Lichfieldianis, 1659.

WALLIS, JOHN: *A Treatise of Algebra, Both Historical and Practical. Shewing, The Original, Progress, and Advancement thereof, from Time to Time; and by what Steps it hath Attained to the Heighth at which now it is. With some Additional Treatises.* London: Printed by John Playford, for Richard Davis, Bookseller, in the University of Oxford, 1685.

WALLIS, JOHN: *A Treatise of Angular Sections.* London: Printed by John Playford for Richard Davis, 1684.

WALTON, BRIAN, ed.: *S. S. Biblia Polyglotta. Complectentia textus originales, Hebraicos, cum Pentateucho Samaritano, Chaldaicos, Graecos, versionumque antiquarum Samaritanae, Graecae sept., Chaldaicae, Syriacae, Vulg. Lat., Arabicae, Aethiopicae, Persicae quicquid comparari poterat ... Ex mss. antiquiss. undique conquisitis optimisque ex impressis summa fide collatis.* 6 vols. London, 1655–57.

WARD, JOHN: *The Lives of the Professors of Gresham College: To which is prefixed the life of the founder, Sir Thomas Gresham.* London, 1740.

WARDHAUGH, BENJAMIN: *Music, Experiment and Mathematics in England, 1653–1705.* Farnham and Burlington, 2008.

WASSENAER, JACOB VAN: *Den on-wissen wis-konstenaer J. Stampioenius ontdeckt door zyne ongegronde weddinge ende mis-lucte solutien van syne eygene questien.* Leiden, 1640.

WBI: *Internationaler Biographischer Index/World Biographical Index*, First CD-ROM edition. Munich, New Providence, London, Paris, 1994.

WEBSTER, CHARLES: *The Great Instauration. Science, Medicine and Reform 1626–1660.* London, 1975.

WEINREB, BEN and CHRISTOPHER HIBBERT: *The London Encyclopaedia.* London, 1983.

WERNER, GEORG CHRISTOPH: *Inventum novum, artis et naturae connubium, in copulatione levitatis cum gravitate & gravitatis cum levitate. Per artificium siphonis machinae aquaticae antliae.* Augsburg, 1670.

Bibliography

WESTFALL, RICHARD: *Database: Catalog of the Scientific Community in the 16th and 17th Centuries.* http://es.rice/ES/humsoc/Galileo/Catalog/catalog.html.

WHITESIDE, DEREK THOMAS: 'Patterns of Mathematical Thought in the Later Seventeenth Century.' *Archive for History of Exact Sciences* 1 (1961), 179–388.

WILKEN, MARTIJN: *Officina algebrae.* Groningen, 1634.

WILKINS, JOHN: *An Essay towards a Real Character and a Philosophical Language.* London, 1668.

WILKINS, JOHN: *The discovery of a world in the moone, or, A discourse tending to prove, that 'tis probable there may be another habitable world in that planet.* London, 1638.

WILLIS, THOMAS: *Pathologiae cerebri.* London, 1667.

WILLIS, THOMAS: *Cerebri anatome: cui accessit, Nervorum descriptio et usus.* London, 1664.

WING, VINCENT: *Astronomia Britannica.* London, 1669.

WING, VINCENT: *Harmonicon coeleste: or, the coelestiall harmony of the visible world.* London, 1651.

WOOD, ANTHONY: *Athenae Oxonienses: an exact history of all the writers and bishops who have had their education in the University of Oxford: to which are added the Fasti, or annals of the said university,* ed. P. BLISS. 5 vols. in 4. London, 1813–20; repr. Hildesheim, 1969.

WOOD, ANTHONY: *History and Antiquities of the University of Oxford; in two books by Anthony à Wood,* ed. J. GUTCH. 2 vols. Oxford, 1792–6.

WOOD, ANTHONY, ed.: *Survey of the Antiquities of the City of Oxford, composed in 1661–6 by Anthony Wood,* ed. A. CLARK. 3 vols. Oxford, 1889–99.

WOOD, ANTHONY: *The Life and Times of Anthony Wood, antiquary, of Oxford, 1632-1695, described by himself,* ed. A. CLARK. 5 vols. Oxford, 1891–1900.

Bibliography

WREN, CHRISTOPHER: 'A Description of Dr. Christopher Wren's Engin, designed for grinding Hyperbolical Glasses.' *Philosophical Transactions* No. 53 (15 November 1669), 1059–60.

WREN, CHRISTOPHER: 'Dr. Christopher Wrens Theory concerning the same Subject; imparted to the R. Society Decemb. 17. last.' *Philosophical Transactions* No. 43 (11 January 1668/9), 867–8.

WREN, CHRISTOPHER: 'Generatio Corporis Cylindroidis Hyperbolici, elaborantis Lentibus Hyperbolicis accomodati.' *Philosophical Transactions* No. 48 (21 June 1669), 961–2.

WREN, CHRISTOPHER: *Parentalia: or, Memoirs of the Family of the Wrens.* London, 1750.

LIST OF LETTERS

? to BROUNCKER (1668/summer), 427
? to WALLIS (1670/III/middle), **343**, 336
? to WALLIS (1670/VIII–IX), 394, **395**

AUBREY to WALLIS (1670/IX/6), **393**, 395, 396

BARROW to COLLINS (1669/III/23), 9
BEALE to OLDENBURG (1668/X/22), 38
BERNARD to COLLINS (1671/IV/13), 431
BERTET to COLLINS (1671/I/26), 529
BERTET to WALLIS (1671/beginning), **426**
BERTET to WALLIS (1671/II–III), 452
BERTET to WALLIS (1671/V–VI), **452**
BERTET to WALLIS (1671/XII/1), 452, **529**, 549
BORELLI to WALLIS (1669/second half (i)), **280**, 294
BORELLI to WALLIS (1669/second half (ii)), **280**, 290, 294
BORELLI to WALLIS (1670/XII/6), **410**
BOYLE, WALLIS, and COLLINS: Catalogue of Books sent to Borelli, 1669/70/I, **297**, 303, 304
BOYLE to OLDENBURG (1663/XI/8), 451
BOYLE to WALLIS (1669/VII/13), **228**, 233
BOYLE to WALLIS (1669/VIII/17), **241**, 245, 247
BRANCKER to PELL (1668/X/17), 16
BROUNCKER to WALLIS (1657/XI/1), 429
BROUNCKER to WALLIS (1671/II/25), **427**

CAVENDISH to MERSENNE (1640/XI/1), 17
CHILDREY to OLDENBURG (1669/IV/1), 164
CHILDREY to OLDENBURG (1670/IV/8), 305
CHILDREY to OLDENBURG (1670/IV/22 and 25), **365**
CHILDREY to WALLIS (1670/III/c.30), **334**, 341
CHILDREY to WARD (1670/III/14), **305**, 327, 336, 341, 400
COLLINS to BAKER (1676/VIII/29), 146
COLLINS to BERNARD (1671/III/26), 427, 430, **431**
COLLINS to BERNARD (1671/XII/early), 543
COLLINS to GREGORY (1669/III/25), 161, 211
COLLINS to GREGORY (1669/XII/5), 210
COLLINS to GREGORY (1670/X/9), 420
COLLINS to GREGORY (1671/I/3), 548
COLLINS to GREGORY (1672/III/24), 544, 548
COLLINS to OLDENBURG (1670/II–III?), 294–297, **303**
COLLINS to PELL (1668/V/1), 8
COLLINS to PELL (1668/VII/28), 8
COLLINS to PELL (1668/X/2), 15
COLLINS to PELL (1668/XI/2), **7**, 15, 17, 24
COLLINS to PELL (1668/XI/24), 17, **44**
COLLINS to VERNON (late 1670), 545
COLLINS to VERNON (1671/II/17), 420, 431

List of letters

COLLINS to VERNON (1671/XI/23), 543
COLLINS to VERNON (1671/XII/24), **543**
COLLINS to VERNON (1672/I/5), 548
COLLINS to WALLIS (1667/II/12), 24
COLLINS to WALLIS (1668/XI/5), **10**, 24
COLLINS to WALLIS (1669/I/22), **99**, 145, 148, 149
COLLINS to WALLIS (1669/late I – II), **154**, 161
COLLINS to WALLIS (1669/III/31?), **160**, 161, 167
COLLINS to WALLIS (1669/VI/27), **210**, 221, 222, 280, 289, 293
COLLINS to WALLIS (1670/I/18), **287**, 289, 290, 297
COLLINS to WALLIS (1670/VII/24), **378**, 385
COLLINS to WALLIS (1670/VII/end), **387**, 388, 392
COLLINS to WALLIS (1671/II/early III?), **430**
COLLINS to WALLIS (1671/III/31), 435, **435**
COLLINS to WALLIS (1672/VIII/22), 293
COSIMO III DE' MEDICI to WALLIS (1670/X/13), **395**, 401

DULAURENS to OLDENBURG (1668/X?), 18, 151

ELLIS to WALLIS (1669/X/17), **251**

FERMAT to DIGBY (1657/IV/20), 506
FLAMSTEED to COLLINS (1670/XII/10), 420
FLAMSTEED to OLDENBURG (1670/III/8), 290

GORNIA to OLDENBURG (1669/VII/10), 171
GOTT to WALLIS (1668/X/29), **3**, 4
GREGORY to COLLINS (1669/I/30), 153
GREGORY to COLLINS (1669/II/25), 99, 100

GREGORY to OLDENBURG (1668/XII/25), 96, 100
GREGORY to WALLIS (1668/IX-X), 12, 28
GREGORY to WALLIS (1668/XI/5), 6, **10**, 43
GREGORY to WALLIS (1668/XII/30?), 67, **86**, 87, 96, 99, 100, 145

HEVELIUS to OLDENBURG (1669/III/21), 158, 167
HEVELIUS to WALLIS (1669/III/21), **158**, 166, 167
HOBBES to the ROYAL SOCIETY (1671/VII/end), **472**, 475, 477, 487, 489, 503, 514, 518
HOBBES to the ROYAL SOCIETY (1671/VIII/first half), **485**, 486, 487, 488, 489, 503, 514
HOBBES to the ROYAL SOCIETY (1671/IX/early), **497**, 503, 514, 518
HOBBES to the ROYAL SOCIETY (1671/IX), **510**, 514, 518
HUET to OLDENBURG (1670/X/30), 391
HUYGENS to DOUBLET (1668/VI/29), 210
HUYGENS to GALLOIS (1668/XI/?), 21, 23, 44, 58, 87
HUYGENS to MORAY (1662/VIII/18), 23, 43, 55
HUYGENS to OLDENBURG (1668/XI/13), 19, 55
HUYGENS to OLDENBURG (166/I/5), 408
HUYGENS to OLDENBURG (1669/II/6), 155
HUYGENS to OLDENBURG (1669/III/30), 99
HUYGENS to OLDENBURG (1669/V/29), 191
HUYGENS to OLDENBURG (1669/VI/26), 228, 238
HUYGENS to OLDENBURG (1669/VIII/10), 241
HUYGENS to OLDENBURG (1670/X/15), 391

List of letters

HUYGENS to OLDENBURG (1670/X/31), 391, 398, 406, 416
HUYGENS to SLUSE (1658/IV/5), 421
HUYGENS to SLUSE (1658/V/28), 424
HUYGENS to WALLIS (1668/IX/6), 421
HUYGENS to WALLIS (1668/VII/1), 400
HUYGENS to WALLIS (1668/XI/13), **19**, 26, 41, 44, 46, 55
HUYGENS to WALLIS (late 1670–late 1671), **420**
HYRNE to WALLIS (1670/II/28), **299**
HYRNE to WALLIS (1670/III/10), 335, 337, 341, 342
HYRNE to WALLIS (1670/IV/12), **343**, 359, 363, 368

JUSTEL to OLDENBURG (1668/V/5), 364
JUSTEL to OLDENBURG (1668/XI/10), 18, 60

LAMPLUGH to WILLIAMSON (1669/V/15), 168, **169**
LAMPLUGH to WILLIAMSON (1669/VII/23), **229**
LEIBNIZ to OLDENBURG (1671/X/25), 521

MORAY to HUYGENS (1662/IX/1), 23, 43
MORAY to HUYGENS (1666/V/24), 19
MORAY to OLDENBURG (1665/X/20), 55
MORAY to WALLIS (1668/II/?), 26

NEILE to OLDENBURG (1668/XII/28), 445, 453
NEILE to OLDENBURG (1669/I/7), 87
NEILE to OLDENBURG (1669/V/23), **181**, 184, 520, 521
NEILE to OLDENBURG (1669/V/30), **186**, 188, 189, 521
NEILE to OLDENBURG (1669/VI/11), **191**, 194, 521
NEILE to OLDENBURG (1669/VI/25), **201**, 213
NEILE to OLDENBURG (1669/VI/26), 228
NEILE to OLDENBURG (1669/VII/3), **219**, 228

OLDENBURG to BERNARD (1670/XII/11), 413
OLDENBURG to BOYLE (1668/III/20), 405
OLDENBURG to CHILDREY (1670/IV/19), 365
OLDENBURG to HEVELIUS (1666/IX/3), 103
OLDENBURG to HEVELIUS (1668/XI/7), 13
OLDENBURG to HEVELIUS (1668/XII/21), 82, 87, 92
OLDENBURG to HEVELIUS (1669/II/2), 103, 75
OLDENBURG to HEVELIUS (1671/XI/19), 534
OLDENBURG to HUYGENS (1666/V/25), 19
OLDENBURG to HUYGENS (1668/XI/28), 23, 41, 44, 58
OLDENBURG to HUYGENS (1670/IX/30), 391
OLDENBURG to HUYGENS (1670/XI/18), 398
OLDENBURG to LEIBNIZ (1670/VIII/20), 386
OLDENBURG to LEIBNIZ (1671/VI/22), 443
OLDENBURG to SLUSE (1669/IX/24), 248, 302
OLDENBURG to SLUSE (1669/X/20), **258**
OLDENBURG to SLUSE (1670/II/5), 248, 302, 303
OLDENBURG to SLUSE (1670/IV/5), 337, 339
OLDENBURG to SLUSE (1670/X/4), 391
OLDENBURG to VERNON (1670/VI/20), 377
OLDENBURG to VERNON (1671/XII/14), 537
OLDENBURG to VOGEL (1671/XI/24), 538
OLDENBURG to WALLIS (1664/X/09), 167
OLDENBURG to WALLIS (1668/VII–VIII ?), 27

List of letters

OLDENBURG to WALLIS (1668/IX–early X), 1
OLDENBURG to WALLIS (1668/early XI), **18**, 37, 39
OLDENBURG to WALLIS (1668/XI/26), **52**, 57
OLDENBURG to WALLIS (1668/XII/6?), 53
OLDENBURG to WALLIS (1668/XI/27), 53, **53**
OLDENBURG to WALLIS (1668/XI/27), 57
OLDENBURG to WALLIS (1668/XII/4), 53, **56**, 58
OLDENBURG to WALLIS (1668/XII/11), **67**, 68, 71
OLDENBURG to WALLIS (1668/XII/12), **68**, 71, 73
OLDENBURG to WALLIS (1668/XII/18), **74**, 80, 82
OLDENBURG to WALLIS (1668/XII/24), **86**, 87, 89
OLDENBURG to WALLIS (1668/XII/25), **86**, 87
OLDENBURG to WALLIS (1669/I/5), 87, **93**, 94
OLDENBURG to WALLIS (1669/I/8), **93**, 94, 95
OLDENBURG to WALLIS (1669/I/10), **93**, 96
OLDENBURG to WALLIS (1669/I/24), **102**
OLDENBURG to WALLIS (1669/I/26), **144**, 150, 152
OLDENBURG to WALLIS (1669/II/5), **153**
OLDENBURG to WALLIS (1669/II/16), **154**, 155
OLDENBURG to WALLIS (1669/II/18), 155, **155**
OLDENBURG to WALLIS (1669/II/26), **158**
OLDENBURG to WALLIS (1669/IV/27), 164, **166**, 167
OLDENBURG to WALLIS (1669/V/11), 166, **169**
OLDENBURG to WALLIS (1669/V/18), 102, **171**, 178
OLDENBURG to WALLIS (1669/V/25?), 181, **183**, 184
OLDENBURG to WALLIS (1669/VI/3), 186, **188**, 189, 191
OLDENBURG to WALLIS (1669/VI/14), 191, 194, **194**, 201
OLDENBURG to WALLIS (1669/VI/22), **200**, 207
OLDENBURG to WALLIS (1669/VII/5), **228**
OLDENBURG to WALLIS (1669/VII/6), 230, 237
OLDENBURG to WALLIS (1669/VII/15), **228**, 230, 237, 239
OLDENBURG to WALLIS (1669/VIII/2), 237, **237**, 239, 241, 242
OLDENBURG to WALLIS (1669/VIII/14), **240**, 241
OLDENBURG to WALLIS (1669/X/24?), **252**, 256
OLDENBURG to WALLIS (1669/XII/18), **265**
OLDENBURG to WALLIS (1669/XII/18, enclosure), **266**
OLDENBURG to WALLIS (1670/I/3), **279**, 288
OLDENBURG to WALLIS (1670/I/16), **281**, 288
OLDENBURG to WALLIS (1670/III/18), **319**, 322
OLDENBURG to WALLIS (1670/III/c.26), 305, **327**, 336, 339
OLDENBURG to WALLIS (1670/IV/5), 320, 335, **337**, 341, 342
OLDENBURG to WALLIS (1670/IV/19), 341, **363**
OLDENBURG to WALLIS (1670/IV/26), 363, **368**
OLDENBURG to WALLIS (1670/VII/28), **380**, 381
OLDENBURG to WALLIS (1670/VIII/7), 378, 381, **387**, 391
OLDENBURG to WALLIS (1670/VIII/16), 391, **393**
OLDENBURG to WALLIS (1670/XI/11), 398, **398**

List of letters

OLDENBURG to WALLIS (1670/XI/18), 398, **401**, 405
OLDENBURG to WALLIS (1670/XII/11), 408, 413, **413**
OLDENBURG to WALLIS (1670/XII/27), 413, **417**
OLDENBURG to WALLIS (1671/IV/9), 439, **442**, 443
OLDENBURG to WALLIS (1671/V/mid–end), **451**, 454
OLDENBURG to WALLIS (1671/V/19), 448, **448**
OLDENBURG to WALLIS (1671/VIII/11), 475, **475**, 477
OLDENBURG to WALLIS (1671/VIII/15), 475, **477**, 479
OLDENBURG to WALLIS (1671/VIII/20), **482**
OLDENBURG to WALLIS (1671/VIII/25), 485, **486**, 488
OLDENBURG to WALLIS (1671/XI/14), **519**, 520, 539
OLDENBURG to WALLIS (1671/XI/21), 520, 522, 535
OLDENBURG to WALLIS (1671/XII/6), 535, 537
OLDENBURG to WALLIS (1671/XII/15), 538, **540**
OLDENBURG to WALLIS (1672/I/9), 544, **554**

PELL to COLLINS (1668/IX/16), 8
PELL to COLLINS (1668/XI/7), **15**, 44
POCOCKE to WALLIS (1671/X/1), 501, **501**

SCHOOTEN, PIETER VAN to WALLIS (1669/XII – 1670/I), 210, **280**, 290
SLUSE to OLDENBURG (1668/XI/2), 8
SLUSE to OLDENBURG (1669/VIII/16), 248, 254
SLUSE to OLDENBURG (1670/III/10), 302, 337, 339
SLUSE to OLDENBURG (1670/XI/22), 413

SLUSE to WALLIS (1670/III/10), 248, **302**, 339
SLUSE to WALLIS (1670/VII/25), 339, 378, 387, 391
SORBIÈRE to OLDENBURG (1663/XII/15), 21

VERNON to COLLINS (1671/XII/14), 543
VERNON to OLDENBURG (1670/VII/19), 377, 398
VERNON to OLDENBURG (1670/IX/4), 393
VERNON to OLDENBURG (1671/XII/3), 543
VERNON to WALLIS (1670/VIII/25?), **393**
VOGEL to OLDENBURG (1671/XI/11), 538

WALLIS to AUBREY (1670/XI/3), **395**, 396, 398
WALLIS to BERTET (late 1670), **420**
WALLIS to BERTET (1671/XII/29), 529, **548**
WALLIS to BORELLI (1670/I/23), 280, 290, **294**, 297, 303, 304, 411, 431
WALLIS to BORELLI (1670/I/23), enclosure: Catalogue of Books for Borelli, **297**, 431
WALLIS to BOYLE (1662/III/24), 378
WALLIS to BOYLE (1666/V/5), 57, 79, 87, 164, 167, 320
WALLIS to BOYLE (1669/VII/second half), **240**, 245
WALLIS to BOYLE (1669/VII/27), 228, 229, **232**, 240, 245, 247
WALLIS to BOYLE (1669/VIII/27), 240, 241, 244, **244**, 247
WALLIS to BOYLE (1671/XII/23), **541**
WALLIS to BROUNCKER (1657/XII/27), 427, 429
WALLIS to BROUNCKER (1668/VIII?), 430, 440, 442
WALLIS to BROUNCKER (1668/XI/1), **6**, 24, 37
WALLIS to BROUNCKER (1668/XI/14), **25**, 42, 44, 59, 66, 69, 100

List of letters

WALLIS to BROUNCKER (1669/X/20), 258

WALLIS to BROUNCKER (1669/XII/31), 270

WALLIS to BROUNCKER (1670/XII/22), 413, 415, **416**

WALLIS to CHILDREY (1670/IV/c.8), 341, **343**

WALLIS to COLLINS (1667/II/15), 24, 222

WALLIS to COLLINS (1668/IX/18), 1, 7, 17, 45

WALLIS to COLLINS (1668/IX/20), 1

WALLIS to COLLINS (1668/X/6), 7, 24

WALLIS to COLLINS (1668/XI/13), 6, 10, 19, **23**

WALLIS to COLLINS (1669/I/18), **99**, 145

WALLIS to COLLINS (1669/I/28), 99, **144**, 149

WALLIS to COLLINS (1669/I/29), 145, **147**, 154

WALLIS to COLLINS (1669/I/30), 99

WALLIS to COLLINS (1669/IV/2), 154, 160, **161**

WALLIS: Corrections to *Mechanica* (1669/VI/4), **227**

WALLIS to COLLINS (1669/VI/27), 280

WALLIS to COLLINS (1669/VII/4), 210, **220**, 223, 227, 280

WALLIS to COLLINS (1669/XI/?), 258, **265**, 270

WALLIS to COLLINS (1670/I/21), 210, 281, 287, **289**, 297

WALLIS to COLLINS (1670/VIII/2), 378, **385**

WALLIS to COLLINS (1670/VIII/14), 388, **388**

WALLIS to COLLINS (1671/III/?), 435

WALLIS to COLLINS (1672/II/4), 548

WALLIS to COSIMO III DE' MEDICI (1670/XI/19), 395, **401**

WALLIS to CREW (1671/XI/16), 519, 520, 520

WALLIS to DIGBY (1657/XII/1), 429

WALLIS to GREGORY (1668/XI/1), **6**, 10, 12, 23, 28, 29, 54

WALLIS to GREGORY (1668/XI/12?), **19**

WALLIS to GREGORY (1668/XI/early XII), **67**, 87

WALLIS to GREGORY (1669/I/18), **98**, 100, 145

WALLIS to HEVELIUS (1651/I/31), 76

WALLIS to HEVELIUS (1663/IV/9), 14, 160

WALLIS to HEVELIUS (1664/IV/15), 75, 158

WALLIS to HEVELIUS (1668/XI/5), **13**

WALLIS to HEVELIUS (1668/XI/22), **41**, 48

WALLIS to HEVELIUS (1669/I/25), **102**, 152, 158, 167

WALLIS to HEVELIUS (1669/I/25, enclosure), 102, **104**

WALLIS to HOBBES (1671/IX/(i)), 472, 485, 498, **503**, 510, 515

WALLIS to HOBBES (1671/IX/(ii)), 485, 498, **514**, 510

WALLIS to HUYGENS (1656/IV/27), 339

WALLIS to HUYGENS (1668/IX/10), 19, 28, 41, 42, 46

WALLIS to HUYGENS (1668/XI/23), 19, 23, **41**, 46, 55, 58

WALLIS to HYRNE (1670/III/19), 299, **320**, 335, 337, 341, 342, 344

WALLIS to HYRNE (1670/IV/14), **359**, 363, 368

WALLIS to LÉOTAUD (1668/II/27), 364

WALLIS to LALOUBÈRE (1668/IX/20), 545

WALLIS to LAMPLUGH (1671/XI/16), **519**, 520

WALLIS to LANGHAM (1669/second half of VII–first half of VIII), **240**

WALLIS to LANGHAM (1669/IX/3), 240, 241, 246, **247**

WALLIS to MORAY? (1668/II/?), 26, 42

WALLIS to NEILE (1670/VIII–IX), **394**, 395

WALLIS to OLDENBURG (1664/IV/16), 79

WALLIS to OLDENBURG (1664/X/1), 79, 158

List of letters

Wallis to Oldenburg (1665/V/18), 158, 168
Wallis to Oldenburg (1666/V/17), 168
Wallis to Oldenburg (1666/VII/28), 57, 87, 164, 320, 335, 400
Wallis to Oldenburg (1667/III/31), 335
Wallis to Oldenburg (1668/II/11), 57
Wallis to Oldenburg (1668/III/17), 39, 320, 364
Wallis to Oldenburg (1668/VII/28), 18, 39, 40, 48
Wallis to Oldenburg (1668/VII/30), 18
Wallis to Oldenburg (1668/VIII/13), 27
Wallis to Oldenburg (1668/X/16), **1**
Wallis to Oldenburg (1668/XI/17), 18, **37**, 48, 54, 75
Wallis to Oldenburg (1668/XI/24), 36, 41, **46**, 53, 66
Wallis to Oldenburg (1668/XI/25), **48**, 54, 58, 69, 71, 73, 92, 152
Wallis to Oldenburg (1668/XI/29), 23, 52, 53, **53**, 56, 66
Wallis to Oldenburg (1668/XII/6?), 25, 49, 52, **56**, 66, 87
Wallis to Oldenburg (1668/XII/10), 60, 63, **66**
Wallis to Oldenburg (1668/XII/13), 61, 67, **68**, 71, 87, 92
Wallis to Oldenburg (1668/XII/15), 67, 68, **70**, 74, 82, 92, 98
Wallis to Oldenburg (1668/XII/19), 74, **74**, 75, 80, 82, 86, 92, 103
Wallis to Oldenburg (1668/XII/19 and 20), 75, **80**
Wallis to Oldenburg (1668/XII/22), 74, 82, **82**, 87, 96
Wallis to Oldenburg (1668/XII/31), 67, 82, 86, 87, **87**, 100
Wallis to Oldenburg (1669/I/10), 93, **94**, 95
Wallis to Oldenburg (1669/I/12), 87, 93, **95**, 100
Wallis to Oldenburg (1669/I/22), 87, 99, **100**, 102, 145, 152
Wallis to Oldenburg (1669/I/29 and 31), 144, **150**, 153
Wallis to Oldenburg (1669/II/21), 144, 154, 155, **155**, 158
Wallis to Oldenburg (1669/IV/25), **164**, 166, 168
Wallis to Oldenburg (1669/V/4), 160, 166, **166**, 169
Wallis to Oldenburg (1669/V/20), 171, **178**, 181
Wallis to Oldenburg (1669/V/27), 181, 184, **184**, 186
Wallis to Oldenburg (1669/VI/8), 186, **188, 189**
Wallis to Oldenburg (1669/VI/17), 194, **194**, 201, 213
Wallis to Oldenburg (1669/VI/25), 200, **207**
Wallis to Oldenburg (1669/VI/29), 201, **213**, 219
Wallis to Oldenburg (1669/VII/26), 228, 229, **230**, 233, 237
Wallis to Oldenburg (1669/VIII/8), 228, 237, **237**
Wallis to Oldenburg (1669/VIII/25), 237, 240, **241**, 244
Wallis to Oldenburg (1669/VIII/25, enclosure), **244**
Wallis to Oldenburg (1669/X/26), 248, 253, **253**, 256
Wallis to Oldenburg (1669/X/27), 248, 253, **256**
Wallis to Oldenburg (1669/XII/19), 265, **269**
Wallis to Oldenburg (1670/I/17), 258, **281**
Wallis to Oldenburg (1670/I/19), 277, 279, 281, **287**, 292
Wallis to Oldenburg (1670/III/20), **322**
Wallis to Oldenburg (1670/III/29), 305, 327, **327**, 338, 364, 365, 400, 406

List of letters

Wallis to Oldenburg (1670/IV/3), 320, **334**, 337, 343, 364
Wallis to Oldenburg (1670/IV/8), 334, 337, 339, **341**, 343, 363
Wallis to Oldenburg (1670/IV/19), **363**, 368
Wallis To Oldenburg (1670/VII/21), **377**
Wallis to Oldenburg (1670/VIII/1), 380, **381**, 387
Wallis to Oldenburg (1670/VIII/14), 387–389, **390**, 393
Wallis to Oldenburg (1670/XI/4), 394, 395, **396**
Wallis to Oldenburg (1670/XI/13), 394, 395, 398, **398**, 401, 406, 407
Wallis to Oldenburg (1670/XI/25), 401, **405**, 409
Wallis to Oldenburg (1670/XII/4), 399, **408**, 413, 414
Wallis to Oldenburg (1670/XII/22), 413, **413**, 416, 417
Wallis to Oldenburg (1671/IV/2), **439**, 441, 442
Wallis to Oldenburg (1671/IV/2, enclosure), **441**
Wallis to Oldenburg (1671/IV/17), **442**, 442, 450
Wallis to Oldenburg (1671/V/19), **448**
Wallis to Oldenburg (1671/V/23), 448, **448**, 453, 454
Wallis to Oldenburg (1671/VI/12), **452**
Wallis to Oldenburg (1671/VI/20), 451, 454
Wallis to Oldenburg (1671/VII/7), **455**
Wallis to Oldenburg (1671/VII/26), 458, **459**, 472, 485, 498
Wallis to Oldenburg (1671/VIII/14), **475**, 477, 479, 482
Wallis to Oldenburg (1671/VIII/15–19), 479, **479**
Wallis to Oldenburg (1671/VIII/20), 477, 479, **479**
Wallis to Oldenburg (1671/VIII/23), 479, **482**, 486
Wallis to Oldenburg (1671/VIII/26), 486, **488**
Wallis to Oldenburg (1671/XI/16), 519, 520, **520**, 522, 538
Wallis to Oldenburg (1671/XII/3), 522, **535**, 538
Wallis to Oldenburg (1671/XII/7), 536, 538, **538**, 540
Wallis to Oldenburg (1672/I/24), 554
Wallis to Oldenburg (1672/I/28), 554
Wallis to Oldenburg (1673/X/14), 394, 395
Wallis to Pococke (1671/X/3), 501, **501**, 537, 538
Wallis to Scarborough (1671/XI/26), 523, **523**
Wallis to Scarborough (1671/XII/1), **523**, 525
Wallis to Scarborough (1671/XII/1, enclosure), **524**
Wallis to Schooten (1669/VI–XII), 210, **279**
Wallis to Sluse (1669/IX/20), **248**, 253, 256, 302
Wallis to Sluse (1670/IV/8), 302, **339**, 341, 378, 379
Wallis to the Royal Society (1671/VIII), **489**, 503
Wallis to Vernon (1670/early VI), **377**
Wallis to Wood (1670/III/20), **324**, 368
Wallis to Wood (1670/V/26), **368**, 389
Wallis to Wood (1670/V/28), **377**, 389
Wallis to Wood (1670/VIII/14), 368, 377, 387, **389**
Wallis: Conjecture on Gott's Proposal of an Artificial Spring (1668/X/29?), 3, **4**
Wallis: Paper on Dulaurens (1668/XII/7), 18, **60**, 66, 69, 88
Wallis: Paper on Hobbes (1668/XII/7), **63**, 66

List of letters

WALLIS: Corrections to Newton's *Art of Practical Gauging* (1669/VII/?), **223**

WALLIS: The Reasonableness of the University's Exemption from Charges (1669/XII/7), **259**

WALLIS: A Brief Account of Mr. Hobbes's Fundamental Mistake, (1669/XII/*c*.31), **277**, 288

WALLIS: Reply to Collins's Question on Algebraic Roots (1670/III/18), **313**

WALLIS, JOHN JR to WALLIS (1669/I/25), **144**, 157

WALLIS, JOHN JR to WALLIS (1669/I/late), **154**, 157

WILLIAMSON to LAMPLUGH (1669/VII/17), 230

WOOD to WALLIS (1670/II/18), **299**, 324

WOOD to WALLIS (1670/IV/25), 368, **368**

WOOD to WALLIS (1670/VIII/2), 377, **387**, 389

INDEX: PERSONS AND SUBJECTS

Aegean Sea, 347
ABUBACER, *see* IBN ṬUFAYL, ĀBU BAKR MUḤAMMAD
ABU'L FEDĀ, ISMAIL (1273–1331), 79
  identified, 536
Abu'l Fedā: Geography, 79, 536
Acts, 487
aether, 3
Africa, 347
air, 449
  pressure, 404
  pump, 451
ALESTREE, RICHARD (1621/2–81) B, 40, 48
algebra, 439, 466
  species, 439
  symbols, 458
ALGERNON PERCY, TENTH EARL OF NORTHUMBERLAND (1602–1668), 9
Alice and Francis, 294, 303
Allhollandtide, 331, 332, 361, 406
ALSOP, NATHANIEL (17th century)
  identified, 233
Amazon, 347
Amsterdam, 8, 13, 58, 211, 212
ANDERSON, ALEXANDER (1581/2–after 1621)
  identified, 211, 293
Anderson, Alexander: Exercitationum mathematicarum deces prima, 293
ANDERSON, ROBERT (*fl.*1666–96)
  identified, 16
ANDERSON, ROBERT (d. 1710), 9, 44

Anderson, Robert: Stereometrical Propositions, 9, 15, 16, 45, 210, 221, 223
angle, 83, 461
  of incidence, 92
  of reflection, 92
  trisection, 11
ANONYMOUS, 395
ANTIATE, TIMAURO, *see* DATI, CARLO
APOLLONIUS of Perga (*c.*260–*c.*190), 78, 531
Apollonius: Conics, 78, 294, 531
approximation, 26, 324
arc
  trisection, 30
Archbishop of Canterbury, 230, 233, 243, 246
Archbishop of Cuckolds (archiepiscopum corniculati), 242
Archbishop of Cuckoos (archiepiscopum corniculati), 245
ARCHIMEDES of Syracus (287?–212), 523, 524
Archimedes: De aequiponderantibus, 525
Archimedes: De lineis spiralibus, 526, 528
Archimedes: Dimensio circuli, 523
architecture, 169, 233, 402
ARISTARCHUS OF SAMOS (310–230), 445
arithmetic, 466, 517
  arithmetical progression, 439, 505, 530
  of infinites, 471, 503
arithmetica speciosa, 471
artificial spring, 3, 5
astronomy, 9, 222

Index: persons and subjects

conjunctions, 434
observations, 13, 14, 79, 89, 158, 160, 402, 433, 434, 535
Astrop Spa, Northamptonshire, 475, 480
asymptotes, 418, 530
AUBREY, JOHN (1626–97) (B), 393, 395, 396, 398
axiom, 12

BACHET DE MÉZIRIAC, CLAUDE-GASPARD (1580–1639), 45
identified, 430, 548
Bachet de Méziriac (ed.): Diophanti Alexandrini arithmeticorum libri sex... Accessit... inventum novum collectum ex... Fermat epistolis, 45, 430
Bachet de Méziriac: Problemes plaisans et delectables, qui se font par les nombres, 548
Bago, 347
BAIK, see ULUG BEG
BALLEVILLE (17th century)), 406
Baltic Sea, 346
Barbary Coast, 347
barometer, 258
experiments, 49, 405, 448, 453
baroscope, 281
BARROW, ISAAC (1630–77), 9, 23, 43, 304, 438, 544
Barrow: Euclidis Elementorum libri XV, 291
Barrow: Lectiones geometricae, 304, 438
Barrow: Lectiones XVIII... in quibus opticorum phaenomenωv genuinae rationes investigantur, 9, 291, 297
BARTHOLIN, ERASMUS (RASMUS BERTHELSEN) (1625–98), 151, 436
Bartholin: Experimenta crystalli Islandici, 398
Bartholin: Historia coelestis, 151
Bartholin: Specimen recognitionis, 151
BARTLET (Oxford carrier) (17th century), 66, 151, 292

BATHURST, RALPH (1620–1704), 3
BAYLIE, THOMAS (?–1699), 170
BEALE, JOHN (1603?–c.1683), 38, 281
BEE,..., see booksellers and printers
BEIG, see ULUG BEG
BEIK, see ULUG BEG
Belgium, Belgian, 211
BENNET, HENRY, first Baron Arlington (1618–85)
identified, 252
BERNARD, EDWARD (1638–96) (B), 78, 103, 144, 151, 413, 415, 431, 544
Bernard: De mensuris et ponderibus antiquis libri tres, 544
BERTET, JEAN (1622–92) (B), 420, 426, 431, 435, 452, 529, 545, 548
BEVERIDGE, WILLIAM (1637–1708), 290, 297
Beveridge: Institutionum chronologicarum libri II, 290, 297
BILLY, JACQUES DE (1602–79), 440
identified, 535
Billy: Diophantus redivivus, 546
Binsey, 263
BISDOMMER, FRANÇOIS (17th century), 293
blood
circulation of, 446
transfusion of, 447
BOBART (BOBERT), JACOB (c.1599–1680), 382
Bodleian Library, see Oxford, University of, Bodleian Library
body, 538, 539
BÖCKLER, GEORG ANDREAS (1617–87), 388
Böckler: Theatrum Machinarum, 388
Bologna, 87, 412
books
auction, 293
binding, 542
exchange, 431
of trade, 8, 17
booksellers and printers, 2, 280, 304, 392

610

Index: persons and subjects

BEE, CORNELIUS (d. 1671/2) (London), 385
DAVIS, RICHARD (1617/18–c.95) (Oxford), 67
MARTYN, JOHN (1617/18–80), 168
MOXON, JOSEPH (1627–91) (London, Russell Street), 222
PITT, MOSES (1639–97) (London, at the 'White Heart', at the 'Angel'), 7, 8, 15, 99, 145, 162, 210, 222, 385, 388, 392, 433
BORELLI, GIOVANNI ALFONSO (1608–79) (B), 2, 57, 87, 97, 98, 280, 287, 294, 304, 399, 404, 406, 409, 410, 414, 431
Borelli: De motionibus naturalibus, 431
Borelli: De vi percussionis, 57, 97, 410, 411
Borelli: Euclides restitutus, 2
Borelli: Historia, et meteorologia, 292, 412
Borelli: Theoricae mediceorum planetarum, 87, 97
BOUILLON, EMMANUEL THÉODOSE (1643–1715)
  identified, 529
BOULLIAU, ISMAËL (1605–94), 103, 439
Boulliau: Astronomia philolaica, 349
Bordeaux, 348
BOYLE, ROBERT (1627–91) (B), 57, 66, 157, 208, 228, 232, 236, 240, 241, 244, 280, 284, 290, 294–541
Boyle: A continuation of New Experiments, 67, 297
Boyle: Certain physiological essays, 297, 445
Boyle: De origine formarum, 434
Boyle: Hydrostatical paradoxes, 297
Boyle: New Experiments, 404
Boyle: Nova Experimenta, 404
Boyle: Some considerations touching the usefulnesse of experimentall naturall philosophy, 297
Boyle: The origine of formes, 297
BRAHE, TYCHO (1546–1601), 57, 68, 73, 79, 80, 87, 92, 151, 445

identified, 535
Brahe: Historia coelestis, 57, 68, 73, 79, 80, 87, 92
BRANCKER, THOMAS (1633–76), 15
BRAZIER, JOHN (d. 1679)
  identified, 245
Brereton, Cheshire, 10
BRERETON, WILLIAM (1631–80), 8, 10, 16, 17, 45
bridges, 262
Bristol, 38, 309, 312, 347
BROUNCKER, WILLIAM (1620?–84) (B), 1, 6, 24, 25, 37, 42, 46, 47, 59, 66, 69, 97, 100, 149, 157, 207, 208, 212, 218, 228, 258, 265, 270, 288–290, 388, 415, 416, 427, 431, 440, 449, 450, 453, 454, 485, 486, 490, 498
BUTLER, JAMES, first Duke of Ormond (1610–88), 245, 246
  identified, 243
BYRD, WILLIAM (d. 1690?)
  identified, 383

calendar, 541
CALVERT, JAMES (1631–98)
  identified, 548
Calvert: Naphtali: seu Colluctationes theologicae cum tribus ingentibus dubiis, 547
Cambaia, 347
Cambridge, University of, 38, 167, 538
Candlemas, 331, 332, 361
Cap Gris-Nez, 309
Capuchins, 431
CARCAVI, PIERRE DE (1600?–1684), 428, 440
CARDANO, GERONIMO (1501–76), 401
carpenter, 435, 509
CARR, ALAN (d. 1676), 543
Cartagena, 357
Cartesians, 521
Caspian Sea, 344
CASSINI, GIOVANNI DOMENICO (1625–1712), 88, 97, 534

611

Index: persons and subjects

Cassini: Ephemerides Bononienses, 87
CAVALIERI, BONAVENTURA (1598?–1647), 294, 401
Cavalieri: Centuria di varii problemi, 295
Cavalieri: Directorium generale Uranometricum, 294
Cavalieri: Geometria indivisibilibus, 402, 473, 490, 506
CELLARIUS, ANDREAS (1596–1665), 350
Cellarius: Harmonia macrocosmica seu Atlas universalis, 350
Charibdes, 347
CHARLES I (1600–49, king of England), 206, 260
CHARLES II (1630–85, king of England), 242, 244–246, 417, 478, 486, 487
Chartes or Carnutes, the metropolis of the Druides, 394
Chatham, 309
Chatham Dockyard, 362
Chepstow, 39, 333
CHERUBIN D'ORLÉANS, PÈRE (1613–97), 431
CHILDREY, JOSHUA (1623–70) (B), 164, 305, 328, 334, 338, 342, 343, 364, 365, 400
Childrey: Britannica Baconia, 164
Childrey: Syzygiasticon instauratum, 164, 310
China, Chinese, 542
　　Almanach, 541, 542
CHRISPINE, ? (17th cent.), 169
CHRYSOCOCCES, GEORGIOS (d. 1336), 103
cinchona, Peruvian bark, Jesuit's powder, quinine, 406, 408
cipher, 155, 157
circle, 462, 500, 554
　　quadrature, 1, 7, 21, 26, 30, 32, 42, 43, 47, 149, 161, 271–278, 370, 438, 507, 517, 531
　　segment of
　　　　table, 145
　　series, 8, 24, 149
cissoid, 2, 421–426

CLARKE, SAMUEL (1625–69), 538
　　identified, 78, 536
CLAVELL, ROBERT (in or before 1633–1711), 548
Clavell: Catalogue of Books, 548
CLENDON, (*fl.*1653–68), 151
clepsydra, 5
coffee, 536
coins, 545
COLBERT, JEAN-BAPTISTE (1619–83), 21
COLLINS, JOHN (1625–83) (B), 1, 7, 10, 15, 23, 25, 26, 42, 44, 99, 144, 147, 154, 160, 161, 167, 208, 210, 220, 227, 265, 287, 289, 294, 297, 303, 313, 378, 385, 387, 388, 392, 400, 411, 412, 430, 431, 435, 534, 535
collision of bodies, *see* mechanics
COLWALL, DANIEL (?–1690), 208
comets, 14, 75
COMMANDINO, FEDERICO (1509–75)
　　identified, 16, 401
Commandino (ed.): Apollonii Pergaei conicorum Libri quattuor, 16
commensurability, 11
COMPTON, HENRY (1631/2–1713)
　　identified, 229
conchoid, 83
　　regula, 418
conic sections, 9, 303
constellations, 335
Conversion of the Jews, 547
COPERNICUS, NICOLAUS (1473–1543), 339, 445
　　hypothesis, 445
Corfu, 347
Cornwall, 309
COSIMO DE' MEDICI (1642–73) (B), 168, 171, 271, 384, 395, 401, 405
COTTEREL, CHARLES (1615–1701), 383
　　identified, 382
COTTON, ROBERT BRUCE (1571–1631)
　　identified, 547
Cotton: Cottoni posthuma, 547
COWART, GEORGE (17th century), 304

Index: persons and subjects

CREW, NATHANIEL (1633–1721) (B), 520
CROMWELL, OLIVER (1599–1658), 231, 234
CROSSE, JOSHUA (1614–76)
  identified, 293
CRUCK, WILLEM VAN DER (17th century), 293
cube, duplication of, 469
CURTZ, ALBERT VON (1600–71), 77
curves, 9
cycloid, 2, 148, 162, 428, 440
cylinder, 250
  section, 428

D'ARÇONS, CÉSAR (d. 1681), 349
  identified, 348
d'Arçons: Le secret decouvert du flux et reflux de la mer et des longitudes, 349
Dalmatia, 347
Danzig, 75
DARY, MICHAEL (17th century)
  identified, 548
DATI, CARLO ROBERTO DE CAMMILLO (1619–76)
  identified, 404, 405
Dati: Lettera a Filaleti di Timauro Antiate, 404, 405
DAVENANT, JAMES (d. 1717)
  identified, 233
DAVIS, RICHARD, *see* booksellers and printers
DE LE BOË, FRANZ (1614–72), 144, 156
DECHALES, CLAUDE FRANÇOIS MILLIET (1621–78), 533
  identified, 529
Dechales: Cursus seu mundus mathematicus, 529, 533
Dechales: l'Art de naviger demontré par principes & confirmé par plusieurs observations tirées de l'experience, 533
deciphering, 252

DEL MONTE, GUIDOBALDO, *see* MONTE, GUIDOBALDO DEL
Denmark, 347
DESCARTES, RENÉ (1596–1650), 30, 90, 95, 242, 401, 529
Descartes: Geometria (ed. Schooten) 1659, 83, 90, 212
DETTONVILLE, A., *see* PASCAL, BLAISE
Devonshire, 309
DIOPHANTUS OF ALEXANDRIA (*fl.* 250), 413
Diophantus: Arithmetica, 413
Dogger Bank, 360
Dokkum, Friesland, 311
Dorset, 171, 309
Dositheus of Pelusium (3rd century BC), 523
Dover, 406
  post, 370
DRING, STEPHEN (17th century)
  commander of *Alice and Francis*, 303
Druides, 394
DU HAMEL, JEAN-BAPTISTE (1623–1706)
  identified, 431
Duke of York's players, 235
DULAURENS, FRANÇOIS (d. c.1675) (B), 18, 21, 60, 63, 65, 69, 87, 92, 151, 436
Dulaurens: Specimina mathematica, 60
Dymchurch wall, 309

EARL OF CLARENDON, *see* HYDE, EDWARD, ... EARL OF CLARENDON
EARL OF NORTHUMBERLAND, *see* ALGERNON PERCY, TENTH EARL OF NORTHUMBERLAND
Earth, 329, 449
  motion, 3, 78, 321, 329, 335, 337, 339–341, 344
eclipse, 160, 534
  lunar, 75, 434
  solar, 76

Index: persons and subjects

EDWARD I (1239–1307, king of England), 260
EDWARD III (1312–77, king of England), 260
elasticity, 68, 69, 71, 91, 92, 94, 96, 97, 101, 102, 157, 404, 405, 444, 450
ELIZABETH OF BOHEMIA (1596–1662)
    identified, 382
ELIZABETH I (1533–1603, queen of England), 260
ellipse, 8, 24, 149, 303, 554
    division, 8
    quadrature, 26
ELLIS, JOHN (c.1646–1738) (B), 251
England, English, 12, 19, 21, 27, 76, 78, 347
English Channel, 309
equations
    analytical, 11
    biquadratic, 212
    coefficients, 315, 317, 319
    cubic, 9, 11, 36, 85, 212, 313, 437
    higher, 7, 8, 11, 436
    homogeneous, 437
    limits of, 437, 439
    quadratic, 8, 36, 313
    roots of, 8, 11, 145, 212, 313, 403, 436, 437, 485, 488, 498, 508–510, 514–516, 554
    solution of, 62, 403
equinox, 38, 306
    vernal, 337
ERPE, THOMAS VAN (ERPENIUS) (1584–1624)
    identified, 538
ESCHINARDI, FRANCESCO (1623–1703), 294
Eschinardi: Centuria problematum opticorum, 294
Essex, 360
ether, 449
Etna, 412
EUCLID of Alexandria (c.325 BC), 12, 484

Euclid: Elements, 8, 12, 456, 460, 462, 480, 490, 498, 504, 506, 511, 517, 543
Euxine (Black) Sea, 344
EVELYN, JOHN (1620–1706), 290
Evelyn: Sylva, or A discourse of forest-trees, 290
experiments, 55, 66, 96, 152, 406, 415, 453
extension, 521

FABRI, HONORÉ (1607–88), 406, 411, 431
Fabri: commentary on Archimedes, 431
Fabri: Dialogi physici, 398, 409, 411, 414
Fabri: Tractatus Physicus de Motu Locali, 414
FAIRTHORNE, WILLIAM (1616–91)
    identified, 363
FALCONIERI, PAOLO (1638–1704), 404
Faversham, 310
fee-farm, 263
FELL, JOHN (1625–86), 38, 75, 82, 169, 230, 233, 243, 244, 246, 382
FERGUSON, JOHAN JACOB (c.1630–91?), 212, 436
Ferguson: Labyrinthus algebrae, 212
FERMAT, CLÉMENT-SAMUEL (c1630–90)
    identified, 430
FERMAT, PIERRE DE (1607/8–65), 413, 427, 429–431, 494, 506, 512
Fermat: Diophanti Alexandrini Arithmeticorum libri sex, 431, 440
Fermat, First Challenge in Number Theory, 427
fixed stars, 102, 103, 104, 434
FLAMSTEED, JOHN (1646–1719), 290, 548
Flamsteed: Lunae ad fixas appulsus visibiles, nec non arctiores juxta eas transitus, observabiles A. 1672, 548
Flanders, 307
Florence, 412
    Accademia dei Lincei, 405
FOGELIUS, MARTIN, *see* VOGEL, MARTIN
force, 3, 238, 242
fortification, 433

Index: persons and subjects

FOSTER, SAMUEL (*c*.1600–52), 160
France, French, 20, 26, 31, 42, 252, 394, 431, 432, 545, 547
FREDERICK III (1609–70, king of Denmark), 151
French gentleman, unidentified (17th century), 258
FRENICLE DE BESSY, BERNARD (1605–1675), 427, 429, 430
Frenicle: Solutio duorum problematum, 427, 429
FROIDMONT, LIBERT (1587–1653), 349
Froidmont: Meteorologicorum libri sex, 349, 358

GALE, THOMAS (1635/6–1702)
    identified, 544
GALILEI, GALILEO (1564–1642), 401, 403, 414
    identified, 446
GALLOIS, JEAN (1632–1707), 21, 23, 58
GASH, MR (17th century), 258
GASSENDI, PIERRE (1592–1655), 379
gauging, gauger, 145, 210, 435
Gentleman's Magazine, 541
geographers, 536
geometrical problems, 535
geometry, *see also* mathematics, 485, 494, 503, 506, 512, 517
    Cartesian, 403
    construction, 460, 461, 473
    of indivisibles, 471
    principles, 12, 417
GERARD, HENRY (17th century), 230, 232, 242
GIRARD, ALBERT (1595–1632), 324, 369
Girard: Invention nouvelle en l'algebre, 324
GLANVILL, JOSEPH (1636–80), 290
Glanvill: Plus ultra: or, The progress and advancement of knowledge, 164, 290

GLISSON, FRANCIS (1599?–1677), 446
GLORIOSO, GIOVANNI CAMILLO (1572–1643), 304
Glorioso: Exercitationum mathematicarum Decas tertia, 304
Gloucestershire, 311
gnomon, 470
GODBID, WILLIAM, *see* booksellers and printers, *see* booksellers and printers
GODDARD, JONATHAN (1616–75), 49
GOLIUS, JACOBUS, *see* GOOL, JAKOB
Golius: De regno Catayo additamentum, 542
GOODWIN, JOHN (*c*.1594–1665), 434
Goodwin: Socianisme discovered, 434
GOOL (GOLIUS), JAKOB VAN (1596–1667), 293
    identified, 210, 542
GOOL, THEODORUS VAN (son of JAKOB VAN GOOL, d. 1679)
    identified, 293
GORNIA, GIOVANNI BATTISTA (17th century), 171
GOTT, SAMUEL (1613–71) (B), 3, 4
Gott: The Divine History of Genesis, 3
GOTTIGNIES, GILLES-FRANÇOIS (1630–89), 304, 431
    identified, 535
    optical manuscripts, 535
Gottignies: Dioptrics, 431
Gottignies: Elementa geometricae planae, 304
GRANDAMI (GRANDAMY) (1588–1672), 349
Grandami: Nova demonstratio immobilitatis terrae petita ex virtute magnetica, 349
gravity, 23, 51, 69, 72, 94, 96, 148, 339, 404, 407, 444, 449
    center of, 30, 306, 329, 472, 493, 503, 504
great seal, 260
GREAVES, EDWARD (1613–80)
    identified, 537, 538

Index: persons and subjects

GREAVES, JOHN (1602–52), 536, 538, 542
   identified, 536
Greaves, John: *A Discourse of the Romane Foot, and Denarius*, 544
Greaves, John: *Chorasmiae et Mawaralnahrae hoc est regionum extra fluvium Oxum descriptio ex tabulis Abulfedae Ismaelis*, 536, 538
Greaves, John: *Epochae Celebriores*, 542
GREAVES, THOMAS (1602–52), 538
GRÉGOIRE DE SAINT-VINCENT, see SAINT-VINCENT, GRÉGOIRE DE
GREGORY, JAMES (1638–75) (B), 6, 7, 10, 19, 21, 23, 25, 28, 36, 37, 42, 45, 46, 54, 56, 58, 59, 67, 86, 88, 96, 98–100, 153, 439
Gregory: *An extract of a letter of Mr. James Gregory*, 36, 101
Gregory: *Answer to the Animadversions of Monsieur Hugenius*, 16, 27, 33, 42
Gregory: *Appendicula ad veram Circuli*, 101
Gregory: *Exercitationes geometricae*, 6, 7, 10, 15, 16, 19, 21, 23, 31, 43, 101, 290, 298
Gregory: *Geometriae pars universalis*, 11, 16, 42, 96
Gregory: *Vera circuli et hyperbolae quadratura*, 6–8, 11, 16, 19, 21, 26, 36, 42
GRIENBERGER, CHRISTOPH (1561–1636), 294, 412
Grienberger: *Catalogus veteres affixarum longitudines, ac latitudines conferens*, 296
Grienberger: *De Luce*, 294, 412
Grienberger: *Speculum ustorium*, 296
GRIMALDI, FRANCESCO MARIA (1618–63), 412
   identified, 410
Grimaldi: *Physico-mathesis de lumine*, 410, 412
GRIMESTON, EDWARD (*fl.*1601–34), 309
Grimeston: *A generall historie of the Netherlands*, 309
GRISIO, SALVATOR (17th century), 294

Grisio: *Antanalisi*, 294
GRYNAEUS, SIMON (1493–1541)
   identified, 104
Guinea, 347

HAAK, THEODOR (1605–90)
   identified, 156, 502
HALL, HENRY, see booksellers and printers
HALLEY, EDMOND (1656–1743)
   identified, 78
Halley (ed.): *Apollonius: Conica*, 78
HARRIOT, THOMAS (1560–1621), 403, 543
Harriot: *Artis analyticae praxis*, 403
HARTS, ROGER, see HORTS, ROGER
HARVEY, WILLIAM (1578–1657), 404
Harvey: *De motu cordis*, 404
HASTINGS, MARYA (d. 1679)
   identified, 235
HAYES, THOMAS (17th century), 232
Hell, 234
HENRY VI (1421–71, king of England), 260
HENRY VIII (1491–1547, king of England), 260
HÉRIGONE, PIERRE (d. c. 1643), 543
Hérigone: *Les six premiers livres des Elements d'Euclide*, 543
HEVELIUS, JOHANNES (1611–87) (B), 13, 37, 38, 41, 48, 54, 57, 66, 75, 78–80, 82, 86, 89, 92, 98, 102, 104, 152, 158, 239
Hevelius: *Cometographia*, 13, 37, 48, 54, 75, 160, 239
Hevelius: *Descriptio Cometae anno 1665*, 75
Hevelius: *Dissertatio de nativa Saturni facie*, 57, 76, 158, 167
Hevelius: *Eclipsis solis observata*, 76
Hevelius: *Epistolae II*, 75
Hevelius: *Machina Coelestis*, 14, 76
Hevelius: *Mercurius in Sole visus*, 75, 79, 326
Hevelius: *Prodromus cometicus*, 75, 89
Hevelius: *Selenographia*, 37, 54, 75, 90

Index: persons and subjects

HICKS, WILLIAM (*c.*1630–82)
   identified, 229
Hicks: Oxford Jests, 229
highways, 262
HOBBES, THOMAS (1588–1679) (B), 60, 62, 63, 69, 73, 87, 174, 222, 270, 277, 280, 281, 459, 464, 471, 472, 485, 497, 503, 506, 510, 514
   as mathematician, 62
   quadrature of the circle, 64, 201
Hobbes: Considerations upon the Answer of Dr. Wallis to the Three Papers of Mr. Hobbes, 510, 518
Hobbes: De corpore, 63, 174
Hobbes: De principiis et ratiocinatione geometrarum, 290, 298
Hobbes: Dialogus physicus, 64
Hobbes: English Works, 485
Hobbes: Examinatio et emendatio mathematicae hodiernae, 63
Hobbes: Opera philosophica, 63, 65, 73, 280
Hobbes: Quadratura circuli, 161, 222, 277
Hobbes: Rosetum Geometricum, 455, 459, 471, 472, 476, 477, 485, 494, 498, 507, 515, 518
Hobbes: Six Lessons to the Professors, 63
Hobbes: Three Papers Presented to the Royal Society Against Dr. Wallis, 485, 497, 510
HODGES, ? (*fl.*1671), 548
HODIERNA or ODIERNA, GIOANBATISTA (1597–1660), 412
Hodierna: De admirandis phasibus, 412
Hodierna: De systemate orbis cometici, 412
Hodierna: Opuscoli, 294, 412
Hodierna: Protei caelestis, 412
HOLINSHEAD, RAPHAEL (d. 1580?), 309, 310
Holinshead: The third volume of Chronicles, 309
Holland, Dutch, 58, 73, 78, 89, 151, 212, 280, 293, 307, 311, 347, 433, 533
   Ambassador at London, 293

Hongroad, near Bristol, 38
HOOKE, ROBERT (1635–1703), 13, 68, 71, 92, 208, 288, 338, 443
Hooke: Micrographia, 290
HOOT (HOOTE), ? (*fl.* 1670–1)
   identified, 545
HORNE, ROBERT (*fl.* 1660–85)
   identified, 545
HORROX, JEREMIAH (1618–41), 14, 75, 79, 167, 290, 326
Horrox, Jeremiah: Opera posthuma (ed. Wallis), 158, 290, 432
Horrox, Jeremiah: Venus in Sole visa, 79, 326
HORROX, JONAS (17th century), 326
HOSKINS, JOHN (1634–1705), 208
HOUGHTON (17th century)
   identified, 293
HOWARD, HENRY (1628–84), 546
HOWES, EDMUND (*fl.* 1602–31), 309
HUDDE, JAN (1628–1704), 293, 435
   identified, 211
Hudde: dioptrica, 211
Hudde: Epistolae duae, 212, 435
HUET, PIERRE DANIEL (1630–1721), 391
HUIPS, FRANS VAN DER (*fl.* 1654), 293
Huips: Algebra, 293
HUME, ALEXANDER (17th century), 211, 293
Hundred of Hoo, 309
HURST, ROGER, *see* HORTS, ROGER
HUYGENS, CHRISTIAAN (1629–95) (B), 7, 19, 21, 26–35, 41, 45–47, 55, 57, 58, 67, 83, 87, 88, 96, 99, 100, 152, 155, 238, 242, 391, 406, 409, 420, 451
   account of motion, 416
   anagram, 155
   solution of Alhazen's problem, 238
Huygens: A summary account of the laws of motion, 157, 408
Huygens: De circuli magnitudine inventa, 21, 83
Huygens: De motu corporum, 101, 408, 409

Index: persons and subjects

Huygens: Examen de Vera Circuli & Hyperboles Quadratura, 21, 27, 42
Huygens: Systema saturnium, 57
Huygens: Theoremata de quadratura hyperboles, ellipsis, et circuli, 21
HUYGENS, CONSTANTIJN (1596–1687), 19, 400
 visit to Oxford, 19, 400
Hven, island of, 535
HYDE, EDWARD, Earl of Clarendon (1609–74)
 identified, 252
HYDE, THOMAS (1636–1703), 76, 77, 103, 169
Hyde, Thomas: Catalogus impressorum, 78
hydrostatics, 5, 454
 experiments, 49, 454
hyperbola, 7, 23, 55, 83, 155, 249, 251, 254, 257, 303, 389, 391, 554
 asymptotes, 24, 83, 250, 379, 418, 530
 quadrature, 1, 7, 23, 26, 43, 149
 series, 24, 149
hyperbolic solid, 155
hyperbolic-cylindroidal solid, 158, 248, 253–256
HYRNE, HENRY (c.1626–after 1672) (B), 299, 320, 338, 341, 343, 359, 363

IBN ṬUFAYL, ĀBU BAKR MUḤAMMAD (ABUBACER) (before 1110–1185), 502
Ibn Ṭufayl: Philosophical Fable, 502
icosahedron, 468
impetus, 50–52, 72, 242, 414
Ireland
 Dublin, 389
Isle of Graine, 309
Isle of St Thomas, 347
Italy, Italian, 12, 295, 347, 401, 412, 432

JEFFREYS, EDWARD (17th century), 288
Jesuits, 394, 435
JOHNSON (17th century)

 identified, 391
Journal des Sçavans, 21, 27, 44, 58, 87, 157, 319, 408
JOYLIFFE, GEORGE (1621–58), 446
JULIUS FRANZ (1641–89, duke of Saxony), 452, 454
Jupiter, 97, 349
 satellites, 79, 402, 405
JUSTEL, HENRI (1620–93), 18, 60, 364
Jutland, 309

KARL I LUDWIG (1617–80, electoral prince of the Palatinate), 382
KARL II (1651–85, electoral prince of the Palatinate), 380, 381
KARL XI (1655–97, king of Sweden), 452, 454
Kent, 307, 312, 331, 360, 390, 406
 Faversham, 361
 Hythe, 336, 343
 Kentish ague, 390
 Romney Marsh, 306, 309, 331, 336, 360
KEPLER, JOHANNES (1571–1630), 535
Kepler, Johannes: Tabulae Rudolphinae, 535
KEPLER, LUDWIG (1607–63), 151
KERSEY, JOHN (1616–77), 436
Kersey: Algebra, 439
Khambhat
 Gulf, 347
KINCKHUYSEN, GERARD (1625–66), 212, 433, 436
Kinckhuysen: Algebra ofte stelkonst, 212
King of England, 243, 263, 264
 purveyors, 262
King's Lynn, 309

LALOUBÈRE (LALOVERA), ANTOINE DE (1600–64), 545
LALOUBÈRE (LALOVERA), ANTOINE DE (1600–64), 1
Laloubère: De cycloide Galilaei et Torricellii propositiones viginti, 148

Index: persons and subjects

Laloubère: Quadratura circuli et hyperbolae, 1
Laloubère: Veterum geometria promota in septem de cycloide libris, 1, 545
LAMPLUGH, KATHERINE (1633–71)
    identified, 171
LAMPLUGH, THOMAS (1615–91) (B), 169, 229, 519, 520
LANGHAM, ELIZABETH (1635–64)
    identified, 235
LANGHAM, JAMES (c.1621–99) (B), 236, 240, 241, 246, 247
LANGHAM, PENELOPE (d. 1684)
    identified, 235
languages, 231
    Arabic, 77, 78, 103, 435, 536
    English, 411, 535
    French, 382, 383, 404, 535
    Latin, 77, 383, 404, 434, 435, 473, 483, 494, 503, 506, 512, 536, 542
    oriental, 77
    Persian, 77, 78, 542
law, lawyer, 246, 542
LE TENNEUR, JACQUES-ALEXANDRE (*fl.* 1639–48), 8
    identified, 17
Le Tenneur: Traité des quantitez incommensurables, 17
LEIBNIZ, GOTTFRIED WILHELM (1646–1716), 443, 538
    identified, 442, 443
Leibniz: Hypothesis physica nova, 442, 443
Leibniz: Theoria motus abstracti, 443, 453
Leicestershire, 537
Leiden, 144, 157, 210, 290, 533
    University of
    Rector Magnificus, 144, 157
LÉOTAUD, VINCENT (1595–1672), 21
Lichfield Cathedral, 396
LICHFIELD, ANN (d. 1671), 414
LICHFIELD, MARGARET (17th century), 222, 414, 416
lightermen, 435

Lincolnshire, 309, 537
LIVY (284?–204?), 357
Livy: Roman History, 357
LOCKEY, THOMAS (c.1602–79)
    identified, 382
logarithm, 8, 23, 43, 437
    curve, 7
logic, 466, 471, 483, 494, 503, 506, 512
logicians, 47
London, 4, 9, 24, 41, 87, 97, 100, 153, 160, 208, 243, 247, 362, 406, 434
    Arundel House, 449
    Arundel Library, 546
    Council of Trade, 8
    Covent Garden, 538
    Drury Lane, 152
    Exeter House, 414
    Great Fire, 304
    Gresham College, 19, 55, 451, 453
    Holborn, 200
    Token House Yard, 304
    watermen, 336, 338
    Woolwich, 228
LONG, WALTER (c.1648–1731), 433
Lords of the Council, 261
Low Countries, the, 82
LOWER, RICHARD (1631–91), 290, 297, 447
Lower: Diatribe Th. Willisii de Febribus vindicatio, 290
Lower: Tractatus de corde, 290, 297
Lowestoft, Suffolk, 360
LOWTHER, JOHN (1642–1706), 208
Lyon, 431, 432, 533, 534

MABB, RALPH (17th century), 545
machines, 50
    simple, 148
MAGALOTTI, LORENZO (1637–1712), 404
Magalotti: Saggi di naturali esperienze, 404
Magellan Straits, 347
MAGHETTI, BENEDETTO (17th century), 294

Index: persons and subjects

magnetic forces, 3
MAIGNAN, EMANUEL (1601?–76), 546
Mainz, 443
MALYNES, GERARD (fl. 1585–1641), 17
Malynes: Consuetudo, vel, lex mercatoria, or The ancient law-merchant, 17
manuscripts, 432, 435
    Arabic, 103, 104
    Greek, 103, 104
    Persian, 103
MARCHETTI, ALESSANDRO (1633–1714), 294
Marchetti: Exercitationes mechanicae, 294, 412
marriage, 236
Mars, 349
Marseilles, 344
MARTINI, CHRISTIAAN (c.1630–76), 212, 293
Martini, Christiaan: Slot en sleutel van de navigation, 212, 293
MARTINI, MARTINO (1614–1661), 542
Martini, Martino: Novus atlas Sinensis, 542
mathematicians, 47
mathematics, 483, 519, 521
    analysis, 62
    application of, 73
    demonstration, 26–29, 31, 35, 36, 44, 59, 73, 82, 101, 208, 409, 457, 461, 486, 501, 508, 518
    differences, 145, 313, 379
    instruments, 155
    mathematical clubs, 436
    operations, 27, 30, 33, 34
    problems, 74, 82, 87, 211, 394, 398, 456, 472, 547
    series, 148
    tables, 146, 161, 210, 221, 223
matter, 521, 539
MAURICE (1567–1625), Prince of Orange, 418

measures, 545
mechanics, see also motion, 435, 466
medicine, 502
Mediterranean Sea, 344, 347
MEMUS, JOHANNES BAPTISTA (15th century), 16
MENGOLI, PIETRO (1625–86), 294, 369, 431
Mengoli: Novae quadraturae arithmeticae, 294, 369, 370
Mengoli: Speculazioni di musica, 431
mercantile affairs, 545
MERCATOR, NICOLAUS (1620–87), 290
    identified, 146
Mercator: Logarithmotechnia, 146, 290, 297
Mercury, 79, 349, 449
MERRET, CHRISTOPHER (1614–95), 290, 297
Merret: Pinax, rerum naturalium Britannicarum continens, 290, 297
MERRY, THOMAS (d. 1682), 435
MERSENNE, MARIN (1588–1648), 17, 408
Messina, 303, 304, 412
methods, 1, 2, 148, 149, 212, 402, 428, 530
    analytical, 6, 7, 11, 12, 27, 28, 30, 32, 33, 42–44, 47, 250
    indivisibles, 149, 402, 403, 458
MEWS, PETER (1619–1706), 382
Mexico, 344
Middleburgh, 212
MITCHELL, JOHN (d. 1556)), 311
Mitchell: A breviat cronicle, 311
MONCEAUX, ANDRÉ DE (17th cent.)
    identified, 416
MONCONYS, BALTHASAR DE (1611–65), 21
money, 2, 48, 385, 392, 546
    payment, 40, 519, 520
Mont Saint-Michel, Normandy, 347
MONTANARI, GEMINIANO (1633–87), 410, 414

Index: persons and subjects

Montanari: Pensieri fisico-matematici, 410, 414
MONTE, GUIDOBALDO DEL (1545–1607), 401
MONTMOR, HENRI LOUIS HABERT DE (*c*.1600–79), 19
Moon, 38, 242, 329, 337, 434
    Lunar months, 541
    motion, 75, 78, 164, 335, 338, 342, 360, 364, 400
MOORE, DOBBINS (Oxford carrier) (17th century), 92, 162, 432
MORAY, ROBERT (1608–73), 19, 23, 26, 27, 42, 43, 46, 55, 77, 208, 288
MORE, HENRY (1614–87), 297
More: Enchiridion ethicum, 297
MORLAND, SAMUEL (1625–95)
    identified, 544
Morland: Tuba Stentoro-Phonica, an instrument of excellent use, as well at sea, as at land, 544
MORRISON, ROBERT (1620–83)
    identified, 536
motion, 522, 538
    experiments, 49, 55, 92, 339–341, 407
    laws of, 52, 71, 73, 148, 444
    theory of, 48, 49, 52–58, 67, 69, 71, 72, 74, 82, 86, 90, 93, 94, 148, 172, 179–207, 209, 213–220, 238, 242, 329, 339, 360, 380, 402, 407, 415, 453, 521
MOXON, JOSEPH, *see* booksellers and printers
music, 231, 234

Nantes, 21
nature
    phenomena, 321, 336, 342, 364
navy
    commissioners, 228
NEALE, WILLIAM, *see* NEILE, WILLIAM
nebula, 534

NEEDHAM, WALTER (1632–91), 290
Needham: Disquisitio anatomica de formato foetu, 290
NEILE (NEALE), WILLIAM (1637–70) (B), 67, 71, 86, 87, 92–94, 102, 209, 213, 394, 445
    hypothesis of motion, 96, 100, 102, 171, 172, 178, 181, 187, 189, 191, 194, 201, 209, 213, 219, 238, 242, 445, 453, 521
    rectification of semi-cubic parabola, 394, 395
Ness Point, 309, 360
Netherlands, *see* Holland, Dutch
newspapers
    London Gazette, edited by Joseph Williamson, 406
    Newes, edited by Henry Muddiman, 406
NEWTON, ISAAC (1643–1727), 212
    identified, 544
    notes on KINCKHUYSEN, 212
NEWTON, JOHN (1622–78), 210
    identified, 145
Newton, John: The Art of Practical Gauging, 210, 221, 223
Newton, John: The Art of natural arithmetick: in whole numbers and fractions, 547
Nile, river, 336
Normandy, 431
NORRIS, EDWARD (*fl.* 1675–1708)
    identified, 229
NORRIS, FRANCIS (1609–69)
    identified, 229
North Foreland, 307
Northampton, 260
Norway, 309, 347
NOTT, ? (17th century), 547
numbers, 212, 439
    commensurable, 30
    infinite, 369, 473, 480, 483, 486, 501, 503–506, 511, 513
    irrational, 532

Index: persons and subjects

negative, 550
rational, 36
surd, 549–554

ODIERNA, GIOANBATISTA, see HODIERNA, GIOVANNI BATTISTA
OLDENBURG, HENRY (1618?–77) (B), 1, 10, 12, 13, 18, 23, 25, 26, 37, 41–43, 46, 48, 52, 53, 56, 60, 63, 66–68, 70, 74, 77, 80, 82, 86, 87, 93–95, 100, 102, 144, 148, 150, 153–155, 158, 160, 164, 166, 169, 171, 178, 181, 183, 184, 186, 188, 189, 191, 194, 200, 201, 207, 212, 213, 219, 222, 228, 230, 237, 240, 241, 244, 249, 252, 253, 256, 265, 266, 269, 279, 281, 287, 294, 303, 319, 322, 327, 334, 337, 341, 343, 363, 365, 368, 377, 380, 381, 387, 388, 390, 393, 396, 398, 401, 405, 408, 413, 417, 439, 441, 442, 448, 451, 452, 454, 455, 459, 475, 477, 479, 482, 486, 488, 502, 519, 520, 522, 535, 537, 538, 540, 554

optics, 9, 402
    lenses, 248
Order of the Garter, 452, 454
ORTENSIUS, MARTINUS, see HORTENSIUS, MARTINUS
Ostend, 310
Oudwoude, Friesland, 311
OUGHTRED, WILLIAM (1575–1669), 24, 40, 61, 62, 402, 543
Oughtred: Clavis mathematicae, 24, 40, 61, 62, 543
OULTON, R. (17th century), 545
Oxford
    experimental philosophy club, 447
Oxford (city, diocese), 246, 434
    aldermen, 262
    assistants, 262
    bishop, 520
    carrier, *see also* Bartlet; Moore; Dobbins, 38, 57, 210, 388
    deputy lieutenants, 261, 263

    high steward, 246
    inns
        The Angel, 169, 397
        The Bear, 383
    local militia, 260
    mayor, 261, 262
    militia, 263
    relations with University, 260
    St Mary the Virgin, 169, 231, 233, 234, 383
    theater, 230
    town hall, 232, 235
    visitors, 1, 19, 21, 68, 169, 241, 380, 381, 401, 405, 415
    wagoner, 89
Oxford, University of, 38, 160, 167, 168, 171, 233, 382
    academical reception, 382
    Act, 231, 243
    All Souls College, 169, 383
    bedels, 383
    Bodleian Library, 37, 54, 57, 75–77, 82, 104, 171
        catalogue, 77
        sub-librarian, 77
    burgesses for parliament, 263
    chancellor, 233, 243, 245, 246, 261
    Christ Church, 169, 233, 382, 502
    colleges and halls, 234, 261
    convocation, 169, 384, 417
    convocation house, 169
    Corpus Christi College, 382
    curators, 38, 234
    heads of houses, 231
    honorary degrees, 384
    local visitors, 171
    Magdalen College, 171, 383, 396
    Magdalen Hall, 396
    Merton College, 382
    orator, 230, 242
    Oriel College, 263, 382
    physic garden, 382
    printers, 9

Index: persons and subjects

privileges, 260
pro-vice-chancellor, 234
proctors, 233
public orator, 233
Queen's College, 171
relations with city, 260
school of moral philosophy, 169
school of music, 171
Sheldonian Theatre, 229, 233, 417
  opening, 230, 237, 240, 244
St Alban Hall, 229
St John's College, 169, 382
taxes, 260
terrae filii, 229, 230, 232–235, 242–244
University College, 383
vice-chancellor, 38, 54, 75, 82, 92, 169, 230, 232–235, 239, 243–246, 261, 382, 383
Wadham College, 231
OZANAM, JACQUES (1640–c.1717)
identified, 535

Pacific Sea, 347
Padua, 25
Palermo, 294, 297, 303, 304
PALMER, JOHN (1612–79), 14, 160, 167
Palmer: Catholique Planisphaer, 14
Panama, 347
parabola, 84, 303, 379, 437, 439, 530, 531, 549, 553
PARDIES, IGNACE GASTON (1636–73), 535, 545
  identified, 535
Pardies: Elémens de géométrie, 535
Pardies: Globi coelestis in tabulas planas redacti descriptio, 535
Paris, 8, 12, 19, 41, 42, 44, 394, 428, 431, 432, 529, 534, 537, 546
  Académie Montmor, 19
  Académie Royale des Sciences, 21
  Observatoire Royale, 535
Parliament, 260

Acts of, 262, 263
PASCAL, BLAISE (1623–1662), 427
Pascal, Historia trochoïdis, 428
Pascal, Lettres de A. Dettonville, 427
Pegu, 347
PELL, JOHN (1611–85) (B), 7, 15, 24, 44, 288, 439
PELZHOFER, JOHANN GEORG (fl. 1666–71), 440, 441
  Pelzhofer's Problem, 439, 441
pendulum, 52, 445
PERCY, HENRY (1564–1632)
  identified, 543
Peripatetics, 521
PERRAULT, CLAUDE (1613–88)
  identified, 535
PERRINCHIEF, RICHARD (1620/1–73), 313
Persian Gulf, 347
pestilence, 260
PETIT, JEAN-FRANÇOIS LE (1546–c.1615), 309
PETTY, WILLIAM (1623–87), 304
Philosophical Transactions, 2, 3, 9, 16, 18, 21, 26, 27, 38, 40, 42, 49, 55–57, 60, 64, 79, 87, 88, 92, 96, 101, 148, 151, 152, 155–157, 207, 253, 256, 277, 290, 366, 378, 401, 406–409, 441, 443, 444, 455, 459, 477, 480, 482, 489, 495, 507, 515
philosophy
  experimental, 168
  natural, 485, 512
  new, 171, 231, 234, 402
  old, 449
physician, 547
physics, 521
physiology
  lymphatic vessels, 446
  parotid duct, 447
PICARD, JEAN (1620–82)
  identified, 535
PICO FONTICULANO, GERONIMO (1541–96), 304
Pico: Tesoro di matematiche, 304

Index: persons and subjects

PIERCE (PEIRSE), Thomas (1621/2–91), 396
  identified, 383
Pisa, 412
PITT, MOSES, *see* booksellers and printers
planets, 78, 360, 434
PLINY THE ELDER (23–79), 309
POCOCKE, EDWARD (1604–91) (B), 79, 103, 104, 501, 502, 536, 538, 544
POCOCKE, EDWARD JR (17th century), 502
Pococke: A Commentary on the Prophecy of Hosea, 544
Pococke: Philosophus autodidactus sive epistola Abi Jaafar, Ebn Tophail, 502, 536, 538
Pococke: The Nature of the drink Kauhi, or Coffe, and the Berry of which it is made, 536, 538
polygon, 11
  regular, 462
POPE, WALTER (c.1627–1714), 288
POTERIUS, ? (17th century), 544
Poterius: De ponderibus ac mensuris, 544
POWLE, HENRY (1630?–92), 39, 364
PRASCH, JOHANN LUDWIG (1637–90), 239
printers, *see* booksellers and printers
printing, 8, 13, 14, 24, 40, 79, 80, 168, 288, 388, 392, 399, 400, 435, 536, 544, 546
printing, *see also* booksellers and printers
PTOLEMY, CLAUDIUS (85?–165?), 102, 435
Ptolemy: Almagest, 103, 104
Ptolemy: Catalogue of Stars, 102, 104, 158, 167

quantities
  infinite, 503
  irrational, 8
quicksilver, 281, 449

RAHN, JOHANN HEINRICH (1622–76), 297

Rahn: Introduction to Algebra, 297
RALLINGSON, RICHARD, *see* RAWLINSON, RICHARD
RANELAGH, KATHARINE (1615–91), 556, 562
  identified, 236
RAWLINSON, RICHARD (d. 1668), 17, 543
  identified, 9
RAYNAUD (REGNAULD), FRANÇOIS (17th century)
  identified, 431
Raynaud: Cogitata de Acceleratione motus gravium, 431
Red Sea, 347
Reggio di Calabria, 347
REINHOLD, ERASMUS (1511–53), 534
Reinhold: Prutenicae tabulae coelestium motuum, 535
RENALDINI, CARLO (1615–98), 294
Renaldini: Geometra promotus, 294, 412
RICCI, MICHELANGELO (1619–82), 294, 412
Ricci: Exercitatio geometrica, 294
RICCIOLI, GIAMBATTISTA (1598–1671), 337–339
Riccioli: Almagestum novum, 339, 379
Riccioli: Astronomiae reformatae tomi duo, 339, 380
ROBERTS, LEWES (1596–1641)
  identified, 545
Roberts: The Merchants Map of Commerce, 545
ROBERVAL, GILLES PERSONNE DE (1602–75), 322, 428
  new balance, 319, 322
Roberval: tract on the cycloid, 428
Rome, 412
  Theatre of Marcellus, 233
ROOKE, LAWRENCE (1619/20–62), 55, 97
roots, *see* equations
Rotherhithe, Cherry Garden Chaires, 411
Royal Academy, 546
Royal Charter, 260–262

Index: persons and subjects

Royal Society
  anatomical operator, 165
Royal Society of London, 13, 19, 20, 23, 25, 31, 36, 38, 39, 43, 44, 46, 47, 49, 53, 67, 71, 73, 74, 79, 82, 101, 153, 164, 167, 172, 208, 212, 231, 234, 242, 244, 288–290, 305, 338, 344, 364, 388, 396, 405, 407, 415, 447, 449, 454, 472, 476–478, 483, 485, 489, 497, 498, 503, 510, 513, 514, 518, 546
  council, 228, 388, 392
  journal book, 451
  president, 66, 288
  register book, 451
  secretary, 66, 474
  St Andrew's day, 416
  treasurer, 208
RYKE, DIRK DE (1640–90), 41
  identified, 19

St Andrews, 45
St Davids & Michaelmas Stream, 39
SAINT-VINCENT, GRÉGOIRE DE (1584–1667), 16, 23, 43, 531, 554
Saint-Vincent, Grégoire de: Opus geometricum, 16, 43, 531, 534
Salisbury, 384
Sandwich, 311
Saturn, 76, 78, 158, 349
SAVILE, HENRY (1549–1622), 524
SAVILE, MR (17th century), 320
SCALIGER, JULIUS CAESAR (1484–1558), 344
Scaliger: Exotericarum exercitationum liber quintus decimus, 344, 348
SCARBOROUGH, CHARLES (1615–1694) (B), 523, 524
SCHÖNBORN, JOHANN PHILIPP VON (1605–73), 443
SCHEINER, CHRISTOPH S.J. (1575–1650)
  identified, 350
Scheiner: Rosa ursina sive sol ex admirando facularum & macularum suarum phaenomeno varius, 350

scholars, 261
scholasticism, 72
SCHOOTEN, FRANS VAN (1615–60), 30, 83, 90, 210
Schooten: Geometria a Renato Descartes, Appendix de cubicarum aequationum resolutione, 30
SCHOOTEN, PIETER VAN (1634–79) (B), 210, 211, 222, 279, 280, 293
SCHOTT, CASPAR (1608–1666)
  identified, 533
Schott: Cursus mathematicus, 533
SCIPIO AFRICANUS (235–183 BC)
  identfied, 357
Scotland, 87, 309, 347
seamen, 435
semi-cubic parabola
  rectification, 394
semicircle
  division, 8
  subtense, 30
  trisection, 11, 30
SENESCHAL, MICHAEL (1606–73)
  identified, 433
Seneschal: Trias evangelica, 433
series, 439, 503, 550
  converging, 8, 26, 27, 29, 32, 33, 56
  differences, 145, 313, 315, 316, 318, 319
  infinite, 532, 551
  proportionals, 8, 315–317, 319, 439, 470, 481, 507–509, 513, 518, 530, 532, 550
  reciprocals, 371
Severn, river, 57, 309, 311, 333
SFONDRATI, PANDOLFO (fl. 1540), 344
Sfondrati: Causa aestus maris, 344
SHARP, JAMES (1618–79), 45
SHELDON, GILBERT (1598–1677), 230, 233, 243
  identified, 233
ship wreck
  weighing of, 238
shipwrights, 435

Index: persons and subjects

Sicily, 347
SLUSE, RENÉ FRANÇOIS DE (1622–85) (B), 9, 15, 45, 69, 248, 253, 302, 337, 339, 378, 391, 413, 439
Sluse: Mesolabum, 9, 15, 45, 69, 297
SNELLIUS, WILLIBRORD (1580–1626)
    identified, 373
Snellius: Cyclometricus, 373
Sobolim, 242, 245
solid, 174, 532, 550
    hyperbolicus acutus, 494, 506
    parabolic, 533
    regular, 469
    ungula, 149
solid problems, *see* equations, cubic
SORBIÈRE, SAMUEL (1615–70)
    identified, 21
Sorbière: Relation d'un voyage en Angleterre, 21
SOUTH, ROBERT (1634–1716), 169, 230, 233, 242
Southampton, 309
space, 521, 539
SPENCER, JOHN (1630–93), 298
Spencer: Dissertatio de Urim & Thummim, 298
sphere
    section, 428
SPRAT, THOMAS (1635–1713), 21, 290
Sprat: The History of the Royal-Society of London, 21, 164, 290, 311
STAG, ? (17th century), 152
Stanford, 260
stella nova, 14
STENSEN, NIELS (1638–86), 447
STEVIN, SIMON (1548–1620), 369, 418
Stevin: Les oeuvres mathematiques, 369, 418
STOW, JOHN (1525?–1605), 309
Stow: Annales, or, A generall Chronicle of England, 309
Straits of Magellan, 346

STREATER, ROBERT (1624–79), 233
    identified, 231
STREETE, THOMAS (1622–89), 290, 297, 433
    astronomical tables, 433
Streete: Astronomia Carolina, 290, 297
STUBBE, HENRY (1632–1676), 164, 168, 396, 397
Stubbe: A censure upon certain passages, 396
Stubbe: An enlargement of the observations, 164
Stubbe: Campanella revived, 396
Stubbe: Legends no Histories, 396
STURMY, SAMUEL (1633–69
    identified), 38
AL SUFI (903–986), 103, 104, 105
Sun, 349, 434
    Solar months, 541
supreme magistrate, 245
surds, 30, 36
Switzerland, Swiss, 401, 406
syllogism, 35, 47

tangents, 83
TARTAGLIA, NICCOLÒ FONTANA (1499?–1557), 548
    identified, 402
tea, 536
telescopes, 535, 544
Thames, river, 263, 306, 309, 332, 358
The Hague, 212
Theatre
    Duke of York's Players, 230
thermoscope, 281
THÉVENOT, MELCHISÉDECH (*c.*1620–92)
    identified, 537
Thévenot: Relations de divers voyages curieux, 537
THIMBLEBY (*alias* ASHBY), RICHARD (1614–80), 394
THRUSTON, MALACHI (*fl.* 1670), 298
Thruston: De respirationis usu primario, 298

Index: persons and subjects

tides, 38, 87, 158, 164, 306, 307, 320–321, 328, 335–336, 341, 344–366
  history, 321
  observations, 38, 56, 57, 156, 328–334, 336, 338, 361, 406
  periods of, 38, 47
Tonbridge School, Kent, 3
TORPORLEY, NATHANIEL (1564–1631) (B), 146
TORRICELLI, EVANGELISTA (1608–47), 49, 403
  barometric experiments, 405, 410
  identified, 446
Torricelli: De solido hyperbolico acuto, 494, 506, 512
Torricelli: De sphaera et solidis sphaeralibus libri duo, 403
trade, 545
trochlea, 250
trochoid, *see* cissoid
Tuscany, 171
TWYSDEN, JOHN (1607–88), 160

ULUG BEG (1394–1449), 103, 542
  astronomical tables, 103
  catalogue of fixed stars, 77, 103, 158, 290
  identified, 75
  translation and printing of his astronomical institutions, 103
Ulug Beg: Tabulae long. ac lat. Stellarum fixarum, ed. Thomas Hyde, 290
UPSALL, ? (17th century), 380–382
Uraniburg, 80, 535

vacuum, 404, 414, 444, 521, 538
van Ceulen: Van den Circkel, 376
VAN DEN HOVE, MARTEEN, *see* HORTENSIUS, MARTINUS
Venice, 309, 344
Venus, 79, 349
VERNON, FRANCIS (1637–77) (B), 97, 377, 393, 431, 537, 538, 543

VIÈTE, FRANÇOIS (1540–1603), 211, 402
Viète: Ad harmonicon coeleste libri quinque priores, 211
Viète: Apollonius Gallus, 402
Viète: In artem analyticem Isagoge, 317, 402
Viète: Opera mathematica, 317
VILLIERS, GEORGE (1628–87)
  identified, 246
VIVIANI, VINCENZO (1622–1703), 404
VOGEL (or FOGELIUS), MARTIN (1634–75)
  identified, 538
VON COCCEJUS, HEINRICH (1644–1719), 440
VOSSIUS, GERARDUS JOANNES (1577–1649), 211
Vossius, Gerardus: De quattuor artibus popularibus, 211
VOSSIUS, ISAAC (1618–89), 211, 347
Vossius, Isaac: De motu marium et ventorum liber, 347, 355

WAGSTAFFE, COMBE (17th century), 396
WAGSTAFFE, THOMAS (1645–1712)
  identified, 396
WALKER, GEORGE (1582?–1651)
  identified, 434
Walker: A Defence of the true sence, 434
WALLIS, ALEXANDER, great-great-grandson of the mathematician (18th century)
  identified, 541
WALLIS, JOHN (1616–1703), 45, 230, 432, 545
  amanuensis, 54
  barometrical observations, 49, 281, 287, 448
  corrections to Newton's *Art of Practical Gauging*, 221, 223
  deciphering, 252
  edition of Horrox's astronomical papers, 79, 432

Index: persons and subjects

hypothesis of motion, 48, 49, 54, 56, 58, 67, 69–73, 92, 94, 148, 178, 184, 191, 360, 416

hypothesis of the tides, 38, 47, 56, 57, 79, 87, 158, 164, 167, 299, 305, 320, 328, 338, 341, 343, 360–362, 364

lecture for the Prince of Orange, 417

mental calculation, 440, 442

model of roof design of interlocking wooden beams, 383

quartan ague, 398, 400, 406, 408, 413, 415, 432, 502, 547

reprinting his works, 73

speech at visit of Cosimo de' Medici, 169

Wallis: *A Brief account of Mr Hobbes's fundamental mistake*, 292

Wallis: *A Discourse of Gravity and Gravitation, grounded on experimental observations*, 449

Wallis: *A Letter... concerning the Variety of the Annual High-Tydes*, 366

Wallis: *A summary account... of the General Laws of Motion*, 157, 416, 444

Wallis: *Adversus Marci Meibomii de proportionibus dialogum tractatus elencticus*, 324

Wallis: *An Accompt of a small Tract, entituled, Thomae Hobbes Quadratura Circuli*, 277

Wallis: *An Answer of Dr. Wallis to Mr. Hobbbes's Rosetum Geometricum*, 459, 477, 480, 495, 513

Wallis: *An Answer to Four Papers of Mr. Hobs, lately published*, 476, 479, 482, 489, 503, 515

Wallis: *An Answer to Three Papers of Mr. Hobs, lately published*, 503, 510, 515

Wallis: *An Appendix... Being an Answer to some Objections*, 301, 335

Wallis: *An Essay... exhibiting his Hypothesis about the Flux and Reflux*, 305, 328, 334

Wallis: *Animadversions... upon Mr Hobs's late Book*, 64

Wallis: *Arithmetica infinitorum*, 12, 29, 31, 36, 44, 326, 369, 376, 403, 421, 473, 494, 506, 511, 530, 532, 549, 553

Wallis: *Commercium epistolicum*, 16, 494, 547

Wallis: *Cono-cuneus*, 9

Wallis: *De aestu maris hypothesis nova*, 158, 335

Wallis: *De curvarum Ἐυϑύνσει*, 550

Wallis: *De cycloide*, 254

Wallis: *De motu*, 409, 494, 503, 549, 550

Wallis: *De postulato quinto*, 12

Wallis: *Defense of the Treatise of the Angle of Contact*, 9

Wallis: *Discourse of Combinations*, 9

Wallis: *Answer... to Mr. Hobbbes's Rosetum Geometricum*, 472

Wallis: *Due Correction for Mr Hobbes*, 63, 281, 297

Wallis: *Elenchus geometriae Hobbianae*, 63, 297

Wallis: *Grammatica linguae anglicanae*, 291

Wallis: *Hobbiani puncti dispunctio*, 63

Wallis: *Hobbius Heauton-timorumenos*, 65, 489

Wallis: *Hypothesis about the Flux and Reflux of the Sea*, 301, 366

Wallis: *Mathesis universalis*, 489

Wallis: *Mechanica*, 9, 24, 49, 72, 91, 148, 162, 385, 388, 398, 458

Wallis: *Mechanica I*, 221, 227, 258, 265, 266, 269, 289, 290, 293, 297, 323, 326

titlepage, 293

Wallis: *Mechanica II*, 304, 326, 386, 387, 391, 393, 398, 476

Wallis: *Mechanica III*, 171, 386, 388, 400, 415, 416, 421, 424, 444, 449, 454, 519, 520

Wallis: *Opera mathematica I*, 233, 249

628

Index: persons and subjects

Wallis: Opera mathematica II, 9, 12, 82, 335
Wallis: Operum mathematicorum pars altera, 297
Wallis: Operum mathematicorum pars prima, 297, 489
Wallis: Thermometrical and Baroscopical Observations, 282
Wallis: Thomae Hobbes quadratura circuli, 201, 207, 222, 270, 277, 288–290, 298
Wallis: Tractatus duo, 2, 156, 162, 249, 421, 427
Wallis: Treatise of Algebra, 9, 24, 525
Wallis: Treatise of Angular Sections, 9
WALLIS, JOHN JR (1650–1717) (B), 30, 144, 151, 154, 235, 389, 452, 454, 519, 565, 520
   severe illness, 398
WALLIS, SUSANNA (1622–87), 30, 235, 565
WALLIS, WILLIAM, great-great-grandson of the mathematician (18th century)
   identified, 541
WALTON, BRIAN (17th century)
   identified, 78
Walton (ed.): Biblia Polyglotta, 79
Wandesburg, 80
war, 260
WARD, SETH (1617–88) (B), 13, 288, 305, 343, 365
WASMER, BENEDICT (17th century), 73
   identified, 68
WASSENAER, JACOB VAN (*c.*1607–?), 74, 82, 86, 87, 90, 96
Wassenaer: Den on-wissen Wis-Konstenaer, 90
weights, 545
WERNER, GEORG CHRISTOPH (*fl.* 1660–70), 386
Werner: Inventum novum, 386
Weymouth, 309, 331, 332
WHARTON, THOMAS (1614–73), 447

WHITE, THOMAS (1593–1676), 429
whores, 232, 234, 243
WILKENS, MARTIJN (17th century), 212, 293
Wilkens: Officina algebrae, 212, 293
WILKINS, JOHN (1614–72), 16, 208, 242, 288, 290
Wilkins: An Essay Towards a Real Character, And a Philosophical Language, 290
Wilkins: The Discovery of a World in the Moone, 242
WILKINS, MARTIN, see WILKENS, MARTIJN
WILLIAM HENRY (1650–1702, prince of Orange and Nassau; 1685 king of England), 417
WILLIAMS, ? (17th century)
   identified, 393
WILLIAMSON, JOSEPH (1633–1701) (B), 169, 229, 232
WILLIS, THOMAS (1621–75), 290, 297
Willis: Cerebri anatome, 297
Willis: Pathologiae cerebri, 290
Winchelsea, 309
winds, 56, 307, 336
WING, VINCENT (1619–68), 222, 290, 297
   identified, 9, 535
Wing: Astronomia Britannica, 222, 290, 297, 535
Wing: Harmonicon coeleste: or, the coelestiall harmony of the visible world, 535
Wisbech, 309
WITHIE, ? (17th century), 169
WOOD, ANTHONY (1632–95), 169
WOOD, ROBERT (1621/2–85) (B), 299, 324, 368, 377, 387, 389
world
   beginning, 483, 491, 506, 512, 519
WREN, CHRISTOPHER (1632–1723), 54, 55, 155, 208, 222, 231, 248–249, 253, 254, 256, 302, 443, 447
   account of motion, 148, 176, 416

Index: persons and subjects

grinding lenses, 249, 253
Sheldonian Theatre, 233
Wren: A Decription of Dr. Christopher Wren's Engin, 256
Wren: Dr. Christopher Wrens Theory, 157

Wren: Generatio Corporis Cylindroidis Hyperbolici, 158, 249, 256

ZANI, ERCOLI (1634–84), 237
zodiac, 331, 361